BOSE–EINSTEIN CONDENSATION OF EXCITONS AND BIEXCITONS
AND COHERENT NONLINEAR OPTICS WITH EXCITONS

Bose–Einstein condensation of excitons is a unique effect in which the electronic states of a solid can self-organize to acquire quantum-phase coherence. The phenomenon is closely linked to Bose–Einstein condensation in other systems such as liquid helium and laser-cooled atomic gases. This is the first book to provide a comprehensive survey of this field, covering theoretical aspects as well as recent experimental work.

After setting out the relevant basic physics of excitons, the authors discuss exciton–phonon interactions as well as the behavior of biexcitons. They cover exciton phase transitions and give particular attention to nonlinear optical effects including the optical Stark effect and chaos in excitonic systems. The thermodynamics of equilibrium, quasi-equilibrium, and nonequilibrium systems are examined in detail.

The authors interweave theoretical and experimental results throughout the book and it will be of great interest to graduate students and researchers in semiconductor and superconductor physics, quantum optics, and atomic physics.

S.A. Moskalenko is Professor of Theoretical and Mathematical Physics of the Institute of Applied Physics of the Academy of Sciences of the Republic of Moldova. He has been a state prize winner of the Moldovan SSR, Laureate of the State Prize of the USSR in the field of Science and Technology, and a full member of the Academy of Sciences of the Republic of Moldova since 1992. He is the author of four books in Russian and many scientific articles.

D.W. Snoke is an Assistant Professor of Physics at the University of Pittsburgh. He is the recipient of a U.S. National Science Foundation early career award, and has been named a Cottrell Scholar by the Research Corporation. He has published many articles on excitons, spectroscopy, and ultrafast processes and was co-editor of the book *Bose–Einstein Condensation*.

BOSE–EINSTEIN CONDENSATION OF EXCITONS AND BIEXCITONS

AND COHERENT NONLINEAR OPTICS WITH EXCITONS

S.A. MOSKALENKO
Academy of Sciences of Moldova

D.W. SNOKE
University of Pittsburgh

CAMBRIDGE
UNIVERSITY PRESS

CAMBRIDGE UNIVERSITY PRESS
Cambridge, New York, Melbourne, Madrid, Cape Town, Singapore, São Paulo

Cambridge University Press
The Edinburgh Building, Cambridge CB2 2RU, UK

Published in the United States of America by Cambridge University Press, New York

www.cambridge.org
Information on this title: www.cambridge.org/9780521580991

First published 2000
This digitally printed first paperback version 2005

A catalogue record for this publication is available from the British Library

Library of Congress Cataloguing in Publication data
Moskalenko, Sviatoslav Anatol evich.
[Boze- ėĭnshteĭnovskaia kondensatsiia ėksitonov l biėksitonov. English]
Bose–Einstein condensation of excitons and biexcitons : and coherent nonlinear optics with excitons / S.A. Moskalenko, D.W. Snoke.
p. cm.
ISBN 0-521-58099-4
1. Exciton theory. 2. Bose–Einstein condensation. 3. Nonlinear optics. I. Snoke. D.W. II. Title.
QC172.8.E9M613 1999
530.4 16--dc21 98–44874
CIP

ISBN-13 978-0-521-58099-1 hardback
ISBN-10 0-521-58099-4 hardback

ISBN-13 978-0-521-02235-4 paperback
ISBN-10 0-521-02235-5 paperback

In memory of Prof. Yulya
S. Boyarskaya.
Soli Deo Gloria

Contents

Preface

The idea for this book grew out of the historic conference on Bose–Einstein Condensation in Trento, Italy, in 1993, which brought together many of the founders of theoretical and experimental research of excitons, as well as experts on Bose–Einstein condensation from many other fields, including astrophysics, nuclear theory, liquid helium, and superconductors. Especially important was the opportunity for Western and Russian researchers to participate together. The authors of this book first met at that time, and it became apparent to both of them that the fundamental works on Bose condensation of excitons remained scattered in the literature, and included many Eastern European publications not readily available in the West. This book is an attempt to bring together for the first time a coherent account of the great amount of theoretical and experimental work in this rapidly growing field.

In the years following that meeting, the theory of weakly interacting Bose systems has taken off, as experiments in new systems have started to pay off, following decades of emphasis on strongly interacting coherent systems. The accompanying table shows the four major areas of interest in coherent condensed matter systems today.

	Atomic systems	Solid systems
Strongly interacting	Liquid Helium-3 and Helium-4	BCS and Hi-T_c superconductors
Weakly interacting	Alkali vapors	Exciton and biexciton Gases

Each has its own appeal. Studies on atoms, including liquid Helium-3 and Helium-4 and the recent, exciting studies of hydrogen and alkali atoms in optical traps, have the appeal that the interactions between the atoms can be easily understood without extensive study of solid-state band structures. On the other hand, atomic condensates are necessarily a low temperature phenomenon, existing only at temperatures from 2 K down to a few nanokelvin. The solid-state phenomena of superconductors and excitonic condensates exist at much higher temperatures, possibly even at room temperature. Two effects exist in solid-state Bose systems which drive much of the research. First, charged bosons (e.g., Cooper pairs) can carry current, so that Bose condensates of charged bosons are superconductors. Several possibilities for novel charged bosons in solid-state systems are discussed in this book. Second, neutral bosons such as excitons typically couple directly to photons, so that coherence in the photon field can be maintained in the electronic states. This leads to a number of nonlinear optical effects, which are discussed at length in this volume. Excitons also couple directly to the crystal phonon states, so that an excitonic condensate can lead to novel acoustic effects.

The idea of the excitonic condensate appeared at the same time when the foundations of the theory of lasers, semiconductors, superconductors, and superfluids were being laid in the 1950s and 1960s. Much of the pioneering work was sponsored by the Academy of Sciences of the USSR, and involved the efforts of academicians such as N.N. Bogoliubov, A.S. Davydov, V.L. Ginzburg, L.V. Keldysh and R.V. Khohlov, and Prof. L.E. Gurevich. In 1982, the division of General Physics and Astronomy of the Academy of Sciences of the USSR listed the theory of high density excitons as one of the major achievements of multinational Soviet physics. S.A. Moskalenko was fortunate to be part of these discussions from the very start, starting with his graduate studies with Professor K.B. Tolpygo of the Kiev Institute of Physics, and wrote some of the earliest papers on the subject. The exciton condensate theory is, in a sense, a triumph of renormalization in quantum field theory. Starting with atoms, one can deduce the band structure of a solid and define a new vacuum equal to the ground state of the solid, in which the only particles are quasiparticles determined by the excitation spectrum of the solid, namely, free electrons, free holes, and lattice phonons. The Coulomb interaction between these electrons and holes leads to pairing into excitons. Starting with this new system of only excitons and phonons, one can move to a new vacuum which is the ground state of the exciton gas, and new quasiparticles appear as the excitations of the interacting exciton system. The beauty of the theory is that one needs to refer only occasionally to the underlying band structure, and in most cases the excitons can be viewed as weakly interacting bosons no different from atoms.

Much of the theoretical work discussed in this book comes from accomplishments of the Department of Theory of Semiconductors and Quantum Electronics of the Institute of Applied Physics (IAP) of the Academy of Sciences of Republic of Moldova, which was headed by S.A. Moskalenko for over 30 years from the time it was organized in 1964. Professor P.I. Khadzhi and Drs. A.I. Bobrysheva, I.V. Beloussov, M.I. Shmiglyuk, and V.R. Misko of this department contributed significantly to this book. Prof. A.H. Rotarv, Drs. M.F. Miglei, S.S. Russu, Yu.M. Shvera, V.A. Zalozh, E.S. Kiseleva, and many other former members of this department made important contributions to the theoretical works cited in the text. The contributions of S.A. Moskalenko to this book are based partially on his earlier books published by the Academy of Sciences of the Republic of Moldova, referred to in the text, and on lecture notes of the special course given from 1980 to 1994 to students of Moldova State University as well as at Uppsala State University in 1995 in connection with his collaboration with Prof. M.A. Liberman.

Our aim throughout the book has been to incorporate discussion of the experimental work with that of the theory. While the theory of excitonic condensates has progressed over the past 40 years, experiments did not begin to make substantial progress until the 1980s, coinciding with the developments of ultrafast laser spectroscopy and high-quality quantum heterostructures. As discussed in this book, there are two realms of experimental effort, the pursuit of spontaneous condensation of excitons in quasiequilibrium and the exploration of coherent effects which occur when a laser field couples directly to the excitonic states. The underlying theory is the same for both, and we have stressed this unity in the text.

Writing of this book would have been impossible without the support of the IAP of Moldova. Supplementary financial support was recieved from INTAS project 94-324. The work of S.A. Moskalenko was generously supported by his family, consisting entirely of physicists. His wife, Professor Yu.S. Boyarskaya, was selflessly devoted to science and especially encouraged him to write this book. For the past three years, S.A. Moskalenko has written this book in her memory, and dedicates his work to her. The efforts of David Snoke have been supported by the National Science Foundation as part of Early Career Award DMR-9722239 and by the Research Corporation through its Cottrell Scholar program. Professors J.P. Wolfe, M. Cardona, G. Baym, Y.C. Chang, and D. Boyanovsky also provided significant support and feedback for this book.

Much remains to be done in this growing field, and we hope that this book will serve as an aid both to those newly entering in and to those already familiar with these topics.

S.A. Moskalenko, Kishinev <exciton@phys.asm.md>
D.W. Snoke, Pittsburgh <snoke@vms.cis.pitt.edu>

of physicists this work. Professor Yu. S. Boyarskaya, was selflessly devoted to science and especially encouraged him to write this book. For the past three years, S. A. Moskalenko has written this book in her memory, and dedicates his work to her. The efforts of David Snoke have been supported by the National Science Foundation as part of an Early Career Award DMR-9722239 and by the Research Corporation through its Cottrell Scholar program. Professors J. P. Wolfe, M. Cardona, G. Baym, Y. C. Chang, and D. Boyanovsky also provided significant support and feedback for this book.

Much remains to be done in this growing field, and we hope that this book will serve as an aid both to those newly entering it and to those already familiar with these topics.

S. A. Moskalenko, Kishinev <exciton@phys.asm.md>
D. W. Snoke, Pittsburgh <snoke@vms.cis.pitt.edu>

I

Introduction

1.1 What is an Exciton?

Many people seem to have trouble with the concept of an exciton. Is it "real" in the same sense that a photon or an atom is? Does the motion of an exciton correspond to the transport of anything real in a solid?

Simply put, an exciton is an electron and a hole held together by Coulomb attraction. Of course, for some people the idea of a "hole" is a difficult concept, so this may not help much. Nevertheless, a hole is a "real" particle and so is an exciton.[a] Modern solid-state theory [1, 2] gives equal footing to both free electrons and holes as charge carriers in a solid, exactly analogous to the way that electrons and positrons are both "real" particles, even though a positron can be seen as the absence of an electron in the negative-energy Dirac sea, i.e., a backwards-in-time-moving electron.

All excitons are *spatially compact*. The strong Coulomb attraction between the negatively charged electron and the positively-charged hole keeps them close together in real space, unlike Cooper pairs, which can have very long correlation lengths because of the weak phonon coupling between them. The sizes of excitons vary from the size of a single atom, e.g., approximately an angstrom up to several hundred angstroms, extending across thousands of lattice sites. Excitons are roughly divided into two categories based on their size. An exciton that is localized to a single lattice site is called a "Frenkel" exciton, after the pioneering work of Frenkel [3] on excitons in molecular crystals. Frenkel excitons appear most commonly in molecular crystals, polymers, and biological molecules, in which they are extremely important for understanding energy transfer. Excitons in the opposite limit, spanning many lattice sites, are known as "Wannier" excitons or "Wannier–Mott" excitons [4, 5]. These are typical in most semiconductors and are the main subject of this book. In between these two limits, there are excitons of intermediate size, which are described by the charge-transfer model.

In general, excitons can *move* through a solid. In the case of Frenkel excitons, this motion is viewed as hopping of both the electron and the hole from one atom to another. If an electron is excited from a valence shell of an atom into an excited state, it may then move

[a] In some of the literature, free electrons, holes, and excitons in a solid are called "quasiparticles," which carry "quasimomentum," to distinguish them from free particles in vacuum. This can lead to the impression that they are not "real," when actually all it means is that they are the fundamental quanta of a field with a renormalized energy spectrum. By this standard, even free electrons and positrons in vacuum may be called quasiparticles, since they have a renormalized energy spectrum compared with the bare electron and positron in quantum electrodynamics. In this volume, we reserve the term "quasiparticle" for elementary excitations of the many-electron and the many-exciton states.

into an unoccupied excited state of a nearby atom. The Coulomb repulsion of the electron on the valence electrons of the new atom will tend to push one of them into the hole left in the valence shell of the original atom when the electron was excited. This effectively causes the hole in the valence shell to follow the excited electron. Equivalently, one may say that the negatively charged excited electron attracts the positively charged hole in the valence band, taking it along to the new atom.

In the case of Wannier excitons, the picture is quite different. The underlying lattice of atoms is treated as a background field in which the electrons and the holes exist as free particles, and an exciton consists of an electron and a hole orbiting each other in this medium. To first order, the entire effect of the underlying atomic lattice on the excitons (or free electrons and holes) is taken into account by means of (1) the renormalized masses of the electron and hole, (2) the dielectric constant of the solid, and (3) scattering with phonons (quanta of vibration) and impurities.

The Wannier exciton is therefore essentially a Rydberg atom analogous to hydrogen or positronium. In the case of the hydrogen atom, the hydrogenic wave function consists of an atomic-orbital part multiplied by a plane-wave factor, $\exp[i(\mathbf{k}\cdot\mathbf{r}-\omega t)]$, where $\mathbf{k}=(m_0+m_P)\mathbf{v}/\hbar$ is the center-of-mass momentum. The energy spectrum for the bound pair is the Rydberg energy plus the kinetic energy associated with the center of mass,

$$E_H = \frac{-\mathrm{Ry}}{n^2} + \frac{\hbar^2 k^2}{2(m_0+m_P)},$$

where $n = 1, 2, \ldots,$ is the principal quantum number,

$$\mathrm{Ry} = \frac{e^2}{2a_0}$$

is the Rydberg energy, and

$$a_0 = \frac{\hbar^2}{e^2 m_r}$$

is the hydrogenic Bohr radius with reduced mass $m_r = m_0 m_P/(m_0+m_P)$. The situation in a semiconductor is analogous: an electron in the conduction band and a hole in the valence band bind together to form an exciton with center-of-mass wave vector $\mathbf{k}$ and total mass $m_{\mathrm{ex}} = m_e + m_h$. The factor e^2 in the definition of the Rydberg in a_0 is replaced by e^2/ϵ_0 (where ϵ_0 is the low-frequency dielectric constant of the solid) and the renormalized band masses m_e and m_h of the electron and the hole are used in place of the free-electron and the proton masses, giving

$$E_{\mathrm{ex}} = \frac{-e^2}{2a_{\mathrm{ex}}\epsilon_0 n^2} + \frac{\hbar^2 k^2}{2(m_e+m_h)}, \tag{1.1}$$

where

$$\mathrm{Ry}_{\mathrm{ex}} = \frac{e^2}{2\epsilon_0 a_{\mathrm{ex}}}, \tag{1.2}$$

$$a_{\mathrm{ex}} = \frac{\hbar^2 \epsilon_0}{e^2 m_r}, \tag{1.3}$$

Table 1.1. *Exciton parameters in various materials*

	ϵ_0	Binding energy (meV)	Approximate radius (Å)
KCl	4.6	580	3
CuCl	5.6	190	7
Cu_2O	7.1	150	7
Si	11.4	12	50
GaAs	13.1	4	150

where $m_r = m_e m_h/(m_e + m_h)$. The rescaling of the Coulomb interaction $e^2 \to e^2/\epsilon_0$ in Eqs. (1.1) to (1.3) implies that the binding energy of the exciton will be several orders of magnitude less than that of hydrogen or positronium. In the semiconductor Cu_2O, for example, which has relatively isolated conduction and valence bands, so that the electron and the hole masses very nearly equal the free-electron mass m_0, these equations imply a binding energy of 13.6 eV/$2\epsilon_0^2$, which for $\epsilon_0 = 7$ in Cu_2O implies $\text{Ry}_{\text{ex}} = 13.6\ \text{eV}/2 \times 49 =$ 0.138 eV, which is very close to the correct value of 0.100 eV, considering that we have used only one material parameter, the dielectric constant, without taking into account the band structure.[b] Table 1.1 gives a list of typical exciton binding energies and excitonic Bohr radii.

Despite the change of energy scale, the Wannier exciton still corresponds to the case of an electron and a hole orbiting each other in the background medium of the solid, exactly like a hydrogen or a positronium atom. The exciton as a whole will move in a straight line at constant momentum until it scatters with a phonon, an impurity, or another exciton, a free electron or a hole. Although the center-of-mass wave vector **k** now corresponds to a crystal momentum that is conserved only *modulo* a reciprocal lattice vector, Umklapp processes that lead to the nonconservation of crystal momentum occur for only high-energy scattering processes, i.e., at temperatures high compared with those of the excitonic Rydberg.

As a complex made of a particle and an antiparticle, excitons are in general *metastable*. Both Frenkel and Wannier excitons have a finite probability for the excited electron to recombine with the hole, leading to the emission of a photon. A typical "life cycle" for an exciton therefore goes as follows: (1) The exciton is created by absorption of a photon, (2) the exciton then moves through the solid, undergoing scattering processes, and (3) finally, the exciton recombines and emits a photon, possibly at some place in the solid quite distant from its creation point. Depending on the symmetries of the recombination processes and the experimental conditions, the lifetime of excitons can range from picoseconds up to milliseconds or longer. In general, the decay rate of an exciton is proportional to the square of the normalized electron–hole orbital wave function at zero separation, which for a hydrogenic Wannier exciton is $\varphi^2(0) = 1/(\pi a_{\text{ex}}^3)$, implying that smaller excitons and Frenkel excitons tend to decay faster.

In one sense, this property is also analogous to the case of positronium, which is a metastable state subject to recombination. Excitons couple to photons in a way different

[b] The excitonic Rydberg for the $n \geq 2$ states in Cu_2O is 100 meV, while the $n = 1$ ground-state binding energy is 150 meV because of central-cell corrections that arise because the size of the $n = 1$ exciton is comparable with the lattice constant. See Ref. 6.

from positronium, however. Figure1.1 illustrates the two cases. By energy and momentum conservation, positronium cannot decay into a single photon; instead, it must decay into two back-to-back photons. The gap energy for excitons is much less than m_0c^2, however. Therefore, although a Wannier exciton can also decay by means of a two-photon process analogous to that of positronium, it can also be created by or decay into a single photon.

Figure 1.2 shows motion of excitons into a crystal of Cu_2O. In these time-resolved images, we can see the above-described characteristics of excitons. Excitons are created at the surface

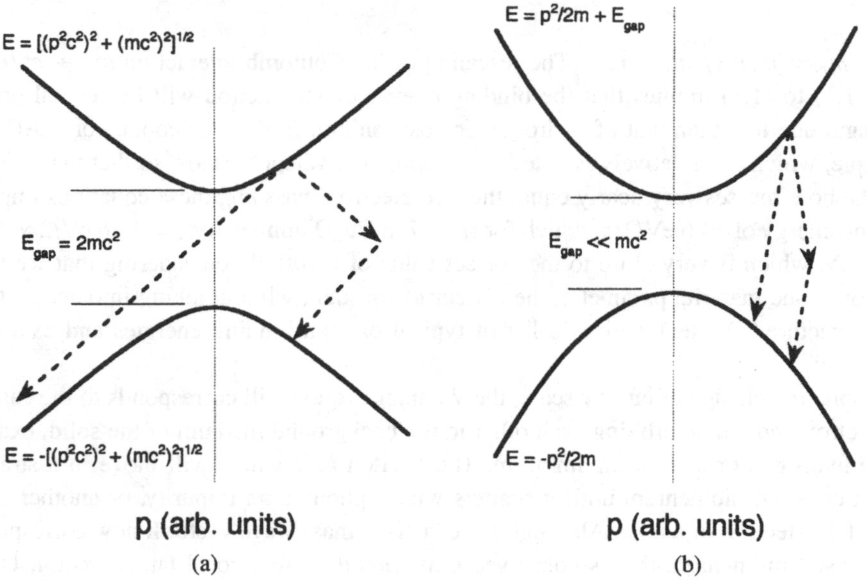

Figure 1.1. (a) Single-versus two-photon electron–positron recombination, (b) single-versus two-photon electron–hole recombination in a direct-gap semiconductor. Since $E_{\text{gap}} \ll mc^2$, these transitions are called "vertical" transitions, since the momentum of the hole is almost the same as the momentum of the electron.

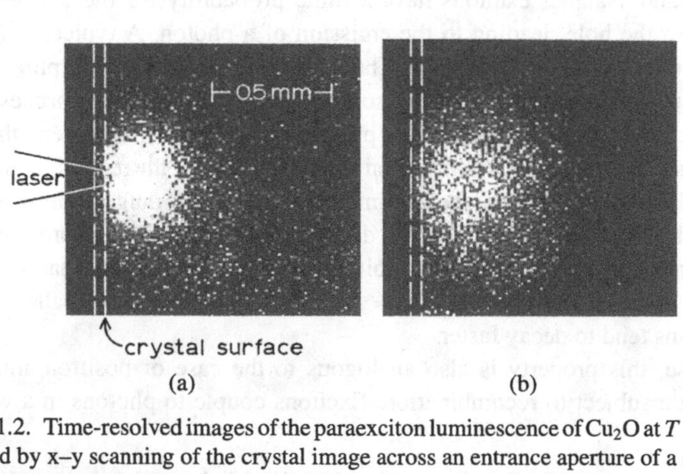

Figure 1.2. Time-resolved images of the paraexciton luminescence of Cu_2O at $T = 2$ K, obtained by x–y scanning of the crystal image across an entrance aperture of a spectrometer: (a) $t = 0.2$ μs after a 100-ns Ar^+ laser pulse is absorbed at the crystal surface, (b) $t = 0.6$ μs (from Ref. 7).

of the crystal by absorption of laser light; they then move into the crystal, scattering with phonons and each other. In general, those scattering processes lead to diffusion of the excitons into the solid. As they go, some of them recombine and emit photons, which are recorded in these images.

We note that the above "life cycle" for excitons has many possible exceptions. Excitons can be created not only by absorption of a photon, but by any process that produces excited electrons and holes, e.g., an electric field. Also, instead of recombining simply into photons, excitons may decay nonradiatively, leading ultimately to the emission of phonons, or by means of a combination of photons and phonons. Also, in some cases excitons may become stable, permanent excitons. The excitonic-insulator state, discussed briefly in Chapters 5 and 10, is one proposed mechanism for this, in which excitons can spontaneously form in a narrow-gap semiconductor. Another example is the case of self-trapped excitons, which can alter the band structure of a solid.

Excitons carry energy and momentum, but not mass or charge. One way of describing an exciton is as a "heavy and slow photon." The exciton can have wavelengths of tens of angstroms, compared with hundreds of nanometers for photons, but it has an energy of a few electron volts, comparable with photons. The coupling of photons to exciton states with very short wavelength leads to many of the potential applications of excitons. For example, one can imagine a photon effectively passing through an aperture of a size much less than the photon wavelength without diffraction – the photon can first be converted into an exciton, which stores the energy in the electronic states with short wavelength while passing through the aperture, and can then convert back into a photon on the other side. In payment for its short wavelength, an exciton moves at speeds far less than the speed of light in the medium, i.e., at speeds of the order of the thermal speeds for free electrons. The photon has been dressed with mass as it travels through the optical medium.

Although one can easily imagine passing light through an aperture smaller than the photon wavelength by converting it into excitons, the problem normally arises that the excitons move diffusively, so that the coherence of the light is lost. In the case of an excitonic Bose condensate, however, the excitons retain coherence or even acquire coherence spontaneously. This is the main topic of this book. As we will see, in many cases there is a continuous transition from coherent states of photons to coherent states of excitons. In other cases, however, coherence in the exciton states is quite distinct from coherence in the photon states.

The exciton is the fundamental quantum of excitation of a solid. This is true even for excitation well above the bandgap, in the electron–hole continuum; the absorption spectrum above the gap is strongly affected by the ionized-exciton wave function [8,9]. Correlations persist between the excited electron and hole even in above-gap excitation of semiconductors until dephasing processes eliminate them [10].

The view of excitons as heavy photons is further illustrated by the "polariton" effect, which mixes the properties of excitons and photons. As seen in Fig. 1.1, the electron and the hole states couple directly by means of a single-photon transition, which means that the dispersion relation for photons in a solid intersects the dispersion relation for Wannier excitons. Depending on the strength of the coupling between photon and exciton states, the region of intersection may be strongly altered, leading to an anticrossing. Figure 1.3 shows a typical polariton dispersion. Polaritons in the region of intersection have mixed character, and, in general, there is no sharp distinction between a photon traveling in the medium and an exciton, other than the group velocity.

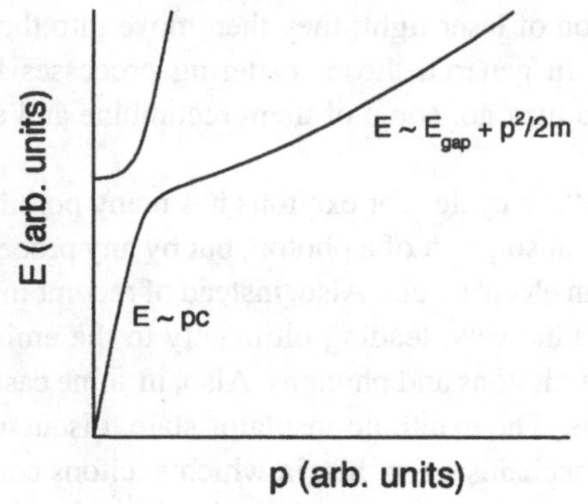

Figure 1.3. Typical polariton dispersion relation for a direct-gap semiconductor, given by $c^2p^2/\epsilon_\infty E^2 = 1 + (E_L^2 - E_T^2)/[E_T^2 + E_T(p^2/m_{ex}) - E^2]$, where E_T and E_L are the transverse and the longitudinal exciton energies, respectively, at zone center [9].

If the medium were infinite, then the coupling of the exciton states to photons in the medium would never lead to a loss of the excitons, since the photons would remain in the medium and couple back to the excitons. For a finite crystal, however, the surface acts to couple polaritons to photons in the external world, which is assumed to be dissipative. For this reason, the lifetime of polaritons depends strongly on the surface-to-volume ratio of a crystal [11].

1.2 The Exciton as a Boson

The premise that excitons act as bosons follows simply from the spin-addition law, $\frac{1}{2} \otimes \frac{1}{2} = 0 \oplus 1$, and similar generalizations for higher-angular-momentum states. In other words, two fermions bound together make a boson. How bound together must they be, however?

In this section, we follow the classic review by Hanamura and Haug [12] to answer this question quantitatively, starting only with the Fermi operators of the underlying electronic states.[c] By writing the Hamiltonian for the electronic system as

$$H = \int d^3x\ \Psi^\dagger(x)H_0(x)\Psi(x) + \frac{1}{2}\int d^3x\ d^3y\ \Psi^\dagger(x)\Psi^\dagger(y)V(x-y)\Psi(y)\Psi(x), \quad (1.4)$$

where $H_0(x)$ is the Hamiltonian for the single electrons that takes into account the crystal-band structure and $V(x) = e^2/\epsilon_0|x|$ is the Coulomb interaction, one can expand the Fermi field operators in terms of the eigenstates of the total self-consistent single-particle Hamiltonian, which includes not only H_0 but also terms due to the interaction of a given band electron with the valence electrons, which will be pointed out below. We write $\Psi(x)$ in terms of the Bloch functions,

$$\Psi(x) = \frac{1}{\sqrt{N}} \sum_{\substack{\mathbf{p} \\ j=c,v}} a_{\mathbf{p}j} u_{\mathbf{p}j}(\mathbf{x}) e^{i\mathbf{p}\cdot\mathbf{x}}, \quad (1.5)$$

in which $u_{\mathbf{p}j}(\mathbf{x})$ is a function with the periodicity of the lattice and $a_{\mathbf{p}j}$ and $a_{\mathbf{p}j}^\dagger$ are electron creation and destruction operators, respectively, that obey the commutation relations $\{a_{\mathbf{p}i}, a_{\mathbf{q}j}\} = 0$ and $\{a_{\mathbf{p}i}, a_{\mathbf{q}j}^\dagger\} = \delta_{i,j}\delta_{\mathbf{p},\mathbf{q}}$.

[c] This derivation is repeated in somewhat more rigorous form in Subsection 7.2.1.

For simplicity, we consider a two-band model with valence band v and conduction band c and assume that the effect of all the other bands has been taken into account in the form of H_0 and the dielectric constant ϵ_0. Introducing the hole operators by a time-reversal transformation K,

$$K a_{\mathbf{p}v} K^\dagger = b_\mathbf{p} \qquad \text{or} \qquad a_{\mathbf{p}v} = K^\dagger b_\mathbf{p} K = b^\dagger_{-\mathbf{p}}, \tag{1.6}$$

and conduction-band operators $a_\mathbf{p} = a_{\mathbf{p}c}$, one can rewrite the Hamiltonian (1.4) as

$$\begin{aligned} H = E_0 &+ \sum E_e(p) a^\dagger_\mathbf{p} a_\mathbf{p} + \sum E_h(p) b^\dagger_\mathbf{p} b_\mathbf{p} + \frac{1}{2}\sum V^{cccc}_{\mathbf{p}_1\mathbf{p}_2\mathbf{p}_3\mathbf{p}_4} a^\dagger_{\mathbf{p}_1} a^\dagger_{\mathbf{p}_2} a_{\mathbf{p}_3} a_{\mathbf{p}_4} \\ &+ \frac{1}{2}\sum V^{vvvv}_{-\mathbf{p}_1-\mathbf{p}_2-\mathbf{p}_3-\mathbf{p}_4} b^\dagger_{\mathbf{p}_1} b^\dagger_{\mathbf{p}_2} b_{\mathbf{p}_3} b_{\mathbf{p}_4} - \sum \left(V^{cvvc}_{\mathbf{p}_1\mathbf{p}_3\mathbf{p}_2\mathbf{p}_4} - V^{cvcv}_{\mathbf{p}_1\mathbf{p}_3\mathbf{p}_4\mathbf{p}_2}\right) a^\dagger_{\mathbf{p}_1} b^\dagger_{\mathbf{p}_2} b_{\mathbf{p}_3} a_{\mathbf{p}_4}, \end{aligned} \tag{1.7}$$

where $V^{ijlm}_{\mathbf{p}_1\mathbf{p}_2\mathbf{p}_3\mathbf{p}_4} = \langle \mathbf{p}_1 i, \mathbf{p}_2 j | V | \mathbf{p}_3 l, \mathbf{p}_4 m\rangle$. Two kinds of terms have been dropped from Eq. (1.7). First, additional terms of the type $a^\dagger_\mathbf{p} a_\mathbf{p}$, $b^\dagger_\mathbf{p} b_\mathbf{p}$, which arise when one rearranges the hole operators into normal order, are combined with the single-particle Hamiltonian H_0 to give the renormalized single-particle energies,

$$\begin{aligned} E_0 &= \sum_\mathbf{p} E^0_v(p) + \sum_{\mathbf{p},\mathbf{q}} \left(V^{vvvv}_{\mathbf{pqqp}} - V^{vvvv}_{\mathbf{pqpq}}\right), \\ E_e(p) &= E^0_c(p) + \sum_\mathbf{q} \left(V^{cvvc}_{\mathbf{pqqp}} - V^{cvcv}_{\mathbf{pqpq}}\right), \\ E_h(p) &= -E^0_v(-p) - \sum_\mathbf{q} \left(V^{vvvv}_{\mathbf{q-p-pq}} - V^{vvvv}_{\mathbf{q-pq-p}}\right), \end{aligned} \tag{1.8}$$

which take into account the contribution of the electrons in the selected valence band to the single-particle band structure. Second, terms that do not conserve the number of pair excitations, e.g., terms with V^{cccv}, have been dropped, since these correspond to polarization of the atomic orbitals and are taken into account in the dielectric function of the crystal. Umklapp processes are also ignored.

Using the effective-mass approximation for the renormalized electron and hole energies,

$$E_e(p) = E_g + \frac{\hbar^2 p^2}{2m_e}, \qquad E_h(p) = \frac{\hbar^2 p^2}{2m_h},$$

one can write an eigenvalue equation for the general pair state,

$$|ex\rangle = \sum A_{\mathbf{p},\mathbf{p}'} a^\dagger_\mathbf{p} b^\dagger_{\mathbf{p}'} |0\rangle,$$

as

$$[E_e(p) + E_h(p') - E] A_{\mathbf{p},\mathbf{p}'} - \sum_{\mathbf{q},\mathbf{q}'} \left(V^{cvvc}_{\mathbf{p}-\mathbf{q}'-\mathbf{p}'\mathbf{q}} - V^{cvcv}_{\mathbf{p}-\mathbf{q}'\mathbf{q}-\mathbf{p}'}\right) A_{\mathbf{q},\mathbf{q}'} = 0. \tag{1.9}$$

The second term in the sum corresponds to an exchange interaction, which is obtained for pair states in which the spins of the missing valence electron and the conduction electron are the same, i.e., in the case of simple spherical bands, the exchange interaction affects only the singlet exciton, leading to a splitting between the "ortho" (triplet) and "para" (singlet) states.

The Feynman diagrams for the Coulomb and exchange interaction of electrons and holes are analogous to the high-energy scattering diagrams

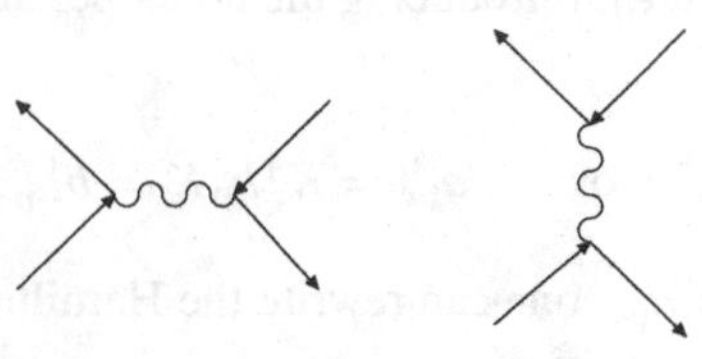

(1.10)

for electron-position Bhabha scattering, as opposed to

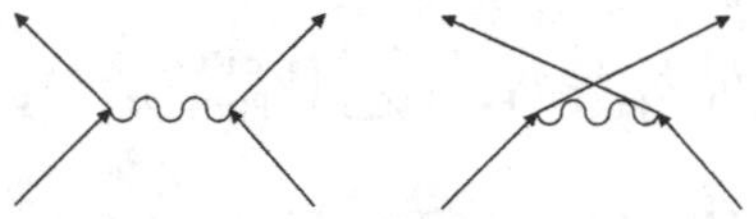

(1.11)

for electron–electron scattering. (For the semiconductor Cu_2O, the valence band is not a pure 1s state, leading to a slightly different picture of the ortho–para splitting. This is discussed in Appendix A.)[d]

For Wannier excitons, the assumption that the wave function of the exciton is slowly varying over the unit cell allows further simplification. Writing

$$V^{c\,v\,v\,c}_{\mathbf{p}-\mathbf{q}'-\mathbf{p}'\mathbf{q}} = \frac{1}{N_{\text{cell}}^2}\int d^3x\, d^3y\; e^{i(\mathbf{q}-\mathbf{p})\cdot\mathbf{x}+i(\mathbf{q}'-\mathbf{p}')\cdot\mathbf{y}}\; V(\mathbf{x}-\mathbf{y})$$
$$\times\, u^*_{\mathbf{p}c}(\mathbf{x})u^*_{-\mathbf{q}'v}(\mathbf{y})u_{-\mathbf{p}'v}(\mathbf{y})u_{\mathbf{q}c}(\mathbf{x})$$

where N_{cell} is the number of unit cells, and a similar term for the exchange energy, one can approximate

$$\frac{1}{v_0}\int_{\text{unit cell}} u^*_{\mathbf{p}\simeq 0,c}(\mathbf{x})u_{\mathbf{p}\simeq 0,c}(\mathbf{x})\,d\mathbf{x} \simeq 1,$$
$$\frac{1}{v_0}\int_{\text{unit cell}} u^*_{\mathbf{p}\simeq 0,c}(\mathbf{x})u_{\mathbf{p}\simeq 0,v}(\mathbf{x})\,d\mathbf{x} \simeq 0,$$

where v_0 is the volume of the unit cell, in which case the exchange integral can be neglected and the Coulomb interaction becomes

$$V^{c\,v\,v\,c}_{\mathbf{p}-\mathbf{q}'-\mathbf{p}'\mathbf{q}} = \frac{1}{N_{\text{cell}}^2}\int d^3x\, d^3y\; e^{i(\mathbf{q}-\mathbf{p})\cdot\mathbf{x}+i(\mathbf{q}'-\mathbf{p}')\cdot\mathbf{y}}\frac{e^2}{\epsilon|\mathbf{x}-\mathbf{y}|}. \tag{1.12}$$

Substituting Eq. (1.12) into Eq. (1.9), one finds that the pair-state eigenfunctions can be

[d] A more complete discussion of electron–hole exchange for Wannier excitons is given in Section 7.2. The "local," or "contact," electron–hole exchange interaction that is described here determines the ortho–para splitting, while the "nonlocal," or "long-range Coulomb" electron–hole exchange interaction, which can be expressed by means of a dipole–dipole interaction, gives rise to the splitting of the exciton into two branches, called "longitudinal" and "transverse" excitons. The latter interaction occurs only in the case of excitons that mix with the photon states by means of an allowed dipole interaction, i.e., a strong polariton effect. These mixed states are the subject of Chapter 7. For a general review of longitudinal and transverse excitons, see Ref. 13.

written as

$$|\text{ex}\rangle = |\mathbf{k}, n\rangle = \frac{1}{\sqrt{V}} \sum_{\mathbf{p},\mathbf{p}'} \delta_{\mathbf{k},\mathbf{p}+\mathbf{p}'}\, \varphi_n(\mathbf{q}) a^\dagger_{\mathbf{p}} b^\dagger_{\mathbf{p}'} |0\rangle = c_{\mathbf{k},n}|0\rangle, \tag{1.13}$$

where $\mathbf{q} = \beta\mathbf{p} - \alpha\mathbf{p}'$, with $\alpha = m_e/(m_e + m_h)$ and $\beta = m_h/(m_e + m_h)$, and $\varphi_n(\mathbf{q}) = \int d^3x\, \varphi_n(\mathbf{x}) e^{-i\mathbf{q}\cdot\mathbf{x}}$ is the Fourier transform of the orbital wave function, which is the solution to the standard hydrogenic equation,

$$\left(-\frac{\hbar^2}{2m_r}\nabla^2 - \frac{e^2}{\epsilon r} + E_n \right) \varphi_n(r) = 0. \tag{1.14}$$

The real-space exciton eigenfunction corresponds to $\phi(\mathbf{x}_e, \mathbf{x}_h) = \frac{1}{\sqrt{V}}\varphi_n(\mathbf{r})e^{i\mathbf{k}\cdot\mathbf{R}}$, where $\mathbf{r} = \mathbf{x}_e - \mathbf{x}_h$ and $\mathbf{R} = \beta\mathbf{x}_e + \alpha\mathbf{x}_h$, with the energy eigenvalues given in Eq. (1.1).

Having now defined the exciton operators in terms of the underlying Fermi operators, we can find the commutation relations of two exciton operators. These are found to be

$$\begin{aligned}
[c_{\mathbf{k},n}, c_{\mathbf{k}',n'}] &= 0, \\
[c^\dagger_{\mathbf{k},n}, c^\dagger_{\mathbf{k}',n'}] &= 0, \\
[c_{\mathbf{k},n}, c^\dagger_{\mathbf{k}',n'}] &= \delta_{\mathbf{k},\mathbf{k}'}\delta_{n,n'} - \frac{1}{V}\sum_{\mathbf{l}} \varphi^*_{n'}(\alpha\mathbf{k}' - \mathbf{l})\varphi_n(\alpha\mathbf{k} - \mathbf{l}) a^\dagger_{\mathbf{k}-\mathbf{l}} a_{\mathbf{k}'-\mathbf{l}} \\
&\quad - \frac{1}{V}\sum_{\mathbf{l}} \varphi^*_{n'}(\mathbf{l} - \beta\mathbf{k}')\varphi_n(\mathbf{l} - \beta\mathbf{k}) b^\dagger_{\mathbf{k}-\mathbf{l}} b_{\mathbf{k}'-\mathbf{l}} \\
&= \delta_{\mathbf{k},\mathbf{k}'}\delta_{n,n'} + \mathcal{O}\big(na^3_{\text{ex}}\big),
\end{aligned} \tag{1.15}$$

where $n = N/V$ is the exciton density. In other words, the excitons will act as bosons as long as $na^3_{\text{ex}} \ll 1$, i.e., the interparticle spacing r_s is large compared with the correlation length a_{ex}.

It is important to note that this commutation relation for "approximate bosons" is true for all composite bosons, not just for excitons. Relations (1.15) follow from the definition of the pair operator $c_{\mathbf{k},n} = \sum A_{\mathbf{p},\mathbf{p}'} a^\dagger_{\mathbf{p}} b^\dagger_{\mathbf{p}'}$ in terms of the anticommuting Fermi operators $a^\dagger_{\mathbf{p}}$ and $b^\dagger_{\mathbf{p}'}$ and the fact that $\varphi_n(r) \to 0$ as $r \to \infty$. The fact that the hole operators are time-reversed electron operators, i.e., antiparticles, does not affect the derivation of this commutation relation; if the $a^\dagger_{\mathbf{p}}$ and $b^\dagger_{\mathbf{p}'}$ operators commute instead of anticommute, relations (1.15) still hold. Therefore the constraints on the density of excitons are no greater or less than those for other composite bosons, such as helium or alkali atoms. This fact gives some sense of the limits of applicability of relations (1.15). In the case of superfluid helium, the condition $na^3 \ll 1$ is manifestly not fulfilled, since the size of the helium atoms is comparable with the interparticle spacing in the liquid. Nevertheless, the Bose properties of the system as a whole still dominate the behavior. What relations (1.15) do indicate is that the picture of independent, uncorrelated bosons will break down at high densities. This issue was addressed in detail by Keldysh and is discussed at length in the second half of Chapter 2.

1.3 Phase Transitions of Excitons

There are two limits in which we can talk about coherent states of excitons, both of which are covered in this book. The first case is that of a nonequilibrium, or "driven," Bose condensate of excitons. In this case, coherent photons from a laser directly couple to some exciton or polariton state. The coherence of the photons is preserved in the excited electronic states by means of the pairing of electrons and holes. This effect is extremely important for nonlinear optical effects, as discussed in Chapters 6–9.

The opposite limit is the case of a spontaneous appearance of coherence. This will occur if the excitons can reach a quasi-equilibrium with a well-defined temperature and chemical potential. A thermodynamic Bose–Einstein condensation (BEC) in this case will lead to a macroscopic number of excitons in the same quantum state. As noted by Nozières and Comte [14], this means that the electronic state of the crystal will not be incoherent, but will have a definite polarization and phase. Luminescence from this coherent state will also be coherent, even in the absence of a coherent pump source. We will return to a fuller treatment of these issues in Chapter 5.

If we treat the excitons as ideal bosons, then basic thermodynamics gives us the phase boundary for Bose condensation [15, 16]. The distribution number of the particles is

$$N_{\mathbf{k}} = \frac{1}{e^{\beta(\hbar^2 k^2/2m_{\mathrm{ex}}-\mu)} - 1}, \tag{1.16}$$

where μ is the chemical potential of the gas, measured from the bottom of the exciton band, and $\beta = 1/k_B T$. In the normal state, the density of particles in the thermodynamic limit is given simply as the integral of $N_{\mathbf{k}}$ over all k:

$$n = \lim_{V\to\infty} \frac{N}{V} = \frac{g}{(2\pi)^3} \int_{-\infty}^{\infty} \frac{d\mathbf{k}}{e^{\beta(\hbar^2 k^2/2m_{\mathrm{ex}}-\mu)} - 1}, \tag{1.17}$$

where g is the spin degeneracy of the excitons. The thermodynamic limit means that N and V both tend to infinity while their ratio remains finite, i.e., a typical macroscopic system. As observed by Einstein [17] (hence the name "Bose–Einstein condensation"), the integral of Eq. (1.17) cannot account for all the particles in a system at all T and μ, since μ must have an upper bound of zero for the distribution function to remain defined at all energies. When $\mu = 0$, the integral of Eq. (1.17) becomes

$$n = 2.612 g \left(\frac{m_{\mathrm{ex}} k_B T}{2\pi\hbar^2} \right)^{3/2}. \tag{1.18}$$

If we want to put more particles into the system without increasing the temperature, what can we do? Does the system prevent more particles from coming in? Einstein had the insight to propose that in thermodynamic equilibrium the extra particles must all "condense" into the ground state, which is treated separately from the rest of the particles. Equation (1.18) therefore defines the critical density for BEC of an ideal Bose gas at a given temperature; alternatively, it can be used to give the critical temperature for Bose condensation at a fixed density. This phase boundary is plotted in Fig. 1.4. BEC occurs when the density is above the density n given by Eq. (1.18). Deviations of the distribution function of the excitons from a Maxwell–Boltzmann form will occur at a density that is approximately one tenth of this density, which corresponds roughly to the density at which λ_D, the thermal de Broglie wavelength of the particles, becomes comparable with the interparticle spacing. For a typical mass of an exciton $m_{\mathrm{ex}} = 2m_0$, Eq. (1.18) implies a critical temperature of 15 K for a typically achievable exciton density of 10^{18} cm^{-3}. Herein lies some of the appeal of experiments on Bose condensation of excitons – the temperatures are easily achievable with standard cryogenic equipment, as opposed to the heroic temperatures of 50 nK or so required for experiments on alkali atoms, because of the $m_{\mathrm{ex}}^{3/2}$ factor in Eq. (1.18).

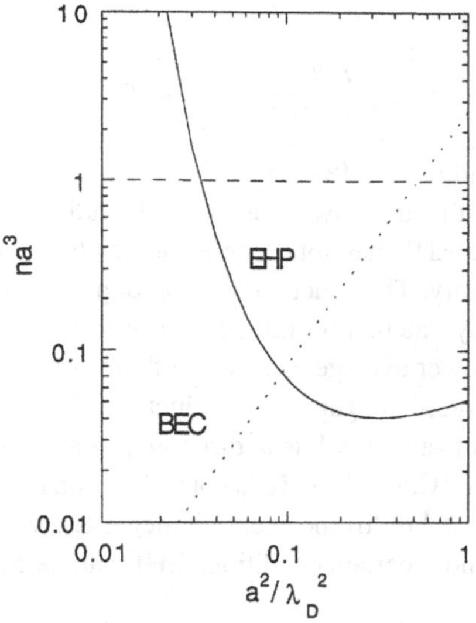

Figure 1.4. Critical densities for excitons. EHP, electron–hole plasma, occurs above the solid line. BEC occurs above the dotted line. When $na^3 > 1$, the electron gas is degenerate.

As indicated by Eqs. (1.15), if the excitons reach too high a density, each exciton will start to see the Fermi nature of the constituent particles of other excitons, instead of the overall Bose nature. Therefore one wants the pair density to remain well below the dashed line in Fig. 1.4, i.e.,

$$n \ll 1/a_{\text{ex}}^3. \tag{1.19}$$

If density approaches this level, phase-space filling will occur and the exciton wave function will depart from its hydrogenic form. The crossover from Bose to Fermi statistics is discussed in detail in the second part of Chapter 2.

The exciton gas will remain a weakly interacting system if [18]

$$nU(0) \ll \frac{\hbar^2(1/a_{\text{ex}}^2)}{2m_{\text{ex}}} \tag{1.20}$$

or

$$na_{\text{ex}}^3 \ll \frac{1}{8\pi} \tag{1.21}$$

for the interaction vertex $U(0) = 4\pi\, a_{\text{ex}}\hbar^2/m_{\text{ex}}$. In other words, since the excitons are neutral, one can assume that the interactions between excitons have a range of the order of the excitonic Bohr radius, so that this condition is essentially the same as that in inequality (1.19) above, except that the density must be somewhat lower.

The situation is complicated somewhat if attractive interactions between the excitons exist. In that case excitonic molecules analogous to the hydrogen molecule, called biexcitons, can exist, and a phase transition to another fermionic phase, called the electron–hole liquid (EHL), can occur. Biexcitons coexist with excitons in the gas phase and do not prevent BEC since they also are bosons that can condense, as discussed in Chapter 4. The occurrence of the EHL presents a greater problem. The condition for nucleation of an EHL is given by [19]

$$n = g\left(\frac{m_{\rm ex}k_BT}{2\pi\hbar^2}\right)^{3/2} e^{-\phi/k_BT}, \tag{1.22}$$

which looks similar to condition (1.18) for BEC, except for the Boltzmann factor, which depends on the strength of the attractive potential ϕ. Condition (1.22) clearly implies that if ϕ is positive, i.e., if the EHL has total energy lower than the exciton gas, BEC will not be possible at any density. The exact value of ϕ depends on a number of parameters that go into the many-body calculation [20], but in general, increasing exciton ground-state degeneracy leads to lower average energy and therefore to greater ϕ. Therefore EHL has been observed in indirect-bandgap semiconductors like Si and Ge [21] with highly degenerate conduction-band valleys, while in direct-gap semiconductors with large exciton binding energy like Cu_2O and CuCl, the EHL is not stable, and BEC of excitons or biexcitons can occur. When stress is used to lift the excitonic degeneracy in Si or Ge, ϕ for the EHL becomes lower [21], and under certain conditions EHL can be completely prevented in Ge [22].

Even in the absence of attractive interactions, a phase transition from the insulating exciton gas to the conducting electron–hole plasma (EHP) will occur when free carriers produced by thermodynamic dissociation of excitons become so numerous that they screen the electron–hole Coulomb attraction and prevent the electron–hole pairs from binding (an "ionization catastrophe" [20]). Excitons themselves will not contribute much to screening, since they are charge neutral. The criterion for this transition is $1/q < a_{\rm ex}$, where $1/q$ is the Debye–Huckel screening length for free electrons coexisting with the excitons, given by $q^2 = 4\pi e^2 n_{\rm free}/\epsilon_0 k_B T$, where $n_{\rm free}$ is the density of free carriers produced by dissociation of excitons. Statistical mechanics [20] gives this density as $n_{\rm free}^2 = n(g_e g_h/g)(m_r/m_{\rm ex})^{3/2}(1/\lambda_D^3)e^{-{\rm Ry}_{\rm ex}/k_BT}$, where ${\rm Ry}_{\rm ex} = \hbar^2/2a_{\rm ex}^2 m_r$ is the exciton binding energy and g_e and g_h are the electron- and the hole-spin degeneracies, respectively. This implies a phase boundary of

$$n = \frac{1}{4ga_{\rm ex}^2\lambda_D}\left(\frac{g}{g_e g_h}\right)\left(\frac{m_r}{m_{\rm ex}}\right)^{1/2} e^{(\lambda_D/a_{\rm ex})^2(m_{\rm ex}/m_r)/4\pi}, \tag{1.23}$$

which is shown as the solid curve in Fig. 1.4. For BEC to occur, the density must remain below this boundary.

The reverse transition, from conducting EHP to insulating exciton gas, called a Mott transition, occurs when the screening length in the plasma becomes greater than the excitonic Bohr radius. A point that has not been well appreciated in the literature is the possibility for hysteresis in the transition from insulating gas to conducting plasma and vice versa. The basic reason for this hysteresis is that neutral bound states will not contribute much to long-range screening. Therefore, if the gas is initially in the insulating phase, a transition to conducting plasma will occur only when the number of free charged particles created by thermal dissociation of bound pairs becomes high enough that the screening due to these particles causes bound states to become unstable. In general, the density of this ionization catastrophe is not the same as the Mott transition in the reverse direction that occurs when the system is initially in the plasma state. This issue is addressed in greater detail in Subsection 5.2.3.

Figure 1.4 shows that, in general, for excitons with repulsive interactions, a region of n–T phase space always exists in which excitonic BEC can occur without crossover to a fermionic EHP phase (assuming that the ground state of the particles is well defined.) In principle, then, an experiment on spontaneous excitonic BEC consists simply of picking a material with repulsive interactions between the excitons, putting it in a cryostat at low temperatures, and creating a large population of excitons with a high-power laser. Determining that a spontaneous equilibrium BEC of excitons has occurred, however, requires careful analysis of the light emitted by the ensemble of excitons.

1.4 Experimental Evidence for Boson Behavior

Is it rigorously established that excitons ever really act as bosons? In other words, although inequality (1.19) says that excitons will be bosonlike for average interparticle distances much greater than the excitonic Bohr radius, how much is "much greater"? Do Bose–Einstein statistics affect the thermodynamics of the excitons when they are only "approximately" bosons?

The answer is yes; in several semiconductors excitons follow the thermodynamic predictions of Bose–Einstein statistics. The most direct evidence of this is found in the phenomenon called "Bose narrowing." Textbook statistics [15] give the equilibrium distribution function of quantum particles as

$$N(E) = \frac{1}{e^{(E-\mu)/k_BT} \pm 1}, \tag{1.24}$$

where the + sign is used for fermions and the − sign is used for bosons. At low density (μ large and negative), the distribution for both fermions and bosons becomes a Maxwell–Boltzmann, $N(E) \propto e^{-E/k_BT}$. The effect of quantum statistics at high density on the shape of the distribution for fermions is well known, namely, the average energy of the gas at a given temperature increases to greater than the classical $\frac{3}{2}k_BT$ because of the appearance of the Fermi sea. The negative sign in the distribution for bosons gives just the opposite effect – the average energy decreases at a given temperature to well below $\frac{3}{2}k_BT$. Figure 1.5 shows the Bose–Einstein distribution function for various values of $\alpha \equiv -(\mu - E_0)/k_BT$, where E_0 is the exciton ground-state energy. As α approaches zero, i.e., as density increases,

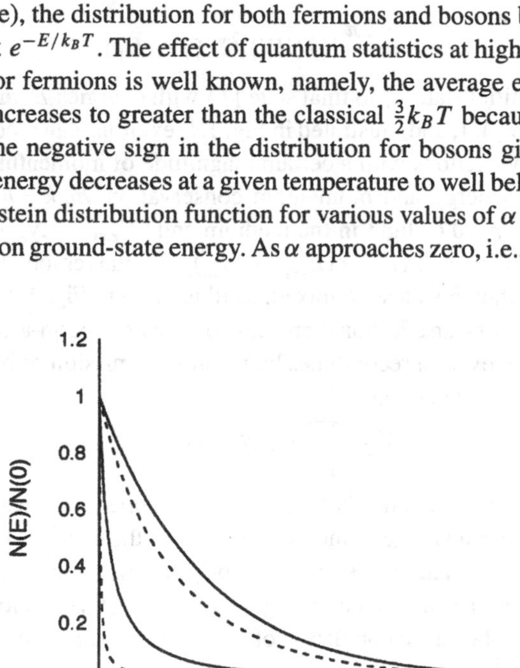

Figure 1.5. Bose occupation number $N(E) = \{\exp[E-\mu/k_BT]-1\}^{-1}$ at various values of chemical potential. The widest distribution is a Maxwell–Boltzmann distribution, and the narrower distributions are, in order, $-\mu/k_BT = 1, 0.1$, and 0.01.

the distribution narrows. The value of μ has an upper bound of zero for the Bose–Einstein distribution function. For density to continue to increase at the same temperature, a δ function must be added at $p = 0$, which is the Bose–Einstein condensate.

Surprisingly, the canonical effect of Bose narrowing is not readily seen in the momentum distribution of He; the distribution in the Bose condensed phase differs by only a few percent [23] from the the Maxwell–Boltzmann distribution. The reason is that the strong interactions in the liquid broaden the states considerably.

Excitons or biexcitons are expected to remain a weakly interacting gas, however, so that narrowing should occur if they really do act as bosons. Can we show that this effect occurs? To prove this, we must first take a detour to understand a common photon emission process of excitons.

1.4.1 Phonon-Assisted Luminescence from Excitons

To present the theory of phonon-assisted luminescence, we begin by summarizing the discussion of Shi, Verechaka, and Griffin [24], which connects the line shape of the phonon-assisted luminescence to the many-body Green's functions of the interacting Bose gas. Although Ref. 24 concentrated on the effect of a Bose condensate of excitons, the theory can also be generalized to discuss the case of collision broadening and nonequilibrium dynamics of the exciton luminescence.

In general, one writes the rate for a luminescence process, based on Fermi's Golden Rule, as

$$W_{if} = \frac{2\pi}{\hbar}|\langle f|V|i\rangle|^2\delta(E_i - E_f), \tag{1.25}$$

for a transition from initial state $|i\rangle$ to final state $|f\rangle$ with energies E_i and E_f, respectively. As discussed in Section, 1.1, and illustrated in Fig. 1.1, excitons can recombine into a single photon. However, only excitons with a certain magnitude of momentum can participate in this process, since by energy and momentum conservation, $\hbar c'k = \hbar^2k^2/2m_{ex} + (E_{gap} - \mathrm{Ry}_{ex})$, where c' is the speed of light in the medium and $(E_{gap} - \mathrm{Ry}_{ex})$ is the ground-state energy of the exciton. The region $\hbar k \sim (E_{gap} - \mathrm{Ry}_{ex})/c'$ is the region of the polariton effect in which exciton and photon states are mixed, as illustrated in Fig. 1.3.

In many semiconductors, an additional process, known as phonon-assisted luminescence, is possible in which the exciton recombines by means of emission of both a photon and an optical phonon, with the interaction

$$V = \sum_{\mathbf{k},\mathbf{q}} L_{\mathbf{k},\mathbf{q}} a_{\mathbf{k}} b^\dagger_{\mathbf{q}} c^\dagger_{\mathbf{k}-\mathbf{q}}, \tag{1.26}$$

where a, b, and c are the exciton, photon, and photon operators, respectively.[e] Because optical phonons with a broad range of momenta couple to the excitons to allow this process, all the excitons can participate in this process, not just those with the special momentum $\hbar k \sim (E_{gap} - \mathrm{Ry}_{ex})/c'$, as illustrated in Fig. 1.6. This fact will allow us to extract the distribution function of the entire population of excitons from the line shape of the phonon-assisted luminescence.

The energy difference between the initial state $|i\rangle$ and the final state $|f\rangle$ in Eq. (1.25) is then

$$E_i - E_f = E_i^{ex} + E_0 - \left(E_f^{ex} + \hbar\omega_{phonon} + \hbar\omega\right),$$

[e] In this volume we will use a, b, and c for the electron, hole, and exciton operators, respectively, when the underlying Fermi components are in view, and a for excitons and b and c for phonons and photons or other particles when they are viewed as pure bosons.

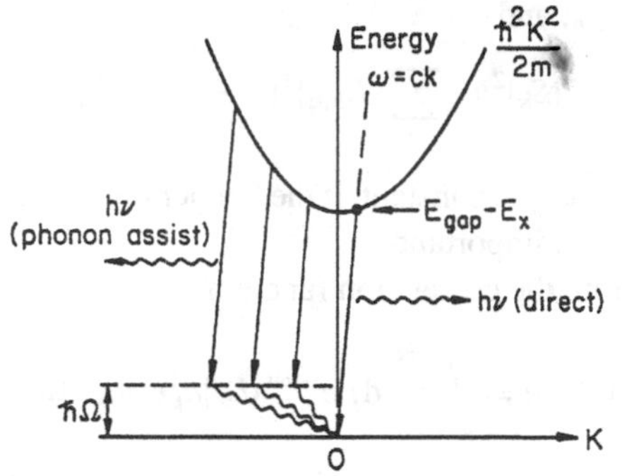

Figure 1.6. Recombination processes of Wannier excitons in a direct band-gap semiconductor (from Ref. 25).

where E_i^{ex} is the initial energy of the exciton gas with N_i excitons, E_f^{ex} is the final energy of the exciton gas with $N_i - 1$ excitons, $E_0 = (E_{gap} - \mathrm{Ry}_{ex})$ is the creation energy of single exciton in its ground state, ω_{phonon} is the phonon energy, and ω is the photon energy. If the excitons act as ideal, noninteracting particles, then $E_i^{ex} - E_f^{ex} = \hbar^2 k^2/2m_{ex}$, the kinetic energy of one exciton. In general, however, the excitons interact with each other as well as with the phonon field of the lattice, so that the energy conservation condition must take into account the possibility that the single-particle states are not the exact eigenstates of the many-body system.

Assuming that the frequency of the optical phonon is a constant independent of $\mathbf{q}$, which is generally true near $\mathbf{k} = 0$ in semiconductors, the Golden Rule implies that the intensity of phonon-assisted luminescence at a given photon frequency ω for a given initial state $|i\rangle$ is proportional to

$$\begin{aligned} I(\omega) &= \sum_f W_{if} \\ &= \sum_f \frac{2\pi}{\hbar} |\langle f|V|i\rangle|^2 \delta(E_i - E_f). \end{aligned} \tag{1.27}$$

Assuming noninteracting photon and phonon states, we can write $|i\rangle = |i_{ex}\rangle|N_{\mathbf{q}}^{phonon}\rangle$ $|N_{\mathbf{q}'}^{photon}\rangle$, where $|N_{\mathbf{q}}^{phonon}\rangle$ and $|N_{\mathbf{q}'}^{photon}\rangle$ are Fock number states and $|i_{ex}\rangle$ is an exact eigenstate of the interacting exciton gas, and therefore

$$|\langle f|V|i\rangle|^2 \cong |\langle f_{ex}|a_{\mathbf{k}}|i_{ex}\rangle|^2 \left(1 + N_{\mathbf{q}}^{phonon}\right)\left(1 + N_{\mathbf{k}-\mathbf{q}}^{photon}\right).$$

Since we will want to treat the excitons as an interacting system, we rewrite the δ-function as

$$\delta(E_i - E_f) = \frac{1}{2\pi\hbar} \int_{-\infty}^{+\infty} dt\, e^{-(i/\hbar)(\hbar\omega_{phonon} + \hbar\omega - E_0)t}\, e^{(i/\hbar)](E_i^{ex} - E_f^{ex})t} \tag{1.28}$$

Assuming that there is no stimulated emission of photons and adopting the Heisenberg representation $a_{\mathbf{k}}(t) = e^{(i/\hbar)Ht}\, a_{\mathbf{k}}\, e^{-(i/\hbar)Ht}$, Eqs. (1.27) and (1.28) imply that

$$I(\omega) = \frac{1}{\hbar^2} \sum_{\mathbf{k}} |S_{\mathbf{k}}|^2 \int_{-\infty}^{+\infty} dt\, e^{-i(\omega - \omega_0)t} \langle i_{ex}|a_{\mathbf{k}}^\dagger(t) a_{\mathbf{k}}(0)|i_{ex}\rangle, \tag{1.29}$$

where $\hbar\omega_0 = E_0 - \hbar\omega_{\text{phonon}}$, and

$$|S_{\mathbf{k}}|^2 = \sum_{\mathbf{q}} |L_{\mathbf{k},\mathbf{q}}|^2 (1 + N_{\mathbf{q}}^{\text{phonon}})$$

can generally be taken as a simple constant, if the temperature is low enough that stimulated optical phonon emission is unimportant.

As seen in this expansion, the correlation function

$$C_i(\mathbf{k}, \omega) = \int_{-\infty}^{+\infty} dt\, e^{-i\omega t} \langle i_{\text{ex}} | a_{\mathbf{k}}^{\dagger}(t) a_{\mathbf{k}}(0) | i_{\text{ex}} \rangle \tag{1.30}$$

determines the spectral line shape of the luminescence. The calculation of this correlation function in the general case of an interacting exciton gas requires the techniques of many-body theory. For a quasi-equilibrium gas, one uses the grand canonical Heisenberg representation $a_{\mathbf{k}}(t) = e^{(i/\hbar)(H-\mu N)t} a_{\mathbf{k}} e^{-(i/\hbar)(H-\mu N)t}$, and defines a thermal average over all possible initial states

$$\langle \ldots \rangle = Z^{-1} \sum_i e^{-\beta(E_i^{\text{ex}} - \mu N_i)} \langle i | \ldots | i \rangle,$$

to obtain the correlation function

$$C(\mathbf{k}, \omega) = \langle C_i(k, \omega) \rangle = Z^{-1} \sum_i e^{-\beta(E_i^{\text{ex}} - \mu N_i)} C_i(k, \omega), \tag{1.31}$$

where Z is the grand canonical partition function, $\beta = 1/k_B T$, and μ is the chemical potential. Then one can show that

$$C(\mathbf{k}, \omega) = N(\omega) A(\mathbf{k}, \omega), \tag{1.32}$$

where $N(\omega)$ is the Bose distribution function

$$N(\omega) = \frac{1}{e^{\beta\hbar\omega} - 1} \tag{1.33}$$

and $A(\mathbf{k}, \omega)$ is the spectral function [18, 26] that gives the effect of the many-body interactions, namely, the distribution of $\mathbf{k}$-states for a single particle in an eigenstate with energy ω. $A(\mathbf{k}, \omega)$ is determined from the quasi-equilibrium Green's functions of the system according to $A(\mathbf{k}, \omega) = -2\,\text{Im}\, G(\mathbf{k}, \omega)$, where

$$G(\mathbf{k}, \omega) = -i \int_0^{\infty} dt\, e^{-i\omega t} \langle [a_{\mathbf{k}}(0), a_{\mathbf{k}}^{\dagger}(t)] \rangle, \tag{1.34}$$

and

$$N_{\mathbf{k}} = \langle c_{\mathbf{k}}^{\dagger}(0) c_{\mathbf{k}}(0) \rangle = \int_{-\infty}^{+\infty} \frac{d\omega}{2\pi} N(\omega) A(\mathbf{k}, \omega). \tag{1.35}$$

The proper form of the quasi-equilibrium Green's functions for the interacting exciton gas will be the subject of much of this book. In the case of weakly interacting, nondegenerate particles, however, the above formalism can be simplified in a straightforward way which applies not only to a gas in equilibrium, but also to a gas far from equilibrium. We start with the correlation function (1.30) and do not perform a thermal average, and assume that the exciton eigenstates are well approximated by Fock number states $|n_1 \cdots n_{\mathbf{k}} \cdots\rangle$, and rewrite (1.30) as

$$C_i(k,\omega) = \int_{-\infty}^{+\infty} dt\, e^{-i\omega t} \langle i_{\rm ex}| e^{(i/\hbar)H^{\rm ex}t} a_{\mathbf{k}}^\dagger \sum_{f_{\rm ex}} |f_{\rm ex}\rangle \langle f_{\rm ex}| e^{-(i/\hbar)H^{\rm ex}t} a_{\mathbf{k}} |i_{\rm ex}\rangle, \qquad (1.36)$$

where $H^{\rm ex} = \sum E_{\mathbf{k}} a_{\mathbf{k}}^\dagger a_{\mathbf{k}} + H_{\rm int}^{\rm ex}$ is the exciton Hamiltonian which includes an interaction term $H_{\rm int}^{\rm ex}$. Then we have

$$C_i(\mathbf{k},\omega) = \int_{-\infty}^{+\infty} dt\ e^{-i\omega t}\ n_{\mathbf{k}} \langle \ldots n_{\mathbf{k}} \ldots | e^{(i/\hbar)H^{\rm ex}t} | \ldots n_{\mathbf{k}} \ldots \rangle$$
$$\times \langle \ldots n_{\mathbf{k}} - 1 \ldots | e^{-(i/\hbar)H^{\rm ex}t} | \ldots n_{\mathbf{k}} - 1 \ldots \rangle.$$

The difference between the energy $H^{\rm ex}$ for the state with $n_{\mathbf{k}}$ particles and the state with $n_{\mathbf{k}} - 1$ particles is just the single-particle energy of state $\mathbf{k}$. Then we can write

$$\begin{aligned} C_i(\mathbf{k},\omega) &= \int_{-\infty}^{+\infty} dt e^{-i\omega t} n_{\mathbf{k}} e^{(i/\hbar)E_{\mathbf{k}}t} \\ &= n_{\mathbf{k}} A_i(\mathbf{k},\omega) \end{aligned} \qquad (1.37)$$

where $A_i(\mathbf{k},\omega) = -2\,{\rm Im}\, G_i(\mathbf{k},\omega)$ and

$$G_i(\mathbf{k},\omega) = \frac{1}{\omega - E_{\mathbf{k}}/\hbar + i\epsilon}. \qquad (1.38)$$

When the particles are interacting, the energy $E_{\mathbf{k}}$ acquires a self-energy correction. In second-order perturbation theory, this is approximated by [29]

$$E^{\rm ex}(\mathbf{k}) = E_{\mathbf{k}} + \langle \mathbf{k} | H_{\rm int}^{\rm ex} | \mathbf{k} \rangle + \sum_{\mathbf{k}'} \frac{|\langle \mathbf{k}' | H_{\rm int}^{\rm ex} | \mathbf{k} \rangle|^2}{E_{\mathbf{k}} - E_{\mathbf{k}'} - i\epsilon} \qquad (1.39)$$

where the latter sum equals

$$\Sigma_{\rm int}(E) = P \int \frac{B(E')}{E - E'} dE' + i\pi B(E) \equiv \Delta + i\Gamma. \qquad (1.40)$$

The first term of Eq. (1.40) gives the real shift in the energy of the exciton states, while the second term gives the line broadening due to the lifetime of the states. When Γ is nonzero, the spectral function takes the Lorentzian form

$$A_i(\mathbf{k},\omega) = \frac{2\hbar\Gamma}{(\hbar\omega - E_{\mathbf{k}} - \Delta)^2 + \Gamma^2}, \qquad (1.41)$$

and the luminescence intensity is given by the convolution of the spectral function with $n_{\mathbf{k}}$,

$$\begin{aligned} I(\omega) &= \frac{1}{\hbar^2} \sum_{\mathbf{k}} |S_{\mathbf{k}}|^2 n_{\mathbf{k}} A(\mathbf{k}, \omega - \omega_0) \\ &= |S|^2 \int d\omega' \rho(\omega') n(\omega') A(\omega', \omega - \omega_0) \end{aligned} \qquad (1.42)$$

where we have made use of the assumption that $|S_{\mathbf{k}}|^2$ is a constant, and $\rho(\omega)$ is the single-particle density of states, proportional to $\sqrt{\omega}$ in three dimensions when $\omega > 0$ and zero otherwise.

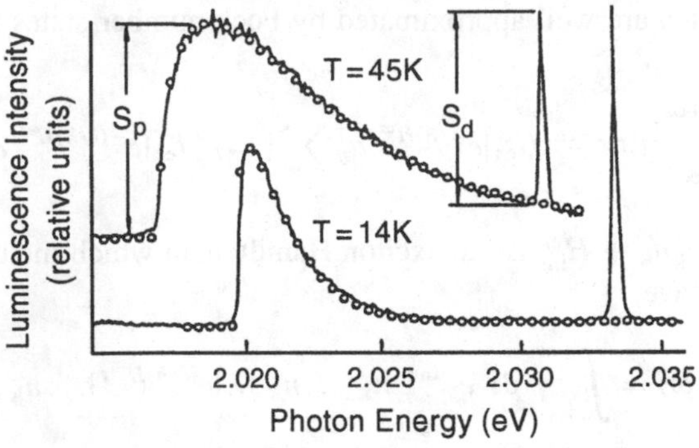

Figure 1.7. Phonon-assisted luminescence from orthoexcitons in the semiconductor Cu_2O at low temperature and low exciton density. The spectral line labeled "S_p" is due to the phonon-assisted process shown in Fig. 1.6; the spectral line labeled "S_d" is due to the direct recombination process. The phonon-assisted line shown is fit (in both cases) to a Maxwell–Boltzmann distribution at the bath temperature (from Ref. 27).

As discussed in Section 2.1.5, this form is incorrect for bosons in the degenerate limit, but when the exciton gas is nondegenerate, i.e., at low exciton density and high temperature, it gives good agreement with experimentally observed luminescence spectra. In the noninteracting limit, $\Sigma_{\rm int} \to 0$, and $A_i(\mathbf{k}, \omega) = 2\pi\delta(\omega - E_{\mathbf{k}\hbar})$, in which case Eq. (1.42) takes the simple form

$$I(\omega) \propto n(\omega - \omega_0)\rho(\omega - \omega_0). \tag{1.43}$$

In other words, the phonon-assisted luminescence line shape gives the kinetic-energy distribution of the excitons directly. This is seen, for example, in Fig. 1.7, in which the phonon-assisted lineshape is well fit by the Maxwell–Boltzmann distribution, $I(\omega) \propto \sqrt{\omega - \omega_0}\, e^{-\beta\hbar(\omega-\omega_0)}$. These data show that the excitons are indeed a gas with the standard free-particle density of states. This kind of line-shape fit to phonon-assisted luminescence is common in semiconductors, including Si, Ge, and CdS.

Since we did not assume a thermalized distribution in deducing Eq. (1.42), it follows that in the weakly interacting limit, the phonon-assisted luminescence spectrum at a given moment in time gives the instantaneous kinetic-energy distribution of the excitons. Figure 1.8 shows luminescence taken from Cu_2O on very short time scales. In this case, at early times the gas does not fit a thermalized distribution because the time resolution of the observations is fast compared with the scattering times of the particles. Only at late times (>50 ps) does the gas fit a Maxwell–Boltzmann distribution, after the excitons have emitted and absorbed several phonons.

At higher temperatures and higher exciton densities, the interactions of the excitons must be taken into account. Figure 1.9 shows phonon-assisted exciton luminescence at high temperatures, when the exciton–phonon interaction gives a measureable contribution to the exciton self energy. As seen in this figure, as well as in Fig. 1.7, the term Δ gives a red shift of the band gap as the temperature increases, since the energy denominator of Eq. (1.38) is negative for the exciton ground state, while the term Γ in the self-energy leads to Lorentzian line broadening via the spectral function. (This kind of broadening is also called "homogeneous" broadening, as opposed to "inhomogeneous" broadening, which is due to random spatial variations of the band-gap energy. The latter usually has a Gaussian shape.) The functions for Δ and Γ due to the exciton–phonon interaction are well understood (see,

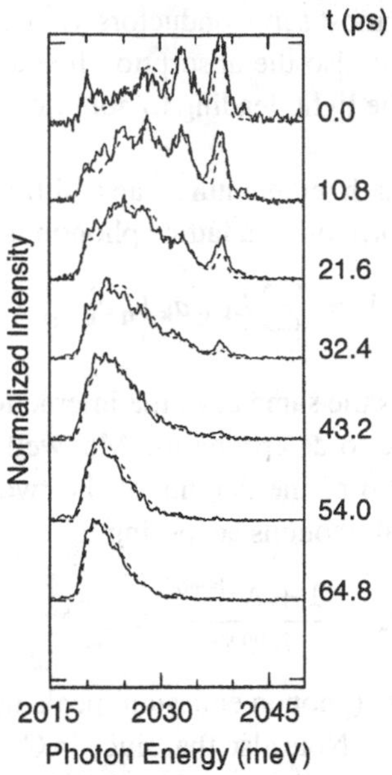

Figure 1.8. Phonon-assisted luminescence from excitons in the semiconductor Cu_2O on time scales that are short when compared with the exciton–phonon and the exciton–exciton scattering times. The dashed curves are a fit to the nonequilibrium theory for the exciton–phonon scattering discussed in Subsection 8.2.3. Time $t = 0$ corresponds to the time of maximum intensity of the short (5 ps) laser pulse which creates the excitons (from Ref. 28).

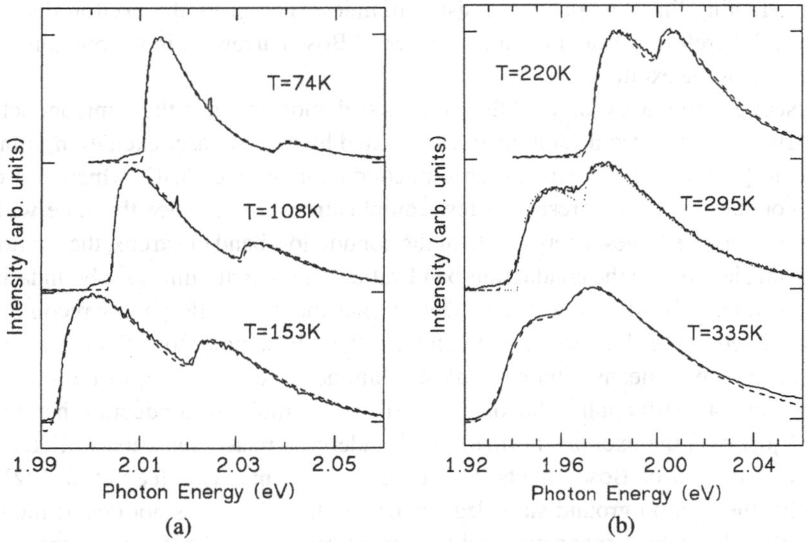

Figure 1.9. Phonon-assisted luminescence from excitons in the semiconductor Cu_2O at low exciton density for several temperatures up to room temperature. The dashed lines are fits to a Maxwell–Boltzmann distribution of the excitons, with the relative height of the Stokes and the Anti-Stokes components given by the phonon population at the same temperature as the excitons, as discussed in the text, with Lorentzian broadening of the luminescence given by the phonon–exciton interaction. The dotted line on the curve for $T = 295$ K is the same theory if this broadening mechanism is ignored (from Ref. 30).

e.g., Ref. 30) and are similar for most semiconductors. Since the homogeneous broadening affects not only the emission but also the absorption line shape, the absorption spectrum can depend on the intensity of the light, leading to nonlinear effects such as bistability and hysteresis [31].

Figure 1.9 also shows that at high temperatures, an additional process of phonon-assisted recombination by means of absorption of a lattice phonon is possible, with the interaction

$$V = \sum_{\mathbf{k},\mathbf{q}} L_{\mathbf{k},\mathbf{q}} a_{\mathbf{k}} b_{\mathbf{q}} c^{\dagger}_{\mathbf{k}-\mathbf{q}}, \tag{1.44}$$

where the matrix element $L_{\mathbf{k},\mathbf{q}}$ is the same as in the interaction (1.26). The data of Fig. 1.9 are fit with a single temperature to determine the Maxwell–Boltzmann line shape of the exciton lines, as well as the ratio of the heights of the two lines, which depends on the occupation number of the optical phonons according to

$$I_S/I_A = \frac{1 + N^{\text{phonon}}}{N^{\text{phonon}}} = e^{E_{\text{phonon}}/k_B T}, \tag{1.45}$$

where I_S is height of the "Stokes" (phonon emission) peak and I_A is the height of the "Anti-Stokes" (phonon absorption) peak. Note also that since in Cu_2O, the exciton binding energy is 150 meV, the excitons exist at room temperature and above. Excitons are not necessarily only a low-temperature phenomenon.

1.4.2 Bose Narrowing

In the case of a degenerate Bose gas, the form of the Green's function becomes complicated; this will be discussed in Chapter 2. One can see from Eq. (1.42), however, that in the weakly interacting limit the phonon-assisted luminescence gives the exciton distribution function, and therefore we can look for evidence of Bose narrowing in the phonon-assisted luminescence of the excitons.

The first experimental example of this effect was demonstrated in the semiconductor Ge. Figure 1.10 shows data from excitons in Ge, created by intense laser excitation, immersed in liquid He [22]. Ge is an indirect-gap semiconductor (unlike Cu_2O, which is a direct-gap semiconductor), which presents a few complications. First, since the state with zero momentum is not the lowest energy state of the conduction-band electrons, the excitons are formed from electrons in the conduction-band valleys at the Brillouin Zone boundaries and holes in the valence band at zone center. This implies that the single-photon recombination process depicted in Fig. 1.1 cannot occur and that there is no polariton effect. The excitons can still recombine by means of phonon-assisted luminescence, however, so the entire theory of Subsection 1.4.1 still applies. Second, since there are multiple conduction-band valleys instead of just one, the excitons normally will undergo a phase transition to EHL, which prevents observation of Bose effects. For these experiments, Timofeev et al. [22] used stress to lift the exciton ground-state degeneracy so that EHL does not occur; the crystal was also placed in a high magnetic field to prevent formation of biexcitons (these effects are discussed in greater detail in Subsection 5.3.2). Under these conditions, the distribution function of the excitons can be observed directly in the phonon-assisted recombination luminescence of the excitons.

At low densities the luminescence has the shape of a Maxwell–Boltzmann distribution, $\sqrt{\tilde{\omega}}e^{-\tilde{\omega}/k_B T}$, at the lattice temperature, as seen in curve 1 of Fig. 1.10 (b); as exciton density is increased when the laser power is turned up, this distribution narrows and fits the ideal-gas

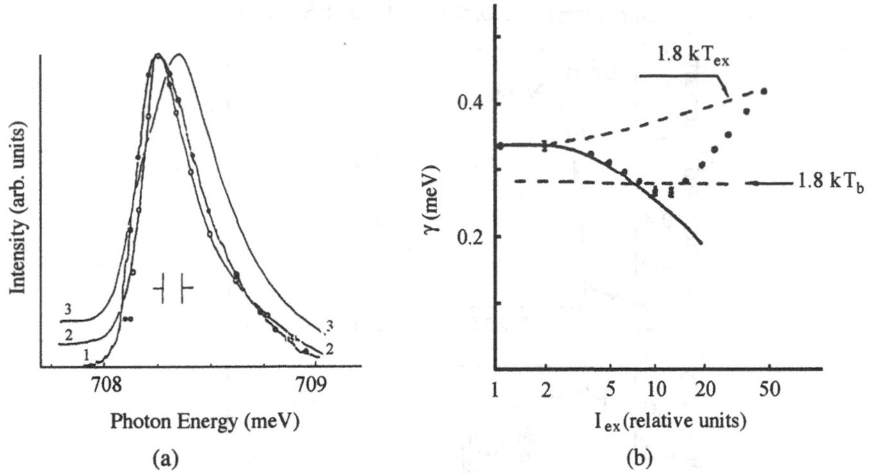

Figure 1.10. (a) Phonon-assisted luminescence from excitons in the semiconductor Ge under stress and magnetic-field conditions that suppress the formation of biexcitons and EHL. Curves 1, 2, and 3 are labeled in order of increasing exciton density. Curve 1 is fit to a Maxwell–Boltzmann distribution, (filled circles) and curve 2 is fit to a Bose–Einstein distribution (open circles). (b) The full width at half maximum γ of the phonon-assisted luminescence shown in (a) as a function of the relative laser excitation density I_{ex}. The distribution initially narrows because of Bose–Einstein effects, then broadens because of heating and interaction effects (from Ref. 22).

Bose–Einstein distribution,

$$I(\omega) \propto \frac{\sqrt{\omega - \omega_0}}{e^{\beta\hbar(\omega-\omega_0-\mu)} - 1} \tag{1.46}$$

shown as the fit in curve 2. As shown in Fig. 1.10 (b), the full width at half maximum (FWHM) drops to below that expected for a classical gas at the bath temperature, $1.8\,k_BT$, even though the exciton gas has an elevated temperature at all times. At even higher excitations, the distribution does not continue to narrow, presumably because exciton–exciton collisions cause extra kinetic energy of the gas as well as spectral broadening.

A similar effect is seen in the case of excitons in Cu_2O. Excitons in Cu_2O are split by electron–hole exchange into a singlet "paraexciton" and a triplet "orthoexciton" at 12-meV higher energy (see Appendix A). When an intense laser pulse excites the crystal well above the bandgap, both orthoexcitons and paraexcitons are created; the number of orthoexcitons versus the number of paraexcitons at any later point in time will depend on the rates of interconversion because of a number of temperature- and density-dependent processes. (For a review of these processes, see Ref. 25.) Fig. 1.11 shows a comparison of phonon-assisted luminescence from paraexcitons and orthoexcitons at the same point in time and in the same region of the crystal after creation by an intense laser pulse [32]. Both are fit to the Bose–Einstein distribution (1.46) for free particles in three dimensions. As seen in this figure, the FWHM of the paraexcitons is less than half of that for the orthoexcitons. Assuming that the orthoexcitons and paraexcitons scatter quickly with each other by means of elastic binary collisions at these densities, the two spectra must have the same temperature. The paraexciton spectrum is then narrower because it has higher quantum degeneracy.

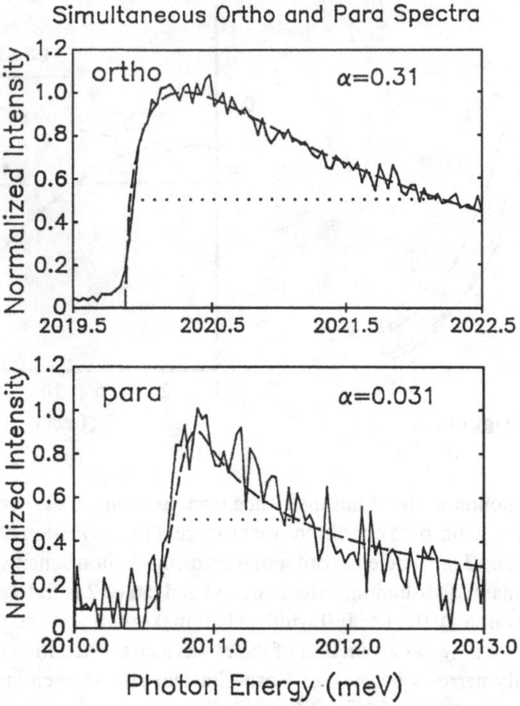

Figure 1.11. Comparison of orthoexciton and paraexciton luminescence from Cu_2O at the same time and place during intense laser excitation. The Bose–Einstein fits to two curves have the same temperature, but different chemical potential (and FWHM), reflecting the different levels of occupation of the two species (from Ref. 32).

The three spin states of the orthoexciton triplet can be split by stress. Figure 1.12 shows orthoexciton phonon-assisted luminescence at high density. In this case, there are three interpenetrating gas species, each with a different chemical potential but the same temperature [33]. The three luminescence spectra for the different species have different widths, consistent with the chemical potential of each component, according to expression (1.46). These spectra can be separated by the detection of different polarizations of the luminescence light.

Unlike the spectra shown for Ge above, the spectra in Figs. 1.11 and 1.12 (as well as the spectra in the original observations of a Bose–Einstein distribution of excitons in Cu_2O [34]) are always broader than a Maxwell–Boltzmann distribution at lattice temperature, which remains near the He bath temperature. This is most likely because a density-dependent heating effect exists in Cu_2O, called Auger recombination [35–37], in which two excitons collide and one recombines, giving its ground-state energy (approximately the bandgap energy) to ionizing the other exciton. With even a very low cross section, this process can still heat the gas considerably, since each event adds an electron–hole pair at 2 eV, or 20,000 K, to the gas. Another possible source of heating at high exciton density is collision-induced conversion of orthoexcitons into paraexcitons, by which two orthoexcitons flip spin in a collision, adding twice the electron–hole exchange energy of the exciton (12 meV) to the kinetic energy of the gas.

Similar Bose narrowing behavior has also been seen for biexcitons in the semiconductor CuCl [38, 39]. Those experiments will be discussed in Chapter 4.

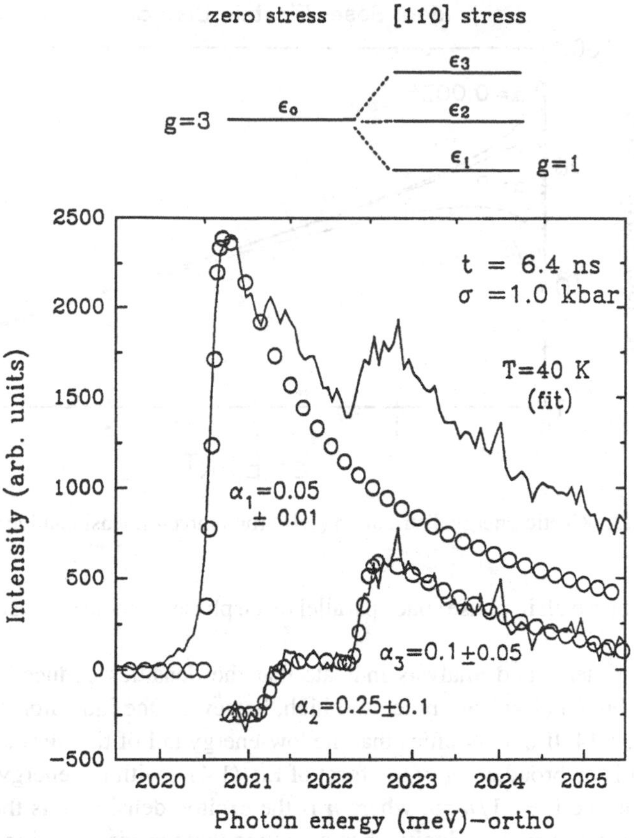

Figure 1.12. Comparison of luminescence from three different spin states of the orthoexciton in Cu_2O, split by applied stress. (Two of the components can be selected by analysis of the luminescence through a polarizer.) The Bose–Einstein fits to the three components have the same temperature, but different chemical potentials (and different FWHM), reflecting the different levels of occupation of the spin states (from Ref. 33).

1.4.3 Ideal-Gas Spectral Fits for Cu_2O

The success of the Maxwell–Boltzmann fit to the phonon-assisted luminescence line shape at low exciton density, e.g., as seen in Fig. 1.7, based on the noninteracting limit of expression (1.43), makes it tempting to fit the phonon-assisted luminescence at higher densities by using the same ideal-gas theory. This approach has been the basis of a large amount of analysis of the experiments on excitons in Cu_2O [27, 32, 33, 36, 37]. Figure 1.13 shows the theoretical ideal Bose–Einstein kinetic-energy distribution (1.46), and Fig. 1.14 shows a fit of this ideal-gas energy distribution to the phonon-assisted luminescence line from orthoexcitons in Cu_2O at high exciton density. As seen in this figure, the fit is quite good, and in fact the orthoexciton phonon-assisted luminescence line from Cu_2O can almost always be well fit by the ideal-gas distribution given in expression (1.46) over a wide range of exciton densities and temperatures. These fits were used in deducing the "Bose saturation" effect [27], in which the density and the temperature of the orthoexciton gas in Cu_2O are found to vary in exactly such a way as to keep the orthoexciton gas on an

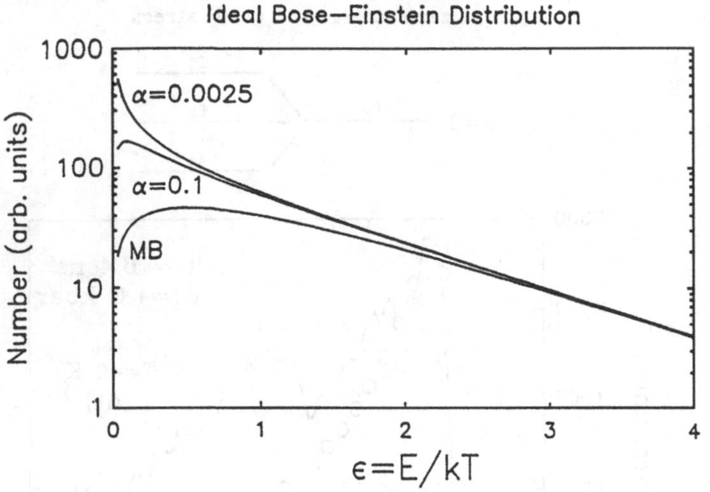

Figure 1.13. Kinetic-energy distribution (1.46) for a three-dimensional ideal Bose gas.

isentrope, i.e., on a path in phase space parallel to the phase boundary for Bose condensation (see Fig. 1.15).

Recent experiments and analysis indicate that the densities deduced from fits of the ideal-gas distribution (1.46) are much too high, however. One indicator, for example, can be seen in Fig. 1.14. If one assumes that the low-energy tail of the luminescence is due to Lorentzian collision broadening of the form of Eq. (1.41), with the energy width $\hbar\Gamma$ given by the classical rate $\Gamma = 1/n\sigma v$, where n is the exciton density, σ is the collision cross section, and v is the average velocity, then assuming that the cross section σ is of the order of the Bohr radius squared gives a collision broadening of 1.3 meV at a density of 10^{19} cm^{-3} [40], which is the density deduced from the fits of the ideal-gas distribution in many experiments. This amount of broadening is approximately ten times the maximum amount seen experimentally. To resolve this, one must either assume that the scattering length of the excitons is much less than the Bohr radius or that the ideal-gas fit gives an incorrect density. Recent measurements of the absolute number of photons emitted by the orthoexcitons in Cu_2O [41] indicate that the latter is more likely and that the exciton density does not exceed 10^{18} cm^{-3} in any of the experiments. To obtain a self-consistent fit of the experimental spectra, it is likely that nonequilibrium kinetics of the type discussed in Chapter 8 will have to be taken into account.

What is learned from the phonon-assisted luminescence from Cu_2O, then? First, we can be fairly certain that the differences in the widths of the luminescence lines from the various spin components reflect the effects of Bose statistics because of the different quantum degeneracies of the components. Classically, there is little to cause a difference in the kinetic-energy distributions of the spin components, even when the gas is not in equilibrium. The exciton–phonon-scattering rates are known for both the paraexcitons [42] and the orthoexcitons [28], and if there is any difference, it is that the orthoexcitons should cool faster than the paraexcitons by means of phonon emission. Second, the interactions of the excitons can be determined by means of the homogeneous broadening of the luminescence, seen on the low-energy side of the spectra. As discussed in Chapter 2, the homogeneous broadening of the luminescence in Cu_2O shows a density-dependent nonlinear behavior strongly indicative of a renormalized excitation spectrum like that of the Bogoliubov model

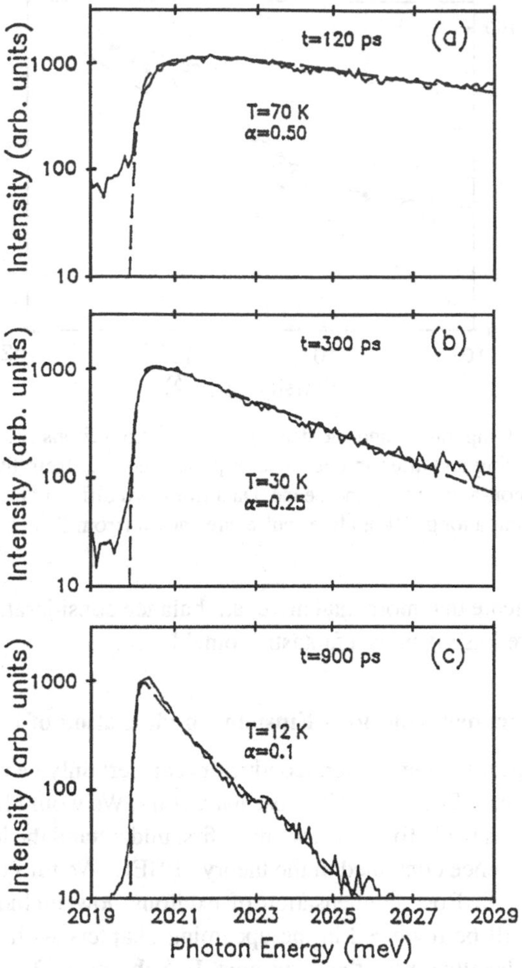

Figure 1.14. Fit of the Bose ideal-gas distribution (1.46) to the phonon-assisted luminescence of orthoexcitons in Cu_2O at various times. The parameter α is the unitless chemical potential $|\mu/k_BT|$ (from Ref. 36).

for the weakly interacting Bose gas. Third, as discussed in Chapter 2, under certain excitation conditions, sharp peaks are observed in the Cu_2O spectra which indicate high quantum degeneracy, albeit in a highly nonequilibrium situation.

The Bose saturation effect seen in Fig. 1.15 primarily corresponds to a shape invariance of the phonon-assisted luminescence line. In other words, as seen in Fig. 1.16, independent of any fit, the orthoexciton phonon-assisted luminescence keeps the same overall shape even while its energy width varies by over a factor of 5. This shape is nominally fit by an ideal Bose distribution with $\alpha = |\beta\mu| = 0.1$. As one can see in Fig. 1.5, a momentum distribution that fits $\alpha = 0.01$–0.1 corresponds to a substantial devation from classical Gaussian statistics, even if the ideal-gas assumption does not apply.

Several authors [37, 43] have suggested that the Bose saturation behavior seen in Fig. 1.15 is simply due to a balance of heating due to the Auger effect and cooling due to phonon emission. The fact that it has exactly the same power law as the phase boundary for Bose condensation, $n = CT^{3/2}$, and the fact that the line shape remains invariant, as shown in

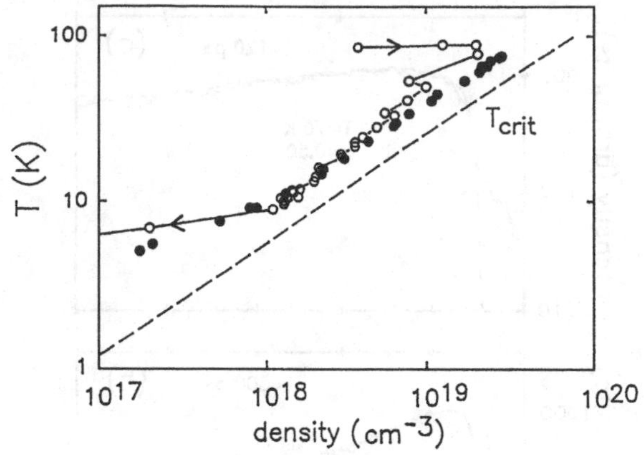

Figure 1.15. The temperature and the density of the orthoexcitons in Cu_2O at several different times during and after an intense laser pulse, deduced from fits of expression (1.46) to the phonon-assisted luminescence. Data from two different experiments, with a short (100-ps) and a long (10-ns) laser pulse, are shown (from Ref. 36).

Fig. 1.16, seem to indicate that more than mere rate-balance considerations come in to play, however, unless nature has given us a fantastic coincidence.

1.5 Experiments on Bose–Einstein Condensation of Excitons

We have seen that excitons in certain semiconductors can certainly be made to demonstrate canonical Bose statistical effects, namely, Bose narrowing. We would like to know whether they can undergo BEC as well. To do this, we must first understand the ideas of spontaneous long-range phase coherence contained in the theory of BEC. We turn to this in Chapter 2.

Experimental work on Bose condensation of excitons now includes several types of experiments. These will be reviewed in the upcoming chapters within the context of the theory that has been developed for these systems. In Subsection 2.1.5 we review the evidence for Bose condensation of excitons in the semiconductor Cu_2O based on spectral analysis. In Section 3.4 we review experiments aimed at detecting superfluidity of excitons in Cu_2O.

As we will see, biexcitons (excitonic molecules) also can undergo Bose condensation, and in Section 4.4 we discuss experiments on Bose condensation of biexcitons in the semiconductor CuCl.

In the above cases, the excitons or biexcitons are assumed to obtain phase coherence spontaneously through a BEC phase transition driven by quasi-equilibrium thermodynamics. In Chapters 6, 7, and 9 we discuss a different kind of coherent exciton system, which can be called a "driven Bose condensate," in which the coherence of laser light is imprinted directly into the excitonic states. Some of these ideas have already been incorporated in devices, such as the optical Stark effect, used in optical switching schemes.[f]

Finally, in Chapter 10 we review some recent work on different kinds of excitonic complexes. In Subsection 10.2.1 we review recent experiments on dipole excitons, also known as

[f] Some people insist that the term "Bose condensation" implies spontaneous coherence and therefore object to the term "driven Bose condensate," preferring instead the term "driven coherent state." There is a fundamental unity to the underlying physics, however, and therefore we use the former term throughout this book.

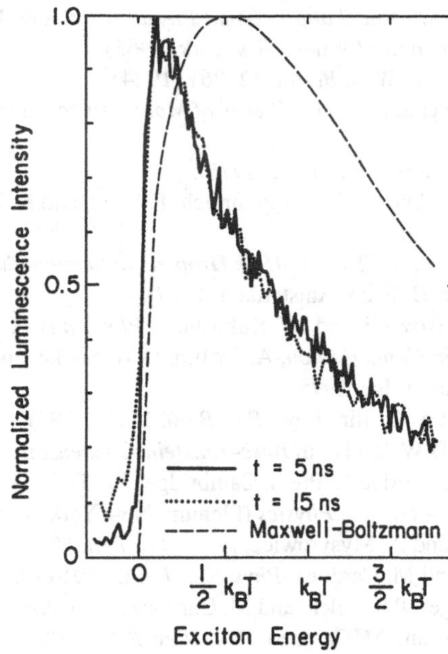

Figure 1.16. The phonon-assisted luminescence from orthoexcitons in Cu_2O at two different times following an intense laser pulse, plotted as a function of k_BT. The temperature at $t = 5$ ns is approximately 50 K, and the temperature at $t = 15$ ns is approximately 10 K (from Ref. 32).

barbell excitons, with electrons and holes in different, parallel two-dimensional planes, and in Section10.3 we review experiments on the excitonic insulator or "permanent excitons."

In the years since Bose condensation of exitons was first proposed [44, 45], it has often been a byword that Bose condensation of excitons is impossible, for some reason. Present theory and experiment show that to be false. There is still much work to be done, however, and many questions remain unanswered.

References

[1] C. Kittel, *Introduction to Solid State Physics*, 7th ed. (Wiley, New York, 1996).
[2] B.K. Ridley, *Quantum Processes in Semiconductors* (Oxford U. Press, Oxford, 1988).
[3] J. Frenkel, *Phys. Rev.* **37**, 17; **37**, 1276 (1931).
[4] G.H. Wannier, *Phys. Rev.* **52**, 191 (1937).
[5] N.F. Mott, *Trans. Faraday Soc.* **34**, 500 (1938).
[6] G.M. Kavoulakis, Y.-C. Chang, and G. Baym, *Phys. Rev. B* **55**, 7593 (1997).
[7] D.P. Trauernicht, J.P. Wolfe, and A. Mysyrowicz, *Phys. Rev. Lett.* **52**, 855 (1984).
[8] R.J. Elliott, *Phys. Rev.* **108**, 1384 (1957).
[9] P.Y. Yu and M. Cardona, *Fundamentals of Semiconductors* (Springer, Berlin, 1996).
[10] See, e.g., A. Cantarero, C. Trallero-Giner, and M. Cardona, *Phys. Rev.* B **39**, 8388 (1989); C. Trallero-Giner, A. Cantarero, and M. Cardona, *Phys. Rev. B* **40**, 4030 (1989).
[11] Ref. 9, p. 351.
[12] E. Hanamura and H. Haug, *Phys. Rep.* **33**, 209 (1977).
[13] M.M. Denisov and V.P. Makarov, *Phys. Status Solidi B* **56**, 9 (1973).
[14] P, Noziéres and C. Comte, *J. Phys. (Paris)* **43**, 1083 (1982).

[15] F. Reif, *Fundamentals of Statistical and Thermal Physics* (McGraw-Hill, New York, 1965).
[16] K. Huang, *Statistical Mechanics* (Wiley, New York, 1963).
[17] A. Einstein, *Absitz. Pr. Akad. Wiss. Berlin*, **22**, 261 (1924).
[18] A.L. Fetter and J.D. Walecka, *Quantum Theory of Many-Particle Systems* (McGraw-Hill, New York, 1971).
[19] G.A. Thomas et al., *Phys. Rev.* B **13**, 1692 (1976).
[20] T.M. Rice, in *Solid State Physics* **32**, H. Ehrenreich, F. Seitz, and D. Turnbull, eds. (Academic, New York, 1977).
[21] J.P. Wolfe and C.D. Jeffries, in *Electron–Hole Droplets in Semiconductors*, C.D. Jeffries and L.V. Keldysh, eds. (North-Holland, Amsterdam, 1987).
[22] V.B. Timofeev, V.D. Kulakovskii, and I.V. Kukushkin, *Physica* B+C **117/118**, 327 (1983).
[23] P. Sokol, in *Bose–Einstein Condensation*, A. Griffin, D.W. Snoke, and S. Stringari, eds. (Cambridge U. Press, Cambridge, 1995).
[24] H. Shi, G. Verechaka and A. Griffin, *Phys. Rev. B* **50**, 1119 (1994).
[25] J.P. Wolfe, J.L. Lin, and D.W. Snoke, in *Bose–Einstein Condensation*, A. Griffin, D.W. Snoke, and S. Stringari, eds. (Cambridge U. Press, Cambridge, 1995).
[26] See, e.g., G. Mahan, *Many-Particle Physics* (Plenum, New York, 1981).
[27] D.W. Snoke, J.P. Wolfe, and A. Mysyrowicz, *Phys. Rev. Lett.* **59**, 827 (1987).
[28] D.W. Snoke, D. Braun, and M. Cardona, *Phys. Rev. B* **44**, 2991 (1991).
[29] See, e.g., P. Lautenschlager, P.B. Allen, and M. Cardona, *Phys. Rev. B* **33**, 5501 (1986).
[30] D.W. Snoke, A.J. Shields, and M. Cardona, *Phys. Rev. B* **45**, 11693 (1992).
[31] Y. Toyozawa, *Solid State Commun.* **28**, 533 (1978); **32**, 13 (1979).
[32] D.W. Snoke, J.P. Wolfe, and A. Mysyrowicz, *Phys. Rev. B* **41**, 11171 (1990).
[33] J.L. Lin and J.P. Wolfe, *Phys. Rev. Lett.* **71**, 1223 (1993).
[34] D. Hulin, A. Mysyrowicz, and C. Benoit a la Guillaume, *Phys. Rev. Lett.* **45**, 1970 (1980).
[35] R.C. Casella, *J. Phys. Chem. Solids* **24**, 19 (1963).
[36] D.W. Snoke and J.P. Wolfe, *Phys. Rev.* B **42**, 7876 (1990).
[37] G.M. Kavoulakis, G. Baym, and J.P. Wolfe, *Phys. Rev. B* **53**, 1 (1996).
[38] L.L. Chase, N. Peyghambarian, G. Grinberg, and A. Mysyrowicz, *Phys. Rev. Lett.* **42**, 1231 (1979).
[39] N. Peyghambarian, L.L. Chase, and A. Mysyrowicz, *Phys. Rev. B* **27**, 2325 (1983).
[40] D.W. Snoke, *Comments Cond. Mater. Phys.* **17**, 325 (1996).
[41] K. O'Hara, L.Ó. Súilleabháin, and J.P. Wolfe, in press.
[42] D.P. Trauernicht and J.P. Wolfe, *Phys. Rev. B* **33**, 8506, (1986).
[43] The suggestion that the power law seen in the Bose saturation effect originates from a balance of the Auger heating and cooling rates was first made by C. Benoit a la Guillaume in 1988.
[44] S.A. Moskalenko, Fiz. Tverd. Tela **4**, 276 (1962).
[45] J.M. Blatt, K.W. Böer, and W. Brandt, Phys. Rev. **126**, 1691 (1962).

2

Basic Theory of Bose–Einstein Condensation of Excitons

2.1 The Bogoliubov Model of the Weakly Nonideal Bose Gas

In this chapter we review the basic theory of Bose–Einstein condensation (BEC) of excitons. To start, this means we must review the basic theory of BEC of any kind of particle, the theory known as the Bogoliubov model, after the foundational contributions made by N. N. Bogoliubov. This model introduces all the strange things associated with BEC: spontaneous symmetry breaking, off-diagonal long-range order (ODLRO), macroscopic occupation of a single quantum state, etc.

It is often said that physicists who spend years studying quantum mechanics eventually warp their intuition so much that things like ODLRO seem normal. It is well worth stepping back every now and then to think about just how strange a Bose condensate is. First, consider the idea of spontaneous symmetry breaking. Many systems exist in which the underlying physics, expressed in the Hamiltonian, do not favor one state over another. The *a priori* probability of occupation of the different states by a particle is equal, i.e., symmetric. Nevertheless, in some cases, thermodynamics requires that a macroscopic number of particles must somehow "choose" one of the states preferentially. If they did not, the system would not be in equilibrium, i.e, could not have a definable temperature. If the particles preferentially choose one state, then the underlying symmetry has been broken.

In the context of magnets, the breaking of the symmetry of the magnet (to have all its constituent dipoles lined up one way and not another) is energetically favored, since there are terms in the Hamiltonian that give lower energy for aligned dipoles, e.g., $\mu \mathbf{S}_1 \cdot \mathbf{S}_2$. In the case of the Bose condensate, there are no such alignment terms in the Hamiltonian. The attraction of the particles to a single state comes from the interference of the quantum-wave functions of the particles. This interference leads to exchange-energy terms that favor symmetry breaking.

If a macroscopic number of particles have all "chosen" to occupy a single quantum state out of many possibilities, does this constitute a "measurement"? How did the particles choose one particular phase and not another? One way of putting this question is to ask whether two different condensates, which have never interacted with each other, possess a definite relative phase [1]. If not, then do they exist in a superposition of all possible phases until they interact, at which point they "collapse" into a definite relative phase? This would seem to be an experimental realization of Schrödinger's Cat, i.e., a superposition of a macroscopic system in different quantum states.

Consider also the idea of long-range phase coherence, or ODLRO. This property means that the wave function of a single particle can extend over macroscopic distances throughout the system; in particular, it penetrates through the barrier of 10^{23} other condensate particles. Therefore, in a very real sense, the particles in the condensate "pass through each other." This is not mere semantics. Quantum mechanically, it is improper to say that one of the particles is "really" on one side of the condensate and not on the other. If the particle were really localized, its wave function would not extend to long range and therefore it could not participate in the condensation. Therefore it is proper to say that a particle in the condensate "passes through" all the other particles. We seem to have no trouble with the idea of "energy bosons," like photons, passing through each other, but this concept seems stranger in the context of macroscopic numbers of massive particles.

Finally, one of the first assumptions of the Bogoliubov model can give us pause. The condensate is characterized as a state of definite phase but not a definite number; the condensate is a coherent state, not a number state. We cannot ask how exactly how many particles are in the condensate; if we try to measure this number, then the coherence will be destroyed and the condensate will disappear, i.e., the temperature will change. As we will see, however, the relative uncertainty in number decreases dramatically as the total size of the condensate increases.

All these fascinating questions can be asked for any condensate, whether liquid He, alkali gases, or excitons. The simplest model that includes all the relevant phenomena is the case of point bosons with infinite lifetime; therefore we begin with this model, and in the second part of this chapter we treat the underlying fermion nature of excitons and other composite bosons. As we will see, all these concepts map directly to the case in which the underlying Fermi particles are considered.

2.1.1 The Quasiaverages

In statistical mechanics, selection rules for the statistical average values of operators appear because of the existence of additive conservation laws. For example, if the Hamiltonian H is invariant with respect to the group of spin rotations, the average value of the total spin operator $\mathbf{S}$ is equal to zero. In the same way, a new selection rule appears if the Hamiltonian commutes with the operator of the total momentum $\mathbf{P}$; namely, the average of the product of a creation operator $a^\dagger_{\mathbf{p}}$ and an annihilation operator $a_{\mathbf{q}}$ of the type $\langle a^\dagger_{\mathbf{p}} a_{\mathbf{q}}\rangle$ is equal to zero, if the momenta $\mathbf{p}$ and $\mathbf{q}$ differ from each other. An additional selection rule also appears when the Hamiltonian commutes with the operator of the total particle number N; this results in the equalities $\langle a_{\mathbf{p}}\rangle = \langle a^\dagger_{\mathbf{p}}\rangle = 0$. These averages are called "regular," and a system with such properties is "nondegenerate." But below the Curie temperature in ferromagnetics or below the critical temperature in the Bose gas, nonzero average values of the type $\langle \mathbf{S}\rangle$ in the former case or of the sort $\langle a_0\rangle$ in the latter case appear spontaneously. These states take some fixed value from among the multitude of possible values. Such states of statistical equilibrium, which arise in systems with spontaneously broken symmetry, are called "degenerate," and the new averages are called "quasiaverages."

Bogoliubov [2] demonstrated his concept of the quasiaverages by using the ideal Bose gas model with the Hamiltonian

$$H = \sum_{\mathbf{k}} \left(\frac{\hbar^2 k^2}{2m} - \mu \right) a^\dagger_{\mathbf{k}} a_{\mathbf{k}}. \tag{2.1}$$

Here $a_{\mathbf{k}}^{\dagger}$ and $a_{\mathbf{k}}$ are the Bose operators of creation and annihilation of particles and μ is their chemical potential. The occupation numbers of the particles are

$$N_0 = \frac{1}{e^{-\beta\mu} - 1}, \qquad N_{\mathbf{k}} = \frac{1}{e^{\beta(\hbar^2 k^2/2m - \mu)} - 1}, \tag{2.2}$$

where $\mu \leq 0$ and $\beta = 1/k_B T$. In the normal state, the density of particles in the thermodynamic limit is given simply as the integral of N_k over all $\mathbf{k}$:

$$n = \lim_{V\to\infty} \frac{N}{V} = \frac{1}{(2\pi)^3} \int_{-\infty}^{\infty} \frac{\mathrm{d}\mathbf{k}}{e^{\beta(\hbar^2 k^2/2m - \mu)} - 1}. \tag{2.3}$$

As discussed in Section 1.3, the integral of Eq. (2.3) cannot account for all the particles in a system at all T and μ, since μ must have an upper bound of zero for the distribution function to remain defined at all energies. When $\mu = 0$, the integral becomes $n = 2.612(mk_B T)^{3/2}/(2\pi\hbar^2)^{3/2}$. At this point, BEC occurs, and a finite value of the density of condensed particles appears in the thermodynamic limit,

$$n_0 = \lim_{V\to\infty} \frac{N_0}{V}, \qquad \mu = -k_B T \ln\left(1 + \frac{1}{N_0}\right), \tag{2.4}$$

and the total particle density contains two parts,

$$n = \lim_{V\to\infty} \frac{N}{V} = n_0 + \frac{1}{(2\pi)^3} \int_{-\infty}^{\infty} \frac{\mathrm{d}^3 k}{e^{\beta\hbar^2 k^2/2m} - 1}. \tag{2.5}$$

It follows from this that the operators $a_0^{\dagger}$ and a_0 asymptotically become c–numbers. To prove this, one may consider the commutator

$$\left[\frac{a_0}{\sqrt{V}}, \frac{a_0^{\dagger}}{\sqrt{V}}\right] = \frac{1}{V},$$

which asymptotically tends to zero. In the thermodynamic limit, the ratio $a_0^{\dagger} a_0/V \sim n_0$ stays constant, however. Therefore the amplitudes $a_0^{\dagger}/\sqrt{V}$ and $a_0/\sqrt{V}$ may be considered as c–numbers, the product of which is equal to n_0. One can then write

$$a_0^{\dagger}/\sqrt{V} \sim \sqrt{n_0} e^{i\alpha}, \qquad a_0/\sqrt{V} \sim \sqrt{n_0} e^{-i\alpha}. \tag{2.6}$$

On the other hand, the regular averages of the operators a_0 and $a_0^{\dagger}$ in the Hamiltonian (2.1) are exactly equal to zero. This is the consequence of the commutativity of the operator H and the operator of the total particle number N, as follows:

$$N = \sum_{\mathbf{k}} a_{\mathbf{k}}^{\dagger} a_{\mathbf{k}}, \qquad [H, N] = 0.$$

As a result, the operator H is invariant with respect to the unitary transformation

$$U = e^{i\phi N} \tag{2.7}$$

with an arbitrary angle ϕ. Such invariance is called gradient invariance of the first kind, or "gauge invariance," and implies that

$$H = U^{\dagger} H U.$$

This property permits the following relation for the condensate amplitude average to be established:

$$\langle a_0\rangle \cong \mathrm{Tr}\,(a_0 e^{-\beta H}) = \mathrm{Tr}\,(a_0 U e^{-\beta H} U^\dagger)$$
$$= \mathrm{Tr}\,(U^\dagger a_0 U e^{-\beta H}) = e^{i\phi}\langle a_0\rangle,$$
$$(1-e^{i\phi})\langle a_0\rangle = 0.$$

Here the relation $U^\dagger a_0 U = e^{i\phi}a_0$ was used. The selection rules

$$\langle a_0\rangle = 0, \qquad \langle a_0^\dagger\rangle = 0, \tag{2.8}$$

then arise, because ϕ is an arbitrary angle. The regular average of Eqs. (2.8) can also be obtained from asymptotical expressions (2.6) if they are integrated over the angle α.

How can this apparent contradiction be resolved? To obtain nonzero average values $\langle a_0\rangle$ and $\langle a_0^\dagger\rangle$ of the type of expressions (2.6), it is necessary to have a Hamiltonian that does not conserve the total particle number. With this aim, an additional term,

$$-\nu(a_0^\dagger e^{i\varphi} + a_0 e^{-i\varphi})\sqrt{V}, \quad \nu > 0, \tag{2.9}$$

is added to the Hamiltonian (2.1), where φ is a fixed angle and ν is an infinitesimal value. The new Hamiltonian has the form

$$H_{\nu,\phi} = \sum_{\mathbf{k}}\left(\frac{\hbar^2k^2}{2m} - \mu\right)a_{\mathbf{k}}^\dagger a_{\mathbf{k}} - \nu(a_0^\dagger e^{i\varphi} + a_0 e^{-i\varphi})\sqrt{V}. \tag{2.10}$$

It does not conserve the condensate total number. Now the regular average values of the operators a_0 and $a_0^\dagger$ over the Hamiltonian $H_{\nu,\varphi}$ differ from zero, i.e., $\langle a_0\rangle_{H_{\nu,\varphi}} \neq 0$ and $\langle a_0^\dagger\rangle_{H_{\nu,\varphi}} \neq 0$. These serve as a basis for determining the quasiaverages. The quasiaverage value of the operator a_0 is designated by $<a_0>$ and is defined as the limit of the regular average $\langle a_0\rangle_{H_{\nu,\varphi}}$ when ν tends to zero:

$$<a_0> = \lim_{\nu\to 0}\langle a_0\rangle_{H_{\nu,\varphi}}. \tag{2.11}$$

It is important to emphasize that the limit $\nu \to 0$ must be effectuated after the thermodynamic limit $V \to \infty$, $N_0 \to \infty$.

To calculate the regular average $\langle a_0\rangle_{H_{\nu,\varphi}}$, one needs to represent the Hamiltonian $H_{\nu,\varphi}$ in a diagonal form. It can be done with the aid of the canonical transformation over the amplitudes a_0 and $a_0^\dagger$ without changing the other amplitudes $a_{\mathbf{k}}^\dagger$ and $a_{\mathbf{k}}$:

$$a_0 = -\frac{\nu}{\mu}e^{i\varphi}\sqrt{V} + \alpha_0, \qquad a_{\mathbf{k}} = \alpha_{\mathbf{k}}. \tag{2.12}$$

In the terms of the new variables, the Hamiltonian $H_{\nu,\varphi}$ takes the form

$$H_{\nu,\varphi} = -\mu\alpha_0^\dagger\alpha_0 + \sum_{\mathbf{k}\neq 0}\left(\frac{\hbar^2k^2}{2m} - \mu\right)\alpha_{\mathbf{k}}^\dagger\alpha_{\mathbf{k}} + \frac{\nu^2 V}{\mu}. \tag{2.13}$$

In the thermodynamic limit, μ is also infinitesimal, and it is possible to choose the ratio of two infinitesimal values μ and $-\nu$ to give the finite value

$$-\frac{\nu}{\mu} = \sqrt{n_0}. \tag{2.14}$$

In the diagonal representation of Eq. (2.13), the regular average value $\langle\alpha_0\rangle_{H_{\nu,\varphi}}$ exactly equals zero, while the value $\langle a_0\rangle_{H_{\nu,\varphi}}$ equals the first term on the right–hand side of formulas (2.12). As a result, the quasiaverage $<a_0>$ is

$$<a_0> = \lim_{\nu\to 0}\langle a_0\rangle_{\hat{H}_{\nu,\varphi}} = \sqrt{N_0}e^{i\varphi}; \tag{2.15}$$

it depends on the fixed angle φ and does not depend on ν. Spontaneous symmetry breaking was implied when we fixed the phase of the condensate amplitude in the Hamiltonian (2.10). When the interaction between the particles is taken into account, such a difference appears for other amplitudes as well. For example, the quasiaverages $<a_{\mathbf{k}}a_{-\mathbf{k}}>$ and $<a^{\dagger}_{\mathbf{k}}a^{\dagger}_{-\mathbf{k}}>$ are nonzero. They give rise to the renormalization of the energy spectrum and to the appearance of a region in the dispersion law that is linear with the wave vector $\mathbf{k}$ when repulsion between particles predominates. In such a way the canonical transformation

$$a_{\mathbf{k}} = \sqrt{N_0}\delta_{\mathbf{k},0}e^{i\varphi} + \alpha_{\mathbf{k}}, \tag{2.16}$$

introduced in the theory of superfluidity, has a quantum-statistical foundation within the framework of the quasiaverage concept. When $T = 0$, the quasiaverage $<a_0>$ coincides with the average over the quantum–mechanical ground state. It is a coherent macroscopic state.

2.1.2 Coherent States

The properties of the Bose condensate ground state are best understood by comparison with the coherent state in the quantum optics, introduced by Glauber [3] (see also Ref [4].) These states are the eigenstates of the photon annihilation operator a. For simplicity the momentum indices of the Bose operators $a^{\dagger}$ and a are omitted.

The coherent state $|z\rangle$ is defined as follows:

$$a|z\rangle = z|z\rangle, \qquad \langle z|a^{\dagger} = z^*\langle z|, \qquad \langle z|z\rangle = 1. \tag{2.17}$$

Different coherent states $|z\rangle$ and $|z'\rangle$ are not orthogonal with each other. Here $z = Ae^{i\phi}$ is a complex number that gives the amplitude and the phase of the state. Thus, while for Fock states with a definite particle number, the operators $a^{\dagger}$ and a have the effect of creation and destruction, more fundamentally these operators are interpreted as measurements of the complex amplitude of the many-particle wave function.

The coherent state can be expanded as a series of Fock states. The Fock states $|n\rangle$ are the eigenfunctions of the particle-number operator $\hat{n} = a^{\dagger}a$, as follows:

$$|n\rangle = \frac{a^{\dagger n}}{\sqrt{n!}}|0\rangle, \qquad a^{\dagger}a|n\rangle = n|n\rangle. \tag{2.18}$$

Here $|0\rangle$ is the wave function of the photon vacuum. The series expansion of the coherent states has the form

$$|z\rangle = e^{-|z|^2/2}\sum_{n=0}^{\infty}\frac{z^n}{\sqrt{n!}}|n\rangle. \tag{2.19}$$

The coherent state can be generated by the action on the ground state $|0\rangle$ of the unitary displacement operator $D(z)$. This operator and its adjoint are defined as

$$D(z) = e^{za^{\dagger}-z^*a}, \qquad D^{\dagger}(z) = D^{-1}(z) = D(-z). \tag{2.20}$$

$D(z)$ can be represented in a factorized form by use of the Baker–Hausdorff formula [4]:

$$D(z) = e^{-|z|^2/2} e^{za^\dagger} e^{-z^* a} = e^{|z|^2/2} e^{-z^* a} e^{za^\dagger}. \tag{2.21}$$

The unitary transformation of the operator a gives rise to its displacement by value z as follows:

$$D^\dagger(z) a D(z) = a + z. \tag{2.22}$$

Now the statement

$$|z\rangle = D(z)|0\rangle \tag{2.23}$$

can be verified. Indeed, using Eq. (2.22), one finds that

$$a|z\rangle = aD(z)|0\rangle = D \underbrace{D^\dagger a D}|0\rangle = z|z\rangle.$$

For this reason, the coherent state is called the ground state of the displaced oscillator, in which the equilibrium position is shifted by value z. When z is a macroscopically large number, i.e., $z \sim \sqrt{N} e^{i\varphi} \sim \sqrt{V}$, the coherent state $|z\rangle$ is called macroscopic.

The theory of BEC of excitons and biexcitons studies coherent macroscopic states. In the homogeneous case, they are characterized by a well-determined wave vector $\mathbf{k}_0$ of the single-particle state, in which the Bose condensation takes place, by a given fixed phase φ, and by a macroscopically large amplitude, proportional to $\sqrt{V}$. The wave vector $\mathbf{k}_0$ can be different from zero, which corresponds to the case of a moving condensate with the superfluid velocity $\mathbf{V}_s = \hbar \mathbf{k}_0/m$, where m is the mass of the particles.

The existence of the coherent macroscopic states involves the appearence of ODLRO [5, 6] in the system. To demonstrate this, one can introduce the operator of the Bose field, $\Psi(\mathbf{r})$,

$$\Psi(\mathbf{r}) = \frac{1}{\sqrt{V}} \sum_{\mathbf{k}} a_{\mathbf{k}} e^{i\mathbf{kr}} = \frac{a_{\mathbf{k}_0}}{\sqrt{V}} e^{i\mathbf{k}_0\mathbf{r}} + \frac{1}{\sqrt{V}} \sum_{\mathbf{k} \neq \mathbf{k}_0} a_{\mathbf{k}} e^{i\mathbf{kr}}, \tag{2.24}$$

the amplitude $a_{\mathbf{k}_0}$ of which is macroscopically large, i.e., $a_{\mathbf{k}_0} \sim \sqrt{N_{\mathbf{k}_0}} \sim \sqrt{V}$. Then the off-diagonal matrix element

$$\langle \Psi^\dagger(\mathbf{r}')\Psi(\mathbf{r})\rangle = \frac{N_{\mathbf{k}_0}}{V} e^{i(\mathbf{k}_0, \mathbf{r}-\mathbf{r}')} + \frac{1}{V} \sum_{\mathbf{k} \neq \mathbf{k}_0} \langle a_{\mathbf{k}}^\dagger a_{\mathbf{k}} \rangle e^{i(\mathbf{k}, \mathbf{r}-\mathbf{r}')} \tag{2.25}$$

remains finite in thermodynamical limit at any distance $|\mathbf{r} - \mathbf{r}'|$ because of the first term on the right-hand side of Eq. (2.25). Since the off-diagonal matrix elements exist only in quantum mechanics and the distance $|\mathbf{r} - \mathbf{r}'|$ can be as large as the size of the sample, the presence of ODLRO means the existence of a macroscopic quantum state in the system [7, 8]. Coherent macroscopic states slowly varying in space and time have been used by London [7] in his attempts to explain the superfluidity, by Ginzburg and Landau in the theory of superconductivity [9, 10], by Ginzburg [11], Pitaevskii [12], and Gross [13] in the theory of quantum vortices, and by Keldysh [14] in the theory of coherent excitons and photons.

In the second half of this chapter, the unitary transformations of Eqs. (2.20) are used to introduce spontaneous symmetry breaking into the Hamiltonian. This allows the possibility of studying BEC of excitons not only in the case in which the excitons and the biexcitons are described by the Bose operators, but also when their underlying electron–hole structure is taken into account [15]. For that case, it will be necessary to introduce, instead of the Bose operators $a_0^\dagger$, the creation operator of the bound electron–hole (e–h) pair with the center-of-mass wave vector equal to zero, as follows:

$$c_0^\dagger = \frac{1}{\sqrt{2V}} \sum_{\mathbf{q},\sigma} \varphi(\mathbf{q}) a_{\mathbf{q},\sigma}^\dagger b_{-\mathbf{q},-\sigma}^\dagger. \tag{2.26}$$

Here $a_{\mathbf{q}\sigma}^\dagger$ and $b_{\mathbf{q},\sigma}^\dagger$ are the Fermi creation operators of the electron and the hole, respectively, with the wave vector $\mathbf{q}$ and the spin projection σ. $\varphi(\mathbf{q})$ is the wave function of the relative e–h motion in the exciton. The new operator of displacement, which breaks the symmetry of the Hamiltonian in the electron–hole representation, can be obtained when the operators $a^\dagger$ and a in expressions (2.20) are substituted by the operators $c_0^\dagger$, c_0. At the same time, the parameter z must be taken as a macroscopically large value, $z \sim \sqrt{N_{\text{ex},0}} \sim \sqrt{V}$. This procedure was followed by Keldysh and Kozlov [15], who studied the collective properties of excitons in the electron–hole description.

This method can also be generalized to the study of coherent biexcitons in the electron–hole representation. With this aim, one can define the biexciton creation operator $C_0^\dagger$ with the center-of-mass wave vector $\mathbf{k} = 0$. It can be constructed as a superposition of the products of two excitonic creation operators of the type of Eq. (2.26). This superposition must describe the bound state of two excitons with the wave function of their relative motion given by $\Phi(\mathbf{q})$. In the most simple case ($m_e = m_h$) it has the form

$$C_0^\dagger = \frac{1}{2V^{3/2}} \sum_{\mathbf{k},\mathbf{p},\mathbf{q}} \sum_{\sigma_1,\sigma_2} \Phi(\mathbf{k})\varphi(\mathbf{p})\varphi(\mathbf{q}) \times a_{\mathbf{k}/2+\mathbf{p},\sigma_1}^\dagger b_{\mathbf{k}/2-\mathbf{p},-\sigma_1}^\dagger a_{-\mathbf{k}/2+\mathbf{q},\sigma_2}^\dagger b_{-\mathbf{k}/2-\mathbf{q},-\sigma_2}^\dagger. \tag{2.27}$$

In the same way as in the case of coherent excitons, to construct the new operators $D^\dagger(z)$ for coherent biexcitons, one needs to substitute the operators $a^\dagger$ and a in formulas (2.20) by the operators $C_0^\dagger$ from Eq. (2.27) and C_0. In addition, it is neccessary to insert, instead of the parameter z, the value $\sqrt{N_{\text{biex},0}} \sim \sqrt{V}$. One can introduce and study "squeezed" coherent macroscopic states of excitons and biexcitons in the same way. Squeezed states are discussed in Subsection 8.3.2.

Before moving on, it is worthwhile to consider the way in which the coherent states introduced here are states of definite phase. As noted above, the coherent states are eigenstates of the operator a, which is the complex amplitude of the particle wave function. One does not normally think of a and $a^\dagger$ as observables because they are non-Hermitian. As discussed in Subsection 2.1.1, however, in the case of a Bose condensate, the operators $a_0^\dagger$ and a_0 asymptotically become c numbers as N_0 becomes macroscopic. As the amplitude becomes large in Eq. (2.19), however, i.e., as $|z| \to \sqrt{N_0}$, the expansion becomes

$$|z\rangle = \sum_{n=0}^{\infty} \left[e^{-(N_0-n)/2}(N_0/n)^{n/2}(2\pi n)^{-1/4}\right] e^{in\phi}|n\rangle, \tag{2.28}$$

in which the quantity in brackets becomes more and more peaked as N_0 becomes large, so that $|z\rangle \to e^{iN_0\phi}|N_0\rangle$. In other words, although the coherent condensed state is not a definite number state, it looks much like one. It also looks like a state of definite phase, since $\langle z|N_0\rangle = e^{iN_0\phi}$. Since an uncertainty relation holds between number and phase, both of these cannot be measured exactly. Nevertheless, since the condensate is macroscopic, they can both be measured extremely well, to within $\hbar$ instead of $N\hbar$, in the same way the position and the momentum x and p can be measured for a macroscopic object like a baseball.

2.1.3 The Excitons as Weakly Interacting, Structureless Bosons

So far, we have considered case of the ideal Bose gas. As Nozières [16] and others have discussed, the ideal Bose gas is actually a pathological case, and even weak interactions between the particles eliminate these pathologies. For example, Nozières showed that the ideal Bose gas is not stable against fragmentation of the condensate into states with effectively the same energy, but differing phase. In other words, the existence of long-range order and superfluidity depend crucially on the existence of interactions between the particles, however weak those interactions may be.

All real systems have interacting particles. Therefore we now move to the Bogoliubov model of the interacting Bose gas. The appeal of exciton gases is that they can, in many cases, be considered weakly interacting bosons, so that much of the Bogoliubov theory, which assumes weak interactions, applies directly. Much of the recent excitement about condensates of alkali gases (e.g., Refs. 17 and 18) stems from the same reason, that in large part they can be treated as weakly interacting point bosons. By contrast, liquid He must be treated as a strongly interacting liquid [19, 20].

Since so many of the calculations depend on the estimation of the exciton–exciton interaction, it is appropiate here to review briefly what we know about this interaction. This interaction is essentially a van der Waals potential between two neutral atoms [21, 22]. Unlike the case of hydrogen, however, the calculation of the exciton–exciton interaction has two additional complications. First, the electron and the hole have comparable masses, so that one cannot generally assume that the hole is motionless like the proton in hydrogen. Second, the electrons and the holes have an exchange interaction, in addition to the electron–electron and the hole-hole exchange terms. Several theoretical works [23–26] have estimated the strength of the interaction between two excitons with various approximations. All these works found an s-wave-scattering length that is positive and of the order of 2–3 times the excitonic Bohr radius for most semiconductors. Experiments measuring the dephasing time of excitons in GaAs found scattering rates consistent with this number [27]; time-resolved measurements of exciton thermalization in Cu_2O [28] yielded an upper bound for the exciton–exciton scattering rate that is also consistent with this number.

Bobrysheva, Miglei, and Shmiglyuk [25, 26] explicitly took into account the different possible spin orientations of the excitons in the case of a simple band structure, i.e., s orbitals for both the conduction and the valence bands. (This work will be reviewed in greater depth in Chapter 4.) Starting with the underlying fermion structure, they found that two paraexcitons, i.e., excitons with electron and hole of opposite spin, have a repulsive interaction constant equal to $13\pi/3\text{Ry}_{\text{ex}}a_{\text{ex}}^3$. Two orthoexcitons, which individually have a total e–h spin equal to unity,[a] interact with different energies, depending on their total spin. If it is equal to 2, the interaction is repulsive and its constant is twice that of two paraexcitons. If the total spin of two orthoexcitons equals zero, the interaction between them is

[a] We note that in the case of Cu_2O, the paraexciton and the orthoexciton states are not pure spin states. See Appendix A for details on the band structure of Cu_2O.

attractive. In this case, bound states are possible, leading to the formation of an excitonic molecule, or biexciton. With the exception of this case, the Pauli exclusion principle leads to a predominantly repulsive interaction between different species of excitons and, as will be shown in Chapter 4, to the stability of the multicomponent exciton–biexciton system [29, 30]. Of course, in any real crystal the particular structure of the conduction and the valence bands will determine the exact interaction between the excitons. Nevertheless, this work shows that for typical direct-gap semiconductors it is a good approximation to assume an overall repulsive, short-range interaction between the excitons. As discussed in Chapter 5, the electron–hole liquid state that was mentioned in Chapter 1, in which the electrons and the holes at high densities become a Fermi liquid, arises only in the case of degenerate bands.

In the model of a weakly nonideal Bose gas, the Hamiltonian of interacting excitons has the form [31–35]

$$H = \sum_{\mathbf{q}} [E_{ex}(\mathbf{q}) - \mu] a_{\mathbf{q}}^{\dagger} a_{\mathbf{q}} + \frac{1}{2V} \sum_{\mathbf{p},\mathbf{q},\mathbf{k}} U(\mathbf{k}) a_{\mathbf{p}}^{\dagger} a_{\mathbf{q}}^{\dagger} a_{\mathbf{q}+\mathbf{k}} a_{\mathbf{p}-\mathbf{k}}. \tag{2.29}$$

Here $a_{\mathbf{p}}^{\dagger}$ and $a_{\mathbf{p}}$ are the Bose creation and annihilation operators, respectively. The electron–hole structure of the excitons in the explicit form is not taken into account – it is reflected only in the properties of exciton–exciton interaction constant $U(\mathbf{k})$, which is assumed to be positive. This corresponds to a repulsive interaction between the excitons, which is realistic for paraexcitons in Cu_2O crystals, for example. $E_{ex}(\mathbf{q})$ is the energy of exciton formation with wave vector $\mathbf{q}$ and with the quadratic dispersion law $E_{ex}(\mathbf{q}) = E_{ex}(0) + T_{\mathbf{q}}$, where $T_{\mathbf{q}} = \hbar^2 \mathbf{q}^2 / 2m_{ex}$. m_{ex} is the exciton translational mass, and μ is the chemical potential of the system.

To introduce spontaneous symmetry breaking and to consider BEC of the excitons in a single-particle state with wave vector $\mathbf{k}_0$, it is necessary to make the canonical transformation

$$a_{\mathbf{q}} = \sqrt{N_{\mathbf{k}_0}} \delta_{\mathbf{k}_0,\mathbf{q}} + \alpha_{\mathbf{q}}. \tag{2.30}$$

Here $N_{\mathbf{k}_0}$ is the macroscopic number $N_{\mathbf{k}_0} = V n_{\mathbf{k}_0}$. As the operators $\alpha_{\mathbf{k}}^{\dagger}$ and $\alpha_{\mathbf{k}}$ are small compared with the condensate amplitude $\sqrt{N_{\mathbf{k}_0}}$, it is possible to develop the perturbation theory with these operators. The Hamiltonian can be represented in the form

$$\begin{aligned} H = {} & [E_{ex}(\mathbf{k}_0) - \mu] N_{\mathbf{k}_0} + L_0 N_{\mathbf{k}_0}/2 + \sqrt{N_{\mathbf{k}_0}} [E_{ex}(\mathbf{k}_0) + L_0 - \mu] (\alpha_{\mathbf{k}_0}^{\dagger} + \alpha_{\mathbf{k}_0}) \\ & + \sum_{\mathbf{q}} [E_{ex}(\mathbf{q}) + L_0 - \mu] \alpha_{\mathbf{q}}^{\dagger} \alpha_{\mathbf{q}} + \sum_{\mathbf{q}} \frac{L_{\mathbf{q}}}{2} (\alpha_{\mathbf{k}_0+\mathbf{q}}^{\dagger} \alpha_{\mathbf{k}_0+\mathbf{q}} + \alpha_{\mathbf{k}_0-\mathbf{q}}^{\dagger} \alpha_{\mathbf{k}_0-\mathbf{q}} \\ & + \alpha_{\mathbf{k}_0+\mathbf{q}}^{\dagger} \alpha_{\mathbf{k}_0-\mathbf{q}}^{\dagger} + \alpha_{\mathbf{k}_0+\mathbf{q}} \alpha_{\mathbf{k}_0-\mathbf{q}}) + \frac{\sqrt{N_{\mathbf{k}_0}}}{V} \sum_{\mathbf{p},\mathbf{k}} U(\mathbf{k}) (\alpha_{\mathbf{p}}^{\dagger} \alpha_{\mathbf{k}_0+\mathbf{k}} \alpha_{\mathbf{p}-\mathbf{k}} \\ & + \alpha_{\mathbf{p}-\mathbf{k}}^{\dagger} \alpha_{\mathbf{k}_0+\mathbf{k}}^{\dagger} \alpha_{\mathbf{p}}) + \frac{1}{2V} \sum_{\mathbf{p},\mathbf{q},\mathbf{k}} U(\mathbf{k}) \alpha_{\mathbf{p}}^{\dagger} \alpha_{\mathbf{q}}^{\dagger} \alpha_{\mathbf{q}+\mathbf{k}} \alpha_{\mathbf{p}-\mathbf{k}}. \end{aligned} \tag{2.31}$$

Here the following notation is introduced:

$$L_{\mathbf{k}} = U(\mathbf{k}) \frac{N_{\mathbf{k}_0}}{V}. \tag{2.32}$$

To eliminate the term that is linear in the operators $\alpha_{\mathbf{k}_0}$ and $\alpha^\dagger_{\mathbf{k}_0}$, one can determine the chemical potential μ as follows:

$$\mu = E_{\text{ex}}(\mathbf{k}_0) + L_0. \tag{2.33}$$

The part of the Hamiltonian that is quadratic in the operators $\alpha^\dagger_{\mathbf{q}}$ and $\alpha_{\mathbf{q}}$ can be diagonalized using Eq. (2.32). This part, which contains the wave vectors $\mathbf{k} \pm \mathbf{k}_0$, has the form

$$\begin{aligned} H_0(\mathbf{k}_0 + \mathbf{k}) + H_0(\mathbf{k}_0 - \mathbf{k}) &= \left(T_{\mathbf{k}} + L_{\mathbf{k}} + \frac{\hbar^2 \mathbf{k}_0\mathbf{k}}{m_{\text{ex}}}\right)\alpha^\dagger_{\mathbf{k}_0+\mathbf{k}}\alpha_{\mathbf{k}_0+\mathbf{k}} \\ &+ \left(T_{\mathbf{k}} + L_{\mathbf{k}} - \frac{\hbar^2 \mathbf{k}_0\mathbf{k}}{m_{\text{ex}}}\right)\alpha^\dagger_{\mathbf{k}_0-\mathbf{k}}\alpha_{\mathbf{k}_0-\mathbf{k}} \\ &+ L_{\mathbf{k}}\left(\alpha^\dagger_{\mathbf{k}_0+\mathbf{k}}\alpha^\dagger_{\mathbf{k}_0-\mathbf{k}} + \alpha_{\mathbf{k}_0+\mathbf{k}}\alpha_{\mathbf{k}_0-\mathbf{k}}\right). \end{aligned} \tag{2.34}$$

The diagonalization can be effectuated by the linear transformation to new Bose operators $\xi^\dagger_{\mathbf{k}}$ and $\xi_{\mathbf{k}}$ as follows:

$$\begin{aligned} \xi_{\mathbf{k}} &= \frac{\alpha_{\mathbf{k}_0+\mathbf{k}} + A_{\mathbf{k}}\alpha^\dagger_{\mathbf{k}_0-\mathbf{k}}}{\sqrt{1 - |A_{\mathbf{k}}|^2}} = u_{\mathbf{k}}\alpha_{\mathbf{k}_0+\mathbf{k}} + v_{\mathbf{k}}\alpha^\dagger_{\mathbf{k}_0-\mathbf{k}}, \\ \alpha_{\mathbf{k}_0+\mathbf{k}} &= \frac{\xi_{\mathbf{k}} - A_{\mathbf{k}}\xi^\dagger_{-\mathbf{k}}}{\sqrt{1 - |A_{\mathbf{k}}|^2}} = u_{\mathbf{k}}\xi_{\mathbf{k}} - v_{\mathbf{k}}\xi^\dagger_{-\mathbf{k}}, \end{aligned} \tag{2.35}$$

where the coefficients of two alternative presentations obey the relations

$$u_{\mathbf{k}} = \frac{1}{\sqrt{1 - |A_{\mathbf{k}}|^2}}, \qquad v_{\mathbf{k}} = \frac{A_{\mathbf{k}}}{\sqrt{1 - |A_{\mathbf{k}}|^2}},$$

$$u^2_{\mathbf{k}} - v^2_{\mathbf{k}} = 1. \tag{2.36}$$

The new operators $\xi^\dagger_{\mathbf{k}}$ and $\xi_{\mathbf{k}}$ have wave vectors $\pm\mathbf{k}$ that are measured relative to the condensate wave vector $\mathbf{k}_0$.

The spectrum of elementary excitations has the well-known form [2, 36]

$$E(\mathbf{k}) = \sqrt{T^2_{\mathbf{k}} + 2T_{\mathbf{k}}L_{\mathbf{k}}}, \tag{2.37}$$

which implies the existence of a linear, phononlike region near $\mathbf{k} = 0$. (We sometimes call these excitations "hydrons" to distinguish them from the lattice phonons.) The transformation coefficients are

$$A_{\mathbf{k}} = \frac{1}{L_{\mathbf{k}}}[T_{\mathbf{k}} + L_{\mathbf{k}} - E(\mathbf{k})],$$

$$u_{\mathbf{k}} = \sqrt{\frac{T_{\mathbf{k}} + L_{\mathbf{k}} + E(\mathbf{k})}{2E(\mathbf{k})}}, \qquad v_{\mathbf{k}} = \sqrt{\frac{T_{\mathbf{k}} + L_{\mathbf{k}} - E(\mathbf{k})}{2E(\mathbf{k})}},$$

$$u^2_{\mathbf{k}} = \frac{1}{2}\left[\frac{T_{\mathbf{k}} + L_{\mathbf{k}}}{E(\mathbf{k})} + 1\right], \qquad v^2_{\mathbf{k}} = \frac{1}{2}\left[\frac{T_{\mathbf{k}} + L_{\mathbf{k}}}{E(\mathbf{k})} - 1\right]. \tag{2.38}$$

The quadratic part of the Hamiltonian obtains the form

$$\sum_{\mathbf{k}>0}[H_0(\mathbf{k}_0 + \mathbf{k}) + H_0(\mathbf{k}_0 - \mathbf{k})] = \frac{1}{2}\sum_{\mathbf{k}}[E(\mathbf{k}) - T_{\mathbf{k}} - L_{\mathbf{k}}] + \sum_{\mathbf{k}}[E(\mathbf{k}) + \hbar\mathbf{V}_s\cdot\mathbf{k}]\xi^\dagger_{\mathbf{k}}\xi_{\mathbf{k}}. \tag{2.39}$$

Here $\mathbf{V}_s$ is the velocity of the superfluid flow and is related to the vector $\mathbf{k}_0$ by the equality

$$\mathbf{V}_s = \frac{\hbar \mathbf{k}_0}{m_{ex}}. \tag{2.40}$$

According to the Landau criterion of superfluidity [10], the velocity $\mathbf{V}_s$ must be less than the critical velocity of superfluidity u^*. The last value is determined from the linear region of the dispersion law at small values of the wave vector $\mathbf{k}$:

$$\lim_{k\to 0} E(\mathbf{k}) = \hbar u^* k, \quad u^* = \sqrt{\frac{U(0)}{m_{ex}}\frac{N_{\mathbf{k}_0}}{V}}. \tag{2.41}$$

Indeed, when $\mathbf{V}_s$ is less than u^*, the energy of the elementary excitation in the presence of a moving condensate $[E(\mathbf{k}) + \hbar \mathbf{V}_s \cdot \mathbf{k}]$ is positive for all values of the vector $\mathbf{k}$. This fact ensures the positive values of the mean distribution numbers of elementary excitations

$$\bar{n}(\mathbf{k}) = \frac{1}{e^{\beta[E(\mathbf{k})+\hbar \mathbf{V}_s\cdot\mathbf{k}]} - 1} \tag{2.42}$$

and the stability of the system [2,36]. The value $U(0)$ can be estimated in the hard-sphere approximation [37,38] as

$$U(0) = \frac{4\pi\hbar^2 a}{m_{ex}}, \tag{2.43}$$

where a has the order of magnitude of the diameter of the exciton.

For excitons in a semiconductor, in which the translational-exciton mass m_{ex} roughly equals the free-electron mass m_0 and the excitonic Bohr radius ranges from 10 to 50 Å, the constant $U(0)$ has values of approximately 10^{-32}–10^{-33} erg cm^3. For the condensate excitons, with density n_0 equal to 10^{16}–10^{17} cm^{-3}, the critical velocity u^* varies from 10^5 to 10^6 cm/s.

Along with the above-mentioned positive branch of the spectrum, there is a second branch with negative energy:

$$E'(\mathbf{k}) = -E(\mathbf{k}) = -\sqrt{T_{\mathbf{k}}^2 + T_{\mathbf{k}} L_{\mathbf{k}}}. \tag{2.44}$$

The coefficients $A'_{\mathbf{k}}$, $u'_{\mathbf{k}}$, and $v'_{\mathbf{k}}$ corresponding to this branch are the following:

$$A'_{\mathbf{k}} = \frac{1}{L_{\mathbf{k}}}[T_{\mathbf{k}} + L_{\mathbf{k}} - E'(\mathbf{k})] = \frac{1}{L_{\mathbf{k}}}[T_{\mathbf{k}} + L_{\mathbf{k}} + E(\mathbf{k})] \geq 1,$$

$$u'_{\mathbf{k}} = \frac{1}{\sqrt{1 - |A'_{\mathbf{k}}|^2}} = \frac{i}{\sqrt{|A'_{\mathbf{k}}|^2 - 1}}, \qquad v'_{\mathbf{k}} = \frac{iA'_{\mathbf{k}}}{\sqrt{|A'_{\mathbf{k}}|^2 - 1}}, \qquad u'^2_{\mathbf{k}} - v'^2_{\mathbf{k}} = 1. \tag{2.45}$$

The new set of operators $\xi'_{\mathbf{k}}$ and $\xi'^{\dagger}_{\mathbf{k}}$ is determined as follows:

$$\xi'_{\mathbf{k}} = \frac{\alpha_{\mathbf{k}_0+\mathbf{k}} + A'_{\mathbf{k}}\alpha^{\dagger}_{\mathbf{k}_0-\mathbf{k}}}{\sqrt{1 - |A'_{\mathbf{k}}|^2}}. \tag{2.46}$$

These commute with the operators $\xi_{\mathbf{k}}$ and $\xi^{\dagger}_{\mathbf{k}}$ defined in Eqs. (2.35) because the orthogonality condition is fulfilled:

$$u_{\mathbf{k}} u'_{\mathbf{k}} - v_{\mathbf{k}} v'_{\mathbf{k}} = 1 - A_{\mathbf{k}} A'_{\mathbf{k}} = 0. \tag{2.47}$$

Other properties of the second solution are discussed in Subsection 6.2.1.

The above results are based on a perturbation theory that takes into account only terms quadratic in the operators $\alpha^{\dagger}_{\mathbf{q}}$ and $\alpha_{\mathbf{q}}$ in the Hamiltonian (2.31). This approximation is valid either at low temperature (in which case $N_0 \sim N$) or at low density. To take into account additional interactions, we need to develop a quantum-field theory for the Bose condensate. The well-known methods of quantum-field theory (e.g., Ref. 39) cannot be directly used in the case of a macroscopically large occupation of one quantum state. Therefore a revised generalization of the quantum-electrodynamics diagram technique to the case of a Bose condensed system was proposed by Beliaev [37]. In Subsection 2.1.4 a short review of this diagram technique is presented, following the original papers of Beliaev and the monograph of Abrikosov, Gor'kov, and Dzyaloshinskii [40]. Primary attention is given to the foundations of the new method, the formulation of the principal rules, and the elucidation of the underlying physics. The concrete application of the Beliaev diagram technique to the exciton–phonon system will be demonstrated in Chapter 3.

2.1.4 The Beliaev Diagram Technique

As noted above, in the ideal Bose gas at $T = 0$, the number of condensed particles coincides with the total particle number N. In this case, the average value of the normal-ordered product of two condensed-particle operators $a_0^\dagger$ and a_0 does not equal zero. Moreover, this value $\langle a_0^\dagger a_0\rangle = N$ is an extensive value proportional to V. In this regard, the Bose gas differs essentially from the vacuum photons in quantum electrodynamics. For example, the application of the Wick theorem [41] to the condensed particles is impossible.

For these reasons, the development of a diagram technique for a degenerate Bose gas needs special investigation. At the same time, however, the single-particle noncondensate states, which have nonzero momenta $\mathbf{p}$ in the ideal Bose gas, are empty at $T = 0$. For that group of particles, the average values of the T-ordered operators $a_{\mathbf{p}}^\dagger$ and $a_{\mathbf{p}}$ can be calculated following the Wick theorem, in full accordance with the methods of quantum field theory.

The Beliaev diagram technique permits one to take into account both of these features. It also shows that interactions between the particles do not completely destroy the condensate – it persists in a changed form. The condensate can be regarded as an external field when the diagram technique for the noncondensate particles is formulated. In the general form, these statements coincide with the main ideas of the Bogoliubov theory of superfluidity [2,36].

Following Refs. 37 and 40, one can introduce the single-particle Green's function by using the wave function and operators in the Heisenberg representation,

$$iG(\mathbf{p}, t - t') = \langle \Phi_H^0 | T[a_{\mathbf{p}H}(t) a_{\mathbf{p}H}^\dagger(t')] | \Phi_H^0 \rangle, \tag{2.48}$$

where $T[\ldots]$ refers to a time-ordered product, as well as in an equivalent interaction representation,

$$iG(\mathbf{p}, t - t') = \frac{\langle T[a_{\mathbf{p}i}(t) a_{\mathbf{p}i}^\dagger(t') S] \rangle_0}{\langle S \rangle_0}, \tag{2.49}$$

where S is the evolution operator and $\langle \ldots \rangle_0$ refers to the mean value in the ground state. The operators $a_{\mathbf{p}H}(t)$ and $a_{\mathbf{p}i}(t)$ are related to the operators $a_{\mathbf{p}}$ in the Schrödinger representation by the unitary transformations

$$\begin{aligned} a_{\mathbf{p}H}(t) &= e^{iHt/\hbar} a_{\mathbf{p}} e^{-iHt/\hbar}, \\ a_{\mathbf{p}i}(t) &= e^{iH_0 t/\hbar} a_{\mathbf{p}} e^{-iH_0 t/\hbar}. \end{aligned} \tag{2.50}$$

Here H is the full Hamiltonian of the system, which consists of the Hamiltonian H_0 of the noninteracting particles and the interaction Hamiltonian H_{int}:

$$H = H_0 + H_{\text{int}}. \tag{2.51}$$

The wave functions in the three representations are connected in the following way:

$$\Phi_H = e^{iHt/\hbar}\Phi_S, \qquad i\hbar\frac{\mathrm{d}\Phi_S(t)}{\mathrm{d}t} = H\Phi_S(t), \qquad \frac{\mathrm{d}\Phi_H}{\mathrm{d}t} = 0,$$

$$\Phi_i = e^{iH_0t/\hbar}\Phi_S, \qquad i\hbar\frac{\mathrm{d}\Phi_i(t)}{\mathrm{d}t} = H_{\text{int},i}(t)\Phi_i(t), \tag{2.52}$$

where

$$H_{\text{int},i}(t) = e^{iH_0t/\hbar}H_{\text{int}}e^{-iH_0t/\hbar}. \tag{2.53}$$

One can introduce the $S(t,0)$ matrix at $t > 0$ as

$$S(t,0) = e^{iH_0t/\hbar}e^{-iHt/\hbar}, \qquad i\hbar\frac{\mathrm{d}S(t,0)}{\mathrm{d}t} = H_{\text{int},i}(t)S(t,0). \tag{2.54}$$

For the arbitrary times $t_1 > t_2$, the $S(t_1,t_2)$ matrix can be expressed through the T product of the interaction operators as follows:

$$S(t_1,t_2) = T\exp\left[-\frac{i}{\hbar}\int_{t_2}^{t_1}\mathrm{d}t'\,H_{\text{int},i}(t')\right]. \tag{2.55}$$

The evolution operator S, which appeared in Eq. (2.49), is nothing but the matrix $S(\infty,-\infty)$.

By using the S matrix, one can establish the relationship between the functions and the operators in the two representations, namely,

$$\Phi_H = S^\dagger(t,0)\Phi_i(t), \qquad \Phi_i(t_1) = S(t_1,t_2)\Phi_i(t_2),$$

$$a_{\mathbf{p}H}(t) = S^\dagger(t,0)a_{\mathbf{p}i}(t)S(t,0). \tag{2.56}$$

The wave function of the ground state in the Heisenberg representation Φ_H^0 coincides with the same wave function in the interaction representation taken at the moment $t=0$, i.e., $\Phi_H^0 = \Phi_i^0(t=0)$, whereas the averaging in expression (2.49) is done with the wave function $\Phi_i^0(t=-\infty) = \Phi_0$. Between the two wave functions there is the relation

$$\Phi_H^0 = S(0,-\infty)\Phi_i^0(t=-\infty). \tag{2.57}$$

Following the adiabatic hypothesis, the interaction between the particles is switched off at $t=-\infty$ and the wave function $\Phi_i^0(t=-\infty)$ coincides with the wave function of the ideal Bose gas at $T=0$. At the same time, the wave function Φ_H^0 of Eq. (2.57) takes into account the interaction between the particles because it is switched on at the moment $t=0$. Taking into account the specific role of the condensed particles, Beliaev proposed to represent the operations of T ordering and $\langle\ \rangle_0$ averaging as the succession of two independent operations, as follows:

$$T = T^0T', \qquad \langle\ \rangle_0 = \langle\langle\ \rangle'\rangle^0.$$

Here the T' ordering acts only on the operators of the noncondensate particles, and the averaging $\langle\ \rangle'$ takes place on their vacuum state.

The operations T^0 and $\langle\ \rangle^0$ deal with only the operators a_0 and $a_0^\dagger$, which commute with all operators $a_{\mathbf{p}}^\dagger$ and $a_{\mathbf{p}}$, with $\mathbf{p} \neq 0$. The Green's function (2.49) can be rewritten in one way for the noncondensate particles,

$$iG(\mathbf{p}, t-t') = \frac{\langle T^0\{\langle T'[a_{\mathbf{p}i}(t)a^\dagger_{\mathbf{p}i}(t')S]\rangle'\}\rangle^0}{\langle S\rangle_0}, \quad \mathbf{p}\neq 0, \tag{2.58}$$

and in another way for the condensed particles,

$$iG(0, t-t') = \frac{\langle T^0[a_{0i}(t)a^\dagger_{0i}(t')\langle S\rangle']\rangle^0}{\langle S\rangle_0}. \tag{2.59}$$

The intermediate matrix element, which appears in the numerator of Eq. (2.58), has the following property:

$$\langle T'[a_{\mathbf{p}i}(t)a^\dagger_{\mathbf{p}i}(t')S]\rangle' = \langle T'[a_{\mathbf{p}i}(t)a^\dagger_{\mathbf{p}i}(t')S]\rangle'_{\text{link}}\langle S\rangle'. \tag{2.60}$$

This equality shows that the matrix element on the left-hand side can be expressed as the product of two factors. One of them is the contribution of only the linked diagrams.

The second factor $\langle S\rangle'$ is the contribution of all the possible unlinked diagrams, called "vacuum loops," which accompany each given linked diagram. In the usual diagram technique, this factor can be reduced by the expression $\langle S\rangle_0$, which is present in the denominator of Eq. (2.58). But in the present case, such reduction is impossible because of the difference between $\langle S\rangle'$ and $\langle S\rangle_0$, which is due to the presence of the condensate. Expression (2.60), after substitution into Eq. (2.58), will give rise to the condensate Green's functions. Indeed, the linked diagrams can contain as parameters the condensed-particle operators $a_{0i}(t_j)$ and $a^\dagger_{0i}(t'_j)$ in different combinations. These operators arise because of the series expansion of the evolution operator $S(\infty, -\infty)$ in terms of the interaction Hamiltonian H_{int}.

For example, one of the linked-diagram contributions can contain the matrix element in the integral form,

$$\mathcal{M}(\mathbf{p}, t-t') = \int\cdots\int \mathrm{d}t_1\cdots\mathrm{d}t_m\mathrm{d}t'_1\cdots\mathrm{d}t'_m M(\mathbf{p}, t-t'; t_1\cdots t_m; t'_1\cdots t'_m) \times a_{0i}(t_1)\cdots a_{0i}(t_m)a^\dagger_{0i}(t'_1)\cdots a^\dagger_{0i}(t'_m). \tag{2.61}$$

After substitution of Eq. (2.61) into Eqs. (2.60) and (2.58), one can observe that the integrand operators of Eq. (2.61) give rise to the m-particle Green's function,

$$iG(0, t_1\cdots t_m; t'_1\cdots t'_m) = \frac{\langle T^0[a_{0i}(t_1)\cdots a_{0i}(t_m)a^\dagger_{0i}(t'_1)\cdots a^\dagger_{0i}(t'_m)\langle S\rangle']\rangle^0}{\langle S\rangle_0}, \tag{2.62}$$

which has the same structure as the one-particle Green's function (2.59).

Here it is supposed that the interaction Hamiltonian has the form

$$H_{\text{int}} = \frac{1}{2V}\sum_{\mathbf{pqk}} U(k)a^\dagger_{\mathbf{p}}a^\dagger_{\mathbf{q}}a_{\mathbf{q}+\mathbf{k}}a_{\mathbf{p}-\mathbf{k}}, \tag{2.63}$$

which conserves the total number of particles. If the number of noncondensate particles is conserved separately, the same takes place for the condensed particles. This fact is reflected in the structure of the Green's function (2.62).

For the sake of simplicity, expression (2.59) is now studied, because this example demonstrates the main properties of condensed systems. At the outset it is necessary to note that the average $\langle H_{\text{int}}\rangle'$ of Eq. (2.63) contains only one term with the condensate-particle operators,

$$\langle H_{\text{int}}\rangle' = U(0)\frac{a_0^{\dagger 2}a_0^2}{2V} \approx V\frac{U(0)n^2}{2}, \quad n = \frac{N}{V}. \tag{2.64}$$

Based on this consideration, the matrix element $\langle S\rangle'$ can be represented in the form

$$\langle S\rangle' = e^{V\sigma}, \tag{2.65}$$

where σ depends on n and does not depend on V. At first sight, it seems to be possible to neglect the noncommutativity of the operators $a_{0i}(t)$ and $a_{0i}^{\dagger}(t)$ because their values in the ideal Bose gas are equal to $\sqrt{N}$, whereas their commutator, equal to unity, is an infinitesimal of the order of $1/V$ compared with the product $a_0^{\dagger}a_0$. But the presence of the matrix element $\langle S\rangle'$ in expression (2.59) does not permit one to do so.

As one can see, the matrix element $\langle S\rangle'$ contains in its series expansion all powers of V. They compensate for the noncommutativity of the operators a_0 and $a_0^{\dagger}$, which must be taken into account carefully. The ordering operation T^0 links the operators $a_{0i}(t)$ and $a_{0i}^{\dagger}(t')$ with the operators $a_{0i}(t_j)$ and $a_{0i}^{\dagger}(t_j')$ entering into the components of $\langle S\rangle'$. In this way the matrix element $\langle S\rangle'$ interferes with the determination of the approximate c-number values of the operators $a_{0i}(t)$ and $a_{0i}^{\dagger}(t')$ and does not permit them to be replaced by the value $\sqrt{N}$, where N is the full number of the particles. Nevertheless, Beliaev showed that such substitution is finally possible, but by c-number $\sqrt{N_0}$, where N_0 is also an extensive value, $N_0 \sim V$. N_0 is the number of condensed particles in the interacting Bose gas and does not coincide with the full particle number N. The contribution of the vacuum-loop diagrams $\langle S\rangle'$ is responsible for transforming the ideal Bose gas condensate into the real condensate.

Beliaev's proof of these statements is complicated and is skipped here. But to obtain a qualitative understanding of the underlying physics, one can return from expression (2.59) to initial formula (2.48).

The Green's function (2.48) for arbitrary $\mathbf{p}$ describes the transfer of a particle from the time t' to another time t. When $\mathbf{p}$ is nonzero and the corresponding single-particle state is empty, the only possibility for making such a transfer is first to create a particle at the time t' and after that to annihilate it at time t. Then the Green's function $G(\mathbf{p}, t-t')$ becomes proportional to $\theta(t-t')$. But if the ground state of the condensed particles contains a large number of participants N_0, another possibility arises, which is first to annihilate one of the existing particles and then to create an additional one. The interchanging of the two operations gives results that differ by the amount $1/N_0$. When N_0 tends to infinity the two operations become independent. This is the physical reason of the factorization of the Green's function $G(0, t-t')$. To determine the time dependence of its two factors, one begins by considering systems with a given total number of the particles N. The ground-state functions are denoted as $\Phi_H^0(N)$. For $t > t'$, the Green's functions $G(0, t-t')$ can be represented as

$$\begin{aligned} iG(0, t-t') &= \langle \Phi_H^0(N)|a_{0H}(t)a_{0H}^{\dagger}(t')|\Phi_H^0(N)\rangle \\ &\approx \langle \Phi_H^0(N)|a_{0H}(t)|\Phi_H^0(N+1)\rangle\langle \Phi_H^0(N+1)|a_{0H}^{\dagger}(t')|\Phi_H^0(N)\rangle \\ &= N_0 e^{-i\mu(t-t')}. \end{aligned} \tag{2.66}$$

Here it was supposed that the action of the operator $a_{0H}(t)$ on the wave function of the ground state $\Phi_H^0(N+1)$ transforms it again into the ground-state function $\Phi_H^0(N)$, but with

N particles, as follows:

$$a_{0H}(t)|\Phi_H^0(N+1)\rangle \approx \sqrt{N_0}|\Phi_H^0(N)\rangle \tag{2.67}$$

Approximation (2.67) can be justified only when N_0 is an extensive value, $N_0 \sim V$. In this case a change of the number N_0 by unity does not result in the essential rearrangement of the gas particles and in the excitation of the system. The time dependence in Eq. (2.66) can be obtained if one takes into account expressions (2.50) and the definition of the chemical potential,

$$\mu = E_0(N+1) - E_0(N). \tag{2.68}$$

The value N_0 that appears in Eq. (2.66) is a mean value because the function $\Phi_H^0(N)$ does not have a well-defined number N_0 of the condensed particles.

The two factors into which the Green's function $G(0, t-t')$ is decomposed are

$$a_{0H}(t) = \sqrt{N_0}e^{-i\mu t}, \qquad a_{0H}^{\dagger}(t) = \sqrt{N_0}e^{i\mu t}. \tag{2.69}$$

If one uses the chemical potential μ instead of N as a thermodynamic variable, then the time dependence of Eqs. (2.69) disappears and the two factors become

$$a_{0H}(t) = \sqrt{N_0}, \qquad a_{0H}^{\dagger}(t) = \sqrt{N_0}. \tag{2.70}$$

As a result, the rules of the Beliaev diagram technique can be formulated.

It is necessary to take into account only the linked diagrams in the perturbation-theory expansion for the Green's function of the noncondensate particles. The operators $a_{0i}(t_j)$ and $a_{0i}^{\dagger}(t_j')$ that appear in the interaction representation must be substituted by the operators $a_{0H}(t_j)$ and $a_{0H}^{\dagger}(t_j')$ that appear in the Heisenberg representation in the forms given in Eqs. (2.69) or (2.70). The density of the condensate n_0 in the interacting Bose gas is not equal to n, even at zero temperature. The operations T' and $\langle\ \rangle'$ act on only the noncondensate particles. After that it is possible to begin the construction of diagrams and graphical equations. The Hamiltonian (2.63) describes the interaction of two incident particles that leads to momentum transfer and their transformation into two scattered particles. The full numbers of incident and scattered particles must coincide, but the particles can belong to different subsystems such as condensate and noncondensate. In the diagrams, the condensed particles are represented by the crooked lines with arrows, whereas the noncondensate particles are represented by the straight thin lines with arrows. Some of the different types of first-order diagrams are represented in Fig. 2.1.

In all the diagrams, the sum of straight and crooked incoming lines is equal to the sum of outgoing lines. An arbitrary diagram belonging to the Green's function $G(\mathbf{p}, \omega)$ can be represented as a chain constructed of irreducible self-energy parts linked by only one straight thin line, which represents the zeroth-order Green's function $G^0(\mathbf{p}, \omega)$. The irreducible self-energy part consists of the elements linked between themselves by two or more thin straight lines. The zeroth-order Green's function $G^0(\mathbf{p}, \omega)$ is represented by a straight thin line with two arrows in the same direction, $\rightarrow\!\!\rightarrow$, while the full Green's function $G(\mathbf{p}, \omega)$ is represented by a thick straight line with two arrows in the same directions $\rightarrow\!\!\rightarrow$.

There are three types of the irreducible self-energy parts, which differ from one another by the number of incoming and outgoing straight and crooked lines. The first type has one incoming and one outgoing straight thin line. In these diagrams the numbers of incoming and outgoing crooked lines are equal. This self-energy part is denoted as $\Sigma_{11}(\mathbf{p}, \omega)$ and is

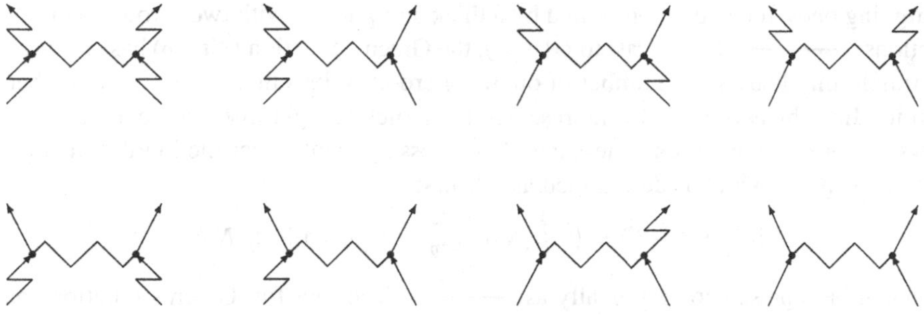

Figure 2.1. The diagrams describing the interaction between two particles either in or out of the condensate (from Ref. 37).

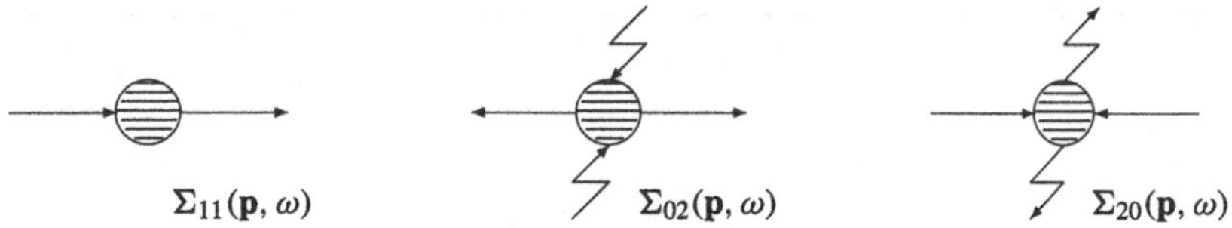

Figure 2.2. Three types of the irreducible self-energy parts (from Ref. 40).

represented in Fig. 2.2 by a dashed circle without crooked lines. There are also two other irreducible self-energy parts, $\Sigma_{02}(\mathbf{p}, \omega)$ and $\Sigma_{20}(\mathbf{p}, \omega)$. $\Sigma_{02}(\mathbf{p}, \omega)$ gathers the diagrams with two outgoing straight thin lines, whereas the number of incoming crooked lines is two more than the number of outgoing crooked lines. This self-energy part is represented by the dashed circle with two outgoing straight lines and two incoming crooked lines. The last self-energy part, $\Sigma_{20}(\mathbf{p}, \omega)$, can be obtained from the previous one by interchanging the incoming and the outgoing lines. It is represented in Fig. 2.2 by the dashed circle with two incoming straight lines and two outgoing crooked lines.

All three types of self-energy parts can be combined in an arbitrary order in the graphical representation of the Green's function $G(\mathbf{p}, \omega)$. The only evident condition is that the matrix elements $\Sigma_{02}(\mathbf{p}, \omega)$ and $\Sigma_{20}(\mathbf{p}, \omega)$ must be present in equal numbers.

The existence of the self-energy parts $\Sigma_{02}(\mathbf{p}, \omega)$ and $\Sigma_{20}(\mathbf{p}, \omega)$ stipulates the introduction of the two anomalous Green's functions $\hat{G}(\mathbf{p}, \omega)$ and $\check{G}(\mathbf{p}, \omega)$ for the noncondensate particles. They are

$$
\begin{aligned}
i\hat{G}(\mathbf{p}, t-t') &= \frac{1}{N_0}\langle \Phi_H^0(N)|T[a_{0H}(0)a_{0H}(0)a^\dagger_{\mathbf{p}H}(t)a^\dagger_{-\mathbf{p}H}(t')]|\Phi_H^0(N)\rangle \\
&\approx \langle \Phi_H^0(N+2)|T[a^\dagger_{\mathbf{p}H}(t)a^\dagger_{-\mathbf{p}H}(t')]|\Phi_H^0(N)\rangle, \\
i\check{G}(\mathbf{p}, t-t') &= \frac{1}{N_0}\langle \Phi_H^0(N)|T[a^\dagger_{0H}(0)a^\dagger_{0H}(0)a_{-\mathbf{p}H}(t)a_{\mathbf{p}H}(t')]|\Phi_H^0(N)\rangle \\
&\approx \langle \Phi_H^0(N-2)|T[a_{-\mathbf{p}H}(t)a_{\mathbf{p}H}(t')]|\Phi_H^0(N)\rangle.
\end{aligned}
\tag{2.71}
$$

The Green's function $\hat{G}(\mathbf{p}, \omega)$ contains the linked diagrams that include two outgoing straight thin lines and a number of incoming crooked lines that exceeds by two the number of outgoing ones. It can be represented by a thick straight line with two arrows in opposite directions: $\longleftrightarrow$. In contrast to $\hat{G}(\mathbf{p}, \omega)$, the Green's function $\check{G}(\mathbf{p}, \omega)$ has two incoming straight thin lines and a number of outgoing crooked lines that exceeds the number of incoming lines by two. It is also represented by a thick straight line, with counterdirected arrows: $\rightarrow\leftarrow$. For completeness it is also necessary to introduce the fourth full Green's function $G'(\mathbf{p}, \omega)$, which is determined as follows:

$$iG'(\mathbf{p}, t - t') = \langle \Phi_H^0(N) | T[a_{\mathbf{p}H}^\dagger(t) a_{\mathbf{p}H}(t')] | \Phi_H^0(N) \rangle,$$

which can be represented graphically as $\longleftarrow$. The four full Green's functions obey two systems of graphical equations, which are equivalent to the Dyson [42] equation in quantum electrodynamics. They are [37, 40]

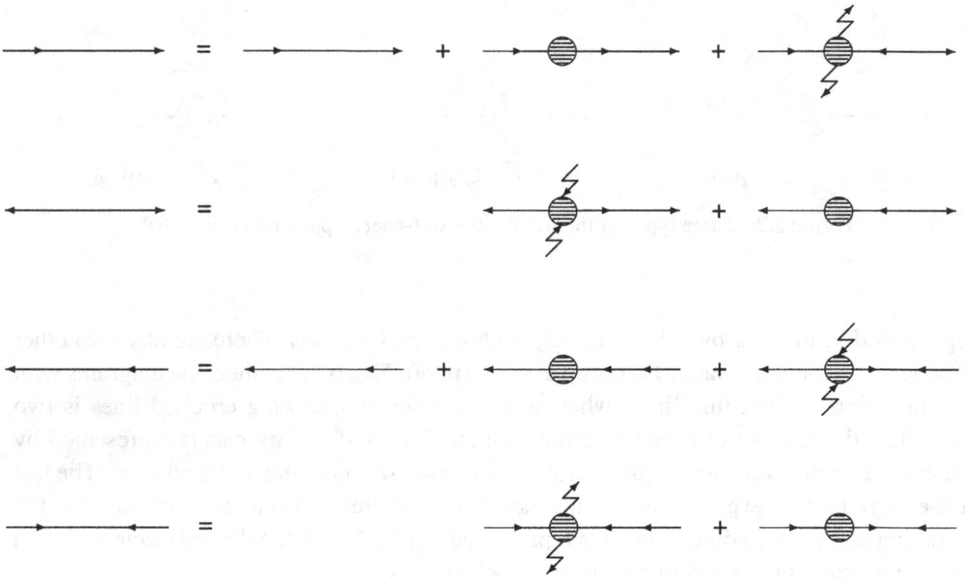

(2.72)

These equations will be studied in detail in Chapter 3, in which the exciton–exciton and the exciton–phonon interactions will be taken into account. In the rest of this section, we apply the Bogoliubov model of the weakly nonideal Bose gas to the system of excitons at high density in semiconductors, neglecting exciton–phonon interactions.

2.1.5 The Excitation Spectrum and Momentum Distribution as Seen in the Phonon-Assisted Luminescence

As shown in Chapter 1, at high excitation levels the phonon-assisted luminescence lines of orthoexcitons and paraexcitons in Cu_2O crystals become narrower than the Maxwell–Boltzmann distribution function at the same temperature. As we mentioned in Subsection 1.4.3, however, the analysis of the line shape of the phonon-assisted luminescence from excitons in Cu_2O in terms of an ideal-gas model breaks down, for example, when one attempt to fit the low-energy tail of the luminescence in a self-consistent way.

The line shape of the phonon-assisted luminescence really should be analyzed in terms of the theory of interacting dilute Bose gas. This was first attempted by Haug and Kranz [43] for the case of $T \cong 0$ and later in the more general case of $T \neq 0$ by Shi, Verechaka, and Griffin [44]. Both of these papers concentrated on the information about the excitation spectrum of the system given by the low-energy tail of the luminescence.

Using the formalism of Subsection 1.4.1, Shi et al. showed that if the phonon-assisted recombination rate is independent of **k**, then one can write the intensity of the phonon-assisted exciton recombination luminescence as

$$I(\omega) = (\text{const.})\frac{N(\tilde{\omega})}{V}\sum_{\mathbf{k}} A(\mathbf{k}, \tilde{\omega}),$$

where $\hbar\tilde{\omega} = \hbar\omega + \hbar\omega_{\text{phonon}} - E_0 - \mu$ and $E_0 = E_{\text{gap}} - \text{Ry}_{\text{ex}}$ is the energy to create an exciton in its ground state. After the separation of the contribution from the Bose condensed excitons with $\mathbf{k} = 0$ from the noncondensed excitons, the luminescence intensity $I(\omega)$ takes the form

$$I(\omega) = 2\pi n_0 \delta(\hbar\tilde{\omega}) + I_{\text{nc}}(\omega), \tag{2.73}$$

in which the constant factor is dropped. In this simplification,

$$I_{\text{nc}}(\omega) = \frac{1}{V}\sum_{\mathbf{k}\neq 0} C(\mathbf{k}, \tilde{\omega}) = \frac{1}{V}\sum_{\mathbf{k}\neq 0} N(\tilde{\omega})A(\mathbf{k}, \tilde{\omega}). \tag{2.74}$$

We obtain the first term in Eq. (2.73) by taking into account the result of Eqs. (2.70) of Beliaev's technique:

$$a_{0H}(t) = \sqrt{N_0}, \qquad a_{0H}(0) = \sqrt{N_0}, \qquad n_0 = (N_0/V).$$

The second term $I_{\text{nc}}(\omega)$ is expressed through the correlation functions $C(\mathbf{k}, \omega)$ and $A(\mathbf{k}, \omega)$:

$$C(\mathbf{k}, \omega) = \int_{-\infty}^{\infty} \mathrm{d}t\, e^{-i\omega t}\langle a^{\dagger}_{\mathbf{k}H}(t)a_{\mathbf{k}H}(0)\rangle,$$

$$A(\mathbf{k}, \omega) = \int_{-\infty}^{\infty} \mathrm{d}t\, e^{-i\omega t}\langle a_{\mathbf{k}H}(0)a^{\dagger}_{\mathbf{k}H}(t) - a^{\dagger}_{\mathbf{k}H}(t)a_{\mathbf{k}H}(0)\rangle, \tag{2.75}$$

which are related to the Green's function of the interacting system, which may in general include not only exciton–exciton interactions but also exciton–phonon interactions, as discussed in Chapter 3. As discussed in Subsection 1.4.1, $C(\mathbf{k}, \omega)$ is related to $A(\mathbf{k}, \omega)$ by the relation

$$C(\mathbf{k}, \omega) = N(\omega)A(\mathbf{k}, \omega), \tag{2.76}$$

where $N(\omega)$ has the Bose–Einstein form,

$$N(\omega) = \frac{1}{e^{\beta\hbar\omega} - 1}, \tag{2.77}$$

and $A(\mathbf{k}, \omega)$ gives rise to the renormalized noncondensate single-particle density of states [44],

$$\rho(\omega) = \frac{1}{V}\sum_{\mathbf{k}\neq 0} A(\mathbf{k}, \omega). \tag{2.78}$$

In the case of an ideal Bose gas described by the Hamiltonian (2.1), one obtains

$$A(\mathbf{k}, \omega) = 2\pi\delta(\hbar\omega + \mu - T_{\mathbf{k}}), \tag{2.79}$$

where $T_{\mathbf{k}} = \hbar^2 k^2/2m$ and

$$\rho(\tilde{\omega}) = \rho(\hbar\omega' - \mu) = \frac{\sqrt{2} m_{\mathrm{ex}}^{3/2}}{\pi\hbar^3}\sqrt{\hbar\omega'}, \tag{2.80}$$

which makes sense for only positive frequencies $\omega' \geq 0$. The line shape for the ideal Bose gas takes the form

$$I(\omega) = 2\pi n_0 \delta(\hbar\omega' - \mu) + \frac{\sqrt{2} m_{\mathrm{ex}}^{3/2}\sqrt{\hbar\omega'}}{\pi\hbar^3[e^{\beta(\hbar\omega' - \mu)} - 1]}, \tag{2.81}$$

where n_0 and μ obey the conditions

$$\begin{aligned} &\mu = 0, \qquad n_0 = n\left[1 - \left(\frac{T}{T_c}\right)^{3/2}\right], \qquad T < T_c, \\ &\mu < 0, \qquad n_0 = 0, \qquad T \geq T_c. \end{aligned} \tag{2.82}$$

Here n is the total density of excitons and T_c is the critical temperature for the ideal Bose gas with the quadratic dispersion law:

$$k_B T_c = \frac{2\pi\hbar^2 n^{2/3}}{m_{\mathrm{ex}}(2.612)^{2/3}} = \frac{3.31\hbar^2 n^{2/3}}{m_{\mathrm{ex}}}. \tag{2.83}$$

To go beyond the ideal-gas approximation, Shi et al. [44] used the finite-temperature version of the Bogoliubov model elaborated by Popov and Faddeev [45,46] to obtain the luminescence line shape for a dilute weakly interacting Bose gas at finite temperatures. Earlier work on the low-temperature behavior of a dilute system of hard spheres was also done by Lee and Yang [38].

In both the work by Haug and Kranz [43] and that by Shi et al. [44], the correlation functions $A(\mathbf{k}, \omega)$ and $C(\mathbf{k}, \omega)$ for the condensed phase were calculated by use of the Bogoliubov transformation and expressions (2.35)–(2.43) for the case $k_0 = 0$. But the formulas for the chemical potentials were different, in accordance with different temperature regions considered in the two papers. The finite-temperature version of Ref. 44 is presented here.

The time dependence of the exciton operator $a_{\mathbf{k}H}(t)$ in the Bogoliubov approximation has the form

$$a_{\mathbf{k}H}(t) = u_{\mathbf{k}} e^{-\frac{i}{\hbar}E_{\mathbf{k}}t}\xi_{\mathbf{k}} - v_{\mathbf{k}} e^{\frac{i}{\hbar}E_{\mathbf{k}}t}\xi_{-\mathbf{k}}^{\dagger}, \tag{2.84}$$

which leads to the expressions for the correlation functions as follows:

$$\begin{aligned} C(\mathbf{k}, \omega) &= 2\pi\left[u_{\mathbf{k}}^2 N_{\mathbf{k}}\delta(\hbar\omega - E_{\mathbf{k}}) + v_{\mathbf{k}}^2(1 + N_{\mathbf{k}})\delta(\hbar\omega + E_{\mathbf{k}})\right], \\ A(\mathbf{k}, \omega) &= 2\pi\left[u_{\mathbf{k}}^2\delta(\hbar\omega - E_{\mathbf{k}}) - v_{\mathbf{k}}^2\delta(\hbar\omega + E_{\mathbf{k}})\right]. \end{aligned} \tag{2.85}$$

Here $N_{\mathbf{k}}$ is the occupation number of the new elementary excitations,

$$N_{\mathbf{k}} = \langle\xi_{\mathbf{k}}^{\dagger}\xi_{\mathbf{k}}\rangle = \frac{1}{e^{\beta E_{\mathbf{k}}} - 1}. \tag{2.86}$$

The density of states $\rho(\omega)$ was calculated with the formulas

$$\rho(\omega) = \frac{1}{\pi}\int_0^\infty k^2 \mathrm{d}\,k\left[u_k^2\delta(\hbar\omega - E_\mathbf{k}) - v_k^2\delta(\hbar\omega + E_\mathbf{k})\right],$$

$$E_\mathbf{k} = \sqrt{T_\mathbf{k}^2 + 2L_\mathbf{k}T_\mathbf{k}}, \qquad u_\mathbf{k}^2 = (T_\mathbf{k} + L_\mathbf{k} + E_\mathbf{k})/2E_\mathbf{k},$$
$$v_\mathbf{k}^2 = (T_\mathbf{k} + L_\mathbf{k} - E_\mathbf{k})/2E_\mathbf{k}, \tag{2.87}$$

where the $\mathbf{k}$ dependence of $L_\mathbf{k}$ was dropped. Then the relations

$$T_\mathbf{k} = -L_0 + \sqrt{L_0^2 + E_\mathbf{k}^2},$$

$$k^2\mathrm{d}k = \sqrt{2}m_\mathrm{ex}^{3/2}\frac{\sqrt{-L_0 + \sqrt{L_0^2 + E_\mathbf{k}^2}}}{\sqrt{L_0^2 + E_\mathbf{k}^2}}E_\mathbf{k}\mathrm{d}E_\mathbf{k} \tag{2.88}$$

can be used.

When all the dust settles, the density of states $\rho(\tilde{\omega})$ and the intensity $I_\mathrm{nc}(\omega)$ in Eq. (2.74) obtain the form

$$\rho(\tilde{\omega}) = \mathrm{sgn}(\tilde{\omega})\frac{m_\mathrm{ex}^{3/2}}{\pi\sqrt{2}}\sqrt{-L_0 + \sqrt{L_0^2 + (\hbar\tilde{\omega})^2}}\left[\sqrt{L_0^2 + (\hbar\tilde{\omega})^2} + \hbar\tilde{\omega}\right],$$

$$I_\mathrm{nc}(\omega) = \rho(\tilde{\omega})N(\tilde{\omega}). \tag{2.89}$$

$\rho(\tilde{\omega})$ changes sign for different signs of the frequency detunings $\tilde{\omega}$ or $(\hbar\omega' - \mu)$. Being multiplied by $N(\tilde{\omega})$, which also changes its sign in dependence on the sign of $\tilde{\omega}$, the resulting luminescence line shape $I_\mathrm{nc}(\omega)$ becomes positive for all values of ω, $\tilde{\omega}$, and ω'. As noted by Shi et al. [44], all the interaction effects go into the renormalized single-particle density of states $\rho(\omega)$. The integrated intensity of the noncondensate contribution to the luminescence line is

$$\int_{-\infty}^{\infty} I_\mathrm{nc}(\omega)\,\mathrm{d}\omega = \frac{2\pi}{V}\sum_{\mathbf{k}\neq 0}\langle a_\mathbf{k}^\dagger a_\mathbf{k}\rangle = 2\pi\,[n - n_0(T)]\,.$$

At $T = 0$ it remains nonzero because of the difference between n and $n_0(T = 0)$ in the interacting Bose gas. As the temperature decreases below T_c, the spectral weight is increasingly shifted to the condensate contribution given by the first term in Eq. (2.73).

As shown in Refs. 43 and 44, expressions (2.73), (2.87), and (2.89) imply that the total luminescence line shape consists of three parts. One of them is the emission of a photon and a phonon due to the decay of the condensed excitons, which remains an unbroadened δ function. The second part is due to the emission of a photon and a phonon due to the decay of an exciton with wave vector $\mathbf{k} \neq 0$, with the simultaneous emission of a collective elementary excitation with wave vector $\mathbf{k}$ and energy E_k. The emission of the elementary excitation diminishes the total energy of the emitted photon and phonon and gives rise to the low-energy tail of the line shape. The third part comes from the emission of the photon and the phonon due to decay of an exciton with wave vector $\mathbf{k} \neq 0$ under the simultaneous absorption of a collective elementary excitation. This contribution occurs on the high-energy side of the exciton–phonon resonance and is especially pronounced at high temperatures, in which case many thermal collective excitations exist. This high-energy wing survives as a

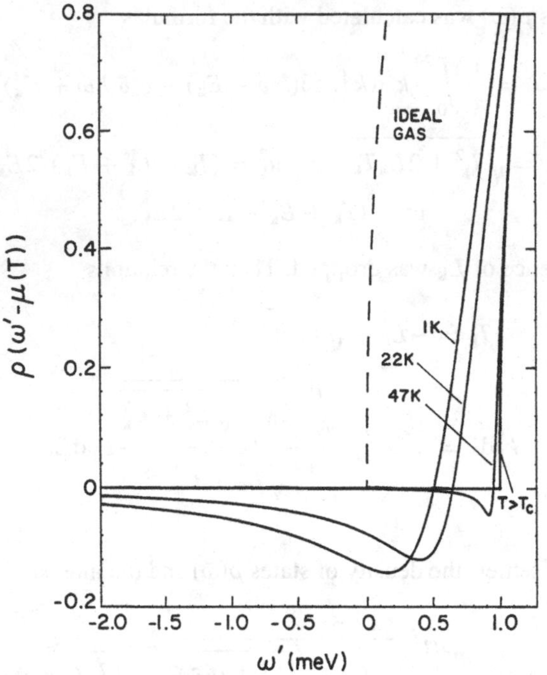

Figure 2.3. Noncondensate single-particle density of states $\rho(\hbar\omega' - \mu)$ as a function of frequency detuning ω' (from Ref. 44).

system goes normal when $u_{\mathbf{k}} = 1$ and $v_{\mathbf{k}} = 0$, but disappears as the temperature approaches zero.

Figure 2.3 shows, from Ref. 44, the noncondensate single-particle density of states $\rho(\hbar\omega' - \mu)$ at different temperatures for the parameters $L_0 = 0.5$ meV, $n_{\text{ex}} = 10^{19}$ cm^{-3}, $m_{\text{ex}} = 2.7m_0$, and $T_c = 51$ K. The value $\rho(\hbar\omega' - \mu)$ is negative for $\hbar\omega' < \mu(T)$. Figure 2.4 shows the decay luminescence spectrum associated with noncondensate excitons for various temperatures above and below T_c. The ideal-gas result at $T = 65$ K is shown for comparison. As seen in Fig. 2.4, the low-energy tail appears only below the critical temperature for condensation. Another result of this model that appears in this figure is a blue shift of the luminescence spectrum due to the interactions in the exciton gas. This blue shift is never seen in the experimental data from Cu_2O; although a low-energy tail appears, the overall spectral position of the phonon-assisted luminescence shifts by less than 0.2 meV, even at the highest exciton densities. As discussed in Section 2.2, however, correlation effects that have been neglected in this model are expected to give a red shift of the ground-state energy, which almost exactly cancels out this blue shift.

Based on this model, the appearance of the low-energy tail in the line shape $I_{\text{nc}}(\omega)$ was taken as a signature of the excitonic BEC by both Haug and Kranz [43] and Shi et al. [44]. In principle, this is much more direct evidence of the condensate than the δ-like condensate peak, since the overall weight of the δ-like peak could be small if the condensate fraction is only a few percent. The low-energy tail, which comes from the term $v_{\mathbf{k}}^2\delta(\hbar\omega + E_{\mathbf{k}})$ in Eqs. (2.87), disappears when $v_{\mathbf{k}} = 0$, i.e., above the critical temperature for Bose condensation in the interacting Bose gas, and persists at all temperatures below T_c because of

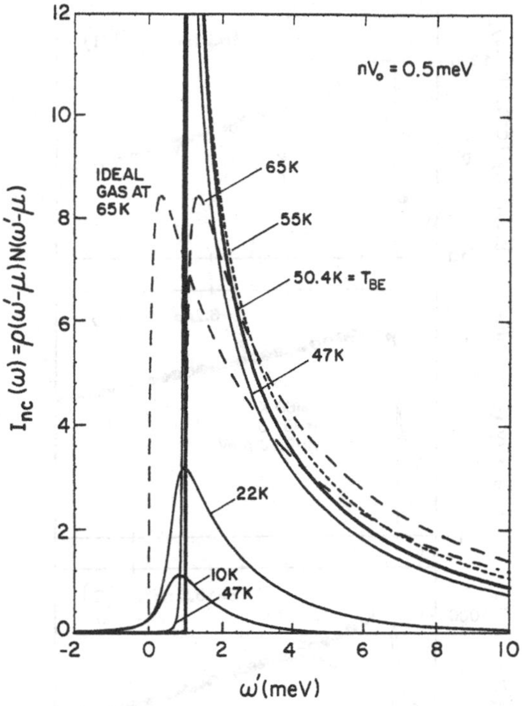

Figure 2.4. Decay luminescence spectrum associated with the noncondensate excitons for various temperatures; $I_{nc}(\omega) = \rho(\hbar\omega' - \mu)N(\hbar\omega' - \mu)$ (from Ref. 44).

the existence of noncondensate excitons with wave vectors $\mathbf{k} \neq 0$ in the ground state of the interacting system, even at $T = 0$. At $T = 0$ these noncondensate particles have the distribution function $\langle a_{\mathbf{k}}^{\dagger} a_{\mathbf{k}} \rangle = v_{\mathbf{k}}^2$; in the condensed phase at finite temperatures $0 < T < T_c$, this occupation number becomes

$$\langle a_{\mathbf{k}}^{\dagger} a_{\mathbf{k}} \rangle = u_k^2 N_k + v_k^2 (1 + N_k), \tag{2.90}$$

which gives rise to the possibility of obtaining luminescence transitions not only at the negative detunings $\tilde{\omega} < 0$, but also at positive ones $\tilde{\omega} > 0$. The high-energy wing survives as the system goes normal when $u_{\mathbf{k}} = 1$ and $v_{\mathbf{k}} = 0$, but disappears when $T = 0$.

We note that the appearance of a low-energy tail only when Bose condensation occurs is the result of an initial assumption of this model, however. In the Bogoliubov approximation, interaction terms that involve only particles in excited states are dropped. The Popov–Fadeev approximation used by Shi et al. goes one step further to include a mean-field term due to interactions with noncondensed particles, but still uses the Bogoliubov transformation for the excitation spectrum. Therefore, by definition, when there is no condensate, the excitation spectrum is unchanged in this model. As discussed in Subsection 1.4.1, however, normal collisions of particles cause broadening of luminescence lines and therefore a low-energy tail, even when the excitons are nondegenerate. Therefore one cannot say that the existence of a low-energy tail alone is a telltale for Bose condensation. On the other hand, normal collision broadening implies a broadening that is directly proportional to the total exciton density, while the width of the low-energy tail predicted by the model presented here depends on only the number of particles in the condensate, which has a strongly nonlinear

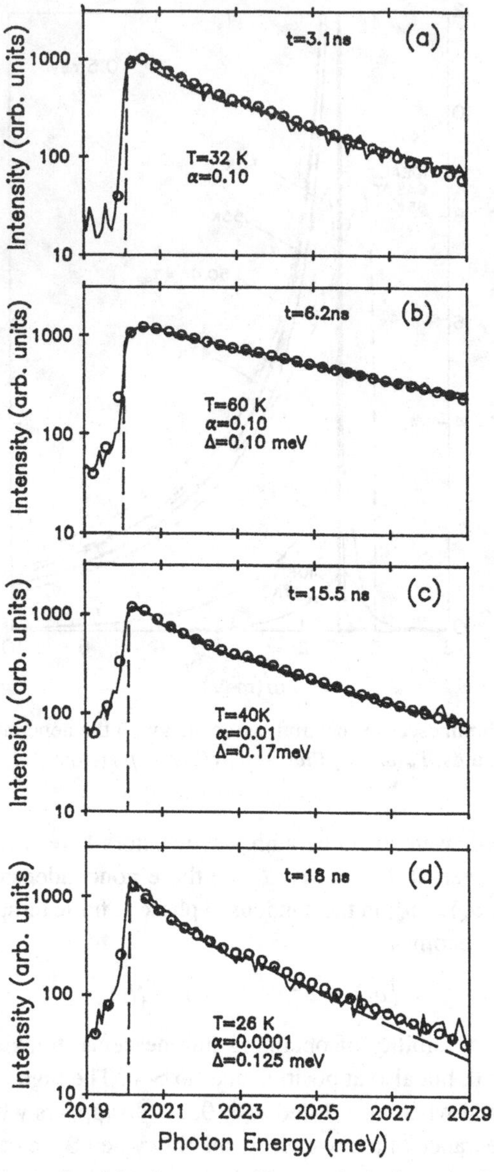

Figure 2.5. Phonon-assisted luminescence from orthoexcitons in Cu_2O taken at four different times following an intense laser pulse. Open circles are a fit to the ideal Bose–Einstein distribution of expression (1.46) convolved with a Lorentzian of width Δ (from Ref. 47).

dependence on the total density. Furthermore, since the interactions with the condensate involve all the condensate particles, while normal collisions involve only two particles at a time, once there is a condensate, these interactions should dominate over the interactions between particles in excited states.

Figure 2.5 shows phonon-assisted luminescence data from Cu_2O at several different densities and temperatures. As seen in this figure, at later times the low-energy tail is much larger, even though the exciton gas is at lower density and lower temperature; for example, the density at time $t = 15.5$ ns is approximately a factor of 2 less than at $t = 6.2$ ns and 50% cooler, but the Lorentzian broadening is almost a factor of 2 greater. This is the opposite of what is expected if the low-energy tail occurs simply because of incoherent collision broadening. In that case, the width of the collision broadening should be proportional to the classical collision rate, $\Gamma = n\sigma v$, i.e., proportional to the total exciton density and average velocity. On the other hand, as the temperature drops, the condensate fraction n_0 can increase, leading to an increase of the Bogoliubov term discussed above. This analysis therefore provides strong evidence for the collective excitations associated with Bose condensation.

Another prediction of this model is that the condensate itself should result in a δ-function spectral peak that is not broadened; this is a direct result of the long-range order in the condensate. If the condensate fraction is not large, this term could be lost in a spectrum taken with finite experimental resolution. Recent results by Lin and Wolfe [48] seem to indicate a large condensate fraction in the paraexciton state, however. Figure 2.6 shows simultaneous paraexciton and orthoexciton luminescence spectra during an intense laser pulse (the energy of the laser photons is much greater than the exciton ground-state energy, so that the laser effectively acts as an incoherent source of excitons). Uniaxial stress was applied to the crystal in this experiment, but this probably did not affect the statistics. As seen in Fig. 2.6, the paraexciton line shape is so narrow that there is no discernible high-energy tail; instead, the spectrum is a symmetric Gaussian with a width of 0.5 meV, which is slightly broader than the experimental spectral resolution. Lin and Wolfe [48] argued that this peak corresponds to symmetrically broadened luminescence from condensate excitons. According to the model presented in this section, however, the condensate luminescence should not be broadened, because of the long-range order. One possibility suggested by

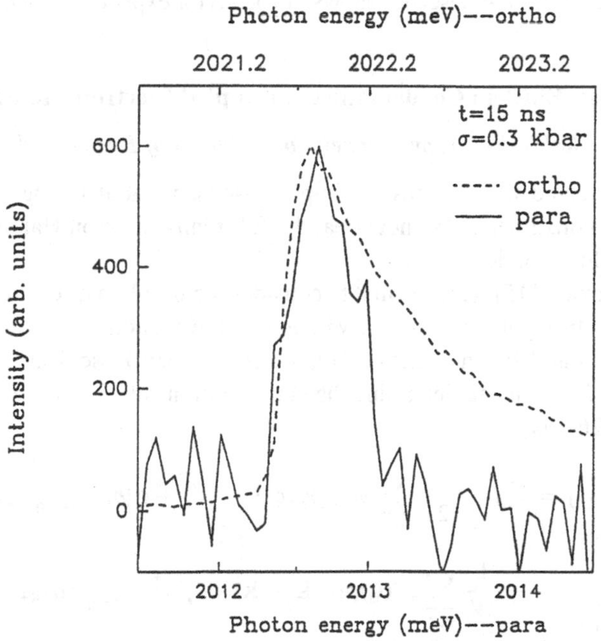

Figure 2.6. Solid curve, paraexciton luminescence; dashed curve, orthoexciton luminescence at the same time and place (from Ref. 48).

Shi et al. [44] and others is that true long-range order is not established in the exciton gas and that a quasi-condensate with topological disorder [49] occurs instead. Another possibility is something analogous to final-state effects in liquid He [19], by which the photons emitted from the condensate suffer broadening that is due to scattering in the medium. In any case, the narrow, symmetric Gaussian form of the paraexciton luminescence seen by Lin and Wolfe is a complete departure from a Maxwellian or ideal-gas distribution.

More recently, Goto et al. [50] used resonant two-photon absorption to create excitons with very low momentum, near the orthoexciton ground state. The main advantage of this method is that high densities of excitons are created at the surface of the crystal, since the two-photon excitation had a very short absorption length (less than 50 μm), just as in the case of the green (5145-Å) laser light used by Lin and Wolfe [48], but the kinetic energy of the excitons generated by resonant two-photon absorption is much lower: $\hbar k \sim (E_{\text{gap}} - \text{Ry}_{\text{ex}})/2c'$. Goto et al. [50] found that the phonon-assisted orthoexciton luminescence spectrum following the excitation pulse fit a two-component distribution, with a δ-like peak with a width of approximately 0.1 meV, equal to the experimental spectral resolution, and an excited-state distribution that fit an ideal-gas distribution [expression (1.46)] with $\mu = 0$, with a negative-energy tail, as discussed above (see Fig. 2.7). The δ peak persisted to late times, after the laser excitation has ceased, which indicates that the distribution represents a thermodynamically stable phase. In recent studies, Shen et al. [51] have shown that the δ peak persists longer when the exciton density is higher; when the laser intensity is reduced to the nonthermal limit, in which the exciton–exciton-scattering time is much longer than the exciton–phonon-scattering time, the δ peak exists only during the laser pulse. It seems likely that because the excitons in these experiments are created directly in states near the orthoexciton ground state, they tend to remain in the orthoexciton level rather than convert into paraexcitons, as in the experiments of Lin and Wolfe [48]. As a result, the orthoexciton luminescence shows the behavior expected for a Bose condensate.

2.2 Bose–Einstein Condensation of Coupled Electron–Hole Pairs

2.2.1 The Keldysh–Kozlov–Kopaev Formulation of Bose Condensation

So far, we have treated the excitons as simple bosons, without taking into account the underlying Fermi constituents. We now treat the full many-electron Hamiltonian. Can we still talk about a Bose condensate?

Keldysh and Kozlov [15] were the first to consider the problem of exciton BEC at $T = 0$ from the many-electron point of view. They noted that the excitons are formed from coupled electrons and holes and for this reason do not exactly obey Bose–Einstein statistics. As shown in Section 1.2, one can determine the commutation rule for two exciton operators, defined in Eq. (2.26), as

$$[\psi_{\text{ex},\mathbf{k}}, \psi^\dagger_{\text{ex},\mathbf{k}'}] = \delta_{\mathbf{k},\mathbf{k}'} - \frac{1}{2V}\sum_{\mathbf{q},\sigma} \varphi^*(\mathbf{q})\varphi[\alpha(\mathbf{k}-\mathbf{k}') + \mathbf{q}] b^\dagger_{\mathbf{k}'-\alpha\mathbf{k}-\mathbf{q},-\sigma} b_{\beta\mathbf{k}-\mathbf{q},-\sigma}$$

$$- \frac{1}{2V}\sum_{\mathbf{q},\sigma} \varphi^*(\mathbf{q})\varphi[\beta(\mathbf{k}'-\mathbf{k}) + \mathbf{q}] a^\dagger_{\mathbf{k}'-\beta\mathbf{k}+\mathbf{q},\sigma} a_{\alpha\mathbf{k}+\mathbf{q},\sigma}, \tag{2.91}$$

where we have neglected the index n for the orbital state (assuming that the excitons are in the ground state) and have explicitly included the spin index σ. Here $\delta_{\mathbf{k},\mathbf{k}'}$ is the Kronecker

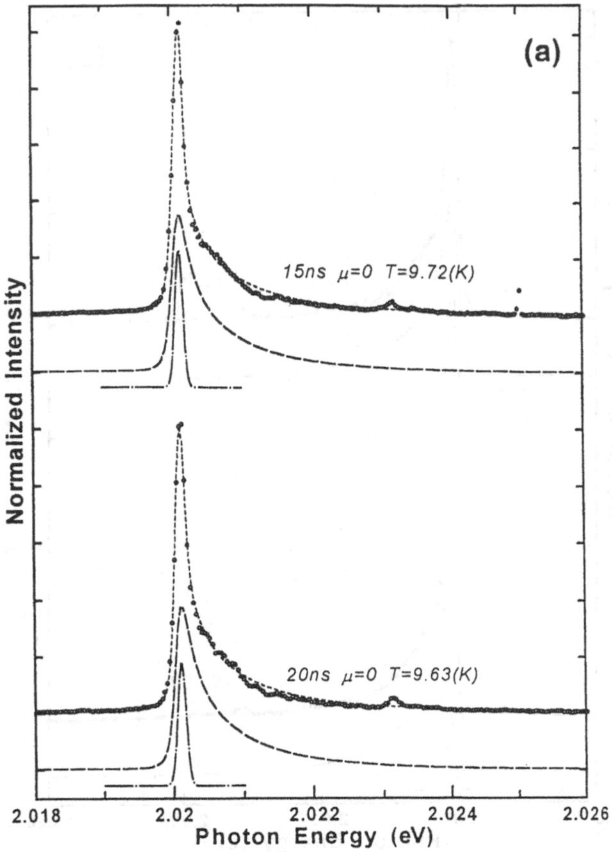

Figure 2.7. Solid curves, orthoexciton luminescence at various times following resonant two-photon excitation of the orthoexciton ground state; long-dashed curves, a fit of the ideal Bose gas distribution of expression (1.46) to the phonon-assisted luminescence at high energies, with broadening to fit the low-energy tail of the spectrum; dashed-dotted curves, the fit of the zero-kinetic-energy peak, with width given by the experimental spectral resolution; short-dashed curves, the sum of the two components (from Ref. 50).

delta symbol; $a^{\dagger}_{\mathbf{p},\sigma}$, $a_{\mathbf{p},\sigma}$, $b^{\dagger}_{\mathbf{p},\sigma}$, and $b_{\mathbf{p},\sigma}$ are the creation and the annihilation Fermi operators of electrons and holes. The deviation of the exciton operators $\hat{\psi}^{\dagger}_{\mathrm{ex},\mathbf{k}}$ and $\hat{\psi}_{\mathrm{ex},\mathbf{k}}$ from Bose statistics increases as the electron–hole density increases. For this reason, as mentioned by Keldysh and Kozlov, the possibility of considering the exciton system as a weakly nonideal Bose gas *a priori* is not evident. For exciton densities of the order of 10^{17}–10^{18} cm^{-3}, the average distance between the excitons starts to become comparable with their radius. In this case, the internal structure begins to play an important role. For example, two electrons belonging to two different excitons cannot approach each other closely if they have the same spin projections. Two holes exhibit the same behavior. In other words, the Pauli exclusion principle leads to a kinematic exciton–exciton interaction, even in the absence of a dynamic interaction. Keldysh and Kozlov [15] pointed out that the effects connected with the difermion exciton structure appear at the same order of the exciton density as that of the effects tied to the Bose gas nonideality. As a result, the systematic investigation of the exciton Bose condensate, even at low exciton densities, cannot be posed as a problem

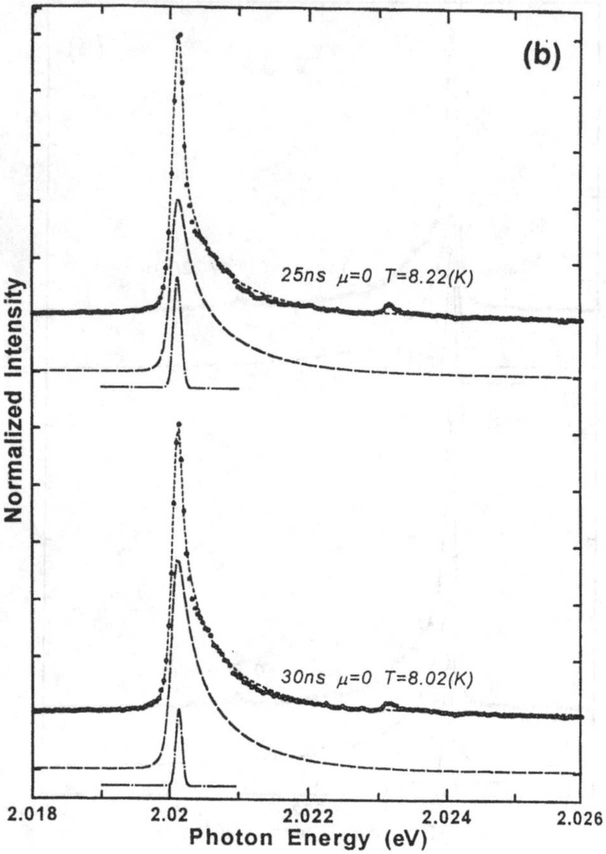

Figure 2.7. *Continued.*

about a weakly nonideal Bose gas. Nevertheless, it was shown that the exciton system with a definite spin structure possesses many similar properties. In particular, at low temperatures, a condensation can take place into a single composite-boson state with wave vector $\mathbf{k} = 0$. In this case, the corrections to the ground-state energy E_0 and to the chemical potential μ_{ex} are quadratic and linear in the exciton density n_{ex}, respectively, as follows:

$$E_0 = N_{ex}\mathrm{Ry}_{ex}\left(1 - 1/2 f n_{ex} a_{ex}^3\right),$$

$$\mu_{ex} = -\mathrm{Ry}_{ex}\left(1 - f n_{ex} a_{ex}^3\right). \tag{2.92}$$

Here N_{ex} is the total number of excitons, Ry_{ex} and a_{ex} are the exciton Rydberg constant and the exciton radius, respectively, and f is a dimensionless parameter of the order of unity, which depends on the scattering amplitudes of two free excitons as well as on the interaction of their Fermi components among themselves. The presence of the excitonic condensate changes the effective interaction between the electrons and holes and therefore their binding energy [15].

One naturally expects that BEC of the excitons will lead to superfluidity, which would be detected as anomalously fast exciton diffusion. This possibility, suggested and discussed in many papers [31,32,35,52] was subjected to criticism by Kohn and Sherrington [53]. They affirmed that the exciton system, while it is a collective of compound *e–h* quasiparticles, can undergo the phenomenon of BEC. But unlike true bosons, in their opinion, composite *e–h* quasi-bosons cannot exhibit the phenomenon of the superfluidity. Keldysh [14], as well as Hanamura and Haug [54], analyzed the latter argument and pointed out that it is based on a misunderstanding. It is necessary from the beginning to determine exactly what exciton superfluidity is and to take into account the exciton features properly. As noted by Keldysh, unlike a flux of atoms, the exciton flux is not accompanied by matter transfer. On this point there is complete accordance with the Kohn and Sherrington paper [53]. But the excitons can transfer their creation energy, their kinetic energy of translational motion, and such things as angular electric or magnetic moments, if they exist. Therefore superfluidity of the exciton gas, which exists only as a nonequilibrium, excited state of the crystal, means the existence of weakly damped fluxes of the energy or of the polarization, but not of the mass or of the charge. The proof of the impossibility of the exciton superfluidity in Ref. 53 was completely based on an investigation of mass transfer.

Furthermore, Keldysh noted that in the description of the nonequilibrium excitons it is necessary to take into account the incomplete equilibrium in the *e–h* system. It is incomplete in the sense that the electrons, holes, and excitons are in equilibrium with each other and with the crystal lattice in all parameters except one, namely, the total number of the excitons and the *e–h* pairs is determined not by the thermodynamical equilibrium condition, but by the external excitation source [14]. The exciton thermalization time is assumed to be much less than its lifetime. In this case, the establishment of a quasi-equilibrium distribution function in the exciton band can be assumed [31]. Following Ref. 14, excitonic BEC means the existence of a coherent wave of the electron density, with definite phase and finite amplitude. Superfluidity of the coherent wave could signify that its scattering processes are completely suppressed by the nonlinear effects. Unlike liquid He, the superfluid exciton flux cannot exist an arbitrarily long time, but only during the exciton lifetime. This means that the flux relaxation time is determined by the exciton lifetime, which is some orders of magnitude greater than the single-exciton relaxation time. These considerations [14] coincide with the main features of the picture of exciton BEC and exciton superfluidity suggested in early works [31,32,35,55–57].

The description of the excitonic BEC as the coherent pairing of electrons and holes was performed in Ref. 15 on the basis of the *e–h* Hamiltonian:

$$
\begin{aligned}
H = &\sum_{\mathbf{p},\sigma}[\mathcal{E}_e(\mathbf{p}) - \mu_e]a^\dagger_{\mathbf{p},\sigma}a_{\mathbf{p},\sigma} + \sum_{\mathbf{p},\sigma}[\mathcal{E}_h(\mathbf{p}) - \mu_h]b^\dagger_{\mathbf{p},\sigma}b_{\mathbf{p},\sigma} \\
&+ 1/2 \sum_{\mathbf{p},\mathbf{q},\mathbf{k},\sigma_1,\sigma_2} V_{\mathbf{k}}\big[a^\dagger_{\mathbf{p},\sigma_1}a^\dagger_{\mathbf{q},\sigma_2}a_{\mathbf{q}+\mathbf{k},\sigma_2}a_{\mathbf{p}-\mathbf{k},\sigma_1} + b^\dagger_{\mathbf{p},\sigma_1}b^\dagger_{\mathbf{q},\sigma_2}b_{\mathbf{q}+\mathbf{k},\sigma_2}b_{\mathbf{p}-\mathbf{k},\sigma_1} \\
&- 2a^\dagger_{\mathbf{p},\sigma_1}b^\dagger_{\mathbf{q},\sigma_2}b_{\mathbf{q}+\mathbf{k},\sigma_2}a_{\mathbf{p}-\mathbf{k},\sigma_1}\big],
\end{aligned} \tag{2.93}
$$

where $V_{\mathbf{k}}$ is the Fourier transform of the Coulomb interactions of two electrons embedded in a continuous medium with dielectric constant ϵ_0 and volume V, given by

$$
V_{\mathbf{k}} = \frac{4\pi e^2}{\epsilon_0 k^2 V}. \tag{2.94}
$$

The Fermi operators $a^\dagger_{\mathbf{p},\sigma}$, $a_{\mathbf{p},\sigma}$, $b^\dagger_{\mathbf{p},\sigma}$, and $b_{\mathbf{p},\sigma}$ describe the creation and the annihilation of the electrons and holes, labeled by the momentum $\mathbf{p}$ and the spin projection σ. The number of the electrons and the holes in the corresponding bands is determined by the choice of experimental conditions. Their quasi-equilibrium states can be characterized by the introduction of two different chemical potentials: μ_e for the electrons and μ_h for the holes. In the case of thermodynamic equilibrium for all quasiparticles, only one common chemical potential exists. The energy of electrons and their chemical potential μ_e are measured relative to the bottom of the conduction band. In the same way, the energy of the holes and their chemical potential μ_h are measured relative to the top of the valence band. The differences $\mathcal{E}_e(\mathbf{p}) - \mu_e$ and $\mathcal{E}_h(\mathbf{p}) - \mu_h$ do not depend on the choice of the energy scale. Assuming parabolic bands, the dispersion laws have the forms

$$\mathcal{E}_e(\mathbf{p}) = \frac{\hbar^2 p^2}{2m_e}, \qquad \mathcal{E}_h(\mathbf{p}) = \frac{\hbar^2 p^2}{2m_h}, \qquad \mathcal{E}_e(\mathbf{p}) + \mathcal{E}_h(\mathbf{p}) = \mathcal{E}(\mathbf{p}) = \frac{\hbar^2 p^2}{2\mu}, \tag{2.95}$$

where μ is the reduced mass of the relative motion; $\mu = m_e m_h/(m_e + m_h)$. In the dilute limit, at $T = 0$ the kinetic energy of the particles will be negligible compared with the Coulomb interaction energy.

As shown in Section 1.2, the Hamiltonian (2.93), which is a rewrite of Eq. (1.7), neglecting electron–hole exchange, leads to paired states of the electrons and the holes, namely, Wannier excitons. The free-exciton energy level is situated below the bottom of the conduction band by a value equal to the exciton Rydberg constant $\mathrm{Ry_{ex}}$, and the chemical potential of the condensed excitons μ_{ex} will coincide with the renormalized exciton level. The chemical reaction

$$e + h \leftrightarrow \mathrm{ex}$$

implies a relation between the chemical potentials,

$$\mu_e + \mu_h = \mu_{\mathrm{ex}}. \tag{2.96}$$

In the following, the finite lifetime of the excitons and the possibilities for their recombination will be neglected.

From physical considerations it is clear that the ground state of the many-body system will be constructed from coupled *e–h* pair states, i.e., excitons. If the excitons form a Bose condensate with wave vector $\mathbf{k} = 0$, the condensate influences the character of the *e–h* coupling, i.e., in this case coherent pairing of electrons and holes with opposite wave vectors occurs. This can be described within the framework of the macroscopic coherent state discussed above. As mentioned in Section 2.1.2, one can introduce spontaneous symmetry breaking into the Hamiltonian (2.93) with the help of the unitary transformation of Eqs. (2.20). In this expression, the Bose operators $a^\dagger$ and a must be replaced by the exciton creation and annihilation operators $c_0^\dagger$ and c_0. For simplicity, the case of a spin-oriented *e–h* system is considered, in which case $c_0^\dagger$ has the form

$$c_0^\dagger = \frac{1}{\sqrt{V}} \sum_{\mathbf{q}} \varphi(\mathbf{q}) a_{\mathbf{q}}^\dagger b_{-\mathbf{q}}^\dagger, \qquad \frac{1}{V} \sum_{\mathbf{q}} |\varphi(\mathbf{q})|^2 = 1, \tag{2.97}$$

and the spin index is omitted.

The unitary transformation is

$$\begin{aligned} \mathcal{D}(\sqrt{N_{ex}}) &= \exp[\sqrt{N_{\mathrm{ex}}}(c_0^\dagger - c_0)] \\ &= \prod_{\mathbf{q}} \exp[\sqrt{n_{\mathrm{ex}}}\varphi(\mathbf{q})(a_{\mathbf{q}}^\dagger b_{-\mathbf{q}}^\dagger - b_{-\mathbf{q}} a_{\mathbf{q}})]. \end{aligned} \tag{2.98}$$

Here N_{ex} is the number of condensed excitons and n_{ex} is their density, where

$$n_{ex} = N_{ex}/V. \tag{2.99}$$

To introduce the Bose condensed state into the Hamiltonian and the spontaneous breaking of its symmetry, one should transform the Hamiltonian as follows:

$$\mathcal{D}^\dagger(\sqrt{N_{ex}})H\mathcal{D}(\sqrt{N_{ex}}). \tag{2.100}$$

It is equivalent to making the transformations

$$\begin{aligned}\mathcal{D}^\dagger a_{\mathbf{q}}\mathcal{D} &= u_{\mathbf{q}}a_{\mathbf{q}} + v_{\mathbf{q}}b^\dagger_{-\mathbf{q}} = \alpha_{\mathbf{q}},\\ \mathcal{D}^\dagger b_{\mathbf{q}}\mathcal{D} &= u_{\mathbf{q}}b_{\mathbf{q}} - v_{\mathbf{q}}a^\dagger_{-\mathbf{q}} = \beta_{\mathbf{q}}.\end{aligned} \tag{2.101}$$

The ground state of the *e–h* Hamiltonian (2.93) is denoted by $|0\rangle$. It is a vacuum state for the electrons and the holes, and it obeys the equalities

$$a_{\mathbf{q}}|0\rangle = b_{\mathbf{q}}|0\rangle = 0, \qquad H|0\rangle = 0. \tag{2.102}$$

The ground state for the transformed Hamiltonian $\mathcal{D}^\dagger H\mathcal{D}$ of expression (2.100) is the macroscopic coherent state $\mathcal{D}^\dagger|0\rangle = |\psi_{cohm}\rangle$. At the same time, it is the vacuum state for the new Fermi operators $\alpha_{\mathbf{q}}$ and $\beta_{\mathbf{q}}$ of Eqs. (2.101). Taking into account the property $\mathcal{D}\mathcal{D}^\dagger = 1$ and equalities (2.102), one obtains

$$\begin{aligned}\mathcal{D}^\dagger H\mathcal{D}|\psi_{cohm}\rangle &= \mathcal{D}^\dagger H\mathcal{D}\mathcal{D}^\dagger|0\rangle = \mathcal{D}^\dagger H|0\rangle = 0,\\ \alpha_{\mathbf{q}}\mathcal{D}^\dagger|0\rangle &= \mathcal{D}^\dagger a_{\mathbf{q}}\mathcal{D}\mathcal{D}^\dagger|0\rangle = \mathcal{D}^\dagger a_{\mathbf{q}}|0\rangle = 0,\\ \beta_{\mathbf{q}}\mathcal{D}^\dagger|0\rangle &= \mathcal{D}^\dagger b_{\mathbf{q}}\mathcal{D}\mathcal{D}^\dagger|0\rangle = \mathcal{D}^\dagger b_{\mathbf{q}}|0\rangle = 0.\end{aligned} \tag{2.103}$$

The coherent macroscopic state $|\psi_{cohm}\rangle$ can be rewritten in the form

$$|\psi_{cohm}\rangle = \mathcal{D}^\dagger(\sqrt{N_{ex}})|0\rangle = \prod_{\mathbf{q}}(u_{\mathbf{q}} - v_{\mathbf{q}}a^\dagger_{\mathbf{q}}b^\dagger_{-\mathbf{q}})|0\rangle. \tag{2.104}$$

The ground state of the coherently paired electrons and holes has the same nature as the ground state in the Bardeen, Cooper, and Shrieffer (BCS) [58] and Bogoliubov, Tolmachev, and Shirkov [59] theories of superconductivity, which can be seen if one compares expression (2.104) with the ground state of the coherently paired electrons studied by Blatt [60]. The coefficients $u_{\mathbf{q}}$ and $v_{\mathbf{q}}$ are determined by expansion of the exponentials $\exp(\pm x_{\mathbf{q}})$ in a series, where

$$x_{\mathbf{q}} = \sqrt{n_{ex}}\varphi(\mathbf{q})(a^\dagger_{\mathbf{q}}b^\dagger_{-\mathbf{q}} - b_{-\mathbf{q}}a_{\mathbf{q}}),$$

and by calculation of the commutators, which are

$$\begin{aligned}u_{\mathbf{q}} &= \cos[\sqrt{n_{ex}}\varphi(\mathbf{q})],\\ v_{\mathbf{q}} &= \sin[\sqrt{n_{ex}}\varphi(\mathbf{q})],\\ u^2_{\mathbf{q}} + v^2_{\mathbf{q}} &= 1.\end{aligned} \tag{2.105}$$

It is easy to see that $u_{\mathbf{q}}$ differs from unity and $v_{\mathbf{q}}$ differs from zero only in the case of a macroscopically large filling of the composite-boson single-particle state with $\mathbf{k} = 0$, when N_{ex} is proportional to V and n_{ex} is finite but not infinitesimal. The wave function $\varphi(\mathbf{q})$ must be determined in a self-consistent way. The wave function is well known for the free-exciton relative motion. For example, the $1s$-type wave function $\varphi_{1s}(\mathbf{q})$ is

$$\varphi_{1s}(\mathbf{q}) = \frac{8\left(\pi a_{\rm ex}^3\right)^{1/2}}{\left(1 + a_{\rm ex}^2 q^2\right)^2}, \qquad \frac{1}{V}\sum_{\mathbf{q}} \varphi_{1s}^2(\mathbf{q}) = 1. \tag{2.106}$$

As a zeroth-order approximation, $\varphi_{1s}(\mathbf{q})$ can be substituted for $\varphi(\mathbf{q})$ in Eqs. (2.105). Then the variable $\sqrt{n_{\rm ex}}\varphi(\mathbf{q})$ of the trigonometric functions (2.105) is less than unity for all wave vectors $\mathbf{q}$, if the inequality

$$64\pi n_{\rm ex} a_{\rm ex}^3 < 1 \tag{2.107}$$

is fulfilled. This is the case of low exciton density. Since this case was considered by Keldysh and Kozlov, we discuss it first. After that, the opposite case of high-density e–h pairs, when $n_{\rm ex}a_{\rm ex}^3 \gg 1$, is also considered, following Refs. 61–63. The function $v_{\mathbf{q}}$ corresponding to this opposite limit is transformed into the steplike Fermi distribution function.

The inverse transformation permits one to express the initial e–h operators $a_{\mathbf{q}}^\dagger$, $a_{\mathbf{q}}$, $b_{\mathbf{q}}^\dagger$, and $b_{\mathbf{q}}$ in terms of the new operators $\alpha_{\mathbf{q}}^\dagger$, $\alpha_{\mathbf{q}}$, $\beta_{\mathbf{q}}^\dagger$, and $\beta_{\mathbf{q}}$ as follows:

$$a_{\mathbf{p}} = u_{\mathbf{p}}\alpha_{\mathbf{p}} - v_{\mathbf{p}}\beta_{-\mathbf{p}}^\dagger,$$
$$b_{\mathbf{p}} = u_{\mathbf{p}}\beta_{\mathbf{p}} + v_{\mathbf{p}}\alpha_{-\mathbf{p}}^\dagger \tag{2.108}$$

In so doing, the Hamiltonian (2.93) can be represented in the form

$$H = U + \tilde{H}_0 + \tilde{H}_{\rm int}. \tag{2.109}$$

Here U plays the role of the ground-state energy and is a functional, which depends on the coefficients $u_{\mathbf{p}}$ and $v_{\mathbf{p}}$ as follows:

$$U = \sum_{\mathbf{p}} [\mathcal{E}(\mathbf{p}) - \mu_{\rm ex}] v_{\mathbf{p}}^2 - \sum_{\mathbf{p},\mathbf{p}'} V_{\mathbf{p}-\mathbf{p}'}\left[v_{\mathbf{p}}^2 v_{\mathbf{p}'}^2 + u_{\mathbf{p}} v_{\mathbf{p}} u_{\mathbf{p}'} v_{\mathbf{p}'}\right]. \tag{2.110}$$

The Hamiltonian $\tilde{H}_0$ contains the terms that are quadratic in the new operators $\alpha_{\mathbf{p}}^\dagger$, $\alpha_{\mathbf{p}}$, $\beta_{\mathbf{p}}^\dagger$, and $\beta_{\mathbf{p}}$:

$$\begin{aligned}\tilde{H}_0 = \sum_{\mathbf{p}} &\left(\left\{1/2[\mathcal{E}(\mathbf{p}) - \mu_{\rm ex}] - \sum_{\mathbf{p}'} V_{\mathbf{p}-\mathbf{p}'} v_{\mathbf{p}'}^2\right\}\left(u_{\mathbf{p}}^2 - v_{\mathbf{p}}^2\right) + 2u_{\mathbf{p}} v_{\mathbf{p}} \sum_{\mathbf{p}'} V_{\mathbf{p}-\mathbf{p}'} u_{\mathbf{p}'} v_{\mathbf{p}'}\right) \\ &\times \left(\alpha_{\mathbf{p}}^\dagger \alpha_{\mathbf{p}} + \beta_{\mathbf{p}}^\dagger \beta_{\mathbf{p}}\right) - \sum_{\mathbf{p}} \left\{\left[\mathcal{E}(\mathbf{p}) - \mu_{\rm ex} - 2\sum_{\mathbf{p}'} V_{\mathbf{p}-\mathbf{p}'} v_{\mathbf{p}'}^2\right] u_{\mathbf{p}} v_{\mathbf{p}} \right. \\ &\left. - \left(u_{\mathbf{p}}^2 - v_{\mathbf{p}}^2\right) \sum_{\mathbf{p}'} V_{\mathbf{p}-\mathbf{p}'} u_{\mathbf{p}'} v_{\mathbf{p}'}\right\}\left(\alpha_{\mathbf{p}}^\dagger \beta_{-\mathbf{p}}^\dagger + \beta_{-\mathbf{p}} \alpha_{\mathbf{p}}\right).\end{aligned} \tag{2.111}$$

The Hamiltonian $\tilde{H}_{\rm int}$ contains the terms with the new operators to the fourth power. They describe the renormalized Coulomb interaction of the new Fermi quasiparticles, which are also referred to as free electrons and holes. The matrix elements described by $\tilde{H}_{\rm int}$ are shown in Fig. 2.8. The solid lines correspond to the electron, the dashed lines correspond to the hole, and the wavy lines indicate the Coulomb interaction.

Along with the usual diagrams that describe the scattering processes with the conservation of the number of participants, there are also diagrams in which the creation or the annihilation of an electron–hole pair during the scattering processes takes place. To each vertex of all these diagrams, it is necessary to ascribe a factor $\gamma_{\mathbf{p},\mathbf{q}}$ or $\tilde{\gamma}_{\mathbf{p},\mathbf{q}}$. These arise because of the u, v canonical transformation of the operators of Eqs. (2.108). The factor $\gamma_{\mathbf{p},\mathbf{q}}$ appears for

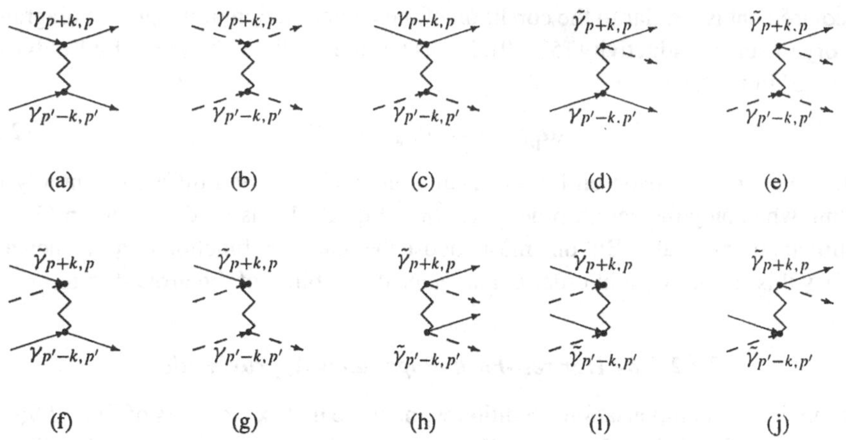

Figure 2.8. The diagrams describing the different terms included in $\tilde{H}_{\text{int}}$ (from Ref. 15).

a vertex in which the scattered particle changes its momentum from $\mathbf{q}$ to $\mathbf{p}$, and the factor $\tilde{\gamma}_{\mathbf{p},\mathbf{q}}$ accompanies the vertices where the creation or the annihilation of an e–h pair with total momentum $\mathbf{p}$–$\mathbf{q}$ takes place. The factors $\gamma_{\mathbf{p},\mathbf{q}}$ and $\tilde{\gamma}_{\mathbf{p},\mathbf{q}}$ can be expressed as follows:

$$\gamma_{\mathbf{p},\mathbf{q}} = u_{\mathbf{p}}u_{\mathbf{q}} + v_{\mathbf{p}}v_{\mathbf{q}} = \cos\{\sqrt{n_{\text{ex}}}[\varphi(\mathbf{q}) - \varphi(\mathbf{p})]\},$$
$$\tilde{\gamma}_{\mathbf{p},\mathbf{q}} = u_{\mathbf{p}}v_{\mathbf{q}} - u_{\mathbf{q}}v_{\mathbf{p}} = \sin\{\sqrt{n_{\text{ex}}}[\varphi(\mathbf{q}) - \varphi(\mathbf{p})]\},$$
$$\gamma^2_{\mathbf{p},\mathbf{q}} + \tilde{\gamma}^2_{\mathbf{p},\mathbf{q}} = 1.$$

In the dilute-density limit, $\tilde{\gamma}_{\mathbf{p},\mathbf{q}} \approx \sqrt{n_{\text{ex}}a^3_{\text{ex}}}$.

The transformed Hamiltonian admits in this way the creation from the ground state, i.e., from the exciton condensate, of a single e–h pair with total momentum equal to zero. This term is present in $\tilde{H}_0$, and its matrix element is described by the term in the last set of braces in Eq. (2.111). But similar matrix elements can be obtained in higher-order perturbation theory. For example, by combining the first-order diagrams of Figs. 2.8(f), 2.8(g), and 2.8(h), one can obtain the second-order diagrams of Figs. 2.9(a) and 2.9(b).

More complicated diagrams can be obtained from Figs. 2.9(a) and 2.9(b) by the introduction of different scattering processes. The summed matrix elements are represented by the diagrams of Figs. 2.9(c) and 2.9(d), in which the block S contains all the possible scattering processes between four particles of two electron–hole pairs. They represent the more general corrections to the term of Eq. (2.111). To avoid the instability of the ground state against the creation of a free e–h pair, it is necessary to require the mutual compensation of all the matrix elements describing these processes in all orders of the perturbation theory.

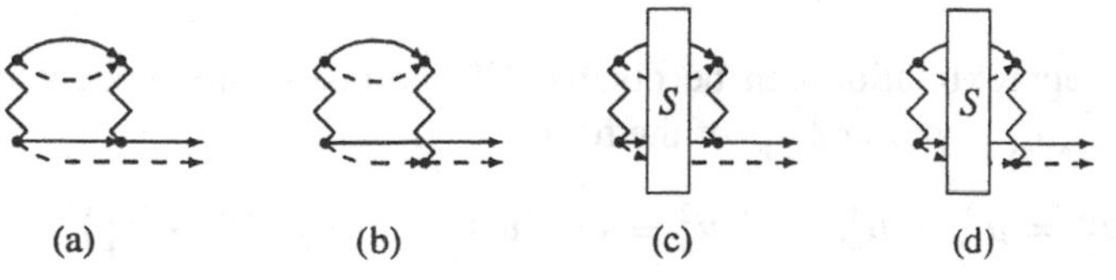

Figure 2.9. The second-order and summed diagrams describing the creation from the condensate of a single e–h pair with total momentum equal to zero (from Ref. 15).

This requirement is similar to the condition of the compensation of dangerous diagrams in the theory of superconductivity [59, 60]. It was written for the *e–h* system by Keldysh and Kozlov [15] in the form

$$\langle \alpha_{\mathbf{p}}^{\dagger} \beta_{-\mathbf{p}}^{\dagger} \rangle = \langle \beta_{-\mathbf{p}} \alpha_{\mathbf{p}} \rangle = 0. \tag{2.112}$$

Here the averaging is made on the full Hamiltonian (2.109) and differs essentially from averaging when only the zeroth-order part $\tilde{H}_0$ of Eq. (2.111) is used. Condition (2.112) is not fulfilled automatically. But one must choose the unknown function $\varphi(\mathbf{q})$ in such a way as to obey this requirement in order to guarantee the stability of the ground state.

2.2.2 The Hartree–Fock–Bogoliubov Approximation

For simplicity, the compensation condition without the matrix elements of Figs. 2.9(c) and 2.9(d) are considered first. It is equivalent to make the averaging in Eq. (2.112) on the Hamiltonian $\tilde{H}_0$ or to the requirement of the disappearance of the last term in Eq. (2.111). It leads to the nonlinear equation

$$\left[\mathcal{E}(\mathbf{p}) - \mu_{\mathrm{ex}} - 2 \sum_{\mathbf{p}'} V_{\mathbf{p}-\mathbf{p}'} v_{\mathbf{p}'}^2 \right] u_{\mathbf{p}} v_{\mathbf{p}} = \left(u_{\mathbf{p}}^2 - v_{\mathbf{p}}^2 \right) \sum_{\mathbf{p}'} V_{\mathbf{p}-\mathbf{p}'} u_{\mathbf{p}'} v_{\mathbf{p}'}. \tag{2.113}$$

This approach is known as a mean-field or Hartree–Fock–Bogoliubov approximation (HFBA). Comte and Nozières dedicated a special paper to this approach [62]. In this subsection the results of Keldysh and Kozlov [15] and Comte and Nozières [62] are reviewed.

Equation (2.113), together with normalization conditions (2.105), gives us the possibility of determining the unknown functions $v_{\mathbf{q}}$ or $\varphi(\mathbf{p})$, as well as of expressing the chemical potential μ_{ex} in terms of the condensed-exciton density n_{ex}. As follows from Eqs. (2.105) and inequality (2.107), in the low-density limit, one can set $v_{\mathbf{q}}$ equal to $\sqrt{n_{\mathrm{ex}}}\varphi(\mathbf{p})$ and develop the perturbation theory in terms of the small parameter $(n_{\mathrm{ex}} a_{\mathrm{ex}}^3)^{1/2}$.

As a zeroth approximation one can write

$$u_{\mathbf{p}}^0 = 1, \qquad v_{\mathbf{p}}^0 = \sqrt{n_{\mathrm{ex}}}\varphi(\mathbf{p}) \ll 1, \qquad \mu_{\mathrm{ex}} = \mu_{\mathrm{ex}}^0 \tag{2.114}$$

and keep in Eq. (2.110) only the terms linear in $v_{\mathbf{p}}^0$. In this approach, Eq. (2.113) transforms into the linear Schrödinger equation for the free-exciton relative motion in the momentum representation,

$$\left[\mathcal{E}(\mathbf{p}) - \mu_{\mathrm{ex}}^0 \right] v_{\mathbf{p}}^0 - \sum_{\mathbf{p}'} V_{\mathbf{p}-\mathbf{p}'} v_{\mathbf{p}'}^0 = 0. \tag{2.115}$$

Its eigenfunction $v_{\mathbf{p}}^0$ and eigenvalue μ_{ex}^0 are

$$v_{\mathbf{p}}^0 = \sqrt{n_{\mathrm{ex}}}\varphi_{1s}(\mathbf{p}), \qquad \mu_{\mathrm{ex}}^0 = -\mathrm{Ry}_{\mathrm{ex}}, \qquad \frac{1}{V} \sum_{\mathbf{p}} \left(v_{\mathbf{p}}^0 \right)^2 = n_{\mathrm{ex}}. \tag{2.116}$$

The first-order approximation can be obtained if one takes into account the first-order corrections in $n_{\mathrm{ex}} a_{\mathrm{ex}}^3$ to μ_{ex} and $u_{\mathbf{p}}$, of the form

$$\mu_{\mathrm{ex}} = \mu_{\mathrm{ex}}^0 + \mu_{\mathrm{ex}}^1, \qquad u_{\mathbf{p}}^2 \cong 1 - \left(v_{\mathbf{p}}^0 \right)^2, \qquad u_{\mathbf{p}} \cong 1 - \left(v_{\mathbf{p}}^0 \right)^2 / 2, \tag{2.117}$$

and keeps the terms proportional to $(v_{\mathbf{p}}^0)^3$ in Eq. (2.113). The terms proportional to $(v_{\mathbf{p}}^0)^5$ are neglected. After multiplication of Eq. (2.113) by $v_{\mathbf{p}}^0$ and summation over the momentum $\mathbf{p}$, one can obtain the equalities

$$\mu_{ex}^{1}\frac{1}{V}\sum_{\mathbf{p}}(v_{\mathbf{p}}^{0})^{2}=\frac{2}{V}\sum_{\mathbf{p},\mathbf{p}'}V_{\mathbf{p}-\mathbf{p}'}\left(v_{\mathbf{p}}^{03}v_{\mathbf{p}'}^{0}-v_{\mathbf{p}}^{02}v_{\mathbf{p}'}^{02}\right),$$

$$\mu_{ex}^{1}=\frac{26\pi}{3}\mathrm{Ry}_{ex}n_{ex}a_{ex}^{3}. \tag{2.118}$$

As a result, in the case of the excitonic condensate, the chemical potential μ_{ex} for the spin-oriented *e–h* pairs in the HFBA is

$$\mu_{ex}=-\mathrm{Ry}_{ex}\left(1-\frac{26\pi}{3}n_{ex}a_{ex}^{3}\right). \tag{2.119}$$

For paraexcitons (spin-zero excitons), the exciton creation operator $\hat{\psi}_{ex,\mathbf{k}}^{\dagger}$ contains the sum over the spin index $\sigma = \pm 1/2$:

$$\psi_{ex,\mathbf{k}}^{\dagger}=\frac{1}{\sqrt{2V}}\sum_{\mathbf{q}\sigma}\varphi(\mathbf{q})a_{\alpha\mathbf{k}+\mathbf{q},\sigma}^{\dagger}b_{\beta\mathbf{k}-\mathbf{q},-\sigma}^{\dagger},$$

$$\frac{1}{V}\sum_{\mathbf{q}}|\varphi(\mathbf{q})|^{2}=1. \tag{2.120}$$

Then the coefficients $u_{\mathbf{q}}$ and $v_{\mathbf{q}}$ of the Bogoliubov canonical transformation in Eqs. (2.105) have an argument equal to $\sqrt{n_{ex}/2}\varphi(\mathbf{q})$, which differs by the factor $1/\sqrt{2}$ from the previous one. For this reason Eqs. (2.118) give for the paraexcitons the following value for μ_{ex}^{1}:

$$\mu_{ex}^{1}=\frac{13\pi}{3}\mathrm{Ry}_{ex}n_{ex}a_{ex}^{3},$$

$$n_{ex}=\frac{1}{V}\sum_{\mathbf{q}}\sum_{\sigma}(v_{\mathbf{q}})^{2}. \tag{2.121}$$

Both values for μ_{ex}^{1} in Eqs. (2.118) and (2.121) are in accordance with the results of Refs. 25 and 26. If condition (2.113) is fulfilled, the Hamiltonian $\tilde{H}_0$ of Eq. (2.111) becomes diagonalized. Then the selection rule for the average values of the Fermi operators in the HFBA is

$$\langle\alpha_{\mathbf{p}}^{\dagger}\beta_{-\mathbf{p}}^{\dagger}\rangle_{0}=\langle\beta_{-\mathbf{p}}\alpha_{\mathbf{p}}\rangle_{0}=0.$$

This is a particular case of the more general requirement of Eq. (2.112) proposed by Keldysh and Kozlov [15]. For the case of equal masses $m_e = m_h$, the new quasiparticles have the positive energy spectrum $\frac{1}{2}E(\mathbf{p})$, as one can see by looking at expression (2.111) and taking into account Eq. (2.113):

$$E(\mathbf{p})=\left\{(\mathcal{E}(\mathbf{p})-\mu_{ex})-2\sum_{\mathbf{p}'}V_{\mathbf{p}-\mathbf{p}'}v_{\mathbf{p}'}^{2}\right\}(u_{\mathbf{p}}^{2}-v_{\mathbf{p}}^{2})$$

$$+4u_{\mathbf{p}}v_{\mathbf{p}}\sum_{\mathbf{p}'}V_{\mathbf{p}-\mathbf{p}'}u_{\mathbf{p}'}v_{\mathbf{p}'}=\frac{1}{u_{\mathbf{p}}v_{\mathbf{p}}}\sum_{\mathbf{p}'}V_{\mathbf{p}-\mathbf{p}'}u_{\mathbf{p}'}v_{\mathbf{p}'}. \tag{2.122}$$

Following the paper of Comte and Nozières [62], it is useful to introduce the notations of the effective chemical potential $\tilde{\mu}_{ex}(\mathbf{p})$ and the ordering parameter $\Delta(\mathbf{p})$, defined as

$$\tilde{\mu}_{ex}(\mathbf{p})=\mu_{ex}+2\sum_{\mathbf{p}'}V_{\mathbf{p}-\mathbf{p}'}v_{\mathbf{p}'}^{2},$$

$$\Delta(\mathbf{p})=\sum_{\mathbf{p}'}V_{\mathbf{p}-\mathbf{p}'}u_{\mathbf{p}'}v_{\mathbf{p}'}. \tag{2.123}$$

In this notation, the Hamiltonian $\tilde{H}_0$, the energy spectrum $E(\mathbf{p})$, Eq. (2.113), and the coefficients $u_\mathbf{p}$ and $v_\mathbf{p}$ take the forms

$$\tilde{H}_0 = \sum_\mathbf{p} \frac{1}{2} E(\mathbf{p})(\alpha_\mathbf{p}^\dagger \alpha_\mathbf{p} + \beta_\mathbf{p}^\dagger \beta_\mathbf{p}),$$

$$E(\mathbf{p}) = \sqrt{[\mathcal{E}(\mathbf{p}) - \tilde{\mu}_{\text{ex}}(\mathbf{p})]^2 + 4\Delta^2(\mathbf{p})},$$

$$[\mathcal{E}(\mathbf{p}) - \tilde{\mu}_{\text{ex}}(\mathbf{p})]u_\mathbf{p} v_\mathbf{p} = (u_\mathbf{p}^2 - v_\mathbf{p}^2)\Delta(\mathbf{p}),$$

$$u_\mathbf{p}^2 = \frac{1}{2}\left[1 + \frac{\mathcal{E}(\mathbf{p}) - \tilde{\mu}_{\text{ex}}(\mathbf{p})}{E(\mathbf{p})}\right], \qquad v_\mathbf{p}^2 = \frac{1}{2}\left[1 - \frac{\mathcal{E}(\mathbf{p}) - \tilde{\mu}_{\text{ex}}(\mathbf{p})}{E(\mathbf{p})}\right], \tag{2.124}$$

where $E(\mathbf{p})$ is the energy needed for the excitation of one free *e–h* pair from the exciton Bose condensate. The minimal energy necessary to transform one of the coherently coupled *e–h* pairs into a free pair approximately equals Ry_{ex}. This is true not only in the HFBA but also in the more general description. Then at $T = 0$ the exact average quasiparticle numbers are equal to zero [15]:

$$\langle \alpha_\mathbf{p}^\dagger \alpha_\mathbf{p} \rangle = \langle \beta_\mathbf{p}^\dagger \beta_\mathbf{p} \rangle = 0. \tag{2.125}$$

Consequently, the average values of the initial electron and hole operators can be determined with formulas (2.108), (2.112), and (2.125), as follows:

$$\langle a_\mathbf{p}^\dagger a_\mathbf{p} \rangle = \langle b_\mathbf{p}^\dagger b_\mathbf{p} \rangle = v_\mathbf{p}^2,$$

$$\langle a_\mathbf{p}^\dagger b_{-\mathbf{p}}^\dagger \rangle = \langle b_{-\mathbf{p}} a_\mathbf{p} \rangle = -u_\mathbf{p} v_\mathbf{p}. \tag{2.126}$$

The average values of the number of electrons N_e and number of holes N_h are equal to the average number of the coupled *e–h* pairs:

$$N_e = N_h = N_{\text{ex}} = \sum_\mathbf{p} v_\mathbf{p}^2. \tag{2.127}$$

The average value of the exciton creation operator $c_0^\dagger$, is a macroscopically large number, $-\sqrt{N_{\text{ex}}}$:

$$\langle c_0^\dagger \rangle = -\frac{1}{\sqrt{V}} \sum_\mathbf{q} \varphi(\mathbf{q}) u_\mathbf{q} v_\mathbf{q} = -\sqrt{N_{\text{ex}}}. \tag{2.128}$$

Using the Keldysh–Kozlov–Kopaev formulation of Bose condensation as a variational ansatz, Comte and Nozières interpolated between dense and dilute systems and showed that even at metallic densities, pairing corrections are very important. Comte and Nozières considered in detail the dilute and the dense limits at which analytic expansions are possible. They briefly discussed the nature of the elementary excitations as well as the critical temperature, whose origin is quite different in the two limits. Their results coincide with the Keldysh and Kopaev [61] effects in the dense limit and with the Keldysh and Kozlov [15] conclusions in the dilute limit. The findings of Comte and Nozières [62] are briefly presented below.

The shape of $v_\mathbf{p}$ gives the internal state of the bound electron–hole pair, while its magnitude $\sqrt{n_{\text{ex}}}$ fixes the density. The behavior of $v_\mathbf{p}^2$ as a function of $\mathcal{E}_\mathbf{p} = p^2/2\mu$ is sketched in Fig. 2.10. Curves a and b in Fig. 2.10 correspond to dilute systems, $n_{\text{ex}} a_{\text{ex}}^3 \ll 1$, in which $v_\mathbf{p}^2$ keeps a constant shape as n_{ex} increases, which is just the Fourier transform of

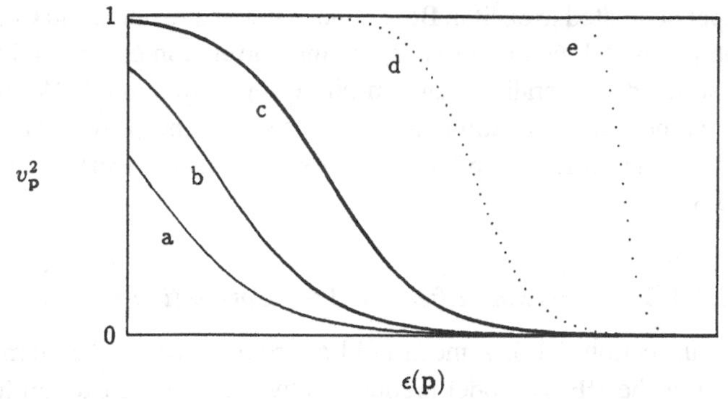

Figure 2.10. The ground-state fermion distribution $v_{\mathbf{p}}^2$ as a function of the electron–hole pair's internal-motion kinetic energy $\mathcal{E}_{\mathbf{p}}$ for various densities $n_{ex}a_{ex}^3$ (from Ref. 62).

the $n=1$ Bohr orbital for the relative electron–hole motion. Curves d and e in Fig. 2.10 correspond to the dense limit. Here the Fermi level edge sharpens and moves to higher energies. The transition between these two limits is smooth, as shown by curve c. As the pair density n_{ex} grows, $v_{\mathbf{p}}^2$ first increases in magnitude without changing its shape. When $v_{\mathbf{p}}^2$ approaches unity, it is stopped by the normalization condition $u_{\mathbf{p}}^2 + v_{\mathbf{p}}^2 = 1$, which is the mathematical statement of the Pauli exclusion principle. Physically, the N_{ex} condensed excitons exhaust the underlying stock of fermion states, because the excitons are nothing but electron–hole bound states. To pack in more particles, $v_{\mathbf{p}}$ must spread further away in $\mathbf{p}$ space. Such a saturation occurs when $n_{ex}a_{ex}^3 \approx 1$ and the excitons start overlapping. Note that Bose condensation also implies phase-space locking and large fluctuations in the number of particles, as well as the macroscopic occupation of the $\mathbf{k}=0$ exciton state.

In the opposite limit, $n_{ex}a_{ex}^3 \gg 1$, as shown by Keldysh and Kopaev [61] and Kohn [64], a dense isotropic *e–h* gas always undergoes an instability analogous to superconductivity. The coherent electron–hole pairing gives rise to the formation of an "excitonic insulator." We briefly discuss this novel phase in Chapters 5 and 10. An explicit solution in this case is equally straightforward. In the zeroth-order approximation, $\Delta(\mathbf{p})$ vanishes and the system is an ordinary Hartree–Fock plasma. In this case, the chemical potential $\tilde{\mu}_{ex}(\mathbf{p})$ plays the role of the Fermi level. In the next order, a small excitonic-insulator instability develops. The ordering parameter $\Delta(\mathbf{p})$ plays the role of the energy gap, which vanishes exponentially when the density n_{ex} goes to infinity.

On the whole, the mean-field approximation, as noted by Comte and Nozières, is a crude, unrealistic model. For example, in the case of the dilute limit, the next-order correction in $n_{ex}a_{ex}^3$ to the chemical potential is negative,

$$\mu_{ex}^{(2)} = -\frac{407\pi^2}{9}\left(n_{ex}a_{ex}^3\right)^2 \mathrm{Ry}_{ex}, \qquad (2.129)$$

and the mean-field approximation breaks down completely [62]. The existence of other contributions to μ_{ex} was recognised by Keldysh and Kozlov [15], who from the very beginning gave the complete expression for compensation condition (2.112). Nevertheless, the results of this model, such as the underlying fermion distribution shown in Fig. 2.10, help to show how the system can move continuously from the dilute, Bose gas limit to the dense fermion-pair limit, and help to illuminate how Bose statistics can arise in a Fermi system.

In a second paper dedicated to exciton Bose condensation, Nozières and Comte [63] tried to improve the above model, taking into account the correlation energy and the screening effects. They constructed a generalized random-phase approximation (RPA), which reduces to the standard treatment of correlations for dense systems and provides an approximate form of the van der Waals attraction between excitons. This and other papers are reviewed in Subsection 2.2.3.

2.2.3 The Screening Effects and the Correlation Energy

As discussed in Subsection 2.1.5, a mean-field approximation in the framework of the Bogoliubov model or the HFBA model predicts a blue shift of the exciton luminescence, i.e., an increase of the ground-state energy due to the repulsive interactions, which is not seen experimentally. It turns out that a more accurate theory that takes into account the correlation effects erases this prediction. Several authors have tackled this problem.

To satisfy the more general condition of Eq. (2.112), Keldysh and Kozlov took into account *ab initio* the matrix elements of Figs. 2.9(c) and 2.9(d) together with the left-hand side expression of Eq. (2.113). The diagrams of Figs. 2.9(c) and 2.9(d) contain at least three vertices in which the creation or annihilation of an e–h pair takes place. Each of them is characterized by a factor $\tilde{\gamma}_{\mathbf{p},\mathbf{q}} \approx \sqrt{n_{\text{ex}}}$ and plays the same role as the condensate line in the Beliaev diagram technique. The lower-order contribution of the matrix elements of Figs. 2.9(c) and 2.9(d) to the chemical potential μ^1_{ex} can be expressed as $\lambda \text{Ry}_{\text{ex}} n_{\text{ex}} a^3_{\text{ex}}$, where λ is a dimensionless parameter. If one adds this contribution to expressions (2.118), the following result will be obtained:

$$\mu_{\text{ex}} = -\text{Ry}_{\text{ex}}\left(1 - f n_{\text{ex}} a^3_{\text{ex}}\right), \quad f = \left(\frac{26\pi}{3} + \lambda\right). \tag{2.130}$$

It contains the mean-field and correlation contributions and determines the density-dependent shift of the exciton-energy level.

Zimmermann [65] investigated this question in more detail. He emphasized the large cancellation of the mean-field and correlation components of the slope coefficient f for bulk crystals. Bose condensation of excitons was studied on the basis of the Hamiltonian proposed by Zimmermann and Rösler [66], which was extended to include correlations, which are essential at elevated densities. Zimmermann noted the importance of a dynamical treatment and established the relative constancy of the exciton ground-state energy. He obtained a mean-field slope coefficient equal to $13\pi/3$, as well as the correlation slope coefficient λ, which turned out to be negative for all mass ratios $\sigma = m_e/m_h$. His numerical calculation for the case $\sigma = 1$ gave the value $\lambda = -14.53$, and therefore $f = -0.92$ [65]. In the high-density limit, the screening works more effectively and forces the appearance of the plasma phase [65]. This work allowed not only a calculation of the density-dependent shift of the exciton ground-state energy, but also the self-energy parts of other exciton levels. In this way Zimmermann et al. [67], as well as Stolz and Zimmermann [68,69], determined the density-dependent shift of the continuum edge of the exciton spectrum, i.e., the shift of the bandgap, ΔE_g. Taking into account the screening effects in a system of e–h pairs, they found that to linear order in the density n_{ex}, for the material parameters of GaAs, the expression ΔE_g is

$$\Delta E_g = -673.4 \text{Ry}_{\text{ex}} n_{\text{ex}} a^3_{\text{ex}}. \tag{2.131}$$

Comparing Refs. 65 and 68, one can conclude that the bandgap shrinks quickly with increasing exciton density, implying that the exciton level remains relatively constant.

The energy per exciton in the Bose condensed state as a function of the exciton density was studied by Henneberger and May [70] in the boson picture. They also arrived at the conclusion that the correlation corrections cancel the blue shift of the exciton level arising in the mean-field approximation and that this blue shift is unrealistic for bulk crystals. In the authors' opinion, the formation of a dielectric liquid of Wannier–Mott excitons could be possible. But the model they used is somewhat restricted because the critical densities are rather large [70]. In another study of the exciton level shift, Hanamura [71] studied the change of the dipole moment of the exciton transition due to phase-space-filling effects.

Nozières and Comte [63] proposed a method that permits one to take into account simultaneously the binding processes and the screening effects in a system of the Bose condensed *e–h* pairs. It can be used both in the dense limit $n_{ex}a_{ex}^3 \gg 1$, in which it essentially coincides with the RPA [72] for a two-component plasma, and in the dilute limit $n_{ex}a_{ex}^3 \ll 1$, in which it is able to describe in part the Kramers polarizability of the neutral atomic excitons and the van der Waals interaction between them.

The formulation of the method, called the generalized RPA (GRPA), was accompanied by detailed explanations of the underlying physics and is very instructive. As a main element it contains the coherence factor $(u_{\mathbf{p}}v_{\mathbf{q}} - u_{\mathbf{q}}v_{\mathbf{p}})^2$, which is typical of a BCS ground state. The main results of the Nozières and Comte paper [63] concerning the Bose condensed excitons are presented here.

Following the Pauli–Feynman theorem [72,73], the ground-state energy E_0 of the system of interacting *e–h* pairs can be expressed in the form

$$E_0 = E_{\text{kin}} + \int_0^{e^2} \frac{d\lambda}{\lambda} E_{\text{int}}(\lambda), \tag{2.132}$$

which contains the kinetic energy E_{kin} of the Bose condensed ideal *e–h* pairs without the Coulonb interaction between them and the mean value of their Coulomb interaction $E_{\text{int}}(\lambda)$ with a variable value of the electric charge squared, λ. The variational parameter λ changes from zero to the value of a real electron charge squared, e^2. To calculate $E_{\text{int}}(\lambda)$, one must also take into account the occurrence of BEC.

The hypothetical *e–h* gas has the wave function $|n(\lambda)\rangle$ that depends on λ and the bare Coulomb interaction,

$$V_{\mathbf{k}}(\lambda) = V_{\mathbf{k}} \frac{\lambda}{e^2}. \tag{2.133}$$

The free-particle energy spectra and their total number do not depend on λ. The Coulomb interaction can be expressed in a factorized form,

$$\frac{1}{2} \sum_{\mathbf{k}} V_{\mathbf{k}} [\rho_{\mathbf{k}} \rho_{\mathbf{k}}^{\dagger} - N_e - N_h], \tag{2.134}$$

where the charge density $\rho_{\mathbf{k}}$ and the full electron and hole number operators N_e and N_h are

$$\rho_{\mathbf{k}} = \sum_{\mathbf{p}} a_{\mathbf{p}+\mathbf{k}}^{\dagger} a_{\mathbf{p}} - \sum_{\mathbf{p}} b_{\mathbf{p}+\mathbf{k}}^{\dagger} b_{\mathbf{p}},$$
$$N_e = \sum_{\mathbf{p}} a_{\mathbf{p}}^{\dagger} a_{\mathbf{p}}, \qquad N_h = \sum_{\mathbf{p}} b_{\mathbf{p}}^{\dagger} b_{\mathbf{p}}. \tag{2.135}$$

Here, as before, the spin-index summation is dropped. The expression $E_{\text{int}}(\lambda)$ can be written as

$$E_{\text{int}}(\lambda) = -N_{\text{ex}} \sum_{\mathbf{k}} V_{\mathbf{k}}(\lambda) + \frac{1}{2} \sum_{\mathbf{k}} V_{\mathbf{k}}(\lambda) \sum_{n(\lambda)} |(\rho_{\mathbf{k}}^{\dagger})_{\text{no}}|^2. \tag{2.136}$$

The first term subtracts the self-interaction of electrons and holes from the factorized part of the Coulomb interaction. N_{ex} is the mean number of e–h pairs or excitons $N_{ex} = N_e = N_h$.

As noted by Nozières and Comte, the screening corrections can act on only real electron transitions allowed by the exclusion principle, leaving the self-interaction term untouched. The matrix elements in Eq. (2.136) can be expressed in terms of the imaginary part of the dynamical dielectric constant $\varepsilon(\mathbf{k}, \omega, \lambda)$ as

$$\frac{1}{2} V_{\mathbf{k}}(\lambda) \sum_{n(\lambda)} |(\rho_{\mathbf{k}}^{\dagger})_{no}|^2 = -\int_0^{\infty} \frac{\hbar d\omega}{2\pi} \mathrm{Im}\left[\frac{1}{\varepsilon(\mathbf{k}, \omega, \lambda)}\right]. \tag{2.137}$$

Relation (2.137) has been deduced in the theory of the linear response of the system to a weak perturbation [72, 73]. Inasmuch as the system is characterized by the variable λ, $\varepsilon(\mathbf{k}, \omega, \lambda)$ also depends on λ.

The ground-state energy of Eq. (2.132) then obtains the form

$$E_0 = E_{\mathrm{kin}} - N_{\mathrm{ex}} \sum_{\mathbf{k}} V_{\mathbf{k}} - \sum_{\mathbf{k}} \int_0^{\infty} \frac{\hbar d\omega}{2\pi} \int_0^{e^2} \frac{d\lambda}{\lambda} \mathrm{Im}\left[\frac{1}{\varepsilon(\mathbf{k}, \omega, \lambda)}\right]. \tag{2.138}$$

The choice of the approximation for $\varepsilon(\mathbf{k}, \omega, \lambda)$ determines the accuracy of the determination of the energy E_0. For the one-component plasma and the electron–hole liquid, the best results were obtained with the RPA together with Hubbard-type corrections [72–74].

In the framework of the RPA, the effective Coulomb interaction between two electrons in a one-component plasma can be represented as a sum of ring diagrams, as in Fig. 2.11. All the wavy lines in Fig. 2.11 have the same transfer wave vector $\mathbf{k}$ and the same contributions $V_{\mathbf{k}}$. The exchange Coulomb diagrams are neglected, which are characterized by the appearance of the contributions $V_{\mathbf{k}_F}$ instead of $V_{\mathbf{k}}$, where $\mathbf{k}_F$ is the Fermi wave vector. In the dense limit, this approximation is valid because the actual values of $\mathbf{k}$ are much less than those of $\mathbf{k}_F$ and $V_{\mathbf{k}_F} \ll V_{\mathbf{k}}$. But in the intermediate metallic range, $n_{\mathrm{ex}} a_{\mathrm{ex}}^3 \approx 1$, the exchange diagrams must be taken into account when $\mathbf{k}$ exceeds $\mathbf{k}_F$. In this case, the direct and the exchange Coulomb interactions of two electrons with the same spin projections cancel each other, and the Hubbard corrections to the usual RPA are needed. Summing up the diagrams in Fig. 2.11, one obtains the effective Coulomb interaction $V_{\mathrm{eff}}(\mathbf{k}, \omega)$ and the dynamical dielectric constant $\varepsilon^{\mathrm{RPA}}(\mathbf{k}, \omega)$ as follows:

$$V_{\mathrm{eff}}(\mathbf{k}, \omega) = \frac{V_{\mathbf{k}}}{\varepsilon^{\mathrm{RPA}}(\mathbf{k}, \omega)} = V_{\mathbf{k}} + V_{\mathbf{k}} \Pi(\mathbf{k}, \omega) V_{\mathbf{k}} + \cdots,$$

$$\varepsilon^{\mathrm{RPA}}(\mathbf{k}, \omega) = 1 - V_{\mathbf{k}} \Pi(\mathbf{k}, \omega) = 1 + 4\pi\alpha(\mathbf{k}, \omega). \tag{2.139}$$

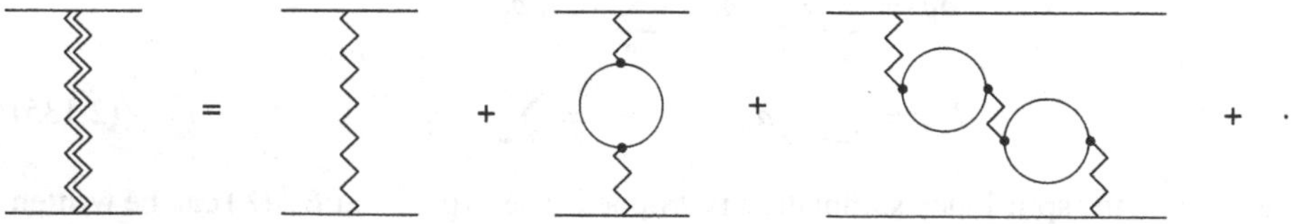

Figure 2.11. The sum diagram containing the polarization loops in the RPA (from Ref. 72).

Here $\alpha(\mathbf{k}, \omega)$ is the polarizability of the electron plasma, whereas $\Pi(\mathbf{k}, \omega)$ is the polarization-loop contribution, which can be expressed through the free-electron Green's functions $G_e^0(\mathbf{p}, \omega')$ in the form

$$\Pi(\mathbf{k}, \omega) = -i \sum_{\mathbf{q}} \int_{-\infty}^{\infty} \frac{\hbar d\omega'}{2\pi} G_e^0(\mathbf{q}, \omega') G_e^0(\mathbf{q} + \mathbf{k}, \omega + \omega'). \tag{2.140}$$

In the case of Bose condensed e–h pairs, as a zeroth-order Hamiltonian one must choose the expression $\tilde{H}_0$ in Eqs. (2.124). It is diagonalized in the new quasiparticle operators $\alpha_{\mathbf{p}}$ and $\beta_{\mathbf{p}}$, which are connected to the operators $a_{\mathbf{p}}$ and $b_{\mathbf{p}}$ of the initial electrons and holes by relations (2.108). In virtue of this, there are four zeroth-order Green's functions constructed from the operators $a_{\mathbf{p}}^\dagger$, $a_{\mathbf{p}}$, $b_{\mathbf{p}}^\dagger$, and $b_{\mathbf{p}}$ as follows:

$$\begin{aligned}
G_a^0(\mathbf{p}, t) &= -i\langle T[a_{\mathbf{p}}^\dagger(t) a_{\mathbf{p}}(0)]\rangle_0 \qquad a \longrightarrow a,\\
G_b^0(\mathbf{p}, t) &= -i\langle T[b_{\mathbf{p}}^\dagger(t) b_{\mathbf{p}}(0)]\rangle_0 \qquad b \longrightarrow b,\\
\hat{G}^0(\mathbf{p}, t) &= -i\langle T[a_{\mathbf{p}}^\dagger(t) b_{-\mathbf{p}}^\dagger(0)]\rangle_0 \qquad a \rightarrow\leftarrow b,\\
\check{G}^0(\mathbf{p}, t) &= -i\langle T[b_{-\mathbf{p}}(t) a_{\mathbf{p}}(0)]\rangle_0 \qquad b \leftarrow\rightarrow a.
\end{aligned} \tag{2.141}$$

Here one must calculate the time dependence of the operators $a_{\mathbf{p}}$ and $b_{\mathbf{p}}$ in the interaction representation and the averaging by using the Hamiltonian $\tilde{H}_0$ in Eqs. (2.124) before going from operators $a_{\mathbf{p}}$ and $b_{\mathbf{p}}$ to the new operators $\alpha_{\mathbf{p}}$ and $\beta_{\mathbf{p}}$. The Fourier transforms of the Green's function (2.2.3) for the case $m_e = m_h$, which is considered here, are

$$G_a^0(\mathbf{p}, \omega) = G_b^0(\mathbf{p}, \omega) = \left[\frac{u_{\mathbf{p}}^2}{\hbar\omega + E(\mathbf{p})/2 - i\delta} + \frac{v_{\mathbf{p}}^2}{\hbar\omega - E(\mathbf{p})/2 + i\delta}\right],$$

$$\hat{G}^0(\mathbf{p}, \omega) = \check{G}^0(\mathbf{p}, \omega) = u_{\mathbf{p}} v_{\mathbf{p}} \left[\frac{1}{\hbar\omega + E(\mathbf{p})/2 - i\delta} - \frac{1}{\hbar\omega - E(\mathbf{p})/2 + i\delta}\right]. \tag{2.142}$$

In this case, the role of the simple polarization loop in Fig. 2.11 is played by a compound polarization loop, the contribution of which consists of three terms,

$$\Pi(\mathbf{k}, \omega) = \Pi_a(\mathbf{k}, \omega) + \Pi_b(\mathbf{k}, \omega) + 2\Pi_{ab}(\mathbf{k}, \omega). \tag{2.143}$$

They correspond to the three simple polarization loops represented in Fig. 2.12. Two of them are determined by the Green's functions $G_a^0(\mathbf{q}, \omega')$ and $G_b^0(\mathbf{q}, \omega')$, whereas the third

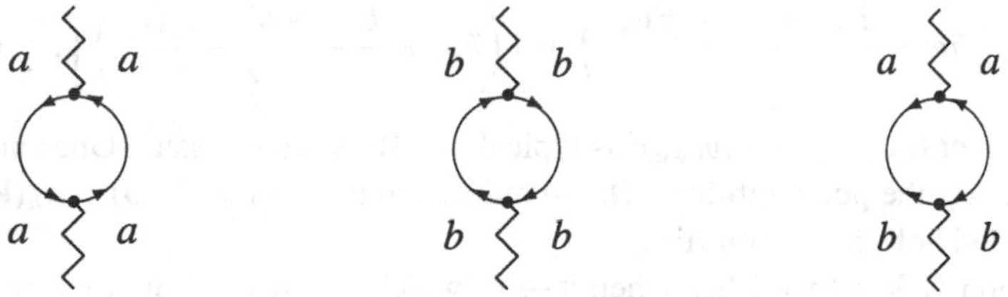

Figure 2.12. Three polarization loops in the case of Bose condensed e–h pairs (from Ref. 63).

one is determined by the anomalous functions $\hat{G}^0(\mathbf{q}, \omega')$ and $\check{G}^0(\mathbf{q}, \omega')$ as follows:

$$\Pi_a(\mathbf{k}, \omega) = \Pi_b(\mathbf{k}, \omega) = -i \sum_{\mathbf{q}} \int_{-\infty}^{\infty} \frac{\hbar d\omega'}{2\pi} G_a^0(\mathbf{q}, \omega') G_a^0(\mathbf{q}+\mathbf{k}, \omega' + \omega),$$

$$\Pi_{ab}(\mathbf{k}, \omega) = -i \sum_{\mathbf{q}} \int_{-\infty}^{\infty} \frac{\hbar d\omega'}{2\pi} \hat{G}^0(\mathbf{q}, \omega') \check{G}^0(\mathbf{q}+\mathbf{k}, \omega' + \omega). \tag{2.144}$$

By comparing the diagrams in Figs. 2.8 and 2.12, one can observe the following. In Fig. 2.12, the wavy lines represent the bare Coulomb interaction of the initial electrons and holes, whereas in Fig. 2.8 they represent the renormalized Coulomb interaction of the new quasiparticles. In the same manner, the origin of the zeroth-order propagators in these two cases is completely different. For these reasons, in the vertices of the diagrams in Fig. 2.8 the creation or the annihilation of a pair of the new quasiparticles can take place, while in Fig. 2.12, at each vertex, only the scattering of the initial electron or hole occurs.

As was determined by Nozières and Comte, the summed polarization contribution is

$$\Pi(\mathbf{k}, \omega) = -\sum_{\mathbf{q}} (u_{\mathbf{k+q}} v_{\mathbf{q}} - u_{\mathbf{q}} v_{\mathbf{k+q}})^2 \times \left[\frac{1}{\hbar\omega + \frac{E(\mathbf{k+q})+E(\mathbf{q})}{2} - i\delta} - \frac{1}{\hbar\omega - \frac{E(\mathbf{k+q})+E(\mathbf{q})}{2} + i\delta} \right]. \tag{2.145}$$

This determined the polarizability of the e–h system,

$$4\pi\alpha(\mathbf{k}, \omega) = -V_{\mathbf{k}} \Pi(\mathbf{k}, \omega). \tag{2.146}$$

After separation of the real and the imaginary parts of the polarizability, one obtains the dispersional $\varepsilon_1(\mathbf{k}, \omega)$ and the absorptional $\varepsilon_2(\mathbf{k}, \omega)$ components of the dynamical dielectric constant,

$$\begin{aligned}
\varepsilon(\mathbf{k}, \omega) &= \varepsilon_1(\mathbf{k}, \omega) + i\varepsilon_2(\mathbf{k}, \omega), \\
\varepsilon_1(\mathbf{k}, \omega) &= 1 + 4\pi\alpha_1(\mathbf{k}, \omega) = 1 - V_{\mathbf{k}} \Pi_1(\mathbf{k}, \omega) \\
&= 1 + V_{\mathbf{k}} \sum_{\mathbf{q}} \frac{(u_{\mathbf{k+q}} v_{\mathbf{q}} - u_{\mathbf{q}} v_{\mathbf{k+q}})^2 (E(\mathbf{k}+\mathbf{q}) + E(\mathbf{q}))}{\left(\frac{E(\mathbf{k+q})+E(\mathbf{q})}{2}\right)^2 - \hbar^2\omega^2}, \\
\varepsilon_2(\mathbf{k}, \omega) &= 4\pi\alpha_2(\mathbf{k}, \omega) = -V_{\mathbf{k}} \Pi_2(\mathbf{k}, \omega) \\
&= \pi V_{\mathbf{k}} \sum_{\mathbf{q}} (u_{\mathbf{k+q}} v_{\mathbf{q}} - u_{\mathbf{q}} v_{\mathbf{k+q}})^2 \\
&\quad \times \left[\delta\left(\hbar\omega - \frac{E(\mathbf{k}+\mathbf{q}) + E(\mathbf{q})}{2}\right) + \delta\left(\hbar\omega + \frac{E(\mathbf{k}+\mathbf{q}) + E(\mathbf{q})}{2}\right) \right].
\end{aligned} \tag{2.147}$$

The coherence factor $(u_{\mathbf{k+q}} v_{\mathbf{q}} - u_{\mathbf{q}} v_{\mathbf{k+q}})^2$ is typical of a BCS ground state. One can verify that for a steplike $v_{\mathbf{q}}$, the polarizability $\alpha(\mathbf{k}, \omega)$ reduces to the sum $\alpha_e(\mathbf{k}, \omega) + \alpha_h(\mathbf{k}, \omega)$ of normal electron and hole polarizabilities.

The contribution $\Pi(\mathbf{k}, \omega)$ vanishes when $\mathbf{k} \to 0$, which ensures a finite dielectric constant at long wavelength. According to Nozières and Comte, physically that feature is a consequence of electron–hole binding, which precludes free-carrier Debye screening. In the dilute limit, the static value $\Pi_1(\mathbf{k},\omega)$ has a maximum of the order of N_{ex}/Ry_{ex} and is nonzero in the range $k \sim 1/a_{ex}$. For this limit to be valid, one must assume that $n_{ex}a_{ex}^3$ is much less than unity but still large enough for molecular biexcitons to be dissociated. In this limit, when $u_{\mathbf{q}} \approx 1$, $v_{\mathbf{q}} \approx \sqrt{n_{ex}}\varphi(\mathbf{q})$, the polarizability $4\pi\alpha_1(\mathbf{k},\omega)$ has the form

$$4\pi\alpha_1(\mathbf{k},\omega) = n_{ex}V_{\mathbf{k}}\sum_{\mathbf{q}}[\varphi(\mathbf{k}+\mathbf{q}) - \varphi(\mathbf{q})]^2 \frac{E(\mathbf{k}+\mathbf{q}) + E(\mathbf{q})}{\left(\frac{E(\mathbf{k}+\mathbf{q})+E(\mathbf{q})}{2}\right)^2 - \hbar^2\omega^2}, \tag{2.148}$$

where

$$\frac{E(\mathbf{k}+\mathbf{q}) + E(\mathbf{q})}{2} \approx \mathrm{Ry}_{ex} + \mathcal{E}_e(\mathbf{k}+\mathbf{q}) + \mathcal{E}_h(\mathbf{q})$$

is the energy needed to transform one exciton into a free e–h pair with the momenta $\mathbf{k}+\mathbf{q}$ and $\mathbf{q}$. The expression $[\varphi(\mathbf{k}+\mathbf{q}) - \varphi(\mathbf{q})]$ is nothing but the matrix element of the charge-density operator $\rho_{\mathbf{k}} = (e^{i\mathbf{k}\cdot\mathbf{r}_e} - e^{i\mathbf{k}\cdot\mathbf{r}_h})$ between the ground $1s$ exciton state $|0\rangle = (1/\sqrt{V})\psi_{1s}(\mathbf{r}_e - \mathbf{r}_h)$ and the excited states $|n\rangle = |\mathbf{k}+\mathbf{q}, -\mathbf{q}\rangle \approx (1/V)e^{i(\mathbf{k}+\mathbf{q})\cdot\mathbf{r}_e - i\mathbf{q}\cdot\mathbf{r}_h}$, which belong to the continuum exciton spectrum. In this way, the e–h correlation is taken into account in only the ground state $|0\rangle$ but not in the excited states $|n\rangle$. If one also takes into consideration the Coulomb interaction in the final states $|n\rangle$, then the Kramers polarizability of the excitons can be obtained. In the present form, the BCS polarizability and the coherence factor describe the dynamic of the electron–hole response incompletely, but this response is not grossly distorted.

To determine the ground-state energy of the Bose condensed e–h pairs, Nozières and Comte substituted into main formula (2.137) the expression for $\mathcal{E}(\mathbf{k},\omega,\lambda)$ found in the framework of the GRPA, of the form

$$\mathcal{E}(\mathbf{k},\omega,\lambda) = 1 - V_{\mathbf{k}}(\lambda)\Pi(\mathbf{k},\omega), \tag{2.149}$$

where the explicit dependence on λ is due to the bare Coulomb interaction constant $V_{\mathbf{k}}(\lambda)$. When Eq. (2.149) is used, the following formula can be obtained:

$$\int_0^{e^2} \frac{d\lambda}{\lambda} \mathrm{Im}\left[\frac{1}{\mathcal{E}(\mathbf{k},\omega,\lambda)}\right] = \arctan\frac{V_{\mathbf{k}}\Pi_2(\mathbf{k},\omega)}{1 - V_{\mathbf{k}}\Pi_1(\mathbf{k},\omega)}. \tag{2.150}$$

If one assumes that $|\mathcal{E}_2(\mathbf{k},\omega)/\mathcal{E}_1(\mathbf{k},\omega)| < 1$ and substitutes $\arctan x$ by x, formula (2.138) takes the form

$$E_0 = E_{kin} - N_{ex}\sum_{\mathbf{k}} V_{\mathbf{k}} - \sum_{\mathbf{k}} V_{\mathbf{k}} \int_0^{\infty} \frac{\hbar\, d\omega}{2\pi} \frac{\Pi_2(\mathbf{k},\omega)}{1 - V_{\mathbf{k}}\Pi_1(\mathbf{k},\omega)}. \tag{2.151}$$

To obtain the ground-state energy in the HFBA, it is enough to neglect the polarizability $V_{\mathbf{k}}\Pi_1(\mathbf{k},\omega)$ compared with unity in the denominator of formula (2.151). Then E_{HFB} is

$$E_{HFB} = E_{kin} - N_{ex}\sum_{\mathbf{k}} V_{\mathbf{k}} + \frac{1}{2}\sum_{\mathbf{k}} V_{\mathbf{k}} \sum_{\mathbf{q}} (u_{\mathbf{k}+\mathbf{q}}v_{\mathbf{q}} - u_{\mathbf{q}}v_{\mathbf{k}+\mathbf{q}})^2. \tag{2.152}$$

The kinetic energy E_{kin} coincides with the first part of the ground-state energy U of Eq. (2.110) in the spin-oriented system. It is easy to prove that the sum of the last two terms in Eq. (2.152) coincides with the second part of U. In this way E_{HFB} completely coincides with U. One can represent the ground-state energy as a sum of two terms. One of them is determined in the HFBA and the second, additional part is called a correlation energy, E_{corr}, where

$$E_0 = E_{HFB} + E_{corr}. \tag{2.153}$$

The correlation energy E_{corr} is the energy in addition to E_{HFB}, obtained by any more precise method than the HFBA. For example, the first-order correction to the polarizability $4\pi\alpha_1(\mathbf{k}, \omega)$ gives the contribution to the correlation energy,

$$\begin{aligned} \delta E_{\text{corr}} &= -\sum_{\mathbf{k}} V_{\mathbf{k}}^2 \int_0^\infty \frac{\hbar\, d\omega}{2\pi} \Pi_2(\mathbf{k}, \omega)\Pi_1(\mathbf{k}, \omega) \\ &= -\sum_{\mathbf{k}} V_{\mathbf{k}}^2 \sum_{\mathbf{p}} \sum_{\mathbf{q}} \frac{(u_{\mathbf{p+k}} v_{\mathbf{p}} - u_{\mathbf{p}} v_{\mathbf{p+k}})^2 (u_{\mathbf{q+k}} v_{\mathbf{q}} - u_{\mathbf{q}} v_{\mathbf{k+q}})^2}{E(\mathbf{p}+\mathbf{k}) + E(\mathbf{p}) + E(\mathbf{q}+\mathbf{k}) + E(\mathbf{q})}, \end{aligned} \tag{2.154}$$

and to the chemical potential

$$\delta\mu_{\text{ex}}^{\text{corr}} = \frac{d\delta E_{\text{corr}}}{dN_{\text{ex}}}. \tag{2.155}$$

Expression (2.154) is the second-order perturbation due to virtual excitation of two quasiparticle pairs with opposite momenta out of the mean-field ground state. This direct-screening correction is negative; it cancels part of the blue shift of the exciton level, in the dilute limit, and acts to depress the chemical potential μ_{ex}. Along with this correction in Eq. (2.154) there is also the exchange-conjugate diagram. Nozières and Comte, discussing the dilute limit, argued on physical grounds that the role of the exchange-screening term will be largely reduced. For the case $m_e = m_h$, they determined the correction in Eq. (2.155) to be

$$\delta\mu_{\text{ex}}^{\text{corr}} = -7.776\, n_{\text{ex}} a_{\text{ex}}^3 \text{Ry}_{\text{ex}}. \tag{2.156}$$

That correction should be combined with the term found in the unscreened mean-field approach. Combining that result with the expression in Eqs. (2.121), $(13\pi/3) n_{\text{ex}} a_{\text{ex}}^3 \text{Ry}_{\text{ex}}$, found for non-spin-polarized *e–h* pairs, the authors [63] obtained the net change of the chemical potential as a function of the exciton density,

$$\mu_{\text{ex}} = \text{Ry}_{\text{ex}}\big(-1 + 5.85\, n_{\text{ex}} a_{\text{ex}}^3\big). \tag{2.157}$$

The net compressibility is therefore positive and the dilute exciton gas is thermodynamically stable, at least when $m_e = m_h$.

In their comments on their results, Nozières and Comte noted that the second-order screening terms that they take into account are nothing but the van der Waals attraction between excitons. These terms describe the virtual polarization of two excitons due to Coulomb interaction. At large distances, the wave functions do not overlap, and only the direct diagrams are important. The exchange-conjugate diagrams come into play when the distance is a_{ex}. But then the exclusion principle is crucial and one can no longer separate the mean-field and the screening effects.

The result of Eq. (2.157) means that the hard-core repulsion supersedes the van der Waals attraction in determining the compressibility of Bose condensed excitons. It is gratifying that the same approximation, which is known to be exact in the limit of high densities, also yields a sensible physical picture in dilute systems. Here it represents an approximation to the van der Waals interaction. The GRPA also provides a reasonable interpolation scheme in the intermediate range of pair densities. Nozières and Comte arrived at the conclusion that the pairing is less efficient in a screened theory, in agreement with the idea of a Mott transition. The tendency to form bound pairs is largely reduced in the presence of screening at intermediate densities. Nevertheless, the electron–hole pairing remains significant, and a joint treatment of screening and pairing seems to be important [63]. One complication is that at a given pair density, there may be more than one possible solution for the equilibrium ratio of excitons to free electrons and holes. This leads to the possibility of hysteresis in the Mott transition, a topic that is discussed in Chapter 5.

2.2.4 Collective Elementary and Macroscopic Excitations

In Subsection 2.2.3 the energy spectrum of one fermionic elementary excitation was derived in the mean-field approximation. For the new *e–h* quasiparticles, when $m_e = m_h$, an expression for the energy spectrum is given by $E(\mathbf{p})/2$, which corresponds to an energy approximately equal to $\mathrm{Ry}_{\mathrm{ex}}/2$. This means that to break a bound exciton state and to create a free *e–h* pair, one needs an energy twice as large.

Keldysh and Kozlov [15] argued that along with the individual one-fermion excitations, there are also collective two-particle excitations, corresponding in the limit $n_{\mathrm{ex}} \to 0$ to the motion of a single exciton as a whole entity, without bond breaking. This kind of excitation changes only the total momentum and the kinetic energy of the translational motion of an exciton. To determine the energy spectrum of that branch of collective excitations, Keldysh and Kozlov introduced the two-particle Green's function $G_2(P, p, p')$, which depends on the four-component momenta $P = (\mathbf{P}, E)$ and $p = (\mathbf{p}, \mathcal{E})$. The first of them describes the translational motion of the pair, whereas p and p' stand for the relative motions of the new quasiparticles. The results of the Keldysh and Kozlov paper [15] on this subject are presented here.

In the presence of the Bose condensate, it is necessary to introduce, along with the usual Green's function G_2, the anomalous two-particle Green's function $\tilde{G}_2(P, p, p')$, which is the sum of the linked diagrams describing the creation of two renormalized *e–h* pairs from the vacuum. Figure 2.13 shows the graphical representation of the two Green's functions. The full and the dashed lines represent the free quasiparticle Green's functions G_e^0 and G_h^0, respectively, which are equal to $[\mathcal{E} - E(\mathbf{p})/2 + i\delta]^{-1}$ in the mean-field approximation when $m_e = m_h$. The Green's functions G_2 and $\tilde{G}_2$, which describe both the collective and the individual excitations, obey the graphical equations represented in Fig. 2.14. The different terms in these equations are grouped in such a way that G_2 and $\tilde{G}_2$ contain only the linked diagrams. The Keldysh–Kozlov equations are to some extent similar to the Beliaev equations for the nonideal Bose gas [37,40]. Unlike the pure algebraic Beliaev equations, however, the Keldysh–Kozlov equations are integral equations over the momenta of the relative motion. For this reason they describe both the exciton motion and the internal structure.

Near the poles corresponding to the oscillations of the exciton-gas density, the main contribution is given by the terms homogeneous in the functions G_2 and $\tilde{G}_2$. There are also inhomogeneous terms, containing the vertices $\Gamma(P, p, p')$ and $\tilde{\Gamma}(P, p, p')$, that have no poles. In the lowest order of perturbation theory, Γ contains the diagram of Fig. 2.8(c), whereas $\tilde{\Gamma}$ is represented by the diagram of Fig. 2.8(h). In higher orders of the perturbation theory, these simplest diagrams must be completed by the blocks that describe all possible scattering processes with the same number of incoming and outgoing quasiparticles as in the

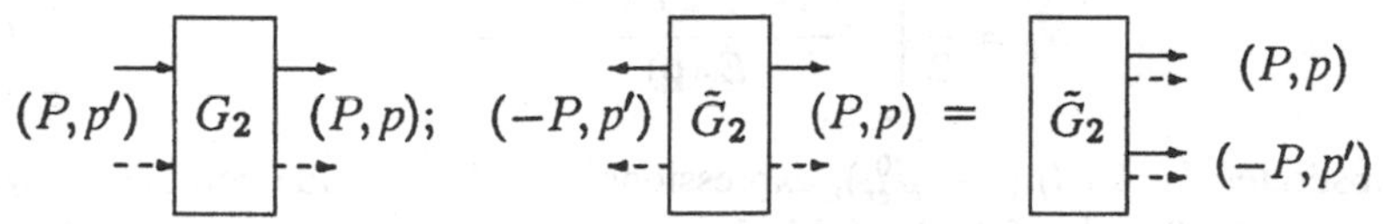

Figure 2.13. The two-particle Green's functions $G_2(P, p, p')$ and $\tilde{G}_2(P, p, p')$ [15].

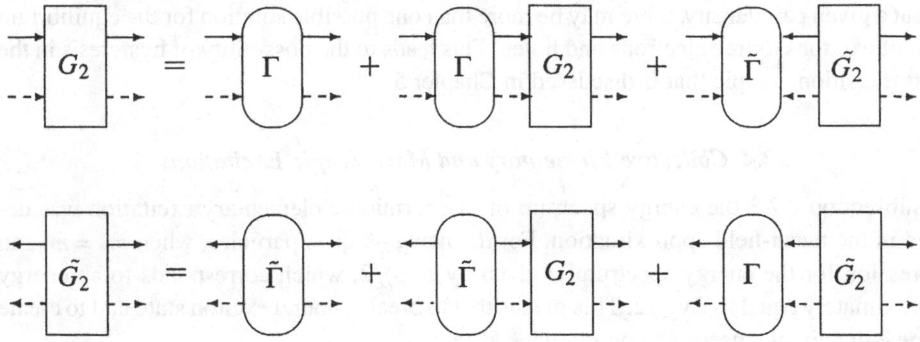

Figure 2.14. The graphical equations for the two-particle Green's functions [15].

simplest diagrams. Keldysh and Kozlov solved the graphical equations by using perturbation theory, taking the values of E and $P^2/2m_{\text{ex}}$ to be small compared with $\mu_{\text{ex}} - \mu_{\text{ex}}^0$, and n_{ex} to be small compared with a_{ex}^{-3}. The pole containing parts of the two-particle Green's functions G_2 and $\tilde{G}_2$ can be presented in the factorized forms

$$G_2(P, p, p') = A(p)g(P)A(p'),$$
$$\tilde{G}_2(P, p, p') = A(p)\tilde{g}(P)A(p'), \tag{2.158}$$

where the multiplier $A(p)$ depends on both exciton parameters and the zeroth-order single-particle Green's functions $G_e^0(p)$, $G_h^0(p)$, as follows:

$$A(p) = G_e^0(p)G_h^0(-p)\left(\text{Ry}_{\text{ex}} + \frac{p^2}{2\mu}\right)\varphi_{1s}(\mathbf{p}). \tag{2.159}$$

The factors $g(P)$ and $\tilde{g}(P)$ of expressions (2.158) coincide in their form with the single-particle Green's functions for the weakly nonideal Bose gas in the Beliaev theory. These functions have the forms

$$g(P) = \frac{1 + N_{\mathbf{p}}}{E - E_c(\mathbf{p})} - \frac{N_{\mathbf{p}}}{E + E_c(\mathbf{p})},$$

$$\tilde{g}(P) = -\frac{\left(\mu_{\text{ex}} - \mu_{\text{ex}}^0\right)}{E^2 - E_c^2(\mathbf{P})}. \tag{2.160}$$

The energy spectrum of the collective elementary excitations $E_c(\mathbf{p})$ and the occupation numbers of the noncondensate excitons $N_{\mathbf{p}}$ are related to the correction of the chemical potential $(\mu_{\text{ex}} - \mu_{\text{ex}}^0)$ in the following way:

$$E_c(\mathbf{p}) = \left[\frac{(\mu_{\text{ex}} - \mu_{\text{ex}}^0)\mathbf{p}^2}{m_{\text{ex}}} + \left(\frac{\mathbf{p}^2}{2m_{\text{ex}}}\right)^2\right]^{1/2},$$

$$N_{\mathbf{p}} = \frac{1}{2}\left[\frac{\left(\mu_{\text{ex}} - \mu_{\text{ex}}^0\right) + \frac{\mathbf{p}^2}{2m_{\text{ex}}}}{E_c(\mathbf{p})} - 1\right]. \tag{2.161}$$

If one substitutes L_0 for $(\mu_{\text{ex}} - \mu_{\text{ex}}^0)$, expressions (2.161) coincide completely with the results of Eqs. (2.37) and (2.38). As Keldysh and Kozlov pointed out, these results are true if a repulsive interaction prevails between the excitons.

Another way to obtain information about the collective elementary excitations was demonstrated by Schmitt-Rink, Chemla, and Haug in papers dedicated to the optical Stark effect in an excitonic system [75, 76]. These papers will be treated in Chapter 6, but the method they used is discussed here. This method was proposed by Anderson in the theory of superconductivity [77] and represents the generalization of the well-known RPA. This method was used to describe the BEC of the electron Cooper pairs that form the ground state of a superconductor. The main idea of this method was developed in reference to the BEC of *e–h* pairs by Schmitt-Rink et al. [75, 76]. Anderson's paper [77] contains many instructive explanations that can help one to have better understanding of similar phenomena in the *e–h* system. The Bose condensed systems of electron pairs in the superconductor on the one hand and of *e–h* pairs in semiconductors on the other exhibit many common properties because of the presence of coherent macroscopic states and therefore can be studied by the same method. Below, we cite some conclusions of the Anderson paper.

In superconductors there are also the collective excitations that coexist alongside the individual elementary excitations. The most important of the collective excitations are the longitudinal waves that have the velocity $v_F/\sqrt{3}$ in the Fröhlich model of superconductor. In the usual case of a high-density electron gas with a singular Coulomb interaction, the collective excitations coincide with the plasma oscillations. The gap in the plasmon spectrum, in a strict sense, is enforced only by the long-range Coulomb effects. Other collective excitations in superconductors resembling higher bound-pair states may or may not exist, but they do not seriously affect the energy gap. The theory proposed by Anderson obeys sum rules and is gauge invariant. The basic Anderson method is a generalized form of both the RPA proposed by Bohm and Pines [78, 72] and the diagram-summing method elaborated by Brueckner and Gell-Mann [79]. Both of these methods lead to a set of linear equations of motion for the elementary excitations in terms of the quantities

$$\rho^{Q}_{\mathbf{k},\sigma} = a^{\dagger}_{\mathbf{k}+\mathbf{Q},\sigma} a_{\mathbf{k},\sigma}. \tag{2.162}$$

These equations of motion were first obtained by physical reasoning and then by a method of direct linearization of the full motion equations. The Anderson method is a natural generalization of this method to the case in which the ground state is not the Fermi sea but the BCS state.

The high-density limit is the only proven domain of validity for the RPA. The RPA gives results in the correlation problem, however, that appear to be satisfactory even in the intermediate range [80]. In any case, it represents the only approach known to be effective in studying collective effects in many-body problems.

The Bohm–Pines RPA technique linearizes the full equations of motion by observing that in the Fermi sea the quantities $n_{\mathbf{k}_0} = a^{\dagger}_{\mathbf{k},\sigma} a_{\mathbf{k},\sigma}$ and $(1 - n_{\mathbf{k}_0}) = a_{\mathbf{k},\sigma} a^{\dagger}_{\mathbf{k},\sigma}$ may have finite c-number average values. The average values are taken as a zeroth-order approximation for these operators, while the $\rho^{\mathbf{Q}}_{\mathbf{k},\sigma}$ act like the first-order infinitesimals.

In the BCS state, not only $n_{\mathbf{k},\sigma}$ but also $b^{\dagger}_{\mathbf{k},\sigma} = a^{\dagger}_{\mathbf{k},\sigma} a^{\dagger}_{-\mathbf{k},-\sigma}$ and $b_{\mathbf{k},\sigma} = a_{-\mathbf{k},-\sigma} a_{\mathbf{k},\sigma}$ have c-number averages, so that one must also derive equations of motion for first-order quantities $b^{\mathbf{Q}\dagger}_{\mathbf{k},\sigma} = a^{\dagger}_{\mathbf{k}+\mathbf{Q},\sigma} a^{\dagger}_{-\mathbf{k},-\sigma}$ etc. The unperturbed BCS state is determined when equations of motion are found for the average values of $\langle n_{\mathbf{k},\sigma}\rangle$ and $\langle b^{\dagger}_{\mathbf{k},\sigma}\rangle$.

The elementary excitations in superconductors fall into two groups. One of them is the individual particlelike excitations, the spectrum of which is practically the same as the BCS–Bogoliubov energy-gap spectrum. The second group contains the collective solutions. One collective excitation of a longitudinal wave resembles a bound pair of electrons of nonzero momentum. This branch of the spectrum has a linear dispersion relation, with the velocity $v = v_F/\sqrt{3}$.

RPA automatically separates the equations of motion of excitations with different total momentum **Q**, because only the zeroth-order quantities have zero momentum. For this reason, the $\mathbf{Q}=0$ equations are decoupled from the rest.

The method of linearization is straightforward in the extreme. Each term in the interaction part of the equations of motion is a product of four fermions and thus a product of two bilinear combinations of b or ρ in a number of ways. Anderson assumes that the state in reference to which he linearizes the equations is the BCS–Bogoliubov product state. This state may have finite zeroth-order values of $b_{\mathbf{k}}^{\dagger}$ or $n_{\mathbf{k}}$ equal to $\langle b_{\mathbf{k}}^{\dagger}\rangle$ and $\langle n_{\mathbf{k}}\rangle$. One keeps only terms that contain one of $\rho_{\mathbf{k},\sigma}^{\mathbf{Q}}$, $b_{\mathbf{k},\sigma}^{\mathbf{Q}}$ quantities.

To better understand the peculiarities of the collective excitations, Anderson [81] studied the coherent excited states generated by the momentum displacement operator, which coincides with the Fourier transform of the electron space-density operator:

$$\rho_{\mathbf{Q}} = \sum_{\mathbf{k},\sigma} a_{\mathbf{k}+\mathbf{Q},\sigma}^{\dagger} a_{\mathbf{k},\sigma}. \tag{2.163}$$

This operator plays a central role in the verification of the gauge invariance of the Hamiltonian used here, as well as in plasma theory. An individual component of $\rho_{\mathbf{Q}}$, namely $a_{\mathbf{k}+\mathbf{Q},\sigma}^{\dagger} a_{\mathbf{k},\sigma}$, when applied to the BCS ground-state wave function $|\psi_g\rangle$, as

$$|\psi_g\rangle = \prod_{\mathbf{q},\sigma} (u_{\mathbf{q}} + v_{\mathbf{q}} a_{\mathbf{q},\sigma}^{\dagger} a_{-\mathbf{q},-\sigma}^{\dagger})|0\rangle, \tag{2.164}$$

creates an excited state $|\psi_{\mathbf{k}+\mathbf{Q},\sigma;-\mathbf{k},-\sigma}\rangle$:

$$|\psi_{\mathbf{k}+\mathbf{Q},\sigma;-\mathbf{k},-\sigma}\rangle = \prod_{\mathbf{q},s} (u_{\mathbf{q}} + v_{\mathbf{q}} a_{\mathbf{q},s}^{\dagger} a_{-\mathbf{q},-s}^{\dagger}) a_{\mathbf{k}+\mathbf{Q},\sigma}^{\dagger} a_{-\mathbf{k},-\sigma}^{\dagger} |0\rangle.$$

$$(\mathbf{q}, s) \neq (\mathbf{k}+\mathbf{Q}, \sigma); (\mathbf{k}, \sigma) \tag{2.165}$$

Here $|0\rangle$ is the vacuum state. The result of the action is

$$a_{\mathbf{k}+\mathbf{Q},\sigma}^{\dagger} a_{\mathbf{k},\sigma} |\psi_g\rangle = u_{\mathbf{k}+\mathbf{Q}} v_{\mathbf{k}} |\psi_{\mathbf{k}+\mathbf{Q},\sigma;-\mathbf{k},-\sigma}\rangle, \tag{2.166}$$

which describes the appearance of one excited pair of electrons $(\mathbf{k}+\mathbf{Q}, \sigma)$ and $(-\mathbf{k}, -\sigma)$ with total momentum pairing equal to **Q** instead of zero. All the other pairs remain in the ground state and have momenta equal to zero. The total operator $\rho_{\mathbf{Q}}$ being applied to $|\psi_g\rangle$ leads to a linear combination of states (2.165). It can be thought of as equivalent to a Cooper bound state of a pair of electrons with nonzero momentum **Q**, superimposed on the BCS ground state:

$$|\psi_{\mathrm{ex},\mathbf{Q}}\rangle = \rho_{\mathbf{Q}}|\psi_g\rangle = \sum_{\mathbf{k},\sigma} u_{\mathbf{k}+\mathbf{Q}} v_{\mathbf{k}} |\psi_{\mathbf{k}+\mathbf{Q},\sigma;-\mathbf{k},-\sigma}\rangle. \tag{2.167}$$

An important conclusion of the Anderson paper [81] is that the longitudinal excitations generated by $\rho_{\mathbf{Q}}$ do not break up the phase coherence of the superconducting state. On the other hand, any transverse excitation of at least one pair in the superconducting ground state involves breaking up the phase coherence over the whole Fermi surface and leads to a lowering of the attractive binding energy.

Another approach to collective elementary excitations was elaborated by Bogoliubov et al. in their book, *New Method in the Theory of Superconductivity* [59]. This method takes into account the profound physical resemblance between the phenomena of superconductivity and superfluidity and represents a generalization of the microscopic theory of superfluidity [2]. Bogoliubov found the dispersion relation of the longitudinal collective branch of the spectrum by using the method of approximate secondary quantization. This method has a virtue in common with the Anderson approach considered above [77], because in both methods one deals with linearized motion equations.

To begin, it is necessary to transform expression (2.163) from the electron operators $a^{\dagger}_{\mathbf{q},\sigma}$ and $a_{\mathbf{q},\sigma}$ to the new particle–hole Fermi-operators $\alpha^{\dagger}_{\mathbf{k},\nu}$ and $\alpha_{\mathbf{k},\nu}$ introduced by Bogoliubov with the aid of the canonical u, v transformation, as follows:

$$\alpha_{\mathbf{k},0} = u_{\mathbf{k}} a_{\mathbf{k},+} - v_{\mathbf{k}} a^{\dagger}_{-\mathbf{k},-},$$
$$\alpha_{\mathbf{k},1} = u_{\mathbf{k}} a_{-\mathbf{k},-} + v_{\mathbf{k}} a^{\dagger}_{\mathbf{k},+}. \tag{2.168}$$

The ground-state wave function (2.164) obeys the conditions

$$\alpha_{\mathbf{k},\nu}|\psi_g\rangle = 0; \quad \nu = 0, 1, \tag{2.169}$$

which means the absence of the new quasiparticles in the superconducting ground state. If one retains only the terms describing the creation or the annihilation of the particle–hole pairs, expression (2.163) becomes

$$\rho_{\mathbf{Q}} = \sum_{\mathbf{k}} (u_{\mathbf{k+Q}} v_{\mathbf{k}} + u_{\mathbf{k}} v_{\mathbf{k+Q}})(\alpha^{\dagger}_{\mathbf{k+Q},0}\alpha^{\dagger}_{\mathbf{k},1} + \alpha_{\mathbf{k},1}\alpha_{\mathbf{k+Q},0}). \tag{2.170}$$

Such pairs of operators do not change the total spin, but do change the total momentum by the value $\mathbf{Q}$.

At sufficiently small values of $\mathbf{Q}$ there is an effective attraction between the fermions $(\mathbf{k+Q}, 0)$ and $(\mathbf{k}, 1)$. This fact encourages one to introduce the particle–hole complexes described by products of two Fermi amplitudes $\alpha^{\dagger}_{\mathbf{k+Q},0}\alpha^{\dagger}_{\mathbf{k},1}$ and $\alpha_{\mathbf{k}1}\alpha_{\mathbf{k+Q},0}$ as inseparable elements of the theory. In the approximate description, one replaces these complexes by the Bose amplitudes $\beta^{\dagger}_{\mathbf{Q}}(\mathbf{k})$, $\beta_{\mathbf{Q}}(\mathbf{k})$, as follows:

$$\alpha^{\dagger}_{\mathbf{k+Q},0}\alpha^{\dagger}_{\mathbf{k},1} \to \beta^{\dagger}_{\mathbf{Q}}(\mathbf{k}), \qquad \alpha_{\mathbf{k}1}\alpha_{\mathbf{k+Q},0} \to \beta_{\mathbf{Q}}(\mathbf{k}). \tag{2.171}$$

After this substitution, the $\rho_{\mathbf{Q}}$ operator becomes a linear operator of the type

$$\rho_{\mathbf{Q}} = \sum_{\mathbf{k}} (u_{\mathbf{k+Q}} v_{\mathbf{k}} + u_{\mathbf{k}} v_{\mathbf{k+Q}})[\beta^{\dagger}_{\mathbf{Q}}(\mathbf{k}) + \beta_{\mathbf{Q}}(\mathbf{k})]. \tag{2.172}$$

This simplification permits one to represent the Coulomb electron interaction,

$$\frac{1}{2}\sum_{\mathbf{q}} V_{\mathbf{q}}\rho_{\mathbf{q}}\rho^{\dagger}_{\mathbf{q}},$$

in a form quadratic in the Bose operators $\beta^{\dagger}_{\mathbf{Q}}(\mathbf{k})$ and $\beta_{\mathbf{Q}}(\mathbf{k})$. For their zeroth-order self-energies, it is necessary to use the sum of the energies of two quasiparticles $(\mathbf{k+Q}, 0)$ and $(\mathbf{k}, 1)$. This method transforms the initial Hamiltonian into a quadratic form. Its energy spectrum can be obtained by the solution of a system of homogeneous linear equations. Bogoliubov showed that the separate roots of the secular equation corresponding to the system of linear equations describe the collective oscillations. The longitudinal collective excitations have the linear dispersion relation

$$E_c(\mathbf{Q}) = \frac{Q v_F}{\sqrt{3}}, \tag{2.173}$$

where v_F is the velocity on the surface of the Fermi sphere. Transverse collective oscillations also exist.

Many of the above-mentioned ideas can be used in the case of the Bose condensed *e–h* pairs. As mentioned above, the ground-state wave function has the form

$$|\psi_g\rangle = \prod_{\mathbf{q},s} (u_\mathbf{q} - v_\mathbf{q} a^\dagger_{\mathbf{q},s} b^\dagger_{-\mathbf{q},-s})|0\rangle. \tag{2.174}$$

The operator $\rho_\mathbf{Q}$ for the electron–hole space density is

$$\rho_\mathbf{Q} = \sum_{\mathbf{k},\sigma} (a^\dagger_{\mathbf{k}+\mathbf{Q},\sigma} a_{\mathbf{k},\sigma} - b^\dagger_{\mathbf{k}+\mathbf{Q},\sigma} b_{\mathbf{k},\sigma}). \tag{2.175}$$

The action of one component of this operator on wave function (2.174) gives the result

$$(a^\dagger_{\mathbf{k}+\mathbf{Q},\sigma} a_{\mathbf{k},\sigma} - b^\dagger_{-\mathbf{k},-\sigma} b_{-\mathbf{k}-\mathbf{Q},-\sigma})|\psi_g\rangle = (u_\mathbf{k} v_{\mathbf{k}+\mathbf{Q}} - u_{\mathbf{k}+\mathbf{Q}} v_\mathbf{k})|\psi_{\mathbf{k}+\mathbf{Q},\sigma;-\mathbf{k},-\sigma}\rangle. \tag{2.176}$$

The functions

$$|\psi_{\mathbf{k}+\mathbf{Q},\sigma;-\mathbf{k},-\sigma}\rangle = \prod_{\mathbf{q},s} (u_\mathbf{q} - v_\mathbf{q} a^\dagger_{\mathbf{q},s} b^\dagger_{-\mathbf{q},-s}) a^\dagger_{\mathbf{k}+\mathbf{Q},\sigma} b^\dagger_{-\mathbf{k},-\sigma}|0\rangle,$$

$$(\mathbf{q}, s) \neq (\mathbf{k}+\mathbf{Q}, \sigma); (\mathbf{k}, \sigma) \tag{2.177}$$

describe the excited state with one excited *e–h* pair with the total momentum **Q** among all the other unperturbed pairs with momenta equal to zero.

As in the case of a superconductor, in the system of Bose condensed *e–h* pairs, the longitudinal excitation generated by the operator $\rho_\mathbf{Q}$ of Eq. (2.175) gives rise to the coherent excited state

$$|\psi_{\mathrm{ex},\mathbf{Q}}\rangle = \rho_\mathbf{Q}|\psi_g\rangle = \sum_{\mathbf{k},\sigma} (u_\mathbf{k} v_{\mathbf{k}+\mathbf{Q}} - u_{\mathbf{k}+\mathbf{Q}} v_\mathbf{k})|\psi_{\mathbf{k}+\mathbf{Q},\sigma;-\mathbf{k},-\sigma}\rangle, \tag{2.178}$$

which is a coherent superposition of wave functions (2.177). It should be mentioned that these states completely differ from the macroscopic coherent state with nonzero wave vector **k**. For true bosons the latter state means the Bose condensation of all particles in the single-particle state with wave vector **k**. The wave function of the metastable ground state in this case has the form

$$|\psi_{g,\mathbf{k}}\rangle = \prod_{\mathbf{q},\sigma} (u_\mathbf{q} - v_\mathbf{q} a^\dagger_{\alpha\mathbf{k}+\mathbf{q},\sigma} b^\dagger_{\beta\mathbf{k}-\mathbf{q},-\sigma})|0\rangle, \tag{2.179}$$

where $\alpha = m_e/m_{\mathrm{ex}}$ and $\beta = m_h/m_{\mathrm{ex}}$. All the bound *e–h* pairs move with the velocity $\mathbf{V}_s = \hbar\mathbf{k}/m_{\mathrm{ex}}$. The total energy of this state is an extensive value proportional to the volume of the system V.

The transition between two coherent macroscopic states $|\psi_{g,\mathbf{k}_1}\rangle$ and $|\psi_{g,\mathbf{k}_2}\rangle$ with $\mathbf{k}_1 \neq \mathbf{k}_2$ needs an extensive amount of energy and cannot be achieved by an elementary excitation. Excitations that accomplish this transition are called macroscopic excitations. These have been discussed, e.g., by Sonin [82] in connection with the relaxation of superfluid He-II. In the 1950s, Feynman [83] suggested that superfluid flow relaxation is due to the formation of quantum vortices as macroscopic excitations, following experiments in which it was found that the critical velocities of superfluidity in He-II turned out to be less than those predicted by the Landau [10] and Bogoliubov [2] theories. The quantum-vortex changes the state of the liquid in a volume equal to $R^2 l$, where R is the radius of the exterior vortex ring and l is its length. This explanation permitted one to determine the values of the critical velocities in better accordance with experiment. Vinen [84] argued that the formation of a quantum vortex with a large radius, being in fact a macroscopic excitation, has a small probability. In spite of this and other difficulties of the theory, there is no doubt about the role of quantum vortices in the superfluid relaxation of He-II. Quantum-vortex formation in the case of Bose condensed excitons has been discussed in Refs. 32, 35, 85, and 86. In the case of dipole active excitons and photons, quantum-vortex formation is accompanied by the production of light filaments [85, 86]. To develop this theory properly, a generalization of the coherent macroscopic states to take into account inhomogeneous space and time evolution is needed.

We have seen that superfluidity of excitons is a fundamental prediction of the proper theory of Bose condensation of the metastable electron–hole system. Excitons do not exist in a container as He-II does, however, in which the superfluid contacts the outside world only by means its surface. Instead, excitons exist inside a viscous medium, which is the phonon field of the crystal. To treat excitonic superfluidity properly, we must understand the exciton–phonon interactions, which are the subject of Chapter 3.

References

[1] A. Leggett, in *Bose-Einstein Condensation*, A. Griffin, D.W. Snoke, and S. Stringari, eds. (Cambridge U. Press, Cambridge, 1995).

[2] N.N. Bogoliubov, *Izv. Akad. Nauk SSSR* Ser. Fiz. **11**, 77 (1947); *Zh. Eksp. Teor. Fiz.* **18**, 7, 622 (1947); Vestn. Mosk. Univ. **7**, 43 (1947); *Lectures on Quantum Statistics* (Radyanska Shkola, Kiev, 1949), (in Ukrainian); *Quasiaverages in Problems of Statistical Mechanics* Preprint R-1451 (Laboratory of Theoretical Physics, Joint Institute for Nuclear Research, Dubna, 1963), (in Russian); *Collection of Papers in Three Volumes* (Naukova Dumka, Kiev, 1971), Vols. 2 and 3 (in Russian).

[3] R.J. Glauber, *Phys. Rev.* **130**, 2529 (1966); **131**, 2776 (1966).

[4] J.R. Klauder and E.C.G. Sudarshan, *Fundamentals of Quantum Optics* (Benjamin, New York, 1968).

[5] O. Penrose, *Philos. Mag.* **42**, 1373 (1951).

[6] C.N. Yang, *Rev. Mod. Phys.* **34**, 694 (1962).

[7] F. London, *Phys. Rev.* **54**, 947 (1938); *Superfluids* (Wiley, New York, 1954), Vols. 1 and 2.

[8] W.F. Vinen, *Rep. Prog. Phys.* **31**, pt. 1, 61 (1968).

[9] V.L. Ginzburg and L.D. Landau, *Zh. Eksp. Teor. Fiz.* **20**, 1064 (1950).

[10] L.D. Landau, *J. Phys. USSR* **5**, 71 (1940); *Zh. Eksp. Teor. Fiz.* **11**, 592 (1941); **14**, 112 (1944); *Collection of Papers in Two Volumes* (Nauka, Moscow, 1969) (in Russian).

[11] V.L. Ginzburg, *Dokl. Akad. Nauk SSSR* **69**, 161 (1949); *Usp. Fiz. Nauk* **97**, 601 (1969).

[12] L.P. Pitaevskii, *Zh. Eksp. Teor. Fiz.* **40**, 646 (1961); *Usp. Fiz. Nauk* **90**, 623 (1966).

[13] E.P. Gross, *Nuovo Cimento* **20**, 454 (1961); *J. Math. Phys.* **4**, 147 (1963); *Ann. Phys.* **9**, 292 (1960).

[14] L.V. Keldysh, in *Problems of Theoretical Physics* (Nauka, Moscow, 1972), pp. 433–444 (in Russian).

[15] L.V. Keldysh and A.N. Kozlov, *Zh. Eksp. Teor. Fiz. Pis'ma* **5**, 238 (1967); *Zh. Eksp. Teor. Fiz.* **54**, 978 (1968) [*Sov. Phys. JETP* **27**, 521 (1968)].

[16] P. Nozières, in *Bose-Einstein Condensation*, A. Griffin, D.W. Snoke, and S. Stringari, eds. (Cambridge U. Press, Cambridge, 1995).

[17] M.H. Anderson et al., *Science* **269** (7), 198 (1995).

[18] K.B. Davis et al., *Phys. Rev. Lett.* **75**, 3969 (1995); M.R. Andrews et al., *Science* **273** (7), 84 (1996).
[19] P. Sokol, in *Bose–Einstein Condensation*, A. Griffin, D.W. Snoke, and S. Stringari, eds., (Cambridge U. Press, Cambridge, 1995).
[20] A. Griffin, *Excitations in a Bose-Condensed Liquid* (Cambridge U. Press, Cambridge, 1993).
[21] M. Kotani, *Handbuch der Physik*, S. Flügge, ed. (Springer, Berlin, 1961), Vol. 37/II.
[22] E. Hanamura and H. Haug, *Phys. Rep. C* **33**, 209 (1977).
[23] G. Manzke, V. May, and K. Henneberger, *Phys. Status Solidi B* **125**, 693 (1984).
[24] S.G. Elkomoss and G. Munschy, *J. Phys. Chem. Solids* **42**, 1 (1981).
[25] A.I. Bobrysheva, M.F. Miglei, and M.I. Shmiglyuk, *Phys. Status Solidi B* **53**, 71 (1972).
[26] A.I. Bobrysheva, *Biexcitons in Semiconductors*, (Shtiintsa, Kishinev, 1979), (in Russian).
[27] L. Schultheis et al., *Phys. Rev. Lett.* **57**, 1635 (1986).
[28] D.W. Snoke, D. Braun, and M. Cardona, *Phys. Rev. B* **44**, 2991 (1991).
[29] A.I. Bobrysheva, S.A. Moskalenko, and Yu.M. Shvera, *Phys. Status Solidi B* **147**, 711 (1988).
[30] A.I. Bobrysheva and S.A. Moskalenko, *Fiz. Tverd. Tela* **25**, 3282 (1983); *Phys. Status Solidi B* **119**, 141 (1983).
[31] S.A. Moskalenko, candidate dissertation, (Institute of Physics of the Academy of Sciences of the Ukrainian SSR, Kiev, 1959); abstracts of the dissertation (Kiev T. G. Shevchenko State University, Kiev, 1960); *Abstracts of the Fourth Workshop on the Theory of Semiconductors* Tbilisi, USSR, 1960 (Academy of Science of the Georgian SSR, Tbilisi, 1960), p. 51; see also E.I.Rashba and K.B. Tolpygo, *Usp. Fiz. Nauk* **74**, 165 (1961); *Abstracts of the All-Union Conference on Spectroscopy Leningrad, 1960* (Acad. Science USSR Moscow, 1960), p. 104; *The Physical Problems of Spectroscopy* Part II, S.E. Frish, ed. (Acad. Sciences USSR, Moscow, 1963). p. 171; *Abstracts of the Second Workshop on the Physics of Alkali-Halide Crystals* Riga, 1961 (Riga State University, Riga, 1961) p. 51; *Physics of the Alkali-Halide Crystals* (Riga State University, Riga, 1962), pp. 77–80.
[32] S.A. Moskalenko, *Fiz. Tverd. Tela* **4**, 276 (1962).
[33] S.A. Moskalenko, P.I. Khadzhi, A.I. Bobrysheva, and A.V. Lelyakov, *Fiz. Tverd. Tela* **5**, 1444 (1963).
[34] S.A. Moskalenko, *Zh. Eksp. Teor. Fiz.* **45**, 1159 (1963).
[35] S.A. Moskalenko, *Bose–Einstein Condensation of Excitons and Biexcitons* (RIO, Academy of Sciences of Moldavian SSR, Kishinev, 1970) (in Russian).
[36] E.M. Lifshitz, and L.P. Pitaevskii, *Statistical Physics, Part* 2 (Pergamon, New York, 1979).
[37] S.T. Beliaev, *Zh. Eksp. Teor. Fiz.* **34**, 417, 433 (1958).
[38] T.D. Lee and C.N. Yang, *Phys. Rev.* **112**, 1419 (1958); **113**, 1406 (1959).
[39] R.P. Feynman, *Phys. Rev.* **76**, 749 (1949).
[40] A.A. Abrikosov, L.P. Gor'kov, and I.E. Dzyaloshinskii, *Methods of Quantum Field Theory in Statistical Physics* (Dover, New York, 1975).
[41] G.C. Wick, *Phys. Rev.* **80**, 268 (1950).
[42] F.J. Dyson, *Phys. Rev.* **75**, 1736 (1949).
[43] H. Haug and H.H. Kranz, *Z. Phys. B* **53**, 151 (1983).
[44] H. Shi, G. Verechaka, and A. Griffin, *Phys. Rev. B* **50**, 1119 (1994).
[45] V.N. Popov and L.D. Fadeev, *Zh. Eksp. Teor. Fiz.* **47**, 1315 (1964).
[46] V.N. Popov, *Functional Integrals and Collective Excitations* (Cambridge U. Press, Cambridge, 1987).
[47] D.W. Snoke, J.P. Wolfe, and A. Mysyrowicz, *Phys. Rev. B* **41**, 11171 (1990).
[48] J.L. Lin and J.P. Wolfe, *Phys. Rev. Lett.* **71**, 1223 (1993).
[49] Yu. Kagan, in *Bose–Einstein Condensation*, A. Griffin, D.W. Snoke, and S. Stringari, eds. (Cambridge U. Press, Cambridge, 1995).
[50] T. Goto, M.Y. Shen, S. Koyama, and T. Yokouchi, *Phys. Rev. B* **55**, 7609 (1997).
[51] M.Y. Shen, T. Yokouchi, S. Koyama, and T. Goto, *Phys. Rev. B* **56**, 13066 (1997).
[52] V.M. Agranovich and B.S. Toshich, *Zh. Eksp. Teor. Fiz.* **53**, 149 (1967).
[53] W. Kohn and D. Sherrington, *Rev. Mod. Phys.* **42**, 1 (1970).

[54] E. Hanamura and H. Haug, *Solid State Commun.* **15**, 1567 (1974).
[55] J.M. Blatt, K.W. Böer, and W. Brandt, *Phys. Rev.* **126**, 1691 (1962).
[56] C. Casella, *J. Phys. Chem. Solids* **24**, 19 (1963); *J. Appl. Phys.* **34**, 1703 (1963); **36**, 2485 (1965).
[57] V.A. Gergel, R.F. Kazarinov, and R.A. Suris, *Zh. Eksp. Teor. Fiz.* **53**, 544 (1967); **54**, 298 (1968); in *Proceedings of the Ninth International Conference on the Physics of Semiconductors* (Nauka, Leningrad, 1969), Vol. 1, p. 472.
[58] J. Bardeen, L.N. Cooper, and J.R. Schrieffer, *Phys. Rev.* **106**, 162 (1957); **108**, 1175 (1957).
[59] N.N. Bogoliubov, V.V. Tolmachev, and D.V. Shirkov, *New Method in the Theory of Superconductivity* (Consultants Bureau, New York, 1959).
[60] J.M. Blatt, *Theory of Superconductivity* (Academic, New York, London, 1964).
[61] L.V. Keldysh and Yu.V. Kopaev, *Fiz. Tverd. Tela* **6**, 2791 (1964) [*Sov. Phys. Solid State* **6**, 2219 (1965)].
[62] C. Comte and P. Nozières, *J. Phys.* **43**, 1069 (1982).
[63] P. Nozières and C. Comte, J. Phys. **43**, 1083 (1982).
[64] W. Kohn, *Phys. Rev. Lett.* **19**, 439 (1967).
[65] R. Zimmermann, *Phys. Status Solidi B* **76**, 191 (1976).
[66] R. Zimmermann and M. Rösler, *Phys. Status Solid B* **75**, 633 (1976); **67**, 525 (1975).
[67] R. Zimmermann, K. Kilimann, W.D. Kraeft, D. Kremp, and G. Röpke, *Phys. Status Solidi B* **90**, 175 (1978).
[68] H. Stolz and R. Zimmermann, *Phys. Status Solidi B* **124**, 201 (1984).
[69] R. Zimmermann and H. Stolz, *Phys. Status Solidi B* **131**, 151 (1985).
[70] F. Henneberger and V. May, *Phys. Status Solidi B* **98**, K77 (1980).
[71] E. Hanamura, *J. Phys. Soc. Jpn.* **29**, 50 (1970).
[72] D. Pines, *Elementary Excitations in Solids* (Benjamin, New York, 1963).
[73] T.M. Rice, in *Solid State Physics* 32, H. Ehrenreich, F. Seitz, D. Turnbull, eds. (Academic, New York, 1977).
[74] C.D. Jeffries and L.V. Keldysh, eds., *Electron–Hole Droplets in Semiconductors* (North-Holland, Amsterdam, 1987).
[75] S. Schmitt-Rink and D.S. Chemla, *Phys. Rev. Lett.* **57**, 2752 (1986).
[76] S. Schmitt-Rink, D.S. Chemla, and H. Haug, *Phys. Rev. B* **37**, 941 (1988).
[77] P.W. Anderson, *Phys. Rev.* **112**, 1900 (1958).
[78] D. Bohm and D. Pines, *Phys. Rev.* **92**, 609 (1953).
[79] K.A. Brueckner and M. Gell-Mann, *Phys. Rev.* **106**, 364 (1957).
[80] P. Nozières and D. Pines, *Phys. Rev.* **109**, 1009 (1958).
[81] P.W. Anderson, *Phys. Rev.* **110**, 827 (1958).
[82] E.B. Sonin, *Usp. Fiz. Nauk* **137**, 267 (1982).
[83] R.P. Feynman, in *Progress in Low Temperature Physics*, C.J. Gorter, ed. (North-Holland, Amsterdam, 1955), Vol. 1, p. 36.
[84] W.F. Vinen, in *Progress in Low Temperature Physics*, C.J. Gorter, ed. (North-Holland, Amsterdam, 1961), Vol. 3, p. 1.
[85] S.A. Moskalenko, M.F. Miglei, M.I. Shmiglyuk, P.I. Khadzhi, and A.V. Lelyakov, *Zh. Eksp. Teor. Fiz.* **64**, 1786 (1973) [*Sov. Phys. JETP* **37**, 902 (1973).]
[86] S.A. Moskalenko, A. I. Bobrysheva, A.V. Lelyakov, M.F. Miglei, P.I. Khadzhi, and M.I. Shmiglyuk, *Interaction of Excitons in Semiconductors* (Shtiintsa, Kishinev, 1974) (in Russian).

3

The Interaction of Condensed Excitons with Lattice Phonons

3.1 Introduction

In this chapter we generalize the theory of the weakly nonideal Bose gas to the case of the interaction of excitons with acoustic or optical phonons, i.e., in the case of interaction with another Bose subsystem, the quasiparticle number of which is not conserved. To simplify the question, we assume that the exciton–phonon interaction leads to only intraband exciton scattering and does not change the total number of excitons in the band. In other words, we consider an isolated exciton band or suppose that the processes of interband scattering are not important because of the symmetries of the exciton and the phonon states and the values of the interaction constants.

The theory of the interaction of condensed excitons with phonons has primary importance in the question of whether excitons will be superfluid. In Chapter 2, we saw that the fact that the excitons are particle–antiparticle pairs does not prevent them from becoming superfluid. At first glance, however, it is not easy to draw an analogy to other Bose condensates such as liquid He. In the case of liquid He, interactions with the external world are limited to the surfaces of the container, while for excitons, the phonon field interpenetrates the same region of space as the exciton gas. The situation is more analogous to the case of a superconductor. Unlike a BCS superconductor, however, in an exciton condensate the lattice phonons have nothing to do with the pairing. We shall see that nevertheless the excitonic condensate remains superfluid in the presence of a phonon field. At the end of this chapter we review recent experiments that may be evidence of excitonic superfluidity.

The theory of the interaction of the excitonic condensate with the phonon field also has other important applications. As discussed in Subsection 3.2.5, the excitonic condensate also has an effect back on the phonon field, leading to a change of the speed of sound of the entire crystal. This leads to predictions for new types of experiments with novel nonlinear effects. In Subsection 3.3.2 we discuss the effect of exciton–phonon interaction on the homogeneous broadening of the absorption and luminescence lines, as mentioned in Subsection 1.4.1.

For the discussion in this chapter we return to the picture of excitons as structureless bosons. Then the Hamiltonian of the exciton–phonon system, taking into account the exciton–exciton and the exciton–phonon interactions, is given by [1, 2]

$$H = \sum_{\mathbf{k}}(\Delta + T_{\mathbf{k}})a^{\dagger}_{\mathbf{k}}a_{\mathbf{k}} + \sum_{\mathbf{k}} B_{\mathbf{k}}b^{\dagger}_{\mathbf{k}}b_{\mathbf{k}} + \frac{1}{2V}\sum_{\mathbf{k}_1,\mathbf{k}_2,\mathbf{k}} U(\mathbf{k})a^{\dagger}_{\mathbf{k}_1}a^{\dagger}_{\mathbf{k}_2}a_{\mathbf{k}_2+\mathbf{k}}a_{\mathbf{k}_1-\mathbf{k}}$$

$$+ \frac{i}{\sqrt{N_a}}\sum_{\mathbf{k},\mathbf{k}'}\Theta(\mathbf{k}-\mathbf{k}')a^{\dagger}_{\mathbf{k}}a_{\mathbf{k}'}(b_{\mathbf{k}-\mathbf{k}'} - b^{\dagger}_{-\mathbf{k}+\mathbf{k}'}). \tag{3.1}$$

Here Δ is the exciton formation energy; $T_{\mathbf{k}}$ is its translational kinetic energy; $a_{\mathbf{k}}^{\dagger}$, $a_{\mathbf{k}}$ and $b_{\mathbf{k}}^{\dagger}$, $b_{\mathbf{k}}$ are the Bose creation and annihilation operators of excitons and phonons with wave vector $\mathbf{k}$, respectively; $B_{\mathbf{k}}$ is the acoustic- or the optical-phonon energy; V is the crystal volume; $U(\mathbf{k})$ is the Fourier transform of the exciton–exciton interaction energy; N_a is the full number of atoms in a crystal; and $\Theta(\mathbf{k})$ is the Fourier transform of the exciton–phonon interaction energy.

As we have remarked previously, one can speak of the exciton–exciton interaction energy only if it is much smaller than the exciton binding energy. If this condition is not fulfilled, it is necessary to start from the electron–hole Hamiltonian. Our model is correct when the excitons are stable formations of a hard-sphere type, between which repulsive forces dominate; in other words, it is possible to speak about the exciton Bose condensate only when an overall repulsion prevails between them. Some types of excitons can bind into biexcitons with a repulsive biexciton–biexciton interaction. In this case, one can consider the biexciton–phonon system by using Hamiltonian (3.1) without loss of generality.

For a deformation-potential interaction of excitons with acoustic phonons and a Fröhlich interaction of excitons with optical phonons, the functions $\Theta(\mathbf{k})$ have the general forms [2–4]

$$\Theta(\mathbf{k}) = \left(\frac{2\hbar k}{Mu}\right)^{1/2} [C_e\phi_e(k) - C_h\phi_h(k)], \tag{3.2}$$

$$\Theta(\mathbf{k}) = \left[2\pi\left(\frac{1}{n_0^2} - \frac{1}{\varepsilon}\right)\hbar\omega_0\frac{e^2}{v_0}\frac{1}{k^2}\right]^{1/2} [\phi_e(k) - \phi_h(k)], \tag{3.3}$$

respectively. Here M is the nucleus mass, u is the acoustic sound velocity in the crystal, n_0 is the refractive index, ε is the static dielectric permeability of the material, ω_0 is the optical-phonon limit frequency, v_0 is the elementary cell volume, and C_e and C_h are the deformation-potential constants for the electrons and the holes, respectively. In the case of simple bands, the functions $\phi_i(k)$ for Wannier–Mott excitons in the $1s$ state are [2]

$$\phi_i(k) = \left(1 + \frac{\mu^2}{m_i^2}\frac{k^2a_{\text{ex}}^2}{4}\right)^{-2}, \tag{3.4}$$

where $m_i = m_e, m_h$ are the effective electron and hole masses. From Eqs. (3.2)–(3.4) one can see that $\Theta(\mathbf{k}) \to 0$ when $k \to 0$; in particular, for acoustic phonons $\Theta(\mathbf{k}) \sim k^{1/2}$ and for optical phonons $\Theta(\mathbf{k}) \sim k$.

The energy spectrum of the system is considered here under the condition that the exciton–exciton collisions take place more often than exciton–phonon scattering. In this case it can be supposed that the excitons form a Bose condensate on time scales that are short compared with those of their interaction with phonons. Then the excitonic Bose condensate acts on the phonon subsystem like an external field, leading to changes in the character of the elementary excitations of the whole system [5]. This implies that the mean-free-path length of the exciton due to the elastic exciton–exciton collisions $l_{\text{ex-ex}}$ is considerably less than the similar values $l_{\text{ex-ph}}^{\text{ac}}$ and $l_{\text{ex-ph}}^{\text{op}}$ for exciton scattering with acoustic and optical phonons. In addition, the impurity concentration in the crystal must be small, so that the free path length $l_{\text{ex-im}}$ of the exciton scattering with impurities is greater than $l_{\text{ex-ex}}$. In other words,

$$l_{\text{ex-ex}} \ll l_{\text{ex-ph}}^{\text{ac}}, l_{\text{ex-ph}}^{\text{op}}, l_{\text{ex-im}}. \tag{3.5}$$

The values l_{ex-ex} and l_{ex-im} can be taken as

$$l_{ex-ex} = \frac{1}{\sigma_{ex-ex} n_{ex}}, \qquad l_{ex-im} = \frac{1}{\sigma_{ex-im} n_{im}},$$

where n_{ex} and n_{im} are the exciton and the impurity concentrations, respectively. The cross sections for elastic exciton–exciton scattering σ_{ex-ex} and exciton–impurity scattering σ_{ex-im} were calculated by Bobrysheva [6, 7] and Kachlishvili [8]. These cross sections are of the order of πa_{ex}^2 and πa_{im}^2, where a_{ex} and a_{im} are the exciton and the impurity radii, respectively. Further discussion of the exciton–exciton-scattering cross section is given in Chapter 4.

The mean free path for the exciton scattering with acoustic phonons has been studied in numerous experimental and theoretical works, e.g., Refs. 3, 4, and 9. Anselm and Firsov [10] found that for temperatures less than the excitation energy from the level $n = 1$ to the level $n = 2$, i.e., for

$$k_B T \ll \frac{3}{4} \frac{\mu e^4}{2\hbar^2 \varepsilon^2},$$

the mean free path is

$$l_{ex-ph}^{ac} = \frac{\pi \rho u^2 \hbar^4}{m_{ex}^2 (C_e - C_h)^2 k_B T},$$

where $\rho = M/v_0$ is the crystal density and μ and m_{ex} are the reduced mass and the total translational mass, respectively. The mean free path for scattering with optical phonons at low temperatures is very long, since the number of these phonons becomes negligible.

For small impurity concentrations, a more strict condition is $l_{ex-ex} \ll l_{ex-ph}^{ac}$, which is fulfilled for

$$T \ll \frac{\pi^2 \rho u^2 \hbar^4 a_{ex}^2 n_{ex}}{m_{ex}^2 (C_e - C_h)^2 k_B}. \tag{3.6}$$

Taking $\rho = 5$ g/cm^3, $u = 3.2 \times 10^5$cm/s, $|C_e - C_h| = 5$ eV, $a_{ex} = 15$ Å, $n_{ex} = 10^{17}$ cm^{-3}, and $m_{ex} = m_0$ (m_0 is the rest mass of the free electron), from inequality (3.6) one concludes that T must be less than 2 K; in other words, cryogenic experiments are required. This does not rule out the possibility of a room-temperature excitonic Bose condensate with strong exciton–phonon interaction, but it means that the simplified theory presented here will not apply. If inequality (3.6) is not fulfilled or if the red shift of the exciton level due to interaction with virtual phonons (discussed in Subsection 1.4.1) is more than the characteristic exciton–exciton interaction energy $n_{ex}U(0)$, then it may be necessary to consider Bose-Einstein condensation (BEC) of the excitons with renormalized translational mass and energy.

If $T = 1$ K, then $l_{ex-ex} = 1.6 \times 10^{-4}$ cm and $l_{ex-ph}^{ac} = 3.2 \times 10^{-4}$ cm. For $n_{im} = 10^{16}$ cm^{-3} and $a_{im} = 10^{-7}$ cm we obtain $l_{ex-im} = 3 \times 10^{-3}$ cm. The scattering time is determined by the ratio of the free path length of the exciton to its thermal velocity u_T. Supposing that $u_T = \sqrt{3k_B T/m_{ex}} \approx 6 \times 10^5$ cm/s, we obtain $\tau_{ex-ex} \approx 2 \times 10^{-10}$ s, $\tau_{ex-ph}^{ac} \simeq 4 \times 10^{-10}$ s, and $\tau_{ex-im} \simeq 5 \times 10^{-9}$ s. In many cases, the lifetime of the excitons is much longer than 10^{-10} s; in some crystals it reaches the values 10^{-8}, 10^{-6}, and even 10^{-4} s [11]. Under these conditions, thermodynamic quasi-equilibrium between the excitons occurs on time scales that are short compared with those of their interaction with the phonons, so that one can initially consider the appearance of exciton BEC in an undeformed lattice.

To consider the excitons as an interacting gas, it is necessary to suppose that their effective free path length due to impurity and phonon scattering is greater than the mean distance $n_{ex}^{-1/3}$ between them. If this is not the case, the excitons will be isolated particles that do not interact with each other. This condition imposes a lower bound on the value of the exciton density. As discussed in Sections 1.3 and 2.2, an upper bound on the value of the exciton density also exists, connected with the electron–hole structure of excitons [12], which leads to screening and metallization at high densities [13, 14].

3.2 The Diagram Technique for the Case $T = 0$. The Energy Spectrum of the System

3.2.1 The Definition of the Green's Functions

The Feynman diagram technique for the special case of interacting Bose particles at $T = 0$ was elaborated by Beliaev [15] and was reviewed in Chapter 2. In Refs. 16 and 17, this technique was expanded for the case $T \neq 0$, taking into account the attractive and the repulsive forces among particles. The Beliaev diagram technique was generalized by Miglei, Moskalenko, and Lelyakov [18] for the case of the excitons interacting with phonons at $T = 0$. The Hamiltonian of such a system is given by Eq. (3.1). The kinematic exciton interaction is taken into account by the introduction of an effective repulsion between excitons at small distances, which, as discussed in Subsection 2.1.3, is a reasonable assumption [19]. In considering the exciton–phonon system, we assume that the exciton number is determined by the pumping, while the number of phonons is not determined.

It is necessary to introduce, besides the exciton Green's functions, the phonon Green's functions and the mixed-type Green's functions, which depend on the four-dimensional quasi-momentum $p = (\mathbf{p}, E)$, where $\mathbf{p}$ is the quasi-momentum. Initially, we use the quasi-particle number N as an independent variable. In this case, the chemical potential $\mu(N)$ is a function of N. One part of the Hamiltonian (3.1) expressed in terms of the operator N, namely $\Delta N = \sum_{\mathbf{k}} \Delta\, a_{\mathbf{k}}^{\dagger} a_{\mathbf{k}}$, can be omitted, since only the intraband scattering for a given number N is considered. Although the ground states of systems with different numbers $(N + i)$ are discussed below, nevertheless the expression $\Delta(N + i)$ remains nearly equal to ΔN because of the extensive value of $N \sim V$.

We introduce the following Green's functions:

$$
\begin{aligned}
G_{11}(p) &= \langle\langle a_{\mathbf{p}} \mid a_{\mathbf{p}}^{\dagger} \rangle\rangle_E = -i \langle N, 0 \mid T(a_{\mathbf{p}} \mid a_{\mathbf{p}}^{\dagger}) \mid N, 0 \rangle_E, \\
G_{33}(p) &= \langle\langle a_{-\mathbf{p}}^{\dagger} \mid a_{-\mathbf{p}} \rangle\rangle_E = -i \langle N, 0 \mid T(a_{-\mathbf{p}}^{\dagger} \mid a_{-\mathbf{p}}) \mid N, 0 \rangle_E, \\
G_{31}(p) &= \langle\langle a_{-\mathbf{p}}^{\dagger} \mid a_{\mathbf{p}}^{\dagger} \rangle\rangle_E = -i \langle N + 2, 0 \mid T(a_{-\mathbf{p}}^{\dagger} \mid a_{\mathbf{p}}^{\dagger}) \mid N, 0 \rangle_E, \\
G_{13}(p) &= \langle\langle a_{\mathbf{p}} \mid a_{-\mathbf{p}} \rangle\rangle_E = -i \langle N, 0 \mid T(a_{\mathbf{p}} \mid a_{-\mathbf{p}}) \mid N + 2, 0 \rangle_E, \\
G_{22}(p) &= \langle\langle b_{\mathbf{p}} - b_{-\mathbf{p}}^{\dagger} \mid b_{\mathbf{p}}^{\dagger} - b_{-\mathbf{p}} \rangle\rangle_E = -i \langle N, 0 \mid T(b_{\mathbf{p}} - b_{-\mathbf{p}}^{\dagger} \mid b_{\mathbf{p}}^{\dagger} - b_{-\mathbf{p}}) \mid N, 0 \rangle_E, \\
G_{21}(p) &= \langle\langle b_{\mathbf{p}} - b_{-\mathbf{p}}^{\dagger} \mid a_{\mathbf{p}}^{\dagger} \rangle\rangle_E = -i \langle N + 1, 0 \mid T(b_{\mathbf{p}} - b_{-\mathbf{p}}^{\dagger} \mid a_{\mathbf{p}}^{\dagger}) \mid N, 0 \rangle_E, \\
G_{23}(p) &= \langle\langle b_{\mathbf{p}} - b_{-\mathbf{p}}^{\dagger} \mid a_{-\mathbf{p}} \rangle\rangle_E = -i \langle N, 0 \mid T(b_{\mathbf{p}} - b_{-\mathbf{p}}^{\dagger} \mid a_{-\mathbf{p}}) \mid N + 1, 0 \rangle_E, \\
G_{12}(p) &= \langle\langle a_{\mathbf{p}} \mid b_{\mathbf{p}}^{\dagger} - b_{-\mathbf{p}} \rangle\rangle_E = -i \langle N, 0 \mid T(a_{\mathbf{p}} \mid b_{\mathbf{p}}^{\dagger} - b_{-\mathbf{p}}) \mid N + 1, 0 \rangle_E, \\
G_{32}(p) &= \langle\langle a_{-\mathbf{p}}^{\dagger} \mid b_{\mathbf{p}}^{\dagger} - b_{-\mathbf{p}} \rangle\rangle_E = -i \langle N + 1, 0 \mid T(a_{-\mathbf{p}}^{\dagger} \mid b_{\mathbf{p}}^{\dagger} - b_{-\mathbf{p}}) \mid N, 0 \rangle_E.
\end{aligned}
\tag{3.7}
$$

The average takes place over the ground state of the system. The first index, $N+i$ (where $i=0,1,2$), indicates the number of excitons in the ground state. The second index (zero) corresponds to the phonons and points out their absence in the ground state. The appearance of matrix elements between the ground states of systems with different quasiparticle numbers $(N+i)$ is connected with the use of quasiaverages instead of usual averages. The fundamental concept of quasiaverages was introduced by Bogoliubov and was discussed in Subsection 2.1.1 in detail. In this procedure, an infinitesimally small term that breaks the particle-number conservation law is introduced in the initial Hamiltonian. The average is calculated with a new gauge-noninvariant Hamiltonian, and afterwards the small term is set to zero.

The Green's function in the coordinate representation [20] is given by

$$\begin{aligned}\langle\langle A(x,t);B(x,\tau)\rangle\rangle &= -i\langle T[A(x,t)B(x,\tau)]\rangle\\ &= -i[\theta(t-\tau)\langle A(x,t)B(x,\tau)\rangle+\theta(\tau-t)\langle B(x,\tau)A(x,t)\rangle], \end{aligned}\tag{3.8}$$

where A and B are either exciton or phonon operators. The transition from the coordinate representation to the momentum representation is done by the Fourier transform (for simplicity, we consider only the time dependence):

$$\langle\langle A(t);B(\tau)\rangle\rangle = \frac{1}{2\pi\hbar}\int_{-\infty}^{\infty}\langle\langle A\mid B\rangle\rangle_E e^{-iE(t-\tau)/\hbar}\,\mathrm{d}E. \tag{3.9}$$

The Green's functions $G_{ik}(p)$ have obvious physical interpretations: $G_{11}(p)$ and $G_{33}(p)$ describe the scattering of the uncondensed quasiparticles with conservation of their number, $G_{22}(p)$ concerns the phonon scattering, and $G_{13}(p)$ and $G_{31}(p)$ reflect the transition processes of two quasiparticles into the condensate and vice versa. The functions $G_{21}(p)$, $G_{23}(p)$, $G_{12}(p)$, and $G_{32}(p)$ have to do with the transition processes of one exciton into and out of the condensate because of interaction with the phonons.

3.2.2 *The Symmetry Properties of the Green's Functions and Their Self-Energy Parts for the Bose System in the Presence of the Condensate*

We now use Bogoliubov's method for the nonideal Bose gas. From definition (3.9) it follows that

$$\langle\langle A(t);B(\tau)\rangle\rangle = \langle\langle B(\tau);A(t)\rangle\rangle.$$

By using Eq. (3.8) one obtains:

$$\langle\langle A\mid B\rangle\rangle_E = \langle\langle B\mid A\rangle\rangle_{-E}.$$

By using the last equality and definitions (3.7), one can easily show that

$$\begin{aligned} G_{33}(\mathbf{p},E) &= G_{11}(-\mathbf{p},-E), & G_{22}(\mathbf{p},E) &= G_{22}(-\mathbf{p},-E),\\ G_{23}(\mathbf{p},E) &= -G_{12}(-\mathbf{p},-E), & G_{13}(\mathbf{p},E) &= G_{13}(-\mathbf{p},-E),\\ G_{32}(\mathbf{p},E) &= -G_{21}(-\mathbf{p},-E), & G_{31}(\mathbf{p},E) &= G_{31}(-\mathbf{p},-E). \end{aligned}\tag{3.10}$$

The other properties of the Green's functions can be found starting from Hamiltonian (3.1).

We note that Hamiltonian (3.1) is invariant under the canonical transformation

$$a_{\mathbf{p}} \to a_{-\mathbf{p}}, \qquad a_{\mathbf{p}}^{\dagger} \to a_{-\mathbf{p}}^{\dagger}, \qquad b_{\mathbf{p}} \to b_{-\mathbf{p}}, \qquad b_{\mathbf{p}}^{\dagger} \to b_{-\mathbf{p}}^{\dagger},$$

which is why the average statistical functions (3.7) must not change under the substitution $\mathbf{p} \to -\mathbf{p}$, i.e.,

$$G_{ik}(\mathbf{p}, E) = G_{ik}(-\mathbf{p}, E), \quad (i, k = 1, 2, 3). \tag{3.11}$$

We further introduce the operators

$$\beta_{\mathbf{p}} = i b_{\mathbf{p}}, \qquad \beta_{\mathbf{p}}^{\dagger} = -i b_{\mathbf{p}}^{\dagger}, \tag{3.12}$$

which are subject to the same commutation rules as the operators $b_{\mathbf{p}}$ and $b_{\mathbf{p}}^{\dagger}$. Then Eq. (3.1) has the form

$$\begin{aligned} H = {} & \sum_{\mathbf{p}} (\Delta + T_{\mathbf{p}}) a_{\mathbf{p}}^{\dagger} a_{\mathbf{p}} + \sum_{\mathbf{p}} B_{\mathbf{p}} \beta_{\mathbf{p}}^{\dagger} \beta_{\mathbf{p}} \\ & + \frac{1}{2V} \sum_{\mathbf{p}_1, \mathbf{p}_2, \mathbf{p}} U(\mathbf{p}) a_{\mathbf{p}_1}^{\dagger} a_{\mathbf{p}_2}^{\dagger} a_{\mathbf{p}_2+\mathbf{p}} a_{\mathbf{p}_1-\mathbf{p}} \\ & + \frac{1}{\sqrt{N_a}} \sum_{\mathbf{p}, \mathbf{p}'} \Theta(\mathbf{p} - \mathbf{p}') a_{\mathbf{p}}^{\dagger} a_{\mathbf{p}'} (\beta_{\mathbf{p}-\mathbf{p}'} + \beta_{-\mathbf{p}+\mathbf{p}'}^{\dagger}). \end{aligned} \tag{3.13}$$

Since all the coefficients in Eq. (3.13) are real, the corresponding motion equations must be invariant in regard to time inversion $t \to -t$ and the substitution $i \to -i$. Therefore the average value of the expression

$$\langle [\beta_{\mathbf{p}}(\tau) + \beta_{-\mathbf{p}}^{\dagger}(\tau)] a_{\mathbf{p}}^{\dagger}(t) \rangle = \int_{-\infty}^{\infty} J_{\mathbf{p}}(E) e^{-iE(t-\tau)/\hbar} \, \mathrm{d}E \tag{3.14}$$

is not changed by the substitutions $t \to -t$, $\tau \to -\tau$, and $i \to -i$. This leads to the equality $J_{\mathbf{p}}^{*}(E) = J_{\mathbf{p}}(E)$. On the basis of Eq. (3.14), one obtains

$$\langle a_{\mathbf{p}}(\tau) [\beta_{\mathbf{p}}^{\dagger}(t) + \beta_{-\mathbf{p}}(t)] \rangle = \langle [\beta_{\mathbf{p}}(\tau) + \beta_{-\mathbf{p}}^{\dagger}(\tau)] a_{\mathbf{p}}^{\dagger}(t) \rangle.$$

Returning to the operators $b_{\mathbf{p}}$ and $b_{\mathbf{p}}^{\dagger}$ leads to the relation

$$-\langle a_{\mathbf{p}}(\tau) [b_{\mathbf{p}}^{\dagger}(t) - b_{-\mathbf{p}}(t)] \rangle = \langle [b_{\mathbf{p}}(\tau) - b_{-\mathbf{p}}^{\dagger}(\tau)] a \dagger_{\mathbf{p}}(t) \rangle$$

or

$$G_{12}(\mathbf{p}, E) = -G_{21}(\mathbf{p}, E). \tag{3.15}$$

Analogously, one can prove the relations

$$G_{23}(\mathbf{p}, E) = -G_{32}(\mathbf{p}, E), \qquad G_{31}(\mathbf{p}, E) = G_{13}(\mathbf{p}, E). \tag{3.16}$$

Now let us introduce the matrix consisting of the Green's functions under consideration,

$$\hat{G}(\mathbf{p}, E) = \begin{Vmatrix} G_{11}(\mathbf{p}, E) & G_{12}(\mathbf{p}, E) & G_{13}(\mathbf{p}, E) \\ G_{21}(\mathbf{p}, E) & G_{22}(\mathbf{p}, E) & G_{23}(\mathbf{p}, E) \\ G_{31}(\mathbf{p}, E) & G_{32}(\mathbf{p}, E) & G_{33}(\mathbf{p}, E) \end{Vmatrix} \tag{3.17}$$

and the analogous self-energy matrix $\tilde{\Sigma}(\mathbf{p}, E)$, consisting of the components $\tilde{\Sigma}_{ik}(\mathbf{p}, E)$, connected with $G(\mathbf{p}, E)$ by the relation

$$\tilde{\Sigma}(\mathbf{p}, E) = G^{-1}(\mathbf{p}, E). \tag{3.18}$$

From definition (3.18) it is clear that the elements $\tilde{\Sigma}_{ik}(\mathbf{p}, E)$ satisfy the same symmetry relations of Eqs. (3.10), (3.11), (3.15), and (3.16) as $G_{ik}(\mathbf{p}, E)$. The symmetry properties $\tilde{\Sigma}_{ik}(\mathbf{p}, E)$ can be proved analytically by use of the inverse matrix $G^{-1}(\mathbf{p}, E)$ and Eqs. (3.10), (3.11), (3.15), (3.16), and (3.18). Therefore

$$\begin{aligned}
&\tilde{\Sigma}_{ik}(\mathbf{p}, E) = \tilde{\Sigma}_{ik}(-\mathbf{p}, E), && \tilde{\Sigma}_{33}(\mathbf{p}, E) = \tilde{\Sigma}_{11}(-\mathbf{p}, -E),\\
&\tilde{\Sigma}_{13}(\mathbf{p}, E) = \tilde{\Sigma}_{31}(\mathbf{p}, E), && \tilde{\Sigma}_{22}(\mathbf{p}, E) = \tilde{\Sigma}_{22}(-\mathbf{p}, -E),\\
&\tilde{\Sigma}_{21}(\mathbf{p}, E) = -\tilde{\Sigma}_{12}(\mathbf{p}, E), && \tilde{\Sigma}_{13}(\mathbf{p}, E) = \tilde{\Sigma}_{13}(-\mathbf{p}, -E),\\
&\tilde{\Sigma}_{32}(\mathbf{p}, E) = -\tilde{\Sigma}_{23}(\mathbf{p}, E), && \tilde{\Sigma}_{23}(\mathbf{p}, E) = -\tilde{\Sigma}_{12}(-\mathbf{p}, -E).
\end{aligned} \tag{3.19}$$

The relations $\tilde{\Sigma}_{31}(\mathbf{p}, E) = \tilde{\Sigma}_{31}(-\mathbf{p}, -E)$ and $\tilde{\Sigma}_{32}(\mathbf{p}, E) = -\tilde{\Sigma}_{21}(-\mathbf{p}, -E)$ follow from the ones given above. From Eq. (3.18), by using Eqs. (3.19), one can express the Green's functions $G_{ik}(\mathbf{p}, E)$ as [the arguments $(\mathbf{p}, E)$ are removed]

$$\begin{aligned}
G_{11} &= \frac{1}{d}\left(\tilde{\Sigma}_{22}\tilde{\Sigma}_{33} + \tilde{\Sigma}_{23}^2\right), & G_{12} &= \frac{1}{d}\left(\tilde{\Sigma}_{12}\tilde{\Sigma}_{33} + \tilde{\Sigma}_{13}\tilde{\Sigma}_{23}\right),\\
G_{13} &= \frac{1}{d}\left(\tilde{\Sigma}_{12}\tilde{\Sigma}_{23} - \tilde{\Sigma}_{13}\tilde{\Sigma}_{22}\right), & G_{22} &= \frac{1}{d}\left(\tilde{\Sigma}_{11}\tilde{\Sigma}_{33} - \tilde{\Sigma}_{13}^2\right),\\
G_{23} &= \frac{1}{d}\left(\tilde{\Sigma}_{11}\tilde{\Sigma}_{23} + \tilde{\Sigma}_{13}\tilde{\Sigma}_{12}\right), & G_{33} &= \frac{1}{d}\left(\tilde{\Sigma}_{11}\tilde{\Sigma}_{22} + \tilde{\Sigma}_{12}^2\right),
\end{aligned} \tag{3.20}$$

where

$$d = \tilde{\Sigma}_{11}\tilde{\Sigma}_{22}\tilde{\Sigma}_{33} + 2\tilde{\Sigma}_{12}\tilde{\Sigma}_{23}\tilde{\Sigma}_{13} + \tilde{\Sigma}_{23}^2\tilde{\Sigma}_{11} + \tilde{\Sigma}_{12}^2\tilde{\Sigma}_{33} - \tilde{\Sigma}_{13}^2\tilde{\Sigma}_{22}.$$

3.2.3 *The Structure of the Equations*

To find the graphical equations, the matrix elements $\Sigma_{ik}(\mathbf{p}, E)$ are introduced. They are connected with $\tilde{\Sigma}_{ik}(\mathbf{p}, E)$ by the relation

$$\tilde{\Sigma}_{ik}(\mathbf{p}, E) = \delta_{ik} G_{ik}^{(0)-1}(\mathbf{p}, E) - \Sigma_{ik}(\mathbf{p}, E),$$

(see below), where $G_{ik}^{0}(\mathbf{p}, E)$ are the Green's functions of the noninteracting particles. The Green's functions (3.7) are represented in the following manner [15, 21]: the solid thick lines with arrows at the ends correspond to the exciton Green's functions, the dashed thick lines correspond to the phonon Green's functions, and the lines that are half solid with one arrow and half dashed represent the Green's functions of mixed type, containing one exciton and one phonon operator:

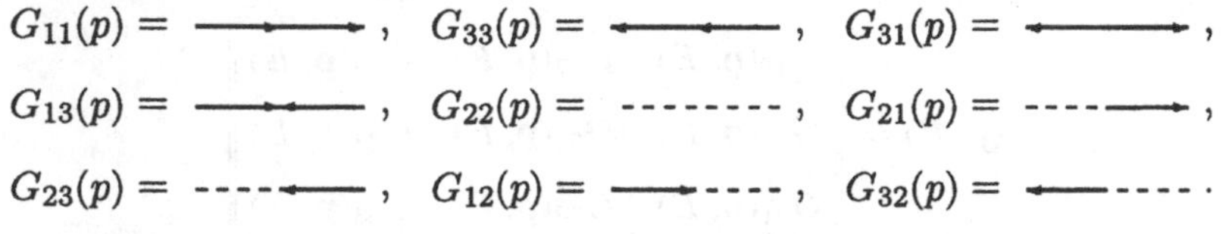
$G_{11}(p) =$,
$G_{33}(p) =$,
$G_{31}(p) =$,
$G_{13}(p) =$,
$G_{22}(p) =$,
$G_{21}(p) =$,
$G_{23}(p) =$,
$G_{12}(p) =$,
$G_{32}(p) =$.

Using a crooked line to represent the condensate particles, as in Subsection 2.1.4, the Green's functions (3.7) satisfy the following three systems of the diagram equations:

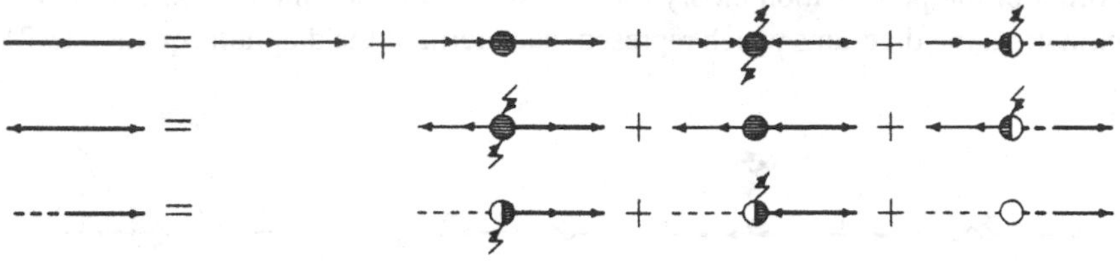

(3.21)

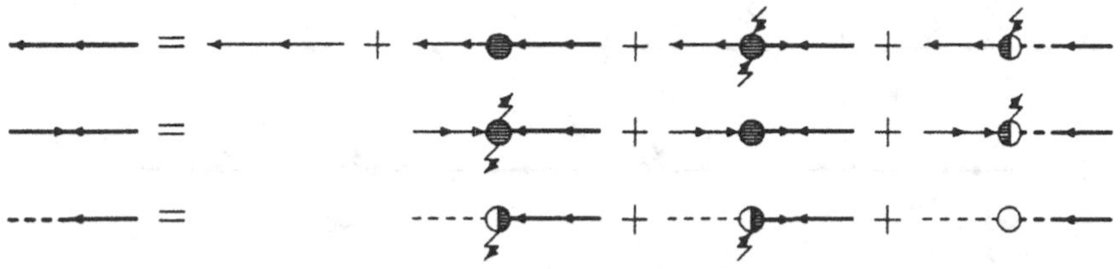

(3.22)

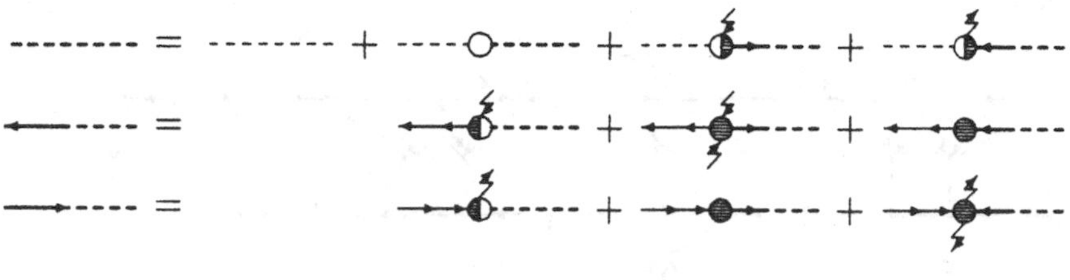

(3.23)

The thin lines correspond to the zeroth-order Green's functions of the excitons and the phonons,

$$G_{11}^0(p) = (E - T_{\mathbf{p}} + i\delta)^{-1}, \qquad G_{22}^0(p) = \frac{2B_{\mathbf{p}}}{E^2 - B_{\mathbf{p}}^2 + i\delta}. \tag{3.24}$$

The self-energy parts are presented in the form of small circles with different shadings:

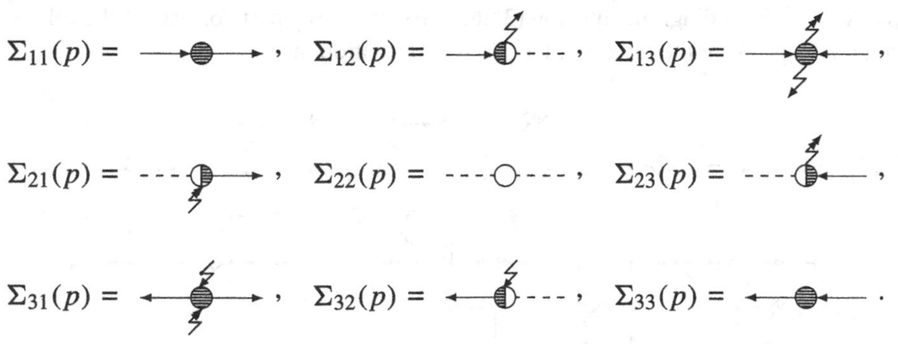

(3.25)

One can be convinced of the correctness of the set of diagram equations (3.21)–(3.23) by enumerating all the diagrams for the Green's functions. We do this below for the first of the diagram equations (3.21); the other eight equations can be verified in the same way. In these diagrams, the wavy line corresponds to the Fourier transform of the interaction energy of two excitons, and the dashed line represents the zeroth-order phonon Green's function. We include only those diagrams which contain a maximum of two wavy or dashed lines, or one wavy and one dashed line. Two inner dashed lines are obtained in the fourth order of the perturbation theory for the exciton–phonon interaction operator. In this approximation, the diagrams which represent each term of the diagram equation (3.21) are

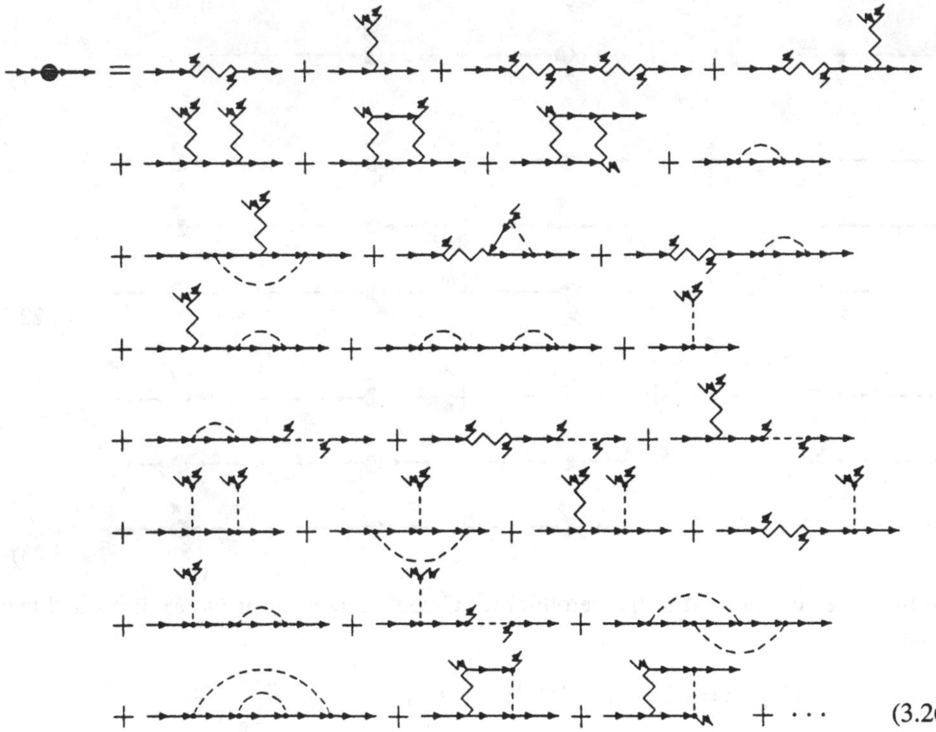

(3.26)

The heavy dots correspond to the Fourier transform of the exciton–phonon energy $\Theta(\mathbf{p})$. As seen from diagram equation (3.26), the self-energy part consists of the following Feynman diagrams in the lowest order of the perturbation theory:

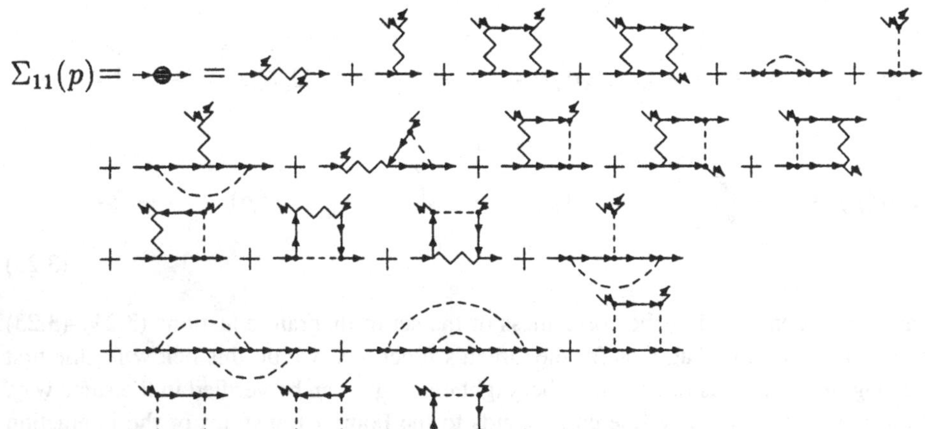

(3.27)

For the third term on the right-hand side of the first equation of diagram equations (3.21) the higher-order diagrams of the perturbation theory are

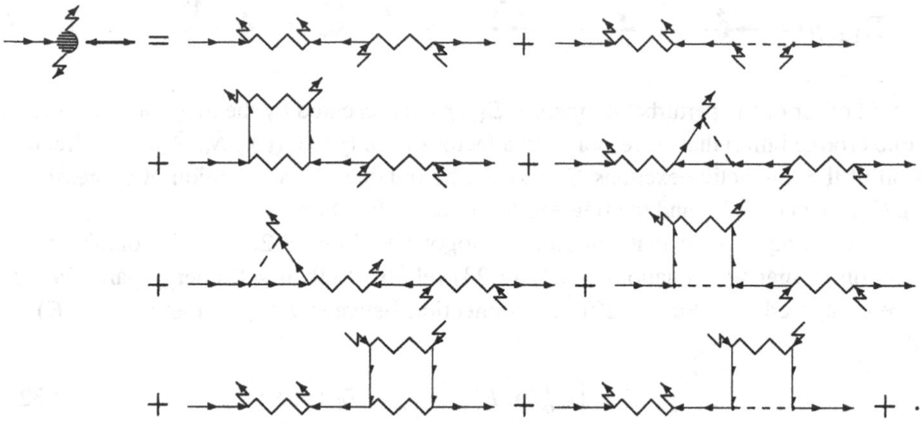

(3.28)

and

$\Sigma_{13}(p) =$ 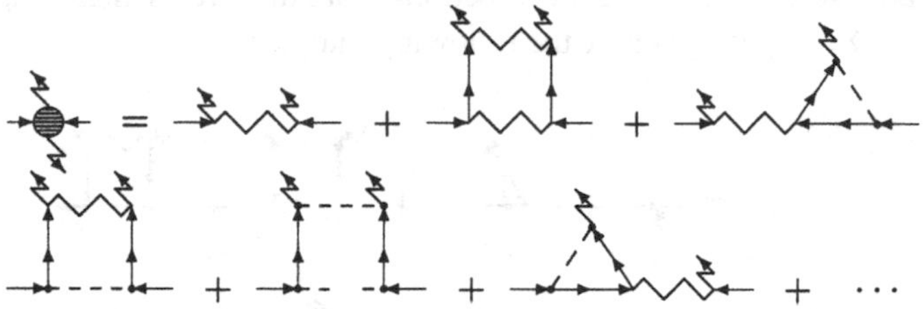

(3.29)

The fourth term of the first equation of diagram equations (3.21) contains the Green's function $G_{21}(p)$,

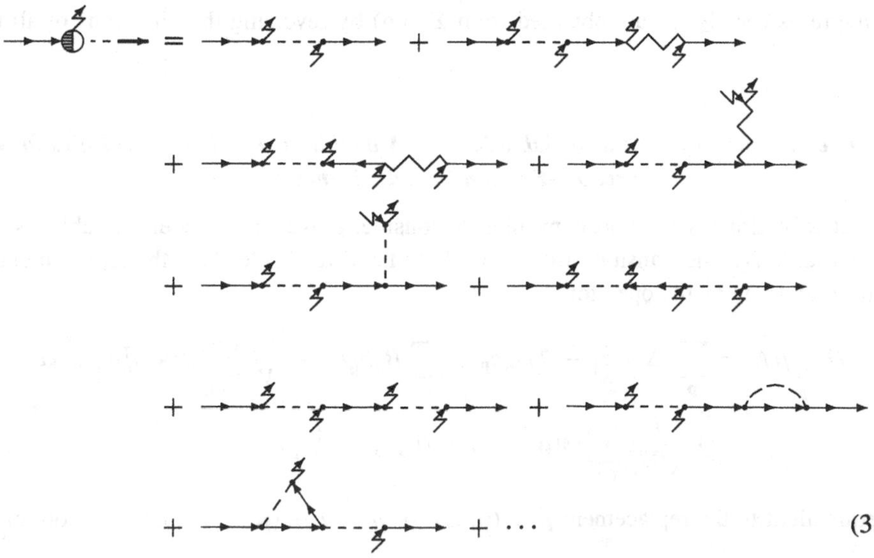

(3.30)

and the self-energy part

$\Sigma_{12}(p)=$  (3.31)

In the first order of the perturbation theory, $\Sigma_{12}(p)$ is presented by the diagram (the vertex with one crooked line) that corresponds to a factor of the type $\Theta(\mathbf{p})\sqrt{N_0/N_a}$. The Green's function of the interacting excitons $G_{11}(p)$ can be obtained by summation of expressions (3.26), (3.28), and (3.30) and the free-exciton Green's function.

It is interesting to note that relations analogous to Eqs. (3.20) can be obtained by the solution of diagram equations (3.21)–(3.23) relative to their self-energy parts. When they are compared with Eqs. (3.20), the connection between $\tilde{\Sigma}_{ik}(\mathbf{p}, E)$ and $\Sigma_{ik}(\mathbf{p}, E)$ is found:

$$\tilde{\Sigma}_{ik}(\mathbf{p}, E) = \left[G^0_{ik}(\mathbf{p}, E)\right]^{-1} \delta_{i,k} - \Sigma_{ik}(\mathbf{p}, E). \tag{3.32}$$

The rest of the diagrams for the self-energy parts contained in diagram equations (3.21)–(3.23) are determined as follows. The diagrams of the lowest order of the perturbation theory for $\Sigma_{33}(p)$, $\Sigma_{31}(p)$, and $\Sigma_{32}(p)$ are obtained from diagram equations (3.27), (3.29), and (3.31), respectively, by reversing the directions of all the arrows, including the crooked ones. $\Sigma_{21}(p)$ and $\Sigma_{22}(p)$ are given by the following diagrams:

$\Sigma_{21}(p)=$ 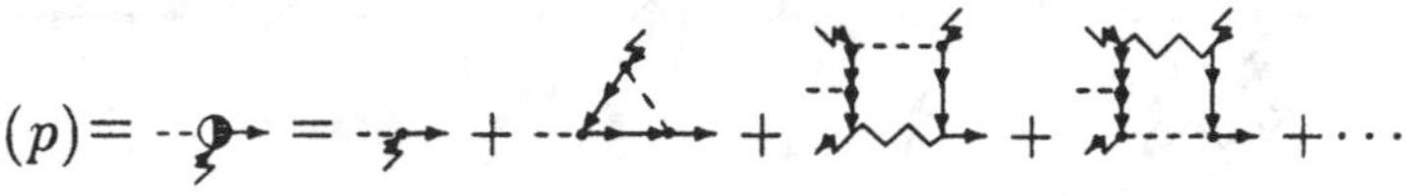 (3.33)

$\Sigma_{22}(p)=$ 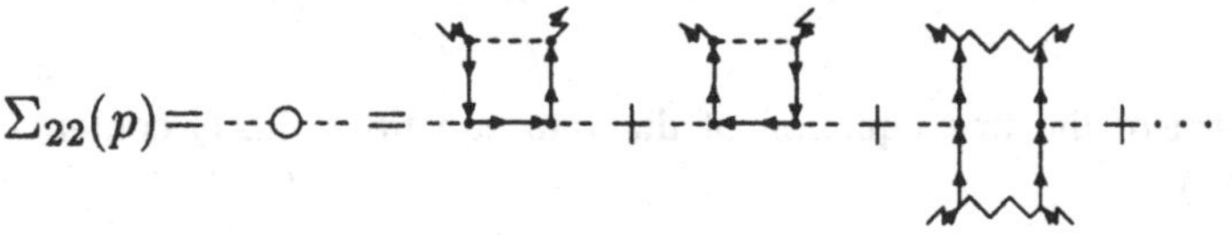 (3.34)

The diagrams for $\Sigma_{23}(p)$ are obtained from $\Sigma_{21}(p)$ by reversing the directions of all the arrows.

3.2.4 *The Transition from the Variable N to the Variable μ. The Feynman Rules for the Exciton–Phonon Green's Functions*

For practical estimates it is more convenient to consider μ as an independent variable instead of the variable N. The transition from fixed N to fixed μ [21] leads to the replacement of Hamiltonian (3.1) by the operator

$$\mathbf{H} = H - \mu N = \sum_{\mathbf{p}}(\Delta + T_{\mathbf{p}} - \bar{\mu})a^\dagger_{\mathbf{p}}a_{\mathbf{p}} + \sum_{\mathbf{p}} B_{\mathbf{p}} b^\dagger_{\mathbf{p}} b_{\mathbf{p}} + \frac{1}{2V}\sum_{\mathbf{p},\mathbf{q},\mathbf{k}} U(\mathbf{k}) a^\dagger_{\mathbf{p}} a^\dagger_{\mathbf{q}} a_{\mathbf{q}+\mathbf{k}} a_{\mathbf{p}-\mathbf{k}}$$
$$+ \frac{i}{\sqrt{N_a}} \sum_{\mathbf{p},\mathbf{q}} \Theta(\mathbf{p}-\mathbf{q}) a^\dagger_{\mathbf{p}} a_{\mathbf{q}} (b_{\mathbf{p}-\mathbf{q}} - b^\dagger_{\mathbf{q}-\mathbf{p}}). \tag{3.35}$$

It is equivalent to the replacement $p \equiv (\mathbf{p}, E) \to p + \mu = (\mathbf{p}, E + \mu)$. The zeroth-order

Green's function has the following form:

$$G_{11}^{0}(p) = (E + \mu - T_{\mathbf{p}} + i\delta)^{-1}.$$

$G_{22}^{0}(p)$ does not change, since the chemical potential of the phonons is equal to zero. We treat μ as an independent variable: $\mu = \bar{\mu} - \Delta$.

We now present the rules by which one can write the expressions corresponding to any diagram for the Green's function of the perturbation theory of the nth order, where $n = m + l$. Here m is the order of the perturbation theory in the interaction $U(\mathbf{p})$, which is equal to the number of wavy lines, and l is the order of the perturbation theory in the interaction $\Theta(\mathbf{p})$ and is equal to the number of thick points. Zeroth-order approximations (3.24) are chosen in such a way that the averages of the exciton and the phonon operators are taken independently. As a result, l is an even number, $l = 2k$, for the homogeneous Green functions G_{11}, G_{33}, G_{31}, G_{13}, and G_{22}, and an odd number, $l = 2k+1$, for the mixed Green functions G_{12}, G_{21}, G_{32}, and G_{23}.

1. Each direct thin solid line from left to right corresponds to the function $G_{11}^{0}(p)$. The lines with opposite direction correspond to the function $G_{33}^{0}(p)$.
2. Each thin dashed line corresponds to the function $G_{22}^{0}(p)$.
3. Each wavy line with momentum $\mathbf{q}$ corresponds to the expression containing the interaction-potential Fourier transform $U(\mathbf{q})/V$.
4. Each thick point corresponds to the Fourier transform of the exciton–phonon interaction-potential $\Theta(\mathbf{q})/\sqrt{N_a}$, where $\mathbf{q}$ is the quasi-momentum of the phonon line connecting with the point.
5. Each crooked line with an arrow that goes into or out of the condensate corresponds to the factor $\sqrt{N_0}$ (N_0 is the mean particle number in the condensate).
6. At every triple vertex the quasi-momentum $\mathbf{q}$ of wavy or dashed lines is equal to the difference of the quasi-momenta of the quasiparticle lines. Integration is performed over each independent four-dimensional quasi-momentum, and the multiplier $(2\pi)^{-4}$ corresponds to each integration.
7. The resulting expressions are multiplied by a factor $(i)^{m+3k-F}$ for the homogeneous Green's functions and by a factor $(i)^{m+3k+1-F}$ for the mixed ones. Here F is an integer that denotes the degree in which the density of condensed excitons $n_0^F = (N_0/V)^F$ appears in the given diagram. It equals the number of pairs of crooked lines. In the case of the Green's functions G_{11}, G_{33}, G_{13}, and G_{31} in the absence of exciton–phonon interaction ($k = 0$), these rules are the same as those in the Beliaev technique [15, 21].

For example, the contributions of two diagrams are

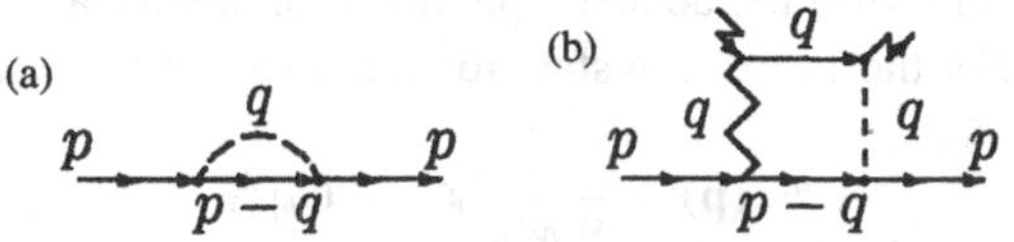

$$M_a^{(2)} = -i\frac{1}{N_a}\left[G_{11}^{(0)}(p)\right]^2 \int \frac{d^4q}{(2\pi)^4} G_{11}^{(0)}(p-q)G_{22}^{(0)}(q)\Theta^2(q),$$

$$M_b^{(3)} = -i\frac{N_0}{N_a}\left[G_{11}^{(0)}(p)\right]^2 \int \frac{d^4q}{(2\pi)^4} U(\mathbf{q})G_{11}^{(0)}(p-q)G_{11}^{(0)}(q)G_{22}^{(0)}(q)\Theta^2(q).$$

3.2.5 The Energy Spectrum and the Change of the Sound Velocity in Crystals due to Bose Condensation of Excitons

We now obtain the energy spectrum in the first order of the perturbation theory. For this purpose it is necessary to select from diagram equations (3.25) the first-order diagrams and to use the rules given above. One finds

$$\Sigma_{11}^{(1)}(p) = \Sigma_{33}^{(1)}(p) = n_0[U(\mathbf{p}) + U(0)], \qquad \Sigma_{31}^{(1)}(p) = \Sigma_{13}^{(1)}(p) = n_0 U(\mathbf{p}),$$
$$\Sigma_{21}^{(1)}(p) = \Sigma_{23}^{(1)}(p) = -\Sigma_{12}^{(1)}(p) = -\Sigma_{32}^{(1)}(p) = i\sqrt{\frac{N_0}{N_a}}\Theta(\mathbf{p}), \qquad \Sigma_{22}^{(1)}(p) = 0. \quad (3.36)$$

After Eqs. (3.36) are inserted into diagram equations (3.21), the condition for a solution is

$$\begin{vmatrix} L_{\mathbf{p}} + T_{\mathbf{p}} - E & L_{\mathbf{p}} & \varphi_{\mathbf{p}} \\ L_{\mathbf{p}} & L_{\mathbf{p}} + T_{\mathbf{p}} + E & \varphi_{\mathbf{p}} \\ 2B_{\mathbf{p}}\varphi_{\mathbf{p}} & 2B_{\mathbf{p}}\varphi_{\mathbf{p}} & E^2 - B_{\mathbf{p}}^2 \end{vmatrix} = 0, \quad (3.37)$$

where

$$\varphi_{\mathbf{p}} = i\sqrt{\frac{N_0}{N_a}}\Theta(\mathbf{p}), \qquad L_{\mathbf{p}} = \frac{N_0}{V}U(\mathbf{p}).$$

The solution is

$$E_{s(s=1,2)} = \left[\frac{1}{2}\left(T_{\mathbf{p}}^2 + 2T_{\mathbf{p}}L_{\mathbf{p}} + B_{\mathbf{p}}^2\right) \pm \frac{1}{2}\sqrt{\left(T_{\mathbf{p}}^2 + 2T_{\mathbf{p}}L_{\mathbf{p}} - B_{\mathbf{p}}^2\right)^2 + 16\frac{N_0}{N_a}B_{\mathbf{p}}T_{\mathbf{p}}\Theta^2(\mathbf{p})}\right]^{1/2}. \quad (3.38)$$

This result is exactly the same as that in Ref. 5, in which the case $T \neq 0$ was discussed. For expression (3.38) to be real it is necessary that

$$B_{\mathbf{p}}\left[T_{\mathbf{p}} + \frac{N_0}{V}U(\mathbf{p})\right] \geq 4\frac{N_0}{N_a}\Theta^2(\mathbf{p}). \quad (3.39)$$

One can see that for standard assumptions about the internal exciton state, the coefficients $\Theta(\mathbf{p})$ are of the type that preserves the linear part of the energy spectrum $E_1(\mathbf{p})$, which is necessary for the existence of excitonic superfluidity. In the case of interaction with optical phonons for small values $|\mathbf{p}|$ we have $B_{\mathbf{p}} = \hbar\omega_0$ and $\Theta^2(\mathbf{p}) \approx p^2$; in this case the stability condition remains the same as in the well-known Bose gas theory, namely, $U(0) > 0$. The interaction of the excitons with the acoustic phonons for small transfer momenta requires a closer examination. For the $1s$ exciton state for small values of p one finds

$$\Theta^2(\mathbf{p}) = \frac{2}{9}\frac{p}{Mu}|C_e - C_h|^2.$$

The other values have the form $U(\mathbf{p}) = U(0)$, $B_{\mathbf{p}} = up$, and $T_{\mathbf{p}} = p^2/2m_{\text{ex}}$. By inserting these expressions and $\rho = N_a M/V$ into relation (3.39), one gets the stability condition

$$U(0) \geq \frac{8}{9}\frac{(C_e - C_h)^2}{\rho u^2}. \quad (3.40)$$

This condition imposes an additional restriction on the value of the repulsive forces between the excitons. If it is violated, as discussed in Ref. 17, it is necessary to take into account the contribution of higher-order terms in the mass operator. Nevertheless, the fact that the linear part of the spectrum remains implies that superfluidity can persist in the presence of a phonon bath. One therefore expects substantial changes in the experimentally measured transport of excitons in the case of an excitonic condensate. Experiments attempting to observe this effect are reviewed in Section 3.4.

As seen in Eq. (3.38), instead of the elementary excitations in the system of degenerate excitons with energy $(T_{\mathbf{p}}^2+2L_{\mathbf{p}}T_{\mathbf{p}})^{1/2}$, found in the Bogoliubov model of Subsection 2.1.3 (which are typically called "phonons" in the context of superfluid He but which are called "hydrons" here to distinguish them from the lattice phonons), and the lattice phonons $\hbar\omega_p$, new elementary excitations appear of mixed type: phonon–hydrons. As a result, both the hydron velocity u^* and the acoustic sound velocity u are changed. The new velocities are denoted by u_1 and u_2. The velocities u^*, u, u_1, and u_2 can be estimated by use of the exciton and the phonon parameters, material characteristics [22–24], and qualitative evaluations of unknown interaction constants.

For example, for a CdS crystal, if one sets $M=4\times 10^{-24}$ g, $u=4.2\times 10^5$ cm/s, $U(0)=10^{-32}$ erg cm^3, $n_0=10^{16}$ cm^{-3}, and $|C_e-C_h|=20$ eV, the new values $u_1=3\times 10^5$ cm/s and $u_2=4.6\times 10^5$ cm/s are found. The upper limit of exciton density in this crystal determined by the metallization condition is of the order of magnitude of 10^{17} cm^{-3}. These calculations show that the change of sound velocity in a crystal with degenerate excitons is considerable for experimentally achievable excitation densities. This model therefore provides a new prediction for experimental evidence of excitonic BEC.

Although we have assumed a weak interaction with phonons, e.g., low-temperature experiments, it is possible to use another approach to the problem when the exciton–phonon interaction is strong. For certain conditions this interaction can be greater than the interaction between the excitons in condensate. In this case one must renormalize the exciton and the phonon spectra and then consider the BEC of the excitons with renormalized mass. One expects, however, that superfluidity may persist to much higher temperatures, as long as the excitons remain condensed.

3.3 In-Depth Study: The Green's Function Method for T > 0

Considering the same problem at $T\neq 0$, one can follow the method of Tserkovnikov [25] and generalize its formalism for the case of the exciton–phonon interaction.

Hamiltonian (3.1) is considered in the thermodynamic limit: $N_a\sim N_{\text{ex}}\to\infty$, $V\to\infty$, but $N_a/V=$ const., $N_{\text{ex}}/V=$const. The exciton energy and the frequencies that appear in the Fourier transforms of the Green's functions are measured relative to the value μ. This value determines the new position of the exciton-band bottom, which is changed because of the interactions in the system.

Following Ref. [25], we single out the condensate with the aid of the Bogoliubov displacement transformation,

$$a_{\mathbf{p}}=\sqrt{N_0}\delta_{\mathbf{p},0}+\alpha_{\mathbf{p}},\qquad b_{\mathbf{p}}=\beta_{\mathbf{p}},\tag{3.41}$$

under the condition $\langle\alpha_0\rangle=\langle\beta_0\rangle=0$. $\delta_{\mathbf{p},0}$ is the Kronecker delta symbol, and N_0 is the exciton number in the condensate. The averaging takes place over the grand canonical ensemble.

For the new operators $\alpha_{\mathbf{p}}$ and $\beta_{\mathbf{p}}$, expression (3.35) is given by

$$H = H_B + H_{\text{int}}, \tag{3.42}$$

$$\begin{aligned} H_B = {} & -\mu N_0 + \frac{N_0^2}{2V}U(0) + \sqrt{N_0}(-\mu + L_0)(\alpha_0 + \alpha_0^\dagger) + \sum_{\mathbf{p}} \xi_{\mathbf{p}} \alpha_{\mathbf{p}}^\dagger \alpha_{\mathbf{p}} \\ & + \sum_{\mathbf{p}} B_{\mathbf{p}} \beta_{\mathbf{p}}^\dagger \beta_{\mathbf{p}} + \frac{1}{2}\sum_{\mathbf{p}} L_{\mathbf{p}}(\alpha_{\mathbf{p}}\alpha_{-\mathbf{p}} + \alpha_{\mathbf{p}}^\dagger \alpha_{-\mathbf{p}}^\dagger) \\ & + \sum_{\mathbf{p}} \varphi_{\mathbf{p}}(\alpha_{\mathbf{p}}\beta_{-\mathbf{p}} - \alpha_{\mathbf{p}}\beta_{\mathbf{p}}^\dagger + \alpha_{\mathbf{p}}^\dagger \beta_{\mathbf{p}} - \alpha_{\mathbf{p}}^\dagger \beta_{-\mathbf{p}}^\dagger), \end{aligned} \tag{3.43}$$

$$\begin{aligned} H_{\text{int}} = {} & \frac{i}{\sqrt{N_a}} \sum_{\mathbf{p}\mathbf{p}'} \Theta(\mathbf{p} - \mathbf{p}')\alpha_{\mathbf{p}}^\dagger \alpha_{\mathbf{p}'}(\beta_{\mathbf{p}-\mathbf{p}'} - \beta_{-\mathbf{p}+\mathbf{p}'}^\dagger) + \frac{\sqrt{N_0}}{V}\sum_{\mathbf{p}_1,\mathbf{p}_2} U(\mathbf{p}_1) \\ & \times \alpha_{\mathbf{p}_1}^\dagger \alpha_{\mathbf{p}_2}^\dagger \alpha_{\mathbf{p}_1+\mathbf{p}_2} + \frac{\sqrt{N_0}}{V}\sum_{\mathbf{p}_1,\mathbf{p}_2} U(\mathbf{p}_1)\alpha_{\mathbf{p}_1+\mathbf{p}_2}^\dagger \alpha_{\mathbf{p}_2}\alpha_{\mathbf{p}_1} \\ & + \frac{1}{2V}\sum_{\mathbf{p}_1,\mathbf{p}_2,\mathbf{p}} U(\mathbf{p})\alpha_{\mathbf{p}_1}^\dagger \alpha_{\mathbf{p}_2}^\dagger \alpha_{\mathbf{p}_2+\mathbf{p}}\alpha_{\mathbf{p}_1-\mathbf{p}}. \end{aligned} \tag{3.44}$$

Here

$$\xi_{\mathbf{p}} = T_{\mathbf{p}} + L_0 + L_{\mathbf{p}} - \mu, \qquad B_{\mathbf{p}} = \hbar\omega_{\mathbf{p}},$$

$$\varphi_{\mathbf{p}} = i\sqrt{\frac{N_0}{N_a}}\Theta(\mathbf{p}), \qquad L_{\mathbf{p}} = \frac{N_0}{V}U(\mathbf{p}). \tag{3.45}$$

Let us determine the Green's functions and their equations of motion. One can introduce the 4×4 matrices as follows:

$$\hat{A}_{\mathbf{p}} = \begin{vmatrix} \alpha_{\mathbf{p}} \\ \beta_{\mathbf{p}} \\ \alpha_{-\mathbf{p}}^\dagger \\ \beta_{-\mathbf{p}}^\dagger \end{vmatrix}, \qquad \hat{A}_{\mathbf{p}}^\dagger = \begin{vmatrix} \alpha_{\mathbf{p}}^\dagger & \beta_{\mathbf{p}}^\dagger & \alpha_{-\mathbf{p}} & \beta_{-\mathbf{p}} \\ 0 & 0 & 0 & 0 \\ 0 & 0 & 0 & 0 \\ 0 & 0 & 0 & 0 \end{vmatrix},$$

$$\hat{U}_{\mathbf{p},\mathbf{p}_1} = \begin{vmatrix} u_{\mathbf{p},\mathbf{p}_1} \\ t_{\mathbf{p},\mathbf{p}_1} \\ u_{-\mathbf{p},-\mathbf{p}_1}^\dagger \\ t_{-\mathbf{p},-\mathbf{p}_1}^\dagger \end{vmatrix}, \qquad \hat{V}_{\mathbf{p},\mathbf{p}_1} = \begin{vmatrix} v_{\mathbf{p},\mathbf{p}_1} \\ 0 \\ v_{\mathbf{p},\mathbf{p}_1}^\dagger \\ 0 \end{vmatrix}, \qquad \hat{\gamma} = \begin{vmatrix} 1 \\ 0 \\ 1 \\ 0 \end{vmatrix},$$

$$\hat{W}_{\mathbf{p},\mathbf{p}_2,\mathbf{p}_2',\mathbf{p}_1'} = \begin{vmatrix} w_{\mathbf{p},\mathbf{p}_2,\mathbf{p}_2',\mathbf{p}_1'} \\ 0 \\ w_{-\mathbf{p},-\mathbf{p}_2,-\mathbf{p}_2',-\mathbf{p}_1'}^\dagger \\ 0 \end{vmatrix}, \qquad \hat{I} = \begin{vmatrix} 1 & 0 & 0 & 0 \\ 0 & 1 & 0 & 0 \\ 0 & 0 & 1 & 0 \\ 0 & 0 & 0 & 1 \end{vmatrix},$$

$$\hat{L}_{\mathbf{p}} = \begin{vmatrix} \xi_{\mathbf{p}} & \varphi_{\mathbf{p}} & L_{\mathbf{p}} & -\varphi_{\mathbf{p}} \\ -\varphi_{\mathbf{p}} & B_{\mathbf{p}} & -\varphi_{\mathbf{p}} & 0 \\ L_{\mathbf{p}} & \varphi_{\mathbf{p}} & \xi_{\mathbf{p}} & -\varphi_{\mathbf{p}} \\ \varphi_{\mathbf{p}} & 0 & \varphi_{\mathbf{p}} & B_{\mathbf{p}} \end{vmatrix}, \qquad \hat{\alpha} = \begin{vmatrix} 1 & 0 & 0 & 0 \\ 0 & 1 & 0 & 0 \\ 0 & 0 & -1 & 0 \\ 0 & 0 & 0 & -1 \end{vmatrix}, \tag{3.46}$$

$$
\begin{aligned}
u_{\mathbf{p},\mathbf{p}_1} &= i\Theta(\mathbf{p}_1)\alpha_{\mathbf{p}_1+\mathbf{p}}\left(\beta_{-\mathbf{p}_1} - \beta^\dagger_{\mathbf{p}_1}\right),\\
t_{\mathbf{p},\mathbf{p}_1} &= -i\Theta(\mathbf{p})\alpha^\dagger_{\mathbf{p}_1}\alpha_{\mathbf{p}_1+\mathbf{p}},\\
v_{\mathbf{p},\mathbf{p}_1} &= [U(\mathbf{p}) + U(\mathbf{p}_1)]\alpha^\dagger_{-\mathbf{p}_1}\alpha_{\mathbf{p}-\mathbf{p}_1} + U(\mathbf{p}_1)\alpha_{\mathbf{p}_1}\alpha_{\mathbf{p}-\mathbf{p}_1},\\
w_{\mathbf{p},\mathbf{p}_2,\mathbf{p}'_2,\mathbf{p}'_1} &= U(\mathbf{p}-\mathbf{p}'_1)\delta_{kr}(\mathbf{p}+\mathbf{p}_2-\mathbf{p}'_1-\mathbf{p}'_2)\alpha^\dagger_{\mathbf{p}_2}\alpha_{\mathbf{p}'_2}\alpha_{\mathbf{p}'_1}.
\end{aligned}
\tag{3.47}
$$

The four-row Green's function matrix is determined as follows:

$$
\hat{G}_{\mathbf{p}}(t) = \langle\langle \hat{A}_{\mathbf{p}}(t); \hat{A}^\dagger_{\mathbf{p}}(0)\rangle\rangle
$$

$$
= \begin{vmatrix}
\langle\langle\alpha_{\mathbf{p}}(t);\alpha^\dagger_{\mathbf{p}}(0)\rangle\rangle & \langle\langle\alpha_{\mathbf{p}}(t);\beta^\dagger_{\mathbf{p}}(0)\rangle\rangle & \langle\langle\alpha_{\mathbf{p}}(t);\alpha_{-\mathbf{p}}(0)\rangle\rangle & \langle\langle\alpha_{\mathbf{p}}(t);\beta_{-\mathbf{p}}(0)\rangle\rangle\\
\langle\langle\beta_{\mathbf{p}}(t);\alpha^\dagger_{\mathbf{p}}(0)\rangle\rangle & \langle\langle\beta_{\mathbf{p}}(t);\beta^\dagger_{\mathbf{p}}(0)\rangle\rangle & \langle\langle\beta_{\mathbf{p}}(t);\alpha_{-\mathbf{p}}(0)\rangle\rangle & \langle\langle\beta_{\mathbf{p}}(t);\beta_{-\mathbf{p}}(0)\rangle\rangle\\
\langle\langle\alpha^\dagger_{-\mathbf{p}}(t);\alpha^\dagger_{\mathbf{p}}(0)\rangle\rangle & \langle\langle\alpha^\dagger_{-\mathbf{p}}(t);\beta^\dagger_{\mathbf{p}}(0)\rangle\rangle & \langle\langle\alpha^\dagger_{-\mathbf{p}}(t);\alpha_{-\mathbf{p}}(0)\rangle\rangle & \langle\langle\alpha^\dagger_{-\mathbf{p}}(t);\beta_{-\mathbf{p}}(0)\rangle\rangle\\
\langle\langle\beta^\dagger_{-\mathbf{p}}(t);\alpha^\dagger_{\mathbf{p}}(0)\rangle\rangle & \langle\langle\beta^\dagger_{-\mathbf{p}}(t);\beta^\dagger_{\mathbf{p}}(0)\rangle\rangle & \langle\langle\beta^\dagger_{-\mathbf{p}}(t);\alpha_{-\mathbf{p}}(0)\rangle\rangle & \langle\langle\beta^\dagger_{-\mathbf{p}}(t);\beta_{-\mathbf{p}}(0)\rangle\rangle
\end{vmatrix}.
\tag{3.48}
$$

The symbol $\langle\langle\ \rangle\rangle$ denotes the retarded Green's function [20]:

$$
\langle\langle A(t); B(0)\rangle\rangle = -i\theta(t)\langle[A(t), B(0)]\rangle; \qquad [A, B] = AB - BA.
$$

Analogously, one can also determine the other Green's functions:

$$
\left\langle\!\left\langle \hat{U}_{\mathbf{p},\mathbf{p}_1}(t); \hat{A}^\dagger_{\mathbf{p}}(0)\right\rangle\!\right\rangle, \qquad \left\langle\!\left\langle \hat{V}_{\mathbf{p},\mathbf{p}_1}(t); \hat{A}^\dagger_{\mathbf{p}}(0)\right\rangle\!\right\rangle, \qquad \left\langle\!\left\langle \hat{W}_{\mathbf{p},\mathbf{p}_2,\mathbf{p}'_2,\mathbf{p}'_1}(t); \hat{A}^\dagger_{\mathbf{p}}(0)\right\rangle\!\right\rangle.
$$

These functions satisfy the following equations of motion:

$$
\begin{aligned}
i\hbar\hat{\alpha}\frac{\mathrm{d}\hat{A}_{\mathbf{p}}}{\mathrm{d}t} &= \sqrt{N_0}(\Delta + L_0 - \hat{\mu})\hat{\gamma}\delta_{kr}(\mathbf{p}) + \hat{L}_{\mathbf{p}}\hat{A}_{\mathbf{p}}\\
&\quad + \frac{1}{\sqrt{N_a}}\sum_{\mathbf{p}_1}\hat{U}_{\mathbf{p},\mathbf{p}_1} + \frac{\sqrt{N_0}}{V}\sum_{\mathbf{p}_1}\hat{V}_{\mathbf{p},\mathbf{p}_1} + \frac{1}{V}\sum_{\mathbf{p}_1,\mathbf{q}_1,\mathbf{p}_2}\hat{W}_{\mathbf{p},\mathbf{p}_2,\mathbf{q}_1,\mathbf{p}_1},
\end{aligned}
\tag{3.49}
$$

$$
\begin{aligned}
&i\hbar\hat{\alpha}\left\langle\!\!\left\langle\frac{\mathrm{d}\hat{A}_{\mathbf{p}}(t)}{\mathrm{d}t}; \hat{A}^\dagger_{\mathbf{p}}(0)\right\rangle\!\!\right\rangle\\
&= \hat{L}_{\mathbf{p}}\langle\langle\hat{A}_{\mathbf{p}}(t); \hat{A}^\dagger_{\mathbf{p}}(0)\rangle\rangle + \frac{1}{\sqrt{N_a}}\sum_{\mathbf{p}_1}\left\langle\!\left\langle\hat{U}_{\mathbf{p},\mathbf{p}_1}(t); \hat{A}^\dagger_{\mathbf{p}}(0)\right\rangle\!\right\rangle\\
&\quad + \frac{\sqrt{N_0}}{V}\sum_{\mathbf{p}_1}\left\langle\!\left\langle\hat{V}_{\mathbf{p},\mathbf{p}_1}(t); \hat{A}^\dagger_{\mathbf{p}}(0)\right\rangle\!\right\rangle + \frac{1}{V}\sum_{\mathbf{p}_2,\mathbf{p}'_2,\mathbf{p}'_1}\left\langle\!\left\langle\hat{W}_{\mathbf{p},\mathbf{p}_2,\mathbf{p}'_2,\mathbf{p}'_1}(t); \hat{A}^\dagger_{\mathbf{p}}(0)\right\rangle\!\right\rangle,
\end{aligned}
\tag{3.50}
$$

$$
\begin{aligned}
&i\hbar\left\langle\!\!\left\langle\frac{\mathrm{d}\hat{U}_{\mathbf{p},\mathbf{p}_1}(t)}{\mathrm{d}t}; \hat{A}^\dagger_{\mathbf{p}}(0)\right\rangle\!\!\right\rangle\hat{\alpha}\\
&= \left\langle\!\left\langle\hat{U}_{\mathbf{p},\mathbf{p}_1}(t); \hat{A}^\dagger_{\mathbf{p}}(0)\right\rangle\!\right\rangle\hat{L}_{\mathbf{p}} + \frac{1}{\sqrt{N_a}}\sum_{\mathbf{q}_1}\left\langle\!\left\langle\hat{U}_{\mathbf{p},\mathbf{p}_1}(t); \hat{U}^\dagger_{\mathbf{p},\mathbf{q}_1}(0)\right\rangle\!\right\rangle\\
&\quad + \frac{\sqrt{N_0}}{V}\sum_{\mathbf{q}_1}\left\langle\!\left\langle\hat{U}_{\mathbf{p},\mathbf{p}_1}(t); \hat{V}^\dagger_{\mathbf{p},\mathbf{q}_1}(0)\right\rangle\!\right\rangle + \frac{1}{V}\sum_{\mathbf{p}_2,\mathbf{p}'_2,\mathbf{p}'_1}\left\langle\!\left\langle\hat{U}_{\mathbf{p},\mathbf{p}_1}(t); \hat{W}^\dagger_{\mathbf{p},\mathbf{p}_2,\mathbf{p}'_2,\mathbf{p}'_1}(0)\right\rangle\!\right\rangle,
\end{aligned}
\tag{3.51}
$$

and so on.

One can remark that

$$i\hbar\left\langle\!\!\left\langle \frac{d\hat{U}_{\mathbf{p},\mathbf{p}_1}(t)}{dt}; \hat{A}^{\dagger}_{\mathbf{p}}(0)\right\rangle\!\!\right\rangle = i\hbar\frac{d}{dt}\langle\!\langle \hat{U}_{\mathbf{p},\mathbf{p}_1}(t); \hat{A}^{\dagger}_{\mathbf{p}}(0)\rangle\!\rangle - \hbar\delta(t)\langle[\hat{U}_{\mathbf{p},\mathbf{p}_1}, \hat{A}^{\dagger}_{\mathbf{p}}]\rangle. \tag{3.52}$$

By using matrix definitions (3.46) and (3.47) and the properties $\langle\alpha_0\rangle = \langle\beta_0\rangle = 0$, one obtains

$$\langle[\hat{A}_{\mathbf{p}}, \hat{A}^{\dagger}_{\mathbf{p}}]\rangle = \hat{\alpha}, \qquad \langle[\hat{V}_{\mathbf{p},\mathbf{p}_1}, \hat{A}^{\dagger}_{\mathbf{p}}]\rangle = \langle[\hat{U}_{\mathbf{p},\mathbf{p}_1}, \hat{A}^{\dagger}_{\mathbf{p}}]\rangle = 0;$$

$$\frac{1}{V}\sum_{\mathbf{p}_2,\mathbf{p}'_2,\mathbf{p}'_1}\langle[\hat{W}_{\mathbf{p},\mathbf{p}_2,\mathbf{p}'_2,\mathbf{p}'_1}, \hat{A}^{\dagger}_{\mathbf{p}}]\rangle\hat{\alpha} = \begin{vmatrix} S_{\mathbf{p}} & 0 & Q_{\mathbf{p}} & 0 \\ 0 & 0 & 0 & 0 \\ Q_{\mathbf{p}} & 0 & S_{\mathbf{p}} & 0 \\ 0 & 0 & 0 & 0 \end{vmatrix}, \tag{3.53}$$

where

$$S_{\mathbf{p}} = \frac{1}{V}\sum_{\mathbf{p}_1}[U(0) + U(\mathbf{p}-\mathbf{p}_1)]\langle\alpha^{\dagger}_{\mathbf{p}_1}\alpha_{\mathbf{p}_1}\rangle,$$

$$Q_{\mathbf{p}} = \frac{1}{V}\sum_{\mathbf{p}_1}U(\mathbf{p}-\mathbf{p}_1)\langle\alpha_{\mathbf{p}_1}\alpha_{-\mathbf{p}_1}\rangle. \tag{3.54}$$

Now the transition from the temporal Green's functions to their Fourier transforms is effectuated:

$$\hat{G}(\mathbf{p}, E) = \langle\!\langle \hat{A}_{\mathbf{p}} \mid \hat{A}^{\dagger}_{\mathbf{p}}\rangle\!\rangle_E,$$

$$\langle\!\langle A \mid B\rangle\!\rangle_E = \int_{-\infty}^{\infty}\langle\!\langle A(t); B(0)\rangle\!\rangle e^{iEt/\hbar}\,dt,$$

$$\langle\!\langle A(t); B(0)\rangle\!\rangle = \frac{1}{2\pi\hbar}\int_{-\infty}^{\infty}\langle\!\langle A \mid B\rangle\!\rangle_E e^{-iEt/\hbar}\,dE,$$

$$\delta(t) = \frac{1}{2\pi\hbar}\int_{-\infty}^{\infty} e^{-iEt/\hbar}\,dE. \tag{3.55}$$

The equations of motion for the four new Green's function matrices, taking into account Eq. (3.52), are the following:

$$(\hat{\alpha}E - \hat{L}_{\mathbf{p}})\hat{G}(\mathbf{p}, E) = \hat{I} + \frac{1}{\sqrt{N_a}}\sum_{\mathbf{p}_1}\langle\!\langle \hat{U}_{\mathbf{p},\mathbf{p}_1} \mid \hat{A}^{\dagger}_{\mathbf{p}}\rangle\!\rangle_E + \frac{\sqrt{N_0}}{V}\sum_{\mathbf{p}_1}\langle\!\langle \hat{V}_{\mathbf{p},\mathbf{p}_1} \mid \hat{A}^{\dagger}_{\mathbf{p}}\rangle\!\rangle_E + \frac{1}{V}\sum_{\mathbf{p}_2,\mathbf{p}'_2,\mathbf{p}'_1}\langle\!\langle \hat{W}_{\mathbf{p},\mathbf{p}_2,\mathbf{p}'_2,\mathbf{p}'_1} \mid \hat{A}^{\dagger}_{\mathbf{p}}\rangle\!\rangle_E, \tag{3.56}$$

$$\langle\!\langle \hat{U}_{\mathbf{p},\mathbf{p}_1} \mid \hat{A}^{\dagger}_{\mathbf{p}}\rangle\!\rangle(\hat{\alpha}E - \hat{L}_{\mathbf{p}}) = \frac{1}{\sqrt{N_a}}\sum_{\mathbf{p}'_1}\langle\!\langle \hat{U}_{\mathbf{p},\mathbf{p}_1} \mid \hat{U}^{\dagger}_{\mathbf{p},\mathbf{p}'_1}\rangle\!\rangle_E + \frac{\sqrt{N_0}}{V}\sum_{\mathbf{p}'_1}\langle\!\langle \hat{U}_{\mathbf{p},\mathbf{p}_1} \mid \hat{V}^{\dagger}_{\mathbf{p},\mathbf{p}'_1}\rangle\!\rangle_E + \frac{1}{V}\sum_{\mathbf{p}_2,\mathbf{p}'_2,\mathbf{p}'_1,}\langle\!\langle \hat{U}_{\mathbf{p},\mathbf{p}_1} \mid \hat{W}^{\dagger}_{\mathbf{p},\mathbf{p}_2,\mathbf{p}'_2,\mathbf{p}'_1}\rangle\!\rangle_E, \tag{3.57}$$

$$\langle\langle \hat{V}_{\mathbf{p},\mathbf{p}_1} \mid \hat{A}^\dagger_{\mathbf{p}} \rangle\rangle_E(\hat{\alpha}E - \hat{L}_{\mathbf{p}}) = \frac{1}{\sqrt{N_a}}\sum_{\mathbf{p}'_1}\langle\langle \hat{V}_{\mathbf{p},\mathbf{p}_1} \mid \hat{U}^\dagger_{\mathbf{p},\mathbf{p}'_1}\rangle\rangle_E + \frac{\sqrt{N_0}}{V}\sum_{\mathbf{p}'_1}\langle\langle \hat{V}_{\mathbf{p},\mathbf{p}_1} \mid \hat{V}^\dagger_{\mathbf{p},\mathbf{p}'_1}\rangle\rangle_E + \frac{1}{V}\sum_{\mathbf{p}_2,\mathbf{p}'_2,\mathbf{p}_1{}'}\langle\langle \hat{V}_{\mathbf{p},\mathbf{p}_1} \mid \hat{W}^\dagger_{\mathbf{p},\mathbf{p}_2,\mathbf{p}'_2,\mathbf{p}'_1}\rangle\rangle_E, \tag{3.58}$$

$$\langle\langle \hat{W}_{\mathbf{p},\mathbf{p}_2,\mathbf{p}'_2,\mathbf{p}'_1} \mid \hat{A}^\dagger_{\mathbf{p}}\rangle\rangle_E(\hat{\alpha}E - \hat{L}_{\mathbf{p}}) = \langle[\hat{W}_{\mathbf{p},\mathbf{p}_2,\mathbf{p}'_2,\mathbf{p}'_1}, \hat{A}^\dagger_{\mathbf{p}}]\rangle\hat{\alpha} + \frac{1}{\sqrt{N_a}}\sum_{\mathbf{q}}\langle\langle \hat{W}_{\mathbf{p},\mathbf{p}_2,\mathbf{p}'_2,\mathbf{p}'_1} \mid \hat{U}^\dagger_{\mathbf{p},\mathbf{q}}\rangle\rangle_E + \frac{\sqrt{N_0}}{V}\sum_{\mathbf{q}}\langle\langle \hat{W}_{\mathbf{p},\mathbf{p}_2,\mathbf{p}'_2,\mathbf{p}'_1} \mid \hat{V}^\dagger_{\mathbf{pq}}\rangle\rangle_E + \frac{1}{V}\sum_{\mathbf{q}_2,\mathbf{q}'_2,\mathbf{q}_1{}'}\langle\langle \hat{W}_{\mathbf{p},\mathbf{p}_2,\mathbf{p}'_2,\mathbf{p}'_1} \mid \hat{W}^\dagger_{\mathbf{p},\mathbf{q}_2,\mathbf{q}'_2,\mathbf{q}'_1}\rangle\rangle_E. \tag{3.59}$$

After inserting Eqs. (3.57) and (3.58) into Eq. (3.56), one finds the Green's function matrix $\hat{G}(\mathbf{p}, E)$:

$$(\hat{\alpha}E - \hat{L}_{\mathbf{p}})\hat{G}(\mathbf{p}, E)(\hat{\alpha}E - \hat{L}_{\mathbf{p}}) = (\hat{\alpha}E - \hat{L}_{\mathbf{p}}) + \hat{K}(\mathbf{p}, E),$$
$$\hat{G}(\mathbf{p}, E) = \hat{G}^B(\mathbf{p}, E) + \hat{G}^B(\mathbf{p}, E)\hat{K}(\mathbf{p}, E)\hat{G}^B(\mathbf{p}, E), \tag{3.60}$$

where

$$\hat{G}^B(\mathbf{p}, E) = (\hat{\alpha}E - \hat{L}_{\mathbf{p}})^{-1}, \tag{3.61}$$

$$\begin{aligned}\hat{K}(\mathbf{p}, E) = {} & \frac{1}{V}\sum_{\mathbf{p}_2,\mathbf{p}'_2,\mathbf{p}_1{}'}\langle[\hat{W}_{\mathbf{p},\mathbf{p}_2,\mathbf{p}'_2,\mathbf{p}'_1}, \hat{A}^\dagger_{\mathbf{p}}]\rangle\hat{\alpha} + \frac{1}{N_a}\sum_{\mathbf{p}_1,\mathbf{q}_1}\langle\langle \hat{U}_{\mathbf{p},\mathbf{p}_1} \mid \hat{U}^\dagger_{\mathbf{p},\mathbf{q}_1}\rangle\rangle_E \\ & + \frac{1}{V}\sqrt{\frac{N_0}{N_a}}\sum_{\mathbf{p}_1,\mathbf{q}_1}\langle\langle \hat{U}_{\mathbf{p},\mathbf{p}_1} \mid \hat{V}^\dagger_{\mathbf{p},\mathbf{q}_1}\rangle\rangle_E + \frac{1}{V}\sqrt{\frac{N_0}{N_a}}\sum_{\mathbf{p}_1,\mathbf{q}_1}\langle\langle \hat{V}_{\mathbf{p},\mathbf{p}_1} \mid \hat{U}^\dagger_{\mathbf{p},\mathbf{q}_1}\rangle\rangle_E \\ & + \frac{N_0}{V^2}\sum_{\mathbf{p}_1,\mathbf{q}_1}\langle\langle \hat{V}_{\mathbf{p},\mathbf{p}_1} \mid \hat{V}^\dagger_{\mathbf{p},\mathbf{q}_1}\rangle\rangle_E \\ & + \frac{1}{V\sqrt{N_a}}\sum_{\mathbf{p}_1,\mathbf{q}_1,\mathbf{p}_2,\mathbf{q}_2}\langle\langle \hat{U}_{\mathbf{p},\mathbf{p}_1} \mid \hat{W}^\dagger_{\mathbf{p},\mathbf{p}_2,\mathbf{q}_2,\mathbf{q}_1}\rangle\rangle_E \\ & + \frac{\sqrt{N_0}}{V^2}\sum_{\mathbf{p}_1,\mathbf{p}_2,\mathbf{q}_1,\mathbf{q}_2}\langle\langle \hat{V}_{\mathbf{p},\mathbf{p}_1} \mid \hat{W}^\dagger_{\mathbf{p},\mathbf{p}_2,\mathbf{q}_2,\mathbf{q}_1}\rangle\rangle_E \\ & + \frac{1}{V\sqrt{N_a}}\sum_{\mathbf{p}_1,\mathbf{p}_2,\mathbf{q}_1,\mathbf{q}_2}\langle\langle \hat{W}_{\mathbf{p},\mathbf{p}_2,\mathbf{q}_2,\mathbf{q}_1} \mid \hat{U}^\dagger_{\mathbf{p},\mathbf{p}_1}\rangle\rangle_E \\ & + \frac{\sqrt{N_0}}{V^2}\sum_{\mathbf{p}_1,\mathbf{p}_2,\mathbf{q}_1,\mathbf{q}_2}\langle\langle \hat{W}_{\mathbf{p},\mathbf{p}_2,\mathbf{q}_2,\mathbf{q}_1} \mid \hat{V}^\dagger_{\mathbf{p},\mathbf{p}_1}\rangle\rangle_E \\ & + \frac{1}{V^2}\sum_{\mathbf{p}_1,\mathbf{p}_2,\mathbf{p}_3,\mathbf{q}_1,\mathbf{q}_2,\mathbf{q}_3}\langle\langle \hat{W}_{\mathbf{p},\mathbf{p}_2,\mathbf{q}_2,\mathbf{p}_1} \mid \hat{W}^\dagger_{\mathbf{p},\mathbf{p}_3,\mathbf{q}_3,\mathbf{q}_1}\rangle\rangle_E.\end{aligned} \tag{3.62}$$

The mass operator or the matrix of the self-energy parts is introduced as

$$\hat{G}^{-1}(\mathbf{p}, E) = \hat{\Sigma}(\mathbf{p}, E) = \hat{G}^{B-1}(\mathbf{p}, E) - \hat{\Sigma}^B(\mathbf{p}, E). \tag{3.63}$$

Starting from Eqs. (3.60) and (3.63), one can obtain

$$[\hat{G}^B(\mathbf{p},E)+\hat{G}^B(\mathbf{p},E)\hat{K}(\mathbf{p},E)\hat{G}^B(\mathbf{p},E)][\hat{G}^{B\,-1}(\mathbf{p},E)-\hat{\Sigma}^B(\mathbf{p},E)]=\hat{I}$$

and the expression for $\hat{\Sigma}^B(\mathbf{p},E)$:

$$\begin{aligned}\hat{\Sigma}^B(\mathbf{p},E)&=[1+\hat{K}(\mathbf{p},E)\hat{G}^B(\mathbf{p},E)]^{-1}\hat{K}(\mathbf{p},E)\\&=\hat{K}(\mathbf{p},E)[1+\hat{G}^B(\mathbf{p},E)\hat{K}(\mathbf{p},E)]^{-1}.\end{aligned}\tag{3.64}$$

Averaging Eq. (3.49) for the operator $\hat{A}_0$ and taking into account the condition $\langle\alpha_0\rangle=0$ and the fact that $\Theta(0)=0$ leads to the following expression for the chemical potential:

$$\begin{aligned}\mu=L_0&+\frac{1}{V}\sum_{\mathbf{p}_1}\{[U(0)+U(\mathbf{p}_1)]\langle\alpha^\dagger_{\mathbf{p}_1}\alpha_{\mathbf{p}_1}\rangle+U(\mathbf{p}_1)\langle\alpha_{\mathbf{p}_1}\alpha_{-\mathbf{p}_1}\rangle\}\\&+\frac{1}{\sqrt{N_0N_a}}\sum_{\mathbf{p}_1}i\Theta(\mathbf{p}_1)[\langle\alpha_{\mathbf{p}_1}\beta_{-\mathbf{p}_1}\rangle-\langle\alpha_{\mathbf{p}_1}\beta^\dagger_{\mathbf{p}_1}\rangle]\\&+\frac{1}{V\sqrt{N_0}}\sum_{\mathbf{p}_1,\mathbf{p}_2}U(\mathbf{p}_1)\langle\alpha^\dagger_{\mathbf{p}_1+\mathbf{p}_2}\alpha_{\mathbf{p}_1}\alpha_{\mathbf{p}_2}\rangle.\end{aligned}\tag{3.65}$$

3.3.1 The Zeroth-Order Approximation

In this approach, one keeps in Hamiltonian (3.42) only the terms H_B quadratic in operators $\alpha_\mathbf{p}$ and $\beta_\mathbf{p}$. Expression (3.65) for μ in lowest order of the interaction constant becomes

$$\mu^{(0)}=L_0,\tag{3.66}$$

which corresponds to a shift of the exciton-band bottom by the value L_0 to the high-energy side. The matrices $\hat{L}_\mathbf{p}$ and $\hat{G}^B(\mathbf{p},E)$ are replaced by $\hat{L}^0_\mathbf{p}$ and $G^{B,(0)}(\mathbf{p},E)$, respectively.

The poles of the Green's function $G^{B,(0)}(\mathbf{p},E)$ are found from the condition

$$\det\left|\hat{\alpha}E-\hat{L}^{(0)}_\mathbf{p}\right|=0,$$

which is equivalent to

$$\begin{vmatrix}E-T_\mathbf{p}-L_\mathbf{p}&-\varphi_\mathbf{p}&-L_\mathbf{p}&\varphi_\mathbf{p}\\\varphi_\mathbf{p}&E-B_\mathbf{p}&\varphi_\mathbf{p}&0\\-L_\mathbf{p}&-\varphi_\mathbf{p}&-E-T_\mathbf{p}-L_\mathbf{p}&\varphi_\mathbf{p}\\-\varphi_\mathbf{p}&0&-\varphi_\mathbf{p}&-E-B_\mathbf{p}\end{vmatrix}=0.\tag{3.67}$$

The energy spectrum has the form

$$\begin{aligned}E_{s(s=1,2)}(\mathbf{p})=\Bigg[&\frac{1}{2}(T^2_\mathbf{p}+2T_\mathbf{p}L_\mathbf{p}+B^2_\mathbf{p})\\&\pm\frac{1}{2}\sqrt{\left(T^2_\mathbf{p}+2T_\mathbf{p}L_\mathbf{p}-B^2_\mathbf{p}\right)^2+16\frac{N_0}{N_a}B_\mathbf{p}T_\mathbf{p}\Theta^2(\mathbf{p})}\,\Bigg]^{1/2}\end{aligned}\tag{3.68}$$

and coincides with expression (3.38) obtained in Subsection 3.2.5. The value $N_0=Nx$ at $T\neq0$ is determined from the condition that the thermodynamic potential has a minimum as a function of the parameter x, which gives the macroscopic fraction of particles in the lowest single-particle state. The evaluations of the thermodynamic potential are presented in Chapter 5.

The Hamiltonian that is quadratic in the operators $\alpha_{\mathbf{p}}$ and $\beta_{\mathbf{p}}$ is diagonalized by the help of the unitary transformation from the original operators to the new ones $\xi_{s\mathbf{p}}$:

$$\alpha_{\mathbf{p}} = \sum_{s=1}^{2} u_{s\mathbf{p}}\xi_{s\mathbf{p}} + \sum_{s=1}^{2} \tilde{u}_{s\mathbf{p}}\xi^{\dagger}_{s,-\mathbf{p}},$$

$$\beta_{\mathbf{p}} = \sum_{s=1}^{2} v_{s\mathbf{p}}\xi_{s\mathbf{p}} + \sum_{s=1}^{2} \tilde{v}_{s\mathbf{p}}\xi^{\dagger}_{s,-\mathbf{p}}; \tag{3.69}$$

$\xi^{\dagger}_{s\mathbf{p}}$ and $\xi_{s\mathbf{p}}$ are the creation and the annihilation operators, respectively, of the quasiparticles with the elementary excitation energies $E_s(\mathbf{p})$. The coefficients of transformations (3.69) $u_{s\mathbf{p}}$, $\tilde{u}_{s\mathbf{p}}$, $v_{s\mathbf{p}}$, and $\tilde{v}_{s\mathbf{p}}$ were determined by Bobrysheva [7]. They have the following properties: $u_{s\mathbf{p}}$ and $\tilde{u}_{s\mathbf{p}}$ are real, but $v_{s\mathbf{p}}$ and $\tilde{v}_{s\mathbf{p}}$ are imaginary. They form two matrices of direct $\hat{C}$ and inverse $\hat{C}^{-1}$ transformations:

$$\hat{C} = \begin{vmatrix} u_{1\mathbf{p}} & u_{2\mathbf{p}} & \tilde{u}_{1\mathbf{p}} & \tilde{u}_{2\mathbf{p}} \\ v_{1\mathbf{p}} & v_{2\mathbf{p}} & \tilde{v}_{1\mathbf{p}} & \tilde{v}_{2\mathbf{p}} \\ \tilde{u}_{1\mathbf{p}} & \tilde{u}_{2\mathbf{p}} & u_{1\mathbf{p}} & u_{2\mathbf{p}} \\ -\tilde{v}_{1\mathbf{p}} & -\tilde{v}_{2\mathbf{p}} & -v_{1\mathbf{p}} & -v_{2\mathbf{p}} \end{vmatrix},$$

$$\hat{C}^{-1} = \begin{vmatrix} u_{1\mathbf{p}} & -v_{1\mathbf{p}} & -\tilde{u}_{1\mathbf{p}} & -\tilde{v}_{1\mathbf{p}} \\ u_{2\mathbf{p}} & -v_{2\mathbf{p}} & -\tilde{u}_{2\mathbf{p}} & -\tilde{v}_{2\mathbf{p}} \\ -\tilde{u}_{1\mathbf{p}} & \tilde{v}_{1\mathbf{p}} & u_{1\mathbf{p}} & v_{1\mathbf{p}} \\ -\tilde{u}_{2\mathbf{p}} & \tilde{v}_{2\mathbf{p}} & u_{2\mathbf{p}} & v_{2\mathbf{p}} \end{vmatrix},$$

$$\hat{C}\hat{C}^{-1} = \hat{I}. \tag{3.70}$$

The commutation properties of the Bose operators $\alpha_{\mathbf{p}}$, $\beta_{\mathbf{p}}$, $\alpha^{\dagger}_{\mathbf{p}}$, $\beta^{\dagger}_{\mathbf{p}}$, $\xi_{s\mathbf{p}}$, and $\xi^{\dagger}_{s\mathbf{p}}$ lead to the following conditions of orthogonality and normalization:

$$\sum_{s=1}^{2}\left(u^2_{s\mathbf{p}} - \tilde{u}^2_{s\mathbf{p}}\right) = 1, \qquad \sum_{s=1}^{2}(|v_{s\mathbf{p}}|^2 - |\tilde{v}_{s\mathbf{p}}|^2) = 1, \tag{3.71}$$

$$\sum_{s=1}^{2}(u_{s\mathbf{p}}\tilde{v}_{s\mathbf{p}} - \tilde{u}_{s\mathbf{p}}v_{s\mathbf{p}}) = 0, \qquad \sum_{s=1}^{2}(u_{s\mathbf{p}}v_{s\mathbf{p}} - \tilde{u}_{s\mathbf{p}}\tilde{v}_{s\mathbf{p}}) = 0,$$

$$u_{s\mathbf{p}} = (-1)^{s-1}\frac{\left[B^2_{\mathbf{p}} - E^2_s(\mathbf{p})\right][T_{\mathbf{p}} + E_s(\mathbf{p})]}{2|T_{\mathbf{p}}E_s(\mathbf{p})|^{1/2}\left\{\left[B^2_{\mathbf{p}} - E^2_s(\mathbf{p})\right]^2 + 4\frac{N_0}{N_a}B_{\mathbf{p}}T_{\mathbf{p}}\Theta^2(\mathbf{p})\right\}^{1/2}},$$

$$v_{s\mathbf{p}} = (-1)^{s-1}\frac{i\sqrt{\frac{N_0}{N_a}}\Theta(\mathbf{p})T_{\mathbf{p}}[B_{\mathbf{p}} + E_s(\mathbf{p})]}{|T_{\mathbf{p}}E_s(\mathbf{p})|^{1/2}\left\{\left[B^2_{\mathbf{p}} - E^2_s(\mathbf{p})\right]^2 + 4\frac{N_0}{N_a}B_{\mathbf{p}}T_{\mathbf{p}}\Theta^2(\mathbf{p})\right\}^{1/2}}. \tag{3.72}$$

$\tilde{u}_{s\mathbf{p}}$ and $u_{s\mathbf{p}}$, as well as $\tilde{v}_{s\mathbf{p}}$ and $v_{s\mathbf{p}}$, are connected by the relations

$$\tilde{u}_{s\mathbf{p}}[E_s(\mathbf{p})] = u_{s\mathbf{p}}[-E_s(\mathbf{p})], \qquad \tilde{v}_{s\mathbf{p}}[E_s(\mathbf{p})] = v_{s\mathbf{p}}[-E_s(\mathbf{p})].$$

The average population number in the approximation of the Hamiltonian H_B is

$$\langle \xi^{\dagger}_{s\mathbf{p}} \xi_{s\mathbf{p}} \rangle_{H_B} = \frac{1}{e^{E_s(\mathbf{p})/k_B T} - 1}. \tag{3.73}$$

Including the main part of the interactions in the Hamiltonian H_B and using the matrix $\hat{G}^{B,0}(\mathbf{p}, E)$ as a zeroth-order Green's function, the poles of which correspond to the collective branches of the spectrum, considerably simplifies further investigation. In this approximation the averages over the full Hamiltonian and the operators in the Heisenberg representation that appeared in Eq. (3.62) can be substituted by the averages over the zeroth-order Hamiltonian and by operators in the interaction representation, respectively. Then

$$\alpha_{\mathbf{p}}(t) = \sum_s u_{s\mathbf{p}} \xi_{s\mathbf{p}} e^{-iE_s(\mathbf{p})t/\hbar} + \sum_s \tilde{u}_{s\mathbf{p}} \xi^{\dagger}_{s,-\mathbf{p}} e^{iE_s(\mathbf{p})t/\hbar},$$
$$\beta_{\mathbf{p}}(t) = \sum_s v_{s\mathbf{p}} \xi_{s\mathbf{p}} e^{-iE_s(\mathbf{p})t/\hbar} + \sum_s \tilde{v}_{s\mathbf{p}} \xi^{\dagger}_{s,-\mathbf{p}} e^{iE_s(\mathbf{p})t/\hbar}. \tag{3.74}$$

In the following, the real and the imaginary parts of the mass operator $\hat{\Sigma}^B(\mathbf{p}, E)$ components are obtained in the first order of the perturbation theory. On this basis, in Subsection 3.3.2 we investigate the alteration of the line shapes of the exciton absorption and luminescence for dipole-active excitons in the presence of a condensate. Since $\hat{K}(\mathbf{p}, E)$ is weak, we can approximate

$$\hat{\Sigma}^B(\mathbf{p}, E) \cong \hat{K}(\mathbf{p}, E). \tag{3.75}$$

As a consequence of Eqs. (3.74), the components $K_{ij}(\mathbf{p}, E)$ can be represented by expressions of the type

$$\sum_{\mathbf{q},s,t} \frac{1 + n_{s\mathbf{q}} + n_{t\mathbf{q}}}{E - [E_s(\mathbf{q}) + E_t(\mathbf{q})] + i\delta} A(\mathbf{q}),$$

$$\sum_{\mathbf{q},s,t} \frac{n_{s\mathbf{q}} - n_{t\mathbf{q}}}{E + [E_s(\mathbf{q}) - E_t(\mathbf{q})] + i\delta} B(\mathbf{q}).$$

The cumbersome coefficients $A(\mathbf{q})$ and $B(\mathbf{q})$ are dropped. Their numerical values are used in Subsection 3.3.2.

From general formula (3.62) and definitions (3.45) and (3.47), because $\varphi_0 = t_{0\mathbf{q}} = 0$, one obtains

$$K_{i2}(0, E) = K_{i4}(0, E) = K_{2i}(0, E) = K_{4i}(0, E) = 0,$$

where $i = 1, 2, 3, 4$. At small values of the quasi-momentum p one can expect that

$$K_{ij}(\mathbf{p}, E) = K_{ij}(0, E) + \text{const}\, p^2. \tag{3.76}$$

At the end of this section it is necessary to make some remarks. In the theory of the weakly interacting Bose gas [15, 21], a general relation was established:

$$\hat{\Sigma}_{11}(0, 0) = \hat{\Sigma}_{13}(0, 0). \tag{3.77}$$

This equality can be used to determine the chemical potential μ in another way. Following Ref. 25, it can be shown that in the absence of phonons, the expression for μ determined from Eq. (3.77) coincides exactly with the one from Eq. (3.65). In the presence of phonons their equivalence was not established. But relation (3.77) is more general than the first-order perturbation-theory results and will be used to determine the exact analytical properties of $\tilde{\Sigma}_{ij}(\mathbf{p}, E)$ in the range of small values of the variables $\mathbf{p}$, E.

In the Beliaev theory of the nonideal Bose gas, it was necessary to replace the interaction constant $U(0)$ with an effective interaction potential equal to $4\pi\hbar^2 a/m_{\text{ex}}$, where a is the scattering amplitude of two quasiparticles with $\mathbf{p} = \mathbf{p}' = 0$, which is of the order of the exciton radius. This effective interaction appears if one takes into account the blocks of ladder diagrams in the self-energy parts.

Having determined the expressions and the properties of the self-energy matrix components, one can proceed to the investigation of the form functions of the exciton absorption and luminescence bands. For the sake of simplicity we restrict ourselves to the case $T = 0$. The absorption coefficient depends on the light frequency. Meanwhile, the values that are finally obtained depend on the energy E, which is measured relative to the chemical-potential value μ. In the former approach with a definite N, the energy was measured relative to the exciton creation energy Δ. In both cases, the light frequency must be measured in the same reference frame as the energy of the elementary excitation. Ultimately, both approaches give rise to the same results.

3.3.2 Phonons and Hydrons in the Dipole-Allowed Single-Photon Transition

For excitonic light absorption it is necessary to have a dissipative subsystem to which the energy of the created excitons can be transmitted [26–36]. In relatively perfect crystals the phonons serve as such a dissipative subsystem for small concentrations of excitons and carriers. In our case, the role of the dissipative subsystem can be played, along with the phonons, by the excitons themselves [37, 38].

We have already discussed in Subsection 1.4.1 the effect of exciton interactions on the luminescence, namely a Lorentzian broadening and a red shift in typical cases. A great deal of work in semiconductor physics has addressed the exciton–phonon interactions and their influence on the form of the absorption and the luminescence bands (e.g., Refs. 2, 37–49). We survey here some of the papers dedicated to this problem.

The qualitative theory of light absorption in molecular crystals was developed by Davydov and Rashba [39]; Davydov and Myasnikov [44], as well as Suna [43], investigated in detail the dependences of the dielectric constant and the extinction coefficient on the light frequency in the case of weak interaction of the molecular-crystal exciton with acoustic and optical phonons. Agranovich and Konobeev [41] studied the light absorption of the crystal, taking into account the polariton effects for the Frenkel exciton interacting with the acoustic phonons. But accounting for the retardation effects for a small exciton oscillator strength is important only if the long-wavelength edge of the exciton absorption band is considered.

The theory of Wannier–Mott exciton light absorption was developed by Toyozawa [2], who took into account the simultaneous creation and annihilation of phonons during the optical transition. In the first variant of the theory [2], the optical and the acoustic phonons were taken into account, but the broadening and the level shift did not depend on the light frequency. It was shown that in the case of a weak exciton–phonon interaction, a small translational exciton mass, and at not very high temperatures, the absorption band has the form of a Lorentzian. If the coupling with the phonons is strong, the translational mass is large and the temperature is high, the band form is Gaussian. Later, Toyozawa generalized his theory and obtained dependences of the exciton level broadening and shift on the frequency ω and the wave vector $\mathbf{k}$ in the case of a weak interaction with the acoustic phonons. Toyozawa found that near the singularity of the derivative of the exciton density of states the multiphonon processes are highly important. The line shape is asymmetric, with a low-energy or high-energy tail, depending on the sign of the exciton translational mass m_{ex}, i.e., whether the point $k = 0$ is the bottom or the top of the exciton band. Further investigations were done by Fedoseev [47], and Lubchenko, Nitsovich, and Tkach [48].

An analogous program for the investigation of excitons weakly interacting with optical phonons was carried out by Moskalenko, Shmiglyuk, and Chinik [45]. It was shown that the band asymmetry may have the opposite sign as that of Ref. [2] and that the optical-phonon contribution to the level shift is significant, even when its contribution to the level broadening is negligible.

As the exciton density increases, the role of exciton–exciton interaction processes becomes more important. For example, inelastic exciton collisions can lead to the appearence of new recombinational emission bands from the exciton state when part of the exciton creation energy is transmitted to a second exciton for its intraband excitation or dissociation [50]. In the case of indirect transitions, a phonon also participates. The position of these new Auger-type emission bands on the energy scale depends on the character of the interaction forces between the excitons and on whether or not biexcitons form. For exciton densities n_{ex} just below the metallization density, the intensities of the new emission bands are proportional to n_{ex}^2. If a second exciton dissociates during the transition process, the low-energy tail of the new emission band must be similar to the high-energy tail of the absorption band because of the exciton photodissociation [50].

Besides the inelastic interaction processes, elastic-scattering processes exist between the excitons that influence the shapes of the exciton absorption and luminescence bands in both the normal and the condensed phases of the exciton system. Besides the classical collision broadening discussed in Subsection 1.4.1, as discussed in Subsection 2.1.3, in the system of degenerate excitons, new phononlike elementary excitations called "hydrons" can appear.[a] These can play the same role in the exciton optical transitions as the acoustic phonons. As shown in Subsection 3.2.5, in the presence of the lattice-phonon field, one finds mixed phonon–hydron excitations.

In the case of a nondegenerate exciton band, which we consider for simplicity, the light absorption coefficient $\mu(\mathbf{k}, w)$ is determined by

$$\mu(\mathbf{k}, \omega) = \pi f \frac{\Omega_p^2}{2c} \frac{f(\mathbf{k}, \omega)}{n(\mathbf{k}, \omega)}, \tag{3.78}$$

where f is the oscillator strength of the optical transition in the exciton state, $\hbar\Omega_p \approx 10$ eV is the plasmon excitation energy for the proper electrons of the crystal, and $n(\mathbf{k}, w)$ is the refractive index.

The form of the absorption band,

$$f(\mathbf{k}, \omega) = \frac{1}{2\pi} \int_{-\infty}^{\infty} \langle a_{\mathbf{k}H}(t) a_{\mathbf{k}H}^\dagger(0) \rangle e^{i\omega t} \, dt, \tag{3.79}$$

is connected with the Fourier transform of the retarded and the advanced exciton Green's functions [20, 21, 42, 51] in the following way:

$$f(\mathbf{k}, \omega) = \frac{i}{2\pi} \frac{1}{1 - e^{-\frac{\hbar\omega}{k_B T}}} [G_R(\mathbf{k}, \omega) - G_A(\mathbf{k}, \omega)].$$

[a] As mentioned in Subsection 3.2.5, these are usually called phonons in the context of liquid He, but are called hydrons here to distinguish them from the lattice phonons.

Here the frequency ω is measured in the laboratory reference frame. For this energy scale, exciton absorption occurs for the energies $\hbar\omega \approx \Delta \gg k_B T$. The value $\exp[-(\hbar\omega/k_B T)]$ may be neglected compared with unity. In the case $T = 0$, G_R and G_A are simply connected with the causal Green's function G [21, 51] and the previous formula is given by ($\omega > 0$)

$$f(\mathbf{k}, \omega) = -\frac{1}{\pi}\mathrm{Im}\, G(\mathbf{k}, \omega). \tag{3.80}$$

In the present case, one must insert the function $G_{11}(\mathbf{k}, w)$ from Eqs. (3.55) in the place of $G(\mathbf{k}, w)$, namely, $G_{11}(\mathbf{k}, w)$ is determined in the first order of the perturbation theory. Now the form function depends on energy E, which is measured relative to the renormalized bottom of the exciton band $\Delta + L_0$. It is shifted by a value L_0 relative to case of a single exciton.

From Eq. (3.63) one obtains

$$G_{11}(\mathbf{k}, E) = \frac{\tilde{\Sigma}_{33}(\mathbf{k}, E)}{\begin{vmatrix} \tilde{\Sigma}_{11}(\mathbf{k}, E) & \tilde{\Sigma}_{13}(\mathbf{k}, E) \\ \tilde{\Sigma}_{31}(\mathbf{k}, E) & \tilde{\Sigma}_{33}(\mathbf{k}, E) \end{vmatrix}}. \tag{3.81}$$

In principle, one should start with a fourth-degree determinant. In the case of small quasi-momenta $\hbar k$ equal to the photon momentum, however, the expressions K_{i2}, K_{i4}, K_{2i}, and K_{4i} are proportional to k^2, as seen in Eq. (3.76). Nondiagonal terms of this type may be accounted for by the perturbation theory and lead to additions in Eq. (3.81) proportional to k^2.

The self-energy part consists of real and imaginary components:

$$\tilde{\Sigma}_{ij}(\mathbf{k}, E) = \sigma_{ij}(\mathbf{k}, E) - i\gamma_{ij}(\mathbf{k}, E). \tag{3.82}$$

Taking into account the symmetry properties of Eqs. (3.19) and definition (3.82), one obtains

$$\begin{aligned} &\sigma_{13}(\mathbf{k}, E) = \sigma_{13}(-\mathbf{k}, -E), && \sigma_{33}(\mathbf{k}, E) = \sigma_{11}(-\mathbf{k}, -E), \\ &\sigma_{13}(\mathbf{k}, E) = \sigma_{31}(\mathbf{k}, E), && \gamma_{13}(\mathbf{k}, E) = -\gamma_{13}(-\mathbf{k}, -E), \\ &\gamma_{33}(\mathbf{k}, E) = -\gamma_{11}(-\mathbf{k}, -E), && \gamma_{13}(\mathbf{k}, E) = \gamma_{31}(\mathbf{k}, E). \end{aligned} \tag{3.83}$$

After inserting Eqs. (3.83) into Eqs. (3.81) and (3.80), one obtains

$$f(\mathbf{k}, E) = \frac{1}{\pi}\frac{M(\mathbf{k}, E)}{N(\mathbf{k}, E)}. \tag{3.84}$$

Here

$$\begin{aligned} M(\mathbf{k}, E) &= 2\sigma_{11}(\mathbf{k}, -E)\sigma_{13}(\mathbf{k}, E)\gamma_{13}(\mathbf{k}, E) - \sigma_{11}^2(\mathbf{k}, -E)\gamma_{11}(\mathbf{k}, E) \\ &\quad + \sigma_{13}^2(\mathbf{k}, E)\gamma_{11}(\mathbf{k}, -E) - \gamma_{11}(\mathbf{k}, -E)\gamma_{13}^2(\mathbf{k}, E) - \gamma_{11}(\mathbf{k}, E)\gamma_{11}^2(\mathbf{k}, -E), \\ N(\mathbf{k}, E) &= [\sigma_{11}(\mathbf{k}, E)\sigma_{11}(\mathbf{k}, -E) - \sigma_{13}^2(\mathbf{k}, E) + \gamma_{11}(\mathbf{k}, E)\gamma_{11}(\mathbf{k}, -E) + \gamma_{13}^2(\mathbf{k}, E)]^2 \\ &\quad + [\gamma_{11}(\mathbf{k}, E)\sigma_{11}(\mathbf{k}, -E) - \sigma_{11}(\mathbf{k}, E)\gamma_{11}(\mathbf{k}, -E) - 2\gamma_{13}(\mathbf{k}, E)\sigma_{13}(\mathbf{k}, E)]^2. \end{aligned} \tag{3.85}$$

The counting of the values $\sigma_{11}(0, E)$, $\sigma_{13}(0, E)$, $\gamma_{11}(0, E)$, and $\gamma_{13}(0, E)$ by use of formulas (3.74) and (3.75) is correct only in the range of values of E that are not too small. For $E \to 0$ the approximate expressions $\sigma_{11}(0, E)$ and $\sigma_{13}(0, E)$ diverge logarithmically, just as $\gamma_{11}(0, E)$ and $\gamma_{13}(0, E)$ meet with breakage at the point $E = 0$. These conclusions are drawn from the analysis of the integrand expressions in Eq. (3.75) for small values of the integrand variable. To avoid these peculiarities, it is necessary to have a more exact accounting of the mass operator, for example as was done in Refs. 2 and 45 for the case of the exciton–phonon interaction.

Following Refs. 15 and 21, we suppose that the exact expressions for the components $\sigma_{ij}(\mathbf{k}, E)$ and $\gamma_{ij}(\mathbf{k}, E)$ are the regular functions for small $\mathbf{k}$ and E. Thus one assumes that

$$\sigma_{11}(\mathbf{k}, E) = \sigma_0 + E + \frac{\hbar^2 k^2}{2m_1} + \cdots,$$
$$\sigma_{13}(\mathbf{k}, E) = \sigma_0 + \frac{\hbar^2 k^2}{2m_3} + \cdots,$$
$$\gamma_{13}(\mathbf{k}, E) = s_1 E + t_1 E^3 + \cdots,$$
$$\gamma_{11}(\mathbf{k}, E) = s_2 E + t_2 E^2 + \cdots. \quad (3.86)$$

The expressions $\gamma_{13}(\mathbf{k}, E)$ and $\gamma_{11}(\mathbf{k}, E)$ in Eqs. (3.86) do not contain free terms, since

$$\gamma_{13}(0, E) = -\gamma_{13}(0, -E), \qquad \gamma_{11}(0, 0) = \gamma_{13}(0, 0).$$

The last equality follows from the condition $\tilde{\Sigma}_{11}(0, 0) = \tilde{\Sigma}_{13}(0, 0)$ [15, 21].

The form function is obtained when Eqs. (3.86) are inserted into Eqs. (3.85) and (3.84). In the range $0 \le |E| \le \hbar^2 k^2/2|\mu_{13}|$, where $\mu_{13}^{-1} = (m_1^{-1} - m_3^{-1})$, one finds

$$f(\mathbf{k}, E) = \frac{2}{\pi} \frac{(s_1 - s_2)}{(\hbar^2 k^2/\mu_{13})^2} E. \quad (3.87)$$

The form function is unique and describes both light absorption and emission. For $s_1 > s_2$ it is positive for $E > 0$, which corresponds to absorption. The form function is negative for $E < 0$, which corresponds to emission. In the neighborhood of the point $E = 0$, each form function has a slope inversely proportional to k^4. This leads to the existence of two sharp central peaks of opposite signs, which came into contact at the point $E = 0$. At the points $E = \pm\hbar^2 k^2/2|\mu_{13}|$, the peaks are broken. For energies E with modulus greater than $\hbar^2 k^2/2|\mu_{13}|$, the slowly varying parts of the form function are stretched. As in the case of one exciton interacting with the phonons, these parts are the wings of the corresponding absorption and luminescence bands. The wing areas may be calculated numerically with Eqs. (3.74) and (3.75). The areas of the central peaks situated in the energy interval 0, $\pm\hbar^2 k^2/2|\mu_{13}|$ are $(s_1 - s_2)/4\pi$. It is essential that they do not depend on k and remain finite for $k \to 0$, when the central-peak widths tend to zero and their slopes become infinite. Therefore they have the properties of $\delta_\pm$ functions.

The width of the central peaks is determined by the value $\hbar^2 k^2/2|\mu_{13}|$. The uncertainty of the wave vector $\mathbf{k}$ is of the same order of magnitude as the absorption coefficient $\mu(\mathbf{k}, w)$, i.e., $k \sim \Delta k \sim \mu(\mathbf{k}, w)$. Taking $\mu_{13} \simeq m_0$, $\mu(\mathbf{k}, w) = 10^5\ \text{cm}^{-1}$, one obtains $\hbar^2 k^2/2|\mu_{13}| \approx 10^{-17}$ erg. Therefore, for the given parameters, the central-peak width is much smaller than the usual width of the separate exciton absorption line.

By analogy with Ref. 45, one can expect that the wing peaks correspond to the energies E of a hydron or a phonon with a wave vector equal to the inverse value of the exciton radius. The calculations by Lelyakov and Moskalenko [37, 49] confirm this. The wings correspond to the creation of a hydron or a phonon in the light absorption or luminescence processes. The integral area of the wing and the central peak of the form function must be equal to unity on both the right and the left sides of the point $E = 0$; from it the value $(s_1 - s_2)$ can be determined.

The estimates of the shapes of the wings and their areas for absorption and luminescence bands were performed in Refs. 49 and 37 for an exciton-condensate density of 10^{17}cm^{-3}, an exciton radius of 50 Å, a hydron velocity of 10^6 cm/s, and $U(0) = 10^{-32}$ erg cm^3;

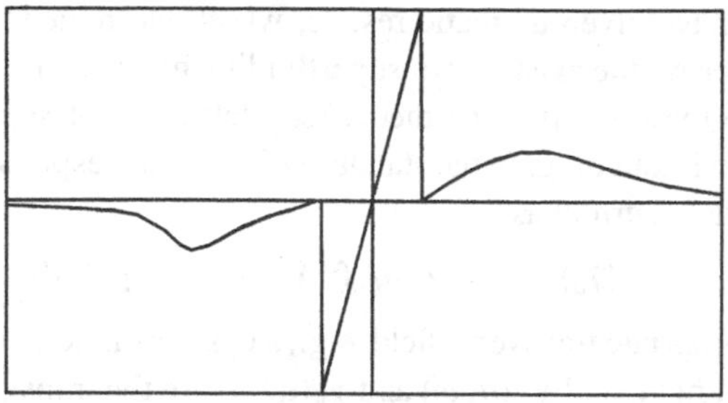

Figure 3.1. The form functions of the absorption and the luminescence bands.

$|C_e - C_h| = 5$ eV, where C_e and C_h are deformation-potential constants. The area of the wing for the absorption band is greater than that of the luminescence band. This follows from the general expression for $\gamma_{11}(\mathbf{k}, E)$, which can be obtained with Eq. (3.75). For simplicity we neglect the phonons. Then the integrand expression for $\gamma_{11}(\mathbf{k}, E > 0)$ contains $u_{\mathbf{q}}^2$, just as $\gamma_{11}(\mathbf{k}, E < 0)$ contains $\tilde{u}_{\mathbf{q}}^2$. As follows from Refs. 37 and 49, $\tilde{u}_{\mathbf{q}}^2 = n_{\mathbf{q}}$ and $\mathbf{u}_{\mathbf{q}}^2 = (1 + n_{\mathbf{q}})$, where $n_{\mathbf{q}}$ is the number of excitons with wave vector $\mathbf{q}$ out of the condensate at $T = 0$. Because of the difference between $(1 + n_{\mathbf{q}})$ and $n_{\mathbf{q}}$, the attenuation for $E > 0$ is more than the gain for $E < 0$. The delta function that appears in the integrand of the expression for $\gamma_{ij}(\mathbf{k}, E)$ chooses a value of $\mathbf{q}$ for which $E = E_{\mathbf{q}}$. Small values of $|E|$ correspond to small values of $\mathbf{q}$, and the difference between $(1 + n_{\mathbf{q}})$ and $n_{\mathbf{q}}$ is small. For large values of $|E|$, this difference is essential and influences the shapes of the wings. It causes a sharper frequency dependence and decreasing of the luminescence-band wing as $|E|$ increases, compared with the behavior of the absorption band. An analogous dependence on the occupation number $n_{\mathbf{q}}$ was found by Gergel', Kazarinov, and Suris [52]. In Ref. 52, however, the attenuation and the gain are δ-like functions, since the authors limited themselves to the zeroth-order approach for the mass operator.

The qualitative picture of the form functions of the absorption and the luminescence bands is presented in Fig. 3.1. When k tends to zero, the form of the central peaks resemble the form of the phononless exciton line obtained by Fedoseev and Hizhnyakov [46]. These results confirm the qualitative considerations about the character of the light absorption and emission by the degenerate excitons suggested earlier by Casella [53], Blatt, Böer, and Brandt [54], and Moskalenko [55, 37].

Petrov [56] discussed the resonance luminescence in liquid He^4 at temperatures below the λ point. Unlike the case of degenerate excitons, the condensate in He-II consists of nonexcited atoms. Therefore the process of the light emission by the excited He atoms and their return to the condensate has a greater probability than the phenomenon of light absorption by the condensate atoms and its excitation out of the condensate.

In conclusion, we note that for small exciton oscillator strength $f < 10^{-4}$, the frequency dependence of the absorption coefficient $\mu(\mathbf{k}, w)$ coincides with the one for the form function. At such a low oscillator strength the refractive index depends very little on ω [45].

3.4 Experiments on Superfluid Exciton Transport and the Phonon Wind

As we have seen in the Section 3.2, an exciton condensate is predicted to be superfluid. Is there any experimental evidence of this? In recent years, measurements of the transport of excitons in Cu_2O have given dramatic results, which we review in this section. Whether these experiments prove the existence of superfluidity, however, remains the subject of some debate. Part of the reason is that the theoretical definition of superfluidity involves more than just transport without drag. In the standard [57] linear-response theory of superfluidity, one writes the particle current as

$$\langle \mathbf{J_q} \rangle = -m\chi_\perp(q, 0)\mathbf{A_q} - m\chi_\parallel(\mathbf{q}, 0)V_\mathbf{q},$$

where $\mathbf{A_q}$ is an appropriate transverse field, e.g., a Coriolis field in the case of superfluid He, $V_\mathbf{q}$ is a longitudinal field, and $\chi_\perp(q, \omega)$ and $\chi_\parallel(\mathbf{q}, \omega)$ are the transverse and the longitudinal current–current response functions, respectively. For a superfluid,

$$\lim_{\mathbf{q}\to 0} \chi_\perp(q, 0) = 0,$$

since the long-range correlations in the superfluid state cause $\mathbf{J_q}$ to be translationally invariant, and

$$\lim_{\mathbf{q}\to 0} \chi_\parallel(q, 0) = -N/m,$$

where N is the number of particles. For a normal fluid,

$$\chi_\perp(q, 0) = \frac{2}{Z}\sum_{nm} \frac{e^{-\beta E_m}|\langle n|\mathbf{J}_{\mathbf{q}\perp}|m\rangle|^2}{\omega_{nm}},$$

which does not vanish in the limit $\mathbf{q} \to 0$, and

$$\lim_{\mathbf{q}\to 0} \chi_\parallel(q, 0) = -N/m.$$

The dramatically different behavior in response to a transverse field leads to the definition of the normal and the superfluid mass densities as

$$\rho_n = -m^2 \lim_{\mathbf{q}\to 0} \chi_\perp(q, 0),$$

$$\rho_s = \rho - \rho_n.$$

In a similar way, the definition of superconductivity involves more than just the measurement of zero resistance. If resistance measurements were all that we had, it would be difficult to tell the difference between truly zero resistance and simply very low resistance. The Meissner effect [58], in which the electromagnetic vector potential plays the role of the transverse field $\mathbf{A_q}$, given by the Coriolis field in the rotating-bucket experiment in He, goes much further toward proving superconductivity. The problem for excitons is that no one has yet found an equivalent transverse field for excitons; since they are charge neutral, a magnetic field causes a change in only their orbital states.

The experiments in Cu_2O show that solitonlike pulses of excitons can propagate for long distances (approximately a centimeter) across a pure crystal following high-intensity laser pulses. In some applications, whether or not these pulses are truly superfluid may be irrelevant, just as in many cases extremely low resistivity is just as good as superconductivity. On the other hand, much of the interest in superfluid excitons stems from the interest in creating a coherent exciton pulse. This would be the case only if the excitons are truly superfluid, i.e., have long-range phase correlation.

If exciton superfluidity could be established under controlled circumstances, it could have significant applications. Since excitons have very short wavelengths compared with those of photons at the same energy, a superfluid exciton pulse would essentially be a coherent wave of optical excitation with a very short wavelength. This could be used, for example, to transmit visible light through tiny semiconductor waveguides with dimensions less than the wavelength of light, without diffraction loss.

3.4.1 Superfluidity of Excitons in Semiconductors

Superfluidity of excitons and biexcitons was predicted in Ref. 55, along with a possible experimental scheme for studying this phenomenon, which is shown in Fig. 3.2 [37, 52].

Gergel' et al. [52] presented a theoretical description of the superfluid flow. They showed that if the excitons undergo the phase transition below the critical temperature and the superfluid component of the exciton gas appears, it will propagate at distances determined by the

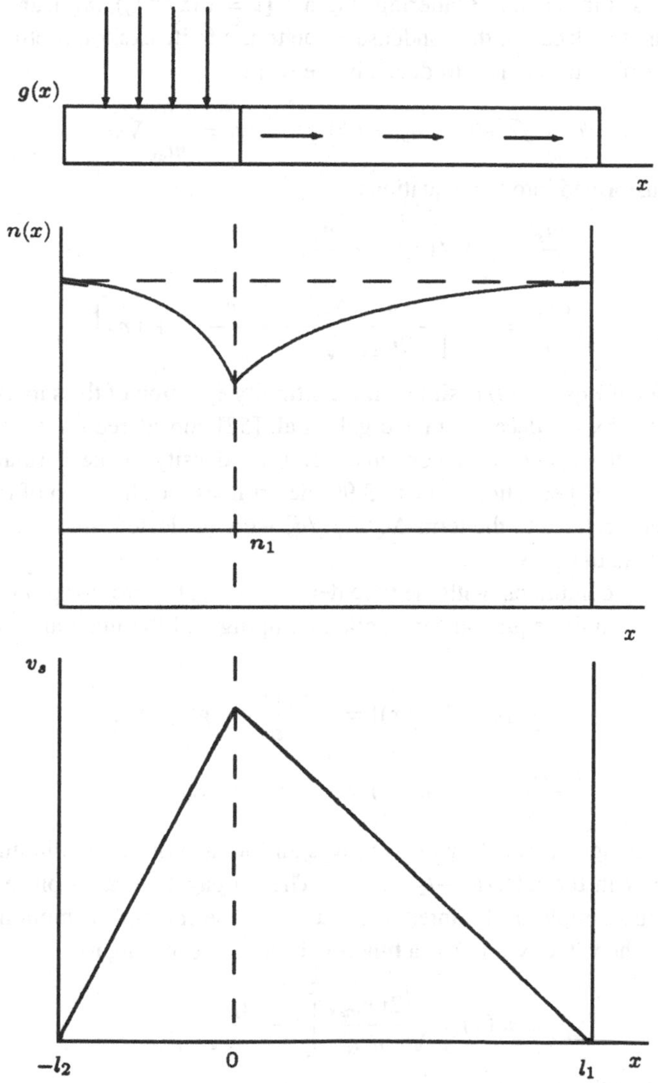

Figure 3.2. The dependences of the density and the velocity of the superfluid component at a steplike pumping [37, 52].

critical velocity of the superfluidity v_{cr} and the exciton lifetime τ ($l = v_{cr}\tau$). The stationary spatial distribution of excitons and the penetration depth of excitons into the nonilluminated region in Fig. 3.2 will considerably differ from the behavior expected for diffusion. In classical diffusive motion, e.g., as seen in Fig. 1.2, the diffusion length $l_D = \sqrt{D\tau}$ is determined by the exciton diffusion coefficient D and the lifetime τ.

The starting point of the paper by Gergel' et al. [52] is the nonlinear Schrödinger equation describing the coherent macroscopic component of the exciton system. The macroscopic wave function ψ obeys the differential equation

$$i\hbar\frac{\mathrm{d}\psi}{\mathrm{d}t} = -\frac{\hbar^2}{2m_{ex}}\Delta\psi + E_{ex}\psi + \nu\,|\psi|^2\,\psi - \frac{i\hbar}{2\tau}\psi, \tag{3.88}$$

where $E_{ex} + \nu\,|\psi|^2 = \mu$ determines the chemical potential, the interaction constant ν is expressed in terms of the exciton-scattering length $f_0[\nu = (4\pi\hbar^2 f_0)/m_{ex}]$, and the last term takes into account the decay of the condensate due to the finite exciton lifetime.

When the wave function ψ is introduced in the form

$$\psi = \sqrt{n_s}e^{i\varphi}, \quad n_s = |\psi|^2, \qquad \mathbf{v}_s = \frac{\hbar}{m_{ex}}\nabla\varphi, \tag{3.89}$$

Eq. (3.88) is transformed into two equations

$$\frac{\partial n_s}{\partial t} = -\nabla\cdot(n_s\mathbf{v}_s) - \frac{n_s}{\tau},$$

$$m_{ex}\frac{\partial \mathbf{v}_s}{\partial t} = -\nabla\left[-\frac{\hbar^2}{2m_{ex}}\frac{\Delta\sqrt{n_s}}{\sqrt{n_s}} + \frac{m_{ex}v_s^2}{2} + \nu n_s\right]. \tag{3.90}$$

The first equation of Eqs. (3.90) is simply the continuity equation of the superfluid component supplemented by the decay term. Gergel' et al. [52] considered only the case $T = 0$, supposing that all the excitons are condensed and the density of the normal component equals zero. The second equation of Eqs. (3.90) determines the alteration of the superfluid velocity $\mathbf{v}_s$. Effects related to the term $\Delta\sqrt{n_s}/\sqrt{n_s}$ were neglected, since the alteration of n_s is much less than that of $\mathbf{v}_s$.

Under stationary conditions, with the time derivatives set to zero, for propagation along the x axis, the continuity equation for steplike pumping and the integral of motion were obtained as

$$\frac{\mathrm{d}}{\mathrm{d}x}\left[n_s(x)\,v_s(x)\right] = -\frac{n_s(x)}{\tau} + g\theta(-x),$$

$$\frac{m_{ex}v_s^2(x)}{2} + \nu n_s(x) = \text{const} = \nu n_{max}. \tag{3.91}$$

The steplike pumping described by $\theta(-x)$ gives an intensity g on the illuminated part of the semiconductor in the interval $-l_2 < x < 0$. The propagation takes place in the darkened region of the sample in the interval $0 < x < l_1$. The integral of motion permits the determination of the velocity $v_s(x)$ as a function of $n_s(x)$ according to

$$v_s(x) = \sqrt{\frac{2\nu n_{max}}{m_{ex}}\left[1 - \frac{n_s(x)}{n_{max}}\right]}. \tag{3.92}$$

The boundary conditions

$$v_s(x = -l_2) = v_s(x = l_1) = 0$$

determine the condensed-exciton densities at the ends of the sample,

$$n_s(x = -l_2) = n_s(x = l_1) = n_{\max}.$$

In the interior part of the sample $-l_2 < x < l_1$, where $v_s \neq 0$, the values of $n_s(x)$ are less than $n_{\max}$. The approximate solutions of Eqs. (3.91) in the interval $0 < x < l_1$ were found [52] to have the forms

$$n_s(x) = n_{\max}\left[1 - \left(\frac{l_1 - x}{L}\right)^2\right], \quad L = \tau\sqrt{\frac{2\nu n_{\max}}{m_{\mathrm{ex}}}},$$

$$v_s(x) = \frac{l_1 - x}{\tau} \qquad 0 < x < l_1. \tag{3.93}$$

This solution is represented in Fig. 3.2.

The maximum value of velocity $v_s(0) = l_1/\tau$ must be less than the critical velocity v_{cr} of superfluidity. It is determined by the Bogoliubov critical velocity $u(x) = \{[\nu n(x)]/m_{\mathrm{ex}}\}^{1/2}$, if quantum vortices are absent. The critical velocity related to the quantum-vortex creation is [37, 52, 55]

$$v_{\mathrm{cr}} = \frac{\hbar}{Rm_{\mathrm{ex}}} \ln\frac{R}{a},$$

where R is the vortex radius and a is the radius of its trunk.

If the length l_1 increases, the necessary stationary velocity $v_s(0)$ needed to maintain the concentration of excitons $n_s(x)$ and to compensate for their decay in the interval l_1 also increases. When l_1 exceeds the maximal admissible value $v_{\mathrm{cr}}\tau$, the condition of superfluidity is broken. The effect of superfluidity can be observed in an explicit way if the length l_1 is less than $v_{\mathrm{cr}}\tau$ but more than the diffusion length $\sqrt{D\tau}$:

$$\sqrt{D\tau} < l_1 \leq v_{\mathrm{cr}}\tau.$$

In this case the transition of the exciton gas from the gaseous phase to the condensed one will lead to a sharp enhancement of the exciton density at the darkened end of the sample.

If the temperature of the exciton gas is less than the critical temperature but nonzero ($0 < T < T_{\mathrm{cr}}$), then the normal component of the two-component liquid model must be taken into account. In this case, the penetration of the superfluid component into the darkened region will be accompanied by the evaporation of the condensed excitons and by the formation of a normal component. The density of the normal component is determined by the temperature of the exciton gas and is given by

$$n_n = 2.612\left(\frac{m_{\mathrm{ex}}k_B T}{2\pi\hbar^2}\right)^{3/2} = n\left(\frac{T}{T_c}\right)^{3/2}, \qquad 0 < T < T_c. \tag{3.94}$$

If the exciton temperature $T < T_c$ remains the same along the interval $0 < x < l_1$, then the density of the normal component will be constant and nonzero. The superfluid component will exist and propagate, evaporating until it forms the normal component [37]. This is also shown in Fig. 3.2.

In principle, this experiment is simple: All one needs is a very powerful cw laser able to flood a semiconductor such as Cu_2O with excitons up to the critical density. To fill the volume of $\sim$1 mm^3 at the critical density for condensation of excitons at $T = 1$ K, however, one needs a red cw laser with power of $\sim$30 W, assuming an exciton lifetime of 1 μs. Since this is experimentally difficult, most recent attempts at seeing excitonic superfluidity have used pulsed lasers with high peak energy. In addition, laser wavelengths with absorption lengths in the medium of the order of microns have been used to reduce the created volume of the excitons and therefore increase their instantaneous density. Nevertheless, this proposal remains conceptually appealing.

3.4.2 Anomalous Exciton Transport in Cu_2O

In Cu_2O crystals, anomalous ballistic transport of excitons over unusually large distances is observed at low lattice temperatures and high initial particle densities, following intense excitation of the surface of the crystal by a pulsed green laser [59–63]. The first results on this topic were obtained by Mysyrowicz, Snoke, and Wolfe [59, 60] in experiments that used time-resolved optical imaging to track the expansion of the excitons away from the surface of the crystal. More recently, papers by Fortin, Fafard, and Mysyrowicz [61], Mysyrowicz et al. [62], and Benson, Fortin, and Mysyrowicz [63] have been dedicated to detailed investigations of this phenomenon, by use of photovoltaic detection of excitons.

The experimental setup is shown in Fig. 3.3. The optical pulses, 10 ns in duration and with peak intensity up to a few megawatts per square centimeter, illuminate the front surface of a sample immersed in superfluid He. The incident radiation of wavelength $\lambda = 532$ nm is absorbed within in a layer of thickness $d = 10^{-4}$ cm. The absorbed photons are converted into electron–hole pairs, which relax toward the $n = 1$ ground state of the yellow excitonic series in less than 1 ps. Cu and Au contacts on the back surface of the sample act as a fast local exciton detector by means of the exciton-mediated photovoltaic effect. At sufficiently high initial particle densities, $n > n_c$, where $n_c \approx 10^{17}$ cm^{-3} at the lattice temperature 2 K, excitons initially created near the front surface of a single crystal are observed to propagate ballistically through the sample in the form of a packet of finite size. The ballistic-exciton packet covers distances approaching 1 cm at velocities near the speed of the sound of the crystal. By contrast, at lower densities, $n < n_c$, the transport is diffusive with the accompanying spread of particle transit times [63]. The stability of the ballistic

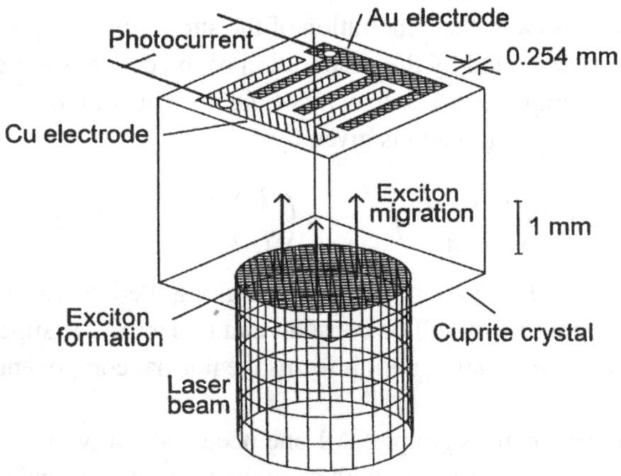

Figure 3.3. Schematic of the experiment to measure exciton transport across a crystal (from Ref. 64).

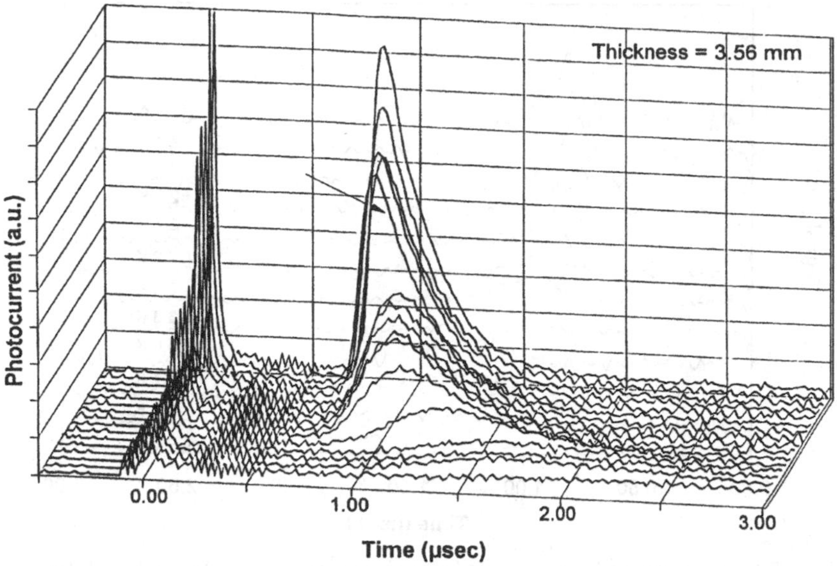

Figure 3.4. Time-resolved photocurrent obtained in a Cu_2O sample following intense laser excitation of the opposite surface for different excitation intensities and a constant temperature of 1.85 K. The intensity increases from the bottom to the top curve from 3.75×10^4 to 1.5×10^6 W/cm^2. The curve marked by the arrow, which shows a dramatic difference from the curve with slightly less intensity, is for an intensity of 6×10^5 W/cm^2 (from Ref. 64).

motion is highly temperature dependent. At a given density near the critical threshold, the packet disintegrates on a small increase of crystal temperature.

The threshold character of the observed events led the authors of Refs. 61–63 to argue that a phase transition must be the physical origin of the effect. Since n_c is close to the critical density required for BEC of excitons in Cu_2O, this anomalous propagation was attributed to excitonic superfluidity. Typical experimental results are represented in Figs. 3.4 and 3.5.

The vertical axis is the photovoltaic signal produced by the dissociation of excitons near the electrode face. The large signal at the times $t = 0.5{-}2\ \mu s$ is due to excitons migrating from the excited front surface to the back surface containing the electrodes. In Ref. 64 it was noted that the striking feature of Figs. 3.4 and 3.5 is the sudden change of the evolution of the excitonic signal over a narrow range of both density and temperature. This suggests that the changes seen in density and temperature are interrelated through the same physical process [63]. Benson et al. also reported that the average exciton velocity increases with increasing intensity and decreasing temperature, saturating just below the speed of sound of the lattice, $v_s = 4.5 \times 10^5$ cm/s. The speed of the exciton packet extracted from various measurements was determined to be 3.9×10^5 cm/s, somewhat below the longitudinal speed of sound in Cu_2O, $v_s = 4.5 \times 10^5$ cm/s.

Benson et al. [63] reported, besides the single-pulse excitation, novel effects from using two-pulse excitation. The excitons created by a weak probe pulse were effectively attracted into the packet formed by the excitons of the first strong pump pulse. In the absence of the pump pulse, the probe excitons propagated diffusively. When the high-density pulse of excitons picked them up, the same excitons propagated ballistically. In this way, the

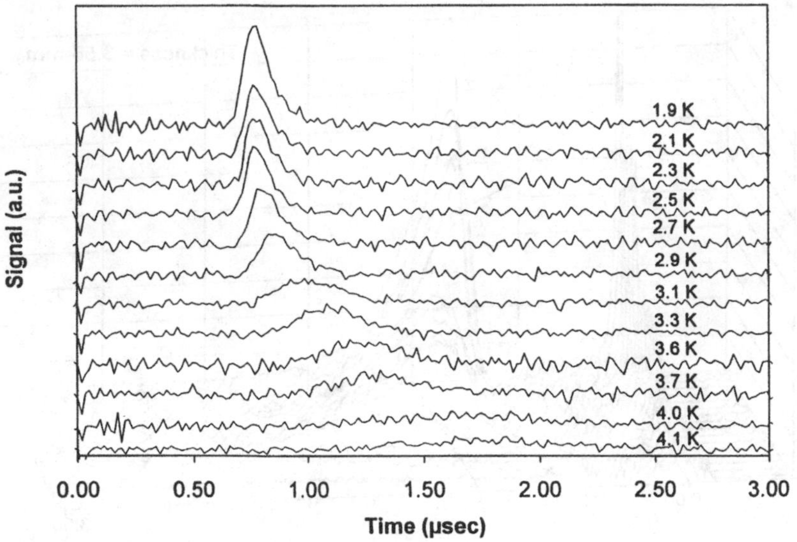

Figure 3.5. Time-resolved photocurrent for the same sample as in Fig. 3.4, at different temperatures and a constant illumination intensity of 1.2×10^6 W/cm^2 (from Ref. 64).

propagating pulse created by the pump could be amplified by passing through an incoherent source of excitons created in the middle of the crystal [65].

3.4.3 The Exciton-Condensate Soliton

The explanation favored by Fortin et al. [61] for the above behavior is that the exciton condensate forms a soliton that propagates across the crystal. Essential to this view is some mechanism by which the condensate brings about a spatial attraction of excitons into the propagating pulse.

Hanamura [66] proposed that solitons of this type could arise from the Gross–Pitaevskii (nonlinear Schrödinger) equation [67–69] for the Bose condensate wave function, $\psi(\mathbf{x}, t)$:

$$i\hbar \frac{\partial \psi}{\partial t} = -\frac{\hbar^2}{2m} \Delta \psi + U \psi^* \psi^2, \tag{3.95}$$

with an attractive potential of exciton–exciton interaction $U < 0$. Starting with the standard Hamiltonian equation (2.29) for the interacting Bose gas and developing a perturbation theory, he found that second-order interaction terms, neglected in the Bogoliubov approximation, could give a negative value of U. However, the influence of exciton–phonon interactions on the dynamics of the condensed excitons was not treated.

Loutsenko and Roubtsov [70] proposed a one-dimensional model that expressly includes the phonons of the lattice and argued that the experiments observed the propagation of a superfluid exciton–phonon condensate. Using a $T = 0$ model, they showed that the exciton–phonon interactions can give rise to solitonlike superfluidity. This kind of soliton is similar to the Davydov soliton in exciton–phonon molecular systems [71–74], introduced to explain the energy transfer along the one-dimensional chains of alpha-helical proteins. Although the energy scales and the interaction constants are quite different in biological systems and semiconductor systems, the theory for exciton–phonon soliton formation is the same as that of Davydov solitons in its general form [75].

While the work of Loutsenko and Roubtsov [70] shows the existence of soliton solutions in the coupled exciton-condensate–phonon fields, the predictions of this model seem to imply that solitons will be unstable in the conditions of the experiments with Cu_2O. For a one-dimensional soliton moving at velocity v, the theory of Loutsenko and Roubtsov gives the amplitude of the exciton condensate as

$$\psi_s(x,t) = e^{ik_0x - i\omega_0 t}\psi_0 \cosh^{-1}[\beta\psi_0(\xi - a)],$$
$$\beta = \sqrt{\frac{m_{ex}U_0}{\hbar^2}\left|\frac{v_0^2 - v^2}{s^2 - v^2}\right|} \tag{3.96}$$

where s is the lattice-phonon sound velocity, $\xi \equiv x - vt$ and a is an arbitrary translation parameter. The critical velocity v_0 is defined by the effective potential between the excitons that occurs by means of the phonon field, given by

$$U_{eff} = U_0\left(\frac{v_0^2 - v^2}{s^2 - v^2}\right), \quad U_0 = \frac{4\pi\hbar^2 f_0}{m_{ex}},$$
$$v_0^2 = s^2 - \frac{\sigma_0^2}{U_0\rho}, \qquad f_0 \simeq 2a_{ex}, \tag{3.97}$$

which is attractive for $v_0 < v < s$. Here U_0 is the hard-sphere interaction potential, σ_0 is the Bardeen–Schockley deformation potential, ρ is the crystal density, and a_{ex} and m_{ex} are the exciton radius and translational mass, respectively, and v_0^2 is assumed to be positive and v_0 is real.

This implies a spatial and a temporal width of the soliton given by

$$l^2 = \frac{\hbar^2}{m_{ex}U_0 n_0}\left|\frac{s^2 - v^2}{v_0^2 - v^2}\right|, \qquad \tau = l/v, \tag{3.98}$$

where $n_0 \equiv \Psi_0^2$ is the density of Bose condensed excitons. For $a_{ex} \approx 10$ Å and $n_0 \approx 10^{17}$ cm^{-3}, this implies a length scale of

$$l \approx 10^{-6}\,\text{cm}\sqrt{\left|\frac{s - v}{v - v_0}\right|}. \tag{3.99}$$

In the experiments with Cu_2O discussed in Subsection 3.4.2 solitonlike wave packets of paraexcitons were observed with $l \geq 10^{-2}$ cm and $\tau \geq 5 \times 10^{-7}$ s. To achieve such enormous values of l, it is necessary to suppose a very small deviation of the velocity v from the critical value v_0, such that $|v - v_0|/v \approx 10^{-8}$. In this case, small fluctuations of the soliton velocity v or of other exciton parameters will lead to arbitrarily large fluctuations of the soliton parameters, which would be quite unstable.

The theory of Hanamura [66] gives similar results. Unlike the theory of Loutsenko and Roubtsov [70], in which the exciton–phonon interaction gives rise to an indirect exciton–exciton attraction, in the case considered by Hanamura, the attraction between the excitons appears in the second order of the perturbation theory because of the exciton–exciton interaction. Hanamura supposed that the effective potential becomes attractive, with an energy scale given by

$$U_{eff} = U_0\varepsilon < 0, \tag{3.100}$$

with U_0 defined as above. In this case,

$$l^2 = \frac{1}{4\pi f_0 n_0 |\varepsilon|}, \tag{3.101}$$

which for the same parameters given above implies that

$$l = 10^{-6}\,\text{cm}\frac{1}{\sqrt{\varepsilon}}. \tag{3.102}$$

To obtain the experimental packet sizes, this implies that $|\varepsilon| \approx 10^{-8}$, which also implies extreme sensitivity to fluctuations. One can conclude that the exciton interactions alone, as well as the interactions between coherent excitons and phonons, in spite of the unusual effective interaction potential, cannot cause a soliton wave packet with $l \geq 10^{-2}$ cm – the expected value is $l \approx 10^{-6}$ cm.

Solitons with spatial length l of the order of the photon wavelength λ ($l > \lambda$) can be formed in the case of polariton wave packets, which are discussed in Chapter 9. Their formation is due to the very small and negative values of the polariton mass for the lower polariton branch in the region of the intersection of the initial photon and exciton branches. Because of the negative polariton mass, the collective bound state of the coherent polaritons and the polariton–soliton formation occurs when the repulsion between the excitons prevails. One can expect that orthoexciton-type polaritons in Cu_2O crystals could form a soliton wave packet with $l \sim \lambda$. Such a theory does not help one to understand the origin of solitons formed by coherent paraexcitons in Cu_2O with length scales of $l \approx 10^{-2}$ cm, however.

3.4.4 The Phonon Wind

In a series of papers [76–79], Tikhodeev and co-workers argued strenuously that the anomalous propagation of excitons in Cu_2O crystals reported in Subsection 3.4.2 can be explained by a purely classical effect known as phonon wind, without invoking superfluidity, except perhaps in the earliest stages.

The phenomenon of phonon wind is well known in semiconductor physics for the case of electron–hole droplets (for a general review, see Ref. 80), although it has never been definitively observed for free excitons, which have a much smaller phonon-absorption cross section. This effect occurs when the energy of the laser photons exciting the crystal is well above that of the bandgap. The electron–hole pairs created by the laser in this case have a large excess energy (often several thousands of degrees Kelvin), which they shed in the form of nonequilibrium phonons as they relax down to the bottom of the band. These nonequilibrium phonons then propagate out into the crystal, "blowing" anything in their path in the same direction.

Before dealing with the specific objections to the interpretation of the experiments of Fortin et al. [61]we first discuss the model of Tikhodeev and co-workers in order to gain a handle on this theory.

Bulatov and Tikhodeev [76], and later Kopelevich, Gippius, and Tikhodeev [77, 78], developed a simplified model for the exciton drag driven by nonequilibrium longitudinal-acoustic phonons, which are generated in the last stage of carrier thermalization. The characteristic rise time of such phonons in Cu_2O crystals τ_{ph} ($\tau_{ph} = 10^{-10}$ s) is assumed to be small compared with the duration of the excitation pulse τ_{pl} ($\tau_{pl} = 10^{-8}$ s) as well as compared with the characteristic observation time, which can be several microseconds. It is assumed that all the excess energy from the generation of electron–hole pairs, equal to ε_{pw} per electron–hole pair, goes into the incoherent ballistic-phonon population. The momentum flux of the acoustic phonons $\mathbf{W}(\mathbf{r},t)$ is determined by the formula

$$\mathbf{W}(\mathbf{r},t) = \frac{\varepsilon_{pw}}{4\pi s}\int \frac{(\mathbf{r}-\mathbf{r}')}{|\mathbf{r}-\mathbf{r}'|^3} g\left(\mathbf{r}', t - \frac{|\mathbf{r}-\mathbf{r}'|}{s}\right) d\mathbf{r}', \tag{3.103}$$

where the integration takes place over the excitation region of the crystal and s is the sound velocity.

The carrier generation rate $g(\mathbf{r},t)$ follows the laser intensity, which is assumed to have a Gaussian time envelope, yielding

$$g(\mathbf{r},t) = g_s e^{-k|z|} e^{-t^2/\tau_{pl}^2}, \quad g_s = \frac{kI}{\hbar\omega}. \tag{3.104}$$

The generation rate is determined by the photon density g_s, expressed in terms of the absorption coefficient k, the intensity of the incident excitation light I, and by the energy of the photon $\hbar\omega$. The transport equation for the exciton gas $n(\mathbf{r},t)$ in the normal, classical state is given by a diffusion equation with drag,

$$\frac{\partial n}{\partial t} + \nabla\cdot(\mathbf{v}n) = D\Delta n - \frac{n}{\tau} + g, \tag{3.105}$$

where $\mathbf{v}$ is the drift velocity of excitons caused by the phonon wind, D is the diffusion coefficient, and g and τ are the exciton pumping rate and the exciton lifetime, respectively. The relation between the drift velocity and the momentum flux of phonons is taken as [76]

$$\mathbf{v}(\mathbf{r},t) = s\frac{\alpha\mathbf{W}(\mathbf{r},t)}{1+\alpha|\mathbf{W}(\mathbf{r},t)|}, \tag{3.106}$$

which incorporates the saturation of the drift force when the excitons are moving at the sound velocity because of the Doppler effect, as seen in phonon wind on electron–hole droplets [80]. Here α is the carrier–phonon-wind coupling constant when $v \ll s$.

The numerical solution [76] of Eq. (3.105), when the parameters for Cu_2O were used, showed approximate agreement with the experimental data [59, 60] during the initial stage of exciton propagation into the crystal bulk at time and distance intervals of the order of 10^{-7} s and 10^{-2} cm. The authors also could reproduce the supersonic transport observed at higher excitation intensities [59, 60] for short periods of time by using this model, assuming that the diffusion constant at $T = 2$ K, $D \approx 600$ cm^2/s, remained unchanged at higher phonon density.

Kopelevich et al. [77, 78] also calculated the long-range transport of excitons as measured in the experiments of Refs. 61–63 and 65. A typical numerical solution of Eq. (3.105) for the exciton flux as a function of time at a distance 3.56 mm from the excitation surface is presented in Fig. 3.6. The parameters chosen for Cu_2O were the same as those in Ref. 76: $\tau_{pl} = 6$ ns, D $= 600$ cm^2/s, $m_{ex} = 2.7m_0$, $\hbar\omega = 2.41$ eV, $k = 1.6\times10^4$ cm^{-1}, $s = 4.5\times10^5$ cm/s, and varying dimensionless excitation intensity I/I_{pw}. The main adjustable parameter for the theory, I_{pw}, can be expressed in terms of other parameters, such as the momentum relaxation time of a e–h pair τ_{rel} and the exciton–phonon cross section σ_{ph}, as

$$I_{pw} = \frac{m_{ex} s\,\hbar\omega}{\tau_{rel}\tau_{pl} k \sigma_{ph} \varepsilon_{pw}}.$$

One can see from Fig. 3.6 that instead of the diffusive signal that appears at lower intensities, at higher intensities a ballistic-exciton signal appears, which arrives with the sound

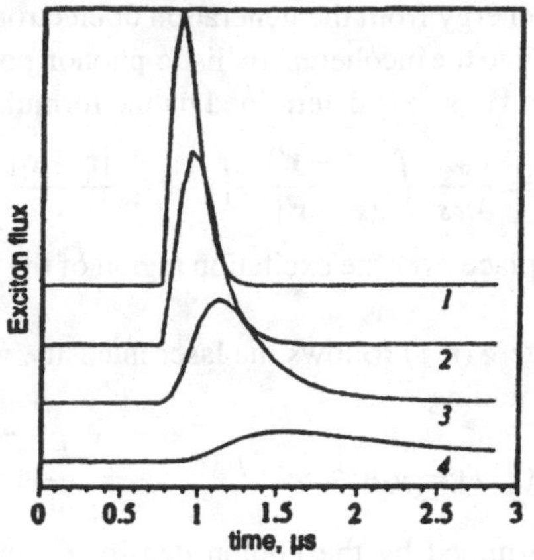

Figure 3.6. Time dependence of the exciton flux at 3.56 mm from the excitation surface, calculated for four dimensionless excitation intensities and D = 600 cm²/s. Labeled curves correspond to 1, $I/I_{pw} = 10^{4.25}$; 2, $I/I_{pw} = 10^{4.00}$; 3, $I/I_{pw} = 10^{3.75}$; 4, $I/I_{pw} = 10^{3.50}$ (from Ref. 78).

velocity, in agreement with the experimental data [61]. The parameter I_{pw} was determined by the fit to be 20 W/cm^{-2} at $T = 1.85$ K.

Since the qualitative agreement of this model is fairly good, one can ask whether all the data can be explained in this model, without invoking a superfluid state. There are several experimental results [61–63, 65] that this model must attempt to explain:

- The ballistic signal due to pulsed surface excitation increases dramatically over a very small temperature range, which is near the expected Bose condensation critical temperature.
- The ballistic signal increases dramatically for laser intensity above a certain threshold. This threshold density depends on the lattice temperature, consistent with the way the critical density for Bose condensation depends on temperature.
- When hot excitons are created by a cw laser in the bulk of the crystal, the ballistic signal due to pulsed surface excitation is decreased, but if the excitons created by the cw laser are cold (i.e., if the cw laser wavelength is tuned near the exciton ground-state resonance) then the ballistic signal is increased. In the superfluid picture, this is because hot excitons in the bulk of the crystal tend to increase the temperature of the exciton gas and therefore put it above the critical temperature for condensation, while cold excitons increase the density without increasing the temperature and therefore increase the superfluid fraction.
- If the pulsed surface excitation is below a critical threshold, then the ballistic signal increases dramatically when the cold cw laser light exceeds a certain threshold. In the superfluid picture, this is because the extra excitons from the cw laser put the exciton density above the critical density for condensation.
- When a second probe laser pulse creates excitons below the critical density at the surface after the first intense laser pulse, the excitons from the probe pulse also move ballistically across the crystal. Excitons created by the weak probe pulse alone propagate diffusively, however.

In the phonon-wind picture, these results have the following explanations:

- To explain the sharp increase in signal as the temperature decreases, Tikhodeev and co-workers argued that the parameter I_{pw} depends on temperature approximately as T^4. This dependence may come about because of the temperature dependence of the e–h pair momentum relaxation time τ_{rel} and the exciton–phonon cross section σ_{ph}. Previous experimental work [9, 81] has shown that the exciton–phonon relaxation time of excitons in Cu_2O increases rapidly with decreasing temperature, with a power law of $\sim T^{-3/2}$. Less is known about the temperature dependence of the relaxation of excited free-e–h pairs.
- An intensity threshold is expected for the phonon-wind effect, since the crossover from diffusive- to ballistic-phonon propagation is a highly nonlinear process.
- Hot excitons created by a cw laser in the bulk of the crystal are expected to increase the number of phonons in the medium and therefore to damp out the ballistic motion. Only phonons created at the surface of the crystal lead to phonon wind, since they have a preferred direction of propagation (away from the surface.) Cold excitons, on the other hand, can be picked up by the phonon wind and added to the ballistic pulse, since they experience the same drift force.
- Kopelevich et al. [78] argued that, in some cases, the excitons created at the surface may actually be outpaced by the phonons they have emitted. The surface-created phonons may then propagate diffusively, while excitons created in the middle of the crystal may be picked up by the phonon wind and travel ballistically.
- The fact that excitons from a second weak pulse are also accelerated can be understood in terms of the phonon "hot spot." It is well known in solid-state physics [82–84] that nonequilibrium phonons can remain in a laser-excitation region for long times, up to microseconds, because of the frequency-dependent phonon–phonon-scattering mechanism. Long-lived phonons generated by the hot spot on the surface of Cu_2O crystal have been discussed by Akimov, Kaplyanskii, and Moskalenko [85]. In the case of a hot spot, the phonon wind can continue to blow for long periods of time, accelerating excitons from the probe pulse as well as from the pump pulse.

Tikhodeev [79] also pointed out an apparent discrepancy in the density calibration of Ref. 65. In that work, the excitons in the middle of the crystal were in some cases able to amplify the incoming ballistic pulse by a factor of 10 or greater. This implies that the number of excitons created by the cw laser was at least ten times greater than the total number of excitons in the condensate at the surface. Given the laser powers actually used, the only way for this to occur would be if the lifetime of the excitons exceeded several milliseconds, i.e., $\sim$100 times the longest recorded lifetime [86]. Recent calculations by O'Hara [87] show that the intrinsic radiative lifetime of paraexcitons in Cu_2O is approximately 8 ms. One expects that surface and impurity recombination effects will give substantially shorter lifetimes, however, even in pure crystals.

The phonon-wind model of Tikhodeev and co-workers contains many simplifications. For example, all the phonons hitting the surface are assumed to be reflected back into the crystal. In fact, typically 50% of phonons hitting the surface are lost by coupling to the liquid-He bath in which the crystal is immersed, depending on the excitation conditions. The model also assumes a diffusion constant of 600 cm^2/s, which is the measured diffusion constant for paraexcitons in Cu_2O at 2 K in the low-density limit. If the creation region at the surface of the crystal is full of hot phonons, however, then one expects that the local lattice temperature is much greater than 2 K, possibly as high as 20 K [88], with consequently a much lower diffusion constant due to exciton–phonon scattering. Tikhodeev and co-workers did not rule out the possibility that for a few nanoseconds after the laser pulse, the exciton gas is truly superfluid, leading to anomalously high diffusion at early times, followed by drift due to the phonon wind at late times. Finally, the model of Tikhodeev and co-workers assumed a single, constant lifetime of 6 ns for exciton recombination. It is well established, however, that the density-dependent Auger recombination process dominates the exciton recombination in Cu_2O at almost all densities, varying from 100 ps at the highest densities [89] to as long as 10 μs at low densities [86].

Despite these simplifications, the qualitative agreement of the model with the data and the issue of the density calibration lead one to examine the role of phonon wind seriously. No single result of the transport of excitons in Cu_2O so far seems to demand the explanation of superfluidity.

We note that the two phenomena can coexist and are not mutually exclusive. Phonon wind may play an important role, even in the case of a superfluid excitonic condensate. For example, the T^4 onset of the ballistic motion discussed above, which is not fully understood in the context of the phonon-wind model, may come about because the adjustable parameter I_{pw} undergoes a considerable change near the critical temperature for the BEC phase transition.

Finally, we note that the recent experiments [90] in which resonant two-photon excitation is used to create a high-density exciton condensate at the surface of a Cu_2O crystal, as discussed in Subsection 2.1.5, may provide a new test for this phenomenon. Since the excitons are created in the ground state, there is no excess energy per exciton and therefore there should be no phonon wind. Measurements of the diffusion of the excitons from this region may provide important new information.

References

[1] A.G. Samoilovich and S.L. Korolyuk, *Fiz. Tverd. Tela* **1**, 1592 (1959).

[2] Y. Toyozawa, *Progr. Theor. Phys.* **20**, 53 (1958); **27**, 89 (1962); *J. Phys. Chem. Solids* **25**, 59 (1964).

[3] B.K. Ridley, *Quantum Processes in Semiconductors* (Oxford U. Press, Oxford, 1988).

[4] K. Seeger, *Semiconductor Physics*, 5th ed. (Springer, Berlin, 1990).

[5] S.A. Moskalenko, P.I. Khadzhi, A.I. Bobrysheva and A.V. Lelyakov, *Fiz. Tverd. Tela* **5**, 1444 (1963).

[6] A.I. Bobrysheva, *Phys. Status Solidi* **16**, 337 (1966).

[7] A.I. Bobrysheva, *Biexcitons in Semiconductors*, (Shtiintsa, Kishinev, 1979) (in Russian).

[8] Z.S. Kachlishvili, *Fiz. Tverd. Tela* **3**, 492 (1961).

[9] D.P. Trauernicht, J.P. Wolfe, and A. Mysyrowicz, *Phys. Rev. Lett.* **52**, 855 (1984).

[10] A.I. Anselm and Yu.A. Firsov, *Zh. Eksp. Teor. Fiz.* **28**, 151 (1955); **30**, 719 (1956).

[11] A. Mysyrowicz, D. Hulin, and A. Antonetti, *Phys. Rev. Lett.* **43**, 1123 (1979).

[12] L.V. Keldysh and A.N. Kozlov, *Zh. Eksp. Teor. Fiz. Pis'ma* **5**, 238 (1967); *Zh. Eksp. Teor. Fiz.* **54**, 978 (1968) [*Sov. Phys. JETP* **27**, 521 (1968)].

[13] V.M. Asnin, A.A. Rogachev, and S.M. Ryvkin, *Fiz. Tekhn. Poluprovodn.* **1**, 1742 (1967).

[14] V.M. Asnin and A.A. Rogachev, *Zh. Eksp. Teor. Fiz. Pis'ma.* **7**, 464 (1968); **9**, 415 (1969).

[15] S.T. Beliaev, *Zh. Eksp. Teor. Fiz.* **34**, 417, 433 (1958).

[16] V.I. Klyatskin, *Izv. Vuzov (Fiz.)*, **1**, 163 (1966).

[17] K.K. Singh and S. Kumar, *Phys. Rev.* **162**, 173 (1967).

[18] M.F. Miglei, S.A. Moskalenko, and A.V. Lelyakov, *Phys. Status Solidi* **35**, 389 (1969).

[19] A.I. Bobrysheva, M.F. Miglei, and M.I. Shmiglyuk, *Phys. Status Solidi B* **53**, 71 (1972).

[20] D.N. Zubarev, *Usp. Fiz. Nauk* **71**, 71 (1960).
[21] A.A. Abrikosov, L.P. Gor'kov, and I.E. Dzyaloshinskii, *Methods of Quantum Field Theory in Statistical Physics* (Dover, New York, 1962).
[22] W.P. Mazon, ed., *Physical Acoustics. Principles and Methods* Vol. 4 of Applications to Quantum and Solid State Physics Series, Part B (Academic, New York, 1968).
[23] E.F. Gross, *Usp. Fiz. Nauk* **63**, 575 (1957); **76**, 433 (1962).
[24] C.A. Hogarth, ed., *Materials Used in Semiconductor Devices* (Wiley and Sons, New York, 1965).
[25] Yu. A. Tserkovnikov, *Dokl. Akad. Nauk SSSR* **143**, 832 (1962).
[26] A.S. Davydov, *Theory of Light Absorption in Molecular Crystals* (Academy of Sciences of the Ukrainian SSR, Kiev, 1951) (in Russian).
[27] R.S. Knox, *Theory of Excitons* (Academic, New York, 1963).
[28] V.M. Agranovich and V.L. Ginzburg, *Crystal Optics with Spatial Dispersion and the Theory of Excitons* (Springer, Berlin, 1984).
[29] A.S. Davydov, *Theory of Molecular Excitons* (Plenum, New York, 1971).
[30] V.M. Agranovich, *Theory of Excitons* (Nauka, Moscow, 1968) (in Russian).
[31] H. Haken and S. Nikitine, eds., *Excitons at High Density* (Springer, Berlin, 1975).
[32] V.L. Broude, E.I. Rashba, and E.F. Sheka, *Spectroscopy of Molecular Excitons* (Energoizdat, Moscow, 1981) (in Russian).
[33] S.I. Pekar, *Crystal Optics and Additional Light Waves* (Naukova Dumka, Kiev, 1982) (in Russian).
[34] S.A. Moskalenko, *Introduction to the Theory of High Density Excitons* (Shtiintsa, Kishinev, 1983) (in Russian).
[35] E.I. Rashba and M.D. Sturge, eds., *Excitons* (North-Holland, Amsterdam, 1982).
[36] M.S. Brodin and I.V. Blonskii, *Exciton Processes in Layered Crystals* (Naukova Dumka, Kiev, 1986) (in Russian).
[37] S.A. Moskalenko, *Bose–Einstein Condensation of Excitons and Biexcitons* (Academy of Sciences of the Moldavian SSR, RIO, Kishinev, 1970) (in Russian).
[38] S.A. Moskalenko, A.I. Bobrysheva, A.V. Lelyakov, M.F. Miglei, P.I. Khadzhi, and M.I. Shmiglyuk, *Interaction of Excitons in Semiconductors* (Shtiintsa, Kishinev, 1974) (in Russian).
[39] A.S. Davydov and E.I. Rashba, *Ukr. Fiz. Zh.* **2**, 226 (1957).
[40] J.J. Hopfield, *J. Phys. Chem. Solids* **22**, 63 (1961).
[41] V.M. Agranovich and Yu.V. Konobeev, *Fiz. Tverd. Tela* **3**, 360 (1961).
[42] I.P. Dzyub, *Fiz. Tverd. Tela* **5**, 1577 (1963).
[43] A. Suna, *Phys. Rev. A* **135**, 111 (1964).
[44] A.S. Davydov and E.N. Myasnikov, *Dokl. Akad. Nauk SSSR* **171**, 1069 (1966); **20**, 143 (1967).
[45] S.A. Moskalenko, M.I. Shmiglyuk, and B.I. Chinik, *Fiz. Tverd. Tela* **10**, 351 (1968).
[46] V. Fedoseev and V. Hiznjakov, *Phys. Status Solidi* **27**, 751 (1968).
[47] V.G. Fedoseev, *Preprints of the Institute of Physics and Astronomy of the Academy of Sciences of the Estonian SSR* (Academy of Sciences, Tartu, Estonian SSR, 1970).
[48] A.F. Lubchenko, V.M. Nitsovich, and N.V. Tkach, *Phys. Status Solidi B* **65**, 609 (1974).
[49] A.V. Lelyakov and S.A. Moskalenko, *Fiz. Tverd. Tela* **11**, 3260 (1969) [*Sov. Phys. Solid State* **11**, 2642 (1969)].
[50] P.I. Khadzhi, *Kinetics of the Recombinational Radiation of Excitons and Biexcitons in Semiconductors* (Shtiintsa, Kishinev, 1977) (in Russian).
[51] V.L. Bonch-Bruevich and S.V. Tyablikov, *The Green's Function Method in Statistical Mechanics* (Fizmatgiz, Moscow, 1961) (in Russian).
[52] V.A. Gergel', R.F. Kazarinov, and R.A. Suris, *Zh. Eksp. Teor. Fiz.* **53**, 544 (1967); **54**, 298 (1968); *Proceedings of the Ninth International Conference on the Physics of Semiconductors, Moscow, 1968* (Nauka, Leningrad, 1969), Vol.1, p. 472.
[53] R.C. Casella, *J. Phys. Chem. Solids* **24**, 19 (1963); *J. Appl. Phys.* **34**, 1703 (1963); **36**, 2485 (1965).

[54] J.M. Blatt, K.W. Böer, and W. Brandt, *Phys. Rev.* **126**, 1691 (1962).
[55] S.A. Moskalenko, *Fiz. Tverd. Tela* **4**, 276 (1962).
[56] Yu.V. Petrov, *Zh. Eksp. Teor. Fiz.* **49**, 1923 (1965).
[57] D. Pines and P. Noziéres, *Theory of Quantum Liquids* (Addison-Wesley, New York, 1989), Vols. 1 and 2.
[58] A.L. Fetter and J.D. Walecka, *Quantum Theory of Many-Particle Systems* (McGraw-Hill, New York, 1971).
[59] D.W. Snoke, J.P. Wolfe, and A. Mysyrowicz, *Phys. Rev. B* **41**, 11171 (1990).
[60] A. Mysyrowicz, D.W. Snoke, and J.P. Wolfe, *Phys. Status Solidi B* **159**, 387 (1990).
[61] E. Fortin, S. Fafard, and A. Mysyrowicz, *Phys. Rev. Lett.* **70**, 3951 (1993).
[62] A. Mysyrowicz, E. Fortin, E. Benson, S. Fafard, and E. Hanamura, *Solid State Commun.* **92**, 957 (1994).
[63] E. Benson, E. Fortin, and A. Mysyrowicz, *Phys. Status Solidi B* **191**, 345 (1995).
[64] E. Fortin, E. Benson, and A. Mysyrowicz, in *Bose-Einstein Condensation*, A. Griffin, D.W. Snoke, and S. Stringari, eds. (Cambridge U. Press, Cambridge, 1995).
[65] A. Mysyrowicz, E. Benson, and E. Fortin, *Phys. Rev. Lett.* **77**, 896 (1996).
[66] E. Hanamura, *Solid State Commun.* **91**, 889 (1994).
[67] E.P. Gross, *Nuovo Cimento* **20**, 454 (1961).
[68] L.P. Pitaevskii, *Zh. Exp. Teor. Fiz.* **40**, 646 (1961).
[69] Z.M. Galasiewicz, *Helium 4*, (Pergamon, New York, 1971).
[70] I. Loutsenko and D. Roubtsov, *Phys. Rev. Lett.* **78**, 3011 (1997).
[71] A.S. Davydov and N.I. Kislukha, *Phys. Status Solidi B* **59**, 465 (1973).
[72] A.S. Davydov, *Solitons in Molecular Systems* (Reidel, New York, 1985), p. 319.
[73] A.S. Davydov, *Biology and Quantum Mechanics* (Pergamon, Oxford, 1982), p. 225.
[74] A.S. Davydov, in *New Horizons of Quantum Chemistry* (Reidel, New York, 1983), pp. 65–75.
[75] S.A. Moskalenko, P.I. Khadzhi, A.H. Rotaru, V.V. Baltaga, E.S. Kiselyova, S.S. Russu, and G.D. Shibarshina, in *The Living State II*, R.K. Mishra, ed. (World Scientific, Singapore, 1985), p. 340.
[76] A.E. Bulatov and S.G. Tikhodeev, *Phys. Rev. B* **46**, 15058 (1992).
[77] G.A. Kopelevich, N.A. Gippius, and S.G. Tikhodeev, in *Proceedings of the Twenty-Second International Conference on the Physics of Semiconductors*, D.J. Lockwood, ed. (World Scientific, Singapore, 1995), p. 61.
[78] G.A. Kopelevich, S.G. Tikhodeev, and N.A. Gippius, *Zh. Eksp. Teor. Fiz.* **109**, 2189 (1996). [*Sov. Phys. JETP* **82**, 1180 (1996)].
[79] S.G. Tikhodeev, *Phys. Rev. Lett.* **78**, 3225 (1997).
[80] L.V. Keldysh and N.N. Sibeldin, in *Nonequilibrium Phonons in Nonmetallic Crystals*, W. Eisenmenger and A.A. Kaplyanskii, eds. (North-Holland, Amsterdam, 1986).
[81] D.W. Snoke, D. Braun, and M. Cardona, *Phys. Rev. B* **44**, 2991 (1991).
[82] D.V. Kazakovtsev and Y.B. Levinson, Sov. Phys. JETP **61**, 1318 (1985); Y.B. Levinson in *Phonons 89*, S. Hunklinger, W. Ludwig, and G. Weiss, eds., (World Scientific, Singapore, 1990).
[83] V.I. Kozub, *Sov. Phys. JETP* **67**, 1191 (1988).
[84] J.A. Shields, M.E. Msall, M.S. Carroll, and J.P. Wolfe, *Phys. Rev. B* **47**, 12510 (1993).
[85] A.V. Akimov, A.A. Kaplyanskii and E.S. Moskalenko, *Fiz. Tverd. Tela* **29**, 509 (1987).
[86] D. Hulin, A. Mysyrowicz, and C. Benoit a la Guillaume, *Phys. Rev. Lett.* **45**, 1970 (1980).
[87] K. O'Hara, Ph.D. dissertation (University of Illinois, Urbana-Champaign, 1999).
[88] D.W. Snoke, J.P. Wolfe, and A. Mysyrowicz, *Phys. Rev. B* **41**, 11171 (1990).
[89] D.W. Snoke and J.P. Wolfe, *Phys. Rev. B* **42**, 7876 (1990).
[90] T. Goto, M.Y. Shen, S. Koyama, and T. Yokouchi, *Phys. Rev. B* **55**, 7609 (1997).

4

Bose–Einstein Condensation of Biexcitons

4.1 Biexcitons and Exciton–Exciton Interactions in Semiconductors

The concept of the excitonic molecule was introduced independently by Moskalenko [1] and Lampert [2]; later, the excitonic molecule was called the biexciton [3]. The biexciton represents the bound state of four Fermi quasiparticles, namely two electrons and two holes. More simply, it can be regarded as a bound state of two excitons.

Besides biexcitons, one can in general talk about "trions" and "quaternions," which are bound states of any combination of three or four electrons and holes, respectively. A trion always carries charge; a quaternion may or may not have charge, biexcitons being one kind of quaternion. Several different types of trions and quaternions are expected to be stable in semiconductors in both bulk and two-dimensional (2D) structures; later in this chapter we discuss a proposal [4] for superconductivity based on charged quaternionic bound states of electrons and holes.

Since biexcitons are also integer-spin bosons, they are expected to obey the same Bose statistics as excitons, including Bose narrowing of the energy distribution, discussed in Chapter 1. This has been seen in the semiconductor CuCl. We review experiments on Bose effects of biexcitons in CuCl at the end of this chapter.

We have already briefly discussed the exciton–exciton interaction in Subsection 2.1.3 in terms of the estimate of the s-wave-scattering length that goes into models of the weakly interacting Bose gas. As we mentioned there, this interaction is essentially a van der Waals force, but with the additional complications of comparable electron and hole masses and electron–hole exchange. In this section, we treat this interaction in greater detail, since it determines whether excitonic molecules and/or electron–hole liquids (EHLs) can form.

4.1.1 Electron–Hole Structure of the Excitonic Molecule

The excitonic molecule is similar to the molecule of hydrogen H_2 when the ratio σ of the electron and the hole masses ($\sigma = m_e/m_h$) tends to zero, and it resembles the molecule of positronium predicted by Wheeler [5, 6] when σ tends to unity. The biexciton is a neutral-compound quasiparticle that is also effectively a boson. It has an integer spin when the spin-orbit coupling can be neglected. Many efforts have been initiated to calculate the biexciton ground-state energy. In the usual notation, it is measured relative to the minimum energy of two free excitons and is denoted as $W(\sigma)$. Its absolute value determines the biexciton dissociation energy. The wave function of the biexciton ground state must be antisymmetric regarding the transpositions of the spin projections of two electrons as well as of two holes separately, when the electron–hole exchange Coulomb interaction is negligibly small compared with the biexciton binding energy. The different spin structures of biexcitons are discussed in Subsection 4.1.2.

A trial wave function for the biexcitons can be chosen that depends on the six distances between the four constituent quasiparticles and that takes into account the different electron–hole configurations. The most probable configuration for the biexciton structure is made of two virtual interacting excitons. There are other possible stable configurations, however, such as a virtual negative trion (*eeh*) interacting with the second hole and a positive trion (*hhe*) interacting with the second electron. As mentioned in Ref. 7, the contribution of the latter configurations in the structure of the H_2 molecule does not exceed 5% of the main atomic configuration.

The best results concerning the biexciton ground-state energy in bulk crystals with a simple band structure were obtained by Akimoto and Hanamura [8, 11], Brinkman, Rice, and Bell [12], and Huang [9]. The results of Refs. 8, 11, and 12, which are the same, are represented by curve 1 in Fig. 4.1(a). Here the ground-state energy $W(\sigma)$ is given as a fraction of the bulk-exciton Rydberg constant Ry_{ex} and is plotted as a function of σ. Curve 2 represents the results of Ref. 9. Both functions $W(\sigma)$ undergo the strong mathematical constraints that result from the symmetry of the electron–hole Hamiltonian with respect to the substitution of the parameter σ by $1/\sigma$. Wehner [13] and Adamowski, Bednarek, and

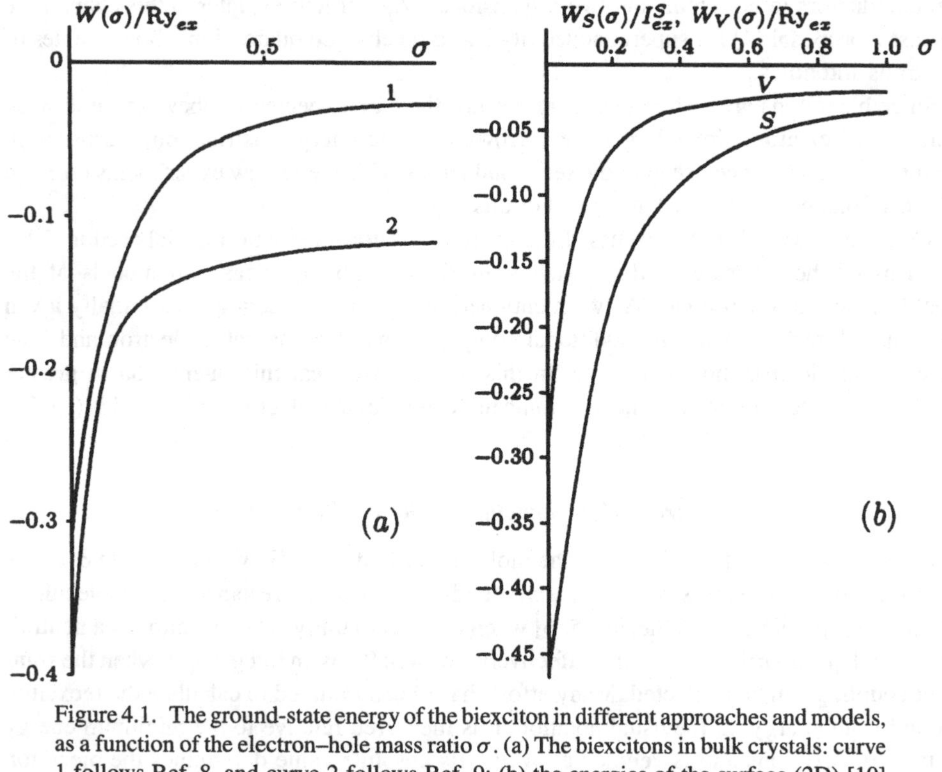

Figure 4.1. The ground-state energy of the biexciton in different approaches and models, as a function of the electron–hole mass ratio σ. (a) The biexcitons in bulk crystals: curve 1 follows Ref. 8, and curve 2 follows Ref. 9; (b) the energies of the surface (2D) [10] and bulk three-dimensional [8] biexcitons.

Suffczynski [14] deduced two differential relations,

$$\left.\frac{\partial W(\sigma)}{\partial \sigma}\right|_{\sigma=1} = 0, \qquad \frac{\partial^2 W(\sigma)}{\partial^2 \sigma} \le -\frac{2}{(1+\sigma)}\frac{\partial W(\sigma)}{\partial \sigma}. \tag{4.1}$$

One can conclude that $W(\sigma)$ is a monotonically increasing function of σ, attaining its maximum value at the point $\sigma = 1$. Both curves 1 and 2 in Fig. 4.1 obey these requirements. The biexciton is stable and can exist at all values of σ in the interval (0, 1). This work also helped determine the results for a molecule of positronium more precisely.

The ground-state energy of a biexciton at the surface of the crystal, as well as in 2D systems with simple band structures was obtained by Bobrysheva et al. [10] and by Kleinmann [15]. The surface, 2D biexciton also exists at all values of σ and has a binding energy larger than that of the bulk biexciton. In Fig. 4.1(b) the dependences $W_s(\sigma)$ and $W_v(\sigma)$ of the surface and the bulk biexcitons, respectively, are given in terms of the ionization energies of the corresponding excitons I_{ex}^S and Ry_{ex}.

In crystals with a complicated band structure, such as CuBr, the ground-state level is multiply degenerate or is split by the crystal field into several closely situated levels. In addition to this structure, excited levels can also exist because of the vibrational and the rotational motions inside the exciton molecule. These levels are characterized by the quantum numbers n and J, which can take the values $0, 1, 2, \ldots$. In bulk crystals with a simple band structure, the discrete levels of the finite motion, including the ground-state level with $n = J = 0$, can be approximated by the formula [16]

$$E_{nJ}(\sigma) = E_{n0}(\sigma) + K_J^R(\sigma), \tag{4.2}$$

where $E_{n0}(\sigma)$ and $K_J^R(\sigma)$ determine the vibrational levels and the quanta of rotational motion, respectively. They have the forms

$$\frac{E_{n0}(\sigma)}{\mathrm{Ry}_{ex}} = -0.35\frac{(1+\sigma^2)^2}{(1+\sigma)^4}\left[1 - 2.46\frac{\sqrt{\sigma(1+\sigma^2)}}{(1+\sigma)^2}\left(n+\frac{1}{2}\right)\right]^2,$$

$$\frac{K_J^R(\sigma)}{\mathrm{Ry}_{ex}} = \frac{\sigma(1+\sigma^2)^2}{(1+\sigma)^6}J(J+1)\left[1 - 1.37\frac{\sigma}{(1+\sigma^2)}J(J+1)\right], \tag{4.3}$$

which are valid only in a region of σ where the expressions in the square brackets are positive at the given values of the quantum numbers n and J. The full energy levels $E_{nJ}(\sigma)$ must be negative. The ground-state level $E_{00}(\sigma)$ coincides with the function $W(\sigma)$ calculated by Akimoto and Hanamura [8, 11]. The discrete biexciton levels obtained in a simple model for the parameters close to the crystals CuCl, CuBr and CdS are represented in Fig. 4.2 [17]. The excited states of the surface and 2D biexcitons are described by formulas similar to Eqs. (4.2) and (4.3) proposed in Ref. 18.

One exciton in the biexciton structure can annihilate radiatively, giving rise to a photon and a remaining exciton. This possibility, suggested by Bobrysheva [19, 20], follows the reaction

$$\text{biexciton}\,(\mathbf{k}) = \text{exciton}\,(\mathbf{k}') + \text{photon}\,(\mathbf{q}).$$

When the photon wave vector $\mathbf{q}$ can be neglected compared with $\mathbf{k}$ and $\mathbf{k}'$, the quantum transitions can be considered as vertical ($\mathbf{k} \cong \mathbf{k}'$). This new luminescence line is called the molecular or M line.

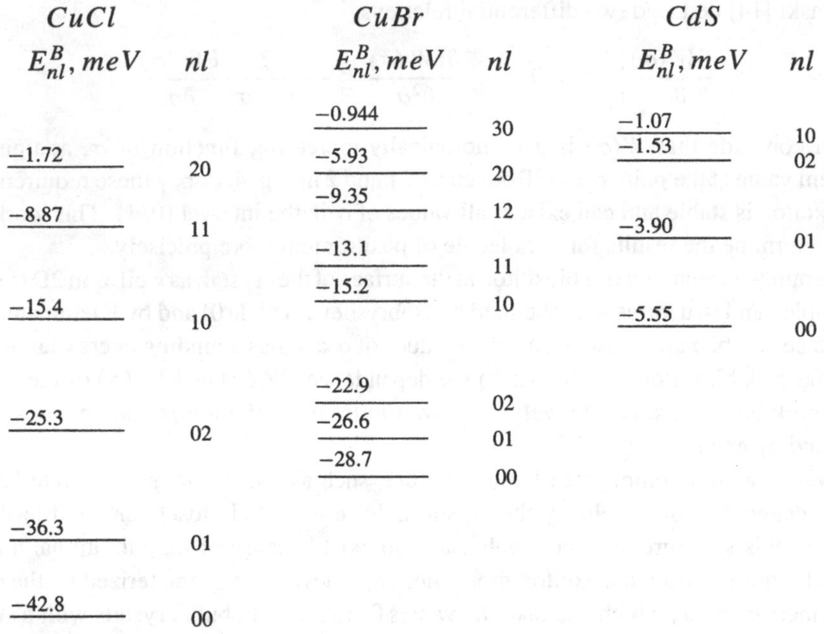

Figure 4.2. The discrete energy levels of the biexcitons in a three-dimensional model with a simple band structure, but with the parameters close to those of the crystals CuCl, CuBr and CdS. For CuCl the values $\sigma = 0.02$ and $\mathrm{Ry}_{\mathrm{ex}} = 190$ meV were used, for CuBr $\sigma = 0.01$ and $\mathrm{Ry}_{\mathrm{ex}} = 110$ meV were used; and for the CdS crystal the values $\sigma = 0.034$ and $\mathrm{Ry}_{\mathrm{ex}} = 30$ meV were used (from Ref. [17]).

As established in a series of papers, the probabilities of quantum transitions with the participation of biexcitons are characterized by anomalously large values. Gogolin and Rashba [21] showed that the absorption processes accompanied by the conversion of an exciton into a biexciton has a gigantic oscillator strength. Later, Hanamura [22] established a similar anomaly in the case of the two-photon transition from the ground state of the crystal to the ground state of the biexciton. Khadzhi and Petrashku [23] noted that the photodissociation process of the biexciton into two free excitons has the same property as the exciton–biexciton conversion. The origin of the gigantic oscillator strength of the exciton–biexciton conversion is related to the existence of the many vertical band-to-band transitions allowed by the momentum conservation law. This case differs essentially from the exciton single-photon quantum transition discussed in Chapter 1, which can occur for only one exciton state, whose wave vector is equal to that of the photon.

Following Gogolin and Rashba, we find that the relation between the two oscillator strengths f_{conv} and f_{ex} is

$$\frac{f_{\mathrm{conv}}}{n_{\mathrm{ex}}} = 2^7\pi\left(\frac{a_m^3}{v_0}\right)\frac{f_{\mathrm{ex}}}{n_a} \approx \frac{f_{\mathrm{ex}}}{n_a}(10^5\text{–}10^6). \tag{4.4}$$

Here n_{ex} and n_a are the densities of the excitons and the proper atoms of the lattice, respectively, a_m is the biexciton radius, and v_0 is the volume of the elementary cell. This

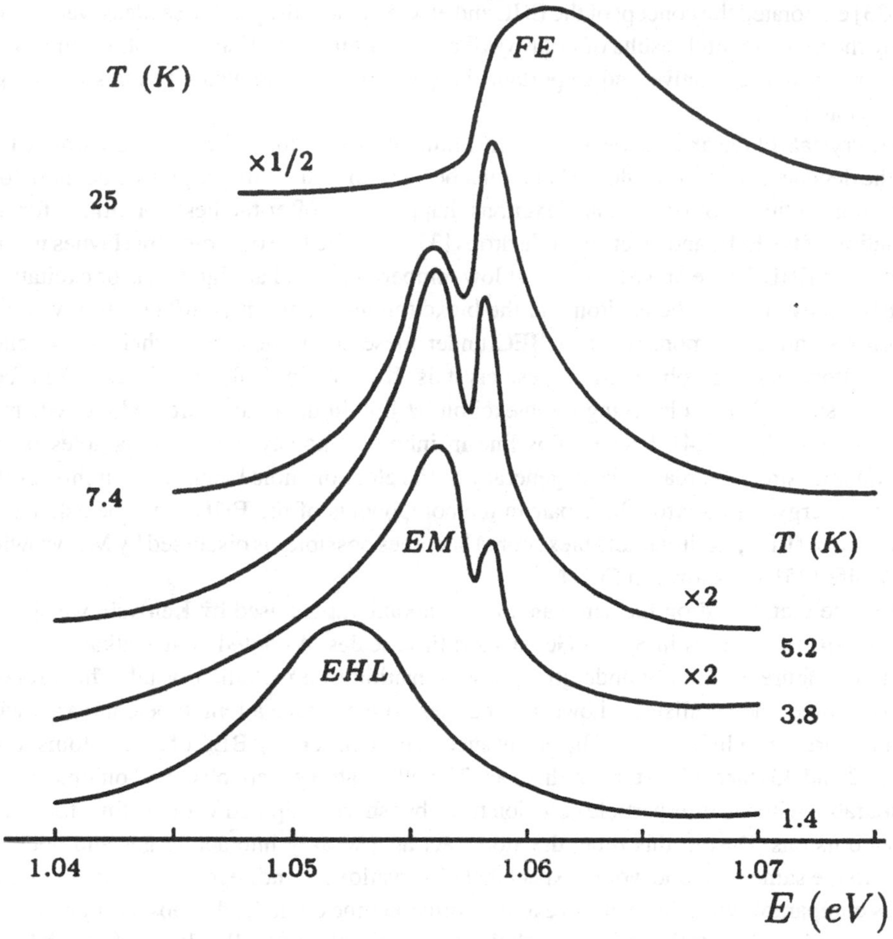

Figure 4.3. The luminescence spectra of the electron–hole liquid and exciton and biexciton gases in stressed Si (from Refs. 24 and 25).

implies that if the density n_{ex} could be the same as n_a, then the oscillator strength f_{conv} would be larger than f_{ex} by the factor (10^5–10^6).

Biexcitons were discovered in the appearance of the new luminescence lines at low temperatures and high levels of excitation (see Fig. 4.3). In a series of papers by Mysyrowicz et al. [26], Nikitine, Mysyrowicz and Grun [27], and Goto, Souma and Ueta [28, 29], the new luminescence lines were reported in CuCl and CuBr crystals. Earlier, Haynes [30] detected a new luminescence line in Si and attributed it to excitonic molecule recombination, but later it was understood that this line has another explanation related to a phase transition of high-density excitons unexpected at that time. Keldysh noted the similarity that exists between the alkali atoms and excitons with large radii and small binding energies of the excitonic molecules. He concluded that just as the alkali atoms at high density and low temperatures undergo metallization, so large-radius excitons can form a metallic-type phase rather different from that of liquid H_2 or liquid He. Starting from these assumptions, Keldysh

[31–33] elaborated the concept of the EHL and electron–hole drops. These ideas were borne out by the experimental results of Pokrovskii and Svistunova [34] and in a short time were transformed into an active and important direction of semiconductor physics (see, e.g., Refs. 35 and 36).

The crystals of Ge and Si, because of the multivalley structure of their conduction bands and the degeneracy of their valence bands and because they are indirect-gap semiconductors with long lifetimes for carriers and excitons, happened to offer the best conditions for the formation of the EHL and electron–hole drops [37, 38]. The line observed by Haynes is now seen as the EHL. In Ge and Si crystals at low temperatures and at high levels of excitation, the EHL prevails over the exciton and the biexciton gases, and it is difficult to reveal the biexcitons and, even more so, their BEC under these conditions. Nevertheless, evidence of biexcitons has been observed in these crystals. Kukushkin, Kulakovskii, and Timofeev [39–41] succeeded in observing the biexciton M line in uniaxially stressed Ge, whereas Gourley and Wolfe [24] detected this line in inhomogenously deformed samples of Si. The uniaxial stress decreases the degeneracy of the electron–hole bands and diminishes the binding energy per electron–hole pair in the components of the EHL. The coexistence of three phases (EHL, excitons, and biexcitons) becomes possible, as discussed by Mysyrowicz and Wolfe [25] and shown in Fig. 4.3.

We note that, based on the same analogy to alkaki vapors used by Keldysh, we should expect BEC of excitons in Si and Ge on short time scales. As noted above, alkali atoms at low temperature eventually undergo a phase transition to a metallic crystal. This process requires three-body collisions, however, and therefore can take a long time compared with the interparticle collision time. The recent successes in observing BEC of alkali atoms (e.g., Refs. 42 and 43) take advantage of this fact. The alkali atoms were observed on time scales comparable with the interparticle collision time, but short compared with the time for three-body collisions. During this time, the atoms act as a weakly interacting gas and undergo BEC. In the same way, one would expect that observations of indirect excitons in Si on time scales comparable with the exciton–exciton collision time could lead to observation of BEC of excitons in this material, even though the excitons will eventually all transform to EHL.

2D biexcitons have been observed experimentally by Miller et al. [44] as well as by Bar-Ad and Bar-Joseph [45] and Kim and Wolfe [46] and others. The excited states of biexcitons were used by Y. Nozue et al. [47, 48] to explain the results of the two-photon Raman scattering in CuBr crystals. Many aspects of the biexcitons are discussed in review articles [49–58], collections of papers [59–63], and monographs [17, 64–69]. New trends in the theory of biexcitons have taken into account the polariton effect, its influence on the radiative decay of biexcitons, and the electronic structure of biexcitons. The Coulomb matrix element of the process ex + ex $\rightarrow$ biex governs the biexciton formation rather than the optical matrix element ex + phot $\rightarrow$ biex discussed above. These investigations were initiated by Ivanov and Haug [70–72] and Ivanov et al. [73], supported by the experimental results of Hasuo, Kawano, and Nagasawa, [74, 75] and of Hasuo et al. [76] on biexcitons in CuCl. Many fascinating experiments have been done with biexcitons in CuCl, and some of these are reviewed at the end of this chapter.

As mentioned above, the ground state of the biexciton is stable when its wave function has a special spin structure of two electrons and of two holes. But the diversity of the wave functions of two excitons is not at all exhausted by this special case, and the theory of the exciton–exciton interaction does not include only biexciton formation. There are other possibilities, which are discussed in Subsection 4.1.2.

4.1.2 Exciton–Exciton Interactions

Čulik [77] investigated the wave functions of two excitons starting from the model of a crystal with a simple band structure that was degenerate in only the electron- and the hole-spin projections $\sigma = \pm 1/2$. The spin-orbit interaction was neglected. The wave functions can be constructed starting from the spin structure of two electrons and two holes separately or can be based on the spin structure of the electron–hole pairs or two excitons. Both of these approaches are equivalent. The latter approach is preferred in Refs. 78–81 and is used below.

Čulik [77] used the second quantization representation and expressed the one- and the two-exciton wave functions through the pairs and the tetrads of creation operators of electrons and holes acting on the vacuum state $|0\rangle$ as follows:

$$a^{\dagger}_{\mathbf{k}_1,\sigma_1} b^{\dagger}_{\mathbf{k}_2,\sigma_2} |0\rangle, \qquad a^{\dagger}_{\mathbf{k}_1,\sigma_1} b^{\dagger}_{\mathbf{k}_2,\sigma_2} a^{\dagger}_{\mathbf{k}_3,\sigma_3} b^{\dagger}_{\mathbf{k}_4,\sigma_4} |0\rangle. \tag{4.5}$$

The spin indices σ_1 and σ_2 of the electron–hole pair form four different combinations and give rise to four exciton functions. They are characterized by the irreducible representations Γ_1^+ and Γ_4^+ in a cubic crystal with a center of inversion. One of them, with the representation Γ_1^+, describes the state of the paraexciton with total spin equal to zero ($s = 0$). The other three wave functions belong to the threefold degenerate irreducible representation Γ_4^+, which describes the orthoexciton states with total spin $s = 1$.[a] The four wave functions can be written as [78, 79]

$$\begin{aligned}
|\psi_{\Gamma_1^+}\rangle &= \frac{1}{\sqrt{2}} \sum_{\mathbf{p}_1,\mathbf{p}_2} \varphi^{n,\mathbf{k}}_{\mathbf{p}_1,\mathbf{p}_2} \sum_{\sigma} a^{\dagger}_{\mathbf{p}_1,\sigma} b^{\dagger}_{\mathbf{p}_2,-\sigma} |0\rangle, \\
|\psi_{\Gamma_4^+,1}\rangle &= \sum_{\mathbf{p}_1,\mathbf{p}_2} \varphi^{n,\mathbf{k}}_{\mathbf{p}_1,\mathbf{p}_2} a^{\dagger}_{\mathbf{p}_1,1/2} b^{\dagger}_{\mathbf{p}_2,1/2} |0\rangle, \\
|\psi_{\Gamma_4^+,-1}\rangle &= \sum_{\mathbf{p}_1,\mathbf{p}_2} \varphi^{n,\mathbf{k}}_{\mathbf{p}_1,\mathbf{p}_2} a^{\dagger}_{\mathbf{p}_1,-1/2} b^{\dagger}_{\mathbf{p}_2,-1/2} |0\rangle, \\
|\psi_{\Gamma_4^+,0}\rangle &= \frac{1}{\sqrt{2}} \sum_{\mathbf{p}_1,\mathbf{p}_2} \varphi^{n,\mathbf{k}}_{\mathbf{p}_1,\mathbf{p}_2} \sum_{\sigma} (-1)^{\sigma-1/2} a^{\dagger}_{\mathbf{p}_1,\sigma} b^{\dagger}_{\mathbf{p}_2,-\sigma} |0\rangle.
\end{aligned} \tag{4.6}$$

The coefficients $\varphi^{n,\mathbf{k}}_{\mathbf{p}_1,\mathbf{p}_2}$ characterize the translational and the relative motions of the e–h pair forming the exciton, respectively, and depend on the corresponding quantum numbers n and $\mathbf{k}$. They have the form

$$\begin{aligned}
\varphi^{n,\mathbf{k}}_{\mathbf{p}_1,\mathbf{p}_2} &= \frac{1}{\sqrt{V}} \delta_{\mathrm{kr}}(\mathbf{p}_1 + \mathbf{p}_2, \mathbf{k}) \varphi_n(\beta \mathbf{p}_1 - \alpha \mathbf{p}_2), \\
&\frac{1}{V} \sum_{\mathbf{q}} |\varphi_n(\mathbf{q})|^2 = 1,
\end{aligned} \tag{4.7}$$

where $\delta_{\mathrm{kr}}(x, y)$ is the Kronecker delta function. The relations among the wave vectors $\mathbf{k}$, $\mathbf{q}$, $\mathbf{p}_1$, and $\mathbf{p}_2$ are

$$\mathbf{p}_1 = \alpha \mathbf{k} + \mathbf{q}, \quad \mathbf{p}_2 = \beta \mathbf{k} - \mathbf{q}, \quad \mathbf{q} = \beta \mathbf{p}_1 - \alpha \mathbf{p}_2, \quad \alpha = m_e/m_{\mathrm{ex}}, \quad \beta = m_h/m_{\mathrm{ex}}.$$

[a] We note that the orthoexciton and the paraexciton in the semiconductor Cu_2O do not have this simple symmetry. See Appendix A for a discussion of the exciton states in Cu_2O.

The 16 wave functions of two excitons are determined by the tetrads of e–h operators (4.5) and are classified according to the irreducible representations resulting from the direct product of the representations describing each of two excitons:

$$(\Gamma_1^+ \oplus \Gamma_4^+) \otimes (\Gamma_1^+ \oplus \Gamma_4^+) = 2\Gamma_1^+ \oplus \Gamma_3^+ \oplus 3\Gamma_4^+ \oplus \Gamma_5^+.$$

The two wave functions of two excitons that hold the most interest for us are those with the spin structure characterized by the irreducible representation Γ_1^+. One of them can be regarded as the product of the wave functions of two paraexcitons and the other one as the product of the wave functions of two orthoexcitons with total spin equal to zero. The structures of these functions are

$$\psi_A(\Gamma_1^+ \in \Gamma_1^+ \otimes \Gamma_1^+) = \psi_{\Gamma_1^+}\psi_{\Gamma_1^+},$$

$$\psi_B(\Gamma_1^+ \in \Gamma_4^+ \otimes \Gamma_4^+) = \frac{1}{\sqrt{3}}\big(\psi_{\Gamma_4^+,1}\psi_{\Gamma_4^+,-1} + \psi_{\Gamma_4^+,-1}\psi_{\Gamma_4^+,1} + \psi_{\Gamma_4^+,0}\psi_{\Gamma_4^+,0}\big). \quad (4.8)$$

In the tetrad representation, the functions have the forms

$$|\psi_A\rangle = \frac{1}{2}\sum_{\substack{\mathbf{p}_1,\mathbf{p}_2,\\ \mathbf{p}_3,\mathbf{p}_4}} F(\mathbf{p}_1,\mathbf{p}_2,\mathbf{p}_3,\mathbf{p}_4)\sum_{s,\sigma} a^\dagger_{\mathbf{p}_1,\sigma} b^\dagger_{\mathbf{p}_2,-\sigma} a^\dagger_{\mathbf{p}_3,s} b^\dagger_{\mathbf{p}_4,-s}|0\rangle,$$

$$|\psi_B\rangle = \frac{1}{2\sqrt{3}}\sum_{\substack{\mathbf{p}_1,\mathbf{p}_2,\\ \mathbf{p}_3,\mathbf{p}_4}} F(\mathbf{p}_1,\mathbf{p}_2,\mathbf{p}_3,\mathbf{p}_4)\Bigg[2\sum_{\sigma} a^\dagger_{\mathbf{p}_1,\sigma} b^\dagger_{\mathbf{p}_2,\sigma} a^\dagger_{\mathbf{p}_3,-\sigma} b^\dagger_{\mathbf{p}_4,-\sigma}$$

$$+\sum_{s,\sigma}(-1)^{\sigma+s} a^\dagger_{\mathbf{p}_1,\sigma} b^\dagger_{\mathbf{p}_2,-\sigma} a^\dagger_{\mathbf{p}_3,s} b^\dagger_{\mathbf{p}_4,-s}\Bigg]|0\rangle. \quad (4.9)$$

When the orthoexciton–paraexciton splitting Δ is much greater than the biexciton dissociation energy $|W|$ ($\Delta \gg |W|$), these functions are interesting in and of themselves [78, 79]. In the opposite limit ($\Delta \ll |W|$), they play an intermediary role. By using them as a basis, we can construct two new functions:

$$\psi_a = \frac{1}{2}(\psi_A - \sqrt{3}\psi_B), \qquad \psi_s = \frac{1}{2}(\psi_B + \sqrt{3}\psi_A).$$

The first function ψ_a is an antisymmetric wave function that takes the transpositions of the spins of two electrons and two holes separately. The second function ψ_s is symmetric regarding these transpositions. The new functions in the tetrad representation become

$$|\psi_a\rangle = \frac{1}{2C_a}\sum_{\substack{\mathbf{p}_1,\mathbf{p}_2,\\ \mathbf{p}_3,\mathbf{p}_4}} F(\mathbf{p}_1,\mathbf{p}_2,\mathbf{p}_3,\mathbf{p}_4)\sum_{s,\sigma}(-1)^{\sigma+s} a^\dagger_{\mathbf{p}_1,\sigma} b^\dagger_{\mathbf{p}_2,s} a^\dagger_{\mathbf{p}_3,-\sigma} b^\dagger_{\mathbf{p}_4,-s}|0\rangle,$$

$$|\psi_s\rangle = \frac{1}{2\sqrt{3}C_s}\sum_{\substack{\mathbf{p}_1,\mathbf{p}_2,\\ \mathbf{p}_3,\mathbf{p}_4}} F(\mathbf{p}_1,\mathbf{p}_2,\mathbf{p}_3,\mathbf{p}_4)\Bigg(\sum_{\sigma} a^\dagger_{\mathbf{p}_1,\sigma} b^\dagger_{\mathbf{p}_2,\sigma} a^\dagger_{\mathbf{p}_3,-\sigma} b^\dagger_{\mathbf{p}_4,-\sigma}$$

$$+\sum_{s,\sigma} a^\dagger_{\mathbf{p}_1,\sigma} b^\dagger_{\mathbf{p}_2,s} a^\dagger_{\mathbf{p}_3,-\sigma} b^\dagger_{\mathbf{p}_4,-s}\Bigg)|0\rangle. \quad (4.10)$$

The normalization constants C_a and C_s obey the equalities

$$\begin{vmatrix} C_a^2 \\ C_s^2 \end{vmatrix} = \sum_{\substack{\mathbf{p}_1,\mathbf{p}_2,\\ \mathbf{p}_3,\mathbf{p}_4}} F(\mathbf{p}_1,\mathbf{p}_2,\mathbf{p}_3,\mathbf{p}_4)\,[F(\mathbf{p}_1,\mathbf{p}_2,\mathbf{p}_3,\mathbf{p}_4) + F(\mathbf{p}_3,\mathbf{p}_4,\mathbf{p}_1,\mathbf{p}_2)$$
$$\pm F(\mathbf{p}_3,\mathbf{p}_2,\mathbf{p}_1,\mathbf{p}_4) \pm F(\mathbf{p}_1,\mathbf{p}_4,\mathbf{p}_3,\mathbf{p}_2)]\,, \quad (4.11)$$

where the plus sign + refers to the coefficient C_a^2 and the minus sign − refers to the coefficient C_s^2. The coefficients $F(\mathbf{p}_1,\mathbf{p}_2,\mathbf{p}_3,\mathbf{p}_4)$ are assumed to be real.

Looking at those expressions, one sees that the antisymmetric spin function is a linear combination of the coefficients $F(\mathbf{p}_1,\mathbf{p}_2,\mathbf{p}_3,\mathbf{p}_4)$ that is symmetric regarding the exchange of the wave vectors describing two electrons $\mathbf{p}_1$ and $\mathbf{p}_3$ and the wave vectors $\mathbf{p}_2$ and $\mathbf{p}_4$ describing two holes. By contrast, the symmetric spin function of two excitons is comprised of an antisymmetric linear combination of the coefficients $F(\mathbf{p}_1,\mathbf{p}_2,\mathbf{p}_3,\mathbf{p}_4)$ in regard to the same transpositions. The same properties can be displayed in the structure of the matrix elements of the Hamiltonian H_{int} describing the Coulomb interactions of the electrons and holes. Chiefly, we consider the diagonal matrix elements of the type $\langle\psi_i|\,H_{\text{int}}\,|\psi_i\rangle$, where ψ_i equals ψ_a, ψ_s, ψ_A, and ψ_B. These matrix elements contain the interactions between the electrons and the holes belonging to two different excitons as well as the interactions between the electrons and holes belonging to the same excitons. Excluding the intraexciton electron–hole Coulomb interactions and keeping only the interexciton electron–hole interactions, one can obtain expressions for the matrix elements $\langle\psi_i|\,H_{\text{int}}\,|\psi_i\rangle'$ describing the exciton–exciton interactions. These matrix elements depend on not only the spin structure of each of the two excitons taking part in the interaction, but also on the total spin of the two excitons. Two of these matrix elements can be represented in a single form as

$$\left\langle \begin{matrix}\psi_a\\ \psi_s\end{matrix} \middle| H_{\text{int}} \middle| \begin{matrix}\psi_a\\ \psi_s\end{matrix} \right\rangle' = \begin{vmatrix} C_a^{-2} \\ C_s^{-2} \end{vmatrix} \sum_{\substack{\mathbf{p}_1,\mathbf{p}_2,\\ \mathbf{p}_3,\mathbf{p}_4}} \sum_{\mathbf{p}} V_{\mathbf{p}} F(\mathbf{p}_1,\mathbf{p}_2,\mathbf{p}_3,\mathbf{p}_4)$$
$$\times \{[F(\mathbf{p}_1+\mathbf{p},\mathbf{p}_2,\mathbf{p}_3-\mathbf{p},\mathbf{p}_4) + F(\mathbf{p}_3-\mathbf{p},\mathbf{p}_4,\mathbf{p}_1+\mathbf{p},\mathbf{p}_2)$$
$$\pm F(\mathbf{p}_1+\mathbf{p},\mathbf{p}_4,\mathbf{p}_3-\mathbf{p},\mathbf{p}_2) \pm F(\mathbf{p}_3-\mathbf{p},\mathbf{p}_2,\mathbf{p}_1+\mathbf{p},\mathbf{p}_4)$$
$$+ F(\mathbf{p}_1,\mathbf{p}_2+\mathbf{p},\mathbf{p}_3,\mathbf{p}_4-\mathbf{p}) + F(\mathbf{p}_3,\mathbf{p}_4-\mathbf{p},\mathbf{p}_1,\mathbf{p}_2+\mathbf{p})$$
$$\pm F(\mathbf{p}_1,\mathbf{p}_4-\mathbf{p},\mathbf{p}_3,\mathbf{p}_2+\mathbf{p}) \pm F(\mathbf{p}_3,\mathbf{p}_2+\mathbf{p},\mathbf{p}_1,\mathbf{p}_4-\mathbf{p})]$$
$$- [F(\mathbf{p}_1+\mathbf{p},\mathbf{p}_2,\mathbf{p}_3,\mathbf{p}_4-\mathbf{p}) + F(\mathbf{p}_3,\mathbf{p}_4-\mathbf{p},\mathbf{p}_1+\mathbf{p},\mathbf{p}_2)$$
$$\pm F(\mathbf{p}_1+\mathbf{p},\mathbf{p}_4-\mathbf{p},\mathbf{p}_3,\mathbf{p}_2) \pm F(\mathbf{p}_3,\mathbf{p}_2,\mathbf{p}_1+\mathbf{p},\mathbf{p}_4-\mathbf{p})$$
$$+ F(\mathbf{p}_1,\mathbf{p}_2-\mathbf{p},\mathbf{p}_3+\mathbf{p},\mathbf{p}_4) + F(\mathbf{p}_3+\mathbf{p},\mathbf{p}_4,\mathbf{p}_1,\mathbf{p}_2-\mathbf{p})$$
$$\pm F(\mathbf{p}_1,\mathbf{p}_4,\mathbf{p}_3+\mathbf{p},\mathbf{p}_2-\mathbf{p}) \pm F(\mathbf{p}_3+\mathbf{p},\mathbf{p}_2-\mathbf{p},\mathbf{p}_1,\mathbf{p}_4)]\}\,. \quad (4.12)$$

These interaction terms are depicted in Fig. 4.4. The diagrams contain two solid and two dashed lines that represent the electrons and the holes, respectively. The wavy lines symbolize the Coulomb interactions. The intersections of two electron lines or two hole lines indicate the exchanges that accompany the Coulomb interactions of the same pair or of another pair of carriers. Initially it is assumed that the first two operators in tetrads (4.5) create the first *e–h* pair and the first exciton and the second pair of operators creates the second exciton. In accordance with this assumption, the first two variables of the coefficients

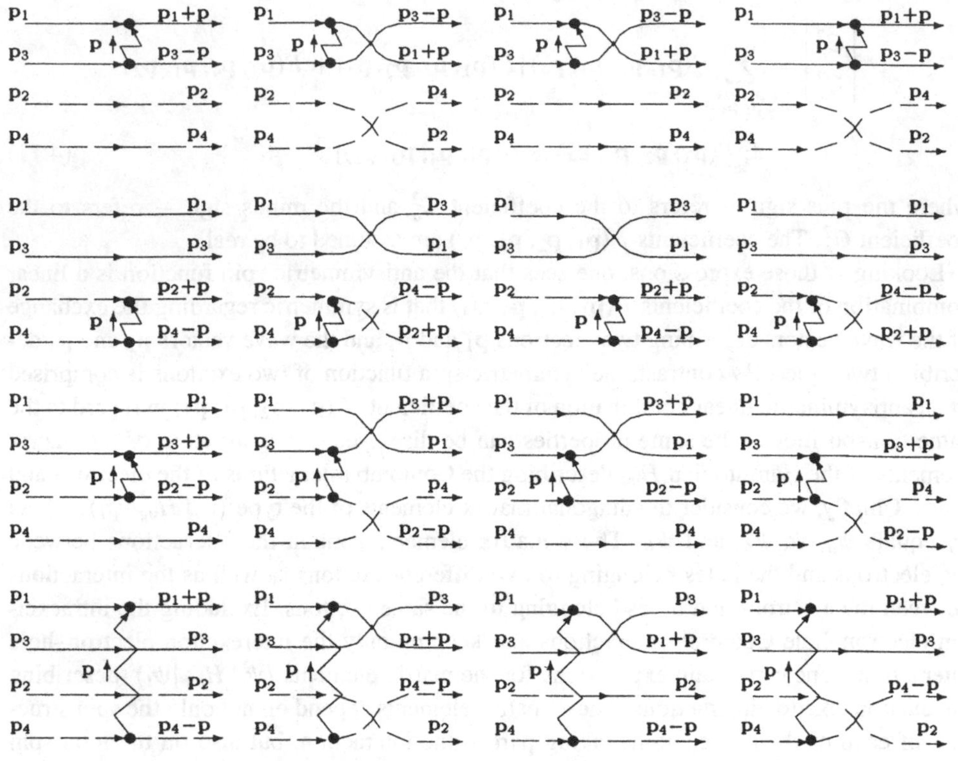

Figure 4.4. The diagrams representing the interactions between two electron–hole pairs that give rise to the exiton–exciton interactions.

$F(\mathbf{p}_1, \mathbf{p}_2, \mathbf{p}_3, \mathbf{p}_4)$ describe the first exciton and the last two variables describe the second exciton. Noting this allows us to select the Coulomb interaction of the electron and the hole belonging to the same exciton and to separate it from the interaction of the electron and the hole belonging to different excitons. The former types of terms are

$$\begin{aligned}
&-[F(\mathbf{p}_1+\mathbf{p}, \mathbf{p}_2-\mathbf{p}, \mathbf{p}_3, \mathbf{p}_4) + F(\mathbf{p}_3, \mathbf{p}_4, \mathbf{p}_1+\mathbf{p}, \mathbf{p}_2-\mathbf{p})\\
&+F(\mathbf{p}_3+\mathbf{p}, \mathbf{p}_4-\mathbf{p}, \mathbf{p}_1, \mathbf{p}_2) + F(\mathbf{p}_1, \mathbf{p}_2, \mathbf{p}_3+\mathbf{p}, \mathbf{p}_4-\mathbf{p})\\
&\pm F(\mathbf{p}_1+\mathbf{p}, \mathbf{p}_4, \mathbf{p}_3, \mathbf{p}_2-\mathbf{p}) \pm F(\mathbf{p}_3, \mathbf{p}_2-\mathbf{p}, \mathbf{p}_1+\mathbf{p}, \mathbf{p}_4)\\
&\pm F(\mathbf{p}_1, \mathbf{p}_4-\mathbf{p}, \mathbf{p}_3+\mathbf{p}, \mathbf{p}_2) \pm F(\mathbf{p}_3+\mathbf{p}, \mathbf{p}_2, \mathbf{p}_1, \mathbf{p}_4-\mathbf{p})]\,.
\end{aligned} \tag{4.13}$$

These terms have been extracted from the full matrix elements $\langle\psi_{a(s)}|H_{\text{int}}|\psi_{a(s)}\rangle$ that were used to obtain the final expressions for matrix elements (4.12) describing the net exciton–exciton interaction.

In an alternative method developed by Hanamura [82] in the theory of high-density excitons, the boson creation and annihilation operators of excitons are introduced and the fermion Hamiltonian of high-density electrons and holes is expanded in terms of these boson operators. An equivalent-diagram representation of the electrostatic interaction between excitons, as well as of the terms that come from the Pauli exclusion principle for the electrons and the holes comprising the excitons, was proposed in Ref. 82.

Further calculations in Refs. 78 and 79 were performed in the Heitler–London approximation with the factorized representation for the coefficients $F(\mathbf{p}_1, \mathbf{p}_2, \mathbf{p}_3, \mathbf{p}_4)$ in the form of a product of the wave functions of two excitons of the type of Eqs. (4.7):

$$F(\mathbf{p}_1, \mathbf{p}_2, \mathbf{p}_3, \mathbf{p}_4) \cong \varphi^{n_1,\mathbf{k}_1}_{\mathbf{p}_1,\mathbf{p}_2} \varphi^{n_2,\mathbf{k}_2}_{\mathbf{p}_3,\mathbf{p}_4}. \tag{4.14}$$

In what follows, it is assumed that the initial and the final states of two excitons are characterized by the wave vectors $\mathbf{k}_1, \mathbf{k}_2$ and $\mathbf{k}_1', \mathbf{k}_2'$, respectively, whereas the discrete quantum numbers n_i and n_i' correspond to the lowest exciton level $n = 1s$. In this case the two exciton wave functions are denoted as $|\psi_i\,(\mathbf{k}_1, \mathbf{k}_2)\rangle$. The matrix elements to be found have the form

$$\langle\psi_i(\mathbf{k}_1, \mathbf{k}_2)|H_{\text{int}}|\psi_i(\mathbf{k}_1', \mathbf{k}_2')\rangle' = \frac{1}{V}\delta_{\text{kr}}(\mathbf{k}_1 + \mathbf{k}_2, \mathbf{k}_1' + \mathbf{k}_2')U_{i,i}(\mathbf{k}_1 - \mathbf{k}_1', \mathbf{k}_2 - \mathbf{k}_1'), \tag{4.15}$$

which demonstrates the momentum conservation law and the noncentral character of the forces acting between two excitons.

Eight terms in the curly braces of Eq. (4.12) do not contain the alternative signs (±). Half of these represent the direct Coulomb interactions of two electrons and two holes belonging to different excitons. The second group contains the Coulomb interactions accompanied by double or bosonic-type exchanges. Because of the neutrality of the excitons, these eight terms nearly compensate one another for arbitrary values of the exciton wave vectors $\mathbf{k}_1, \mathbf{k}_2, \mathbf{k}_1'$, and $\mathbf{k}_2'$. Their total contribution is exactly zero in the random-phase approximation when one can set $\mathbf{k}_1 = \mathbf{k}_2 = \mathbf{k}_1' = \mathbf{k}_2'$. In this case, the resulting contributions to the interexciton interactions are determined exclusively by the Coulomb interaction terms accompanied by one exchange. These are characterized by the alternative signs (±) in formula (4.12). These terms describe the exchange Coulomb interaction of two electrons or two holes, the direct Coulomb interaction for the example of two electrons accompanied by the exchange of two holes and vice versa, and, finally, the direct Coulomb electron–hole interactions accompanied by the exchanges between two electrons or between two holes.

With these approximations, following Refs. 78 and 79, one can obtain

$$\begin{aligned} U_{a,a}(0) &= -\frac{26\pi}{3}\text{Ry}_{\text{ex}}a_{\text{ex}}^3, \\ U_{s,s}(0) &= \frac{26\pi}{3}\text{Ry}_{\text{ex}}a_{\text{ex}}^3. \end{aligned} \tag{4.16}$$

The nondiagonal matrix elements $U_{a,s}(0)$ and $U_{s,a}(0)$ are equal to zero because of the different spin structures of the functions $|\psi_a\rangle$ and $|\psi_s\rangle$. In another limit, $\Delta \gg |W|$, the wave functions of two excitons of the types $|\psi_A\rangle$ and $|\psi_B\rangle$ are available. In this limit the matrix elements were determined with the orthoexciton–paraexciton splitting Δ taken into account [78, 80, 81]. The following results were obtained:

$$\begin{aligned} U_{A,A}(0) &= U_{pp,pp}(0) = \pi a_{\text{ex}}^3\left(\frac{13}{3}\text{Ry}_{\text{ex}} - \frac{9}{4}\Delta\right), \\ U_{B,B}(0) &= U_{oo,oo}(0) = \pi a_{\text{ex}}^3\left(\frac{-13}{3}\text{Ry}_{\text{ex}} + \frac{3}{4}\Delta\right), \\ U_{op,op}(0) &= \pi a_{\text{ex}}^3\left(\frac{26}{3}\text{Ry}_{\text{ex}} - \frac{5}{2}\Delta\right). \end{aligned} \tag{4.17}$$

The last matrix element was calculated on the functions of two excitons, one of which was an orthoexciton and the other a paraexciton. These two functions have spin equal to unity ($s = 1$).

The results expressed by formulas (4.16) and (4.17) lead to the conclusion that two excitons can form bound states like molecules in both limiting cases $\Delta \ll |W|$ and $\Delta \gg |W|$. In the first limiting case, only the function $|\psi_a\rangle$, which is antisymmetric regarding the transpositions of the spin projections of two electrons and two holes separately, can ensure biexciton stability. In the opposite limit, the biexciton can be formed by any two orthoexcitons with total spin equal to zero [80].

Other than these two matrix elements, all the other matrix elements of Eqs. (4.16) and (4.17) are positive, yielding repulsion between the excitons. The different types of excitons interact among themselves with forces of different signs. These investigations showed that even in crystals with a simple band structure, there is a variety of different exciton–exciton interactions. This result allows us to consider the coherent pairing of excitons and to obtain a Bose condensate of biexcitons that is stable against collapse.

These results concerning bulk materials were enlarged by Bobrysheva, Zyukov, and Beryl [83], who investigated exciton–exciton interactions in 2D structures and on the surface of the crystals. Their results in this case were

$$\langle\psi_i|H_{\text{int}}|\psi_i\rangle' = \frac{1}{S}\delta_{\text{kr}}(\mathbf{k}_1 + \mathbf{k}_2, \mathbf{k}_1' + \mathbf{k}_2')U^s_{i,i}(\mathbf{k}_1 - \mathbf{k}_1', \mathbf{k}_2 - \mathbf{k}_1'). \tag{4.18}$$

Here, instead of the volume V of the bulk crystal, the surface area S of the 2D system appears. The matrix elements were expressed in terms of the 2D exciton ionization potential I^s_{ex} and the 2D exciton radius a^s_{ex},

$$I^s_{\text{ex}} = \frac{2\mu e^4}{\hbar^2\varepsilon^2_{\text{eff}}}, \qquad a^s_{\text{ex}} = \frac{\hbar^2\varepsilon_{\text{eff}}}{2\mu e^2},$$

which depend on the effective dielectric constant ε_{eff}. In the limit $\Delta \ll |W|$, the authors obtained

$$U^s_{a,a}(0) = -2\pi I^s_{\text{ex}}\left(a^s_{\text{ex}}\right)^2, \qquad U^s_{s,s}(0) = 2\pi I^s_{\text{ex}}\left(a^s_{\text{ex}}\right)^2. \tag{4.19}$$

In the opposite limit, $\Delta \gg |W|$, the results differ by one half for the first two values:

$$U^s_{pp,pp}(0) = \pi I^s_{\text{ex}}\left(a^s_{\text{ex}}\right)^2, \qquad U^s_{oo,oo}(0) = -\pi I^s_{\text{ex}}\left(a^s_{\text{ex}}\right)^2,$$
$$U^s_{op,op}(0) = 2\pi I^s_{\text{ex}}\left(a^s_{\text{ex}}\right)^2. \tag{4.20}$$

Similar results were obtained independently by Schmitt-Rink, Chemla, and Miller [84]. Having the matrix elements of the exciton–exciton interactions, one can obtain the density-dependent shift of the exciton level in the mean-field approximation. This question was investigated by Bobrysheva, Moskalenko, and Vybornov [85].

There are several similar methods that permit the transformation of the Hamiltonian of the fermion system to the boson representation. One of them is based on the Usui transformation [86], which was used by Hanamura [82] and Hiroshima [87]. To this end, in Ref. 85, the exciton creation operators $c^\dagger_{\mathbf{k},\sigma_1,\sigma_2}$, which have the form

$$c^\dagger_{\mathbf{k},\sigma_1,\sigma_2} = \frac{1}{\sqrt{V}}\sum_{\mathbf{q}}\varphi(\mathbf{q})a^\dagger_{\alpha\mathbf{k}+\mathbf{q},\sigma_1}b^\dagger_{\beta\mathbf{k}-\mathbf{q},\sigma_2}, \tag{4.21}$$

were substituted approximately by the pure boson operators $a^\dagger_{\mathbf{k},i}$ as follows:

$$a^\dagger_{\mathbf{k},1} = c^\dagger_{\mathbf{k},1/2,1/2}, \qquad a^\dagger_{\mathbf{k},3} = c^\dagger_{\mathbf{k},1/2,-1/2},$$
$$a^\dagger_{\mathbf{k},2} = c^\dagger_{\mathbf{k},-1/2,-1/2}, \qquad a^\dagger_{\mathbf{k},4} = c^\dagger_{\mathbf{k},-1/2,1/2}. \tag{4.22}$$

After this substitution, the internal structure of the exciton can be forgotten. The excitons are treated as structureless quasiparticles forming a weakly interacting Bose gas.

In the boson representation, the Hamiltonian describing the exciton–exciton interaction has the form

$$\begin{aligned}
H_{\text{int}} = \frac{1}{2V} \sum_{\substack{\mathbf{k}_1,\mathbf{k}_2,\\ \mathbf{k}_1',\mathbf{k}_2'}} \Bigg[& U_{ss}(\mathbf{k}_1, \mathbf{k}_2; \mathbf{k}_1', \mathbf{k}_2') \Bigg(\sum_{i=1}^{4} a^\dagger_{\mathbf{k}_1,i} a^\dagger_{\mathbf{k}_2,i} a_{\mathbf{k}_2',i} a_{\mathbf{k}_1',i} \\
& + 2 \sum_{i=3,4} \sum_{j=1,2} a^\dagger_{\mathbf{k}_1,i} a^\dagger_{\mathbf{k}_2,j} a_{\mathbf{k}_2',j} a_{\mathbf{k}_1',i} + a^\dagger_{\mathbf{k}_1,3} a^\dagger_{\mathbf{k}_2,4} a_{\mathbf{k}_2',4} a_{\mathbf{k}_1',3} \\
& + a^\dagger_{\mathbf{k}_1,1} a^\dagger_{\mathbf{k}_2,2} a_{\mathbf{k}_2',1} a_{\mathbf{k}_1',2} + a^\dagger_{\mathbf{k}_1,3} a^\dagger_{\mathbf{k}_2,4} a_{\mathbf{k}_2',1} a_{\mathbf{k}_1',2} + a^\dagger_{\mathbf{k}_1,1} a^\dagger_{\mathbf{k}_2,2} a_{\mathbf{k}_2',3} a_{\mathbf{k}_1',4} \Bigg) \\
& + U_{aa}(\mathbf{k}_1, \mathbf{k}_2; \mathbf{k}_1', \mathbf{k}_2') \big(a^\dagger_{\mathbf{k}_1,3} a^\dagger_{\mathbf{k}_2,4} a_{\mathbf{k}_2',4} a_{\mathbf{k}_1',3} + a^\dagger_{\mathbf{k}_1,1} a^\dagger_{\mathbf{k}_2,2} a_{\mathbf{k}_2',2} a_{\mathbf{k}_1',1} \\
& - a^\dagger_{\mathbf{k}_1,3} a^\dagger_{\mathbf{k}_2,4} a_{\mathbf{k}_2',2} a_{\mathbf{k}_1',1} - a^\dagger_{\mathbf{k}_1,2} a^\dagger_{\mathbf{k}_2,1} a_{\mathbf{k}_2',4} a_{\mathbf{k}_1',3} \big) \Bigg].
\end{aligned} \tag{4.23}$$

In the mean-field approximation, the tetradic Hamiltonian (4.23) is transformed into a quadratic one. The coefficients $U_{aa(ss)}(\mathbf{k}_1, \mathbf{k}_2; \mathbf{k}_1', \mathbf{k}_2')$ in the random-phase approximation are substituted by $U_{aa(ss)}(0)$. In these approximations expression (4.23) obtains the form

$$\sum_i \sum_{\mathbf{k}} U_{ss}(0) n_{\text{ex}} a^\dagger_{\mathbf{k},i} a_{\mathbf{k},i}, \tag{4.24}$$

where the exciton density n_{ex} consists of four partial densities n_i,

$$n_i = \frac{1}{V} \sum_{\mathbf{k}} \langle a^\dagger_{\mathbf{k},i} a_{\mathbf{k},i} \rangle, \qquad n_{\text{ex}} = \sum_i n_i,$$

which are assumed to be equal to one another.

The resulting shift of the exciton level is positive and equals

$$U_{ss}(0) n_{\text{ex}}. \tag{4.25}$$

The shifts of the orthoexciton and the paraexciton levels were obtained in the same way. They are equal and depend on the densities of the paraexcitons and the orthoexcitons n_p and n_{or} as follows:

$$U_{ss}(0)\,(n_p + n_{\text{or}}). \tag{4.26}$$

Although there are four types of excitons with either repulsive or attractive forces among them and the special case of attraction leads to biexciton formation, nevertheless the overall exciton–exciton interaction is repulsive. Therefore the exciton levels undergo a blue shift that depends on the exciton density. This fact plays an important role in the stability of the exciton–biexciton system.

The goal in this chapter is the study of the coherent macroscopic states of biexcitons. In the case of excitons, we introduced their coherent macroscopic state in Chapter 2 by means of two methods. One of them was the coherent pairing of electrons and holes, with the Keldysh–Kopaev–Kozlov method to take into account the electron–hole structure of the excitons. The other approach regards the excitons as structureless bosons and is based on the Bogoliubov displacement operation or a Glauber unitary transformation. In the case of biexcitons, one can propose three different methods that describe their coherent macroscopic states with various degrees of accuracy. One method is the coherent binding of four-particle electron–hole complexes or "quaternions," proposed by Moskalenko and developed by Moskalenko and Bobrysheva [88] and by Bobrysheva et al. [63]. This is simply the generalization of the Keldysh–Kopaev–Kozlov method for the case of more complicated, compound electron–hole quasiparticles. It permits one, in principle, to take into account in a complete manner the electron–hole structure of biexcitons. The second approach regards the excitons as structureless pure bosons and the biexciton coherent state is found as a coherent pairing of excitons. The third approximation considers the biexcitons as pure bosons, neglecting their internal structure. It is the roughest approximation, but it permits one to simplify many difficult questions.

4.2 Coherent Binding of Electron–Hole Quaternions

4.2.1 Formulation of the Method

As in Section 4.1 we consider a crystal with a simple band structure with only spin degeneracy. The biexciton creation operator is constructed with the structure of the function $|\Psi_a\rangle$ of Eqs. (4.10) taken into account. The tetrads of Fermi operators are chosen in such a way that the spins of two electron operators separately and of two hole operators individually are counterdirected. Only this spin configuration, with definite phase factors, will guarantee in the case $\Delta \ll |W|$ that the biexciton ground state is stable. The total spin of the biexciton in the ground state equals zero. When the wave vector $\mathbf{k}$ is used for the biexciton center of mass, the biexciton creation operator $C_{\mathbf{k}}^{\dagger}$ can be represented in the form [63, 88]

$$C_{\mathbf{k}}^{\dagger} = \frac{1}{2CV^{3/2}} \sum_{\substack{\mathbf{k}_1,\mathbf{k}_2,\\ \mathbf{k}_3,\mathbf{k}_4}} \sum_{s_1,s_2} (-1)^{s_1+s_2} F(\mathbf{k}_1,\mathbf{k}_2,\mathbf{k}_3,\mathbf{k}_4) a_{\mathbf{k}_1,s_1}^{\dagger} b_{\mathbf{k}_2,s_2}^{\dagger} a_{\mathbf{k}_3,-s_1}^{\dagger} b_{\mathbf{k}_4,-s_2}^{\dagger}, \tag{4.27}$$

where the four running variables $\mathbf{k}_i$ obey the momentum conservation law:

$$\sum_{i=1}^{4} \mathbf{k}_i = \mathbf{k}. \tag{4.28}$$

The spin indices take two values $s_1, s_2 = \pm 1/2$. The coefficients $F(\cdots \mathbf{k}_i \cdots)$ depend on the three independent variables $\mathbf{Q}$, $\mathbf{p}$, and $\mathbf{q}$ as follows:

$$F(\mathbf{k}_1,\mathbf{k}_2,\mathbf{k}_3,\mathbf{k}_4) = \delta_{\mathrm{kr}}\left(\sum_i \mathbf{k}_i, \mathbf{k}\right) F_{\mathbf{k}}(\mathbf{Q},\mathbf{p},\mathbf{q}). \tag{4.29}$$

Two different representations of the coefficients $F(\cdots \mathbf{k}_i \cdots)$ are used, depending on the object in view. The normalization constant C is determined by an equality of the type of

Eq. (4.11):

$$C^2 = \frac{1}{V^3} \sum_{\cdots \mathbf{k}_i \cdots} F(\mathbf{k}_1, \mathbf{k}_2, \mathbf{k}_3, \mathbf{k}_4) \times [F(\mathbf{k}_1, \mathbf{k}_2, \mathbf{k}_3, \mathbf{k}_4) + F(\mathbf{k}_3, \mathbf{k}_4, \mathbf{k}_1, \mathbf{k}_2) + F(\mathbf{k}_3, \mathbf{k}_2, \mathbf{k}_1, \mathbf{k}_4) + F(\mathbf{k}_1, \mathbf{k}_4, \mathbf{k}_3, \mathbf{k}_2)]. \tag{4.30}$$

The three last terms in the square brackets can be obtained from the first one by different permutations of the four variables. The coherent macroscopic state of the biexcitons with wave vector $\mathbf{k} = 0$ can be introduced into the electron–hole Hamiltonian equation (2.93) by means of the unitary transformation:

$$D(\sqrt{N_{b,0}}) = \exp X, \qquad X = \sqrt{N_{b,0}}(C_0^\dagger - C_0). \tag{4.31}$$

It is similar to the transformation of Eq. (2.98) introduced in Subsection 2.2.1 in the case of coherent excitons. $N_{b,0}$, as well as $N_{ex,0}$, is an extensive value proportional to the volume V of the system. The density of the coherent biexcitons is denoted by $n_{b,0}$, where

$$n_{b,0} = \frac{N_{b,0}}{V}. \tag{4.32}$$

In what follows below, the sum of the variables $\mathbf{k}_i$ in all the functions $F(\cdots \mathbf{k}_i \cdots)$ are set equal to zero ($\sum_i \mathbf{k}_i = 0$). The biexciton creation and annihilation operators obey the Bose commutation relation only approximately:

$$[C_{\mathbf{k}}, C_{\mathbf{k}'}^\dagger] = \delta_{\mathbf{k},\mathbf{k}'} + \text{terms of the type}$$
$$a^\dagger a + b^\dagger b + a^\dagger b^\dagger ab + a^\dagger a^\dagger aa + b^\dagger b^\dagger bb + a^\dagger a^\dagger b^\dagger aab + b^\dagger b^\dagger a^\dagger bba. \tag{4.33}$$

The deviation from the case of true bosons is much more significant than in the case of excitons, as given in Eqs. (1.15) or (2.91).

In the limit of low excitation density, the biexciton operators tend to become those of pure bosons. In this limit, the average values of the additional terms in Eq. (4.33) vanish, since in the vacuum state the occupation numbers of electrons and holes vanish. The increase of the electron and the hole densities leads at first to the disappearance of the biexcitons as bound compound quasiparticles. With a subsequent increase in the excitation level, the excitons begin to dissociate, too. This destruction of the compound quasiparticles at a high excitation level is reflected in the commutation relations of the corresponding operators.

The unitary transformation of the electron–hole Hamiltonian (2.93) by the operator $D(\sqrt{N_{b,0}})$ of Eq. (4.31) leads to the calculation of the expressions

$$Da_{\mathbf{p}\sigma}D^+ = a_{\mathbf{p}\sigma} + \frac{1}{1!}[X, a_{\mathbf{p}\sigma}] + \frac{1}{2!}[X, [X, a_{\mathbf{p}\sigma}]] + \cdots,$$
$$Db_{\mathbf{p}\sigma}D^+ = b_{\mathbf{p}\sigma} + \frac{1}{1!}[X, b_{\mathbf{p}\sigma}] + \frac{1}{2!}[X, [X, b_{\mathbf{p}\sigma}]] + \cdots. \tag{4.34}$$

In the case of coherent excitons, these calculations were performed exactly, leading to Bogoliubov u, v transformations (2.100). In the case of coherent biexcitons, it is possible to obtain only approximate results. There are several possible approximations. One of them is to confine the calculation to the first order of a perturbation theory, expanding the transformed Hamiltonian $\mathcal{H}$ in series on the square root of the biexciton density $\sqrt{n_{b,0}}$. Another possibility is to simplify the second-order commutators $[X, [X, a_{\mathbf{p}\sigma}]]$ and $[X, [X, b_{\mathbf{p}\sigma}]]$ by reducing them in the mean-field approximation to the operators $a_{\mathbf{p}\sigma}$ and $b_{\mathbf{p}\sigma}$, respectively. Both approximations are used below.

With the accuracy of third-order perturbation theory, the transformed Hamiltonian $\mathcal{H}$ has the form

$$\begin{aligned}\mathcal{H} &= D(\sqrt{N_{b,0}})HD^{+}(\sqrt{N_{b,0}}\,) \\ &\approx H + \frac{1}{1!}[X, H] + \frac{1}{2!}[X, [X, H]] + \frac{1}{3!}[X, [X, [X, H]]]. \end{aligned} \tag{4.35}$$

The first commutator $[X, H]$ gives rise to the expression

$$\begin{aligned}[X, H] = &-\frac{\sqrt{n_{b,0}}}{2CV} \sum_{\substack{\mathbf{k}_1,\mathbf{k}_2,\\ \mathbf{k}_3,\mathbf{k}_4}} \sum_{s_1,s_2} (-1)^{s_1+s_2} \\ &\times \{[\varepsilon_e(\mathbf{k}_1) + \varepsilon_e(\mathbf{k}_3) + \varepsilon_h(\mathbf{k}_2) + \varepsilon_h(\mathbf{k}_4) - 2\mu_e - 2\mu_h]F(\mathbf{k}_1, \mathbf{k}_2, \mathbf{k}_3, \mathbf{k}_4) \\ &+ \sum_{\mathbf{k}} V(\mathbf{k})[F(\mathbf{k}_1 - \mathbf{k}, \mathbf{k}_2, \mathbf{k}_3 + \mathbf{k}, \mathbf{k}_4) + F(\mathbf{k}_1, \mathbf{k}_2 - \mathbf{k}, \mathbf{k}_3, \mathbf{k}_4 + \mathbf{k}) \\ &- F(\mathbf{k}_1 - \mathbf{k}, \mathbf{k}_2 + \mathbf{k}, \mathbf{k}_3, \mathbf{k}_4) - F(\mathbf{k}_1, \mathbf{k}_2, \mathbf{k}_3 - \mathbf{k}, \mathbf{k}_4 + \mathbf{k}) \\ &- F(\mathbf{k}_1 - \mathbf{k}, \mathbf{k}_2, \mathbf{k}_3, \mathbf{k}_4 + \mathbf{k}) - F(\mathbf{k}_1, \mathbf{k}_2 + \mathbf{k}, \mathbf{k}_3 - \mathbf{k}, \mathbf{k}_4)]\} \\ &\times \left(a^{\dagger}_{\mathbf{k}_1,s_1} b^{\dagger}_{\mathbf{k}_2,s_2} a^{\dagger}_{\mathbf{k}_3,-s_1} b^{\dagger}_{\mathbf{k}_4,-s_2} + b_{\mathbf{k}_4,-s_2} a_{\mathbf{k}_3,-s_1} b_{\mathbf{k}_2,s_2} a_{\mathbf{k}_1,s_1}\right) \\ &- \frac{\sqrt{n_{b,0}}}{2CV} \sum_{\substack{\mathbf{k}_1,\mathbf{k}_2,\\ \mathbf{k}_3,\mathbf{k}_4}} \sum_{s_1,s_2} (-1)^{s_1+s_2} \\ &\times \sum_{\mathbf{k}} V(\mathbf{k})[F(\mathbf{k}_1 - \mathbf{k}, \mathbf{k}_2, \mathbf{k}_3, \mathbf{k}_4) - F(\mathbf{k}_1, \mathbf{k}_2 - \mathbf{k}, \mathbf{k}_3, \mathbf{k}_4) \\ &+ F(\mathbf{k}_1, \mathbf{k}_2, \mathbf{k}_3 - \mathbf{k}, \mathbf{k}_4) - F(\mathbf{k}_1, \mathbf{k}_2, \mathbf{k}_3, \mathbf{k}_4 - \mathbf{k})] \\ &\times \left(a^{\dagger}_{\mathbf{k}_1,s_1} b^{\dagger}_{\mathbf{k}_2,s_2} a^{\dagger}_{\mathbf{k}_3,-s_1} b^{\dagger}_{\mathbf{k}_4,-s_2} \rho_{\mathbf{k}} + \rho^{+}_{\mathbf{k}} b_{\mathbf{k}_4,-s_2} a_{\mathbf{k}_3,-s_1} b_{\mathbf{k}_2,s_2} a_{\mathbf{k}_1,s_1}\right). \end{aligned} \tag{4.36}$$

Here $\rho_{\mathbf{k}}$ is the Fourier transform of the electron–hole (e–h) spatial charge-density operator

$$\begin{aligned}\rho_{\mathbf{k}} &= \sum_{\mathbf{p},\sigma} a^{\dagger}_{\mathbf{p}\sigma} a_{\mathbf{p}+\mathbf{k},\sigma} - \sum_{\mathbf{p},\sigma} b^{\dagger}_{\mathbf{p}\sigma} b_{\mathbf{p}+\mathbf{k},\sigma} \\ \rho^{+}_{\mathbf{k}} &= \rho_{-\mathbf{k}}, \end{aligned} \tag{4.37}$$

which determines the spatial fluctuation of the e–h charge density.

Expression (4.36) consists of two parts. One of them contains the four-particle complexes, or quaternions, with creation or annihilation operators with total wave vector equal to zero, as follows from the coefficients $F(\cdots \mathbf{k}_i \cdots)$ accompanying them. This group of operators describes the creation from the vacuum and the return into it of the quaternions consisting of two electrons and two holes with total wave vector and total spin equal to zero. These terms are similar to the creation or the annihilation of e–h pairs in the theory of excitonic BEC elaborated by Keldysh and Kozlov [89], discussed in Chapter 2. In the Bogoliubov theory of superconductivity [90], the pairs of electrons with opposite wave vectors and spin projections take part in similar processes. The presence of these terms in the Hamiltonians for the coherent excitons and superconductors, which would lead to instabilities of the respective ground states, was removed by means the Bogoliubov principle of compensation of the dangerous diagrams [90]. Keldysh and Kozlov [89] introduced a similar requirement [Eq. (2.112)], which leads to the compensation of the matrix elements that describe these processes in all orders of the perturbation theory. The method proposed in this section, as noted above, is the generalization of the Keldysh–Kozlov method to the case of more complicated compound quasiparticles such as biexcitons. In this case, instability of the ground state can be removed in the same way, when the exact average value of the group of four operators is set to zero, with total wave vector equal to zero, i.e.,

$$\langle a^{\dagger}_{\mathbf{k}_1,s_1} b^{\dagger}_{\mathbf{k}_2,s_2} a^{\dagger}_{\mathbf{k}_3,-s_1} b^{\dagger}_{\mathbf{k}_4,-s_2} \rangle = 0, \qquad \sum_i \mathbf{k}_i = 0. \tag{4.38}$$

The average value must be calculated with the fully transformed Hamiltonian $\mathcal{H}$.

In the first order of the perturbation theory, condition (4.38) is equivalent to the requirement that the sum of the matrix elements enclosed in the curly braces of expression (4.36) must be set to zero. This approximation is discussed below. Before that, it is necessary to pay attention to the latter part of Eq. (4.36). It contains, for example, a group of four operators of the type $a^{\dagger}_{\mathbf{k}_1,s_1} b^{\dagger}_{\mathbf{k}_2,s_2} a^{\dagger}_{\mathbf{k}_3,-s_1} a^{\dagger}_{\mathbf{k}_4,-s_2}$ multiplied by the operator $\rho_{\mathbf{k}}$. The total wave vector of these four operators is nonzero because of the functions $F(\cdots \mathbf{k}_i \cdots)$ accompanying this group. These terms describe the transformations of e–h quaternions with a nonzero total wave vector into spatial fluctuations of the e–h charge density, or plasmons, with the same wave vector. These transformations are due to the Coulomb interaction and do not appear in the case of coherent excitons. These terms differ essentially from the ones that are present in the first part of expression (4.36) and do not directly influence the ground-state stability condition, if it is determined in the first order of the perturbation theory. This condition is

$$\begin{aligned}
&\left[\varepsilon_e(\mathbf{k}_1) + \varepsilon_e(\mathbf{k}_3) + \varepsilon_h(\mathbf{k}_2) + \varepsilon_h(\mathbf{k}_4) - 2\mu_e^0 - 2\mu_h^0\right] F^0(\mathbf{k}_1, \mathbf{k}_2, \mathbf{k}_3, \mathbf{k}_4) \\
&\quad + \sum_{\mathbf{k}} V(\mathbf{k}) [F^0(\mathbf{k}_1 - \mathbf{k}, \mathbf{k}_2, \mathbf{k}_3 + \mathbf{k}, \mathbf{k}_4) + F^0(\mathbf{k}_1, \mathbf{k}_2 - \mathbf{k}, \mathbf{k}_3, \mathbf{k}_4 + \mathbf{k}) \\
&\quad - F^0(\mathbf{k}_1 - \mathbf{k}, \mathbf{k}_2 + \mathbf{k}, \mathbf{k}_3, \mathbf{k}_4) - F^0(\mathbf{k}_1, \mathbf{k}_2, \mathbf{k}_3 - \mathbf{k}, \mathbf{k}_4 + \mathbf{k}) \\
&\quad - F^0(\mathbf{k}_1 - \mathbf{k}, \mathbf{k}_2, \mathbf{k}_3, \mathbf{k}_4 + \mathbf{k}) - F^0(\mathbf{k}_1, \mathbf{k}_2 + \mathbf{k}, \mathbf{k}_3 - \mathbf{k}, \mathbf{k}_4)] = 0.
\end{aligned} \tag{4.39}$$

It determines the total chemical potential $2\mu_e^0 + 2\mu_h^0$ in the zeroth-order approximation, which coincides with the energy of the biexciton level. The wave function $F^0(\cdots \mathbf{k}_i \cdots)$ and the biexciton energy in the zeroth-order approximation do not depend on the density-dependent corrections. Taking into account the Schrödinger equation in the momentum representation (4.39) and the normalization condition (4.30), one can determine the biexciton energy level $E_b^0(0)$ at the bottom of the biexciton band in the zeroth-order approximation:

$$\begin{aligned}E_b^0(0) &= 2\mu_e^0 + 2\mu_h^0\\ &= \frac{1}{C^2V^3}\sum_{\substack{\mathbf{k}_1,\mathbf{k}_2,\\ \mathbf{k}_3,\mathbf{k}_4}}\{[\varepsilon_e(\mathbf{k}_1)+\varepsilon_e(\mathbf{k}_3)+\varepsilon_h(\mathbf{k}_2)+\varepsilon_h(\mathbf{k}_4)]\,F^0(\mathbf{k}_1,\mathbf{k}_2,\mathbf{k}_3,\mathbf{k}_4)\\ &+\sum_{\mathbf{k}} V(\mathbf{k})\big[F^0(\mathbf{k}_1-\mathbf{k},\mathbf{k}_2,\mathbf{k}_3+\mathbf{k},\mathbf{k}_4)+F(\mathbf{k}_1,\mathbf{k}_2-\mathbf{k},\mathbf{k}_3,\mathbf{k}_4+\mathbf{k})\\ &-F^0(\mathbf{k}_1-\mathbf{k},\mathbf{k}_2+\mathbf{k},\mathbf{k}_3,\mathbf{k}_4)-F^0(\mathbf{k}_1,\mathbf{k}_2,\mathbf{k}_3-\mathbf{k},\mathbf{k}_4+\mathbf{k})\\ &-F^0(\mathbf{k}_1-\mathbf{k},\mathbf{k}_2,\mathbf{k}_3,\mathbf{k}_4+\mathbf{k})-F^0(\mathbf{k}_1,\mathbf{k}_2+\mathbf{k},\mathbf{k}_3-\mathbf{k},\mathbf{k}_4)\big]\}\\ &\times[F^0(\mathbf{k}_1,\mathbf{k}_2,\mathbf{k}_3,\mathbf{k}_4)+F^0(\mathbf{k}_3,\mathbf{k}_4,\mathbf{k}_1,\mathbf{k}_2)+F^0(\mathbf{k}_3,\mathbf{k}_2,\mathbf{k}_1,\mathbf{k}_4)\\ &+F^0(\mathbf{k}_1,\mathbf{k}_4,\mathbf{k}_3,\mathbf{k}_2)]. \end{aligned} \tag{4.40}$$

This expression contains the mean kinetic energy and the Coulomb interactions of four charged quasiparticles and coincides with results obtained earlier in the theory of biexcitons [11], which confirms the validity of the proposed method. This makes it possible to obtain the density-dependent corrections to the biexciton ground-state energy in the third order of the perturbation theory.

4.2.2 Coherent Pairing of Electrons and Holes with Opposite-Sign Trions

Now let us consider the unitary transformation (4.34). In the case of coherent excitons, X_{ex} has the form $(a^\dagger b^\dagger - ba)$. Its first-order commutators $[X, a_{\mathbf{p}\sigma}]$ and $[X, b_{\mathbf{p}\sigma}]$ give rise to the operators $b^\dagger_{-\mathbf{p},-\sigma}$ and $a^\dagger_{-\mathbf{p},-\sigma}$, whereas the second-order commutators lead to the initial operators $a_{\mathbf{p},\sigma}$ and $b_{\mathbf{p},\sigma}$. This fact permits one to obtain a closed form of the exact u, v transformation of Eqs. (2.108). By contrast, the commutators $[X, a_{\mathbf{p}\sigma}]$ and $[X, b_{\mathbf{p}\sigma}]$, when X is determined by Eq. (4.31), generate the creation operators of the "trions" $A^\dagger_{-\mathbf{p},-\sigma}$ and $B^\dagger_{-\mathbf{p},-\sigma}$ with resulting wave vectors $-\mathbf{p}$ and resulting spin projections $-\sigma$. They are defined as

$$B^\dagger_{-\mathbf{p},-\sigma} = \frac{[X, a_{\mathbf{p}\sigma}]}{\sqrt{n_{b,0}W_\alpha(\mathbf{p})}}, \qquad A^\dagger_{-\mathbf{p},-\sigma} = -\frac{[X, b_{\mathbf{p}\sigma}]}{\sqrt{n_{b,0}W_\beta(\mathbf{p})}}. \tag{4.41}$$

The constants $W_\alpha(\mathbf{p})$ and $W_\beta(\mathbf{p})$ are specified below.

A few words on "trions" are in order here. Trions are bound states of three charged quasiparticles. The negative trion, also called a negative biexcitonic ion (X^-), consists of two electrons and one hole (*eeh*), whereas the positive trion, also known as the positive biexcitonic ion (X_2^+), consists of two holes and one electron (*hhe*). Their existence was suggested by Lampert [2]. The stability of both types of trions in bulk crystals with simple band structure was proved via direct calculations by Bobrysheva (reported in Ref. 91), Gerlach [92], Munschy and Stébé [93, 94], and Insepov and Norman [95]. The stability of surface and 2D trions was established in Refs. 96 and 97. Experimental evidence for trions has been seen in studies of the luminescence lines of various semiconductor materials, e.g., evidence for trions in quantum-well structures was seen by Finkelstein, Shtrikman, and Bar-Joseph [98–101].

Trions, as well as excitons, electrons and holes, can be considered as the fragments and the constituent parts of biexcitons. For this reason, the appearance of the trion creation operators in transformation (4.41) is not surprising. The commutator $\left[X, a_{\mathbf{p}\sigma}\right]$ has the form

$$[X, a_{\mathbf{p}\sigma}] = \frac{\sqrt{n_{b,0}}}{2CV} \sum_{\mathbf{k}_2,\mathbf{k}_4} \left(b^\dagger_{\mathbf{k}_2,-\sigma} b^\dagger_{\mathbf{k}_4,\sigma} - b^\dagger_{\mathbf{k}_2,\sigma} b^\dagger_{\mathbf{k}_4,-\sigma}\right)$$
$$\times \left[\sum_{\mathbf{k}_1} F(\mathbf{k}_1, \mathbf{k}_2, \mathbf{p}, \mathbf{k}_4) a^\dagger_{\mathbf{k}_1,-\sigma} + \sum_{\mathbf{k}_3} F(\mathbf{p}, \mathbf{k}_2, \mathbf{k}_3, \mathbf{k}_4) a^\dagger_{\mathbf{k}_3,-\sigma}\right]. \tag{4.42}$$

Two hole operators and one electron operator are bounded by the functions $F\left(\cdots \mathbf{k}'_i \cdots\right)$. This binding can be made more evident if the biexciton creation operator $C_0^\dagger$ is represented in a particular case as being constructed from two exciton creation operators $c^\dagger_{\mathbf{Q},s_1,s_2}$ as follows:

$$C_0^\dagger = \frac{1}{2C\sqrt{V}} \sum_{\mathbf{Q}} \sum_{s_1,s_2} (-1)^{s_1+s_2}\, \Phi(\mathbf{Q})\, c^\dagger_{\mathbf{Q},s_1,s_2} c^\dagger_{-\mathbf{Q},-s_1,-s_2}. \tag{4.43}$$

The exciton creation operators in their turn are constructed from the electron–hole creation operators as determined earlier in Eq. (1.13):

$$c^\dagger_{\mathbf{Q},s_1,s_2} = \frac{1}{\sqrt{V}} \sum_{\mathbf{p}} \varphi(\mathbf{p})\, a^\dagger_{\alpha\mathbf{Q}+\mathbf{p},s_1} b^\dagger_{\beta\mathbf{Q}-\mathbf{p},s_2}. \tag{4.44}$$

The functions $\Phi(\mathbf{Q})$ and $\varphi(\mathbf{p})$ determine the relative motions of the two excitons that make up the biexciton and of the e–h pair that makes up the exciton, respectively. These functions are normalized according to

$$\frac{1}{V} \sum_{\mathbf{Q}} \Phi^2(\mathbf{Q}) = 1, \qquad \frac{1}{V} \sum_{\mathbf{p}} \varphi^2(\mathbf{p}) = 1. \tag{4.45}$$

In this particular model, the four variables $\mathbf{k}_i$ are expressed in terms of three independent variables $\mathbf{Q}$, $\mathbf{p}$, and $\mathbf{q}$ as

$$\begin{aligned}
&\mathbf{k}_1 = \alpha\mathbf{Q} + \mathbf{p}, \qquad \mathbf{k}_3 = -\alpha\mathbf{Q} + \mathbf{q},\\
&\mathbf{k}_2 = \beta\mathbf{Q} - \mathbf{p}, \qquad \mathbf{k}_4 = -\beta\mathbf{Q} - \mathbf{q},\\
&\mathbf{p} = \beta\mathbf{k}_1 - \alpha\mathbf{k}_2,\ \mathbf{q} = \beta\mathbf{k}_3 - \alpha\mathbf{k}_4, \qquad \mathbf{Q} = \mathbf{k}_1 + \mathbf{k}_2 = -\left(\mathbf{k}_3 + \mathbf{k}_4\right),\\
&\alpha = m_e/m_{\text{ex}}, \qquad \beta = m_h/m_{\text{ex}}.
\end{aligned} \tag{4.46}$$

The function $F\left(\cdots \mathbf{k}_i \cdots\right)$ takes the form

$$F(\mathbf{k}_1, \mathbf{k}_2, \mathbf{k}_3, \mathbf{k}_4) = \Phi(\mathbf{Q})\varphi(\mathbf{p})\varphi(\mathbf{q}). \tag{4.47}$$

The commutators to be found have the forms

$$\begin{aligned}
[X, a_{\mathbf{p}\sigma}] &= \frac{\sqrt{n_{b,0}}}{C\sqrt{V}} \sum_{\mathbf{Q}} \Phi(\mathbf{Q})\varphi(\alpha\mathbf{Q} + \mathbf{p})\\
&\quad\times [c^\dagger_{\mathbf{Q},-\sigma,-\sigma} b^\dagger_{-\mathbf{p}-\mathbf{Q},\sigma} - c^\dagger_{\mathbf{Q},-\sigma,\sigma} b^\dagger_{-\mathbf{p}-\mathbf{Q},-\sigma}],\\
[X, b_{\mathbf{p}\sigma}] &= \frac{\sqrt{n_{b,0}}}{C\sqrt{V}} \sum_{\mathbf{Q}} \Phi(\mathbf{Q})\varphi(\beta\mathbf{Q} + \mathbf{p})\\
&\quad\times [c^\dagger_{\mathbf{Q},\sigma,-\sigma} a^\dagger_{-\mathbf{p}-\mathbf{Q},-\sigma} - c^\dagger_{\mathbf{Q},-\sigma,-\sigma} a^\dagger_{-\mathbf{p}-\mathbf{Q},\sigma}].
\end{aligned} \tag{4.48}$$

The second-order commutators $[X, [X, a_{\mathbf{p}\sigma}]]$ and $[X, [X, b_{\mathbf{p}\sigma}]]$ contain expressions proportional to the operators $a_{\mathbf{p},\sigma}$ and $b_{\mathbf{p},\sigma}$, respectively, multiplied by factors of the same type as appeared in the commutator of Eq. (2.91), namely

$$[X, [X, a_{\mathbf{p}\sigma}]] \cong a_{\mathbf{p}\sigma}[1 + a^\dagger a + b^\dagger b + a^\dagger b^\dagger ab + b^\dagger b^\dagger bb],$$
$$[X, [X, b_{\mathbf{p}\sigma}]] \cong b_{\mathbf{p}\sigma}[1 + a^\dagger a + b^\dagger b + a^\dagger b^\dagger ab + a^\dagger a^\dagger aa]. \qquad (4.49)$$

The operators in the square brackets can be taken into account in the mean-field approximation, being substituted by their mean values. They can be neglected in only the low excitation limit. In both approximations, the commutators of relations (4.49) become proportional to the initial operators $a_{\mathbf{p}\sigma}$ and $b_{\mathbf{p}\sigma}$. In doing so, one can represent transformations (4.34) approximately by the forms

$$Da_{\mathbf{p}\sigma}D^\dagger \cong u_\alpha(\mathbf{p})a_{\mathbf{p}\sigma} + v_\alpha(\mathbf{p})B^\dagger_{-\mathbf{p},-\sigma},$$
$$Db_{\mathbf{p}\sigma}D^\dagger \cong u_\beta(\mathbf{p})b_{\mathbf{p}\sigma} - v_\beta(\mathbf{p})A^\dagger_{-\mathbf{p},-\sigma}, \qquad (4.50)$$

where

$$u_\alpha(\mathbf{p}) = \cos\left[\sqrt{n_{b,0}}W_\alpha(\mathbf{p})\right],$$
$$v_\alpha(\mathbf{p}) = \sin\left[\sqrt{n_{b,0}}W_\alpha(\mathbf{p})\right]. \qquad (4.51)$$

When the density-dependent corrections and the overlap integrals are neglected, the parameter $W_\alpha(\mathbf{p})$ and the normalization constant C receive the simplified expressions

$$W_\alpha(\mathbf{p}) = \frac{1}{V}\sum_{\mathbf{Q}} \Phi^2(\mathbf{Q})\varphi^2(\alpha\mathbf{Q} + \mathbf{p}), \qquad C = \sqrt{2}. \qquad (4.52)$$

The parameter $W_\alpha(\mathbf{p})$ is the average value of the product of two probabilities, $\Phi^2(\mathbf{Q})$ and $\varphi^2(\alpha\mathbf{Q} + \mathbf{p})$. One of them is the probability of finding an exciton with a center-of-mass wave vector $\mathbf{Q}$ as a constituent of the biexciton. The second one is the probability of finding a hole with the wave vector $-(\mathbf{p} + \mathbf{Q})$ as a constituent of the exciton with wave vector $-\mathbf{Q}$. In this case, the wave vector of the relative motion in the exciton will be equal to $\alpha\mathbf{Q} + \mathbf{p}$, which appears in the second factor. The parameter $W_\beta(\mathbf{p})$ differs from $W_\alpha(\mathbf{p})$ because of the substitution of α by β. The singlet-exciton creation operator has the form

$$c^\dagger_{\mathbf{Q}} = \frac{1}{\sqrt{2}}(c^\dagger_{\mathbf{Q},\sigma,-\sigma} + c^\dagger_{\mathbf{Q},-\sigma,\sigma}). \qquad (4.53)$$

The transformation $DC^\dagger_{\mathbf{Q}}D^\dagger$ can be obtained approximately in the same way and has the form

$$DC_{\mathbf{Q}}D^\dagger \simeq u_{\text{ex}}(\mathbf{Q})c_{\mathbf{Q}} - v_{\text{ex}}(\mathbf{Q})c^\dagger_{-\mathbf{Q}}, \qquad (4.54)$$

where the coefficients $u_{\text{ex}}(\mathbf{Q})$ and $v_{\text{ex}}(\mathbf{Q})$ are

$$u_{\text{ex}}(\mathbf{Q}) = \cosh\left[\sqrt{n_{b,0}/2}\Phi(\mathbf{Q})\right],$$
$$v_{\text{ex}}(\mathbf{Q}) = \sinh\left[\sqrt{n_{b,0}/2}\Phi(\mathbf{Q})\right]. \qquad (4.55)$$

The Coulomb interaction in the initial Hamiltonian is represented by the matrix element $V_{\mathbf{k}}$. In the transformed Hamiltonian $\widetilde{\mathcal{H}}$, this matrix element is multiplied by coefficients of the types $\gamma_{\mathbf{p+k,p}}$ and $\tilde{\gamma}_{\mathbf{p+k,p}}$, which appeared in Fig. 2.8. But in the present case, this task is much more cumbersome than in the case of coherent excitons and has not yet been performed. Nevertheless, certain approximations are permitted, which are necessary if one wants to introduce the coherent pairing of excitons of relation (4.54). This approach is used below.

Besides the case of BEC of neutral bound *e–h* quaternions discussed above, Yudson [4] suggested the possibility of BEC of charged bosonic complexes such as bound quaternions consisting of three electrons and one spatially separated hole (*eee–h*) in a double quantum well (DQW) near a metal plate. Since transport of these complexes would be accompanied by a nonzero total current along the DQW plane, superfluidity would therefore also correspond to superconductivity.

Yudson paid attention to the special role of image charges near a conducting surface. When the metal plate is close to the *e* layer, for example, the *e–e* Coulomb repulsion is suppressed considerably by the image-charge polarization of the metal plate. This suppression can hinder, for example, the crystallization of the 2D electron gas, giving rise to the cold melting of the 2D electron Wigner crystal [102]. Yudson used the advantage of suppression of the *e–e* repulsion in systems of spatially separated *e* and *h* and demonstrated the possibility of the formation of stable charged complexes of several electrons bound to a spatially separated hole. Complexes of either Fermi (*eeh*) or Bose (*eeeh*) statistics may coexist in the ground state. Their stability may be increased by an external magnetic field perpendicular to the plane of the layers. This possibility is discussed further in Subsection 10.2.2.

BEC of the charged bound quaternions in Yudson's model can be considered by two methods. One of them is to regard these new quaternions as true charged bosons. The second possibility is to take into account the Fermi nature of their constituent fragments. Following this method, BEC of the charged bound quaternions will be similar to the coherent pairing of negative trions (*ee–h*) with the electrons that have opposite momenta and spin projections.

4.3 Coherent Pairing of Excitons

The coherent pairing of bosons, which is formally similar to the Cooper pairing of electrons in superconductors, was first investigated by Valatin and Butler [103]. Later, Coniglio and Marinaro [104] pointed out that in a system of bosons, depending on the character of the interaction between them, BEC in a single-particle state or in a two-particle state is possible. For certain forms of the interaction potential, condensations of both types can occur [105]. These ideas, including the coherent pairing of the bosons, were used by Evans and Imri [106] to explain the superfluidity of liquid He.

Mavroyannis [107, 108] was the first to try to apply these ideas to the Frenkel excitons in molecular crystals and to the electron–hole pairs that form Wannier excitons in semiconductors. Mavroyannis suggested that the coherent pairing of excitons with wave vectors $\mathbf{k}$ and $-\mathbf{k}$ is simply the Bose condensation of biexcitons in a single-particle state with the center-of-mass wave vector equal to zero. But the realization of this program was inadequate, first, because the wave function of the relative motion of the electron and the hole forming the Wannier exciton was not introduced. Also, Fermi-type commutation relations were used for the creation and annihilation operators of the excitons and *e–h* pairs to account for the fact that two Frenkel excitons cannot occupy the same site, but this approach does not account for the fact that the exciton operators for different sites commute, instead of anticommuting, leading to a mixed commutation relation called the Pauli type. In addition, it was supposed that the biexcitons can be formed even with an infinitesimal attractive interaction between excitons. In spite of these limitations, the main idea of the Mavroyannis papers [107, 108] is correct, that is, the coherent pairing of excitons can lead to BEC of biexcitons. The proof of this statement and the reexamination of this problem were initiated independently by Nozières and Saint James [109] and by Bobrysheva, Moskalenko and Shvera [110–112]. The coherent pairing of excitons due to their indirect attractive interaction by means of phonons was investigated by Nandakumaran and Sinha [113]. If the attractive interaction is insufficient for the formation of biexcitons in bulk crystals, then the coherent pairing of excitons does not correspond to the minimum of the energy and BEC of excitons in the single-particle state will be favored [114].

Although an attractive interaction is necessary for the formation of biexcitons, it can also lead to the collapse of the biexciton condensate [109, 112] because of terms that give an unlimited lowering of the biexciton energy with increasing density. For this reason the question of the coherent pairing is somewhat complicated. The solution of this question was found [112] on the basis of results concerning the multicomponent exciton system [85]; the same solution was proposed in Ref. 109. The coherent pairing of excitons in the multicomponent system leads to the possibility of achieving BEC of biexcitons and at the same time allows the stabilization of the condensed biexcitons. In Subsection 4.3.1 we discuss the models investigated by Bobrysheva et al. [110, 111].

4.3.1 Coherent Pairing in the Multicomponent Exciton System

As mentioned above, in semiconductors with simple band structures, any one of four species of excitons with different orientations of the electron-spin and the hole-spin projections can be considered. We consider two limits: in the first case, when the orthoexciton–paraexciton splitting Δ is small compared with the biexciton binding energy $|W|$, and second, when the opposite is true. The exciton–exciton interaction Hamiltonians for these two limiting cases were derived in Refs. 80 and 85; these include terms that describe attraction as well as repulsion between different species of excitons. The interaction of each type of exciton with all the other types of excitons is repulsive on the average. As a result, in the Hartree–Fock approximation the exciton energy level is shifted to higher energy when the exciton density increases. On the other hand, the attractive terms lead to the biexciton formation. Therefore one can expect that the internal spin structure of the excitons will prevent the system from collapsing.

The condition $\Delta < |W|$ is realized in CuCl, CuBr, and CdS crystals, in which biexcitons have been detected experimentally. Moreover, induced nonequilibrium BEC of excitons was observed [118, 119], which is reviewed in Section 4.4. The condition $\Delta > |W|$ concerns, for example, the Cu_2O crystal, in which a biexciton can be formed by two orthoexcitons with the resulting spin $s = 0$ [80]. No biexciton recombination line occurs in Cu_2O, in agreement with calculations [78, 115–117] that predict biexcitons are unstable or, at best, very weakly bound when the electron and hole masses are nearly equal, as is the case in Cu_2O. In this subsection we show that in both limiting cases, the coherent pairing of excitons coincides with the BEC of biexcitons formed by excitons with suitable spin structures. The exciton–exciton interaction in a multicomponent system that is repulsive on the average leads to a positive compressibility and thus prevents the system from collapsing. The coexistence of a single-particle condensate of paraexcitons and the condensate of biexcitons formed by two orthoexcitons with total spin $s = 0$ is also discussed.

It is useful to recall that in the case $\Delta < |W|$ the four species of excitons can be described by the boson operators $a_{\mathbf{k},i}$ of Eqs. (4.22), which correspond to the real exciton operators $c_{\mathbf{k},\sigma_1,\sigma_2}$ as follows:

$$a_{\mathbf{k},1} = c_{\mathbf{k},1/2,1/2}, \qquad a_{\mathbf{k},2} = c_{\mathbf{k},-1/2,-1/2},$$
$$a_{\mathbf{k},3} = c_{\mathbf{k},1/2,-1/2}, \qquad a_{\mathbf{k},4} = c_{\mathbf{k},-1/2,1/2}.$$

The interaction Hamiltonian of the four species of excitons was discussed in Subsection 4.1.2 and is represented by formula (4.23). The full Hamiltonian has the form

$$H = H_0 + H_{\text{int}},$$

$$H_0 = \sum_{\mathbf{k}} \sum_{i=1}^{4} [E_{\text{ex}}(\mathbf{k}) - \mu]\, a^{\dagger}_{\mathbf{k},i} a_{\mathbf{k},i},$$

$$\begin{aligned} H_{\text{int}} = \frac{1}{2V} \sum_{\substack{\mathbf{k}_1,\mathbf{k}_2,\\ \mathbf{k}_1',\mathbf{k}_2'}} \delta_{\text{kr}}(\mathbf{k}_1 + \mathbf{k}_2; \mathbf{k}_1' + \mathbf{k}_2') \Bigg\{ & U_{ss}(\mathbf{k}_1, \mathbf{k}_2; \mathbf{k}_1', \mathbf{k}_2') \\ & \times \left(\sum_{i=1}^{4} a^{\dagger}_{\mathbf{k}_1,i} a^{\dagger}_{\mathbf{k}_2,i} a_{\mathbf{k}_2',i} a_{\mathbf{k}_1',i} + 2 \sum_{i=3,4} \sum_{j=1,2} a^{\dagger}_{\mathbf{k}_1,i} a^{\dagger}_{\mathbf{k}_2,j} a_{\mathbf{k}_2',j} a_{\mathbf{k}_1',i} \right) \\ & + [U_{ss}(\mathbf{k}_1, \mathbf{k}_2; \mathbf{k}_1', \mathbf{k}_2') + U_{aa}(\mathbf{k}_1, \mathbf{k}_2; \mathbf{k}_1', \mathbf{k}_2')] \\ & \times \left(a^{\dagger}_{\mathbf{k}_1,3} a^{\dagger}_{\mathbf{k}_2,4} a_{\mathbf{k}_2',4} a_{\mathbf{k}_1',3} + a^{\dagger}_{\mathbf{k}_1,1} a^{\dagger}_{\mathbf{k}_2,2} a_{\mathbf{k}_2',2} a_{\mathbf{k}_1',1} \right) \\ & + [U_{ss}(\mathbf{k}_1, \mathbf{k}_2; \mathbf{k}_1', \mathbf{k}_2') - U_{aa}(\mathbf{k}_1, \mathbf{k}_2; \mathbf{k}_1', \mathbf{k}_2')] \\ & \times \left(a^{\dagger}_{\mathbf{k}_1,3} a^{\dagger}_{\mathbf{k}_2,4} a_{\mathbf{k}_2',1} a_{\mathbf{k}_1',2} + a^{\dagger}_{\mathbf{k}_1,1} a^{\dagger}_{\mathbf{k}_2,2} a_{\mathbf{k}_2',3} a_{\mathbf{k}_1',4} \right) \Bigg\}, \end{aligned}$$

$$U(\mathbf{k}_1, \mathbf{k}_2; \mathbf{k}_1', \mathbf{k}_2') = U(\mathbf{k}_1', \mathbf{k}_2'; \mathbf{k}_1, \mathbf{k}_2) = U(\mathbf{k}_2, \mathbf{k}_1; \mathbf{k}_2', \mathbf{k}_1') = U(-\mathbf{k}_1, -\mathbf{k}_2; -\mathbf{k}_1', -\mathbf{k}_2'). \tag{4.56}$$

Here $E_{\text{ex}}(\mathbf{k})$ is the exciton creation energy and μ is the chemical potential that determines the exciton density n_{ex},

$$n_{\text{ex}} = \frac{1}{V} \sum_{\mathbf{k}} \sum_{i=1}^{4} \langle a^{\dagger}_{\mathbf{k},i} a_{\mathbf{k},i} \rangle = \sum_i n_i. \tag{4.57}$$

The Fourier transforms of the exciton–exciton interaction energy $U(\mathbf{k}_1, \mathbf{k}_2; \mathbf{k}_1', \mathbf{k}_2')$ were calculated in Subsection 4.1.2. In the random-phase approximation when $\mathbf{k}_1 = \mathbf{k}_2 = \mathbf{k}_1' = \mathbf{k}_2'$, they are determined by expressions (4.16). The Bogoliubov canonical transformations giving rise to the coherent pairing of excitons of different species and opposite wave vectors have the forms

$$a_{\mathbf{k},1} = u_{\mathbf{k}} \alpha_{\mathbf{k},1} + v_{\mathbf{k}} \alpha^{\dagger}_{-\mathbf{k},2}, \qquad a_{\mathbf{k},3} = u_{\mathbf{k}} \alpha_{\mathbf{k},3} - v_{\mathbf{k}} \alpha^{\dagger}_{-\mathbf{k},4},$$
$$a_{\mathbf{k},2} = u_{\mathbf{k}} \alpha_{\mathbf{k},2} + v_{\mathbf{k}} \alpha^{\dagger}_{-\mathbf{k},1}, \qquad a_{\mathbf{k},4} = u_{\mathbf{k}} \alpha_{\mathbf{k},4} - v_{\mathbf{k}} \alpha^{\dagger}_{-\mathbf{k},3}. \tag{4.58}$$

Here the normalization condition must be added,

$$u_{\mathbf{k}}^2 - v_{\mathbf{k}}^2 = 1. \tag{4.59}$$

As one can see, the excitons of type 1 with the total spin projection $s_z = 1$ are correlated with the excitons of type 2 with $s_z = -1$. Correlated pairs of such excitons have the resultant spin projection and the resultant wave vector both equal to zero.

Two other species of excitons with $i = 3$ and $i = 4$ can also form correlated pairs with a net spin projection equal to zero. But the signs before the coefficients $v_{\mathbf{k}}$ in the canonical transformation for these species of excitons are the opposite of those of the first two species. These signs are in accordance with the spin structure of the function $|\psi_a\rangle$ of Eqs. (4.10), which guarantees biexciton formation only in the limit $\Delta < |W|$. If the signs before the coefficients $v_{\mathbf{k}}$ in Eqs. (4.58) are chosen to be the same, a biexciton bound state is impossible. By using transformations (4.58), one can select in the transformed Hamiltonian $\tilde{H}$ the quadratic part of the operators $\alpha_{\mathbf{k},i}^{\dagger}\alpha_{\mathbf{k},i}$. This quadratic form $\tilde{H}_2$ has the form

$$\begin{aligned}\tilde{H}_2 = &\sum_{\mathbf{k}}\sum_{i=1}^{4}\left[\tilde{\mathcal{E}}(\mathbf{k})\left(u_{\mathbf{k}}^2 + v_{\mathbf{k}}^2\right) + 2u_{\mathbf{k}}v_{\mathbf{k}}\Delta(\mathbf{k})\right]\alpha_{\mathbf{k},i}^{\dagger}\alpha_{\mathbf{k},i}\\ &+\sum_{\mathbf{k}}\left[2\tilde{\mathcal{E}}(\mathbf{k})\,u_{\mathbf{k}}v_{\mathbf{k}} + \Delta(\mathbf{k})\left(u_{\mathbf{k}}^2 + v_{\mathbf{k}}^2\right)\right]D_{\mathbf{k}}.\end{aligned} \tag{4.60}$$

Here the following notations were introduced:

$$\begin{aligned}\tilde{\mathcal{E}}(\mathbf{k}) &= E_{\mathrm{ex}}(\mathbf{k}) - \mu + \frac{1}{2V}\sum_{\mathbf{p}}\left[9U_{ss}(\mathbf{k},\mathbf{p};\mathbf{k},\mathbf{p}) + U_{aa}(\mathbf{k},\mathbf{p};\mathbf{k},\mathbf{p})\right]v_{\mathbf{p}}^2,\\ \Delta(\mathbf{k}) &= \frac{1}{V}\sum_{\mathbf{p}}U_{aa}(\mathbf{k},-\mathbf{k};\mathbf{p},-\mathbf{p})\,u_{\mathbf{p}}v_{\mathbf{p}},\\ D_{\mathbf{k}} &= \left(\alpha_{\mathbf{k},1}^{\dagger}\alpha_{-\mathbf{k},2}^{\dagger} + \alpha_{-\mathbf{k},2}\alpha_{\mathbf{k},1}\right) - \left(\alpha_{\mathbf{k},3}^{\dagger}\alpha_{-\mathbf{k},4}^{\dagger} + \alpha_{-\mathbf{k},4}\alpha_{\mathbf{k},3}\right).\end{aligned} \tag{4.61}$$

The dangerous diagrams arising from the operator $D_{\mathbf{k}}$ describe the creation from vacuum and the annihilation of pairs of quasiparticles with opposite wave vectors and spin projections. The Bogoliubov compensation condition for these diagrams has the form

$$2\tilde{\mathcal{E}}(\mathbf{k})u_{\mathbf{k}}v_{\mathbf{k}} + \Delta(\mathbf{k})\left(u_{\mathbf{k}}^2 + v_{\mathbf{k}}^2\right) = 0. \tag{4.62}$$

Equation (4.62), together with normalization condition (4.59), permits the determination of the coefficients $u_{\mathbf{k}}$ and $v_{\mathbf{k}}$ and the energy spectrum of the single-particle elementary excitation $E(\mathbf{k})$ through the coherence parameter $\Delta(\mathbf{k})$. These expressions correspond to the mean-field approximation. They are

$$\begin{aligned}u_{\mathbf{k}}^2 = \frac{1}{2}\left[\frac{\tilde{\mathcal{E}}(\mathbf{k})}{E(\mathbf{k})} + 1\right], \qquad v_{\mathbf{k}}^2 = \frac{1}{2}\left[\frac{\tilde{\mathcal{E}}(\mathbf{k})}{E(\mathbf{k})} - 1\right],\\ E(\mathbf{k}) = \sqrt{\tilde{\mathcal{E}}^2(\mathbf{k}) - \Delta^2(\mathbf{k})}.\end{aligned} \tag{4.63}$$

The quadratic Hamiltonian has the diagonalized form

$$\tilde{H}_2 = \sum_{\mathbf{k}}\sum_{i=1}^{4} E(\mathbf{k})\alpha_{\mathbf{k},i}^{\dagger}\alpha_{\mathbf{k},i}. \tag{4.64}$$

The ground state $|0\rangle$ is determined as

$$\alpha_{\mathbf{k},i}|0\rangle = 0. \tag{4.65}$$

The density of excitons in the ground state is

$$n_{\text{ex}} = \frac{1}{V}\sum_{\mathbf{k}}\sum_{i=1}^{4}\langle a^{\dagger}_{\mathbf{k},i}a_{\mathbf{k},i}\rangle = \frac{4}{V}\sum_{\mathbf{k}} v^2_{\mathbf{k}},$$

$$\langle a^{\dagger}_{\mathbf{k},i}a_{\mathbf{k},i}\rangle = v^2_{\mathbf{k}}, \qquad v_{\mathbf{k}} \sim \sqrt{n_{\text{ex}}}. \tag{4.66}$$

The chemical potential μ can be expanded in a series on the square root of the density $\sqrt{n_{\text{ex}}}$. In the lowest order of the perturbation theory, one can set $\mu = \mu^0$ and $u^0_{\mathbf{k}} = 1$ and keep only the first-order coefficients $v^0_{\mathbf{k}}$, neglecting the higher-order terms $(v^0_{\mathbf{k}})^n$ with $n \geq 2$. Then condition (4.62) takes the form

$$2[E_{\text{ex}}(\mathbf{k}) - \mu^0]v^0_{\mathbf{k}} + \frac{1}{V}\sum_{\mathbf{p}} U_{aa}(\mathbf{k}, -\mathbf{k}; \mathbf{p}, -\mathbf{p})v^0_{\mathbf{p}} = 0, \tag{4.67}$$

which is the Schrödinger equation for the biexciton in the momentum representation. The bound state of the biexciton is possible because of the attractive potential $U_{aa} < 0$. As a result, one can write

$$2\mu^0 = E_b(0) < 2E_{\text{ex}}(0). \tag{4.68}$$

One obtains the density correction to the chemical potential by taking into account the expressions $\mu = \mu^0 + \delta\mu$ and $u_{\mathbf{k}} = 1 + (v^0_{\mathbf{k}})^2/2$ and keeping the terms $(v^0_{\mathbf{k}})^3$ inclusively. This procedure leads to the result

$$\delta\mu \cong U_{ss}(0)\, n_{\text{ex}}, \tag{4.69}$$

which means that the system has a positive compressibility $[U_{ss}(0) > 0]$. This last condition implies a multicomponent exciton composition with an overall repulsive interaction.

In the opposite limit, $\Delta > |W|$, one can consider the two gases of orthoexcitons and paraexcitons in quasi-eqilibrium with two different chemical potentials μ_{or} and μ_{p}. Instead of the boson operators $a_{\mathbf{k},i}$ with $i = 1, 2, 3, 4$, one must use the operators for the paraexciton $a_{\mathbf{k}\,\text{p}}$ and the orthoexciton $a_{\mathbf{k},m\ \text{or}}$, with $m = 1, 2, 3$:

$$a_{\mathbf{k}\,\text{p}} = \frac{1}{\sqrt{2}}(a_{\mathbf{k},3} + a_{\mathbf{k},4}) = b_{\mathbf{k}},$$

$$a_{\mathbf{k},1\ \text{or}} = a_{\mathbf{k},1}, \qquad a_{\mathbf{k},2\ \text{or}} = a_{\mathbf{k},2},$$

$$a_{\mathbf{k},3\ \text{or}} = \frac{1}{\sqrt{2}}(a_{\mathbf{k},3} - a_{\mathbf{k},4}). \tag{4.70}$$

In what follows, the subscript or is dropped and the orthoexcitons are denoted simply as $a_{\mathbf{k},m}$ with $m = 1, 2, 3$.

The Hamiltonian of the two-component system in the new representation has the form

$$H = H_0 + H_{\text{int}},$$

$$H_0 = \sum_{\mathbf{k}}[E_{\text{p}}(\mathbf{k}) - \mu_{\text{p}}]b^{\dagger}_{\mathbf{k}}b_{\mathbf{k}} + \sum_{\mathbf{k}}\sum_{m=1}^{3}[E_{\text{or}}(\mathbf{k}) - \mu_{\text{or}}]a^{\dagger}_{\mathbf{k},m}a_{\mathbf{k},m},$$

$$H_{\text{int}} = \frac{1}{2V} \sum_{\substack{\mathbf{k}_1,\mathbf{k}_2,\\ \mathbf{k}_1',\mathbf{k}_2'}} \delta_{\text{kr}}(\mathbf{k}_1+\mathbf{k}_2;\mathbf{k}_1'+\mathbf{k}_2') \Bigg\{ \frac{1}{4}[U_{aa}(\mathbf{k}_1,\mathbf{k}_2;\mathbf{k}_1',\mathbf{k}_2') + 3U_{ss}(\mathbf{k}_1,\mathbf{k}_2;\mathbf{k}_1',\mathbf{k}_2')]$$

$$\times b^\dagger_{\mathbf{k}_1} b^\dagger_{\mathbf{k}_2} b_{\mathbf{k}_2'} b_{\mathbf{k}_1'} + U_{ss}(\mathbf{k}_1,\mathbf{k}_2;\mathbf{k}_1',\mathbf{k}_2') \sum_{m=1}^{3} \left(a^\dagger_{\mathbf{k}_1,m} b^\dagger_{\mathbf{k}_2} b_{\mathbf{k}_2'} a_{\mathbf{k}_1',m} + b^\dagger_{\mathbf{k}_1} a^\dagger_{\mathbf{k}_2,m} a_{\mathbf{k}_2',m} b_{\mathbf{k}_1'} \right)$$

$$+ \sum_{i,j,l,m=1}^{3} U_{ijlm}(\mathbf{k}_1,\mathbf{k}_2;\mathbf{k}_1',\mathbf{k}_2') a^\dagger_{\mathbf{k}_1,i} a^\dagger_{\mathbf{k}_2,j} a_{\mathbf{k}_2',l} a_{\mathbf{k}_1',m} \Bigg\}. \tag{4.71}$$

Here $E_{\text{p}}(\mathbf{k})$ and $E_{\text{or}}(\mathbf{k})$ denote the paraexciton and the orthoexciton bands, respectively.

The coefficients U_{ijlm} were determined in Refs. 80 and 120. They are

$$U_{3333} = \frac{1}{4}(U_{aa} + 3U_{ss}), \qquad U_{1111} = U_{2222} = U_{ss},$$

$$U_{1212} = U_{1221} = U_{2112} = U_{2121} = \frac{1}{4}(U_{aa} + U_{ss}),$$

$$U_{1233} = U_{2133} = U_{3312} = U_{3321} = \frac{1}{4}(U_{aa} - U_{ss}),$$

$$U_{2323} = U_{2332} = U_{3223} = U_{3232} = U_{1313} = U_{1331} = U_{3113} = U_{3131} = \frac{U_{ss}}{2}. \tag{4.72}$$

All the other coefficients are equal to zero.

The canonical transformation corresponding to the coherent pairing of the orthoexcitons with the total spin equal to zero has the form

$$a_{\mathbf{k},1} = u_{\mathbf{k}}\alpha_{\mathbf{k},1} - v_{\mathbf{k}}\alpha^\dagger_{-\mathbf{k},2},$$

$$a_{\mathbf{k},2} = u_{\mathbf{k}}\alpha_{\mathbf{k},2} - v_{\mathbf{k}}\alpha^\dagger_{-\mathbf{k},1},$$

$$a_{\mathbf{k},3} = u_{\mathbf{k}}\alpha_{\mathbf{k},3} - v_{\mathbf{k}}\alpha^\dagger_{-\mathbf{k},3}. \tag{4.73}$$

In regard to the paraexcitons, two different possibilities of the collective properties of paraexcitons were studied in Refs. 110 and 111. One of them was the coherent pairing of paraexcitons simultaneously with the coherent pairing of orthoexcitons. But the paraexcitons cannot form biexcitons without the help of the indirect exciton–phonon interaction [19], because the direct interaction between the paraexcitons is repulsive. The only remaining possibility is the BEC of paraexcitons in the single-particle state $\mathbf{k}=0$. This possibility was taken into account by the Bogoliubov displacement operation

$$b_{\mathbf{k}} = \sqrt{N_{0,\text{p}}}\delta_{\mathbf{k},0} + \beta_{\mathbf{k}}, \qquad n_{0,\text{p}} = \frac{N_{0,\text{p}}}{V}, \tag{4.74}$$

which was introduced into Hamiltonian equations (4.71) together with canonical transformations (4.73). As in the previous case ($\Delta < |W|$), the compensation condition of the dangerous diagrams with the participation of the orthoexitons has the form of Eqs. (4.61). But the renormalized energy $\tilde{\mathcal{E}}_{\text{or}}(\mathbf{k})$ and the coherence parameter $\Delta_{\text{or}}(\mathbf{k})$ are slightly different:

$$\tilde{\mathcal{E}}_{\text{or}}(\mathbf{k}) = E_{\text{or}}(\mathbf{k}) - \mu_{\text{or}} + n_{0,\text{p}} U_{ss}(\mathbf{k},0;\mathbf{k},0)$$

$$+ \frac{1}{2V}\sum_{\mathbf{p}} [7U_{ss}(\mathbf{k},\mathbf{p};\mathbf{k},\mathbf{p}) + U_{aa}(\mathbf{k},\mathbf{p};\mathbf{k},\mathbf{p}) v_{\mathbf{p}}^2],$$

$$\Delta_{\text{or}}(\mathbf{k}) = \frac{1}{V}\sum_{\mathbf{p}} \left[\frac{3}{4} U_{aa}(\mathbf{k},-\mathbf{k};\mathbf{p},-\mathbf{p}) + \frac{1}{4} U_{ss}(\mathbf{k},-\mathbf{k};\mathbf{p},-\mathbf{p}) \right] u_{\mathbf{p}} v_{\mathbf{p}}. \tag{4.75}$$

By comparing $\Delta_{or}(\mathbf{k})$ with $\Delta(\mathbf{k})$ of Eqs.(4.61), one can conclude that the attractive potential between two orthoexcitons with resulting spin $s = 0$ is approximately two times lower than in the case $\Delta < |W|$. Nevertheless, in the case $\Delta > |W|$, the biexciton exists for all values of $\sigma = m_e/m_h$ [80]. The value $2\mu_{or}^0$ coincides with the biexciton level $E_b(0)$. The correction $\delta\mu_{or}$ to the chemical potential μ_{or} is positive:

$$\delta\mu_{or} = U_{ss}(0)(n_{0,p} + n_{or}), \qquad n_{or} = \frac{3}{V}\sum_{\mathbf{p}} v_{\mathbf{p}}^2. \tag{4.76}$$

One finds the chemical potential μ_p for the paraexcitons by setting the coefficients of the operators $\beta_0^\dagger$ and β_0 to zero. It has the form

$$\mu_p = E_p(0) + \frac{U_{ss}(0)}{2} n_{0,p} + \frac{3}{V}\sum_{\mathbf{p}} U_{ss}(\mathbf{p}, 0; \mathbf{p}, 0) v_{\mathbf{p}}^2. \tag{4.77}$$

The corrections to both chemical potentials happen to be positive, depending only on $U_{ss}(0) > 0$. The condition for thermodynamic stability of both gases of biexcitons and paraexcitons is satisfied because of the attractive and the repulsive forces acting in the multicomponent exciton system.

4.3.2 In-Depth Study: *Coherent Pairing of Excitons Induced by Laser Radiation*

Bobrysheva, Moskalenko, and Kam [121] studied the possibility of inducing coherent pairing of excitons under the influence of coherent laser radiation. In this case, induced single-particle BEC of excitons is created with the same wave vector and polarization as those of laser radiation in the crystal. Therefore, two condensates induced simultaneously are considered, both an exciton condensate and a biexciton condensate. A crystal like CuCl is chosen as the model, in which there are two lower exciton levels characterized by the irreducible representations Γ_5 and Γ_2 of the crystallographic group T_d at the point $\mathbf{k} = 0$. The separation between these energy levels is less than the biexciton dissociation energy, and all four states of these two levels are involved in the formation of the biexciton. Because of the spin-orbit coupling, the hole states of the valence band Γ_7 are characterized by the angular momentum with two projections that resemble the two spin projections in the Čulik band model.

The electron-spin – hole-angular-momentum structure of the threefold degenerate level Γ_5 is the same as the electron-spin – hole-spin structure of the orthoexcitons discussed above. The spin–angular-momentum structure of the excitons of symmetry Γ_2 is the same as that in the case of paraexcitons. Unlike the orthoexcitons considered above, however, the Γ_5 excitons are dipole active. Instead of the orthoexciton states described by the operators $a_{\mathbf{k},i}$ introduced in the Čulik model it is more useful to use their linear combinations, as follows:

$$a_{\mathbf{k},\Gamma_5,x} = \frac{1}{\sqrt{2}}(a_{\mathbf{k},1} + a_{\mathbf{k},2}), \qquad a_{\mathbf{k},\Gamma_5,y} = \frac{1}{\sqrt{2}}(a_{\mathbf{k},1} - a_{\mathbf{k},2}). \tag{4.78}$$

The operators $a_{\mathbf{k},\Gamma_5,z}$ and $a_{\mathbf{k},\Gamma_2}$ remain the same as in the case of orthoexcitons and paraexcitons introduced in Eqs. (4.70), namely

$$a_{\mathbf{k},\Gamma_5,z} = \frac{1}{\sqrt{2}}(a_{\mathbf{k},3} - a_{\mathbf{k},4}), \qquad a_{\mathbf{k},\Gamma_2} = \frac{1}{\sqrt{2}}(a_{\mathbf{k},3} + a_{\mathbf{k},4}).$$

Although the Bloch functions in CuCl crystals guarantee dipole-active transitions in the Γ_5 exciton states, nevertheless the exciton–exciton interactions remain the same as in a more simple band model in which the orthoexciton–paraexciton splitting and the transverse–longitudinal splitting of the dipole-active excitons are not taken into account. In what follows the operators $a_{\mathbf{k},\Gamma_5,x}$, $a_{\mathbf{k},\Gamma_5,y}$, and $a_{\mathbf{k},\Gamma_5,z}$ are denoted as $\alpha_{\mathbf{k},i}$, where $i = x, y, z$.

The biexciton wave function can be constructed from the wave functions of two excitons in a CuCl crystal in the form

$$\psi_{\Gamma_1} = \psi_{\Gamma_{5,x}}\psi_{\Gamma_{5,x}} + \psi_{\Gamma_{5,y}}\psi_{\Gamma_{5,y}} + \psi_{\Gamma_{5,z}}\psi_{\Gamma_{5,z}} - \Psi_{\Gamma_2}\psi_{\Gamma_2}.$$

This kind of expression is justified by the fact that the two quadratic forms constructed by the corresponding operators,

$$a_{\Gamma_{5,x}}a_{\Gamma_{5,x}} + a_{\Gamma_{5,y}}a_{\Gamma_{5,y}} + a_{\Gamma_{5,z}}a_{\Gamma_{5,z}} - a_{\Gamma_2}a_{\Gamma_2} = a_1a_2 + a_2a_1 - a_3a_4 - a_4a_3, \tag{4.79}$$

are equivalent. Only the last type of spin structure guarantees the stabiity of the biexciton.

The Hamiltonian of the excitons of the four types $i = 1, 2, 3, 4$, which we now use to number the states $\Gamma_{5,x}$, $\Gamma_{5,y}$, $\Gamma_{5,z}$, and Γ_2, includes the interaction with the laser radiation and the exciton–exciton interaction. This Hamiltonian has the form

$$H = H_0 + H_1 + H_{\text{int}}, \tag{4.80}$$

$$H_0 = \sum_{\mathbf{k}}\sum_{j=1}^{3} E_{\Gamma_5}(\mathbf{k})\alpha^{\dagger}_{\mathbf{k},j}\alpha_{\mathbf{k},j} + \sum_{\mathbf{k}} E_{\Gamma_2}(\mathbf{k})\alpha^{\dagger}_{\mathbf{k},4}\alpha_{\mathbf{k},4}, \tag{4.81}$$

$$H_1 = -\mu_{\text{ex}}F^{1/2}_{\mathbf{k}_0}\left(e^{-i\omega_L t}\alpha^{\dagger}_{\mathbf{k}_0,1} + e^{i\omega_L t}\alpha_{\mathbf{k}_0,1}\right), \tag{4.82}$$

$$\begin{aligned} H_{\text{int}} = \frac{1}{2V}\sum_{\mathbf{p},\mathbf{q},\mathbf{k}} \Bigg\{ & \frac{1}{4}[U_{aa} + 3U_{ss}]\sum_{j=1}^{4} \alpha^{\dagger}_{\mathbf{p},i}\alpha^{\dagger}_{\mathbf{q},i}\alpha_{\mathbf{q}+\mathbf{k},i}\alpha_{\mathbf{p}-\mathbf{k},i} \\ & + \sum_{\substack{i,j=1\\ i\neq j}}^{4} U_{ss}\alpha^{\dagger}_{\mathbf{p},i}\alpha^{\dagger}_{\mathbf{q},j}\alpha_{\mathbf{q}+\mathbf{k},j}\alpha_{\mathbf{p}-\mathbf{k},i} + \frac{1}{4}\sum_{\substack{i,j=1\\ i\neq j}}^{3}[U_{aa} - U_{ss}]\alpha^{\dagger}_{\mathbf{p},i}\alpha^{\dagger}_{\mathbf{q},i}\alpha_{\mathbf{q}+\mathbf{k},j}\alpha_{\mathbf{p}-\mathbf{k},j} \\ & + \frac{1}{4}\sum_{j=1}^{3}[U_{ss} - U_{aa}]\left(\alpha^{\dagger}_{\mathbf{p},j}\alpha^{\dagger}_{\mathbf{q},j}\alpha_{\mathbf{q}+\mathbf{k},4}\alpha_{\mathbf{p}-\mathbf{k},4} + \alpha^{\dagger}_{\mathbf{p},4}\alpha^{\dagger}_{\mathbf{q},4}\alpha_{\mathbf{q}+\mathbf{k},j}\alpha_{\mathbf{p}-\mathbf{k},j}\right)\Bigg\}. \end{aligned} \tag{4.83}$$

Here μ_{ex} is a coupling constant proportional to the matrix element of the dipole transition from the ground state of the crystal to a Γ_5 excitonic state and ω_L and $\mathbf{k}_0$ are the laser radiation frequency and wave vector, respectively. The constants U_{ss} $(\mathbf{p}, \mathbf{q}; \mathbf{p}-\mathbf{k}, \mathbf{q}+\mathbf{k})$ and U_{aa} $(\mathbf{p}, \mathbf{q}; \mathbf{p}-\mathbf{k}, \mathbf{q}+\mathbf{k})$ are the Fourier transforms of the exciton interaction energy. They are calculated with the wave functions of two excitons that are either symmetric or antisymmetric under permutations of the spin projections of the two electrons as well as under permutations of the angular-momentum projections of the two holes. For small exciton momenta, as mentioned above,

$$U_{ss}(0) = -U_{aa}(0) = \frac{26\pi}{3}\text{Ry}_{\text{ex}}a^3_{\text{ex}},$$

where Ry_{ex} and a_{ex} denote the ionization potential and the exciton radius, respectively. It was assumed that the laser radiation is polarized in the direction $e\|x$ and excites only one exciton state with the same wave vector and polarization. The value $F^{1/2}_{\mathbf{k}_0}$ is the amplitude of the laser radiation $\left(F_{\mathbf{k}_0} \sim V\right)$.

The polariton effect was not taken into account since it does not have a significant influence on the internal structure of the biexciton when its radius a_{biex} remains equal to 2–3 a_{ex}. In this case, the excitons taking part in the biexciton formation have wave vectors of the order of magnitude of a_{biex}^{-1}. They belong to the excitonlike part of the lower polariton branch, where the polariton effects are negligible. This was first noted by Agranovich and was discussed in the recent paper by Agranovich, La Rocca and Bassani [122]. The polariton effect can be important, however, as an alternative mechanism of biexciton generation and recombination.

As shown above, the optical properties of crystals mediated by the excitonic molecules are related to the one- and two-photon transitions involving the biexciton, exciton and polariton states as well as the ground state of the crystal. The biexciton states were changed only by strong laser radiation, whereas the weak probe electromagnetic field, i.e., the polariton effect, did not change the internal structure of the biexciton. This assumption merits a closer look. How great is the influence of the polariton effect on the excitonic structure of the biexciton and on its dispersion law?

These aspects of biexciton physics were discussed in detail in a comprehensive review article by Ivanov, Haug and Keldysh [123]. The traditional point of view is based on the reasoning that the wave vectors of the exciton relative motion inside the biexciton, being of the order of magnitude of $1/a_m$, are much greater than the optical wave vector $k_{\text{opt}} = (w_t\sqrt{\varepsilon_0})/c$ at which the polariton effect is important. Here ω_t is the transverse frequency of the exciton, c is the light velocity in the vacuum and ε_0 is the static dielectric constant. Because the resonant photon wave length is incommensurate with the biexciton radius a_m and the inequality $k_{\text{opt}}a_m \ll 1$ holds, the influence of the polariton effect on the structure of biexciton was considered to be a small perturbation. As an alternative to this model, Ivanov and Haug [70–72] proposed a bipolariton model of the excitonic molecule, taking into account simultaneously the exciton–exciton interaction in the adiabatic approximation and the resonant exciton–photon interaction. In these approximations, the exact Schrödinger equation containing the contribution of both polariton branches was obtained. To simplify this complicated equation, the lower polariton branch (LPB) approximation was used. In the framework of this approximation, a Mexican-hat structure of the bipolariton dispersion law was obtained, i.e., a maximum at $\mathbf{k} = 0$. This means that one cannot expect spontaneous BEC of excitonic molecules at the point $\mathbf{k} = 0$.

Here it is useful to remember the properties of the LPB dispersion curve, which has a minimum of the group velocity $v_g(\mathbf{k}) = d\omega^{\text{LPB}}(\mathbf{k})/d\mathbf{k}$ and a singularity of the polariton effective mass $\hbar[d^2\omega^{\text{LPB}}(\mathbf{k})/d\mathbf{k}^2]^{-1}$ in the vicinity of this minimum. Their influence on the structure of the LPB polariton soliton is discussed in Chapter 9; possibly these properties imply the Mexican hat structure of the bipolariton dispersion curve. As mentioned in Ref. 123, further investigations are needed for more definite conclusions. There is also experimental evidence in favor of this point of view [73–76]. These aspects are not taken into account below, however, since the main topic is to provide evidence of the polarization of the internal biexciton structure under the influence of laser radiation.

The time dependence in Eq. (4.82) can be eliminated by application of the unitary transformation $U = \exp(-i\omega t N)$, where $N = \sum_{\mathbf{q}} \sum_{j=1}^{4} \alpha_{\mathbf{q},i}^{\dagger}\alpha_{\mathbf{q},i}$. We consider, instead of the operator H, the operator $\mathcal{H} = U^{\dagger}HU - \hbar\omega_L N$. Two Bose condensates of the excitons and the biexcitons are introduced into the Hamiltonian $\hat{\mathcal{H}}$ by means of the unitary transformation

$$D = D_1\left(\sqrt{N_{\mathbf{k}_0,x}^{\text{ex}}}\right) D_2\left(\sqrt{N_{2\mathbf{k}_0}^{\text{biex}}}\right). \tag{4.84}$$

The unitary transformations D_1 and D_2 give rise to the displacements of the exciton and the biexciton operators, respectively. Since the biexciton operators are constructed from products of the exciton operators, the unitary transformation D_2 involves the coherent pairing of the excitons. The displacement operators D_1 and D_2 depend on the macroscopic values $N^{\mathrm{ex}}_{\mathbf{k}_0,x}$ and $N^{\mathrm{biex}}_{2\mathbf{k}_0}$ ($N^{\mathrm{ex}}_{\mathbf{k}_0,x} \sim N^{\mathrm{biex}}_{2\mathbf{k}_0} \sim V$). The operator D_1 has the form

$$D_1\left(\sqrt{N^{\mathrm{ex}}_{\mathbf{k}_0,x}}\right) = \exp\left[\sqrt{N^{\mathrm{ex}}_{\mathbf{k}_0,x}}\left(\alpha^{\dagger}_{\mathbf{k}_0,1} - \alpha_{\mathbf{k}_0,1}\right)\right]. \tag{4.85}$$

In a similar way, the operator D_2 can be determined by use of the biexciton creation and annihilation operators $C^{\dagger}_{2\mathbf{k}_0}$ and $C_{2\mathbf{k}_0}$. The former has the form

$$C^{\dagger}_{2\mathbf{k}_0} = \frac{1}{\sqrt{V}}\sum_{\mathbf{q}}\sum_{i=1}^{4}\Phi(\mathbf{q})\, C_i \alpha^{\dagger}_{\mathbf{k}_0+\mathbf{q},i}\alpha^{\dagger}_{\mathbf{k}_0-\mathbf{q},i}, \tag{4.86}$$

where the function of the relative motion of a biexciton $\Phi(\mathbf{q})$ and the coefficients C_i obey the normalization conditions

$$\frac{1}{V}\sum_{\mathbf{q}}\Phi^2(\mathbf{q}) = 1, \qquad 2\sum_{i=1}^{4} C_i^2 = 1. \tag{4.87}$$

If one neglects the difference between the levels E_{Γ_5} and E_{Γ_2}, comparing expressions (4.86) and (4.79), one can conclude that in the absence of the laser radiation the coefficients of Eq. (4.86), C_i^0, are equal to

$$C_1^0 = C_2^0 = C_3^0 = -C_4^0 = \frac{1}{2\sqrt{2}}.$$

The displacement operator D_2,

$$D_2 = \exp\left[\sqrt{N^{\mathrm{biex}}_{2\mathbf{k}_0}}\left(\psi^{\dagger}_{2\mathbf{k}_0} - \psi_{2\mathbf{k}_0}\right)\right], \tag{4.88}$$

is equivalent to the u_i, v_i transformations of the exciton operators of Eqs. (4.55):

$$D_2 a_{\mathbf{k}_0+\mathbf{q},i} D_2^{\dagger} = u_i(\mathbf{q}) a_{\mathbf{k}_0+\mathbf{q},i} - v_i(\mathbf{q}) a^{\dagger}_{\mathbf{k}_0-\mathbf{q},i}, \tag{4.89}$$

where

$$u_i(\mathbf{q}) = \cosh\left[2C_i\sqrt{n^{\mathrm{biex}}_{2\mathbf{k}_0}}\,\Phi(\mathbf{q})\right],$$

$$v_i(\mathbf{q}) = \sinh\left[2C_i\sqrt{n^{\mathrm{biex}}_{2\mathbf{k}_0}}\,\Phi(\mathbf{q})\right],$$

$$u_i^2(\mathbf{q}) - v_i^2(\mathbf{q}) = 1, \qquad n^{\mathrm{biex}}_{2\mathbf{k}_0} = N^{\mathrm{biex}}_{2\mathbf{k}_0}/V. \tag{4.90}$$

In the limit of low densities, $n^{\mathrm{biex}}_{2\mathbf{k}_0} a^3_{\mathrm{biex}} \ll 1$, one has

$$v_i(\mathbf{q}) \approx 2C_i\sqrt{n^{\mathrm{biex}}_{2\mathbf{k}_0}}\,\Phi(\mathbf{q}). \tag{4.91}$$

As in the case of the spontaneous coherent pairing considered above, u_i, v_i transformations (4.89) lead to the appearence of terms in the transformed Hamiltonian $D\mathcal{H}D^{\dagger}$ of the type $\alpha_{\mathbf{k}_0+\mathbf{q},i}\alpha_{\mathbf{k}_0-\mathbf{q},i}$, which give rise to the dangerous diagrams. The compensation condition of these diagrams makes it possible to conclude that $C_2 = C_3 = -C_4 \neq C_1$. The equations arising from the compensation conditions, together with normalization conditions (4.90), allow the possibility of determining the coefficients C_1 and C_2 as well as the biexciton density $n^{\text{biex}}_{2\mathbf{k}_0}$. One can find the density of the Bose condensed excitons $n^{\text{ex}}_{\mathbf{k}_0,x} = N^{\text{ex}}_{\mathbf{k}_0,x}/V$ in the single-particle state characterized by the quantum numbers $(\Gamma_{5,x}, \mathbf{k}_0)$ in the same way as in Eq. (2.31), by setting to zero the terms containing the operators $\alpha^\dagger_{\mathbf{k}_0}$ and $\alpha_{\mathbf{k}_0}$, which appeared after the action of displacement operation (2.30). In doing so, one can obtain the exciton-condensate density $n^{\text{ex}}_{\mathbf{k}_0,x}$:

$$n^{\text{ex}}_{\mathbf{k}_0,x} = \frac{|\mu_{\text{ex}}|^2 f_{\mathbf{k}_0}}{\left[E_{\Gamma_5}(0) - \hbar\omega_L + L_0/2 + \mathcal{L}(\mathbf{k}_0) + M(\mathbf{k}_0)\right]^2 + \gamma^2_{\text{ex}}}, \tag{4.92}$$

where

$$f_{\mathbf{k}_0} = F_{\mathbf{k}_0}/V, \qquad L_0 = n^{\text{ex}}_{\mathbf{k}_0,x} U_{ss}(0),$$

$$\mathcal{L}(\mathbf{k}_0) = \frac{1}{V}\sum_{\mathbf{q}}\sum_{j=1}^{4} U_{ss}(\mathbf{k}_0, \mathbf{q}; \mathbf{k}_0, \mathbf{q}) v_j^2(2\mathbf{k}_0 - \mathbf{q}),$$

$$M(\mathbf{k}_0) = \frac{1}{2V}\sum_{\mathbf{q}} U_{ss}(\mathbf{k}_0, \mathbf{k}_0; 2\mathbf{k}_0 - \mathbf{q}, \mathbf{q})\left[\sum_{i=1}^{4} u_i(\mathbf{q})v_i(\mathbf{q}) - 2u_1(\mathbf{q})v_1(\mathbf{q})\right], \tag{4.93}$$

and γ_{ex} has been introduced phenomenologically. The Schrödinger equations for the functions $v_i(\mathbf{k})$ are

$$\begin{aligned}
&\left\{2\left[E_{\Gamma_5}(0) - \hbar\omega_L + L_0 + \mathcal{L}(\mathbf{k}_0)\right] + \frac{\hbar^2 k^2}{m_{\text{ex}}}\right\} u_1(\mathbf{k})v_1(\mathbf{k}) \\
&\quad + \frac{1}{2V}\sum_{\mathbf{q}} U_{aa}(\mathbf{k}, 2\mathbf{k}_0 - \mathbf{k}; \mathbf{q}, 2\mathbf{k}_0 - \mathbf{q})[3u_2(\mathbf{q})v_2(\mathbf{q}) - u_1(\mathbf{q})v_1(\mathbf{q})] \\
&\quad \times [u_1(\mathbf{k})u_1(2\mathbf{k}_0 - \mathbf{k}) + v_1(\mathbf{k})v_1(2\mathbf{k}_0 - \mathbf{k})] \\
&\quad = \frac{L_0}{2}[u_1(\mathbf{k})u_1(2\mathbf{k}_0 - \mathbf{k}) + v_1(\mathbf{k})v_1(2\mathbf{k}_0 - \mathbf{k})],
\end{aligned} \tag{4.94}$$

$$\begin{aligned}
&\left\{2\left[E_{\Gamma_5}(0) - \hbar\omega_L + L_0 + \mathcal{L}(\mathbf{k}_0)\right] + \frac{\hbar^2 k^2}{m_{\text{ex}}}\right\} u_2(\mathbf{k})v_2(\mathbf{k}) \\
&\quad + \frac{1}{2V}\sum_{\mathbf{q}} U_{aa}(\mathbf{k}, 2\mathbf{k}_0 - \mathbf{k}; \mathbf{q}, 2\mathbf{k}_0 - \mathbf{q})[u_2(\mathbf{q})v_2(\mathbf{q}) + u_1(\mathbf{q})v_1(\mathbf{q})] \\
&\quad \times [u_2(\mathbf{k})u_2(2\mathbf{k}_0 - \mathbf{k}) + v_2(\mathbf{k})v_2(2\mathbf{k}_0 - \mathbf{k})] \\
&\quad = -\frac{L_0}{2}[u_2(\mathbf{k})u_2(2\mathbf{k}_0 - \mathbf{k}) + v_2(\mathbf{k})v_2(2\mathbf{k}_0 - \mathbf{k})].
\end{aligned} \tag{4.95}$$

In the absence of laser radiation, $\Phi(\mathbf{q})$ satisfies the Schrödinger equation for relative biexciton motion:

$$\begin{aligned}
&\frac{1}{V}\sum_{\mathbf{k}} \frac{\hbar^2 k^2}{m_{\text{ex}}}\Phi^2(\mathbf{k}) + \frac{1}{V^2}\sum_{\mathbf{kq}} U_{aa}(\mathbf{k}, 2\mathbf{k}_0 - \mathbf{k}; \mathbf{q}, 2\mathbf{k}_0 - \mathbf{q})\Phi(\mathbf{k})\Phi(\mathbf{q}) \\
&\quad = \langle T\rangle + \langle V\rangle = -|W|.
\end{aligned} \tag{4.96}$$

Here the attractive interaction potential $U_{aa} < 0$ gives rise to the negative average potential energy $\langle V\rangle = -|\langle V\rangle|$, which together with the average kinetic energy $\langle T\rangle$ leads to the biexciton bound-state formation and determines the biexciton dissociation energy $|W|$ as

$$E_{\text{biex}}(0) = 2E_{\Gamma_5}(0) - |W|.$$

To the lowest order of the biexciton density, by using the approach of approximation (4.91), we arrive at the following expressions of the coefficients C_1 and C_2:

$$C_{\frac{1}{2}} = \frac{|\langle V\rangle| \mp \Delta_{\text{biex}}}{\sqrt{2[3\,(|\langle V\rangle| + \Delta_{\text{biex}})^2 + (|\langle V\rangle| - \Delta_{\text{biex}})^2]}}, \tag{4.97}$$

$$C_2 - C_1 = \frac{\sqrt{2}\Delta_{\text{biex}}}{\sqrt{[3\,(|\langle V\rangle| + \Delta_{\text{biex}})^2 + (|\langle V\rangle| - \Delta_{\text{biex}})^2]}}. \tag{4.98}$$

Here the detuning from resonance is defined as

$$\begin{aligned} \Delta_{\text{biex}} &= 2\left[E_{\Gamma_5}(0) - \hbar\omega_L + L_0 + \mathcal{L}(\mathbf{k}_0)\right] - |W| \\ &= E_{\text{biex}}(0) - 2\hbar\omega_L + 2\,[L_0 + \mathcal{L}(\mathbf{k}_0)]\,. \end{aligned} \tag{4.99}$$

In this approximation the biexciton-condensate density is

$$n_{2\mathbf{k}_0}^{\text{biex}} = \frac{\left[n_{\mathbf{k}_0,x}^{\text{ex}}\,U_{ss}\,(0)\right]^2}{\pi a_b^3\left[4\,(C_1 - C_2)^2\,(\Delta_{\text{biex}} + 2\,|\langle V\rangle|)^2 + \gamma_{\text{biex}}^2\right]}. \tag{4.100}$$

Here a phenomenological decay constant for biexcitons, γ_{biex}, has also been introduced.

When the frequency detuning is small compared with the average potential energy of the interaction between two excitons $\Delta_{\text{biex}} < |\langle V\rangle|$, then the excitons convert quickly one into another because collisions between themselves and the difference $C_1 - C_2$ tends to zero. The densities of the different types of excitons taking part in the coherent pairing are determined by the expression

$$n_i = \frac{1}{V}\sum_{\mathbf{q}}{}'\langle 0|\alpha_{\mathbf{q},i}^{\dagger}\alpha_{\mathbf{q},i}|0\rangle = \frac{1}{V}\sum_{\mathbf{q}}{}' v_i^2(\mathbf{q}) = 4C_i^2 n_{2\mathbf{k}_0}^{\text{biex}}, \tag{4.101}$$

where the average is taken over the ground state $|0\rangle$ of the transformed Hamiltonian $D\mathcal{H}D^{+}$.

The prime on the sum in Eq. (4.101) means that the number of Bose condensed excitons $N_{\mathbf{k}_0,x}^{\text{ex}}$ in the single-particle state with $i = 1$ and $\mathbf{q} = \mathbf{k}_0$ is excluded, since these excitons do not take part in the coherent pairing. When the frequency detuning Δ_{biex} tends to zero, the densities of the four types of excitons taking part in the coherent pairing are practically the same, even though the laser radiation directly excites only one exciton mode because of the interactions and the conversions among the different types of excitons. When the absolute value of the detuning $|\Delta_{\text{biex}}|$ exceeds the average potential energy $|\langle V\rangle|$, then the exciton–exciton interactions cannot equalize the densities of the different types of the nonequilibrium excitons that form the biexcitons. The relative numbers of these excitons become unequal, and the biexcitons formed by them are anisotropic. The same thing happens when a new biexciton is excited by the test light beam in the presence of laser radiation that has already generated the nonequilibrium multicomponent exciton system. The density of the newly created excitons is infinitesimal compared with the density of the nonequilibrium excitons, and the structure of the new biexciton will be the same as that discussed above.

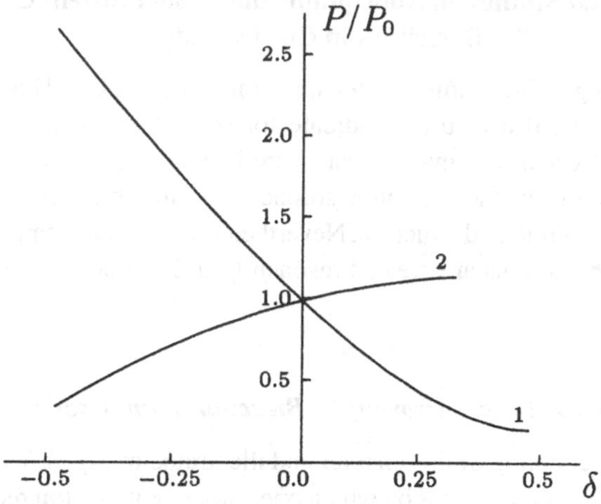

Figure 4.5. The relative probabilities of two-photon transitions for two different orientations of the polarization vector of the probing photons: curve 1, $\mathbf{e}_t \parallel X$, curve 2, $\mathbf{e}_t \perp X$; $\delta = \Delta_{\text{biex}} / |\langle V \rangle|$ (from Ref. 121).

For these reasons the wave function of the biexciton with an arbitrary wave vector $2\mathbf{k}$ in the coherently polarized crystal has the form

$$|\psi_{2\mathbf{k}}^{\text{biex}}\rangle = \frac{1}{\sqrt{V}} \sum_{\mathbf{q}} \Phi(\mathbf{q})[C_1 \alpha^\dagger_{\mathbf{k}+\mathbf{q},1} \alpha^\dagger_{\mathbf{k}-\mathbf{q},1} + C_2(\alpha^\dagger_{\mathbf{k}+\mathbf{q},2}\alpha^\dagger_{\mathbf{k}-\mathbf{q},2} + \alpha^\dagger_{\mathbf{k}+\mathbf{q},3}\alpha^\dagger_{\mathbf{k}-\mathbf{q},3} - \alpha^\dagger_{\mathbf{k}+\mathbf{q},4}\alpha^\dagger_{\mathbf{k}-\mathbf{q},4})]|0\rangle, \qquad (4.102)$$

where $C_1$and C_2 are determined by Eq. (4.97) and $|0\rangle$ is the ground state of the transformed Hamiltonian $D\mathcal{H}D^\dagger$.

Let us consider the probability of the two-photon transition from the ground state of the coherently polarized crystal $|0\rangle$ to the biexciton state of Eq. (4.102). If we assume that the polarizations of the two probing photons are both equal to $\mathbf{e}_t$, the transition probability is found to depend on the orientation of $\mathbf{e}_t$ with respect to the X axis along which the pump radiation is polarized. The fractional value of the probability, P/P_0, is given by

$$\frac{P}{P_0} = \begin{cases} 8|C_1|^2 & \mathbf{e}_t \parallel X \\ 8|C_2|^2 & \mathbf{e}_t \perp X \end{cases},$$

where P_0 is the probability of a transition in the absence of biexciton polarization and $|C_1^0|^2 = |C_2^0|^2 = 1/8$. The dependence of P/P_0 on the detuning Δ_{biex} in units of $|\langle V \rangle|$ is depicted in Fig. 4.5. Sometimes it is more convenient experimentally to use, in the two-photon transition, one photon from the pump laser proper. Then the second probing photon must have a polarization $(\mathbf{e}_t \cdot \mathbf{e}_x) \neq 0$ and an energy equal to $\hbar\omega_L + \Delta_{\text{biex}}$.

To conclude, one can note that as the detuning from resonance grows and the deviation of curves 1 and 2 (Fig. 4.5) from unity increases, the biexciton binding energy decreases and its radius increases. These factors are not taken into account in this description. For this reason it is valid in only a small region of δ. But even in this case, the effect is significant and can be observed in experiments. In this way, the probabilities of hyper-Raman-scattering processes in the quantum wells have been determined [120, 124].

4.4 Experimental Studies on Nonequilibrium Bose–Einstein Condensation of Biexcitons in CuCl Crystals

As discussed at the beginning of this chapter, the semiconductor CuCl has very tightly bound biexcitons, which make it a natural candidate for observing Bose effects of biexcitons. One complication of CuCl, however, is that there is a strong polariton effect; this means that the luminescence from the biexciton ground state, in which one would expect Bose condensation, has a complicated structure. Nevertheless, many fascinating effects have been seen in CuCl when biexcitons are created resonantly in low energy states near the ground state.

4.4.1 Bose Narrowing of Biexciton Luminescence

As discussed at the beginning of this chapter and illustrated in Fig. 4.6, biexcitons can emit luminescence by means of a process by which one constituent exciton recombines, emitting a photon and leaving behind a free exciton. As in the case of phonon-assisted luminescence discussed in Chapter 1, this process has the virtue that biexcitons at all kinetic energies can participate, since the remaining exciton can take up the momentum of the annihilated exciton in the same way that the optical phonon does in phonon-assisted recombination. Therefore one can deduce the momentum distribution of the biexcitons directly from the luminescence. Ignoring the effect of the polariton on low energy states, energy and momentum conservation imply that the luminescence intensity will be given by

$$I(\omega) \propto \rho(\tilde{\omega})N_{\tilde{\omega}}, \qquad \hbar\tilde{\omega} = (\hbar\omega - E_{\text{gap}} + \text{Ry}_{\text{ex}} + |W|)\left(1 - \frac{m_{\text{biex}}}{m_{\text{ex}}}\right)^{-1}, \tag{4.103}$$

where $|W|$ is the biexciton binding energy; the biexciton mass m_{biex} is nominally twice the exciton mass m_{ex}. This implies that the multiplicative factor that depends on the masses will be negative, and biexcitons with higher kinetic energy will appear at lower photon energy, the reverse of the case for luminescence from single excitons by means of phonon-assisted recombination. Otherwise the luminescence line shape gives the kinetic-energy distribution just as the phonon-assisted luminescence does for Cu_2O or Ge; at low density the line shape fits a Maxwell–Boltzmann distribution at bath temperature. In the dilute limit, far from the polariton region, the density of states $\rho(\tilde{\omega})$ has the standard form for a three-dimensional gas, proportional to $\sqrt{\tilde{\omega}}$ for $\tilde{\omega} > 0$ and zero otherwise.

Figure 4.7 shows luminescence from CuCl when biexcitons are created by two-photon absorption. As illustrated in Fig. 4.6, there are two spectral lines, labeled M_T and M_L, because the biexciton can decay into either a transverse or a longitudinal exciton. (For a general review of the properties of transverse and longitudinal excitons, see Ref. 125.) As seen when the two curves of lowest density in Fig. 4.7 are compared, the energy distribution narrows with increasing density as the bath temperature is kept constant. The canonical effect of Bose narrowing, dicussed in Chapter 1, is clearly seen.

Since the density of states is distorted near $E = 0$ by the coupling of the photon to the exciton states by means of the polariton effect, as the density continues to increase, the distribution does not narrow continuously; instead, a new line, labeled N_T, appears on the M_T line, which corresponds to the energy of the biexcitons created by the laser pulse, with nearly zero kinetic energy. This line is actually well below the laser photon energy, however, because of the polariton dispersion, as illustrated in Fig. 4.6.

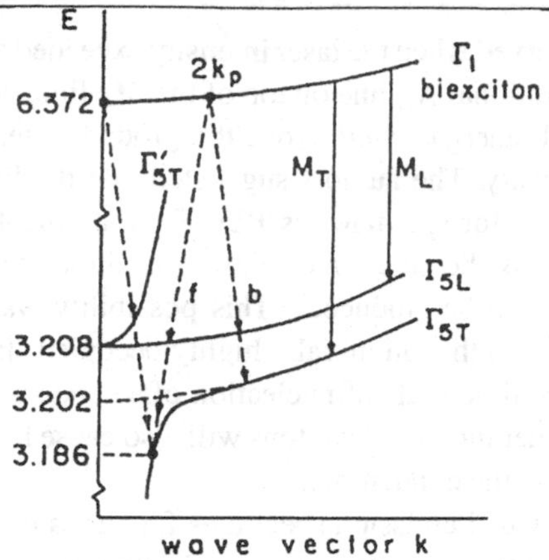

Figure 4.6. Schematic of the dispersion curves of the biexciton and the polariton branches of CuCl (from Ref. 119). "f" and "b" indicate the decay processes with emission of a photon in the forward and backward direction, respectively, relative to the pump laser photons.

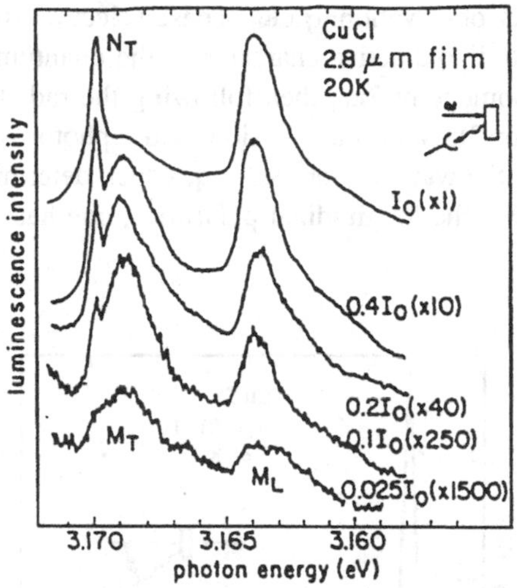

Figure 4.7. Luminescence from CuCl at several excitation levels. The bottom curve corresponds to the lowest excitation level and the top curve to the highest (from Ref. 119).

4.4.2 The Nonequilibrium Bose–Einstein Condensate

The simplest interpretation of the new luminescence line at high biexciton density in Fig. 4.7 is to view it as luminescence from a Bose–Einstein condensate of biexcitons. Peyghambarian, Chase, and Mysyrowicz [119] and Chase et al. [118] performed a comprehensive investigation of the biexciton luminescence *M* lines in CuCl films with resonant two-photon excitation and concluded that this luminescence indeed represents a real, nonequilibrium ("induced") Bose–Einstein condensate of biexcitons in CuCl crystals.

The authors used two–photon excitation from a single laser beam, similar to the excitation used by Goto et al. [126] for experiments in Cu_2O, discussed in Chapter 2. When the laser wavelength was tuned to almost exactly half the biexciton ground-state energy, the sharp

N_T line in Fig. 4.7 was observed when the laser intensity exceeded a temperature-dependent threshold, as well as an additional N_L line on top of the M_L line. When the laser was tuned to generate biexcitons with energies well above the ground state, however, no such lines appeared at any laser intensity. The authors suggested that the lifetime of the biexcitons in CuCl crystals is too short for spontaneous BEC from an incoherent source, but when the condensate is "seeded" by the creation of biexcitons directly in states near the ground state, then the condensate can be "induced." This possibility was considered by Yakhot and Levich [127], who argued that an initially highly occupied state can act as a center of nucleation. The issue of the time scale of nucleation of a condensate will be considered in Chapter 8. Nonresonant generation of biexcitons will also cause heating because of phonon emission, which can destroy the condensate.

The authors studied the CuCl emission spectra as functions of the sample temperature, excitation intensity, laser frequency, polarization, and detection geometry. Figure 4.8 shows the dependence of the intensities and the polarizations of the N_T and the N_L lines on the angle θ between the wave vector $\mathbf{k}_0$ of the laser photon and the wave vector $\mathbf{q}$ of the emitted photon. One can see that the N_L and the N_T lines are greatly reduced for emission in the forward direction. The N_L line disappears in the collinear geometry of observation and is polarized for perpendicular observation, $\mathbf{q}\perp\mathbf{k}_0$. These selection rules can be explained if one takes into account the dipole-active character of the quantum transition. If the laser creates biexcitons with momentum $2\mathbf{k}_0$, then following the radiative recombination of a biexciton with wave vector $2\mathbf{k}_0$ and the emission of the photon with wave vector $\mathbf{q}$, the polariton that remains has the wave vector $2\mathbf{k}_0 - \mathbf{q}$, which determines the dipole moment of the polariton, $\mathbf{d}_{2\mathbf{k}_0-\mathbf{q}}$. For the longitudinal polaritons, one has $\mathbf{d}^L_{2\mathbf{k}_0-\mathbf{q}} \| (2\mathbf{k}_0 - \mathbf{q})$, and

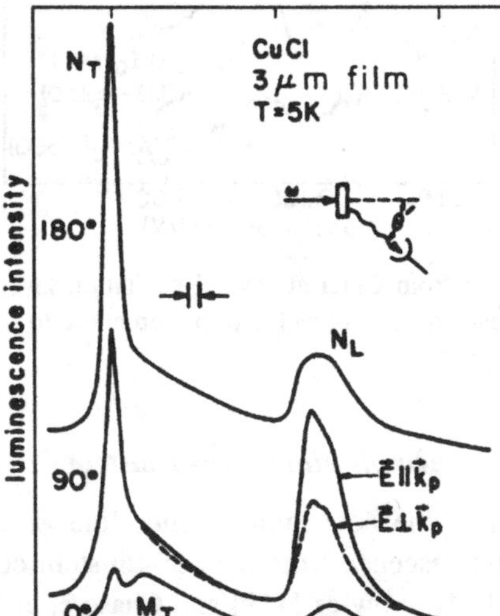

Figure 4.8. Anisotropy and polarization of the biexciton luminescence for $2\mathbf{k}_0$ excitation. Values θ shown on the left are the angles of propagation of the emitted photons relative to the direction of the incident laser beam outside the film (from Ref. 119).

for the transverse polaritons there are two possible perpendicular orientations of the dipole moments, $\mathbf{d}^{T^i}_{2\mathbf{k}_0-\mathbf{q}} \perp (2\mathbf{k}_0 - \mathbf{q})$, where $i = 1, 2$. Because the ground state of the biexciton at the point $\mathbf{k} \approx 0$ is a fully symmetric state, the selection rule for the quantum transition from the biexciton ground state to the polariton state is the same as for the quantum transition from the ground state of the crystal to the polariton state. The probability of this transition has the general form $W \approx (\mathbf{e_q} \cdot \mathbf{d}_{2\mathbf{k}_0-\mathbf{q}})^2$, where $\mathbf{e_q}$ is the polarization of the emitted photon. In a concrete form, for the spikes N_T and N_L one has

$$W_T \approx \left(\mathbf{e_q} \cdot \mathbf{d}^{T^1}_{2\mathbf{k}_0-\mathbf{q}}\right)^2 + \left(\mathbf{e_q} \cdot \mathbf{d}^{T^2}_{2\mathbf{k}_0-\mathbf{q}}\right)^2, \qquad W_L \approx \left(\mathbf{e_q} \cdot \mathbf{d}^{L}_{2\mathbf{k}_0-\mathbf{q}}\right)^2 \approx \left(\mathbf{e_q} \cdot \mathbf{k}_0\right)^2. \quad (4.104)$$

One can see that in the collinear geometry $\mathbf{q} \| \mathbf{k}_0$, when $(\mathbf{e_q} \cdot \mathbf{q}) = 0$, as well as in the case $\mathbf{q} \perp \mathbf{k}_0 \perp \mathbf{e_q}$, the N_L component is forbidden, while the N_T components, which depend on the two transverse orientations of the dipole moments $\mathbf{d}^{T^i}_{2\mathbf{k}_0-\mathbf{q}}$, are allowed. The forward-scattered luminescence from the N_T component is energetically quite far from the backward-scattered N_T luminescence, however, since it must equal the laser photon energy, as illustrated in Fig. 4.6. These selection rules agree with the experimental data. Bobrysheva, Lelyakov, and Vybornov [128] also arrived at the same conclusion.

These polarization studies show that the additional luminescence features maintain a memory of the wave vector of the original laser light. In other words, the coherence of the laser has been maintained in the biexciton gas. Is it possible that this occurs simply because the gas is not thermalized, i.e., because the relaxation time out of the states excited by the laser is long compared with the radiative lifetime? Peyghambarian et al. [119] were well aware of this possible explanation and carefully studied the issues of thermalization and relaxation time. Several experimental results argue against the interpretation of the data as nonthermalized. First, as seen in Fig. 4.7, at low density, the biexciton gas is well thermalized by phonon scattering; even though the biexcitons are all created with nearly zero kinetic energy, phonon scattering brings them into a Maxwell–Boltzmann distribution within their lifetime. The extra peak appears especially at high density, when particle–particle collisions would normally cause even faster thermalization. If the lifetime of the biexcitons decreased dramatically as density increased, then possibly the gas could become less thermalized at high density. The lifetime of the biexcitons does not decrease substantially, however, as indicated, for example, by measurements of the total luminescence intensity versus laser intensity. A second argument is based on estimates of the actual collision time versus measurements of the lifetime. By examining the Lorentzian collision broadening of the two-photon absorption (discussed in Chapter 1), Peyghambarian et al. deduced that the collision rate was $\sim 10^{12}\ \mathrm{s}^{-1}$ at the experimental biexciton densities in the range 10^{18}–$10^{19}\mathrm{cm}^{-3}$. By comparison, the biexciton radiative lifetime is $\sim 10^{-9}$ s. In addition, it was shown that particles added to the system in different parts of momentum space were preferentially attracted into the condensate. It seems reasonable to conclude, then, that the coherence is maintained in the biexciton states and that this "induced Bose condensate" is stable against collisions with phonons and other particles. This is not too different from the original prediction of Ref. 129 of the possibility of obtaining an induced BEC of excitons by means of direct resonant excitation of the exciton band.

We note that in these experiments, as well as in the experiments on free excitons in Ge [130] and in Cu_2O [126, 131, 132] discussed in Chapters 1 and 2, the apparent temperature of the gas increases as density increases. This may be caused by a nonlinear heating process, or it may be an intrinsic effect of interactions in the system.

4.4.3 Optical Phase Conjugation Based on Stimulated Scattering

The standard equilibrium distribution of quantum-statistical mechanics,

$$N_{\mathbf{k}} = \frac{1}{e^{\beta(E_{\mathbf{k}}-\mu)} \pm 1},$$

is actually just the steady-state implication [133] of a more general rule for scattering processes of quantum particles, which is that all scattering processes must be multiplied by a factor that depends on the number of particles already in the final scattering state:

$$\Gamma_{i\to f} \propto n_i(1 \pm n_f). \tag{4.105}$$

The $-$ sign for fermions yields the familiar result of Pauli exclusion, which gives zero scattering probability to an occupied state, while the $+$ sign for bosons causes occupied states to "suck in" other bosons, leading to all the strange thermodynamic effects discussed in this book. This stimulated scattering is the basis of the laser, but is also a fundamental property of any nonequilibrium Bose system. In particular, even on very short time scales when no temperature is definable, highly occupied states should attract particles from other states.

Figure 4.9 shows data from a four-wave-mixing experiment with biexcitons in CuCl [134]. In this experiment, a "pump" creates biexcitons in random-spin states with energy well above the ground state. A "probe" consists of three lasers arranged for a four-wave-mixing measurement called optical phase conjugation (illustrated in the inset of Fig. 4.9). Two counterpropagating lasers, tuned to half the biexciton ground-state energy, place biexcitons directly into states very near the ground state by means of an energy- and momentum-conserving two-photon absorption process. A third laser at approximately half the ground-state energy also hits the crystal at the same time. If states near the ground state are highly occupied, the probe beam will stimulate the emission of a fourth photon in order to complete an energy- and momentum-conserving two-photon emission process. This process, called optical phase conjugation, provides a way of probing states very near to the ground state, because momentum is translated into a measurement of the angle of the generated fourth light beam, which can be selected sensitively.

When the pump pulse is present, creating a high density of biexcitons at high kinetic energy, the phase-matched signal (the stimulated emission of the fourth photon) increases relative to the signal from the probe alone. This is the opposite of what is expected and observed in experiments on four-wave mixing with electrons, which are Fermi particles. In that case, the high density of particles created by the pump pulse serves mainly to knock particles out of the probed state and decrease the phase-matched signal [135, 136]. The biexcitons created in the ground state by the probe pulses suck in the high-energy, incoherent biexcitons from the pump pulse.

When only the high-power pump laser excites the sample, no two-photon emission from states near the ground state is observed, no matter how high the biexciton density. As discussed in Subsection 4.4.2, the lifetime of the biexcitons does not appear to be long enough for thermodynamic BEC. Nevertheless, the action of biexcitons created by the probe to suck extra particles into such a small volume of phase space is a remarkable demonstration of Bose–Einstein statistics.

Space does not allow a full discussion of many of the other interesting experiments with biexcitons. The types of semiconductor structures in which biexcitons are studied continue to increase. These now include superlattices [137], quantum wells [46, 138–141], quantum wires and quantum dots [142–144], and processes such as superradiance [145], the optical

Stark effect and Authler–Townes splitting [146], four-wave mixing and phase conjugation [147], and squeezed states [148]. The autodissociated states of biexcitons [149], the biexcitons in molecular crystals discovered by Gaididei et al. [150], and polyexcitons [151] in semiconductors have added to the diversity of biexciton studies. Some coherent nonlinear phenomena with the participation of biexcitons will be discussed in Chapters 6 and 9.

The intense investigation of the biexcitons in the past 25 years has led to revived interest in the positronium molecule (see, e.g. the review article by Abdel-Raouf [6].) The concept of the positronium molecule suggested by Wheeler [5], together with the knowledge accumulated by molecular physics, led to the formulation of the concept of excitonic molecules or biexcitons [1, 2]. One hopes that the modern theory of the positronium molecule will be

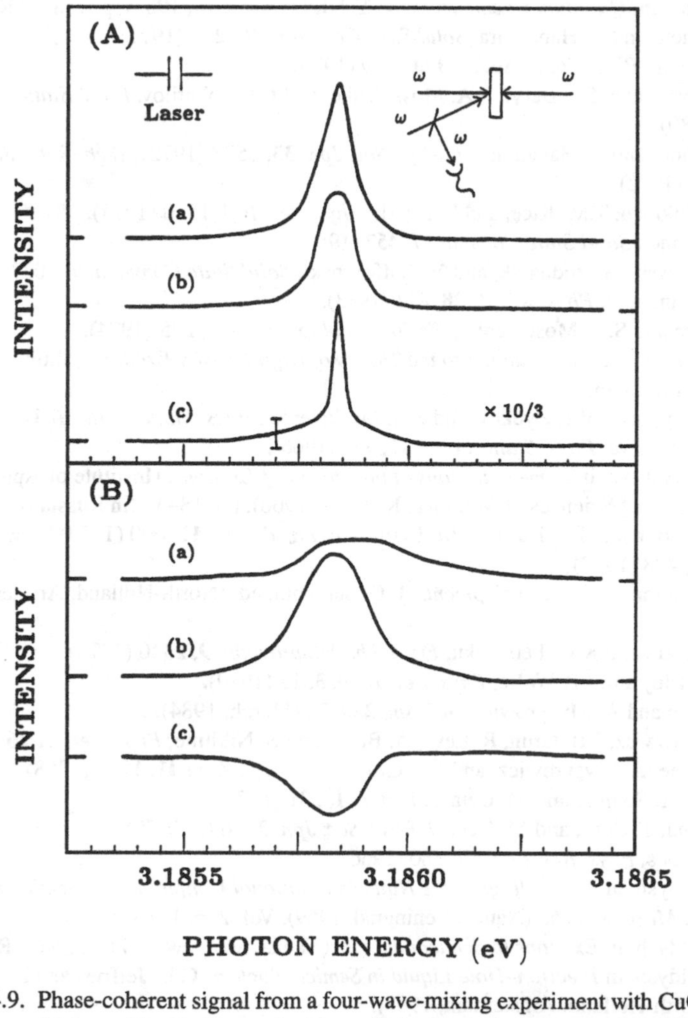

Figure 4.9. Phase-coherent signal from a four-wave-mixing experiment with CuCl. (A) Data from a high-quality CuCl sample. Curve (b) is the phase-coherent signal from the probe alone, curve (a) is the phase-coherent signal from the pump + probe, and curve (c) is the difference. The phase-coherent signal is increased in the presence of strong scattering of biexcitons at high density. (B) the same data for a low-quality sample, with a short biexciton lifetime (from Ref. 134).

enhanced by the best achievements of the theory of large-radius biexcitons in semiconductors, which may help in the experimental search for the positronium molecule.

References

[1] S.A. Moskalenko, *Zh. Opt. Spektrosk.* **5**, 147 (1958).
[2] M.A. Lampert, *Phys. Rev. Lett.* **1**, 450 (1958).
[3] S.A. Moskalenko, candidate dissertation (Institute of Physics of the Academy of Sciences of the Ukrainian SSR, Kiev, 1959).
[4] V.I. Yudson, *Phys. Rev. Lett.* **77**, 1564 (1996).
[5] J.A. Wheeler, *Ann. N.Y. Acad. Sci.* **48**, 219 (1946).
[6] M.A. Abdel-Raouf, *Fortschr. Phys.* **36**, 521 (1988).
[7] G.G. Gelman, *Quantum Chemistry* (ONTY, NKTP, Moscow, 1937), p. 138 (in Russian).
[8] O. Akimoto and E. Hanamura, *Solid State Commun.* **10**, 253 (1972).
[9] W.T. Huang, *Phys. Status Solidi B* **60**, 309 (1973).
[10] A.I. Bobrysheva, S.I. Beryl, S.A. Moskalenko, and E.P. Pokatilov, *Phys. Status Solidi B* **100**, 281 (1980).
[11] O. Akimoto and E. Hanamura, *J. Phys. Soc. Jpn.* **33**, 1537 (1972); *Tech. Rep. ISSP*, Series A, No. 526 (1972).
[12] W.F. Brinkman, T.M. Rice, and B.T. Bell, *Phys. Rev. B* **8**, 1570 (1973).
[13] R.R. Wehner, *Solid State Commun.* **7**, 457 (1969).
[14] J. Adamowski, S. Bednarek, and M. Suffczynski, *Solid State Commun.* **9**, 2037 (1971).
[15] D.A. Kleinmann, *Phys. Rev. B* **28**, 871 (1983).
[16] I.A. Karp and S.A. Moskalenko, *Fiz. Tekh. Poluprovodn.* **8**, 285 (1974).
[17] S.A. Moskalenko, *Introduction to the Theory of High Density Excitons* (Shtiintsa, Kishinev, 1983) (in Russian).
[18] A.I. Bobrysheva, V.T. Zyukov, and S.A. Moskalenko, *Phys. Status Solidi B* **105 K**, 45 (1981).
[19] A.I. Bobrysheva, *Phys. Status Solidi* **16**, 337 (1966).
[20] A.I. Bobrysheva, in *Some Questions of the Theory of Excitons*, (Institute of Applied Physics of Academy of Sciences of Moldova, Kishinev, 1966), pp. 18–38 (in Russian).
[21] A.A. Gogolin and E.I. Rashba, *Zh. Eksp. Teor. Fiz. Pis'ma* **17**, 690 (1973) [*Sov. Phys. JETP Lett.* **17**, 478 (1973)].
[22] E. Hanamura, in *New Developments*, E.O. Seraphin, ed. (North-Holland, Amsterdam, 1976), p. 81.
[23] P.I. Khadzhi and K.G. Petrashku, *Fiz. Tekh. Poluprovodn.* **9**, 2340 (1975).
[24] P.L. Gourley and J.P. Wolfe, *Phys. Rev. B* **20**, 3319 (1979).
[25] J.P. Wolfe and A. Mysyrowicz, *Sci. Am.* **250**, 70 (March, 1984).
[26] A. Mysyrowicz, J.B. Grun, R. Levy, A. Bivas, and S. Nikitine, *Phys. Lett. A* **26**, 615 (1968).
[27] S. Nikitine, A. Mysyrowicz, and J.B. Grun, *Helv. Phys. Acta* **41**, 1058 (1968).
[28] T. Goto, H. Souma, and M. Ueta, *J. Lumin.* **1**, 231 (1970).
[29] H. Souma, T. Goto, and M. Ueta, *J. Phys. Soc. Jpn.* **29**, 697 (1970).
[30] J.R. Haynes, *Phys. Rev. Lett.* **17**, 860 (1966).
[31] L.V. Keldysh, in *Proceedings of the Ninth International Conference on Semiconductor Physics, Moscow, 1968* (Nauka, Leningrad, 1969), Vol. 2, p. 1384.
[32] L.V. Keldysh, in *Excitons in Semiconductors* (Nauka, Moscow, 1971), p. 5 (in Russian).
[33] L.V. Keldysh, in *Electron–Hole Liquid in Semiconductors*, C.D. Jeffries and L.V. Keldysh, eds. (North-Holland, Amsterdam, 1987).
[34] Ya.E. Pokrovskii, *Phys. Status Solidi B* **11**, 385 (1972).
[35] T.M. Rice, in *Solid State Physics* **32**, H. Ehrenreich, F. Seitz, and D. Turnbull, eds. (Academic, New York, 1977).
[36] D. Hensel, T. Phillips, and G. Thomas, in *Solid State Physics* **32**, H. Ehrenreich, F. Seitz, and D. Turnbull, eds. (Academic, New York, 1977).
[37] W.T. Brinkman and T.M. Rice, *Phys. Rev. B* **7**, 1508 (1973).

[38] M. Combescot and P. Nozieres, *J. Phys. C* **5**, 2369 (1972).
[39] I.V. Kukushkin, V.D. Kulakovskii, and V.B. Timofeev, *Zh. Eksp. Teor. Fiz.* **32**, 304 (1980).
[40] V.D. Kulakovskii, I.V. Kukushkin, and V.B. Timofeev, *Zh. Eksp. Teor. Fiz.* **81**, 684 (1981).
[41] V.B.Timofeev, in *Excitons*, E.I. Rashba and M.D. Sturge, eds. (North-Holland, Amsterdam, 1982), p. 254.
[42] M.H. Anderson et al., *Science* **269 (7)**, 198 (1995).
[43] K.B. Davis et al., *Phys. Rev. Lett.* **75**, 3969 (1995); M.R. Andrews et al., *Science* **273 (7)**, 84 (1996).
[44] R.C. Miller, D.A. Kleinmann, A.C. Gossard, and O. Munteanu, *Phys. Rev. B* **25**, 6545 (1982).
[45] S. Bar-Ad and I. Bar-Joseph, *Phys. Rev. Lett.* **68**, 349 (1992).
[46] J.C. Kim and J.P. Wolfe, *Phys. Rev. B* **57**, 9861 (1998).
[47] Y. Nozue, N. Miyahara, S. Takagi, and M. Ueta, *Solid State Commun.* **38**, 1199 (1981).
[48] Y. Nozue, N. Miyahara, S. Takagi, and M. Ueta, *J. Lumin.* **24/25**, 429 (1981).
[49] S. Nikitine, in *Excitons at High Density*, H. Haken and S. Nikitine, eds. (Springer, Berlin, 1975), p. 18.
[50] E. Hanamura, in *Excitons at High Density*, H. Haken and S. Nikitine, eds. (Springer, Berlin, 1975), p. 43.
[51] R. Levy and J.B. Grun, *Phys. Status Solidi B* **22**, 11 (1974).
[52] S.A. Moskalenko, in *Molecular Spectroscopy of Dense Phases*, M. Grosmann, S.G. Elkomoss, and J. Ringeissen, eds. (Elsevier, New York, 1976), p. 45.
[53] J.B. Grun, *Nuovo Cimento B* **39**, 579 (1977).
[54] E. Hanamura and H. Haug, *Phys. Rep. C* **33**, 210 (1977).
[55] H. Haug and C. Klingshirn, *Phys. Rep.* **70**, 315 (1981).
[56] J.B. Grun, B. Hönerlage, and R. Levy, in *Excitons*, E.I. Rashba and M.D. Sturge, eds. (Nauka, Moscow, 1985), p. 333.
[57] A.I. Bobrysheva, S.A. Moskalenko, and P.I. Khadzhi, *Bul. Acad. Shtiintse Repub. Moldova, (Seria fizica si technica)* **2**, 77 (1990).
[58] S.A. Moskalenko and P.I. Khadzhi, in *Interaction of Excitons with Laser Radiation* (Shtiintsa, Kishinev, 1991), pp. 3–31 (in Russian).
[59] A. Griffin, D.W. Snoke, and S. Stringari, eds., *Bose–Einstein Condensation* (Cambridge U. Press, Cambridge, 1995).
[60] *Nonlinear Optical Properties of Excitons in Semiconductors of Different Dimensionalities*, S.A. Moskalenko and P.I. Khadzhi, eds. (Shtiintsa, Kishinev, 1992) (in Russian).
[61] *Excitons and Biexcitons in Size-Confined Systems* S.A. Moskalenko and P.I. Khadzhi, eds. (Shtiintsa, Kishinev, 1990) (in Russian).
[62] *Quantum Statistical Properties of High-Density Excitons*, V.A. Moskalenko and P.I. Khadzhi, eds. (Shtiintsa, Kishinev, 1988) (in Russian).
[63] A.I. Bobrysheva, S.A. Moskalenko, V.R. Misko, and S.I. Negru, *Proceedings of the International Semiconductor Conference*, 18th ed. (IEEE Romanian Section, Sinaia, Romania, October 1995), pp. 263–266.
[64] P.I. Khadzhi, *Kinetics of the Recombinational Radiation of Excitons and Biexcitons in Semiconductors* (Shtiintsa, Kishinev, 1977) (in Russian).
[65] A.I. Bobrysheva, *Biexcitons in Semiconductors* (Shtiintsa, Kishinev, 1979) (in Russian).
[66] M.I. Shmiglyuk and V.N. Pitei, *Coherent polaritons in Semiconductors* (Shtiintsa, Kishinev, 1989) (in Russian).
[67] I.I. Jeru, *Low-Frequency Resonances of Excitons and Impurity Centers* (Shtiintsa, Kishinev, 1976) (in Russian).
[68] S.A. Moskalenko, A.I. Bobrysheva, A.V. Lelyakov, M.F. Miglei, P.I. Khadzhi, and M.I. Shmiglyuk, *Interaction of Excitons in Semiconductors* (Shtiintsa, Kishinev, 1974) (in Russian).
[69] S.S. Russu and K.G. Petrashku, *Kinetics of High-Density Excitons in Semiconductors* (Shtiintsa, Kishinev, 1986) (in Russian).
[70] A.L. Ivanov and H. Haug, *Phys. Rev. B* **48**, 1490 (1993).
[71] A.L. Ivanov and H. Haug, *Phys. Rev. Lett.* **74**, 438 (1995).

[72] A.L. Ivanov and H. Haug, *Phys. Status Solidi B* **188**, 61 (1995).
[73] A.L. Ivanov, M. Hasuo, N. Nagasawa, and H. Haug, *Phys. Rev. B* **52**, 11017 (1995).
[74] M. Hasuo, H. Kawano, and N. Nagasawa, *Phys. Status Solidi B* **188**, 77 (1995).
[75] M. Hasuo, H. Kawano, and N. Nagasawa, *J. Lumin.* **60/61**, 672 (1994).
[76] M. Hasuo, M. Nishino, T. Itoh, and A. Mysyrowicz, *Phys. Rev. Lett.* **70**, 1303 (1993).
[77] F. Čulik, *Czech. J. Phys. B* **70**, 194 (1966).
[78] A.I. Bobrysheva, M.F. Miglei, and M.I. Shmiglyuk, *Phys. Status Solidi B* **53**, 71 (1972).
[79] A.I. Bobrysheva, M.F. Miglei, S.A. Moskalenko, and M.I. Shmiglyuk, in *Compound Semiconductors and Their Physical Properties* (Shtiintsa, Kishinev, 1971), pp. 25–34 (in Russian).
[80] A.I. Bobrysheva and S.A. Moskalenko, *Phys. Status Solidi B* **119**, 141 (1983).
[81] A.I. Bobrysheva, S.A. Moskalenko, M.I. Shmigluk, and S.S. Russu, *Phys. Status Solidi B* **190**, 481 (1995).
[82] E. Hanamura, *J. Phys. Soc. Jpn.* **29**, 50 (1970).
[83] A.I. Bobrysheva, V.T. Zyukov, and S.I. Beryl, *Phys. Status Solidi B* **101**, 69 (1980).
[84] S. Schmitt-Rink, D.S. Chemla, and D.A.B. Miller, *Phys. Rev. B* **32**, 6601 (1985).
[85] A.I. Bobrysheva, S.A. Moskalenko, and V.I. Vybornov, *Phys. Status Solidi B* **76**, K51 (1976).
[86] T. Usui, *Prog. Theor. Phys.* **23**, 787 (1960).
[87] T. Hiroshima, *Phys. Rev. B* **40**, 3862 (1989).
[88] S.A. Moskalenko and A.I. Bobrysheva, in *Abstracts of the First International Conference on Excitonic Processes in Condensed Matter (EXCON-94)* (North Territory University, Darwin, Australia, 1994) in *Abstracts of the Second International Conference on Excitonic Processes in Condensed Matter (EXCON-96)* (Dresden U. Press, Bad Schandau, Germany, 1996); in *Proceedings of the Seventeenth Annual Semiconductor Conference, (CAS-94) Sinaia, Romania, 1994*, Vol. 2, (Romanian Academy, Sinaia, Romania, 1994) pp. 555–558.
[89] L.V. Keldysh and A.N. Kozlov, *Pis'ma Zh. Eksp. Teor. Fiz.* **5**, 238 (1967); *Zh. Eksp. Teor. Fiz.* **54**, 978 (1968) [*Sov. Phys. JETP* **27**, 521 (1968)].
[90] N.N. Bogoliubov, V.V. Tolmachev, and D.V. Shirkov, *New Method in the Theory of Superconductivity* (Consultants Bureau, New York, 1959).
[91] S.A. Moskalenko, *Bose–Einstein Condensation of Excitons and Biexcitons* (Academy of Sciences of the MSSR, Kishinev, 1970) (in Russian).
[92] B. Gerlach, *Phys. Status Solidi B* **63**, 459 (1974).
[93] G. Munschy and B. Stébé, *Phys. Status Solidi B* **64**, 213 (1974); **72**, 135 (1975).
[94] B. Stébé and G. Munschy, *Solid State Commun.* **17**, 1051 (1975).
[95] Z.A. Insepov and H.E. Norman, *Zh. Eksp. Teor. Fiz.* **69**, 1321 (1975).
[96] A.I. Bobrysheva, V.T. Zyukov, and P.G. Bilinkis, *Fiz. Tekh. Poluprovodn.* **15**, 1400 (1981).
[97] A.I. Bobrysheva, M.V. Grodetskii, and V.T. Zyukov, *J. Phys. C* **16**, 5723 (1983).
[98] G. Finkelstein and I. Bar-Joseph, *Nuovo Cimento D* **17**, 1239 (1995).
[99] G. Finkelstein, H. Shtrikman, and I. Bar-Joseph, *Phys. Rev. Lett.* **74**, 976 (1995).
[100] G. Finkelstein, H. Shtrikman, and I. Bar-Joseph, *Phys. Rev. B* **53**, R1709 (1996).
[101] G. Finkelstein, H. Shtrikman, and I. Bar-Joseph, *Phys. Rev. B* **53**, 12593 (1996).
[102] Yu.E. Lozovik and V.I. Yudson, *Pis'ma Zh. Eksp. Teor. Fiz.* **22**, 26 (1975) [*Sov. Phys. JETP Lett.* **22**, 11 (1975)].
[103] J.G. Valatin and D. Butler, *Nuovo Cimento* **10**, 37 (1958).
[104] A. Coniglio and M. Marinaro, *Nuovo Cimento B* **48**, 249 (1967).
[105] A. Coniglio, F. Manchini, and M. Maturi, *Nuovo Cimento B* **63**, 227 (1969).
[106] W.A.B. Evans and Y. Imri, *Nuovo Cimento B* **63**, 155 (1969).
[107] C. Mavroyannis, *Phys. Rev. B* **10**, 1741 (1974).
[108] C. Mavroyannis, *J. Low Temp. Phys.* **25**, 501 (1976).
[109] P. Noziéres and D. Saint James, *J. Phys. (Paris) B* **43**, 1133 (1982).
[110] A.I. Bobrysheva, S.A. Moskalenko, and Yu.M. Shvera, *Fiz. Tekh. Poluprovodn.* **21**, 366 (1987).
[111] A.I. Bobrysheva, S.A. Moskalenko, and Yu.M. Shvera, *Phys. Status Solidi B* **147**, 711 (1988).

[112] A.I. Bobrysheva and S.A. Moskalenko, *Fiz. Tverd. Tela* **25**, 3282 (1983).
[113] V.N. Nandakumaran and K.P. Sinha, *Z. Phys. B* **22**, 173 (1975).
[114] A.I. Bobrysheva, S.A. Moskalenko, M.I. Shmigliuk, and S.S. Russu, *Phys. Status Solidi B* **190**, 481 (1995).
[115] F. Bassani and M. Rovere, *Solid State Commun.* **19**, 887 (1976).
[116] A.I. Bobrysheva and S.A. Moskalenko, *Fiz. Tverd. Tela* **25**, 3282 (1983); *Phys. Status Solidi B* **119**, 141 (1983).
[117] A.I. Bobrysheva, S.A. Moskalenko, and S. Russu, *Phys. Status Solidi B* **167**, 625 (1991).
[118] L.L. Chase, N. Peyghambarian, G. Grinberg, and A. Mysyrowicz, *Phys. Rev. Lett.* **42**, 1231 (1979).
[119] N. Peyghambarian, L.L. Chase, and A. Mysyrowicz, *Phys. Rev. B* **27**, 2325 (1983).
[120] A.I. Bobrysheva and S.A. Moskalenko, *Bul. acad. shtiintse Repub. Moldova Ser. Fiz. Teh.* **2** (14), 3 (1994).
[121] A.I. Bobrysheva, S.A. Moskalenko, and H.N. Kam, *Zh. Eksp. Teor. Fiz.* **103**, 301 (1993); [*Sov. Phys. JETP* **76**, 163 (1993)].
[122] V.M. Agranovich, G.C. La Rocca, and F. Bassani, *J. Lum.* **76/77**, 161 (1998).
[123] A.L. Ivanov, H. Haug, and L.V. Keldysh, *Phys. Reports* **296**, 237 (1998).
[124] A.I. Bobrysheva and S.A. Moskalenko, in *Bose-Einstein Condensation*, A. Griffin, D.W. Snoke, and S. Stringari, eds. (Cambridge U. Press, Cambridge, 1995), pp. 507–512.
[125] M.M. Denisov and V.P. Makarov, *Phys. Status Solidi B* **56**, 9 (1973).
[126] T. Goto, M.Y. Shen, S. Koyama, and T. Yokouchi, *Phys. Rev. B* **55**, 7609 (1997).
[127] Y. Yakhot and E. Levich, *Phys. Lett. A* **80**, 301 (1980).
[128] A.I. Bobrysheva, A.V. Lelyakov, and V.I. Vybornov, *Fiz. Tverd. Tela* **20**, 1620 (1978).
[129] S.A. Moskalenko, *Fiz. Tverd. Tela* **4**, 276 (1962).
[130] V.B. Timofeev, V.D. Kulakovskii, and I.V. Kukushkin, *Physica B+C* **117/118**, 327 (1983).
[131] D.W. Snoke, J.P. Wolfe, and A. Mysyrowicz, *Phys. Rev. Lett.* **59**, 827 (1987).
[132] J.L. Lin and J.P. Wolfe, *Phys. Rev. Lett.* **71**, 1223 (1993).
[133] D.I. Blokhintsev, *Quantum Mechanics* (Dordrecht-Holland, Amsterdam, 1964), p. 493ff.
[134] M. Hasuo, N. Nagasawa, T. Itoh, and A. Mysyrowicz, *Phys. Rev. Lett.* **70**, 1303 (1993).
[135] P.C. Becker, H.L. Fragnito, C.H. Brito Fork, J.E. Cunningham, J.E. Henry, and C.V. Shank, *Phys. Rev. Lett.* **61**, 1647 (1988).
[136] J.-Y. Bigot, M.T. Portella, R.W. Schoenlein, J.E. Cunningham, and C.V. Shank, *Phys. Rev. Lett.* **67**, 636 (1991).
[137] M. Nakayama, K. Suyama, and H. Nishimura, *Phys. Rev. B* **51**, 7870 (1995).
[138] K.H. Pantke, D. Oberhauser, V.G. Lyssenko, and J.M. Hvan, *Phys. Rev. B* **47**, 2413 (1993).
[139] J. Zhang, T. Pang, and C. Chen, *Phys. Lett. A* **206**, 101 (1995).
[140] F. Kreller, M. Lowich, J. Puls, and F. Henneberger, *Phys. Rev. Lett.* **75**, 2420 (1995).
[141] Y. Yamada, T. Mishina, Y. Masumoto, Y. Kawakami, and S. Yamaguchi, *Phys. Rev. B* **51**, 2596 (1995).
[142] L. Banyai, *Phys. Rev. B* **39**, 8022 (1989).
[143] Y.Z. Hu, S.W. Koch, M. Lindberg, N. Peyghambarian, E.L. Pollock, and F.F. Abraham, *Phys. Rev. Lett.* **64**, 1805 (1990).
[144] Y. Masumoto, S. Okamoto, and S. Katayanagi, *Phys. Rev. B* **50**, 18658 (1994).
[145] V.M. Agranovich and S. Mukamel, *Phys. Lett. A* **147**, 155 (1990).
[146] Z.G. Koinov, *J. Phys. C* **8**, L391 (1996).
[147] B.F. Feuerbacher, J. Kuhl, and K. Ploog, *Phys. Rev. B* **43**, 2439 (1991).
[148] N. Ba An, *Phys. Rev. B* **48**, 11732 (1993).
[149] D. Labrie, V.A. Karasyuk, M.K. Nissen, Ya.E. Pokrovskii, and M.L.W. Thevalt, *Phys. Rev. B* **47**, 1628 (1993).
[150] Yu.B. Gaididei, V.M. Loktev, A.F. Prikhot'ko, and L.I. Shanskii, *Pis'ma Zh. Eksp. Teor. Fiz.* **18**, 164 (1973).
[151] A.G. Steele, W.G. McHullin, and M.L. Thewalt, *Phys. Rev. Lett.* **59**, 2899 (1987).

5

Phase Transitions and Thermodynamics of High-Density Excitons

5.1 Introduction

As discussed briefly in Chapter 1, the general theory of phase transitions in the electron–hole system predicts that in a system with a simple band structure, when annihilation and polariton effects are neglected, there will always be a region of phase space in which BEC of electron–hole pairs occurs. In this chapter we examine some of the more complicated features of the electron–hole phase diagram.

The many-body system of electrons and holes appears deceptively simple. If one neglects the possibility for electron–hole annihilation, one can consider the following apparently simple system: two types of Fermi particles with equal mass and opposite charge, interacting only by means of Coulomb interaction. As we will see in this chapter, however, the phase diagram for this system is extraordinarily complex, and the full theory for the phase diagram is not yet complete. We emphasize that most of these complexities arise solely from the above simple model and not from the other complicated features of solids such as band structure, band-to-band recombination, etc. Therefore much of this theory applies equally well to the case of electron–positron plasma or electron–ion plasma. The similarity of the *e–h* system to the electron–proton system, for example, suggested the existence of excitonic molecules [1, 2] and, moreover, the possibility of a gas–liquid phase transition of exciton and biexciton gas into a Fermi electron–hole liquid (EHL). The EHL is the bound state of a macroscopically large number of electrons and holes [3–6], similar to the many-electron–proton system.

As stressed by Keldysh, however, this similarity is not complete. In a comprehensive and instructive review [7], he pointed out the peculiar properties distinguishing the *e–h* system from any other, which we summarize here. One of them is the greatly reduced Coulomb interaction due to dielectric screening in the host crystal. Another peculiarity is the absence in the *e–h* system of any really heavy particles such as nuclei, since the electron and hole masses have the same order of magnitude or differ by, at most, one order. This fact results in a dominant role for quantum effects at all temperatures $k_B T < \mathrm{Ry}_{\mathrm{ex}}$ and leads to very large zero-energy vibrations of excitons in the excitonic molecules. The amplitude of these vibrations is of the order of, or even larger than, the excitonic radius a_{ex} itself, so that the biexcitons are always very loosely bound complexes with binding energies not exceeding $0.1\mathrm{Ry}_{\mathrm{ex}}$. For the same reason, when the electron and hole masses are comparable, nothing like an "excitonic crystal," analogous to solid hydrogen, can exist. The weak van der Waals attraction between the excitons is not able to confine light particles such as excitons or biexcitons.

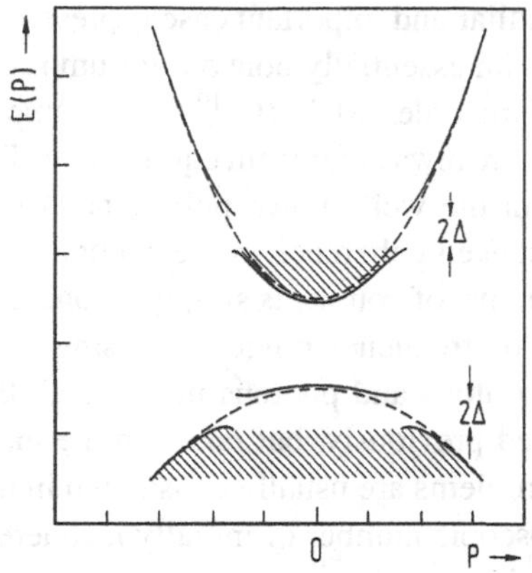

Figure 5.1. Energy spectrum in the nonequilibrium exciton-insulator phase. The hatched regions are the energy ranges occupied by electrons (from Ref. 7).

A condensed phase of biexcitonic molecules, i.e., a molecular liquid similar to liquid hydrogen, also cannot exist because the s-wave-scattering length of two biexcitons is large and positive. But an EHL, similar to metallic hydrogen or alkali metals, does exist [3–6]. The EHL transforms into a nonequilibrium excitonic-insulator phase at low temperatures [8–10] if the effective masses of both electrons and holes are more or less isotropic. As illustrated in Fig. 5.1, the collective pairing of electrons and holes in the vicinity of the Fermi surfaces is quite similar to Bardeen-Cooper-Shrieffer (BCS) pairing in superconductors and is manifested in the appearance of energy gaps near the Fermi surfaces. These gaps may be considered as a remnant of the binding energy of a single exciton and decrease quickly with increasing particle density. In this sense, then, the excitonic-insulator state in the nonequilibrium e–h system is a coherent BEC state of high-density excitons, with $n_{ex}a_{ex}^{-3} \geq 1$. Recent experiments [11] in which short, intense laser pulses were used to generate high nonequilibrium densities of electrons and holes have been interpreted as observations of this excitonic-insulator state [12].

Another important difference of the e–h system from ordinary matter is its essentially nonequilibrium nature, produced by some external action, usually illumination. Then the total number N of e–h pairs becomes an independent variable with a value controlled by an external source. As mentioned above, in certain semiconductors the lifetime of the nonequilibrium e–h system may be much longer than the thermalization time, so that conservation of N is not broken by the recombination processes on short time scales. The system appears to be in quasi-equilibrium, the only nonequilibrium parameter being the total number of particles N.

As Keldysh noted [7], it is also possible to create an equilibrium excitonic-insulator state, i.e., a population of permanent excitons. This can occur if the binding energy per pair in the excitonic molecule or the EHL is larger than the initial energy band gap for free-e–h pair creation, E_{gap}. In this case, spontaneous reconstruction of the electronic structure of the crystal occurs. Nothing like superfluidity can arise, however, because the Hamiltonian does not conserve the total number of excitons or e–h pairs [8, 13]. In this case, the interband scattering matrix elements of the Coulomb interaction are the sources of the creation of e–h pairs, lifting the gauge invariance of the Hamiltonian [14]. We discuss experimental and theoretical work on this possibility in Chapter 10.

In the opposite limit, a peculiar and important case is presented by BEC of dipole-active excitons and photons. This is an essentially nonequilibrium problem, since the lifetime of dipole-allowed excitons is of the order 10^{-9}–10^{-10} s, i.e., comparable with the thermalization times, and therefore they are always far from equilibrium. The quantum superpositions of photons and excitons form the well-known polaritons [15]. In this case, there is not always a clear distinction between coherence in the excitonic states and coherence in the photon states. A BEC of photons, of course, is simply a coherent electromagnetic wave of finite amplitude in the polariton frequency range, i.e., lasing.

BEC of mixed states of excitons and photons includes all the many-body phenomena of BEC of excitons developed previously, but takes on the meaning of nonlinear optical phenomena. Two different problems are usually considered in this context. One of them is the accumulation of a macroscopic number of initially incoherent-excitation quanta into a single photon mode. This lasing transition, i.e., the appearance of a coherent mode from multimode intense noise in a pumped system, is a typical example of a nonequilibrium phase transition [16–21]. When this theory is applied to the lower polariton branch in semiconductors, one gets the "polariton-bottleneck" picture. Because of the small density of states in the polariton–exciton-crossing region, the polariton occupation number can reach large values even if the total exciton density is small. Then the mechanism of line narrowing by stimulated emission starts. Under the condition of sufficiently strong stationary pumping, the stationary incoherent-radiation regime becomes instable and a finite-amplitude coherent mode appears. In a finite time interval, the exciton interaction synchronizes all the modes in the region of the bottleneck and a coherent state with a macroscopically large number of polaritons arises. This is simply a BEC of polaritons. We will treat this problem in Chapter 7.

Another direction of research is nonlinear optics proper, i.e., the propagation of intense electromagnetic waves or pulses and their interaction with a medium under the condition of polariton formation [21,22]. In this case, the coherence is induced by the external source of radiation [7]. We will address some of these cases in Chapter 9. In the present chapter we restrict our attention to systems in which quasi-equilibrium has been established, i.e., in which scattering times are much shorter than the particle recombination lifetimes.

5.2 Excitons and Electron–Hole Plasma

5.2.1 Electron–Hole Liquid and Electron–Hole Plasma

As mentioned in Chapter 4, Keldysh originated the idea of the EHL [3,4,23]. He pointed out for the first time that the excitons and the excitonic molecules of the types similar to alkali metal atoms and molecules, with comparatively small binding energies, will become EHLs and electron–hole droplets (EHDs) in the case of high-intensity excitation. These suggestions were justified completely by the subsequent theoretical and experimental investigations of high-density excitons in such crystals as Ge and Si [5,24–31]. Brinkman and Rice [27] and Combescot and Nozières [28] showed that the many-valley structure of the conduction bands in Ge and Si, the fourfold degenerate structure of their valence bands, and the anisotropy of the corresponding masses not only facilitate but play a decisive role in the formation of EHLs and EHDs. The EHL happens to be much more stable than the exciton and biexciton gases in these crystals.

The theory of electron–hole plasma (EHP) and the more complicated case of EHL begins with the determination of the energy per *e–h* pair in the Hartree–Fock approximation. As usual, this energy E_{HF}^{eh} is measured from the bottom of the conduction band. A short review of the theory at $T = 0$ is presented here.

The energies E_{HF}^{eh} for the degenerate EHP at $T = 0$ in Ge and Si were determined in Refs. 27 and 28, taking into account their real band structures, and are listed below together with the well-known result for the crystal with a simple band structure [32]:

$$E_{\mathrm{HF}}^{eh} = \left(\frac{0.468}{r_s^2} - \frac{1.136}{r_s}\right) \mathrm{Ry_{ex}} \qquad \text{(Ge)},$$

$$E_{\mathrm{HF}}^{eh} = \left(\frac{0.727}{r_s^2} - \frac{1.157}{r_s}\right) \mathrm{Ry_{ex}} \qquad \text{(Si)},$$

$$E_{\mathrm{HF}}^{eh} = \left(\frac{2.21}{r_s^2} - \frac{1.832}{r_s}\right) \mathrm{Ry_{ex}} \qquad \text{(simple model)}. \tag{5.1}$$

Here r_s is the dimensionless mean distance between the particles expressed in units of the exciton Bohr radius. For the electron–hole densities $n_e = n_h = n$, one has

$$r_s = \frac{r_0}{a_{\mathrm{ex}}}, \qquad r_0 = \left(\frac{3}{4\pi n}\right)^{1/3}. \tag{5.2}$$

The total energy per e–h pair is defined in terms of the correlation energy E_{corr}^{eh} and the energy E_{HF}^{eh} deduced in Eqs. (5.1) as

$$E^{eh} = E_{\mathrm{HF}}^{eh} + E_{\mathrm{corr}}^{eh}. \tag{5.3}$$

The estimation of the correlation energy, which takes into account the screening effects, is the most complicated part of the theory. The calculation of the screening effects requires knowledge of the dielectric constant $\varepsilon(\mathbf{k}, \omega)$. In the theory of the EHL, $\varepsilon(\mathbf{k}, \omega)$ is determined in the random-phase approximation (RPA) with the Hubbard [33] or Nozières and Pines [34] corrections taken in account, which is necessary when r_s is near unity.

In this way Brinkman and Rice [27] found that in the case of a simple band structure, the energy per e–h pair has the minimal value $E^{eh} = -0.86\,\mathrm{Ry_{ex}}$, corresponding to the point $r_s = 1.95$, when the ratio σ of the electron–hole masses ($\sigma = m_e/m_h$) equals unity. The value E^{eh} ($\sigma = 1$) lies above the exciton level $E_{\mathrm{ex}}^{eh} = -\mathrm{Ry_{ex}}$. As a result, the exciton gas is a more stable state than that of the EHL. This is the primary reason why Cu_2O is so attractive for studies of excitonic condensation, since it has a simple band structure and an electron–hole mass ratio of nearly unity.

In the case $\sigma \to 0$, the value $E^{eh}(\sigma = 0)$ lies a little lower on the energy scale than the exciton level, but is situated higher than the energy per e–h pair in the biexciton complex. This last value equals $E_{\mathrm{biex}}^{eh} = -\mathrm{Ry_{ex}} - |W|/2$, where $|W|$ is the dissociation energy of the biexciton. Brinkman and Rice concluded that the biexciton gas is more stable than the EHL in the case of a simple band structure at arbitrary values of σ.

A completely different situation is realized in the crystals Ge and Si, in which the polarizabilities of the electrons and the holes are greater, essentially because of the multiple degeneracy of the conduction-band minima. This fact leads to the important role played by the screening effects and the correlation energies in both crystals. For example, in the case of the Ge crystal, the position of the energy E^{eh} on the energy scale is much lower than the exciton level and has the value $E^{eh} \cong -2\mathrm{Ry_{ex}}$. The results obtained for Ge by Brinkman

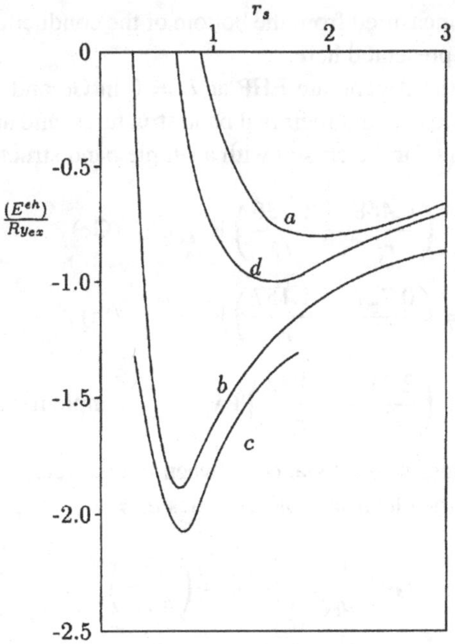

Figure 5.2. The ground-state energy per e–h pair in a Ge crystal as a function of the dimensionless distance r_s: curve a, the isotropic simplified model; curve b, Brinkman and Rice results [27]; curve c, Combescot and Nozières results [28]; curve d, uniaxially stressed Ge [27].

and Rice [27], as well as by Combescot and Nozières [28], are presented in Fig. 5.2 Curve a concerns the isotropic simple model with equal electron–hole masses ($\sigma = 1$). Curves b and c represent results obtained by Brinkman and Rice [27] and Combescot and Nozières [28], respectively. Curve d refers to the crystal Ge subject to a uniaxial stress along the $\langle 111\rangle$ direction, as calculated by Brinkman and Rice [27]. The $\langle 111\rangle$ uniaxial stress removes the fourfold degeneracy of the conduction-band valleys, since only one of them remains as the lowest electron band. As one can see, the minimum of the energy E^{eh} in the unstressed Ge approximately equals $E^{eh} \simeq -2\mathrm{R}_{\mathrm{y}}^{\mathrm{ex}}$ and is achieved at the point $r_s = 0.63$, which corresponds to the equilibrium density of the e–h pairs $n_{\mathrm{eq}} = 1.8 \times 10^{17}$ cm^{-3}.

If the mean density of the e–h pairs excited in the Ge crystal is less than the equilibrium value n_{eq}, then the e–h pairs, having enough time to join together because of their long lifetime, will form EHDs, and a condensation of the exciton gas in real space takes place. This represents the opposite limit of BEC in momentum space, which depends on the repulsive interaction between the excitons. It is generally the case that if repulsive interactions prevent real-space ordering, then a reduction of entropy must lead to momentum-space ordering, i.e., Bose condensation. In the case of EHL, the transition from the gaseous exciton phase to the EHL phase at $T \neq 0$ is a first-order phase transition.

EHL differs from EHP in the same way that metallic Hg differs from homogeneous electron–ion plasma. As one can see from Fig. 5.2, EHL in a uniaxially stressed Ge crystal becomes much less stable – its ground-state energy per e–h pair is not far from the exciton level. It was under these conditions that Kukushkin, Kulakovskii, and Timofeev [29–31,35] saw the experimental evidence for Bose effects in the recombination luminescence line shape of the excitons, discussed in Chapter 1. Since the biexciton dissociation energy in Ge is small, if one applies, in addition to the unaxial stress, a constant external magnetic field H such that the energy of the electron or hole spin in the magnetic field $\mu_B H$ (μ_B is the Bohr magneton) is comparable with the biexciton dissociation energy, then the biexcitons will become unstable. Another way of looking at this is to note that although the biexciton is stable when the spins of two electrons and two holes separately are oriented antiparallel each other, as discussed in Chapter 4, orienting the spins of the particles parallel to each other by means of an external magnetic field will eliminate the unique spin orientations favorable for the existence of biexcitons. Kukushkin et al. succeeded in removing the biexciton luminescence line by means of an applied magnetic field in exactly this way [29,30].

This experimental situation, with spin-aligned electrons, holes, and excitons, is the basis of the theoretical model discussed below. In this three-component system, the biexcitons and the difference between EHL and EHP can be neglected. The phase transition in this system is studied in the mean-field approximation by use of the Bogoliubov variational principle. The theory is similar to that of spin-aligned hydrogen, in which the magnetic field is used to prevent the formation of hydrogen molecules and metallic hydrogen at high densities in order to see Bose effects of the individual atoms [36].

5.2.2 First-Order and Second-Order Phase Transitions in the Degenerate Electron–Hole System

Early work was done by Silver [37] on the first-order phase transition from the classical to quantum-degenerate EHP. The Fermi nature of the electrons comes into play in two ways, by means of the exchange correction to the Coulomb interaction between electrons, which gives an effective long-range attraction, and by means of the phase-space filling (PSF) effect, which gives an effective short-range repulsion.

Haug [38] generalized Silver's paper by introducing the excitons explicitly. Haug elaborated a simplified but effective method for treating the phase transitions in the three-component system consisting of excitons and EHP, based on the Hartree–Fock approximation and the Bogoliubov variational principle. The EHP Hamiltonian was complemented by the introduction of the excitons and their interactions among themselves and with the EHP. These interactions are determined by the electron–hole structure of the excitons and by their spin orientation.

The work by Haug [38] was later supplemented by that of Moskalenko, Russu and Zalozh [39] and that of Moskalenko et al. [40] in some aspects. Special attention was given to the second-order phase transition between the classical and quantum-degenerate states of excitons (BEC) taking place in the presence of the first-order phase transition from the exciton gas to the degenerate EHP. Such a possibility was pointed out by Haug [38], but the coexistence conditions of the two phase transitions were not elucidated. This program was carried out in Ref. 39 for different values of σ. The phase transitions in the three-component system localized in two-dimensional (2D) structures were also studied in Ref. 39. In what follows, we review the results obtained by Haug [38] as well as the results of Refs. 39 and 40.

Following Ref. 38, the model Hamiltonian of the three-component, electron–hole–exciton system with definite spin orientation has the form

$$\begin{aligned}\mathcal{H} = &\sum_{\mathbf{k}} [E_e(\mathbf{k}) - \mu_e]\, a_{\mathbf{k}}^{\dagger} a_{\mathbf{k}} + \sum_{\mathbf{k}} [E_h(\mathbf{k}) - \mu_h]\, b_{\mathbf{k}}^{\dagger} b_{\mathbf{k}} + \sum_{\mathbf{k}} [E_{\text{ex}}(\mathbf{k}) - \mu_{\text{ex}}]\, c_{\mathbf{k}}^{\dagger} c_{\mathbf{k}} \\ &+ \sum_{\mathbf{kqp}} V_{\mathbf{k}} (a_{\mathbf{p}}^{\dagger} a_{\mathbf{q}}^{\dagger} a_{\mathbf{q+k}} a_{\mathbf{p-k}} + b_{\mathbf{p}}^{\dagger} b_{\mathbf{q}}^{\dagger} b_{\mathbf{q+k}} b_{\mathbf{p-k}} - 2 a_{\mathbf{p}}^{\dagger} b_{\mathbf{q}}^{\dagger} b_{\mathbf{q+k}} a_{\mathbf{p-k}}) \\ &+ \frac{1}{V} \sum_{\mathbf{kqp}} \nu_{e-\text{ex}}(\mathbf{p}, \mathbf{q}, \mathbf{k}) a_{\mathbf{p}}^{\dagger} c_{\mathbf{q}}^{\dagger} c_{\mathbf{q+k}} a_{\mathbf{p-k}} + \frac{1}{V} \sum_{\mathbf{kqp}} \nu_{h-\text{ex}}(\mathbf{p}, \mathbf{q}, \mathbf{k}) b_{\mathbf{p}}^{\dagger} c_{\mathbf{q}}^{\dagger} c_{\mathbf{q+k}} b_{\mathbf{p-k}} \\ &+ \frac{1}{2V} \sum_{\mathbf{kqp}} U_{\text{ex}-\text{ex}}(\mathbf{p}, \mathbf{q}, \mathbf{k}) c_{\mathbf{p}}^{\dagger} c_{\mathbf{q}}^{\dagger} c_{\mathbf{q+k}} c_{\mathbf{p-k}}. \end{aligned} \tag{5.4}$$

Here the chemical potentials of electrons μ_e, holes μ_h, and excitons μ_{ex} are interconnected by the relation

$$\mu_e + \mu_h = \mu_{\text{ex}}, \tag{5.5}$$

which reflects the e–h pair–exciton mutual transformations represented by the chemical reaction

$$e + h \rightleftharpoons \text{ex}.$$

The energy spectra of the electrons, holes, and excitons are

$$E_e = E_g + \frac{\hbar^2 k^2}{2m_e}, \qquad E_h = \frac{\hbar^2 k^2}{2m_h}, \qquad E_{\text{ex}} = E_g - \text{Ry}_{\text{ex}} + \frac{\hbar^2 k^2}{2m_{\text{ex}}}. \tag{5.6}$$

The creation and annihilation operators $a_{\mathbf{k}}^{\dagger}$ and $a_{\mathbf{k}}$, $b_{\mathbf{k}}^{\dagger}$ and $b_{\mathbf{k}}$, and $c_{\mathbf{k}}^{\dagger}$ and $c_{\mathbf{k}}$ describe the electrons, holes, and excitons, respectively. The interaction constants were determined to be

$$\begin{aligned} &\nu_{e-\text{ex}}(\mathbf{p}, \mathbf{q}, \mathbf{k}) = \nu_{h-\text{ex}}(\mathbf{p}, \mathbf{q}, \mathbf{k}) = \nu(0) = 24\pi a_{\text{ex}}^3 \text{Ry}_{\text{ex}}, \\ &U_{\text{ex}-\text{ex}}(\mathbf{p}, \mathbf{q}, \mathbf{k}) = U(0) = \frac{26\pi}{3} a_{\text{ex}}^3 \text{Ry}_{\text{ex}}, \qquad \text{Ry}_{\text{ex}} = \frac{m^* e^4}{2\hbar^2 \varepsilon_0^2}, \\ &V_q = \frac{4\pi e^2}{\varepsilon_0 V q^2}, \qquad m^* = \frac{m_e m_h}{m_{\text{ex}}}, \qquad m_{\text{ex}} = m_e + m_h. \end{aligned} \tag{5.7}$$

The Bogoliubov variational principle [41] consists of the introduction of the trial quadratic Hamiltonian $\mathcal{H}_{\text{tr}}$ and the corresponding thermodynamic potential ψ_{tr}, by which the upper limit $\bar{\psi}$ of the true thermodynamic potential ψ can be determined. The trial Hamiltonian

$$\begin{aligned}\mathcal{H}_{\text{tr}} = &\sum_{\mathbf{k}} [E_e(\mathbf{k}) - \mu_e + \Sigma_e(\mathbf{k})]\, a_{\mathbf{k}}^{\dagger} a_{\mathbf{k}} + \sum_{\mathbf{k}} [E_h(\mathbf{k}) - \mu_h + \Sigma_h(\mathbf{k})]\, b_{\mathbf{k}}^{\dagger} b_{\mathbf{k}} \\ &+ \sum_{\mathbf{k}} [E_{\text{ex}}(\mathbf{k}) - \mu_{\text{ex}} + \Sigma_{\text{ex}}(\mathbf{k})]\, c_{\mathbf{k}}^{\dagger} c_{\mathbf{k}} \end{aligned} \tag{5.8}$$

and the thermodynamic potential

$$\psi_{\text{tr}} = -k_B T \ln Z_{\text{tr}}, \qquad Z_{\text{tr}} = \text{Tr}\big(e^{-\beta \mathcal{H}_{\text{tr}}}\big) \tag{5.9}$$

contain the unknown self-energy parts $\Sigma_i(\mathbf{k})$, where $i = e, h, \text{ex}$, which have to be determined. As Bogoliubov established [41], the upper limit $\bar{\psi}$ of the desired thermodynamic potential ψ is

$$\bar{\psi} = \psi_{\text{tr}} + \langle \mathcal{H} - H_{\text{tr}} \rangle_{\mathcal{H}_{\text{tr}}}. \tag{5.10}$$

The unknown self-energy parts $\Sigma_i(\mathbf{k})$ can be determined from the additional conditions that they ensure the minimum of the thermodynamic potential $\bar{\psi}$,

$$\frac{d\bar{\psi}}{d\Sigma_i(\mathbf{k})} = 0, \qquad i = e, h, \text{ex}. \tag{5.11}$$

The expressions for $\bar{\psi}$ and $\Sigma_i(\mathbf{k})$ were found to be

$$\bar{\psi} = -k_B T \sum_{\mathbf{k}} (\ln\{1 + \exp[-\beta \tilde{E}_e(\mathbf{k})]\} + \ln\{1 + \exp[-\beta \tilde{E}_h(\mathbf{k})]\}$$
$$- \ln\{1 - \exp[-\beta \tilde{E}_{\text{ex}}(\mathbf{k})]\}) - \frac{1}{2} \sum_{\mathbf{k}} [\Sigma_e(\mathbf{k})\bar{n}_e(\mathbf{k}) + \Sigma_h(\mathbf{k})\bar{n}_h(\mathbf{k}) + \Sigma_{\text{ex}}(\mathbf{k})\bar{n}_{\text{ex}}(\mathbf{k})], \tag{5.12}$$

$$\tilde{E}_i(\mathbf{k}) = E_i(\mathbf{k}) - \mu_i + \Sigma_i(\mathbf{k}), \qquad i = e, h,$$
$$\Sigma_i(\mathbf{k}) = \nu(0) n_{\text{ex}} + \sigma_i(\mathbf{k}),$$
$$\sigma_i(\mathbf{k}) = -\sum_{\mathbf{q}} V_{\mathbf{k}-\mathbf{q}} \bar{n}_i(\mathbf{q}), \tag{5.13}$$

$$\tilde{E}_{\text{ex}}(\mathbf{k}) = E_{\text{ex}}(\mathbf{k}) - \mu_{\text{ex}} + \Sigma_{\text{ex}}(\mathbf{k}),$$
$$\Sigma_{\text{ex}}(\mathbf{k}) = \nu(0)\,(n_e + n_h) + 2U(0) n_{\text{ex}}. \tag{5.14}$$

Here the values of $\bar{n}_i(\mathbf{k})$ are determined by

$$\bar{n}_i(\mathbf{k}) = \{\exp[\beta \tilde{E}_i(\mathbf{k})] + 1\}^{-1}, \quad i = e, h,$$
$$\bar{n}_{\text{ex}}(\mathbf{k}) = \{\exp[\beta \tilde{E}_{\text{ex}}(\mathbf{k})] - 1\}^{-1}. \tag{5.15}$$

The densities of the electrons, holes, and excitons are

$$n_i = \frac{1}{V} \sum_{\mathbf{k}} \bar{n}_i\,(\mathbf{k}), \qquad i = e, h, \text{ex}. \tag{5.16}$$

To make the question more transparent and to find analytical expressions for the values n_i, one can neglect the dependences of the self-energy parts $\sigma_i(\mathbf{k})$ on $\mathbf{k}$, substituting them by $\bar{\sigma}_i$ [40]. In this case, the only dependences of the effective energy spectra $\tilde{E}_i(\mathbf{k})$ on $\mathbf{k}$ would be their quadratic laws, which lead to the formulas

$$n_i = \left(\frac{m_i k_B T}{2\pi\hbar^2}\right)^{3/2} f_{3/2}(e^{-Z_i}), \qquad Z_i = \beta \tilde{E}_i(0), \qquad i = e, h,$$
$$n_{\text{ex}} = \left(\frac{m_{\text{ex}} k_B T}{2\pi\hbar^2}\right)^{3/2} g_{3/2}\left(e^{-Z_{\text{ex}}}\right), \qquad Z_{\text{ex}} = \beta \tilde{E}_{\text{ex}}(0),$$

with

$$Z_{\text{ex}} \geq 0, \qquad \mu_{\text{ex}} \leq E_{\text{ex}}(0) + \Sigma_{\text{ex}}. \tag{5.17}$$

Here $f_{3/2}(x)$ is the Fermi integral and $g_{3/2}(x)$ is the generalized Riemann zeta function [42]. In the case of excitonic BEC, one needs to introduce the condensate density n_{ex}^0 along with the thermal density n_{ex}^T according to

$$n_{\text{ex}} = n_{\text{ex}}^0 + n_{\text{ex}}^T, \qquad Z_{\text{ex}} = 0, \qquad \mu_{\text{ex}} = E_{\text{ex}}(0) + \Sigma_{\text{ex}}. \tag{5.18}$$

In this case, the variation of μ_{ex} is due only to the variation of n_{ex}^0 and vice versa. Although this model was used in Ref. 40, below we present a much more accurate model, taking into account the explicit dependences $\sigma_i(\mathbf{k})$. Skipping ahead, one can remark that the two approaches of Refs. 39 and 40, as well as the approximation for $\sigma_i(\mathbf{k})$ proposed by Silver [37] and used by Haug [38], give similar results.

Up to now formulas (5.4)–(5.16) follow exactly the theory developed by Haug [38]. Some differences, except for formulas (5.17), now arise because of an alternative estimate of $\sigma_i(\mathbf{k})$ proposed in Ref. 40, instead of the Silver approximation [37]:

$$V_{\mathbf{k}-\mathbf{q}} = \frac{4\pi e^2}{\varepsilon_0 V\,|\mathbf{k}-\mathbf{q}|^2} \approx \frac{4\pi e^2}{\varepsilon_0 V k q}. \tag{5.19}$$

The self-energy parts $\sigma_i(\mathbf{k})$ can be calculated exactly at $T = 0$:

$$\sigma_i(\mathbf{k})|_{T=0} = -\frac{e^2}{\pi\varepsilon_0} k_F^i \left[1 + \frac{1}{2}\left(\frac{k_F^i}{k} - \frac{k}{k_F^i}\right) \ln\left|\frac{k + k_F^i}{k - k_F^i}\right|\right]. \tag{5.20}$$

Here k_F^i are the Fermi wave vectors for the electrons and the holes [32].

This formula was generalized for the case $T \neq 0$ by substituting for k_F^i the equivalent values [39]

$$k_F^i \to \sqrt{\frac{12\pi e^2 n_i}{\varepsilon_0\,|\sigma_i(0)|}}. \tag{5.21}$$

Haug [38], in approximation (5.19), calculated the density n_i and the diagrams representing the dependences of pressure $P = -(\mathrm{d}\bar{\psi}/\mathrm{d}V)_T$ on the mean distance between the particles, $d = n^{-1/3}$. The total density of e–h pairs and excitons under the condition of electroneutrality is determined as

$$n = n_e + n_{\text{ex}}, \qquad n_e = n_h. \tag{5.22}$$

The dimensionless pressure $\bar{P}$, distance $\bar{d}$, inverse temperature $\bar{\beta}$, and chemical potential $\bar{\mu}$ were introduced:

$$\bar{P} = \frac{P a_{\text{ex}}^3}{\text{Ry}_{\text{ex}}}, \qquad \bar{d} = \left(n a_{\text{ex}}^3\right)^{-1/3}, \qquad \bar{\beta} = \beta \text{Ry}_{\text{ex}}, \qquad \bar{\mu} = \frac{\mu_{\text{ex}}}{\text{Ry}_{\text{ex}}}. \tag{5.23}$$

The condition of electroneutrality $n_e = n_h$ and chemical reaction (5.5) establish two relations among the three chemical potentials μ_i and lead to only one independent variable, $\bar{\mu}$.

Despite the simplicity of the system considered in this model, which consists of only equal-mass, spin-aligned fermions of opposite charge, complicated properties of nonlinear systems are manifested in the phase diagram, which can lead to optical bistability effects [43, 44]. All the values to be found, such as the densities n_i and the thermodynamic potential $\bar{\psi}$ and its derivatives, for example $\bar{P}$, considered as functions of the variable $\bar{d}$ or the related variable $\bar{\mu}$, possess not just one but three branches of solutions in some regions of phase space. In these regions, hysteresis loops appear on the corresponding curves [44]. The first

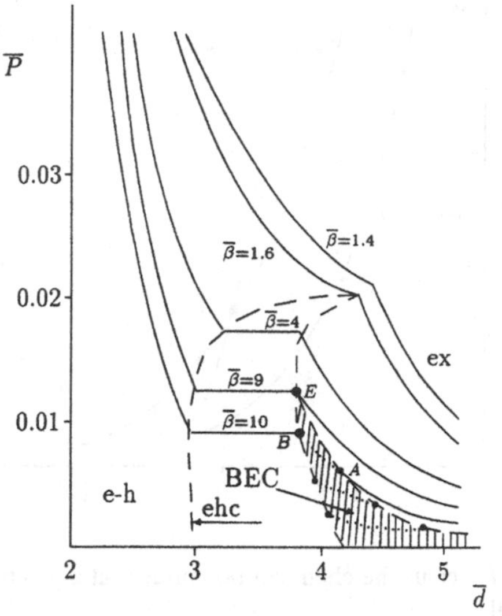

Figure 5.3. The dependences $\bar{P}(\bar{d})$ for different inverse temperatures $\bar{\beta}$ (from Ref. 38).

branch, which corresponds to the exciton gas, spreads from the region of large $\bar{d} \to \infty$ up to the region of the hysteresis. This consists of a mixture of excitons and *e–h* pairs in which the number of excitons exceeds the number of *e–h* pairs. The third branch extends from the region of the hysteresis loop to the region of small $\bar{d}$ ($\bar{d} < 1$). This phase, in contrast with the first one, consists of a high density of degenerate *e–h* pairs and small densities of excitons and is therefore called degenerate EHP. These two phases are stable. In the region of the hysteresis loop there is also one unstable phase, which consists of high and comparable densities of excitons and *e–h* pairs.

Haug constructed the dependences $\bar{P}(\bar{d})$ for different $\bar{\beta}$ by using the stable solutions of the obtained branches. These are presented in Fig. 5.3. Haug noted that they resemble the diagrams for van der Waals gas and are similar to the phase diagrams describing the phase transition from classical to quantum-degenerate EHP studied by Silver [37]. When the electron and the hole masses are equal ($\sigma = m_e/m_h = 1$), the critical temperature of the first-order phase transition from the exciton phase to the degenerate EHP phase was determined as

$$k_B T_c^{ehc} = 0.63\text{Ry}_{\text{ex}}, \qquad T_c^{\text{BEC}} \simeq T_c^{ehc}/6. \tag{5.24}$$

That phase transition is characterized by a coexistence region in which drops of degenerate EHP exist in an exciton-rich gaseous phase. At temperatures lower than approximately $T_c^{ehc}/6$, a second-order phase transition from the gaseous phase to the degenerate exciton phase occurs.

As noted by Haug [38], the value of the critical temperature T_c^{ehc} is unrealistically high because of the overestimation of the exciton stability in the mean-field approximation, which does not take into account the screening of the Coulomb potential. This was confirmed by the calculations of Moskalenko et al. [39], based on the alternative approach expressed by formulas (5.46) and (5.47). For example, the critical temperatures $k_B T_c^{ehc} = 0.7\text{Ry}_{\text{ex}}$ and $k_B T_c^{\text{BEC}} = 0.11\text{Ry}_{\text{ex}}$ at $\sigma = 1$ were obtained [39]. For $\sigma = 0.2$, the value $k_B T_c^{\text{BEC}} = 0.07\text{Ry}_{\text{ex}}$ was found.

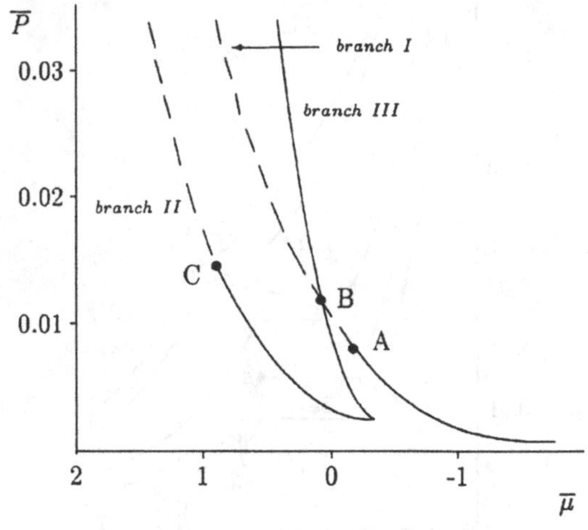

Figure 5.4. Pressure $\bar{P}$ versus the chemical potential $\bar{\mu}$ at a given inverse temperature $\bar{\beta} = 10$ [34, 37, 39, 40].

The dependence $\bar{P}(\bar{\mu})$ is represented in Fig. 5.4. The results obtained earlier by Haug are designated by the solid curves, whereas the supplementary results of Refs. 39 and 40 are designated by the dashed curves. The exciton gas is represented by the first branch. It contains a segment AB where the degenerate exciton phase exists. At point A, excitonic BEC occurs. Up to point B, this phase is more stable than the third branch representing the degenerate EHP, because the first branch corresponds to lower values of $\bar{\mu}$ at the same values of $\bar{P}$ and $\bar{\beta}$. The segment AB can be extended inside the two-phase region, and the observation of excitonic BEC on this segment is also possible. In spite of the proximity of the first-order phase transition due to the condensation of the EHP in the real space, the main condition of excitonic BEC remains the same. For this to be true, it is necessary to have repulsion between the excitons [$U(0) > 0$], which is physically reasonable for this model, which assumes spin-aligned particles. Moreover, a repulsion between the excitons and the free e, h carriers must also be assumed ($\nu_0 > 0$). The remark of Haug concerning the overestimation of the exciton stability applies to the results of Refs. 39 and 40 shown here.

In Fig. 5.3, the isotherms obtained by Haug are supplemented by details in the hatched region, where the second-order phase transition related to the excitonic BEC takes place. The dashed curve in this figure designates the boundary of the two-phase region, in which drops of degenerate EHP coexist with the gaseous or degenerate exciton phases. The appearance of excitonic BEC can be expected not only in the hatched region of Fig. 5.3, but also in the two-phase region. At upper point E of the hatched region, two phase transitions, both first and second order, coexist simultaneously. At this point three phases can appear: the exciton gaseous and degenerate phases, as well as the degenerate EHP phase. At lower temperatures, two consecutive phase transitions will occur: excitonic BEC at lower densities and EHP condensation at higher densities. The phase diagram obtained by Haug [38] is represented in Fig. 5.5, which shows the temperature and density regions where different phases exist, designated as the degenerate EHP, gaseous and degenerate exciton phases, and the two-phase region.

Since screening of the Coulomb interaction and the resulting decrease of the exciton binding energy were neglected in this model, the Mott transition [45, 46] in the exciton–EHP

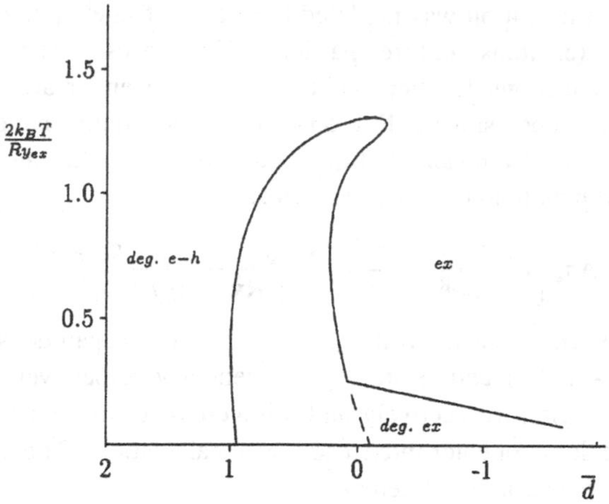

Figure 5.5. Phase diagram: the dependence of k_BT versus the the distance $\bar{d} = 10$ (from Ref. 38).

system could not be studied. In Subsection 5.2.3 we discuss a different model, which explicitly takes into account the decrease of the exciton binding energy due to screening.

Many others have also looked at this problem, e.g., Vashishta, Kalia, and Singwi [47], Insepov and Norman [48], Rice [5], Ebeling et al. [49], Combescot and Benoit à la Guillaume [50], and Rogachev [51].

As mentioned at the beginning of this chapter, Keldysh paid attention to the possible coherent *e–h* pairing of the electrons and the holes near the corresponding Fermi surfaces taking part in the formation of EHL [4]. The coherent pairing of electrons and holes at high density ($n_{e,h}a_{\text{ex}}^3 > 1$) in semimetals and the formation of BCS-type *e–h* pairs was studied by Keldysh and Kopaev [8], who introduced the concept of the excitonic insulator. Keldysh proposed some variants of the new phase diagrams, including the EHL and excitonic-insulator phases side by side with the exciton gaseous and excitonic degenerate phases [4, 7]. Keldysh also pointed out that the formation of a three-dimensional (3D) "excitonic crystal" is impossible if the excitons have the translational masses m_{ex} of the order of the free-electron mass m_0. The energy criteria of solid melting show that it usually occurs when the amplitudes of the thermal oscillations of the crystal-lattice atoms achieve values of the order of one tenth of the interatomic distance. For the hypothetical excitonic crystal, this means that the kinetic energy of the excitonic localization would be more than the exciton ionization energy, and the excitonic crystal would be melted even for $T = 0$ [4, 23]. The case of the excitonic crystal in one-dimensional (1D) quantum wires was discussed by Ivanov and Haug [52].

5.2.3 Hysteresis due to Screening Effects

In Subsection 5.2.2 a model was considered in which the quantum statistics of the constituent particles were taken into account, but the effect of screening on the stability of the bound pairs was ignored. In this section, we discuss a model [53] that takes the opposite approach, taking into account the screening effects explicitly, but ignoring the quantum-statistical effects. This model also shows hysteresis by means of a completely different mechanism.

In Ref. 53 the Mott transition was modeled by means of rate equations for conversion between bound states (excitons) and free particles. This process was assumed to occur by means of three-body collisions. If other mechanisms for conversion are used, e.g., phonon-assisted ionization, the conclusions of Ref. 53 do not substantially change.

For species conversion by means of three-body collisions, the rate of change of the bound-state (exciton) population $n_{\rm ex}$ was written as

$$\frac{\partial n_{\rm ex}}{\partial t} = An_e\left[n_e^2 - \frac{n_{\rm ex}n_Q}{e^{{\rm Ry}_{\rm ex}/k_BT} - (1 + {\rm Ry}_{\rm ex}/k_BT)}\frac{g_e g_h}{g_{\rm ex}}\left(\frac{m_r}{m}\right)^{3/2}\right] \tag{5.25}$$

where n_e is the free-electron density, with $n_{\rm ex} + n_e = n$, the total pair density; g_e, g_h, and $g_{\rm ex}$ are the electron-, ion- (hole-), and exciton-spin degeneracy, respectively, m and m_r are the total and the reduced masses, respectively, and A is a cross-section constant that determines the absolute rate and does not enter into steady-state calculations. The quantum density of states that enters this equation is defined as

$$n_Q = \left(\frac{mk_BT}{2\pi\hbar^2}\right)^{3/2} \equiv 1/\lambda_D^3,$$

where λ_D is the de Broglie wavelength, which when compared with Eq. (1.18) is approximately just the same as the critical density for Bose condensation. Setting $\partial n_{\rm ex}/\partial t = 0$ in Eq. (5.25) with ${\rm Ry}_{\rm ex} \gg k_BT$ gives the standard Saha equation [54] $(n_e^2/n_Q) = n_{\rm ex}e^{-{\rm Ry}_{\rm ex}/k_BT}$ $(g_eg_h/g_{\rm ex})(m_r/m)^{3/2}$, but at high temperature the Saha result is altered by the use of the Planck–Larkin partition function [54, 55] $Z = e^{{\rm Ry}_{\rm ex}/k_BT} - (1 + {\rm Ry}_{\rm ex}/k_BT)$, which takes into account the fact that states with ${\rm Ry}_{\rm ex} \ll k_BT$ are quasi-free.

The interesting dynamics in this system come from the dependence of ${\rm Ry}_{\rm ex}$ on n_e. When screening is present, so that the particles do not interact with pure Coulomb attraction but with a Yukawa-type interaction, the binding energy decreases approximately according to [56]

$${\rm Ry}_{\rm ex}(n_e, T) = \begin{cases} {\rm Ry}_{\rm ex}(0)\left[1 - \frac{2}{1+(qa)^{-1}}\right], & qa < 1 \\ 0, & qa \geq 1 \end{cases}. \tag{5.26}$$

The screening constant is given by the Debye–Huckel formula [57],

$$q^2 = \frac{4\pi e^2 n_e}{\epsilon_0 k_BT}. \tag{5.27}$$

The excitons are assumed to give no contribution to screening, which is physically reasonable since they are electrically neutral, with only a weak polarizability.

Figure 5.6 shows points of equilibrium $\partial n_{\rm ex}/\partial t = 0$ found with a self-consistent numerical approach in Ref. 53. The solid curve marks the boundary between the region with one steady-state solution and the region with three steady-state solutions, one of which is unstable. Figure 5.7 shows the ionized fraction as a function of total pair density at a fixed temperature, as density is varied. In Fig. 5.7(d), a classic hysteresis curve appears. The upper and the lower solutions for the ionized fraction are stable against fluctuations, while the middle one is unstable.

By simple arguments, one can see how hysteresis occurs in this system. Figure 5.8 shows the relevant crossovers for the case $m = 4m_r$ and $g_e = g_h = 2$, $g_{\rm ex} = 1$ of Figs. 5.6 and 5.7. If we assume that the system is in the plasma state, then we expect a Mott transition when $1/a = q$, where q is given by Debye formula (5.27) with $n_e = n$. In terms of the unitless

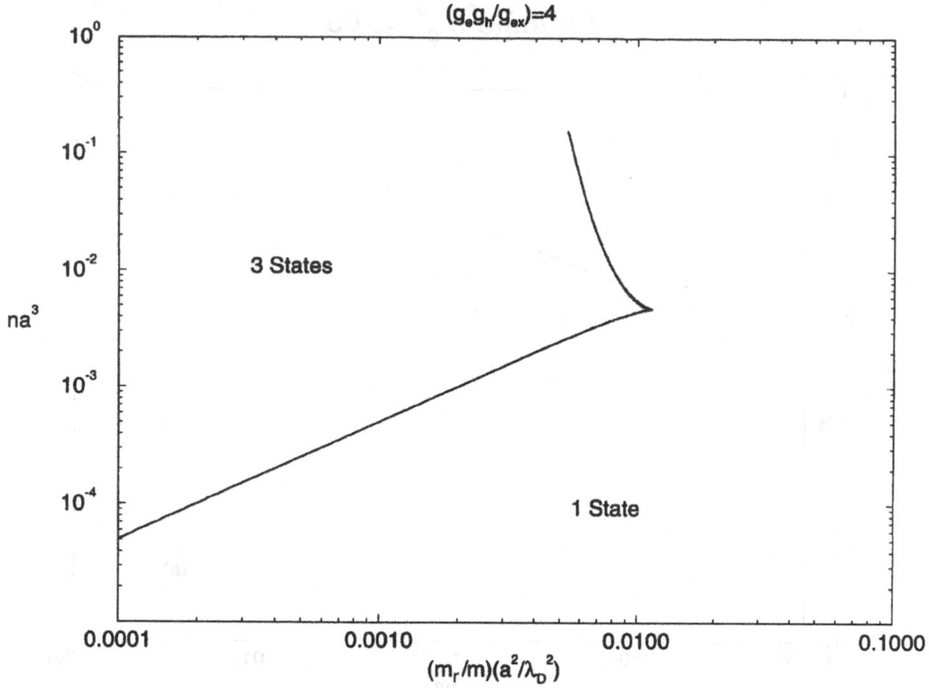

Figure 5.6. The boundary between the regions of phase space with one steady-state solution and three steady-state solutions of Eq. (5.25) for the case $g_e = g_h = 2$, $g = 1$ (from Ref. 53).

parameters na^3 (unitless pair density) and a^2/λ_D^2 (unitless temperature), this becomes

$$na^3 = \frac{1}{2}\left(\frac{m_r}{m}\right) a^2/\lambda_D^2, \tag{5.28}$$

shown as the heavy solid line in Fig. 5.8. On the other hand, if we assume that the system is in the insulating-gas (excitonic) phase, we would expect a transition when $1/a = q$, where q is given by Eq. (5.27), but instead of setting n_e equal to n, we would use the equilibrium value from Eq. (5.25), assuming that $\text{Ry}_{\text{ex}} = \text{Ry}_{\text{ex}}(0)$ and $n_e \ll n$. This yields

$$na^3 = \frac{1}{4}\left(\frac{a^2}{\lambda_D^2}\right)^{1/2}\left(\frac{g}{g_e g_h}\right)\left(\frac{m_r}{m}\right)^{1/2} e^{(\lambda_D^2/a^2)(m/m_r)/4\pi}, \tag{5.29}$$

shown as the light solid curve in Fig. 5.8. Since at very low temperatures the number of thermally ionized particles becomes extremely small, the "ionization-catastrophe" density becomes exponentially large at low temperature. For comparison, the conditions for a transition to a Fermi degenerate gas ($na^3 \sim 1$) and for a transition to a weakly interacting Bose condensate [$na^3 = 2.612(a^2/\lambda_D^2)^{3/2}$] are shown as the dashed and the dotted lines in Fig. 5.8, respectively. Since the model presented here involves only Maxwellian statistics, it is valid for only densities well below both of these curves.

At low temperature and low density, the system is an insulating excitonic (bound-state) gas, assuming repulsive interactions between excitons. As density is raised, in the low-temperature limit the system remains an insulating gas until an ionization catastrophe [5] occurs. As seen in Fig. 5.8, this can occur at substantially higher density than that of the Mott

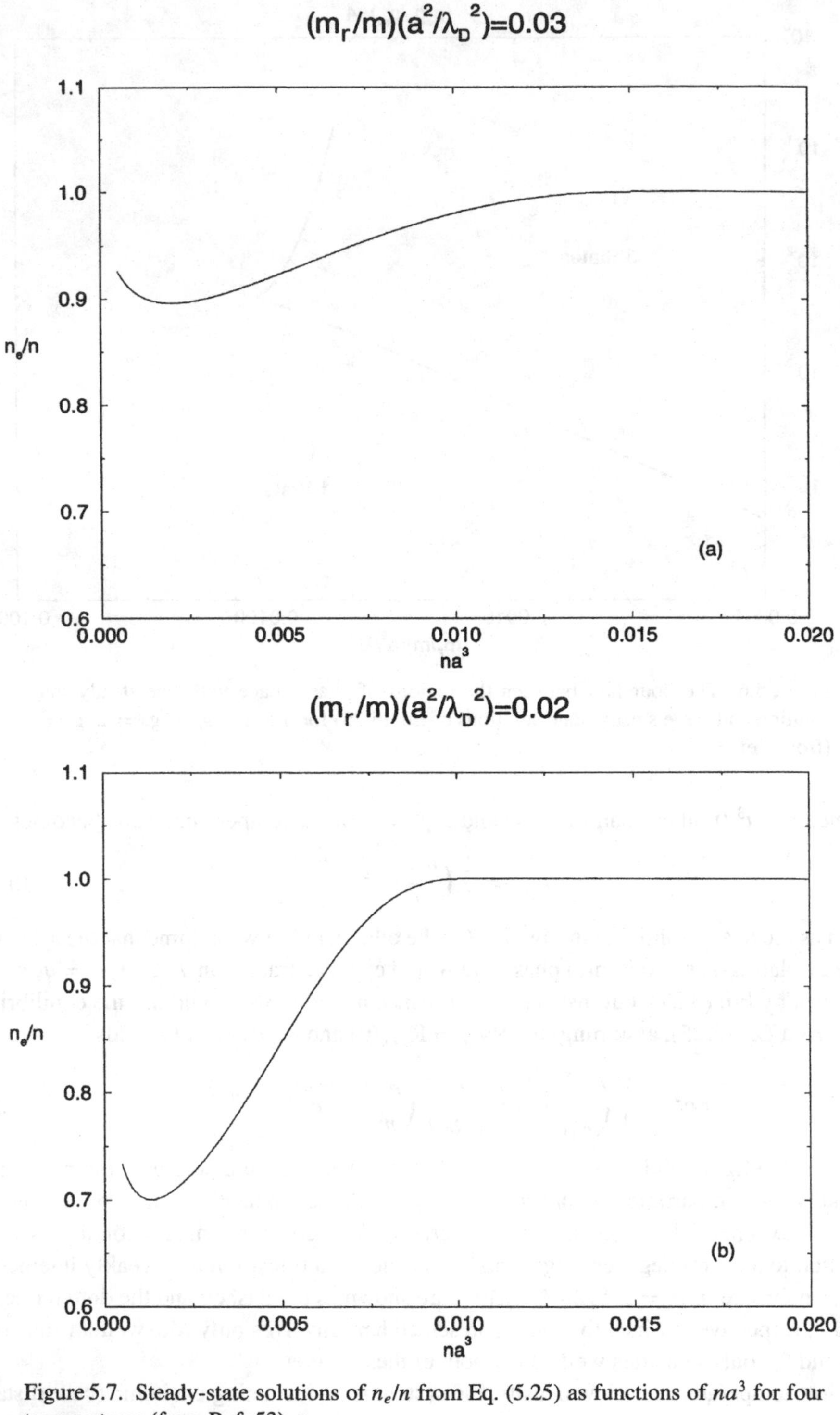

Figure 5.7. Steady-state solutions of n_e/n from Eq. (5.25) as functions of na^3 for four temperatures (from Ref. 53).

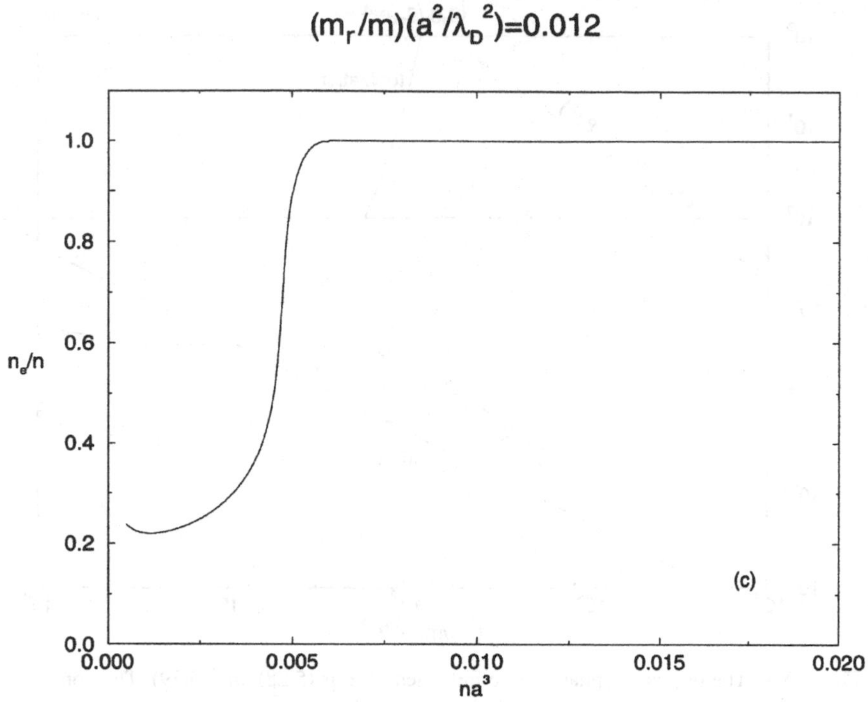

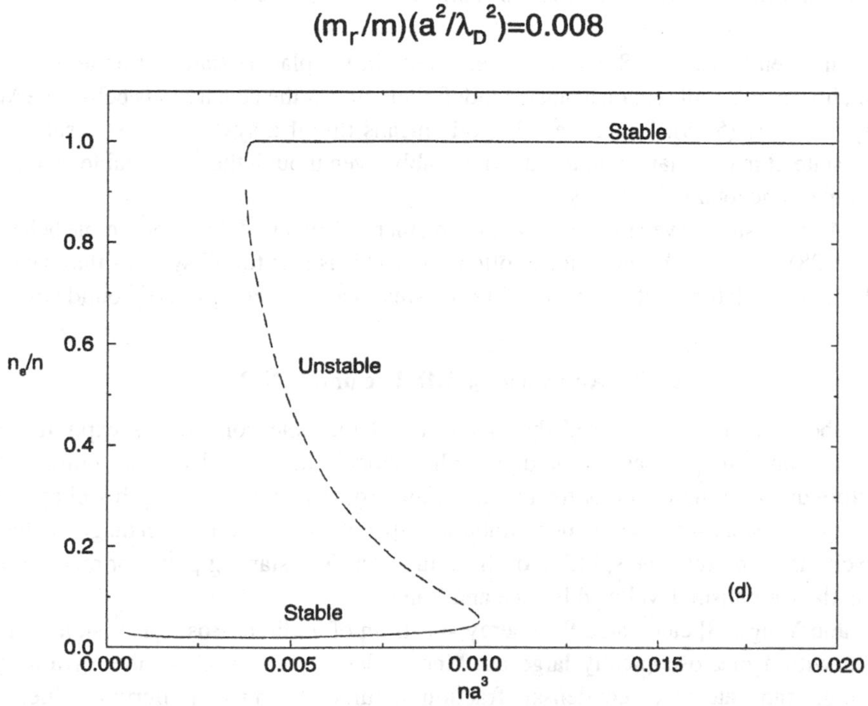

Figure 5.7. *Continued.*

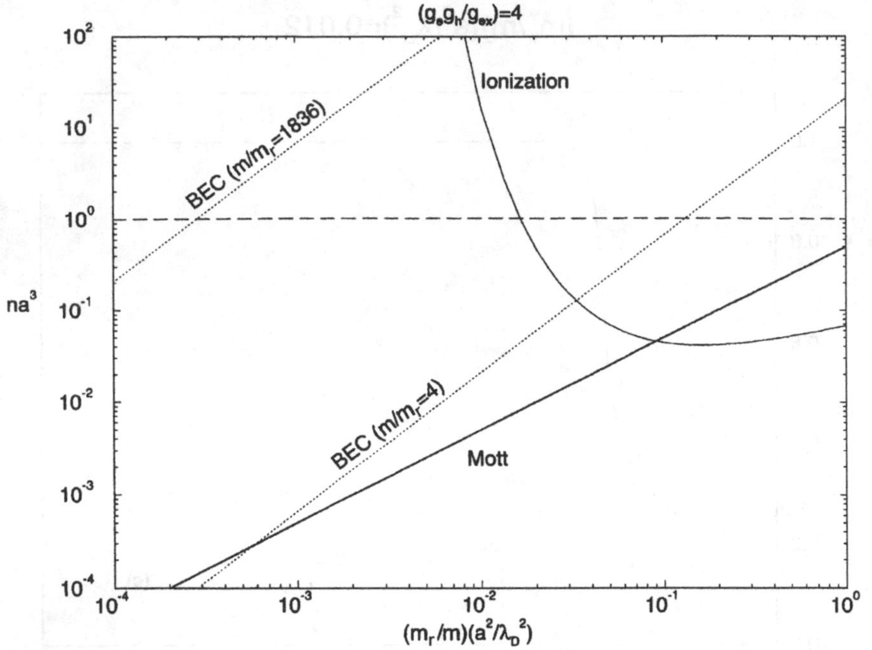

Figure 5.8. The regions of phase space indicated by Eqs.(5.28) and (5.29). The conditions for BEC for two different values of m/m_r are shown for comparison.

transition given by Eq. (5.28). If the system starts in the plasma state in this temperature regime, however, it will remain a plasma indefinitely unless the density falls below the Mott density given by (5.28). Experimentally, this means that if a system is first created in a plasma state, it may remain in that state metastably, even though there is a stable insulating state at the same total pair density.

The Mott transition, even in a classical system, therefore does not have the simple behavior of Eq. (5.28), although this equation is often used. This is true for all systems that have the possibility of a Mott transition from neutral plasma to an insulating gas of bound pairs.

5.3 The Approach of T.D. Lee and C.N. Yang

In the above models, we treated the electron and the hole constituents explicitly, but considered only simple band structures. Other models have tackled more complicated Hamiltonians with the excitons treated as simple bosons. In the rest of this chapter we consider models of excitons with nonparabolic dispersion laws, excitons in magnetic fields, and excitons with exchange splitting of the spin states. The starting point for much of this work is the formalism developed by Lee and Yang.

Lee and Yang [58] calculated the energy spectrum of a dilute Bose gas system of hard spheres with a macroscopically large but "incomplete" occupation number of a single-particle ground state, i.e., a condensate fraction perturbed from its equilibrium value, with the method proposed by Lee, Huang, and Yang [59]. These energy levels were used by Lee and Yang to compute the statistical properties of the system. The thermodynamic functions of the system in both gaseous and degenerate phases were obtained to the lowest order of the interaction constant, and the phase transition was investigated on this basis.

The energy spectrum of the weakly interacting Bose gas with a condensate in the single-particle ground state was found to be [58]

$$E(x,\ldots,n_{\mathbf{k}},\ldots) = E_0(x) + \sum_{\mathbf{k}} E(x,\mathbf{k})\, n_{\mathbf{k}}, \tag{5.30}$$

where the $n_{\mathbf{k}}$ are positive integers satisfying the condition

$$\sum_{\mathbf{k}} n_{\mathbf{k}} = N(1-x)\,. \tag{5.31}$$

The fraction x gives the ratio of the total number N of particles in the system and the number N_0 of the Bose condensed particles,

$$\begin{gathered} N_0 = xN, \qquad 0 \le x \le 1, \\ n_0 = \frac{N_0}{V}, \qquad n = \frac{N}{V}, \\ N \sim N_0 \sim V. \end{gathered} \tag{5.32}$$

$E_0(x)$ is the energy of the ground state of the system with the incomplete condensate and $E(x,\mathbf{k})$ is the energy of the elementary excitation arising in the system.

Formula (5.31) expresses the conservation law of the total number of quasiparticles, which equals the total number of the noncondensed fundamental particles of the gas. The energy spectrum of the Bogoliubov-type elementary excitations was obtained in Chapter 2 and has the form of Eq. (2.37):

$$E(x,\mathbf{k}) = \sqrt{T_{\mathbf{k}}^2 + 2T_{\mathbf{k}} n_0 U(0)}. \tag{5.33}$$

Here $U(0)$ is the interaction constant, which in the model of hard spheres does not depend on the momentum transferred during the interaction of the particles. In real space the interaction represents a contact repulsion.

The phase when the Bose condensate appears was called the "degenerate" phase by Lee and Yang and was described in detail in Ref. 58. To this end, the temperature region below the critical point T_c of BEC was divided into three intervals. One of them contains moderately low temperatures not so far from T_c, when the mean thermal value of the kinetic energy $\langle T_p \rangle$ is much more than the characteristic interaction energy,

$$L_0 = U(0)n_0 = U(0)nx,$$

$$\langle T_p \rangle \gg L_0 = U(0)nx, \qquad x \ll 1.$$

At the same time, these moderate temperatures are not so close to the critical point that it is necessary to take into account the interaction between the elementary excitations in the degenerate phase.

Lee and Yang also investigated two other temperature intervals of low and very low temperatures. In discussing the phase transition, however, we restrict ourselves to the results of Ref. 58 by Lee and Yang concerning the degenerate phase at moderately low temperatures. The phase above the critical point is called the "gaseous" phase. At moderately low temperatures, the dispersion law can be simplified and represented in the form

$$E(x,\mathbf{k}) \simeq T_{\mathbf{k}} + U(0)nx. \tag{5.34}$$

This distorts the true dispersion law (5.33) in the region of small values of the wave vectors. But this region is not so important in the case of moderate temperatures. Instead, approximation (5.34) permits one to calculate analytically the thermodynamic values in the moderate degenerate phase. Expression (5.34) also describes the energy spectrum in a gaseous phase if one sets $x = 0$. Approximation (5.34) was also used by Huang, Yang, and Luttinger [60].

The thermodynamic functions obtained by Lee and Yang [58] for the degenerate phase at moderate temperatures and for the gaseous phase were extrapolated to the critical point from below and from above. Comparing the thermodynamic functions at this point led them to conclude that the transition is a second-order phase transition.

At the same time, Lee and Yang [58] cautioned that their results are true only in the lowest order of the interaction and that taking into account the higher-order corrections could lead to discontinuities of the thermodynamic values in the critical point. Hanamura and Haug [61] showed that by applying the Bogoliubov theory beyond its range of validity, i.e., to the high-temperature regime near the critical point, one can arrive at the conclusion that it is a first-order phase transition. These remarks testify to the fact that the real situation is much more complicated than the approach discussed below. To study this question in depth, one can consult several monographs [62–71] and articles [72–76].

In the approach of Lee and Yang, the ground-state energy of the system with a macroscopically large but incomplete condensate has the form, to the lowest order on the interaction constant,

$$E_0(x) = U(0)nN[1 + (1 - x)^2]. \tag{5.35}$$

Lee and Yang [58] and Hugenholtz and Pines [77] deduced the corrections in the next order of the interaction parameter $(L_0)^{1/2} = [U(0)n_0]^{1/2}$. These corrections depend on the form of the dispersion law T_p and even on the specific form of the interaction forces. In what follows, it is important to note that expression (5.35) does not depend on the specific form of the dispersion law T_p.

The partition function Z for a dilute gas of hard spheres obeying Bose statistics was determined as

$$Z = \sum_{\cdots n_{\mathbf{k}} \cdots} \exp\left[-\beta E\left(x \cdots n_{\mathbf{k}} \cdots\right)\right], \qquad \beta = (k_B T)^{-1}. \tag{5.36}$$

Here, as in formula (5.30), the positive integer values $n_{\mathbf{k}}$ must obey particle conservation law (5.31). The quantity $\ln Z$ was evaluated by means of the method of steepest descent. The thermodynamic potential was determined as

$$\psi = -k_B T \ln Z \tag{5.37}$$

and has the form

$$\frac{\psi}{V} = \frac{E_0(x)}{V} + nk_B T(1 - x)\ln \xi + \frac{k_B T}{V} \sum_{\mathbf{k}} \ln\left[1 - \xi e^{-\beta E(x,\mathbf{k})}\right]. \tag{5.38}$$

The saddle point ξ obeys the extremum condition, which is the condition of steepest descent,

$$\left.\frac{\mathrm{d}\psi(x, \xi)}{\mathrm{d}\xi}\right|_x = 0, \tag{5.39}$$

and gives rise to the equation

$$\frac{1}{V} \sum_{\mathbf{k}} \frac{1}{\frac{1}{\xi} e^{\beta E(x,\mathbf{k})} - 1} = n(1 - x), \qquad 0 \le \xi \le 1. \tag{5.40}$$

Equation (5.40) is the same as conservation law (5.31), in which the true integer values $n_{\mathbf{k}}$ are substituted by their mean thermal values:

$$\bar{n}_{\mathbf{k}} = \frac{1}{\frac{1}{\xi} e^{\beta E(x,\mathbf{k})} - 1}. \tag{5.41}$$

The parameters x and ξ, as well the critical temperature T_c, are chosen to be the same as those in the case of ideal Bose gas, namely,

$$x = 0, \qquad 0 \le \xi \le 1 \text{ at } T > T_c,$$
$$0 \le x \le 1, \qquad \xi = 1 \text{ at } T \le T_c. \tag{5.42}$$

Formulas (5.34), (5.35), (5.38), (5.40), and (5.42) permit one to calculate the thermodynamic properties of the gaseous and moderately degenerate phases. We emphasize again that the approach elaborated by Lee and Yang is based on the model of hard spheres, in the lowest order of the perturbation theory on the interaction. The interaction between elementary excitations is neglected. In what follows, these results are applied to the case of high-density excitons. In addition, we take into account some factors intrinsic to the excitons. Among them are the more general form of the dispersion law, which can differ from the isotropic quadratic one, the existence of multiply degenerate exciton levels, which can be split by external constant fields and uniaxial deformations, and the exciton–phonon interaction. All these aspects and their influence on the phase transition are discussed below within the framework of the Lee and Yang formalism.

5.3.1 In-Depth Study: *Excitons with Different Dispersion Laws*

In this subsection we review the work of Khadzhi and Moskalenko, which was published in Ref. 78. In anisotropic crystals in the vicinity of the point $\mathbf{k} = 0$, as well as in cubic crystals in the vicinities of some symmetric points in the Brillouin zone, the dispersion law in the excitonic band is quadratic but anisotropic, of the type

$$T_{\mathbf{p}} = \frac{p_{\parallel}^2}{2m_{\parallel}} + \frac{p_{\perp}^2}{2m_{\perp}}, \qquad p_{\perp} = \sqrt{p_x^2 + p_y^2}, \qquad p_{\parallel} = p_z. \tag{5.43}$$

Here $m_{\parallel}$ and $m_{\perp}$ are the longitudinal and the transverse masses, respectively. In general, the energy to create one exciton $E_{\text{ex}}(p) = E_{\text{ex}}(0) + T_{\mathbf{p}}$ must also be considered. Because the total number of excitons in the exciton band will be kept constant, however, one can drop the energy of the exciton-band bottom.

As shown by Rashba and Sheka [79], in crystals of the same type as CdS, without a center of inversion, the dispersion laws of electrons and holes may contain the terms linear in $p_{\perp}$. This can happen in the bands with Γ_7 and Γ_8 irreducible representations, when the spin-orbital coupling is taken into account. Besides quadratic terms of the type of Eq. (5.43), one can expect the existence of the terms $cp_{\perp}$, where c is the velocity. In CdS and CdSe crystals, the velocity c for the carriers has been estimated as 4×10^5–10^6 cm/s and the transverse mass is of the order of 0.5–0.7 m_0 [80].

The dispersion law for the exciton translational motion in these crystals can be found by solution of the Schrödinger equation in the matrix form. Although it was not deduced explicitly, one can expect the appearance of a similar linear term in the exciton dispersion law too. The expected expression is

$$T_{\mathbf{p}} = \frac{p_{\parallel}^2}{2m_{\parallel}} + \frac{p_{\perp}^2}{2m_{\perp}} + cp_{\perp}. \tag{5.44}$$

For the parameters c and $m_{\perp}$, the same as for the electrons and holes in CdS and CdSe crystals, one can see that the $cp_{\perp}$ term is more important than the $p_{\perp}^2/2m_{\perp}$ term at temperatures below 2–5 K. In what follows below, the thermodynamics of the exciton gas are considered with the general dispersion law:

$$T_{\mathbf{p}} = \frac{p_{\parallel}^2}{2m_{\parallel}} + cp_{\perp}^s, \quad s = 1, 2. \tag{5.45}$$

Within the framework of the Lee and Yang approach and dispersion law (5.45), the thermodynamic potential of the exciton gas describing both phases simultaneously has the form

$$\begin{aligned}\frac{\psi}{V} &= \frac{U(0)n^2}{2}[1 + (1 - x)^2] + nk_BT(1 - x)\ln\xi \\ &- \frac{sc}{4\pi^2\hbar^3}\int_0^{\infty} \mathrm{d}p_{\parallel}\int_0^{\infty} \frac{p_{\perp}^{s+1}\mathrm{d}p_{\perp}}{\frac{1}{\xi}e^{\beta\frac{p_{\parallel}^2}{2m_{\parallel}k_BT}+\beta cp_{\perp}^s+\beta nU(0)x} - 1}.\end{aligned} \tag{5.46}$$

By using the integral value [42]

$$\int_0^{\infty} \frac{x^{p-1}\,\mathrm{d}x}{\frac{1}{q}e^x - 1} = \Gamma(p)g_p(q), \tag{5.47}$$

where $\Gamma(p)$ is the gamma function and $g_p(q)$ is the generalized Riemann zeta function [42, 81],

$$g_p(q) = \sum_{k=1}^{\infty} \frac{q^k}{k^p}, \qquad \Gamma(p + 1) = p!, \tag{5.48}$$

one can obtain the expression for the thermodynamic potential and the condition of the steepest descent in the forms [78, 82]

$$\begin{aligned}\frac{\psi}{V} &= \frac{U(0)n^2}{2}[1 + (1 - x)^2] + nk_BT(1 - x)\ln\xi \\ &- k_BT\lambda^{-3}g_{\frac{3s+4}{2s}}\{\xi\exp[-\beta nU(0)x]\},\end{aligned} \tag{5.49}$$

$$n\,(1 - x) = \lambda^{-3}g_{\frac{s+4}{2s}}\{\xi\,\exp[-\beta nU\,(0)\,x]\}, \tag{5.50}$$

where the thermal de Broglie wavelength λ and the critical temperature of the ideal Bose gas with dispersion law (5.45) are

$$\lambda^{-3} = \sqrt{\frac{\pi m_{\parallel}}{2}}\frac{\Gamma\left(1 + \frac{2}{s}\right)(k_BT)^{\frac{s+4}{2s}}}{4\pi^2\hbar^3c^{2/s}}, \tag{5.51}$$

$$kT_c = \left[\frac{2^{2/5}\pi^{3/2}\hbar^3nc^{2/s}}{g_{\frac{s+4}{2s}}(1)\,\Gamma\left(1 + \frac{2}{s}\right)m_{\parallel}^{1/2}}\right]^{\frac{2s}{s+4}}, \tag{5.52}$$

$$x = \left[1 - \left(\frac{T}{T_c}\right)^{\frac{s+4}{2s}}\right], \quad T \leq T_c. \tag{5.53}$$

Here n is the exciton density. The finite fraction x determines the condensate density $n_0 = nx$. As follows from Eq. (5.52) in the case of a quadratic but anisotropic dispersion law, taking $c^{-1} = 2m_\perp$, one will obtain

$$T_c \sim \left(m_\parallel m_\perp^2\right)^{-1/3} n^{2/3}, \tag{5.54}$$

whereas in the case $s = 1$, the result will be quite different:

$$T_c \sim (m_\parallel)^{-1/5} n^{2/5} c^{2/5}. \tag{5.55}$$

For the parameters $m_\parallel = m_0$, $c = 4 \times 10^5$ cm/s, and $n = 10^{16}\,\text{cm}^{-3}$, the critical temperature equals 1.2 K.

Let us proceed to the calculation of the entropy S, the total energy E, the pressure P, the chemical potential per particle μ, the heat capacity C_v, and the compressibility $(\partial P/\partial n)_T$. These thermodynamic values are determined by the formulas

$$S = -\left(\frac{\partial \psi}{\partial T}\right)_V, \qquad P = -\left(\frac{\partial \psi}{\partial V}\right)_T, \qquad C_V = \left(\frac{\partial E}{\partial T}\right)_V,$$

$$E = \psi + TS, \qquad n\mu = \frac{\psi}{V} + P. \tag{5.56}$$

In the gaseous phase, as mentioned above, one sets $x = 0$ and considers $0 \le \xi \le 1$. The thermodynamic values are then

$$SV^{-1} = -nk_B \ln \xi + \left(\frac{3s+4}{2s}\right) k_B \lambda^{-3} g_{\frac{3s+4}{2s}}(\xi),$$

$$EV^{-1} = U(0)n^2 + \left(\frac{s+4}{2s}\right) k_B T \lambda^{-3} g_{\frac{3s+4}{2s}}(\xi),$$

$$P = U(0)n^2 + k_B T \lambda^{-3} g_{\frac{3s+4}{2s}}(\xi),$$

$$\mu = 2U(0)n + k_B T \ln \xi,$$

$$C_V V^{-1} = \frac{(s+4)(3s+4)}{4s^2} \frac{nk_B g_{\frac{3s+4}{2s}}(\xi)}{g_{\frac{s+4}{2s}}(\xi)} - \frac{(s+4)^2}{4s^2} \frac{nk_B g_{\frac{s+4}{2s}}(\xi)}{g_{\frac{4-s}{2s}}(\xi)},$$

$$\left(\frac{\partial P}{\partial n}\right)_T = 2U(0)n + \frac{k_B T g_{\frac{s+4}{2s}}(\xi)}{g_{\frac{4-s}{2s}}(\xi)}, \qquad n = \lambda^{-3} g_{\frac{s+4}{2s}}(\xi). \tag{5.57}$$

Studying the gaseous phase and the degenerate phase at moderate temperatures, where the value $y = \beta n U(0) x$ is less than unity, one deals with the functions $g_p(\xi)$ for $0 < \xi \le 1$ and $g_p(e^{-y})$ when $0 < y \ll 1$. The functions $g_p(\xi)$ in the vicinity of the point $\xi = 1$ have some peculiarities. For example, the function $g_{3/2}(\xi)$ has a finite value $g_{3/2}(1)$ but its first derivative diverges in the point $\xi = 1$:

$$\left.\frac{\mathrm{d}g_{3/2}(\xi)}{\mathrm{d}\xi}\right|_{\xi=1} = g_{1/2}(1) \to \infty.$$

Nevertheless one can propose the approximate analytical expressions for some frequently encountered functions $g_p(e^{-y})$ when y tends to zero, which can serve to illustrate the origin of the difficulties that appear. For example, one can use the formulas [83]

$$g_{5/2}(e^{-y}) \underset{y\to 0}{\approx} g_{5/2}(1) - g_{3/2}(1)y + \frac{4\sqrt{\pi}}{3}y^{3/2},$$

$$g_{3/2}(e^{-y}) \underset{y\to 0}{\approx} g_{3/2}(1) - 2\sqrt{\pi y}, \qquad g_{1/2}(e^{-y}) \underset{y\to 0}{\approx} \sqrt{\frac{\pi}{y}}. \tag{5.58}$$

In the opposite limit, ($y \to \infty$), these functions equal e^{-y}, which is the first term in their series representations (5.48).

Taking into account the singularities of functions (5.58) and their derivatives, one can nevertheless formulate a regular self-consistent description of the phase transition within the framework of the approach discussed above. This can be done with a first-order approximation in the interaction constant for the thermodynamic values and with a zeroth-order approximation in the interaction constant for the parameters ξ and x, which means that the saddle point ξ and the condensate fraction x are taken to be the same as those for the ideal Bose gas. Consideration of the higher-order corrections in the interaction constant is beyond the frame of this simplified approach and represents an unreliable procedure.

In what follows, a series expansion of the type

$$g_p(e^{-y}) \underset{y\to 0}{\approx} g_p(1) - g_{p-1}(1)y, \qquad p = \frac{3s+4}{2s} \tag{5.59}$$

is used, taking into account the general requirement that the resulting corrections that are obtained must not exceed first order in the interaction constant. In contrast to these approximations, the parameter x in the degenerate phase at moderate temperatures is calculated on the basis of the most simplified equation,

$$n\,(1-x) = \lambda^{-3} g_{\frac{s+4}{2s}}\,(1)\,. \tag{5.60}$$

In these approximations, the following thermodynamic values for the degenerate phase at moderate temperatures were obtained:

$$\psi V^{-1} = U(0)n^2[1 + (1-x)^2] - k_B T\lambda^{-3} g_{\frac{3s+4}{2s}}(1) + \lambda^{-3} U(0) n x g_{\frac{s+4}{2s}}\,(1)\,,$$

$$SV^{-1} = -\frac{U(0)n^2}{T}\left(\frac{s+4}{2s}\right) x\,(1-x) + \left(\frac{3s+4}{2s}\right) k_B \lambda^{-3} g_{\frac{3s+4}{2s}}(1),$$

$$EV^{-1} = U(0)n^2\left[1 - \left(\frac{s+4}{2s}\right)x + \frac{2}{s}x^2\right] + \left(\frac{s+4}{2s}\right) k_B T\lambda^{-3} g_{\frac{3s+4}{2s}}(1),$$

$$P = U(0)n^2\left(1 - x + \frac{x^2}{2}\right) + k_B T\lambda^{-3} g_{\frac{3s+4}{2s}}(1),$$

$$\mu = 2U(0)n\left(1 - \frac{x}{2}\right), \qquad \left(\frac{\partial P}{\partial n}\right)_T = U(0)n,$$

$$C_V V^{-1} = \frac{U(0)n^2}{T}\left[\left(\frac{s+4}{2s}\right)^2 (1-x) + \frac{2\,(s+4)}{s^2} x(1-x)\right]$$

$$+ \frac{(s+4)\,(3s+4)}{4s^2} n k_B (1-x) \frac{g_{\frac{3s+4}{2s}}\,(1)}{g_{\frac{s+4}{2s}}\,(1)}. \tag{5.61}$$

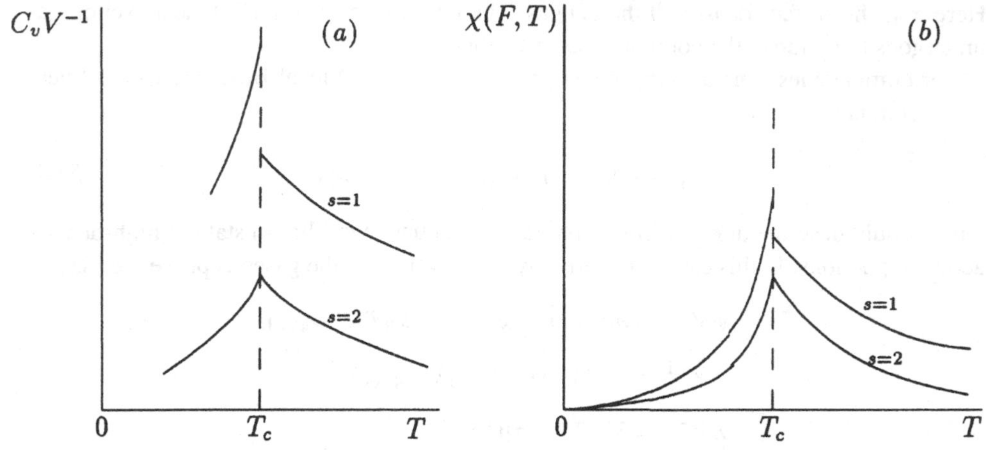

Figure 5.9. (a) The jumps of the heat capacity, (b) susceptibility for the ideal Bose gas with different dispersion laws: $s = 1$ and $s = 2$ (from Ref. 83).

The thermodynamic functions are continuous at the point of the phase transition – the discontinuities appear in only the heat capacity and the compressibility. The corresponding jumps are

$$V^{-1}C_V|_{\rm deg} - V^{-1}C_V|_{\rm gas} = \frac{U(0)n^2}{T_c}\left(\frac{s+4}{2s}\right)^2 + \frac{(s+4)^2}{4s^2}nk_B\frac{g_{\frac{s+4}{2s}}(1)}{g_{\frac{4-s}{2s}}(1)},$$

$$\left(\frac{\partial P}{\partial n}\right)_T\bigg|_{\rm deg} - \left(\frac{\partial P}{\partial n}\right)_T\bigg|_{\rm gas} = -U(0)n - k_BT_c\frac{g_{\frac{s+4}{2s}}(1)}{g_{\frac{4-s}{2s}}(1)}. \tag{5.62}$$

The jumps depend on not only the interaction constant, but also on the exciton dispersion law. For the ideal Bose gas $[U(0) = 0]$ and the quadratic dispersion law ($s = 2$), the jumps disappear $[g_{1/2}(1) \to \infty]$. Nevertheless, for the ideal Bose gas, but with a linear term in the dispersion law ($s = 1$), the jumps are nonzero. These two cases are illustrated in Fig. 5.9(a).

As one can see, consideration of the different dispersion laws of the 3D exciton gas leads to new qualitative results. Yet the linear term plays a more important role in the case of 2D exciton gas. If the crystal slab is cut perpendicularly to the crystallographic axis, then to calculate the partition function, it is necessary to integrate over only the plane $p_\perp$. The energy of elementary excitations with wave vectors lying in this plane has the dispersion law $cp_\perp$. This dispersion law at small values of $p_\perp$ gives much higher energy than the quadratic dispersion law $p_\perp^2/2m_\perp$. This fact leads to a decrease of the total number of the noncondensate particles, which facilitates condensate formation in the case of a 2D system.

In contrast to the case of the quadratic dispersion law, the critical temperature T_c for the 2D exciton gas with linear dispersion law $T_p = cp_\perp$ in the plane of the slab is nonzero and is given as [83]

$$k_BT_c = \left[\frac{2\pi\hbar^2c^2\sigma}{g_2(1)}\right]^{1/2}, \qquad \sigma = \frac{N}{S}. \tag{5.63}$$

Here σ is the surface density of the 2D gas. The heat capacity of the 2D ideal exciton gas undergoes the jump at the point of phase transition.

For completeness one also can consider the case of a 3D ideal Bose gas with a linear dispersion law,

$$T_{\mathbf{p}} = cp, \qquad p = \sqrt{p_x^2 + p_y^2 + p_z^2}, \tag{5.64}$$

which could describe nonequilibrium photons or a quasi-equilibrium state of high-density acoustic phonons. In this case, the thermodynamic values in the gaseous phase are [83]

$$\begin{aligned}
\psi V^{-1} &= nk_BT(1-x)\ln\xi - k_BT\lambda^{-3}g_4(\xi),\\
SV^{-1} &= -nk_B\ln\xi + 4k_B\lambda^{-3}g_4(\xi),\\
EV^{-1} &= 3k_BT\lambda^{-3}g_4(\xi),\\
P &= k_BT\lambda^{-3}g_4(\xi), \qquad \mu = k_BT\ln\xi,\\
C_VV^{-1} &= 12nk_B\frac{g_4(\xi)}{g_3(\xi)} - 9nk_B\frac{g_3(\xi)}{g_2(\xi)},\\
\left(\frac{\partial P}{\partial n}\right)_T &= k_BT\frac{g_3(\xi)}{g_2(\xi)}.
\end{aligned} \tag{5.65}$$

In the above, the saddle point ξ was determined by the equation

$$n = \lambda^{-3}g_3(\xi), \qquad \lambda^{-3} = \left(\frac{k_BT}{c}\right)^3\frac{1}{\pi^2\hbar^3}. \tag{5.66}$$

In the degenerate phase when $0 \le T \le T_c$, the condensate fraction x obeys the equation

$$n(1-x) = \lambda^{-3}g_3(1). \tag{5.67}$$

The thermodynamic values are

$$\begin{aligned}
\psi V^{-1} &= -k_BT\lambda^{-3}g_4(1), & SV^{-1} &= 4k_B\lambda^{-3}g_4(1),\\
EV^{-1} &= 3k_BT\lambda^{-3}g_4(1), & P &= k_BT\lambda^{-3}g_4(1),\\
C_VV^{-1} &= 12k_B\lambda^{-3}g_4(1), & \left(\frac{\partial P}{\partial n}\right)_T &= 0, \quad \mu = 0.
\end{aligned} \tag{5.68}$$

The jumps of the heat capacity and compressibility are

$$\begin{aligned}
V^{-1}C_V|_{\text{deg}} - V^{-1}C_V|_{\text{gas}} &= 9nk_B\frac{g_3(1)}{g_2(1)},\\
\left(\frac{\partial P}{\partial n}\right)_T\Big|_{\text{deg}} - \left(\frac{\partial P}{\partial n}\right)_T\Big|_{\text{gas}} &= -k_BT_c\frac{g_3(1)}{g_2(1)}.
\end{aligned} \tag{5.69}$$

Up to now, we have not considered the influence of the exciton–phonon interaction on the phase transition of high-density excitons. We now turn to this question, using the results of Refs. 82 and 84. Omitting the intermediate calculations, we write the contribution of the exciton–phonon interaction to the jump of the heat capacity of the exciton gas created in a deformable crystal. In addition to expressions (5.62) one must add the term

$$\frac{s+4}{2s}\frac{1}{k_B T_c^2}\sum_{\mathbf{p}}\frac{2T_{\mathbf{p}}B_{\mathbf{p}}\theta^2(p)}{T_{\mathbf{p}}^2-B_{\mathbf{p}}^2}\left(\bar{n}_{\mathbf{p}}^2 e^{\beta T_{\mathbf{p}}}-\bar{m}_{\mathbf{p}}^2 e^{\beta B_{\mathbf{p}}}\right)\frac{N}{N_a}. \tag{5.70}$$

Here the notations are the same as those used in Chapter 3. $B_{\mathbf{p}}$ is the dispersion law of the phonon branch, and $\bar{n}_{\mathbf{p}}$ and $\bar{m}_{\mathbf{p}}$ are the mean occupation numbers of the excitons and phonons at $T = T_c$, where

$$\bar{n}_{\mathbf{p}} = \frac{1}{e^{\beta T_{\mathbf{p}}}-1}, \qquad \bar{m}_{\mathbf{p}} = \frac{1}{e^{\beta B_{\mathbf{p}}}-1},$$

$$\sum_{\mathbf{p}}\bar{n}_{\mathbf{p}} = N,\ \xi = 1, \qquad x = 0, \qquad V = N_a v_0. \tag{5.71}$$

$\theta(p)$ is the constant of the exciton–phonon interaction, N_a is the number of the proper atoms of the lattice, and v_0 is the volume of the elementary cell. One can see that additional term (5.70) to the heat-capacity jump is negative. It may prevail or not over the first two contributions given by formulas (5.62).

It is important to note that the thermodynamic potential ψ, energy E, entropy S, and pressure P remain continuous functions at the point of the phase transition. As before, we are dealing with a second-order phase transition.

To estimate the influence of the phonons on the heat-capacity jump, we use the following realistic parameters: the exciton density $n = 10^{16}$ cm^{-3}, the sound velocity in crystal $u = 5\times10^5$ cm/s, and the exciton radius $a_{ex} = 20$ Å, with a quadratic dispersion law ($s = 2$). At the same time, we choose the dispersion law of the acoustic phonon $B_p = up$, the exciton translational mass $m_{ex} = m_0$, the constant $|C_e - C_h| = 5$ eV, and $U(0) = 10^{-34}$ erg cm^3. In this case, the resulting jump of the heat capacity is negative and equals -2.2×10^{-3} erg cm^{-3} K^{-1}. At larger values of $U(0)$, the jump will be positive, but lower than in the absence of the exciton–phonon interaction. A similar influence of the phonons on the heat capacity of ferromagnetics was described by Becker [85]. In this case, the spin waves play the role of the exciton system below the critical point $T \le T_c$.

Another characteristic of the exciton dispersion law is its degeneracy. Besides its anisotropy, which was discussed above in the context of deformable and rigid lattices, the degenerate exciton states can be characterized by the proper dipole moments and by the linear splitting under the influence of constant external electric and magnetic fields as well as that of the uniaxial deformation. Some aspects of the BEC of excitons with such an energy spectrum are discussed below.

5.3.2 In-Depth Study: *Bose Condensation of Excitons in an External Field – Ferromagnetism and Ferroelectricity*

The ferromagnetism of orthoexcitons as a result of their Bose condensation was first discussed by Blatt, Böer, and Brandt [86]. In this section, the results obtained by Moskalenko [83, 87] are reported. We begin with a review of the symmetry rules.

Besides the magnetic moment of the orthoexciton due to the spins of electron–hole pair, there is also the orbital magnetic moment associated with the electron–hole motion. One can formulate the selection rules that determine those irreducible representations of the exciton states that allow the existence of a corresponding dipole moment. The direct product of the representations $\Gamma\times\Gamma$ must contain the irreducible representations formed by the components of the desired dipole moment. In the case of a cubic crystal with a center of inversion, the components of the orbital magnetic moment form the irreducible representation Γ_4^+. Then all 3D representations $\Gamma_4^\pm$ and $\Gamma_5^\pm$ of the cubic group O_h can possess a magnetic moment and can undergo the linear Zeeman effect. The components of the electric dipole moment form the same representations as the coordinates x, y, and z. All 3D representations of the groups without a center of inversion, T, T_d, and O, as well as the 2D and some 1D representations of the groups D_3 and C_{3V}, allow the existence of a proper electric dipole moment. The splittings of the exciton levels under the influence of the external constant fields often depend on the orientation of these fields with regard to the crystallographic axis. For some symmetry directions, the splitting is incomplete.

In exactly the same way, one can determine the splitting of a degenerate exciton level under the action of a uniaxial deformation. In this case, one has to consider the irreducible representations formed by the components of the second-order symmetric tensor. In the cubic group O_h, these irreducible representations are Γ_1^+, Γ_3^+, and Γ_5^+. Then the corresponding selection rule has the form of the requirement $\Gamma \times \Gamma \in \Gamma_1^+, \Gamma_3^+, \Gamma_5^+$.

In what follows below, it is assumed that there is an external constant field or uniaxial stress F and a proper dipole moment f of the degenerate exciton level, which splits linearly in the external field. The split components have energies relative to the bottom of the lower exciton band equal to

$$T_{\mathbf{k}} + tfF, \tag{5.72}$$

where t equals 0, 1, 2 in the case of a threefold degenerate level or 0, 2 in the case of a twofold level. The energy of the ideal exciton gas in the presence of the external constant field F can be written as

$$E = \sum_{\mathbf{k},t} (T_{\mathbf{k}} + tfF)\,\bar{n}_{\mathbf{k},t}. \tag{5.73}$$

The full number of the noncondensate excitons in the split-off exciton bands, assumed to be constant, is

$$\frac{1}{V}\sum_{\mathbf{k},t} \bar{n}_{\mathbf{k},t} = n\,(1-x). \tag{5.74}$$

The mean value $\bar{n}_{\mathbf{k},t}$ of the exciton occupation number in the state characterized by the wave vector $\mathbf{k}$ and the subband number t has the form

$$\bar{n}_{\mathbf{k},t} = \frac{1}{\frac{1}{\xi}e^{\beta(T_k + tfF)} - 1}. \tag{5.75}$$

Here T_k has the general anisotropic form introduced in Eq. (5.45), whereas the saddle point ξ and the fraction x, which determines the macroscopic but incomplete occupation of the lowest state of the lowest subband, are determined as usual for the ideal Bose gas:

$$\begin{aligned} &x = 0,\ 0 \le \xi \le 1 \quad \text{at } T > T_c,\\ &0 \le x \le 1,\ \xi = 1 \quad \text{at } T \le T_c. \end{aligned} \tag{5.76}$$

Here the interaction between the excitons is neglected. This is justified when the splittings in the external field tfF are much more than the characteristic interaction energy $nU(0)x$.

The thermodynamic potential ψ and the total number of noncondensate particles were determined [83, 87] by the generalization of the formulas obtained above:

$$\psi V^{-1} = nk_BT\,(1-x)\ln\xi - k_BT\lambda^{-3}\sum_t g_{\frac{3s+4}{2s}}[\xi\,\exp(-\beta tfF)], \tag{5.77}$$

$$n\,(1-x) = \lambda^{-3}\sum_t g_{\frac{s+4}{2s}}[\xi\,\exp(-\beta tfF)], \qquad \beta^{-1} = k_BT,$$

$$\lambda^{-3} = \sqrt{\frac{\pi m_\parallel}{2}}\frac{\Gamma\left(1+\frac{2}{s}\right)(k_BT)^{\frac{s+4}{2s}}}{4\pi^2\hbar^3c^{2/s}}. \tag{5.78}$$

The critical temperature T_c depends on the field strength F and results from Eqs. (5.78) on substitution of the values $x = 0$ and $\xi = 1$. One has

$$kT_c(F) = \left\{\frac{4\sqrt{2}\pi^2\hbar^3nc^{2/s}}{(\pi m_\parallel)^{1/2}\Gamma\left(1+\frac{2}{s}\right)\sum_t g_{\frac{s+4}{2s}}[\exp(-\beta tfF)]}\right\}^{\frac{2s}{s+4}}. \tag{5.79}$$

When F tends to zero, all the subbands give the same contributions, whereas in the limit $F \to \infty$ only one term with $t = 0$ gives a nonzero contribution. The critical temperature increases with the external field:

$$\frac{T_c(F\to\infty)}{T_c(F=0)} = \left(\sum_t 1\right)^{\frac{2s}{s+4}}. \tag{5.80}$$

In addition to the thermodynamic values considered above, one can introduce the specific moment $M(F,T)$ and the susceptibility $\chi(F,T)$, determined as

$$M(F,T) = -\frac{1}{V}\left(\frac{d\psi}{dF}\right)_{T,V}, \qquad \chi(F,T) = \left[\frac{dM\,(F,T)}{dF}\right]_{T,V}. \tag{5.81}$$

In the gaseous phase,

$$M_{\text{gas}}(F,T) = f\lambda^{-3}\sum_t(1-t)g_{\frac{s+4}{2s}}(\xi e^{-\beta tfF}), \tag{5.82}$$

$$\chi_{\text{gas}}(F,T) = \frac{f^2n}{k_BT}\frac{\sum_{tt'}(t-1)(t-t')g_{\frac{4-s}{2s}}(\xi e^{-\beta tfF})g_{\frac{s+4}{2s}}(\xi e^{-\beta t'fF})}{\sum_{tt'}g_{\frac{4-s}{2s}}(\xi e^{-\beta tfF})g_{\frac{s+4}{2s}}(\xi e^{-\beta t'fF})}. \tag{5.83}$$

In the case $F = 0$, the specific moment $M_{\text{gas}}(0,T)$ equals zero because it is induced by the external field F. Mathematically it results from the equality

$$\sum_t(1-t) = 0,$$

in the both threefold and twofold degenerate cases. The susceptibility $\chi(F,T)$ is nonzero.

In the analysis of the expression $\chi_{\text{gas}}(F,T)$ of Eq. (5.83), one can begin with the case $s = 2$ for a small external field, such that βfF is much less than unity ($\beta fF \ll 1$), and consider the temperature region approaching the critical point $T_c (T - T_c \ll T_c)$ from above. An important role in these conditions is played by the singular function $g_{\frac{1}{2}}(\xi e^{-\beta fF})$, written in the form $g_{\frac{1}{2}}[e^{-(y+\beta fF)}]$, which diverges when the exponent $(y + \beta fF)$ tends to zero. Its analytical approximation (5.58) gives rise to two alternative expressions,

$$g_{\frac{1}{2}}(\xi e^{-\beta f F})_{\xi\to 1,\,\beta f F\ll 1} = g_{\frac{1}{2}}[e^{-(y+\beta f F)}]_{(y+\beta f F)\ll 1}$$

$$\approx \frac{\sqrt{\pi}}{\sqrt{y+\beta f F}} = \begin{cases} \frac{\sqrt{\pi}}{\sqrt{\beta f F}}, & y < \beta f F \ll 1 \\ \sqrt{\frac{\pi}{y}}, & \beta f F < y \ll 1 \end{cases}. \tag{5.84}$$

The small exponent y can be expressed in terms of the temperature difference $(T - T_c)/T_c \ll 1$, starting from Eq. (5.78) and approximations (5.58). Indeed, in the case $s = 2$, Eq. (5.78) becomes

$$n = \lambda^{-3} \sum_t g_{\frac{3}{2}}[\xi \exp(-\beta t f F)], \tag{5.85}$$

which in the limit $\beta f F < y \ll 1$ can be rewritten as

$$n \simeq \lambda^{-3}\left(\sum_t 1\right) g_{\frac{3}{2}}(e^{-y}) \simeq \lambda^{-3}\left(\sum_t 1\right)[g_{\frac{3}{2}}(1) - 2\sqrt{\pi y}]. \tag{5.86}$$

The parameter $\lambda^{-3} \cong T^{3/2}$ can be approximated as

$$\lambda^{-3}(T) \simeq \lambda^{-3}(T_c)\left(1 + \frac{3}{2}\frac{T - T_c}{T_c}\right), \qquad 0 < \frac{T - T_c}{T_c} \ll 1,$$

$$\lambda^{-3}(T_c)\left(\sum_t 1\right) g_{\frac{3}{2}}(1) = n. \tag{5.87}$$

Here is the critical temperature $T_c(F = 0)$ is used because $\beta f F$ is assumed to be less than y. Substituting relation (5.87) into relation (5.86), one has

$$\sqrt{\pi y} \approx \frac{3}{4} g_{\frac{3}{2}}(1) \frac{T - T_c}{T_c}. \tag{5.88}$$

On the basis of these relations and other inequalities of the type

$$g_{\frac{1}{2}}(e^{-y}) > g_{\frac{1}{2}}(e^{-\beta f F}) \qquad \text{at } y < \beta f F \ll 1, \tag{5.89}$$

and vice versa, one can obtain for the threefold degenerate exciton level ($t = 0, 1, 2$) the following final expressions for the susceptibility of the exciton system with $s = 2$ in the gaseous phase:

$$\chi_{\text{gas}}(F, T) = \frac{0.87 f^2 n}{\sqrt{k_B T_c f F}}, \qquad \left(\frac{T - T_c}{T_c}\right)^2 < \beta f F \ll 1, \tag{5.90}$$

$$\chi_{\text{gas}}(F, T) = \frac{C}{T - T_c}, \qquad \beta f F < \left(\frac{T - T_c}{T_c}\right)^2 \ll 1. \tag{5.91}$$

Here the Curie–Weiss constant C is

$$C = \frac{8\pi f^2 n}{9\,(2.612)^2\, k_B}. \tag{5.92}$$

The Curie–Weiss law (5.92) determines the temperature dependence of the susceptibility $\chi_{\text{gas}}(F, T)$ in the limit of small fields $\beta f F \ll 1$ and in the temperature interval not too close to the critical point, so that the inequality $(T - T_c)^2/T_c^2 > \beta f F$ holds. When the temperature T approaches T_c from above and this inequality is inverted, then the Curie–Weiss law is changed by expression (5.90). In the case $s = 1$, the susceptibility of the gaseous phase does not obey the Curie–Weiss law anywhere. This is related to the fact that the kinetic energy of the quasiparticles with small wave vectors is much more than in the case of a quadratic dispersion law ($s = 2$) and the influence of the external field is not so important. Assuming that $s = 1$, one deals with the functions $g_{5/2}(z)$ and $g_{3/2}(z)$, which converge in the point $z = 1$,

$$g_{5/2}(1) = 1.341, \qquad g_{3/2}(1) = 2.612,$$

unlike $g_{1/2}(z)$, which diverges.

This fact simplifies the calculation of the susceptibility $\chi_{\text{gas}}(F, T)$. For a threefold degenerate exciton level ($t = 0, 1, 2$), neglecting the influence of the external field, one can obtain

$$\chi_{\text{gas}}(T_c) = \frac{2}{3}\frac{2.612}{1.341}\frac{f^2 n}{k_B T_c}, \qquad s = 1. \tag{5.93}$$

Estimating this expression for the parameters $n = 10^{17}\text{cm}^{-3}$, $f = 10^{-17}$cgs units, and $T_c = 2$ K gives the value $\chi_{\text{gas}} = 0.04$ and the contribution $4\pi\chi_{\text{gas}} \cong 0.5$ to the dielectric constant.

Summarizing the results of Eqs. (5.90)–(5.93), one can conclude that there are two types of ferromagnetic and ferroelectric Bose gases, either those that are undergoing the Curie–Weiss regularity or those that are not.

At temperatures much higher than the critical temperature, the difference between the Bose gases with $s = 1$ and $s = 2$ disappears, and they both have a susceptibility that does not depend on the external field,

$$\chi_{\text{gas}}(T \to \infty) = \frac{f^2 n}{k_B T}\frac{\sum_{tt'}(t - t')(t - 1)}{\sum_{tt'} 1}. \tag{5.94}$$

This result follows from the fact that in this case all the functions $g_p(\xi e^{-\beta t f F})$ can be substituted by the same expression $\xi e^{-\beta t f F} \approx \xi$.

Now let us consider the degenerate phase, which exists at all temperatures $0 < T < T_c$, when the interaction between the quasiparticles is neglected. As usual, in this phase one has $\xi = 1$ and $0 < x < 1$. The specific moment $M(F, T)$ and susceptibility $\chi(F, T)$ in the degenerate phase are

$$M_{\text{deg}}(F, T) = nfx + f\lambda^{-3}\sum_t (1 - t)\, g_{\frac{s+4}{2s}}(e^{-\beta t f F}),$$

$$\chi_{\text{deg}}(F, T) = \frac{f^2\lambda^{-3}}{k_B T}\sum_t t^2 g_{\frac{4-s}{2s}}(e^{-\beta t f F}). \tag{5.95}$$

In the vicinity of the critical temperature and small values of the external field $\beta f F \ll 1$, assuming $t = 0, 1, 2$, one has

$$\chi_{\text{deg}}(F, T) = \begin{cases} \frac{0.87 f^2 n}{\sqrt{k_B T_c f F}}, & s = 2 \\ \frac{5}{3}\frac{2.612}{1.341}\frac{f^2 n}{k_B T_c}, & s = 1 \end{cases}. \tag{5.96}$$

By comparing these expressions with corresponding formulas (5.90) and (5.93), one can conclude that the susceptibility is a continuous function at the transition point for the quadratic dispersion law ($s = 2$). It undergoes a discontinuity in the case $s = 1$ of the same form and origin as observed for the heat capacity. At lower temperatures $kT \ll fF < kT_c$, the susceptibility of the degenerate phase tends to zero, decreasing exponentially:

$$\chi_{\deg}(F, T) = \frac{f^2 \lambda^{-3}}{k_B T} e^{-\beta f F} \approx T^{\frac{4-s}{2s}} e^{-\frac{fF}{k_B T}}. \tag{5.97}$$

The behavior of the susceptibility is shown in Fig. 5.9(b).

The condensate fraction x determines the spontaneous specific dipole moment M_s and the corresponding field amplitude $F_s = 4\pi M_s$. They are

$$M_s = nfx, \qquad F_s = 4\pi nfx, \tag{5.98}$$

where

$$x = \left\{1 - \left[\frac{T}{T_c(F)}\right]^{\frac{s+4}{2s}}\right\}. \tag{5.99}$$

Instead of $T_c(F)$, one can use $T_c(F \to \infty)$ if the value fF exceeds $k_B T$ or $T_c(F \to 0)$ in the opposite case. The amplitude of the spontaneous field plays the role of the proper Weiss field. For the parameters $n = 10^{17}$ cm^{-3}, $f = 10^{-17}$ cgs units, and $x = 0, 1$ one obtains the value $F_s = 1$ cgs unit $= 300$ V/cm.

We conclude this section with some general remarks. First, the model of an ideal Bose gas placed in an external, constant field can be supplemented by the interaction between the excitons. This can be done by combining the results of Subsections 5.3.1 and 5.3.2. It is possible to neglect the exciton–exciton interaction if the splitting in the field fF exceeds the characteristic value $U(0)nx$ of the exciton-level shift. If $U(0) = 10^{-33}$ erg cm^3, the minimal external field will be of the order of $F_{\min} = U(0)nx/f$ and will achieve the value F_s, mentioned above, at the same values of n, x, and f. It can also be interesting to study exciton states split simultaneously by two different fields. By analogy with Ref. 88, one can expect the appearance of mixed-type ferromagnetoelectricity.

As one can see, an interconnection exists between apparently different phenomena such as BEC, on one hand, and ferromagnetism and ferroelectricity, on the other hand. This fact stimulates one to apply the obtained formulas to real samples. By doing so and applying formula (5.92) to the ferroelectrics of the type KH_2PO_4 and triglycine sulfate with a constant electric dipole moment $f = 2\times10^{-18}$ cgs units, with two orientations $t = 0, 2$ and the concentration of the dipoles $n = 10^{23}$ cm^{-3}, one obtains the Curie–Weiss constant $C \cong$ 2000 K. It coincides with the order of magnitude of the experimental values, 2200–3300 K.

The restricted scope of this chapter does not allow the possibility of discussing many interesting papers and results. One of them is the BEC and superfluidity of excitons in a high magnetic field, or "magnetoexcitons," investigated by Liberman and Korolev [89–91]. They paid attention to the fact that a high magnetic field, such that the distance between the Landau levels exceeds the binding energy of an exciton, gives the opportunity to create various new states of matter, i.e., the exciton crystal, molecular complexes, and BEC of an exciton gas in a semiconductor, depending on the dimensionality of the system [91–93]. Korolev and Liberman considered the problem of excitonic interaction in its multielectron formulation, starting from the second-quantized representation of the Hamiltonian of interacting electrons and holes in a high magnetic field, and studied the exciton ground state in a high magnetic field [94–97]. A system of a such excitons is similar to that of a weakly nonideal Bose gas. The theoretical basis for the existence and the stability of the Bose condensate in a bulk semiconductor due to an overall decrease of the interaction between excitons and an increase of their binding energy in a high magnetic field is therefore well established for a high density of excitons [89, 90]. The existence of a built-in condensate of excitons over a broad density range significantly changes the excitation spectrum of coupled excitons and photons in a high magnetic field [91] and results in a number of interesting optical phenomena. It was established that the interaction of two excitons, one of which is in an excited state, can lead to strong coupling. Different types of exciton complexes in a quasi-1D semiconductor quantum wire, including excitonic crystals and molecular chains, were investigated, obtained by variation of both the direction and the intensity of the magnetic field and by variation of the exciton density [91–93].

Earlier work on magnetoexcitons in two dimensions [98, 99] showed that superfluid excitons in these structures would move in a direction perpendicular to both the magnetic field and an applied electric field, i.e., that the magnetoexciton superfluid would be a superconductor. Subsequent experimental investigations [100–104] of these structures, made from GaAs quantum wells, showed strong nonlinear optical effects, in agreement with theory. Recent theoretical and experimental work on the special case of spatially-separated two-dimensional electrons and holes is discussed in section 10.2.

Finally, we note that the Lee and Yang formalism has also been applied to the case of a two-component Bose gas, i.e., a gas made of two different types of bosons, which may convert into each other [105, 106]. In particular, this method has been applied to the case of Cu_2O, in which the lowest exciton level is split by electron–hole exchange into two sublevels, the singlet paraexcitons, which are the lowest level, and the triplet orthoexcitons, which lie 12 meV higher (see Appendix A). Both of these levels are highly occupied in most experiments, as discussed in Subsection 1.4.2. Starting with Hamiltonian (4.71), which describes the interaction between the orthoexcitons and the paraexcitons, assuming a common chemical potential for the two species, and using the Bogoliubov variational principle [41] introduced in Subsection 5.2.2 to determine an upper limit for the thermodynamic potential, Moskalenko et al. [105, 106] predicted that the mean-field blue shift of the paraexciton ground-state energy, because of interactions, should be greater than that of the orthoexcitons and that at high density the paraexcitons can actually undergo a mean-field blue shift so large that they move to a higher energy than the orthoexcitons, leaving the orthoexcitons as the lowest exciton state. This has never been seen experimentally in Cu_2O, and probably will not be, since the ortho–para splitting is so large in Cu_2O, and the orthoexcitons and paraexcitons almost never reach chemical equilibrium at low temperature because the interconversion processes become slow. Nevertheless, this effect could be the basis of fascinating nonlinear optical effects in other semiconductor systems with overall repulsive interactions between the excitons.

References

[1] S.A. Moskalenko, *Zh. Opt. Spektrosk.* **5**, 147 (1958).

[2] M.A. Lampert, *Phys. Rev. Lett.* **1**, 450 (1958).

[3] L.V. Keldysh, in *Proceedings of the Ninth International Conference on Semiconductor Physics, Moscow, 1968*, (Nauka, Leningrad, 1969), Vol. 2.

[4] L.V. Keldysh, in *Electron-Hole Droplets in Semiconductors*, C.D. Jeffries and L.V. Keldysh, eds. (North-Holland, Amsterdam, 1987).
[5] T.M. Rice, in *Solid State Physics*, H. Ehrenreich, F. Seitz, and D. Turnbull, eds. (Academic, New York, 1977), Vol. 32.
[6] L.V. Keldysh, *Contemp. Phys.* **27**, 395 (1986).
[7] L.V. Keldysh, in *Bose-Einstein Condensation*, A. Griffin, D.W. Snoke and S. Stringari, eds. (Cambridge U. Press, Cambridge, 1995).
[8] L.V. Keldysh and Yu. V. Kopaev, *Fiz. Tverd. Tela* **6**, 2791 (1964) [*Sov. Phys. Solid State* **6**, 2219 (1965)].
[9] J. des Cloizeaux, *J. Phys. Chem. Solids* **26**, 259 (1965).
[10] B.I. Halperin and T.M. Rice, *Solid State Phys.* **21**, 115 (1968).
[11] J. Ding et al., *Phys. Rev. Lett.* **69**, 1707 (1992).
[12] M.E. Flatte, E. Runge, and H. Ehrenreich, *Appl. Phys. Lett.* **66**, 1313 (1995).
[13] E.A. Andryushin, L.V. Keldysh and A.P. Silin, *Sov. Phys. JETP* **46**, 616 (1977).
[14] R.R. Guseinov and L.V. Keldysh, *Sov. Phys. JETP* **36**, 1193 (1973).
[15] J.J. Hopfield, *Phys. Rev.* **112**, 1555 (1958).
[16] H. Haug, *Z. Phys.* **200**, 57 (1967).
[17] H.Haug, *Phys. Rev.* **184**, 338 (1969).
[18] R. Graham and H. Haken, *Z. Phys.* **237**, 31 (1970).
[19] V. Degiorgio and M.O. Scully, *Phys. Rev. A* **2**, 1170 (1970).
[20] H. Haken, *Rev. Mod. Phys.* **47**, 67 (1975).
[21] H. Haug and S.W. Koch, *Quantum Theory of Optical and Electronic Properties of Semiconductors* (World Scientific, London, 1993).
[22] A. Stahl and I. Balslev, *Electrodynamics of Semiconductors Band Edge* (Springer, Berlin, 1987).
[23] L.V. Keldysh, in *Excitons in Semiconductors* (Nauka, Moscow, 1971), p. 5 (in Russian).
[24] C.D. Jeffries and L.V. Keldysh, eds., *Electron-Hole Droplets in Semiconductors* (North-Holland, Amsterdam, 1987).
[25] Ya.E. Pokrovskii, *Phys. Status Solidi B* **11**, 385 (1972).
[26] D. Hensel, T. Phillips and G. Thomas, in *Solid State Physics* **32**, H. Ehrenreich, F. Seitz, and D. Turnbull, eds. (Academic, New York, 1977).
[27] W.F. Brinkman and T.M. Rice, *Phys. Rev. B* **7**, 1508 (1973).
[28] M. Combescot and P.Nozieres, *J. Phys. C* **5**, 2369 (1972).
[29] I.V. Kukushkin, V.D. Kulakovskii, and V.B. Timofeev, *Zh. Eksp. Teor. Fiz.* **32**, 304 (1980).
[30] V.D. Kulakovskii, I.V. Kukushkin, and V.B. Timofeev, *Zh. Eksp. Teor. Fiz.* **81**, 684 (1981).
[31] V.B. Timofeev, in *Excitons*, E.I. Rashba and M.D. Sturge, eds. (North-Holland, Amsterdam, 1982).
[32] D. Pines, *Elementary Excitations in Solids* (Benjamin, New York, 1963).
[33] J. Hubbard, *Proc. R. Soc. London Ser. A* **243**, 336 (1957).
[34] P. Nozières and D. Pines, *Phys. Rev.* **111** , 442 (1958).
[35] V.B. Timofeev, V.D. Kulakovskii, and I.V. Kukushkin, *Physica B + C* **117/118**, 327 (1983).
[36] T.J. Greytak and D. Kleppner, in *New Trends in Atomic Physics*, G. Grynberg and R. Stora, eds. (North-Holland, Amsterdam, 1984).
[37] R.N. Silver, *Phys. Rev. B* **8**, 2403 (1973).
[38] H. Haug, *Z. Phys. B* **24**, 351 (1976).
[39] S.A. Moskalenko, S.S. Russu, and V.A. Zalozh, in *Quantum Statistical Properties of High-Density Excitons* (Shtiintsa, Kishinev, 1988), p.143 (in Russian).
[40] S.A. Moskalenko, S.S. Russu, V.V. Baltaga, A.I. Bobrysheva, A.V. Lelyakov, and V.A. Zalozh, in *Proceedings of the International Conference "Exciton 84," Güstrow, Germany*, 22–26 October 1984, part II (Pedagogishe Hohschule "Liselotte Hermann", Güstrow, Germany, 1984), p.162.
[41] N.N. Bogoliubov, *Nuovo Cimento* **7**, 794 (1958).
[42] I.S. Gradshtein and I.M. Ryzhik, *Tables of Integrals, Series and Products* A. Jeffrey, ed. (Academic, Boston, 1994).

[43] P.I. Khadzhi, *Nonlinear Optical Processes in the System of Excitons and Biexcitons in Semiconductors* (Shtiintsa, Kishinev, 1985) (in Russian).

[44] P.I. Khadzhi, G.D. Shibarshina, and A.H. Rotaru, *Optical Bistability in the System of Coherent Excitons and Biexcitons in Semiconductors* (Shtiintsa, Kishinev, 1988) (in Russian).

[45] N.F. Mott, *Philos. Mag.* **62**, 287 (1961).

[46] N.F. Mott, *Metal-Insulator Transitions* (Barnes and Noble, New York, 1974).

[47] P. Vashishta, R.K. Kalia, and K.S. Singwi, in *Electron-Hole Droplets in Semiconductors*, C.D. Jeffries and L.V. Keldysh, eds. (North-Holland, Amsterdam, 1987).

[48] Z.A. Insepov and H.E. Norman, *Zh. Eksp. Teor. Fiz.* **62**, 2290 (1972).

[49] W. Ebeling, W.D. Kraeft, D. Kremp, and K. Kilimann, *Phys. Status Solidi B* **78**, 241 (1976).

[50] M. Combescot and C. Benoit à la Guillaume, *Phys. Rev. Lett.* **44**, 182 (1980).

[51] A.A. Rogachev, *Ann. Phys.* **20**, 389 (1995).

[52] A.L. Ivanov and H. Haug, *Phys. Rev. Lett.* **71**, 3182 (1993).

[53] D.W. Snoke and J.D. Crawford, *Phys. Rev. E* **52**, 5796 (1995).

[54] W. Ebeling, W.-D. Kraft, and D. Kremp, *Theory of Bound States and Ionization Equilibrium in Plasmas and Solids* (Akademie-Verlag, Berlin, 1976).

[55] W.-D. Kraft, D. Kremp, W. Ebeling, and G. Röpke, *Quantum Statistics of Charged Particle Systems* (Plenum, New York, 1986).

[56] C.R. Smith, *Phys. Rev.* A **134**, 1235 (1964).

[57] E.g., B.K. Ridley, *Quantum Processes in Semiconductors* (Oxford U. Press, Oxford, 1988).

[58] T.D. Lee and C.N. Yang, *Phys. Rev.* **112**, 1419 (1958); **113**, 1406 (1959).

[59] T.D. Lee, K. Huang, and C.N. Yang, *phys. Rev.* **106**, 1135 (1957).

[60] K. Huang, C.N. Yang, and J.M. Luttinger, *Phys. Rev.* **105**, 776 (1957).

[61] E. Hanamura and H. Haug, *Phys. Rep.* **33**, 209 (1977).

[62] V.V. Tolmachev, *Theory of the Bose Gas* (Moscow State University, Moscow, 1969) (in Russian).

[63] K. Huang, *Statistical Mechanics* (Wiley, New York, 1963).

[64] R.P. Feynman, *Statistical Mechanics* (Benjamin, New York, 1972).

[65] Studies of the V.A. Steklova Mathematical Institute, *Statistical Mechanics and the Theory.* Statisticheskaya mekhanika i *of Dynamic Systems* (Nauka, Moscow, 1989), Vol. CLXXXXI (in Russian).

[66] N.N. Bogoliubov and N.N. Bogoliubov Jr., *Introduction to Quantum Statistical Mechanics* (Nauka, Moscow, 1984) (in Russian).

[67] P.N. Brusov and V.N. Popov, *Superfluidity and Collective Properties of Quantum Liquids* (Nauka, Moscow, 1988) (in Russian).

[68] F. London, *Superfluids II* (Wiley, New York, 1954).

[69] I.M. Khalatnikov, *Theory of Superfluidity* (Nauka, Moscow, 1971) (in Russian).

[70] K.G. Wilson and J. Kogut, *Phys. Rep.* **12**, 75 (1974).

[71] S.-K. Ma, *Modern Theory of Critical Phenomena* (Benjamin, Reading, MA, 1980).

[72] K. Huang, *Phys. Rev.* **115**, 765 (1959); **119**, 1129 (1960); Physica **26**, Suppl. 58 (1960).

[73] A.E. Glassgold, A.N. Kaufman, and K.M. Watson, *Phys. Rev.* **120**, 660 (1960).

[74] F. Mohling, *Phys. Rev.* **135**, 881 (1964); **139**, 664 (1965).

[75] D.N. Zubarev, *Zh. Eksp. Teor. Fiz.* **27**, 129 (1955); *Usp. Fiz. Nauk* **71**, 71 (1960).

[76] Yu.A. Tserkovnikov, *Dok. Akad. Nauk SSSR* **143**, 832 (1962).

[77] N. Hugenholtz and D.Pines, *Phys. Rev.* **116**, 489 (1959).

[78] S.A. Moskalenko, P.I. Khadzhi, A.I. Bobrysheva, and A.V. Lelyakov, *Fiz. Tverd. Tela* **5**, 1444 (1963).

[79] E.I. Rashba and V.I. Sheka, *Fiz. Tverd. Tela* **2**, 162 (1959).

[80] G.D. Mahan and J.J. Hopfield, *Phys. Rev.* **135**, 428 (1964).

[81] M. Abramowitz and I.A. Stegun, eds., *Handbook of Mathematical Functions* (National Bureau of Standards, Washington, D.C., 1964).

[82] S.A. Moskalenko, *Izv. Akad. Nauk MSSR* **7**, 79 (1963).

[83] S.A. Moskalenko, *Bose-Einstein Condensation of Excitons and Biexcitons* (RIO, Academy of Sciences of MSSR, Kishinev, 1970), (in Russian).

[84] S.A. Moskalenko, *Zh. Eksp. Teor. Fiz.* **45**, 1159 (1963).
[85] E. Becker, *Phys. Lett. A* **28**, 361 (1968).
[86] J.M. Blatt, K.W. Böer, and W. Brandt, *Phys. Rev.* **126**, 1691 (1962).
[87] S.A. Moskalenko, *Izv. Akad. Nauk MSSR* **12**, 89 (1966).
[88] V.G. Bar'yakhtar and I.E. Chunis, *Fiz. Tverd. Tela* **10**, 3546 (1968).
[89] M.A. Liberman and A.V. Korolev, *Phys. Rev. Lett.* **72**, 270 (1994).
[90] M.A. Liberman and A.V. Korolev, *Phys. Rev. B* **50**, 14077 (1994).
[91] A.V. Korolev and M.A. Liberman, *Solid State Commun.* **98**, 49 (1995).
[92] A.V. Korolev and M.A. Liberman, *Phys. Lett. A* **209**, 201 (1995).
[93] A.V. Korolev and M.A. Liberman, *Int. J. Mod. Phys. B* **10**, 729 (1996).
[94] R.J. Elliott and R. Loudon, *J. Phys. Chem. Solids* **15**, 196 (1960).
[95] H. Hasegava and R.E. Howard, *J. Phys. Chem. Solids* **21**, 179 (1961).
[96] L.P. Gor'kov and I.E. Dzyaloshinskii, *Zh. Eksp. Teor. Fiz.* **53**, 717 (1967).
[97] A.V. Korolev and M.A. Liberman, *Phys. Rev. A* **45**, 1762 (1992).
[98] D. Paquet, T.M. Rice, and K. Ueda, *Phys. Rev. B* **32**, 5208 (1985).
[99] I.V. Lerner and Yu.E. Lozovik, *Sov. Phys. JETP* **53**, 763 (1981).
[100] J.B. Stark, W.H. Knox, D.S. Chemla, W. Schäfer, S. Schmitt-Rink, and C. Stafford, *Phys. Rev. Lett.* **65**, 3033 (1990).
[101] S. Schmitt-Rink, J.B. Stark, W.H. Knox, D.S. Chemla, and W. Schäfer, *Appl. Phys. A* **53**, 491 (1991).
[102] J.B. Stark, W.H. Knox, and D.S. Chemla, *Phys. Rev. Lett.* **68**, 3080 (1992).
[103] C. Stafford, S. Schmitt-Rink, and W. Schäfer, *Phys. Rev. B* **41**, 10000 (1990).
[104] D.S. Chemla, J.B. Stark, and W.H. Knox, *Ultrafast Phenomona VIII* (Springer, Berlin, 1993), p. 21.
[105] S.A. Moskalenko, A.I. Bobrysheva, S.S. Russu, V.V. Baltaga, and A.V. Lelyakov, *J. Phys. C* **18**, 989 (1985).
[106] S.A. Moskalenko, S.S. Russu, V.V. Baltaga, A.I. Bobrysheva, and A.V. Lelyakov, *Phys. Status Solidi B* **129**, 657 (1985).

6

The Optical Stark Effect and the Virtual Bose Condensate

6.1 Nonequilibrium Theory of the Optical Stark Effect in the Excitonic Range of the Spectrum

One of the most important applications of the theory of Bose condensation of excitons is the optical Stark effect, or "AC Stark effect," first demonstrated for excitons in Cu_2O [1] and since seen in several semiconductors and semiconductor heterostructures (e.g., Refs. 2–5). This effect is a promising tool for the field of optical communications. One of the major interests in this field is the development of all-optical-switching methods, by which light signals are switched on or off directly by other light signals, just as in electronics, electrical signals switch other electrical signals. At the present, most optical communications systems use electrical signals (e.g., electro-optic or acousto-optic devices) to switch the light signals, which means that the limiting bandwidth of the system is controlled by that of the electrical signals, not the optical bandwidth. The optical Stark effect offers this possibility. When one laser beam impinges on a sample, it can drastically alter the excitonic absorption line shape. Therefore a second beam can see either a transparent or an absorbing medium, depending on the presence of the control beam. This is the optical equivalent of a transistor.

The basic effect is shown in Fig. 6.1. When an intense laser is tuned to a photon energy that is just below the excitonic ground state, the exciton ground state is shifted in frequency and the oscillator strength is altered. This leads to a strong reduction in the optical absorption at the original wavelength of the exciton ground state. Therefore the pump laser can turn the absorption on or off, controlling the transmission of a second laser at a different wavelength. As seen in this figure, at low energies there is also the appearance of optical gain, in the absence of population inversion.

This effect differs considerably from experiments we have considered so far, in which real excitons or biexcitons were created by a laser tuned to photon energies at or above the exciton ground state. In the optical Stark effect, the coherent excitons are virtual excitons, which exist only when the laser field is present. When the pump pulse ceases, the effect vanishes, as seen in the curve at the 1.2-ps delay in Fig. 6.1. As we will see, much of the theory directly connects to the theory of the quasi-equilibrium real condensate, however.

BEC of excitons in a strong electromagnetic field was studied for the first time by Elesin and Kopaev [6]. In this chapter, we present the more general nonequilibrium theory of the optical Stark effect in the excitonic range of the spectrum developed by Schmitt-Rink, Chemla, and Haug (SCH). Their basic papers [7, 8] contain not only many rigorous results, but in addition are instructive because of the plausible justifications of the method they

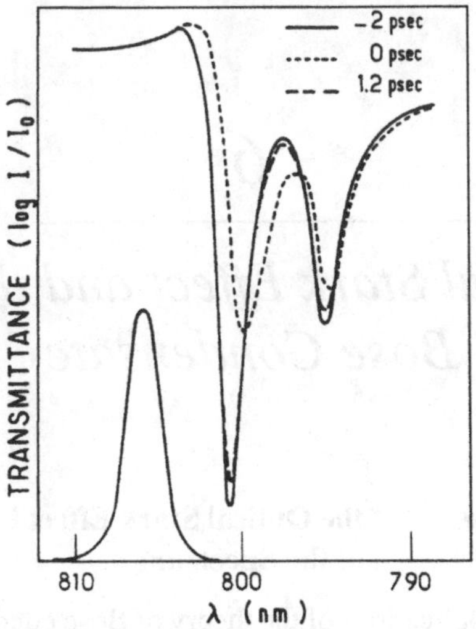

Figure 6.1. Subpicosecond time-resolved absorption spectrum of a GaAs–AlGaAs multiple-quantum-well sample with a well and barrier thickness of 100 Å, recorded at different delays relative to an intense, nonresonant optical pump pulse. The spectral position of the pump pulse is shown at the bottom of the figure (from Ref. 2).

choose and the explanations of the underlying physics. The results of other papers [6, 9] are discussed as they are related to the BEC of the excitons.

For better understanding of the methods SCH used, we begin with a short review of the Keldysh diagram technique for nonequilibrium processes. In particular, we outline the chief differences from the usual technique elaborated for equilibrium statistics.

6.1.1 The Keldysh Diagram Technique

The diagram technique developed by Keldysh [10] is similar to the usual Feynman technique in the field theory. It was elaborated for calculating the Green's functions of statistical systems in states far from thermodynamic equilibrium. As discussed by Lifshitz and Pitaevskii [11], the construction of the diagram technique for the nonequilibrium processes begins in the same way as in the equilibrium case at $T = 0$. The averaging of the electron operators, for example, $a^{\dagger}_{\mathbf{p}\sigma}$ and $a_{\mathbf{p}\sigma}$, in the interaction representation takes place also over the state of the noninteracting particles. But this state can be either a nonstationary or a stationary-but-excited state $|\Phi_i\rangle$, different from the ground state $|\Phi_0\rangle$. In this case it is impossible to apply the adiabatic hypothesis. That hypothesis is applicable only to the ground state $|\Phi_0\rangle$, which, being unique and nondegenerate, remains the same under the action of slowly increasing or decreasing interactions. It obtains a phase factor $e^{i\alpha}$ under the effect of the evolution operator $S(\infty, -\infty)$ as follows:

$$S(\infty, -\infty)|\Phi_0\rangle = e^{i\alpha}|\Phi_0\rangle.$$

As usual, one needs to calculate the average of the time-ordered operators in the interaction representation, of the type

$$\langle\Phi_n|S^{\dagger}(\infty, -\infty)T[S(\infty, -\infty)a_{\mathbf{p}\sigma,i}(t)a^{\dagger}_{\mathbf{p}\sigma,i}(t')]|\Phi_n\rangle. \tag{6.1}$$

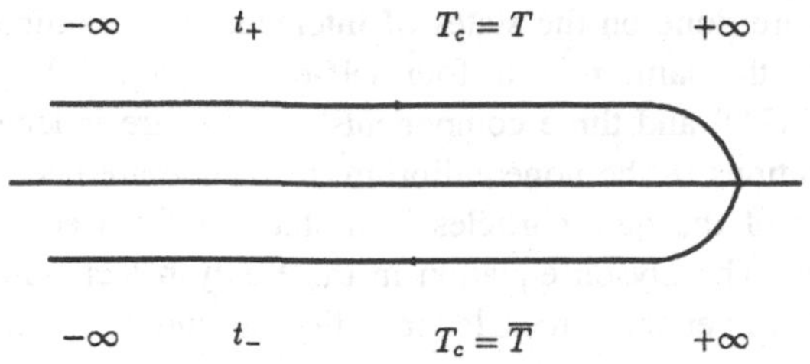

Figure 6.2. The contour C of two real times.

Here T is the time-ordering operator. The Latin letter indices i and H of the operators denote the interaction or the Heisenberg representations, respectively. Along with the evolution operator $S(\infty, -\infty)$ in expression (6.1), the Hermitian conjugate operator $S^{\dagger}(\infty, -\infty)$ is also presented. The adiabatic hypothesis considerably facilitates calculations when $\langle\Phi_n| = \langle\Phi_0|$, allowing the possibility of transforming expression (6.1) into the well-known formula [12, 13]

$$\frac{\langle\Phi_0|T[S(\infty, -\infty)a_{\mathbf{p}\sigma,i}(t)a^{\dagger}_{\mathbf{p}\sigma,i}(t')]|\Phi_0\rangle}{\langle\Phi_0|S(\infty,-\infty)|\Phi_0\rangle}. \tag{6.2}$$

When $\langle\Phi_n| = \langle\Phi_i|$, however, the operator $S^{\dagger}(\infty, -\infty)$ cannot be extracted. This gives rise to a new type of diagram. In contrast to the operator $S(\infty, -\infty)$, which is time ordered, the operator $S^{\dagger}(\infty, -\infty)$ is reverse time ordered. All its terms, when it is series expanded, must be kept on the left-hand side of the other operators in expression (6.1). These requirements can be taken into account properly by introduction of the contour C (see Fig. 6.2.), which goes along the upper side of the real time axis from $t_+ = -\infty$ to $t_+ = +\infty$ and back again in reverse on the lower, negative side from $t_- = +\infty$ to $t_- = -\infty$.

The time-ordering operator along the contour C is denoted by T_C. It is identical to the usual time-ordering operator $T_C = T$ along the upper side and to the reverse-time-ordering operator $T_C = \bar{T}$ on the lower side. The later time is $t_- = -\infty$ and the earlier time is $t_+ = -\infty$. After that it is necessary to ascribe the operator $S(\infty, -\infty)$ to the upper side and the operator $S^{\dagger}(\infty, -\infty)$ to the lower side. This assignment allows the possibility of gathering them without mixing in a new evolution operator S_C, as follows:

$$S_C = T_C \exp\left[-\frac{i}{\hbar}\int_{-\infty}^{\infty} H_{\text{int},i}(t'_+)\,\mathrm{d}t'_+ + \frac{i}{\hbar}\int_{-\infty}^{\infty} H_{\text{int},i}(t'_-)\,\mathrm{d}t'_-\right], \tag{6.3}$$

and it guarantees the arrangement of the operator $S^{\dagger}$ on the left-hand side with reference to the operator S as is required by expression (6.1). The interaction operators $H_{\text{int},i}(t'_{\pm})$, which are present in Eq. (6.3) with different signs before the integral, cannot be reduced because of the different signs of the times $t'_{\pm}$. They could be made different if different independent external sources are introduced, as proposed in Refs. 14–16.

Because of the existence of two real times $t_{\pm}$, there are four possible combinations of $t_{1\pm}$ and $t_{2\pm}$ when the Green's functions and the self-energy parts are constructed. The full and zero-order Green's functions $G^{\alpha,\beta}(t_1, t_2)$ and $G_0^{\alpha,\beta}(t_1, t_2)$ are determined as follows:

$$G^{\alpha,\beta}(t_1, t_2) = -i\langle T_C[a_{\mathbf{p}\sigma,H}(t_{1\alpha})a^{\dagger}_{\mathbf{p}\sigma,H}(t_{2\beta})]\rangle,$$

$$G_0^{\alpha,\beta}(t_1, t_2) = -i\langle T_C[a_{\mathbf{p}\sigma,i}(t_{1\alpha})a^{\dagger}_{\mathbf{p}\sigma,i}(t_{2\beta})]\rangle_0. \tag{6.4}$$

Here the averagings are done on the states of interacting and noninteracting quasiparticles, respectively. For the same reason, four self-energy parts $\Sigma^{\alpha,\beta}(t_1, t_2)$ appear. Only three components of $G^{\alpha,\beta}$ and three components of $\Sigma^{\alpha,\beta}$ are independent. The supplementary Green's functions in the nonequilibrium technique are needed to determine the distribution functions of the quasiparticles in a state far from equilibrium, along with their energy spectrum. The Dyson equation in the Keldysh technique contains the supplementary summation over the Greek letter indices γ and δ, which are labeled by $\pm$, namely,

$$\begin{aligned} G^{\alpha,\beta}&(t_1, t_2) \\ &= G_0^{\alpha,\beta}(t_1, t_2) + \sum_\gamma \sum_\delta \int dt_3 \int dt_4\, G_0^{\alpha,\gamma}(t_1, t_3)\Sigma^{\gamma,\delta}(t_3, t_4)G^{\delta,\beta}(t_4, t_2) \\ &= G_0^{\alpha,\beta}(t_1, t_2) + \sum_\gamma \sum_\delta \int dt_3 \int dt_4\, G^{\alpha,\gamma}(t_1, t_3)\Sigma^{\gamma,\delta}(t_3, t_4)G_0^{\delta,\beta}(t_4, t_2). \end{aligned} \tag{6.5}$$

As the system forgets the initial state of noninteracting particles $|\Phi_i\rangle$ under the influence of the interaction H_{int}, it is reasonable to exclude from the Dyson equation the memory of the initial distribution function and, in the same manner, the zeroth-order Green's function. This aim can be achieved by the help of the differential operator $G_0^{-1}(t) = [i\hbar\frac{\partial}{\partial t} - \mathcal{E}(\mathbf{p})]$, where $\mathcal{E}(\mathbf{p})$ is a free-electron energy spectrum. Taking into account the relations

$$G_0^{-1}(t_1)G_0^{\pm\pm}(t_1, t_2) = \pm\hbar\delta(t_1 - t_2), \qquad G_0^{-1}(t_1)G_0^{\pm\mp}(t_1, t_2) = 0, \tag{6.6}$$

one can obtain the integrodifferential Dyson equation,

$$\begin{aligned} \left[G_0^{-1}(t_1) - G_0^{-1*}(t_2)\right] G^{-+}(t_1, t_2) &= i\hbar\left(\frac{\partial}{\partial t_1} + \frac{\partial}{\partial t_2}\right) G^{-+}(t_1, t_2) \\ &= \hbar \sum_\alpha \int dt' [\Sigma^{-,\alpha}(t_1, t')G^{\alpha,+}(t', t_2) + G^{-,\alpha}(t_1, t')\Sigma^{\alpha,+}(t', t_2)]. \end{aligned} \tag{6.7}$$

As mentioned in Ref. 10, it is often advantageous to introduce the relative time $t_a = t_1 - t_2$ and the center time $t = (t_1 + t_2)/2$ as microscopic and macroscopic scales, respectively, and to represent the Green's functions and the self-energy parts as follows:

$$\begin{gathered} G^{\alpha,\beta}(t_1, t_2) = G^{\alpha,\beta}(t \mid t_a), \qquad \Sigma^{\alpha,\beta}(t_1, t_2) = \Sigma^{\alpha,\beta}(t \mid t_a), \\ G^{\alpha,\beta}(t \mid \omega) = \int_{-\infty}^{\infty} dt_a e^{i\omega t_a} G^{\alpha,\beta}(t \mid t_a), \qquad \Sigma^{\alpha,\beta}(t \mid \omega) = \int_{-\infty}^{\infty} dt_a e^{i\omega t_a} \Sigma^{\alpha,\beta}(t \mid t_a). \end{gathered} \tag{6.8}$$

Two times describe the dynamic and the kinetic stages of the evolution. If the variations on the two scales are sufficiently well separated, in particular for slow variation on the macroscopic scale, then one can Fourier transform Eq. (6.7) with respect to the relative time t_a:

$$i\frac{\partial}{\partial t} G^{-,+}(t \mid \omega) = \sum_\alpha [\Sigma^{-,\alpha}(t \mid \omega)G^{\alpha,+}(t \mid \omega) + G^{-,\alpha}(t \mid \omega)\Sigma^{\alpha,+}(t \mid \omega)]. \tag{6.9}$$

The nonequilibrium theory of the optical Stark effect in the excitonic range of the spectrum was elaborated by SCH [7, 8] within the framework of these methods and approximations.

Below, their results are presented, but in a form slightly different from the original, because of our use of the hole operators instead of valence electron operators. The *e–h* operators were introduced in Chapter 2.

6.1.2 Nonequilibrium Theory of Virtual Excitonic Bose–Einstein Condensation

As mentioned in Refs. 7 and 8, in experiments on the excitonic optical Stark effect, the coherent *e–h* pairs responsible for the shift of the absorption line and the change of the oscillator strength are generated only virtually, and they persist only as long as a pump beam is present. The optical Stark shift is the manifestation of the light-induced renormalization of the excitonic states, which closely resembles the case in superconductors and Bose condensed systems. They can be described with one and the same theory.

In the Green's function formalism, the coherent polarization induced by a coherent laser field is determined by the off-diagonal interband Green's function. It corresponds to the anomalous Green's function in a superconductor or to the condensate wave function in a Bose gas.

The coherent *e–h* pairing is not spontaneous but is externally enforced by the symmetry-breaking pump field. SCH theory gives the possibility of recovering the transition from real-space pairing to momentum-space pairing as the pump intensity increases. Such investigations in the case of spontaneous BEC were done by Comte and Nozières [17, 18] and were discussed in Chapter 2.

The Hamiltonian of the *e–h* system in the presence of laser radiation consists of two parts. One of them is Hamiltonian (2.93) of Chapter 2, but without the chemical potential. The second term describes the interaction of the semiconductor with laser radiation characterized by frequency ω_L and electric-field strength E_L. Only the interband transitions with the dipole moment d in the rotating-wave approximation are taken into account. This term has the form

$$H_r = \sum_{\mathbf{p}\sigma} \left[(dE_L)e^{-i\omega_L t} a^\dagger_{\mathbf{p}\sigma} b^\dagger_{-\mathbf{p},-\sigma} + (dE_L)^* e^{i\omega_L t} b_{-\mathbf{p},-\sigma} a_{\mathbf{p}\sigma}\right]. \tag{6.10}$$

The interaction of the *e–h* pairs with vacuum photons is not taken into account.

The explicit time dependence can be eliminated by the help of the unitary transformation,

$$V = \exp\left(-\frac{i\omega_L t}{2} N\right), \qquad N = \sum_{\mathbf{p}\sigma} (a^\dagger_{\mathbf{p}\sigma} a_{\mathbf{p}\sigma} + b^\dagger_{\mathbf{p}\sigma} b_{\mathbf{p}\sigma}). \tag{6.11}$$

It means the transition from the laboratory frame to a reference frame rotating with the frequency ω_L. In this reference frame, the full Hamiltonian has the form

$$\begin{aligned}
\mathcal{H} &= V^\dagger (H + H_r) V - i\hbar V^\dagger \frac{\mathrm{d}V}{\mathrm{d}t} \\
&= \sum_{\mathbf{p}\sigma} [E_e(\mathbf{p}) a^\dagger_{\mathbf{p}\sigma} a_{\mathbf{p}\sigma} + E_h(\mathbf{p}) b^\dagger_{\mathbf{p}\sigma} b_{\mathbf{p}\sigma}] + \sum_{\mathbf{p}\sigma} [(dE_L) a^\dagger_{\mathbf{p}\sigma} b^\dagger_{-\mathbf{p},-\sigma} + (dE_L)^* b_{-\mathbf{p},-\sigma} a_{\mathbf{p}\sigma}] \\
&\quad + \frac{1}{2} \sum_{\mathbf{p},\mathbf{q},\mathbf{k}} \sum_{\sigma_1\sigma_2} V_{\mathbf{k}} \big(a^\dagger_{\mathbf{p}\sigma_1} a^\dagger_{\mathbf{q}\sigma_2} a_{\mathbf{q}+\mathbf{k},\sigma_2} a_{\mathbf{p}-\mathbf{k},\sigma_1} + b^\dagger_{\mathbf{p}\sigma_1} b^\dagger_{\mathbf{q}\sigma_2} b_{\mathbf{q}+\mathbf{k},\sigma_2} b_{\mathbf{p}-\mathbf{k},\sigma_1} \\
&\quad - 2 a^\dagger_{\mathbf{p}\sigma_1} b^\dagger_{\mathbf{q}\sigma_2} b_{\mathbf{q}+\mathbf{k},\sigma_2} a_{\mathbf{p}-\mathbf{k},\sigma_1} \big).
\end{aligned} \tag{6.12}$$

The free e and h energies, measured relative to the middle of the semiconductor energy gap, are

$$E_i(\mathbf{p}) = \frac{E_g - \hbar\omega_L}{2} + \frac{\hbar^2 p^2}{2m_i}, \qquad i = e, h. \tag{6.13}$$

Here, only a simple model of a two-band semiconductor with spin degeneracy is considered.

As mentioned above, the dense system of e–h pairs generated by laser radiation requires the use of the many-body nonequilibrium technique. The Green's functions of the one-component system of Eqs. (6.4) must be generalized for the case of a two-component system in the form of a 2×2 matrix,

$$\hat{G}^{\alpha,\beta}(\mathbf{p}; t_1, t_2) = \left\| G_{ij}^{\alpha,\beta}(\mathbf{p}; t_1, t_2) \right\|$$
$$= \begin{vmatrix} -i\langle T_C[a^{\dagger}_{\mathbf{p}\sigma,H}(t_{1\alpha})a_{\mathbf{p}\sigma,H}(t_{2\beta})]\rangle & -i\langle T_C[a^{\dagger}_{\mathbf{p}\sigma,H}(t_{1\alpha})b^{\dagger}_{-\mathbf{p},-\sigma,H}(t_{2\beta})]\rangle \\ -i\langle T_C[b_{-\mathbf{p},-\sigma,H}(t_{1\alpha})a_{\mathbf{p}\sigma,H}(t_{2\beta})]\rangle & -i\langle T_C[b_{-\mathbf{p},-\sigma,H}(t_{1\alpha})b^{\dagger}_{-\mathbf{p},-\sigma,H}(t_{2\beta})]\rangle \end{vmatrix}. \tag{6.14}$$

The zeroth-order matrix $\hat{G}_0^{\alpha,\beta}(\mathbf{p}; t_1, t_2)$ consists of the zeroth-order Green's function of the same composition as that in matrix (6.14). The diagonal elements of the matrix $\hat{G}^{\alpha,\beta}(\mathbf{p}; t_1, t_2)$ describe the propagation of an electron or a hole within the corresponding bands, while the off-diagonal elements describe the creation and the annihilation of an e–h pair. The components of matrix (6.14) can be determined in different ways with the accuracy of the operator transpositions. When the electron component is chosen in the form $-i\langle T[a^{\dagger}(t)a(t')]\rangle$, the hole component appears in the alternative form $-i\langle T[b(t)b^{\dagger}(t')]\rangle$, and vice versa. As a result, the self-energy parts $\Sigma_{ii}^{\alpha,\beta}$ for two diagonal Green's functions will be determined with different signs and therefore will also appear in the motion equations with different signs. The Dyson equation for the component $G_{ij}^{-,+}(\mathbf{p}; t_1, t_2)$ will contain the summation over the Latin and Greek letter indices,

$$\begin{aligned} G_{ij}^{-,+}(\mathbf{p}; t_1, t_2) &= G_{0,ij}^{-,+}(\mathbf{p}; t_1, t_2) \\ &\quad + \sum_{l,m}\sum_{\alpha,\beta} \int dt_3 \int dt_4 G_{0,il}^{-,\alpha}(\mathbf{p}; t_1, t_3)\Sigma_{lm}^{\alpha,\beta}(\mathbf{p}; t_3, t_4)G_{mj}^{\beta,+}(\mathbf{p}; t_4, t_2) \\ &= G_{0,ij}^{-,+}(\mathbf{p}; t_1, t_2) \\ &\quad + \sum_{l,m}\sum_{\alpha,\beta} \int dt_3 \int dt_4 G_{il}^{-,\alpha}(\mathbf{p}; t_1, t_3)\Sigma_{lm}^{\alpha,\beta}(\mathbf{p}; t_3, t_4)G_{0,mj}^{\beta,+}(\mathbf{p}; t_4, t_2). \end{aligned} \tag{6.15}$$

The zeroth-order Green's functions $G_{0,ij}^{\alpha,\beta}(\mathbf{p}; t_1, t_2)$ can be excluded by the help of differential operators,

$$G_{01}^{-1}(t) = \left[i\hbar\frac{\partial}{\partial t} - E_e(\mathbf{p}) \right], \qquad G_{02}^{-1}(t) = \left[i\hbar\frac{\partial}{\partial t} - E_h(\mathbf{p}) \right],$$

and their complex-conjugate expressions. They can be used in different combinations, taking into account the relations of the type of Eqs. (6.6).

Equation (6.15) can be simplified by neglect of the self-energy matrices $\hat{\Sigma}^{\pm\mp}$, compared with $\hat{\Sigma}^{\pm\pm}$, because the first matrices cannot appear in the first order of the perturbation theory, while the second can. In this approximation, only the self-energy matrix remains:

$$\hat{\Sigma} = \hat{\Sigma}^{++} \approx -\hat{\Sigma}^{--}.$$

The next transformation of Eq. (6.15) is related to the Fourier transform of the type of Eqs. (6.8) and (6.9). As a result one obtains four motion equations for the components $G_{\bar{i}\bar{j}}^{-+}(\mathbf{p}, t \mid \omega)$. They are

$$
\begin{aligned}
i\frac{\partial}{\partial t}G_{\bar{i}\bar{i}}^{-+}(\mathbf{p}, t \mid \omega) &= \sum_l [\Sigma_{il}(\mathbf{p}, t \mid \omega)G_{\bar{l}\bar{i}}^{-+}(\mathbf{p}, t \mid \omega) - G_{\bar{i}\bar{l}}^{-+}(\mathbf{p}, t \mid \omega)\Sigma_{li}(\mathbf{p}, t \mid \omega)],\\
\left\{i\frac{\partial}{\partial t} + [E_e(\mathbf{p}) + E_h(\mathbf{p})]/\hbar\right\} G_{\bar{1}\bar{2}}^{-+}(\mathbf{p}, t \mid \omega) &= \sum_l [\Sigma_{1l}(\mathbf{p}, t \mid \omega)G_{\bar{l}\bar{2}}^{-+}(\mathbf{p}, t \mid \omega)\\
&\quad - G_{\bar{1}\bar{l}}^{-+}(\mathbf{p}, t \mid \omega)\Sigma_{l2}(\mathbf{p}, t \mid \omega)],\\
\left\{i\frac{\partial}{\partial t} - [E_e(\mathbf{p}) + E_h(\mathbf{p})]/\hbar\right\} G_{\bar{2}\bar{1}}^{-+}(\mathbf{p}, t \mid \omega) &= \sum_l [\Sigma_{2l}(\mathbf{p}, t \mid \omega)G_{\bar{l}\bar{1}}^{-+}(\mathbf{p}, t \mid \omega)\\
&\quad - G_{\bar{2}\bar{l}}^{-+}(\mathbf{p}, t \mid \omega)\Sigma_{l1}(\mathbf{p}, t \mid \omega)]. \quad (6.16)
\end{aligned}
$$

The next approximation is related to the neglect of the ω dependence of the self-energy matrix components. As SCH [7, 8] emphasized, such a possibility can be achieved in the collisionless regime. Experimentally this situation can be realized, for example, under nonresonant excitation well below the exciton level of virtual e–h pairs, when no energy is deposited in the sample. The nonresonant field drives coherent valence charge fluctuations that can be viewed as virtual e–h pairs. The coherence implies that these pairs experience no real collisions. The virtual transitions occur in a time shorter than $|E_{\rm ex}/\hbar - \omega_L|^{-1}$, which in its turn must be shorter than any relaxation time in a medium. The other possibility is the transient stage that appears for resonant excitation with an ultrashort laser pulse, on short time scales before the first collision takes place. SCH pointed out that the affirmation "no relaxation" does not imply "no correlation."

The influence of the Coulomb interaction was taken into account within the Hartree–Fock–Bogoliubov approximation or mean-field approach. It allows for virtual collisions, which will renormalize the system parameters. Hamiltonian (6.12) in the mean-field approximation is

$$\mathcal{H}_{m-f} = \sum_{\mathbf{p}\sigma}[\tilde{E}_e(\mathbf{p})a^\dagger_{\mathbf{p}\sigma}a_{\mathbf{p}\sigma} + \tilde{E}_h(\mathbf{p})b^\dagger_{\mathbf{p}\sigma}b_{\mathbf{p}\sigma} + \Delta(\mathbf{p})a^\dagger_{\mathbf{p}\sigma}b^\dagger_{-\mathbf{p},-\sigma} + \Delta^*(\mathbf{p})b_{-\mathbf{p},-\sigma}a_{\mathbf{p},\sigma}], \quad (6.17)$$

where the renormalized e–h energies and the Rabi frequency are

$$
\begin{gathered}
\tilde{E}_i(\mathbf{p}) = E_i(\mathbf{p}) + \sigma_i(\mathbf{p}),\\
\sigma_e(\mathbf{p}) = -\sum_{\mathbf{q}} V_{\mathbf{p}-\mathbf{q}}\langle a^\dagger_{\mathbf{q}\sigma}a_{\mathbf{q}\sigma}\rangle, \qquad \sigma_h(\mathbf{p}) = -\sum_{\mathbf{q}} V_{\mathbf{p}-\mathbf{q}}\langle b^\dagger_{\mathbf{q}\sigma}b_{\mathbf{q}\sigma}\rangle,\\
\Delta(\mathbf{p}) = (dE_L) - \sum_{\mathbf{q}} V_{\mathbf{p}-\mathbf{q}}\langle b_{-\mathbf{q},-\sigma}a_{\mathbf{q}\sigma}\rangle. \quad (6.18)
\end{gathered}
$$

When $\Sigma_{ij}(\mathbf{p}, t \mid \omega)$ do not depend on ω, Eqs. (6.16) can be transformed into the motion equations for the reduced density matrix $\hat{n}(\mathbf{p}, t)$, which is determined as follows:

$$\hat{n}(\mathbf{p}, t) = i \int_{-\infty}^{\infty} \frac{d\omega}{2\pi} G^{-+}(\mathbf{p}, t \mid \omega) = i G^{-+}(\mathbf{p}; t, t),$$

$$\hat{n}(\mathbf{p}, t) = \begin{vmatrix} \langle a^{\dagger}_{\mathbf{p}\sigma,H}(t) a_{\mathbf{p}\sigma,H}(t)\rangle & \langle a^{\dagger}_{\mathbf{p}\sigma,H}(t) b^{\dagger}_{-\mathbf{p},-\sigma,H}(t)\rangle \\ \langle b_{-\mathbf{p},-\sigma,H}(t) a_{\mathbf{p}\sigma,H}(t)\rangle & \langle b_{-\mathbf{p},-\sigma,H}(t) b^{\dagger}_{-\mathbf{p},-\sigma,H}(t)\rangle \end{vmatrix}. \quad (6.19)$$

In this way, for $t_1 = t_2 = t$ the 2×2 distribution-function matrix $i\hat{G}^{-+}(\mathbf{p}; t, t)$ generalizes the concept of the distribution function in classical statistical mechanics and equals the reduced density matrix $\hat{n}(\mathbf{p}, t)$. Its diagonal components are related to the e–h distribution functions:

$$n_{11}(\mathbf{p}, t) = n_e(\mathbf{p}, t), \qquad n_{22}(\mathbf{p}, t) = 1 - n_h(\mathbf{p}, t). \quad (6.20)$$

The off-diagonal components play the role of wave functions $n_{21}(\mathbf{p}, t) = \Psi(\mathbf{p}, t)$ and $n_{12}(\mathbf{p}, t) = \Psi^*(\mathbf{p}, t)$. The motion equations for the components of the reduced density matrix have the form

$$i\frac{\partial n_{ii}(\mathbf{p}, t)}{\partial t} = \sum_l \left[\Sigma_{il}(\mathbf{p}, t) n_{li}(\mathbf{p}, t) - n_{il}(\mathbf{p}, t)\Sigma_{li}(\mathbf{p}, t)\right], \qquad i = 1, 2,$$

$$\left[i\frac{\partial}{\partial t} + \frac{E_e(p) + E_h(p)}{\hbar}\right] n_{12}(\mathbf{p}, t) = \sum_l \left[\Sigma_{1l}(\mathbf{p}, t) n_{l2}(\mathbf{p}, t) - n_{1l}(\mathbf{p}, t)\Sigma_{l2}(\mathbf{p}, t)\right],$$

$$\left[i\frac{\partial}{\partial t} - \frac{E_e(p) + E_h(p)}{\hbar}\right] n_{21}(\mathbf{p}, t) = \sum_l \left[\Sigma_{2l}(\mathbf{p}, t) n_{l1}(\mathbf{p}, t) - n_{2l}(\mathbf{p}, t)\Sigma_{l1}(\mathbf{p}, t)\right]. \quad (6.21)$$

They can be rewritten as

$$\begin{aligned} i\frac{\partial n_{11}(\mathbf{p}, t)}{\partial t} &= \Sigma_{12}(\mathbf{p}, t) n_{21}(\mathbf{p}, t) - n_{12}(\mathbf{p}, t)\Sigma_{21}(\mathbf{p}, t), \\ i\frac{\partial n_{22}(\mathbf{p}, t)}{\partial t} &= \Sigma_{21}(\mathbf{p}, t) n_{12}(\mathbf{p}, t) - n_{21}(\mathbf{p}, t)\Sigma_{12}(\mathbf{p}, t), \\ i\frac{\partial n_{12}(\mathbf{p}, t)}{\partial t} &= -\left[\frac{E_e(p) + E_h(p)}{\hbar} + \Sigma_{22}(\mathbf{p}, t) - \Sigma_{11}(\mathbf{p}, t)\right] n_{12}(\mathbf{p}, t) \\ &\quad + \Sigma_{12}(\mathbf{p}, t)\left[n_{22}(\mathbf{p}, t) - n_{11}(\mathbf{p}, t)\right] \\ i\frac{\partial n_{21}(\mathbf{p}, t)}{\partial t} &= \left[\frac{E_e(p) + E_h(p)}{\hbar} + \Sigma_{22}(\mathbf{p}, t) - \Sigma_{11}(\mathbf{p}, t)\right] n_{21}(\mathbf{p}, t) \\ &\quad + \Sigma_{21}(\mathbf{p}, t)\left[n_{11}(\mathbf{p}, t) - n_{22}(\mathbf{p}, t)\right]. \end{aligned} \quad (6.22)$$

Comparing the combinations $[E_e(p) - \hbar\Sigma_{11}(\mathbf{p}, t)]$ and $[E_h(p) + \hbar\Sigma_{22}(\mathbf{p}, t)]$ with corresponding expressions (6.18), one can conclude that $\hbar\Sigma_{11}(\mathbf{p}, t) = -\sigma_e(\mathbf{p}, t)$ and $\hbar\Sigma_{22}(\mathbf{p}, t) = \sigma_h(\mathbf{p}, t)$. At the same time $\hbar\Sigma_{12}(\mathbf{p}, t) = -\Delta^*(\mathbf{p}, t)$. Commenting on the expression for $\Sigma_{12}(\mathbf{p}, t)$, SCH mentioned that the Coulomb interaction couples transitions with different wave vectors. An e–h pair with given $\mathbf{p}$ of internal motion not only experiences the external field alone, it also feels a significant internal field associated with e–h pairs with other wave vectors $\mathbf{q}$, which is similar to a molecular field. The external and the Coulomb fields combine to give an effective self-consistent local field. For small pump intensities the local field corrections dominate and transform the free-e–h pairs into excitons. For large pump intensities the external field dominates.

Equations (6.22) possess two integrals of motion,

$$n_{11}(\mathbf{p},t)+n_{22}(\mathbf{p},t)=1,$$
$$4n_{12}(\mathbf{p},t)n_{21}(\mathbf{p},t)+[n_{11}(\mathbf{p},t)-n_{22}(\mathbf{p},t)]^2=1, \tag{6.23}$$

which drive the equalities

$$n_e(\mathbf{p},t)=n_h(\mathbf{p},t), \qquad 4|\psi(\mathbf{p},t)|^2\le 1;$$
$$n_e(\mathbf{p},t)=\frac{1}{2}\big[1-\sqrt{1-4|\psi(\mathbf{p},t)|^2}\big]. \tag{6.24}$$

In the stationary case SCH obtained a self-consistent equation for the the light-induced pair amplitude:

$$\left[E_e(\mathbf{p})+E_h(\mathbf{p})-\sum_{\mathbf{q}}V_{\mathbf{p}-\mathbf{q}}n_e(\mathbf{q})-\sum_{\mathbf{q}}V_{\mathbf{p}-\mathbf{q}}n_h(\mathbf{q})\right]\psi(\mathbf{p})$$
$$=[1-2n_e(\mathbf{p})]\left[-(dE_L)+\sum_{\mathbf{q}}V_{\mathbf{p}-\mathbf{q}}\psi(\mathbf{q})\right]. \tag{6.25}$$

Without the nonlinear corrections, Eq. (6.25) corresponds to an inhomogeneous Wannier equation in momentum space driven by the pump field E_L. The self-consistent nonlinear corrections describe the effects on the excitons of the phase-space filling (PSF) through the factor $[1-2n_e(\mathbf{p})]$ and fermion-exchange self-energies expressed by $\sigma_i(\mathbf{p})$.

In the case of nonresonant excitation $\hbar\omega_L<E_{\text{ex}}$ and small intensities $|E_L|^2$, the nonlinear correction term can be treated by means of a perturbation theory. In this case,

$$\psi(\mathbf{p})\sim E_L, \qquad n_i(\mathbf{p})=|\psi(\mathbf{p})|^2\approx|E_L|^2. \tag{6.26}$$

The fermion distribution functions $n_i(\mathbf{p})$ are thus determined by the probability of finding the virtually excited e and h in the bosonic-exciton states. The coupling constant with the driving external field is also reduced by the factor $[1-2n_e(\mathbf{p})]$ because of the PSF effect. This fact can be suitably reinterpreted as a correction that is due to the proper normalization of the exciton wave function in the presence of other e–h pairs.

Equation (6.25) can be rewritten in the low-density limit in the form

$$\frac{1}{V}\sum_{\mathbf{p}}\left(E_g-\hbar\omega_L+\frac{\hbar^2\mathbf{p}^2}{2\mu}\right)|\psi(\mathbf{p})|^2-\frac{1}{V}\sum_{\mathbf{p},\mathbf{q}}V_{\mathbf{p}-\mathbf{q}}\psi(\mathbf{q})\psi^*(\mathbf{p})$$
$$-\frac{2}{V}\sum_{\mathbf{p},\mathbf{q}}V_{\mathbf{p}-\mathbf{q}}|\psi(\mathbf{q})|^2|\psi(\mathbf{p})|^2+\frac{2}{V}\sum_{\mathbf{p},\mathbf{q}}V_{\mathbf{p}-\mathbf{q}}\psi(\mathbf{q})\psi^*(\mathbf{p})|\psi(\mathbf{p})|^2$$
$$=-\frac{(dE_L)}{V}\sum_{\mathbf{p}}\psi^*(\mathbf{p})[1-2|\psi(\mathbf{p})|^2], \tag{6.27}$$

with the normalization condition

$$\frac{1}{V}\sum_{\mathbf{p},\sigma}|\psi(\mathbf{p})|^2=n_{\text{ex}}, \tag{6.28}$$

where the summation over the spin index is dropped in the case of a spin-aligned system. The solution of Eqs. (6.27) and (6.28) can be chosen in a form proportional to the $1s$ exciton wave function of the relative motion in a momentum space,

$$\psi(\mathbf{p}) = \sqrt{\frac{n_{\mathrm{ex}}}{2}}\varphi_{1s}(\mathbf{p})$$

in the case of a spin-degenerate system, and

$$\psi(\mathbf{p}) = \sqrt{n_{\mathrm{ex}}}\varphi_{1s}(\mathbf{p}) \tag{6.29}$$

in the case of a spin-aligned system, where

$$\varphi_{1s}(\mathbf{p}) = \frac{8\left(\pi a_{\mathrm{ex}}^3\right)^{1/2}}{\left(1 + p^2 a_{\mathrm{ex}}^2\right)^2}, \tag{6.30}$$

which coincides with the Keldysh and Kozlov result of Eqs. (2.116) reported in Chapter 2. Here n_{ex} and a_{ex} are the exciton density and radius, respectively. The shift of the exciton energy is determined by the self-energy correction due to exchange interaction,

$$-\frac{2}{V}n_{\mathrm{ex}}\sum_{\mathbf{p},\mathbf{q}} V_{\mathbf{p}-\mathbf{q}}\varphi_{1s}^2(\mathbf{q})\varphi_{1s}^2(\mathbf{p}),$$

and by the vertex correction due to the PSF effect,

$$\frac{2}{V}n_{\mathrm{ex}}\sum_{\mathbf{p},\mathbf{q}} V_{\mathbf{p}-\mathbf{q}}\varphi_{1s}^3(\mathbf{q})\varphi_{1s}(\mathbf{p}).$$

They combine simply to give an effective exciton–exciton interaction,

$$\frac{2}{V}n_{\mathrm{ex}}\sum_{\mathbf{p},\mathbf{q}}\left[V_{\mathbf{p}-\mathbf{q}}\varphi_{1s}^3(\mathbf{p})\varphi_{1s}(\mathbf{q}) - V_{\mathbf{p}-\mathbf{q}}\varphi_{1s}^2(\mathbf{p})\varphi_{1s}^2(\mathbf{q})\right] = \frac{26\pi}{3}n_{\mathrm{ex}}a_{\mathrm{ex}}^3\mathrm{Ry}_{\mathrm{ex}}, \tag{6.31}$$

which coincides with case (6.27) in the spin-aligned case and differs by a factor of 2 from the case of paraexcitons in a spin-degenerate system.

Expression (6.27) can be represented in the form well known in the theory of optical bistability,

$$n_{\mathrm{ex}} = \frac{|dE_L\psi_{1s}(0)|^2\left(1 - 7\pi n_{\mathrm{ex}}a_{\mathrm{ex}}^3\right)^2}{\left[E_{\mathrm{ex}}(0) - \hbar\omega_L + 26\pi/3n_{\mathrm{ex}}a_{\mathrm{ex}}^3\mathrm{Ry}_{\mathrm{ex}}\right]^2 + \gamma_{\mathrm{ex}}^2}, \tag{6.32}$$

if the decay rate γ_{ex} is added in the denominator phenomenologically.

Here it is appropriate to give an account of the investigation by Elesin and Kopaev[6] of the excitonic Bose condensate in a strong electromagnetic field. This was done in the mean-field approximation and can be presented with Hamiltonian (6.17) and formula (6.25) after some transformations. From the beginning, Elesin and Kopaev realized that the laser radiation frequency ω_L and amplitude E_L play the roles of the chemical potential and the external condensate source, respectively. These conclusions are common to all subsequent papers [7–9]. The frequency ω_L was supposed to be less than the interband transition threshold E_g but above the ground-state exciton level $E_{\mathrm{ex}}(0) = E_g - \mathrm{Ry}_{\mathrm{ex}}$. But this last restriction is not essential and can be dropped. If one writes formula (6.25) in the form

$$[\tilde{E}_e(\mathbf{p}) + \tilde{E}_h(\mathbf{p})]u_{\mathbf{p}}v_{\mathbf{p}} = \left(u_{\mathbf{p}}^2 - v_{\mathbf{p}}^2\right)\Delta(\mathbf{p}) \tag{6.33}$$

and uses its solutions

$$u_{\mathbf{p}}^2 = \frac{1}{2}\left[1 + \frac{\tilde{E}_e(\mathbf{p}) + \tilde{E}_h(\mathbf{p})}{E(\mathbf{p})}\right], \qquad v_{\mathbf{p}}^2 = \frac{1}{2}\left[1 - \frac{\tilde{E}_e(\mathbf{p}) + \tilde{E}_h(\mathbf{p})}{E(\mathbf{p})}\right],$$

$$E(\mathbf{p}) = \sqrt{[\tilde{E}_e(\mathbf{p}) + \tilde{E}_h(\mathbf{p})]^2 + 4\Delta^2(\mathbf{p})}, \tag{6.34}$$

one can obtain the relations between the ordering parameter $\Delta(\mathbf{p})$ of Eq. (6.18) and the wave function $\psi(\mathbf{p})$:

$$-\psi(\mathbf{p}) = u_{\mathbf{p}} v_{\mathbf{p}} = \frac{\Delta(\mathbf{p})}{E(\mathbf{p})},$$

$$E(\mathbf{p})\psi\mathbf{p} = -(dE_L) + \textstyle\sum_{\mathbf{q}} V_{\mathbf{p}-\mathbf{q}}\psi(\mathbf{q}). \tag{6.35}$$

This nonlinear integral equation forms the starting point of the discussions by Elesin and Kopaev [6] as well as those by Comte and Mahler [9].

6.1.3 The Coulomb Green's Function Method

Elesin and Kopaev linearized Eq. (6.35) by using the approximation

$$E(\mathbf{p}) = E_g' + \mathcal{E}(\mathbf{p}) + \frac{2\Delta^2(0)}{E_g'}, \qquad E_g' = E_g - \hbar\omega_L > 0. \tag{6.36}$$

In this approach Eq. (6.35) becomes

$$[E + \mathcal{E}(\mathbf{p})]\psi(\mathbf{p}) - \sum_{\mathbf{q}} V_{\mathbf{p}-\mathbf{q}}\psi(\mathbf{q}) = \lambda, \tag{6.37}$$

where the notations

$$E = E_g - \hbar\omega + \frac{2\Delta^2(0)}{E_g'} > 0, \qquad \lambda = -(dE_L) \tag{6.38}$$

are introduced.

In the real-space representation, one obtains the linear inhomogeneous Schrödinger equation with the Coulomb potential and coherent $\delta(\mathbf{r} - \mathbf{r}')$ external source as follows:

$$\left(E - \frac{\hbar^2\Delta}{2\mu} - \frac{e^2}{\epsilon_0 r}\right)\psi_E(r) = \lambda\delta(\mathbf{r}). \tag{6.39}$$

Its solution is the Coulomb Green's function, which has the form [19–21]

$$\psi_E(\mathbf{r}, \mathbf{r}') = \lambda \sum_{l,m,n} \frac{\psi_{nlm}^*(\mathbf{r}')\psi_{nlm}(\mathbf{r})}{E + E_{nl}}. \tag{6.40}$$

Here $\psi_{nlm}(\mathbf{r})$ and E_{nl} are the eigenfunctions and the eigenvalues, respectively, of the homogeneous Schrödinger equation that belong to discrete and continuum states. The Coulomb Green's function has a closed-form expression derived by Hostler and Pratt [20, 21]. It was used in several papers [6, 9, 22, 23] and has the eigenfunction expansion

$$\psi_E(\mathbf{r}, \mathbf{r}') = \lambda \sum_{l,m} g_l(E, r, r')Y_{lm}^*(\theta', \varphi')Y_{lm}(\theta, \varphi) \tag{6.41}$$

as well as

$$\delta(\mathbf{r}-\mathbf{r}') = \frac{\delta(r-r')}{rr'}\sum_{l,m} Y^*_{lm}(\theta',\varphi')Y_{lm}(\theta,\varphi). \tag{6.42}$$

The radial Schrödinger equation for the radial Coulomb Green's function [23] has the form

$$\left[E - \frac{\hbar^2}{2\mu}\frac{1}{r^2}\frac{\mathrm{d}}{\mathrm{d}r}\left(r^2\frac{\mathrm{d}}{\mathrm{d}r}\right) + \frac{\hbar^2 l(l+1)}{2\mu r^2} - \frac{e^2}{\epsilon_0 r}\right] g_l(E,r) = \frac{\delta(r)}{r^2}. \tag{6.43}$$

Elesin and Kopaev found the solution for the case $l = 0$:

$$g_0(E,r) = \frac{\mu}{\pi\hbar^2}\Gamma(1-\nu)\frac{1}{r}W_{\nu,1/2}\left(\frac{2r}{a_{\mathrm{ex}}\nu}\right), \tag{6.44}$$

$$\nu = \sqrt{\frac{\mathrm{Ry}_{\mathrm{ex}}}{E}}, \quad E > 0.$$

Here $W_{\nu,1/2}(x)$ are the Whittaker functions and ν plays the role of a fundamental quantum number of the virtual state. The gamma function $\Gamma(1-\nu)$ determines the dependence on ν.

When the laser frequency ω_L coincides with one of the discrete exciton levels shifted by the value $2\Delta^2(0)/E'_g$, then ν equals an integer quantum number n and the gamma function becomes divergent [24]:

$$\hbar\omega_L = E_g - \frac{\mathrm{Ry}_{\mathrm{ex}}}{n^2} + \frac{2\Delta^2(0)}{E'_g}, \qquad \nu = n,$$

$$\Gamma(1-n) = (-n)! = \infty, \quad n = 1, 2, \ldots. \tag{6.45}$$

In the case of resonances, the amplitudes of the linearized Coulomb–Schrödinger field without damping expressed by the Coulomb Green's functions are divergent.

This result concerning the discrete resonances was pointed out by Comte and Mahler [9]. Elesin and Kopaev [6] confined themselves to the ground-state exciton level $n = 1$. In the vicinity of this resonance they used the approximations

$$\Gamma(1-\nu) = \frac{1}{\nu-1} \approx \frac{2}{\nu^2-1},$$

$$n_{\mathrm{ex}} = \int \psi_E^2(\mathbf{r})\,\mathrm{d}\mathbf{r} = \lambda^2 A\Gamma^2(1-\nu) \approx \frac{4A\lambda^2}{(\nu^2-1)^2},$$

$$\frac{\Delta(0)}{E'_g} = \psi(p=0) = \int \psi_E(\mathbf{r})\,\mathrm{d}\mathbf{r} = \lambda B\Gamma(1-\nu),$$

$$\frac{\Delta^2(0)}{E'_g} = \alpha n_{\mathrm{ex}}\mathrm{Ry}_{\mathrm{ex}}, \quad \alpha = \frac{B^2 E'_g}{A\mathrm{Ry}_{\mathrm{ex}}} \approx 16\sqrt{2}\pi a^3_{\mathrm{ex}}. \tag{6.46}$$

If one introduces phenomenologically the damping γ of the exciton level, the nonlinear dependence of the exciton density n_{ex} on the laser intensity λ^2 takes the form

$$n_{\mathrm{ex}} = \frac{4A(\mathrm{Ry}_{\mathrm{ex}})^2\lambda^2}{[E_{\mathrm{ex}}(0) + \alpha n_{\mathrm{ex}}\mathrm{Ry}_{\mathrm{ex}} - \hbar\omega_L]^2 + \gamma^2}, \tag{6.47}$$

which coincides in general form with expression (6.32).

Equations (6.32) and (6.47) are cubic. When the intensity of the source exceeds a critical value, there is a range of values of ω_L in which all three roots of the equations are real.

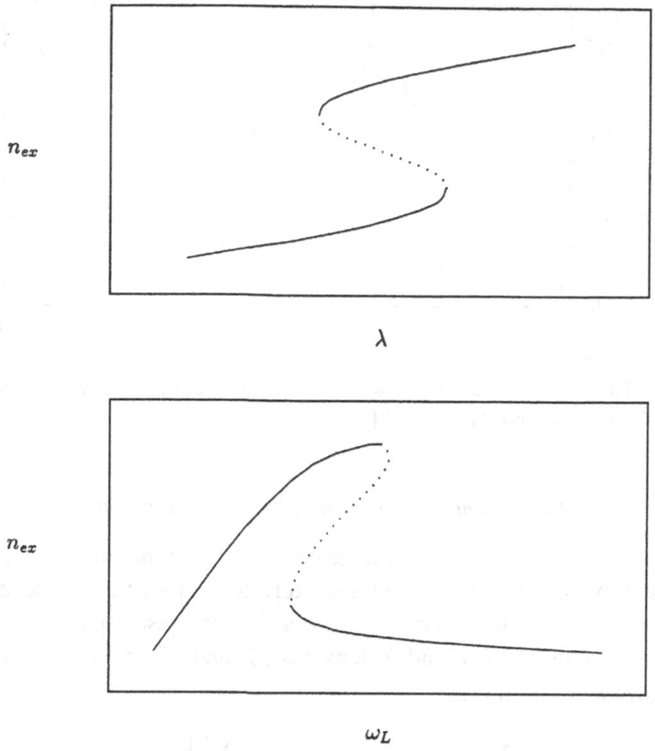

Figure 6.3. The amplitude hysteresis and the frequency hysteresis of the condensed exciton density n_{ex} in the vicinity of the ground exciton level [6].

A discontinuous change of Bose condensate density occurs on variation of either the frequency ω_L or the intensity λ^2. Correspondingly, either frequency or amplitude hysteresis appears. These are represented in Fig. 6.3.

In the concluding remarks of their paper, Elesin and Kopaev noted that excitonic Bose condensation in a strong electromagnetic field takes place at an arbitrary temperature, as in the case of magnetization of a ferromagnet in the presence of an external magnetic field. Along with the above-mentioned hysteresis, one can also expect temperature hysteresis. In this case the system will possess a phase transition of the first kind.

Comte and Mahler [9] established that the linearized Coulomb–Schrödinger field with coherent source exhibits resonances at all points $\nu = n = 1, 2, 3 \ldots$, but allows for solutions at arbitrary energy $E \geq 0$ and parameter $\nu > 0$. For these off-resonant states, n_{ex} goes to zero with the source strength. These states can be characterized as a cloud of virtual excitons dressing the source. Under off-resonance conditions, the electron–hole pairs may be called virtual in the same sense that a cloud of virtual phonons dresses the electron charge in the polar semiconductor [9], giving rise to polaron formation. The resonances are represented in Fig. 6.4. Because of the interaction between the identical fermions, the solution changes its character, evoking real exciton branches for all exciton levels. The possibility of a novel hysteresis effect appears [9]. Such effects are known as optical-bistability phenomena and will be discussed in detail in Chapter 9.

The last part of the SCH paper is dedicated to determining the collective excitations of the coherently pumped system and its linear response to a weak test-beam perturbation.

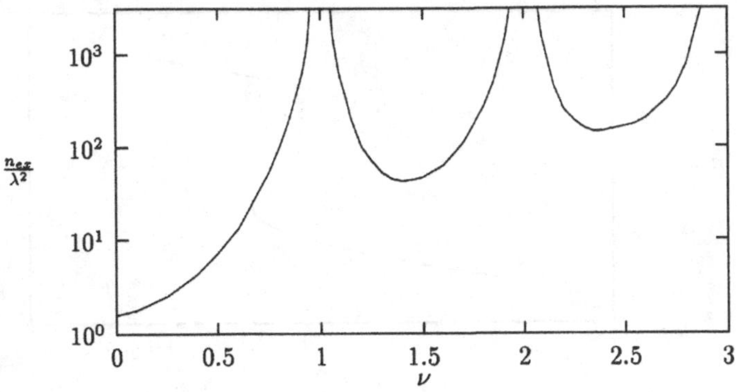

Figure 6.4. The density of excitons as a function of the quantum number ν for the linearized Coulomb–Schrödinger field [9].

6.1.4 Linear Response to a Weak Perturbation

The main results of SCH in this part of their paper can be reproduced if one follows the method proposed by Anderson [25] (see also Ref. 26). Anderson proposed to consider bilinear forms constructed from Fermi operators. In the present case, it is necessary to consider the fluctuations of the e and h densities $\rho^e_{\mathbf{Q}}$ and $\rho^h_{\mathbf{Q}}$ and the polarization of the system $\rho^{eh}_{\mathbf{Q}}$ determined as follows:

$$\rho^e_{\mathbf{Q}} = \sum_{\mathbf{k},s} a^\dagger_{\mathbf{k}+\mathbf{Q},s} a_{\mathbf{k}s}, \qquad \rho^h_{\mathbf{Q}} = \sum_{\mathbf{k},s} b^\dagger_{\mathbf{k}+\mathbf{Q},s} b_{\mathbf{k}s},$$
$$\rho^{eh\dagger}_{\mathbf{Q}} = \sum_{\mathbf{k},s} a^\dagger_{\mathbf{k}+\mathbf{Q},s} b^\dagger_{-\mathbf{k},-s}, \qquad \rho^{eh}_{\mathbf{Q}} = \sum_{\mathbf{k},s} b_{-\mathbf{k},-s} a_{\mathbf{k}+\mathbf{Q},s}. \tag{6.48}$$

To begin, it is necessary to consider the partial components of the fluctuation operators such as $a^\dagger_{\alpha\mathbf{Q}+\mathbf{q},s} a_{\mathbf{q}-\beta\mathbf{Q},s}$, $b_{-(\alpha\mathbf{Q}+\mathbf{q}),-s} b^\dagger_{\beta\mathbf{Q}-\mathbf{q},-s}$, $a^\dagger_{\alpha\mathbf{Q}+\mathbf{q},s} b^\dagger_{\beta\mathbf{Q}-\mathbf{q},-s}$, and $b_{-(\alpha\mathbf{Q}+\mathbf{q}),-s} a_{-\beta\mathbf{Q}+\mathbf{q},s}$ for which the motion equations can be obtained with the help of the full Hamiltonian (6.12). Here $\alpha = m_e/m_{\mathrm{ex}}$ and $\beta = m_h/m_{\mathrm{ex}}$.

On this basis the motion equation can be derived for the matrix $\hat{R}^{\mathbf{Q}}(\mathbf{q}, s)$ constructed from the averages of the partial fluctuations. It has the form

$$\hat{R}^{\mathbf{Q}}(\mathbf{q}, s) = \begin{vmatrix} \langle a^\dagger_{\alpha\mathbf{Q}+\mathbf{q},s} a_{-\beta\mathbf{Q}+\mathbf{q},s} \rangle & \langle a^\dagger_{\alpha\mathbf{Q}+\mathbf{q},s} b^\dagger_{\beta\mathbf{Q}-\mathbf{q},-s} \rangle \\ \langle b_{-(\alpha\mathbf{Q}+\mathbf{q}),-s} a_{-\beta\mathbf{Q}+\mathbf{q},s} \rangle & \langle b_{-(\alpha\mathbf{Q}+\mathbf{q}),-s} b^\dagger_{\beta\mathbf{Q}-\mathbf{q},-s} \rangle \end{vmatrix}. \tag{6.49}$$

The motion equation can be obtained in the random-phase approximation (RPA), which means that only the terms containing the components $\hat{R}^{\mathbf{Q}}_{ij}(\mathbf{q}, s)$ in the first order of the perturbation theory are taken into account. It is supposed that $\hat{R}^{\mathbf{Q}}_{ij}(\mathbf{q}, s)$ are much smaller than the reduced density matrix components

$$\left|\hat{R}^{\mathbf{Q}}_{ij}(\mathbf{q}, s)\right| \ll |n_{ij}(\mathbf{q}, s)|. \tag{6.50}$$

To elucidate better the changes of the system properties due to the action of the coherent pump field, as usual, a measurement of the linear response to a weak perturbation is considered, and so a weak, external, test electric field with the strength $E_t(\mathbf{Q})$, depending on the selected wave vector $\mathbf{Q}$ and the frequency $\Delta\omega$ in the rotating reference frame, is chosen. Its influence on the system can be described by the matrix $\delta\hat{\epsilon}_{\text{ext}}(\mathbf{Q}, \mathbf{q}, t)$. The motion equations within the framework of the RPA in a matrix form were obtained by SCH for the particular case $\alpha = \beta = 1/2$ and for arbitrary values of α and β by Moskalenko and Misko. The equations have the form

$$\begin{aligned} i\hbar\frac{\mathrm{d}}{\mathrm{d}t}\hat{R}^{\mathbf{Q}}(\mathbf{q}, s) &= \hat{R}^{\mathbf{Q}}(\mathbf{q}, s)\hat{\epsilon}(\mathbf{q} - \beta\mathbf{Q}) - \hat{\epsilon}(\mathbf{q} + \alpha\mathbf{Q})\hat{R}^{\mathbf{Q}}(\mathbf{q}, s) \\ &+ [\hat{n}(\mathbf{q} + \alpha\mathbf{Q}) - \hat{n}(\mathbf{q} - \beta\mathbf{Q})]\, V_{\mathbf{Q}}\langle\rho^e_{\mathbf{Q}} - \rho^h_{\mathbf{Q}}\rangle \\ &- \hat{n}(\mathbf{q} + \alpha\mathbf{Q})\sum_{\kappa} V_\kappa \hat{R}^{\mathbf{Q}}(\mathbf{q} + \kappa, s) + \sum_{\kappa} V_\kappa \hat{R}^{\mathbf{Q}}(\mathbf{q} + \kappa, s)\hat{n}(\mathbf{q} - \beta\mathbf{Q}) \\ &+ \hat{n}(\mathbf{q} + \alpha\mathbf{Q})\delta\hat{\epsilon}_{\text{ext}}(\mathbf{q}, \mathbf{Q}, t) - \delta\hat{\epsilon}_{\text{ext}}(\mathbf{q}, \mathbf{Q}, t)\hat{n}(\mathbf{q} - \beta\mathbf{Q}). \end{aligned} \tag{6.51}$$

Here the matrices $\hat{\epsilon}(\mathbf{q})$ and $\delta\hat{\epsilon}_{\text{ext}}(\mathbf{q}, \mathbf{Q}t)$ have the forms

$$\hat{\epsilon}(\mathbf{q}) = \begin{vmatrix} \tilde{E}_e(\mathbf{q}) & \Delta^*(\mathbf{q}) \\ \Delta(\mathbf{q}) & -\tilde{E}_h(\mathbf{q}) \end{vmatrix}, \tag{6.52}$$

$$\delta\hat{\epsilon}_{\text{ext}} = \begin{vmatrix} 0 & -(\mathrm{d}E_t)^* e^{i\Delta\omega t} \\ -(\mathrm{d}E_t)e^{-i\Delta\omega t} & 0 \end{vmatrix}. \tag{6.53}$$

In their comments on Eq. (6.51), SCH stressed the fact that it cannot be obtained on the basis of the Hamiltonian $\mathcal{H}_{m-f}$ of Eq. (6.17) taken in the mean-field approximation. Indeed, in this approximation only the two first terms on the right-hand side of Eq. (6.51), which contain the matrix $\hat{\epsilon}(\mathbf{q})$ of Eq. (6.52), can be obtained. They determine the self-energy corrections and are due to Fock and Bogoliubov self-energy terms in $\mathcal{H}_{m-f}$.

Along with them in Eq. (6.51) there are also three other terms that constitute the vertex corrections. One of them is proportional to $V_{\mathbf{Q}}\langle\rho^e_{\mathbf{Q}} - \rho^h_{\mathbf{Q}}\rangle$. It is due to the Hartree term of the Coulomb interaction, and it can be represented by the sum of bubble diagrams. The last two vertex corrections are due to Fock and Bogoliubov terms, which can be represented by ladder diagrams. In conclusion, SCH noticed that the self-energy and the vertex corrections are treated on equal footing, so that the Ward identities are fulfilled exactly and the theory is gauge invariant.

The linear response of the system to this perturbation, as was shown by SCH, is determined by two elementary nonlinear processes. One of them is the renormalization of *e–h* pairs excited out of the coherent ground state (condensate). It is different from the renormalization of the condensate *e–h* pairs themselves, although both renormalizations are due to the anharmonic interaction with pump photons and with other *e–h* pairs in the condensate. Another nonlinear process is related to the *e–h* pair–pair creation and annihilation, when two *e–h* pairs out of the condensate are destroyed and two renormalized excited *e–h* pairs are created; one transforms into a test photon but the other remains as a collective excitation, and vice versa. Both anharmonic interactions contribute to the "depletion of the condensate" stimulated by the presence of test photons. It leads to optical gain even though there is no population inversion.

The theory of SCH recovers exactly the Bogoliubov–Beliaev equations for the excitation spectrum of a weakly nonideal Bose gas, including a driven term. They obtained the spectrum of collective excitations with $\mathbf{Q} = 0$ as

$$\hbar\omega_n = \left[\left(E_n^{HF} - \hbar\omega_L\right)^2 - \left|\Sigma_{nn}^B\right|^2\right]^{\frac{1}{2}}, \tag{6.54}$$

where E_n^{HF} is the renormalized transition energy to the nth exciton level of the relative motion and Σ_{nn}^B is the corresponding Bogoliubov self-energy. Expression (6.54) is identical to the Bogoliubov–Beliaev expression for the excitation spectrum of a weakly nonideal Bose gas, but it has a gap. The laser photon energy $\hbar\omega_L$ can be thought of as a chemical potential. Optical absorption into renormalized exciton states occurs for $\omega_t = \omega_L + \omega_n$ and optical gain for $\omega_t = \omega_L - \omega_n$, i.e., at a frequency symmetric about ω_L. If $\hbar\omega_L$ is replaced by the chemical potential of the condensed system in the low-density limit and the symmetry-breaking pump field goes to zero, the excitation spectrum is gapless, and for small momentum $\mathbf{Q}$ has the dependence $\omega(Q) = cQ$. The sound velocity is density dependent and interpolates smoothly between the Anderson–Bogoliubov mode at high densities and the Bogoliubov mode at low densities.

In their concluding remarks, SCH emphasized that their theory can describe both the ground state of Bose condensed *e–h* pairs and their excitation spectrum within the framework of one and the same model, which reduces to the Bogoliubov–Beliaev theory of a weakly nonideal Bose gas in one limit and an ordinary weakly coupling BCS theory in the other. Virtual excitons behave like a driven condensate of interacting weakly nonideal bosons. The excitons have an anharmonic interaction among themselves that can be interpreted as arising from mutual polarization and hard-core repulsion. The interaction of excitons with photons is anharmonic as well. This anharmonicity is directly determined by the PSF effect and has its roots in the underlying Fermi statistics obeyed by the components of composite bosons [7, 8].

6.2 The Energy Spectrum and the Exciton Absorption and Gain Bands in the Presence of Laser Radiation

As mentioned in Section 6.1, SCH determined the energy spectrum of the elementary excitations of the exciton system in the presence of nonresonant laser radiation only near the point $\mathbf{k} = 0$ and in the case in which the laser frequency was well below the exciton-band bottom. One aim in this section is to study this question in a more general case, namely, the energy spectrum at arbitrary values of wave vectors and frequency detunings are presented, following Refs. 27–29. On this basis, the exciton absorption and emission bands in the presence of laser radiation are obtained [30, 31]. It is shown that under certain conditions, instabilities of the energy spectrum can appear. For this purpose the Bose description of excitons is used, as is the Hamiltonian (2.48), supplemented by additional terms. One of them describes the interaction of the excitons with wave vector $\mathbf{k}_0$ with coherent laser radiation with the same wave vector. This introduces coherent pumping of the exciton system.

The laser radiation acts as a source of coherent excitons. The system is therefore not in equilibrium, and there is no reason to keep the chemical potential μ, as was done in Eq. (2.48). Instead, the laser frequency ω_L will play its role. For the laser to generate excitons directly, the excitons must couple to the light field, which nominally implies a

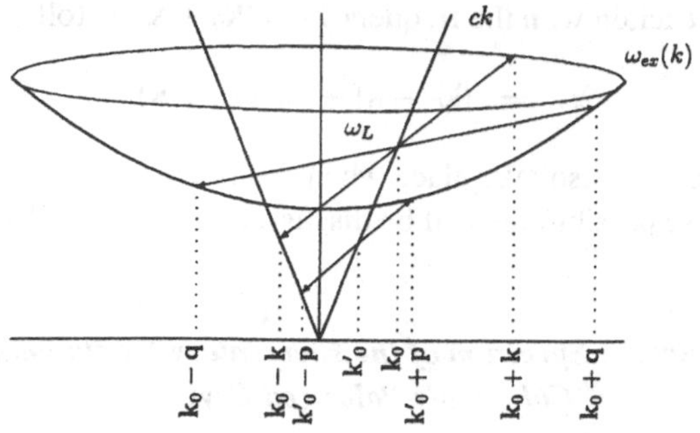

Figure 6.5. The real transformations of two laser photons with wave vectors $\mathbf{k}_0$ ($\mathbf{k}'_0$) into two scattered excitons with wave vectors $\mathbf{k}_0 \pm \mathbf{q}$ or into one vacuum photon with wave vector $\mathbf{k}_0 - \mathbf{k}$ ($\mathbf{k}'_0 - \mathbf{p}$) and one scattered exciton with wave vector $\mathbf{k}_0 + \mathbf{k}$ ($\mathbf{k}'_0 + \mathbf{p}$).

dipole or possibly a quadrupole interaction. The general case of BEC of dipole-active excitons will be discussed in Chapter 7. In what follows, the interaction of the excitons with the light source is taken into account only as a cause of quantum transitions. Keeping in mind our intention to consider the absorption and the emission exciton bands, we add to our Hamiltonian terms for the free photons belonging to the probe-light source and to the vacuum electromagnetic field, as well as for the corresponding exciton–photon interactions.

The main physical cause of the new processes studied below is the fact that the laser radiation induces the appearance of coherent excitons, which cause induced oscillations with the frequency ω_L in the laboratory reference frame, for both positive and negative detunings of the laser.

When the laser frequency is greater than the bottom of the exciton band, the real possibility arises of two laser photons transforming into two scattered excitons. The sum frequency of two photons $2\omega_L$ and their full wave vector $2\mathbf{k}_0$ is equal to the corresponding values of two scattered excitons as follows:

$$\tilde{\omega}_{\text{ex}}(\mathbf{k}_0 + \mathbf{k}) + \tilde{\omega}_{\text{ex}}(\mathbf{k}_0 - \mathbf{k}) = 2\omega_L. \tag{6.55}$$

Here the frequencies $\tilde{\omega}_{\text{ex}}(\mathbf{k}_0 \pm \mathbf{k})$ are shifted with regard to the bare values $\omega_{\text{ex}}(\mathbf{k}_0 \pm \mathbf{k})$ by the density-dependent corrections. This transformation does not take place directly, but rather indirectly through the intermediate states. This role is played by the excitons of the coherently driven mode with the wave vector $\mathbf{k}_0$. Two photons of the laser radiation virtually transform themselves into two excitons with the same wave vector $\mathbf{k}_0$. For this virtual transition the energy conservation law is not needed. Because of the exciton–exciton interaction, two excitons with wave vector $\mathbf{k}_0$ collide with each other and transform into two scattered excitons with wave vectors $\mathbf{k}_0 \pm \mathbf{k}$. The energy and momentum conservation laws (6.55) take place between the initial and the final states. This equality plays the role of the phase matching condition. This process is demonstrated in Fig. 6.5. When the laser frequency lies below the bottom of the exciton band, then condition (6.55) is not fulfilled and the real transition of two photons into two scattered excitons is impossible. The situation changes if one takes into account the vacuum photons, the dispersion branch of which spreads above as well as below the bottom of the exciton band. In this case there is the possibility of two laser photons transforming into one vacuum photon with the frequency $c|\mathbf{k}_0 - \mathbf{k}|$ lower than

ω_L and one scattered exciton with the frequency $\tilde{\omega}_{\text{ex}}(\mathbf{k}_0 + \mathbf{k})$ as follows:

$$2\omega_L = c|\mathbf{k}_0 - \mathbf{k}| + \tilde{\omega}_{\text{ex}}(\mathbf{k}_0 + \mathbf{k}). \tag{6.56}$$

Of course, this process can also take place when the frequency ω_L exceeds the bottom of the exciton band. These possibilities will be discussed in detail in Chapter 7.

6.2.1 Energy Spectrum of the Elementary Excitations in a Coherently Polarized Crystal

Following Refs. 27–29 and 32, we assume that the Hamiltonian of excitons interacting between themselves, with the laser radiation, and with photons of a broadband, weak light source, has the form

$$H = \sum_{\mathbf{p}} \hbar\omega_{\text{ex}}(p) a_{\mathbf{p}}^{\dagger} a_{\mathbf{p}} + \sum_{\mathbf{p}} \hbar c p c_{\mathbf{p}}^{\dagger} c_{\mathbf{p}}$$
$$+ \sum_{\mathbf{p}} \lambda_{\mathbf{p}} (a_{\mathbf{p}}^{\dagger} c_{\mathbf{p}} + c_{\mathbf{p}}^{\dagger} a_{\mathbf{p}}) + \lambda_{\mathbf{k}_0} \left(a_{\mathbf{k}_0}^{\dagger} c_{\mathbf{k}_0} + c_{\mathbf{k}_0}^{\dagger} a_{\mathbf{k}_0}\right) + \frac{1}{2V} \sum_{\mathbf{p},\mathbf{q},\mathbf{k}} U(\mathbf{k}) a_{\mathbf{p}}^{\dagger} a_{\mathbf{q}}^{\dagger} a_{\mathbf{q}+\mathbf{k}} a_{\mathbf{p}-\mathbf{k}}, \tag{6.57}$$

where $a_{\mathbf{p}}^{\dagger}$, $a_{\mathbf{p}}$ and $c_{\mathbf{p}}^{\dagger}$, $c_{\mathbf{p}}$ are the creation and the annihilation operators of excitons and photons, respectively, and $U(\mathbf{k})$ and $\lambda_{\mathbf{p}}$ are the constants of the exciton–exciton and the exciton–photon interactions. It is assumed that the laser radiation has the wave vector $\mathbf{k}_0$ and the photon frequency $\omega_L = ck_0$. The antiresonance terms of the type $\exp(-2i\omega_L t)$ are neglected. The coherent laser radiation is introduced into Hamiltonian (6.57), with the operators $c_{\mathbf{k}_0}^{\dagger}$ and $c_{\mathbf{k}_0}$ substituted by expressions of the type

$$c_{\mathbf{k}_0} = \sqrt{F_{\mathbf{k}_0}} e^{-i\omega t} e^{-i\psi}, \quad F_{\mathbf{k}_0} \sim V. \tag{6.58}$$

The quantum single-particle states of the photons with $\mathbf{p} \neq \mathbf{k}_0$ describe the electromagnetic field of the vacuum and the probe signal. The explicit time dependence of Hamiltonian (6.57), arising after the substitution of expressions (6.58), can be removed if one goes to the reference frame that is rotating with the laser frequency ω_L. This change is achieved with the help of the unitary transformation

$$V = e^{-i\omega_L t N}, \qquad N = \sum_{\mathbf{p}} (a_{\mathbf{p}}^{\dagger} a_{\mathbf{p}} + c_{\mathbf{p}}^{\dagger} c_{\mathbf{p}}), \tag{6.59}$$

which gives rise to a new Hamiltonian

$$\mathcal{H} = V^{\dagger} H V - \hbar\omega_L N. \tag{6.60}$$

Hamiltonian (6.60) does not depend on time, but the energies of the quasiparticles are measured relative to the laser photon energy $\hbar\omega_L$ as follows:

$$\mathcal{H} = \sum_{\mathbf{p}} \hbar \left[\omega_{\text{ex}}(p) - \omega_L\right] a_{\mathbf{p}}^{\dagger} a_{\mathbf{p}} - \sum_{\mathbf{p}} \hbar (cp - \omega_L) c_{\mathbf{p}}^{\dagger} c_{\mathbf{p}}$$
$$+ \sum_{\mathbf{p}} \lambda_{\mathbf{p}} (a_{\mathbf{p}}^{\dagger} c_{\mathbf{p}} + c_{\mathbf{p}}^{\dagger} a_{\mathbf{p}}) + \sqrt{F_{\mathbf{k}_0}} (a_{\mathbf{k}_0}^{\dagger} + a_{\mathbf{k}_0}) + \frac{1}{2V} \sum_{\mathbf{p},\mathbf{q},\mathbf{k}} U(\mathbf{k}) a_{\mathbf{p}}^{\dagger} a_{\mathbf{q}}^{\dagger} a_{\mathbf{q}+\mathbf{k}} a_{\mathbf{p}-\mathbf{k}}. \tag{6.61}$$

The terms that were linear in the operators $a^{\dagger}_{\mathbf{k}_0}$ and $a_{\mathbf{k}_0}$ were excluded after the Bogoliubov displacement operation,

$$a_{\mathbf{p}} = \sqrt{N_{\mathbf{k}_0}} e^{-i\phi} \delta_{\mathbf{p},\mathbf{k}_0} + \alpha_{\mathbf{p}}. \tag{6.62}$$

The requirement of the disappearance of the terms linear in the operators $\alpha^{\dagger}_{\mathbf{k}_0}$ and $\alpha_{\mathbf{k}_0}$ after the substitution of Eq. (6.62) into Hamiltonian (6.61) gives the possibility of establishing a relation between the amplitudes $\sqrt{N_{\mathbf{k}_0}}$ and $\sqrt{F_{\mathbf{k}_0}}$. Formerly, in Section 6.1, this requirement allowed us to determine the chemical potential μ through the condensate-particle density $n_{\mathbf{k}_0} = N_{\mathbf{k}_0}/V$. In the present variant the unknown parameter is $n_{\mathbf{k}_0}$, because the laser frequency, which plays the role of the chemical potential, is determined by the experiment. As a result, one obtains

$$\sqrt{N_{\mathbf{k}_0}}\,[\hbar\omega_{\perp}(\mathbf{k}_0) - \hbar\omega_L + L_0] = -\lambda_{\mathbf{k}_0} e^{i(\varphi-\phi)} \sqrt{F_{\mathbf{k}_0}}. \tag{6.63}$$

The phase difference $(\varphi - \phi)$ can be chosen, taking into account the frequency detuning $\tilde{\Delta}$, determined as

$$\tilde{\Delta} = \Delta + L_0, \qquad \Delta = \hbar[\omega_{\perp}(\mathbf{k}_0) - \omega_L], \qquad L_{\mathbf{k}} = U(\mathbf{k})\frac{N_{\mathbf{k}_0}}{V}. \tag{6.64}$$

The macroscopic number $N_{\mathbf{k}_0} \sim V$ determines the filling of the selected mode $\mathbf{k}_0$. It is related to the macroscopic number $F_{\mathbf{k}_0}$ of coherent photons of laser radiation by the formulas [6]

$$n_{\mathbf{k}_0} = \frac{\lambda^2_{\mathbf{k}_0} f_{\mathbf{k}_0}}{\Delta^2 + \gamma^2_{\text{ex}}}, \qquad n_{\mathbf{k}_0} = \frac{N_{\mathbf{k}_0}}{V}, \qquad f_{\mathbf{k}_0} = \frac{F_{\mathbf{k}_0}}{V}. \tag{6.65}$$

It is supposed that between the excitons repulsion prevails and $L_{\mathbf{k}} > 0$. The damping constant γ_{ex} was introduced phenomenologically. The relation $n_{\mathbf{k}_0}$ as a function of $f_{\mathbf{k}_0}$ describes the phenomenon of optical bistability in the excitonic range of the spectrum [6]. Expanding Hamiltonian (6.61) on the small operators $\alpha^{\dagger}_{\mathbf{k}_0+\mathbf{k}}$ and $\alpha_{\mathbf{k}_0+\mathbf{k}}$, with $\mathbf{k} \neq 0$, one can separate out the additive constant, the quadratic part, and the terms of higher orders, which are infinitesimal. Only the quadratic part will be kept below:

$$\begin{aligned}
\mathcal{H}^{(2)} &= \sum_{\mathbf{k}} \{\hbar[\omega_{\text{ex}}(\mathbf{k}_0+\mathbf{k}) - \omega_L] + L_0 + L_{\mathbf{k}}\}\alpha^{\dagger}_{\mathbf{k}_0+\mathbf{k}}\alpha_{\mathbf{k}_0+\mathbf{k}} \\
&+ \frac{1}{2}\sum_{\mathbf{k}} L_{\mathbf{k}}\left(e^{-2i\phi}\alpha^{\dagger}_{\mathbf{k}_0+\mathbf{k}}\alpha^{\dagger}_{\mathbf{k}_0-\mathbf{k}} + e^{2i\phi}\alpha_{\mathbf{k}_0+\mathbf{k}}\alpha_{\mathbf{k}_0-\mathbf{k}}\right) \\
&+ \sum_{\mathbf{k}} \hbar\left(c\,|\mathbf{k}_0+\mathbf{k}| - \omega_L\right) c^{\dagger}_{\mathbf{k}_0+\mathbf{k}} c_{\mathbf{k}_0+\mathbf{k}} \\
&+ \sum_{\mathbf{k}} \lambda_{\mathbf{k}_0+\mathbf{k}}\left(c^{\dagger}_{\mathbf{k}_0+\mathbf{k}}\alpha_{\mathbf{k}_0+\mathbf{k}} + \alpha^{\dagger}_{\mathbf{k}_0+\mathbf{k}} c_{\mathbf{k}_0+\mathbf{k}}\right).
\end{aligned} \tag{6.66}$$

The photons of the broadband light source will be considered as a cause of the quantum transitions. For this reason only the exciton part of Hamiltonian (6.66) will be diagonalized. In this case one can put $\phi = 0$ without loss of generality. The diagonalization is achieved by the introduction of new operators $\xi^{\dagger}_{\mathbf{k}}$ and $\xi_{\mathbf{k}}$, with the Bogoliubov canonical transformation:

$$\xi_{\mathbf{k}} = \frac{\alpha_{\mathbf{k}_0+\mathbf{k}} + A_{\mathbf{k}}\alpha^{\dagger}_{\mathbf{k}_0-\mathbf{k}}}{\sqrt{1-|A_{\mathbf{k}}|^2}} = u_{\mathbf{k}}\alpha_{\mathbf{k}_0+\mathbf{k}} + v_{\mathbf{k}}\alpha^{\dagger}_{\mathbf{k}_0-\mathbf{k}},$$

$$\alpha_{\mathbf{k}_0+\mathbf{k}} = \frac{\xi_{\mathbf{k}} - A_{\mathbf{k}}\xi^{\dagger}_{-\mathbf{k}}}{\sqrt{1-|A_{\mathbf{k}}|^2}} = u_{\mathbf{k}}\xi_{\mathbf{k}} - v_{\mathbf{k}}\xi^{\dagger}_{-\mathbf{k}},$$

$$u_{\mathbf{k}} = \frac{1}{\sqrt{1-|A_{\mathbf{k}}|^2}}, \qquad v_{\mathbf{k}} = \frac{A_{\mathbf{k}}}{\sqrt{1-|A_{\mathbf{k}}|^2}}. \tag{6.67}$$

The coefficients $A_{\mathbf{k}}$ take two possible values $A_{\mathbf{k}i}$ as follows:

$$A_{\mathbf{k}i} = \frac{(\tilde{\Delta} + T_{\mathbf{k}} + L_{\mathbf{k}}) - \varepsilon_i(\mathbf{k})}{L_{\mathbf{k}}}, \qquad \varepsilon_i(\mathbf{k}) = \pm\varepsilon(\mathbf{k}), \quad i = 1, 2. \tag{6.68}$$

Depending on the $\pm$ sign before the square root, $\varepsilon(\mathbf{k})$ is

$$\varepsilon(\mathbf{k}) = \sqrt{(\tilde{\Delta} + T_{\mathbf{k}} + L_{\mathbf{k}})^2 - L^2_{\mathbf{k}}}. \tag{6.69}$$

The values $\varepsilon_i(\mathbf{k})$ determine the two branches of the energy spectrum in a system with a motionless condensate. One of them determines the ordinary elementary excitation branch, whereas the second one describes the quasi-energy states that appear in the presence of the Bose condensate because of processes involving the joint participation of condensate and noncondensate particles. The moving condensate induced by the laser radiation has the velocity $\mathbf{V}_s$, which depends on the photon momentum $\hbar\mathbf{k}_0$ and the exciton translational mass m_{ex}. The energy-spectrum branches $E_i(\mathbf{k})$ in this case differ from the values $\varepsilon_i(\mathbf{k})$ by an additional term $\hbar\mathbf{V}_s\mathbf{k}$:

$$E_i(\mathbf{k}) = \varepsilon_i(\mathbf{k}) + \hbar\mathbf{V}_s\mathbf{k}, \quad \mathbf{V}_s = \frac{\hbar\mathbf{k}_0}{m_{\text{ex}}}. \tag{6.70}$$

The quasi-exciton elementary excitation branch is denoted as $E_1(\mathbf{k})$. It is chosen so that the sign of $\varepsilon_1(\mathbf{k})$ will coincide with the sign of the expression $(\tilde{\Delta} + T_{\mathbf{k}} + L_{\mathbf{k}})$, which determines the original exciton branch,

$$\hbar\tilde{\omega}_{\text{ex}}(\mathbf{k}_0 + \mathbf{k}) = \hbar\omega_{\text{ex}}(\mathbf{k}_0 + \mathbf{k}) - \hbar\omega_L + L_0 + L_{\mathbf{k}} = (\tilde{\Delta} + T_{\mathbf{k}} + L_{\mathbf{k}}) + \hbar\mathbf{V}_s\mathbf{k}. \tag{6.71}$$

Therefore $\varepsilon_i(\mathbf{k})$ are determined as follows:

$$\varepsilon_1(\mathbf{k}) = \operatorname{sgn}(\tilde{\Delta} + T_{\mathbf{k}} + L_{\mathbf{k}})\varepsilon(\mathbf{k}),$$

$$\varepsilon_2(\mathbf{k}) = -\varepsilon_1(\mathbf{k}). \tag{6.72}$$

Between the quasi-exciton branch $E_1(\mathbf{k})$ and the quasi-energy branch $E_2(\mathbf{k})$ there is the relation

$$E_2(\mathbf{k}) = \varepsilon_2(\mathbf{k}) + \hbar\mathbf{V}_s\mathbf{k} = -\varepsilon_1(\mathbf{k}) + \hbar\mathbf{V}_s\mathbf{k} = -E_1(-\mathbf{k}). \tag{6.73}$$

The quasi-energy branch $E_2(\mathbf{k})$ recalls, in general, features of the starting quasi-energy branch,

$$-\hbar\tilde{\omega}_{\text{ex}}(\mathbf{k}_0 - \mathbf{k}) = \hbar\omega_L - \hbar\omega_{\text{ex}}(\mathbf{k}_0 - \mathbf{k}) - L_0 - L_{\mathbf{k}} = -(\tilde{\Delta} + T_{\mathbf{k}} + L_{\mathbf{k}}) + \hbar\mathbf{V}_s\mathbf{k}, \tag{6.74}$$

which appear because of the existence of the Bose condensate. In consequence of these definitions, the coefficients $A_{\mathbf{k}i}$ have the properties

$$|A_{\mathbf{k}1}| \leq 1, \qquad |A_{\mathbf{k}2}| \geq 1, \qquad A_{\mathbf{k}1}A_{\mathbf{k}2} = 1. \tag{6.75}$$

When the expression $(\tilde{\Delta} + T_{\mathbf{k}} + L_{\mathbf{k}})^2$ exceeds $L_{\mathbf{k}}^2$, the coefficients $A_{\mathbf{k}i}$ are real values, whose moduli tend to converge in the limit $\varepsilon(\mathbf{k}) \to 0$. In the opposite case $(\tilde{\Delta} + T_{\mathbf{k}} + L_{\mathbf{k}})^2 < L_{\mathbf{k}}^2$, the energy $\varepsilon(\mathbf{k})$ of Eq. (6.69) becomes purely imaginary and the coefficients turn out to be complex values with the moduli equal to unity ($|A_{\mathbf{k}i}| = 1$). In these cases unitary transformation (6.67) makes no sense. Nevertheless it pays to study the imaginary part of the energy spectrum because it indicates the appearance of the new processes in the system. The imaginary parts represent the increments of the growing or damping waves that appear because of the presence of the laser radiation. These questions will be discussed in a detailed way in Chapter 7.

As usual, BEC of a weakly nonideal Bose gas is considered, assuming a repulsive interaction between the particles, which ensures the stability of the energy spectrum in the Bogoliubov theory of superfluidity. In the presence of the induced exciton condensate, the condition $L_{\mathbf{k}} > 0$ remains sufficient for only the positive frequency detunings $\tilde{\Delta} \geq 0$.

The evolution of the energy spectrum $\pm\varepsilon(\mathbf{k})$, depending on the frequency detunings $\tilde{\Delta}$, is represented in the Fig. 6.6. They are depicted together with starting exciton and quasi-energy branches $\pm\tilde{\omega}_{\text{ex}}(\mathbf{k}_0 \pm \mathbf{k})$.

Both branches are calculated in the rotating reference frame and are drawn as functions of the relative wave vector $\mathbf{k}$. The figures show that at $\tilde{\Delta} \geq 0$ the instabilities in the spectrum do not appear. When $\tilde{\Delta}$ is positive and differs from zero, $\tilde{\Delta} > 0$, the energy spectrum has a gap at the point $\mathbf{k} = 0$. This fact was mentioned in Ref. 8. It means that it is necessary to add a finite amount of energy to transform a laser photon into an exciton. When $\tilde{\Delta}$ equals zero, the energy gap disappears. The energy spectrum exhibits a linear section in the range of small wave vectors $\mathbf{k}$, such that $T_{\mathbf{k}} \leq L_{\mathbf{k}}$. It completely coincides with the energy spectrum in the case of quasi-equilibrium, when the chemical potential, following Eqs. (2.52), is $\mu = \hbar\omega_{\text{ex}}(\mathbf{k}_0) + L_0$. In this particular case the virtual Bose condensate turns out to be a real, but nonequilibrium, coherent macroscopic state induced by the resonant laser radiation. The velocity of the exciton superfluid flow $\mathbf{V}_s = \hbar\mathbf{k}_0/m_{\text{ex}}$ must be less than the critical velocity of the superfluidity u^*. This criterion is fulfilled when the exciton-condensate density $n_{\mathbf{k}_0}$ obeys the condition

$$n_{\mathbf{k}_0} > \frac{2T_{\mathbf{k}_0}}{U(0)}. \tag{6.76}$$

When $\tilde{\Delta} = -L_0$, the region of the instability covers the small values of wave vectors, for which the relation

$$T_{\mathbf{k}} \leq L_{\mathbf{k}}$$

takes place.

The wave vectors of the photons that take part in the optical transitions from the ground state of the crystal to the exciton states lie in this region. One can expect considerable changes of the forms of the exciton absorption and luminescence bands. At negative values of the detuning $\tilde{\Delta}$, but a large absolute value, the region of the instability in the exciton

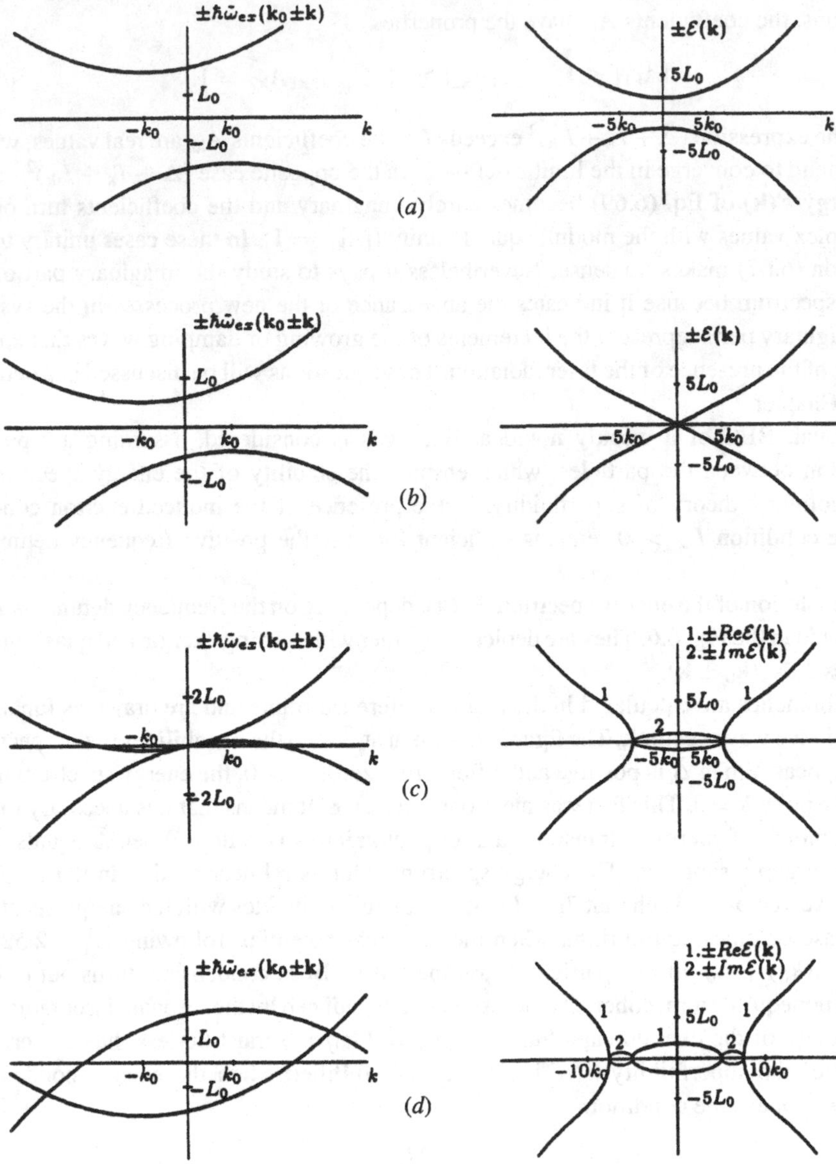

Figure 6.6. The bare exciton and quasi-energy branches of the spectrum $\pm\hbar\tilde{\omega}_{\text{ex}}(\mathbf{k}_0 \pm \mathbf{k})$ and the renormalized branches of the spectrum $\pm\text{Re}\,\varepsilon(k)$ and $\pm\text{Im}\ \varepsilon(k)$ at different frequency detunings: (a) $\tilde{\Delta} = L_0$, (b) 0, (c) $-L_0$, (d) $-3L_0$ [28].

band is shifted toward large values of the wave vectors, where it is difficult to detect by spectroscopic methods. By slowly varying the laser frequency ω_L, it is possible to realize all the above-mentioned variants. These properties, as well as the existence of the induced exciton condensate and the changes of the internal exciton state discussed in Section 6.1, are the distinguishing features of the optical Stark effect in the excitonic range of the spectrum. These optical transitions are discussed in detail below.

Although there are two branches of the energy spectrum $E_i(\mathbf{k})$, nevertheless there is only one set of independent operators $\xi^\dagger_\mathbf{k}$ and $\xi_\mathbf{k}$ with all possible values of $\mathbf{k}$. For such a set of operators, one can choose the operators $\xi^\dagger_{\mathbf{k},1}$ and $\xi_{\mathbf{k},1}$ corresponding to the coefficients $A_{\mathbf{k}1}$ and energy spectrum $E_1(\mathbf{k})$. The second set of operators, determined with the help of coefficients $A_{\mathbf{k}2}$, can be chosen in the form

$$\xi_{\mathbf{k},2} = \frac{\alpha_{\mathbf{k}_0+\mathbf{k}} + A_{\mathbf{k}2}\alpha^\dagger_{\mathbf{k}_0-\mathbf{k}}}{\sqrt{|A_{\mathbf{k}2}|^2 - 1}},$$

$$\alpha_{\mathbf{k}_0+\mathbf{k}} = \frac{A_{\mathbf{k}2}\xi^\dagger_{-\mathbf{k},2} - \xi_{\mathbf{k},2}}{\sqrt{|A_{\mathbf{k}2}|^2 - 1}}.$$

After the substitution $A_{\mathbf{k}2} = 1/A_{\mathbf{k}1}$, one can find the relations

$$\xi_{\mathbf{k},2} = \xi^\dagger_{-\mathbf{k},1}, \qquad \xi^\dagger_{\mathbf{k},2} = \xi_{-\mathbf{k},1}.$$

The operators $\xi^\dagger_{\mathbf{k},2}$ and $\xi_{\mathbf{k},2}$ coincide with the Hermitian conjugate operators $\xi^\dagger_{-\mathbf{k},1}$ and $\xi_{-\mathbf{k},1}$ with opposite wave vectors. The full set of operators $\xi^\dagger_{\mathbf{k},1}$ and $\xi_{\mathbf{k},1}$, together with the coefficients $A_{\mathbf{k}1}$ and energy branch $E_1(\mathbf{k})$, is sufficient to describe both types of elementary excitations as well as the quantum transitions in the system. Toward this end, it is necessary to take into account the exciton–photon interaction. In the following, the index 1 in the operators $\xi^\dagger_{\mathbf{k},1}$ and $\xi_{\mathbf{k},1}$ and in the coefficients $A_{\mathbf{k}1}$ is dropped, but is kept in the energy $E_1(\mathbf{k})$. After the diagonalization of the exciton part of quadratic Hamiltonian (6.66) and the introduction of the new photon operators

$$\eta^\dagger_\mathbf{k} = c^\dagger_{\mathbf{k}_0+\mathbf{k}}, \qquad \eta_\mathbf{k} = c_{\mathbf{k}_0+\mathbf{k}},$$

the Hamiltonian $\mathcal{H}^{(2)}$ will take the form

$$\mathcal{H}^{(2)} = \sum_\mathbf{k} E_1(\mathbf{k})\xi^\dagger_\mathbf{k}\xi_\mathbf{k} + \sum_\mathbf{k} \hbar\,(c\,|\mathbf{k}_0 + \mathbf{k}| - \omega_L)\,\eta^\dagger_\mathbf{k}\eta_\mathbf{k}$$
$$+ \sum_\mathbf{k} \frac{\lambda_{\mathbf{k}_0+\mathbf{k}}}{\sqrt{1 - |A_\mathbf{k}|^2}}[\xi^\dagger_\mathbf{k}\eta_\mathbf{k} + \eta^\dagger_\mathbf{k}\xi_\mathbf{k} - A_\mathbf{k}(\xi_{-\mathbf{k}}\eta_\mathbf{k} + \eta^\dagger_\mathbf{k}\xi^\dagger_{-\mathbf{k}})]. \qquad (6.77)$$

Here the resonant and the antiresonant terms in regard to operators $\eta_\mathbf{k}$ and $\xi_\mathbf{k}$ are important. The ground state of the excitons in the coherently polarized crystal can be chosen as the vacuum state of the annihilation operators $\xi_\mathbf{k}$ as follows:

$$\xi_\mathbf{k}|0\rangle_{\text{ex}} = 0. \qquad (6.78)$$

The choice of the ground state needs to be discussed, because the energy $E_1(\mathbf{k})$ of the elementary excitation in the rotating reference frame can be negative in some cases. Questions about the nonequilibrium distribution function of the quasiparticles under these conditions and about the stability of the chosen ground state of Eq. (6.78) appear. The answers to these questions are discussed in Chapter 8, where the nonequilibrium distribution functions of the scattered polaritons in the presence of one selected coherently excited polariton mode is studied [33]. The distribution function has peaks on the energy scale in the laboratory reference frame symmetrically situated relative to the energy of the selected mode and depends on the absolute values of the differences of the scattered polariton energies and the energy of the selected mode [33]. One can be convinced that this conclusion is justified if one takes into account the character of the conversion of two laser photons into two scattered excitons, as shown in Fig. 6.5. One of two excitons has energy greater than the energy of the photon, and the second exciton has energy lower than the photon energy by the same absolute value. Nevertheless the probability of their appearance is the same because it takes place in a single quantum transition.

These properties are reflected in the fact that the energies $E_1(\mathbf{k})$ of the elementary excitations obey equality (6.73), which can be written as

$$E_1(\mathbf{k}) + E_2(-\mathbf{k}) = 0.$$

At first sight it seems that the system allows the spontaneous creation and annihilation of pairs of quasiparticles. Each pair consists of a quasi-exciton with energy $E_1(\mathbf{k})$ and a quasi-energy complex of quasiparticles with energy $E_2(-\mathbf{k})$. It seems that these processes can take place at arbitrary wave vector $\mathbf{k}$ and frequency detuning $\tilde{\Delta}$. Actually, the nonequilibrium single-particle distribution functions $N(E)$, being symmetric in relation to the substitution $E_1(\mathbf{k})$ by $E_2(-\mathbf{k})$, depend on the moduli $|E_1(\mathbf{k})|$, as follows:

$$N[E_1(\mathbf{k})] = N[E_2(-\mathbf{k})] = N[-E_1(\mathbf{k})] = N[|E_1(\mathbf{k})|]. \tag{6.79}$$

These distribution functions are equal to zero in the ground state of Eq. (6.78) at $T = 0$, which ensures its stability. Real quantum transitions of the type of Eq. (6.55) take place in the vicinity of the points where the expression $(\tilde{\Delta} + T_{\mathbf{k}} + L_{\mathbf{k}})$ tends to zero. In these regions the energy $\varepsilon(\mathbf{k})$ has imaginary values, the coefficients $|A_{\mathbf{k}}|$ equal unity, and the above-quoted formulas are not valid. Although in the ground state of Eq. (6.78) there are no elementary excitations at $T = 0$, nevertheless there are nonzero occupation numbers of the initial exciton states,

$$\left\langle \alpha^\dagger_{\mathbf{k}_0+\mathbf{k}}\alpha_{\mathbf{k}_0+\mathbf{k}}\right\rangle = n^{\mathrm{ex}}_{\mathbf{k}_0+\mathbf{k}} = v^2_{\mathbf{k}} = \frac{A^2_{\mathbf{k}}}{1 - |A_{\mathbf{k}}|^2},$$

$$\left\langle \alpha_{\mathbf{k}_0+\mathbf{k}}\alpha^\dagger_{\mathbf{k}_0+\mathbf{k}}\right\rangle = 1 + n^{\mathrm{ex}}_{\mathbf{k}_0+\mathbf{k}} = u^2_{\mathbf{k}} = \frac{1}{1 - |A_{\mathbf{k}}|^2}. \tag{6.80}$$

Thus, in the presence of resonant or nonresonant laser radiation, a high-density exciton gas appears that consists of either real or virtual excitons. The latter differ from the former in that the virtual excitons exist only during the action of the laser radiation and disappear simultaneously with its disappearance. By contrast, the real excitons will continue to exist during their lifetime.

In the zeroth-order approximation, the states of the polarized crystal and the probe-photon field are independent. The exciton–photon interaction is considered to be weak, serving only as a cause of quantum transitions. The strong interaction between noncondensate excitons and photons in the presence of laser radiation was considered in Ref. 32.

6.2.2 Exciton Absorption and Gain Bands in a Coherently Polarized Crystal

Following Refs. 30, 31, and 34, one can consider quantum transitions under the action of a probe photon with wave vector $\mathbf{Q}$ from the ground state of the coherently polarized crystal to the quasi-exciton state with the wave vector $\mathbf{P}$. The initial and the final states of the two component system in the rotating reference frame are

$$|i\rangle = |0\rangle_{\text{ex}}\eta^{\dagger}_{\mathbf{Q}}|0\rangle_{\text{ph}}, \qquad E_i = \hbar\,(c|\mathbf{k}_0+\mathbf{Q}| - \omega_L)\,,$$
$$|f\rangle = \xi^{\dagger}_{\mathbf{P}}|0\rangle_{\text{ex}}|0\rangle_{\text{ph}}, \qquad E_f = E_1(\mathbf{P}), \tag{6.81}$$

where $|0\rangle_{\text{ph}}$ is the ground state of the probe photons.

The amplitude of the quantum transition in the exciton–photon interaction Hamiltonian (6.77) is

$$\langle i|H_{\text{int}}|f\rangle = \delta_{\mathbf{Q},\mathbf{P}}\frac{\lambda_{\mathbf{k}_0+\mathbf{Q}}}{\sqrt{1-|A_{\mathbf{Q}}|^2}}, \tag{6.82}$$

where $\delta_{\mathbf{Q},\mathbf{P}}$ is the Kronecker symbol.

The transition probability summed over all the final states is

$$P_{\text{abs}}(\mathbf{Q}) = \frac{2\pi}{\hbar}\frac{\lambda^2_{\mathbf{Q}}}{1-|A_{\mathbf{Q}}|^2}\delta[\hbar c|\mathbf{k}_0+\mathbf{Q}| - \hbar\omega_L - E_1(\mathbf{Q})]. \tag{6.83}$$

It depends on the wave vector $\mathbf{Q}$ accounted from the wave vector $\mathbf{k}_0$. The δ function can be represented by the Lorentzian, if one introduces the damping $\Gamma(\mathbf{Q})$ of the elementary excitation:

$$\delta(x) = \frac{1}{\pi}\frac{\Gamma}{x^2+\Gamma^2}, \qquad \Gamma \to 0.$$

The full wave vector of the absorbed photon is $\mathbf{q} = \mathbf{k}_0 + \mathbf{Q}$. Its energy in the laboratory reference frame is $\hbar\omega = \hbar c q = \hbar c|\mathbf{k}_0+\mathbf{Q}|$. The transition probability as a function of wave vector $\mathbf{q}$ can be written in the form

$$P_{\text{abs}}(\mathbf{q}) = \frac{2}{\hbar}\frac{\lambda^2_{\mathbf{q}}}{1-|A_{\mathbf{q}-\mathbf{k}_0}|^2}\frac{\Gamma(\mathbf{q}-\mathbf{k}_0)}{[\hbar\omega - \hbar\omega_L - E_1(\mathbf{q}-\mathbf{k}_0)]^2 + \Gamma^2(\mathbf{q}-\mathbf{k}_0)},$$
$$\hbar c q = \hbar\omega. \tag{6.84}$$

The transition probability depends essentially on the orientation between the vectors $\mathbf{q}$ and $\mathbf{k}_0$. This dependence is reflected in a more evident way by the coefficients $|A_{\mathbf{q}-\mathbf{k}_0}|^2$ and is discussed below. Along with this quantum transition there are also three others. Two of them, related to the absorption of one quasi-exciton elementary excitation, have probability equal to zero because of the lack of such elementary excitations at $T = 0$. Only the Stokes process of the light emission remains, along with the concomitant emission of one quasi-exciton elementary excitation.

This process has the following vacuum initial state and two-particle final state:

$$|i\rangle = |0\rangle_{\text{ex}}|0\rangle_{\text{ph}}, \qquad E_i = 0,$$
$$|f\rangle = \xi^{\dagger}_{\mathbf{P}}|0\rangle_{\text{ex}}\eta^{\dagger}_{\mathbf{Q}}|0\rangle_{\text{ph}}, \qquad E_f = \hbar c|\mathbf{k}_0+\mathbf{Q}| - \hbar\omega_L + E_1(\mathbf{P}). \tag{6.85}$$

This two-quantum transition involves the third term of interaction Hamiltonian (6.77). The transition probability summed over all possible final states with a given photon wave vector $\mathbf{Q}$ and with arbitrary wave vector $\mathbf{P}$ is

$$P_{\text{em}}(\mathbf{Q}) = \frac{2\pi}{\hbar}\frac{\lambda^2_{\mathbf{k}_0+\mathbf{Q}}|A_{\mathbf{Q}}|^2}{1-|A_{\mathbf{Q}}|^2}\delta[\hbar c|\mathbf{k}_0+\mathbf{Q}| - \hbar\omega_L + E_1(-\mathbf{Q})]. \tag{6.86}$$

Here the Lorentzian can also be introduced instead of the δ function. The energy conservation laws resulting from this quantum transition look like

$$\hbar c|\mathbf{k}_0 + \mathbf{Q}| - \hbar\omega_L + E_1(-\mathbf{Q}) = 0,$$
$$\hbar c|\mathbf{k}_0 + \mathbf{Q}| = \hbar\omega_L + E_2(\mathbf{Q}), \tag{6.87}$$

which means the emission of a photon due to the annihilation of the one quasi-energy elementary excitation, the energy of which in the laboratory reference frame is $\hbar\omega_L + E_2(\mathbf{Q})$. If one remembers that $\hbar\omega_L + E_1(-\mathbf{Q})$ approximately equals $\hbar\omega_{\text{ex}}(\mathbf{k}_0 - \mathbf{Q})$, one can represent Eqs. (6.87) in the form

$$\hbar c|\mathbf{k}_0 + \mathbf{Q}| + \hbar\omega_{\text{ex}}(\mathbf{k}_0 - \mathbf{Q}) \approx 2\hbar\omega_L. \tag{6.88}$$

It shows that the transformation of two laser photons into one photon of the weak light source and one noncondensate exciton takes place in the system, obeying the reaction

$$\text{photon}(\mathbf{k}_0) + \text{photon}(\mathbf{k}_0) = \text{photon}(\mathbf{k}_0 + \mathbf{Q}) + \text{exciton}(\mathbf{k}_0 - \mathbf{Q}). \tag{6.89}$$

These considerations show that the only cause of the light emission or of gain of a weak light signal by the system is the external laser radiation. The difference of the two probabilities (6.83) and (6.86) gives rise to the probability of a net absorption in one spectral region and of a gain or net emission in another spectral region. It is determined by the formula

$$\begin{aligned} P_{\text{net}}(\mathbf{q}) &= P_{\text{abs}}(\mathbf{q}) - P_{\text{em}}(\mathbf{Q}) \\ &= \frac{2\lambda^2_{\mathbf{k}_0+\mathbf{Q}}}{\hbar}\left\{\frac{1}{1-|A_{\mathbf{Q}}|^2}\frac{\Gamma(\mathbf{Q})}{[\hbar c|\mathbf{k}_0+\mathbf{Q}| - \hbar\omega_L - E_1(\mathbf{Q})]^2 + \Gamma^2(\mathbf{Q})}\right. \\ &\quad \left. - \frac{|A_{\mathbf{Q}}|^2}{1-|A_{\mathbf{Q}}|^2}\frac{\Gamma(-\mathbf{Q})}{[\hbar c|\mathbf{k}_0+\mathbf{Q}| - \hbar\omega_L + E_1(-\mathbf{Q})]^2 + \Gamma^2(-\mathbf{Q})}\right\} \end{aligned} \tag{6.90}$$

and can be rewritten as a function of $\hbar cq = \hbar c|\mathbf{k}_0 + \mathbf{Q}| = \hbar\omega$, as

$$\begin{aligned} P_{\text{net}}(\hbar\omega) &= \frac{2\lambda^2_{\mathbf{q}}}{\hbar}\left\{\frac{1}{1-|A_{\mathbf{q}-\mathbf{k}_0}|^2}\frac{\Gamma(\mathbf{q}-\mathbf{k}_0)}{[\hbar\omega - \hbar\omega_L - E_1(\mathbf{q}-\mathbf{k}_0)]^2 + \Gamma^2(\mathbf{q}-\mathbf{k}_0)}\right. \\ &\quad \left. - \frac{|A_{\mathbf{q}-\mathbf{k}_0}|^2}{1-|A_{\mathbf{q}-\mathbf{k}_0}|^2}\frac{\Gamma(\mathbf{k}_0-\mathbf{q})}{[\hbar\omega - \hbar\omega_L + E_1(\mathbf{k}_0-\mathbf{q})]^2 + \Gamma^2(\mathbf{k}_0-\mathbf{q})}\right\}. \end{aligned} \tag{6.91}$$

Here it is reasonable to compare these expressions with the line shapes of the absorption and the luminescence exciton bands at $T = 0$ when spontaneous BEC of excitons in the case of an allowed dipole transition is considered [35–37]. As shown in Ref. 35 and discussed in Subsection 3.3.2, these line shapes consist of sharp central peaks at frequencies close to the energy of the condensed excitons, with broad wings. The wing of the absorption band is situated on the side of higher energies relative to the central peak. During this quantum transition the elementary excitation is created. The intensity of this wing is determined by the coefficient $u^2_{\mathbf{q}} = 1 + n^{\text{ex}}_{\mathbf{q}}$. The wing of the luminescence band is situated on the side of the lower energies relative to the central peak, because an elementary excitation is emitted simultaneously with the emission of a light photon. The intensity of this wing depends on the coefficients $v^2_{\mathbf{q}} = n^{\text{ex}}_{\mathbf{q}}$, which are less then the coefficients $u^2_{\mathbf{q}}$. Exactly the same properties arise from expressions (6.83) and (6.86) in the case of induced BEC of excitons.

To conclude this section, we give here an overview of the new phenomena discussed above. The appearance of the luminescence band is due to the transformation of two laser photons into the probe photon and one nonequilibrium scattered exciton. The inverse process means the absorption of the probe photon and the annihilation of one nonequilibrium scattered exciton and their joint transformation into two laser photons. In the presence of the induced BEC of excitons, these processes can be described in the first order of the perturbation theory in the same way as the usual exciton absorption process. Studying these new processes in the absence of coherent virtual excitons requires using higher-order perturbation theory with biexcitons or pairs of excitons as the intermediate states. For example, the description of the hyper-Raman-scattering process of two laser photons by means of the biexciton intermediate state requires third-order perturbation theory [38–41]. The changes of the system and its new properties can be tested by the measurement of the linear response to weak perturbations. The functions $E_1(\mathbf{q}-\mathbf{k}_0)$, $|A(\mathbf{q}-\mathbf{k}_0)|$, and $[1-|A(\mathbf{q}-\mathbf{k}_0)|^2]^{-1}$ depend on the modulus $|\mathbf{q}-\mathbf{k}_0|/k_0$. It has different dependences on the dimensionless frequency parameter $x=\omega/\omega_L=q/k_0\geq 0$ for different orientations between the wave vectors $\mathbf{q}$ and $\mathbf{k}_0$. For example, when $\mathbf{q}$ is parallel ($\mathbf{q}\uparrow\uparrow\mathbf{k}_0$), perpendicular ($\mathbf{q}\perp\mathbf{k}_0$), or antiparallel ($\mathbf{q}\uparrow\downarrow\mathbf{k}_0$) to $\mathbf{k}_0$, these dependences are $|x-1|$, $\sqrt{x^2+1}$, and $(x+1)$ respectively.

The factor $[1-|A(\mathbf{q}-\mathbf{k}_0)|^2]^{-1}$ has a well-defined anisotropy that manifests itself when the coefficients $|A(\mathbf{q}-\mathbf{k}_0)|$ are close to unity. At high positive values of $\tilde{\Delta}$, the coefficients $|A_{\mathbf{k}}|$ for the excitonlike energy spectrum are much lower than unity, and the anisotropy of the factor $[1-|A(\mathbf{q}-\mathbf{k}_0)|^2]^{-1}$ is very small. At $\tilde{\Delta}=L_0$ the anisotropy is of the order of 2%. At the detuning $\tilde{\Delta}=0$, the induced BEC of excitons transforms itself into real but nonequilibrium BEC. In this case the occupation numbers of Eqs. (6.80) $n^{ex}_{\mathbf{k}_0+\mathbf{k}}$ at $\mathbf{k}\to 0$ diverge. The pronounced difference appears between the occupation numbers of $n^{ex}_{\mathbf{q}}$,

$$n^{ex}_{\mathbf{q}}=\frac{|A(\mathbf{q}-\mathbf{k}_0)|^2}{1-|A(\mathbf{q}-\mathbf{k}_0)|^2},\qquad \mathbf{q}=\mathbf{k}_0+\mathbf{k}, \tag{6.92}$$

when $\mathbf{q}$ tends to $\mathbf{k}_0$ or when it tends to $-\mathbf{k}_0$. Because the absorption and the emission probabilities depend on the values $(1+n^{ex}_{\mathbf{q}})$ and $n^{ex}_{\mathbf{q}}$, respectively, these probabilities become anisotropic too.

At negative values $\tilde{\Delta}$, the instabilities in the system occur. In those regions of the wave vectors, where the energy spectrum is complex, the coefficients $|A_{\mathbf{k}}|^2=1$ and the corresponding occupation numbers of Eq. (6.92) are infinite. In these regions canonical transformation (6.67) makes no sense. Here the generation of new waves due to induced exciton combinational scattering takes place. In our case this generation is thresholdless because damping of the exciton levels was neglected from the beginning. The threshold will be studied in Chapter 7. In the neighborhood of the instability regions, the occupation numbers $n^{ex}_{\mathbf{q}}$ are finite but anomalously large. The light absorption and emission probabilities reveal similar behavior. It is interesting to note that the boundaries of the regions with anomalous values of $n^{ex}_{\mathbf{q}}$ also depend on the geometry of the observation. Besides the factor $[1-|A(\mathbf{q}-\mathbf{k}_0)|^2]^{-1}$, the probabilities of net transitions (6.90) and (6.91) also contain other factors, which have the Lorentzian forms. The Lorentzians can also reveal different frequency dependences at different orientations between the vector $\mathbf{q}$ and $\mathbf{k}_0$, but these dependences are less than that of the first factor.

The fact is that the arguments of the δ functions or the Lorentzians mainly represent small differences of the large values $\hbar\omega$ and $\hbar\omega_L$, each of which is much more than the energy $E_1(\mathbf{q}-\mathbf{k}_0)$ of an elementary excitation. In this case it is difficult to reveal the anisotropy of the energy spectrum. Nevertheless, the presence of the elementary excitation energy $E_1(\mathbf{q}-\mathbf{k}_0)$ in the arguments of expressions (6.90) and (6.91) is important. If one neglects these energies $\pm E_1(\mathbf{Q})$ and supposes $\Gamma(\mathbf{Q}) = \Gamma(-\mathbf{Q})$, formulas (6.90) and (6.91) will express isotropic net absorption. At large damping $\Gamma(\mathbf{Q}) \geq |E_1(\mathbf{Q})|$, the anisotropy will be not well defined. One can expect a pronounced anisotropy in the opposite case $\Gamma(\mathbf{Q}) < |E_1(\mathbf{Q})|$ and at small frequency detunings $\tilde{\Delta}$, when the coefficients $|A_{\mathbf{Q}}|^2$ fall to nearly unity. The exciton absorption and gain bands are represented in Fig. 6.7 for two frequency detunings $\tilde{\Delta}$. In these cases, the singularity related to the δ function was removed by introduction of the damping parameter $\Gamma(Q)$. This method turns out to be insufficient in the case $\tilde{\Delta} < 0$, when another singularity related to the factor $[1 - |A(\mathbf{Q}|^2]^{-1}$ appears. To remove both types of singularities, one must introduce the damping parameter from the start in the expressions for the coefficient $A(\mathbf{Q})$ [Eq. (6.68)] and the energy spectrum $E_1(\mathbf{Q})$ [Eq. (6.70)].

Finally, we recall that the anisotropy of the two-photon transition from the ground state of the crystal to the ground state of the biexciton in the presence of laser radiation was studied in Ref. 42 and was discussed in Section 4.4. The degenerate exciton levels in the case of

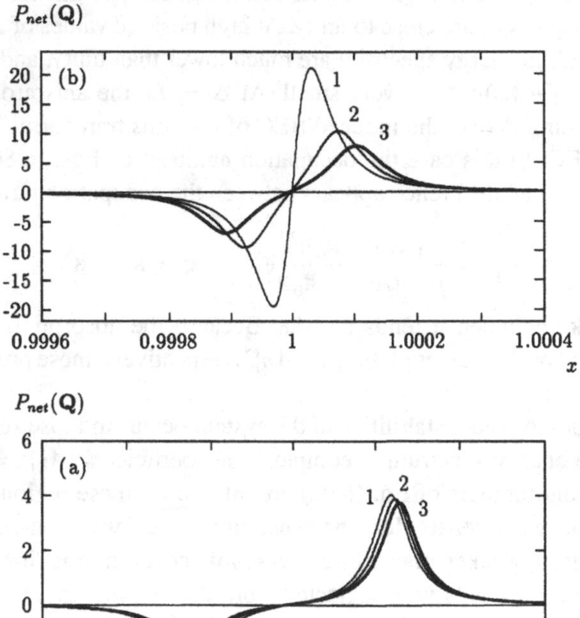

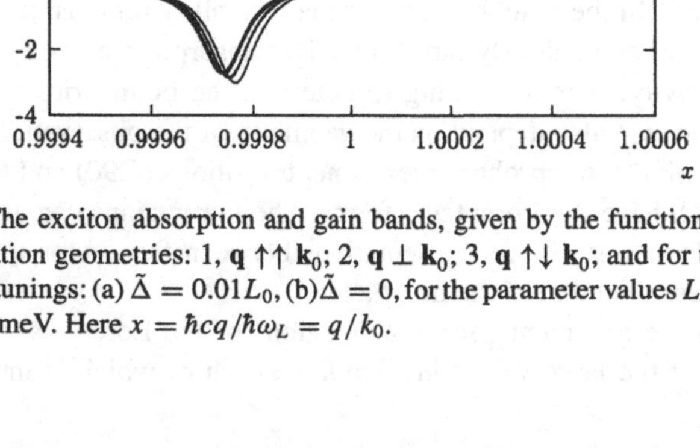

Figure 6.7. The exciton absorption and gain bands, given by the function $P_{\rm net}(\mathbf{Q})$, for three observation geometries: 1, $\mathbf{q} \uparrow\uparrow \mathbf{k}_0$; 2, $\mathbf{q} \perp \mathbf{k}_0$; 3, $\mathbf{q} \uparrow\downarrow \mathbf{k}_0$; and for two different frequency detunings: (a) $\tilde{\Delta} = 0.01 L_0$, (b) $\tilde{\Delta} = 0$, for the parameter values $L_0 = 3.1$ meV and $\Gamma = 0.1$ meV. Here $x = \hbar cq/\hbar\omega_L = q/k_0$.

the optical Stark effect undergo splitting and polarization. These questions were studied by Combescot [43, 44]. In the case considered above, the exciton level is nondegenerate and the anisotropy depends on the propagation direction of the probe-light signal relative to the wave vector of the laser radiation.

6.3 In-Depth Study: Biexcitons in a Coherent Cloud of Virtual Excitons

So far in this chapter we have considered only free excitons. Many fascinating nonlinear optical effects also arise in the case of the mixed exciton–biexciton system. This adds an extra complication, of course. To understand these effects, we need a tractable theoretical model.

As discussed in Chapter 4, three different approaches have been used in the theoretical description of biexcitons and their contribution to physical phenomena. Two of them represent limiting cases: one is the direct microscopic treatment [45, 46], treating the biexciton as a four-particle bound state, and the other is the phenomenological one [47–50], describing the biexciton as an independent boson excitation capable of decaying into an exciton and a photon. Both of these approaches, in many cases successful, have certain intrinsic disadvantages for applications. It seems that the first approach is too complicated to be used as a starting point for a realistic many-body problem. (The first attempt in this direction was discussed in Section 4.2.) The second limiting case is inadequate to account for the changes of the biexciton properties and parameters due to interactions in a dense exciton–biexciton system.

The modern approach to the many-body physics of high-density excitons was formulated in Refs. 8 and 51. This approach makes it possible to describe the continuous destruction of compound quasiparticles such as excitons and biexcitons. The destruction takes place when the densities of the constituent quasiparticles increase. In the case of biexcitons, the excitons can be considered as the constituent parts. Since the radii of the excitons $a_{\rm ex}$ are a few times less than the radii of the biexcitons $a_{\rm biex}$, there is a relatively wide interval of exciton densities over which the excitons are virtually unchanged, whereas the excitonic molecules may transform drastically, i.e., exciton densities of the order of $a_{\rm biex}^{-3}$, but much less than $a_{\rm ex}^{-3}$ ($n_{\rm ex} \leq a_{\rm biex}^{-3} < a_{\rm ex}^{-3}$). The coherent pairing of excitons can be studied in precisely this density range. In the following discussion, the nonlinear optical properties of the coherently paired excitons coherently driven by an external laser field are considered, following the papers of Keldysh [52–54]. Unlike Section 4.3, in what follows the coherent pairing of excitons is considered within the framework of nonequilibrium theory.

In the case of a coherent pumping wave with momentum $\mathbf{p}_0$ and frequency ω_L, a rotating reference frame is used in which the frequency ω_L becomes the origin of the frequency axis. The exciton energy in this reference frame is

$$\tilde{E}_{\rm ex}(\mathbf{p}) = E_g - {\rm Ry}_{\rm ex} + \frac{p^2}{2m_{\rm ex}} - \hbar\omega_L = E_{\rm ex}(\mathbf{p}) - \hbar\omega_L.$$

In the same way, a photon of the probe beam has the energy $\hbar\Omega(\mathbf{p}) = \hbar cp - \hbar\omega_L$. The rotating-wave approximation takes into account only the resonant terms in the vicinity of the one-photon exciton resonance and of the two-photon biexciton resonance. The antiresonant terms in the exciton–photon interaction are neglected.

The excitons are taken as true bosons, and the model proposed by Ivanov, Keldysh, and Panashchenko [45] is used. In this model, two equivalent types of excitons $i = 1, 2$ with the same energy spectrum are introduced. The biexcitons are treated as the bound state of two excitons, which can appear if an attractive interaction exists between them. At low temperatures, however, the system of Bose particles with only attractive interactions is unstable against spontaneous contraction. To avoid this difficulty, it is assumed that excitons of different types attract each other, $V_{12} = V_{21} = -V < 0$, whereas excitons of the same type have a repulsive interaction, $U_{11} = U_{22} = U > V > 0$. Such an approach leads simultaneously to the existence of bound complexes of two excitons of different types and to the stability of the many-body system as a whole. The molecular gas of excitons becomes stable. The interaction in the two-component exciton system is introduced by the 2×2 matrix,

$$\hat{V} = U\hat{I} - V\hat{\sigma}_x,$$
$$\hat{I} = \begin{vmatrix} 1 & 0 \\ 0 & 1 \end{vmatrix}, \qquad \hat{\sigma}_x = \begin{vmatrix} 0 & 1 \\ 1 & 0 \end{vmatrix}. \tag{6.93}$$

The system under consideration is a driven Bose liquid, which arises from the macroscopically occupied mode of excitons coherently created by the external electromagnetic field. As a result of the mutual scattering of excitons in this mode, other modes become populated and biexcitons may also arise.

A consistent description of the system is obtained in terms of the nonequilibrium Green's function technique [8, 10, 55–57]. This picture is similar to the case of the two-component Fermi system considered in detail in the first part of this chapter. For this reason, we simply summarize the results of Keldysh [52–54].

In the two-component Bose system, the normal Green's function $\hat{G}_{ij}(\mathbf{p}, t, t')$ was introduced in the momentum representation, describing the propagation of the two types of excitons $(i, j = 1, 2)$ including their possible change into one another, as well as the anomalous Green's functions $\hat{\hat{G}}_{ij}$ depicting the correlated appearance of two particles i and j [52–54]. The latter functions are the manifestation of the presence of a macroscopically large number of correlated pairs. These functions also contain all the information about the biexcitons. As a result of the transformation of two laser photons with the four-momenta $(\mathbf{p}_0, \omega_L)$ into two scattered excitons, one of them is characterized by four-momentum $(\mathbf{p}, \omega)$, and the second one by $(2\mathbf{p}_0 - \mathbf{p}, 2\omega_L - \omega)$ in the laboratory reference frame.

Similar correlations are well known in the phenomenological description of such nonlinear processes as four-wave mixing and self-phase modulation. According to the nonequilibrium Green's function formalism, the functions $\hat{G}_{ij}$ and $\hat{\hat{G}}_{ij}$, as well as the corresponding self-energy parts $\hat{\Sigma}_{ij}$ and $\hat{\hat{\Sigma}}_{ij}$, are the 2×2 matrices. In the triangular representation one can write two of these matrices in the forms

$$\hat{G}_{ij} = \begin{vmatrix} 0 & G^a_{ij} \\ G^r_{ij} & F^{\text{ex}}_{ij} \end{vmatrix}, \qquad \hat{\Sigma}_{ij} = \begin{vmatrix} \Omega^{\text{ex}}_{ij} & \Sigma^r_{ij} \\ \Sigma^a_{ij} & 0 \end{vmatrix}. \tag{6.94}$$

Here F^{ex}_{ij} are closely related to the distribution functions, whereas Ω^{ex}_{ij} are the noise correlators in the system. The indices r and a denote the retarded and the advanced components, respectively.

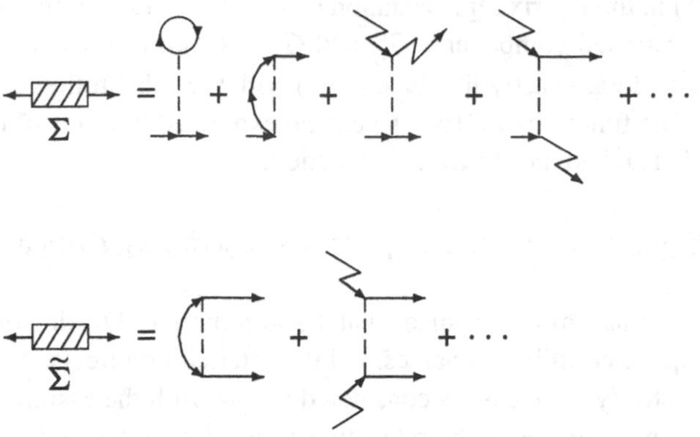

Figure 6.8. Feynman representation of the self-energy parts in the self-consistent field approximation. The normal Green's functions are represented by direct line ⟶, the anomalous ones by the direct line with counteroriented arrows ⟷, and the macroscopically occupied states by the wavy lines.

The Green's functions are determined by the Dyson equations, which are similar to the corresponding equations for the Bose liquid [56–58]:

$$\left[i\hbar\frac{\mathrm{d}}{\mathrm{d}t} - \tilde{E}_{\mathrm{ex}}(\mathbf{p})\right]\hat{G}_{ij}(\mathbf{p},t,t') - \int \mathrm{d}t_1\,\hat{\Sigma}_{il}(\mathbf{p},t,t_1)\hat{G}_{lj}(\mathbf{p},t_1,t')$$
$$- \int \mathrm{d}t_1\,\hat{\hat{\Sigma}}_{il}(\mathbf{p},t,t_1)\hat{\hat{G}}_{lj}(\mathbf{p},t_1,t') = \delta_{ij}\delta(t-t'),$$
$$\left[-i\hbar\frac{\mathrm{d}}{\mathrm{d}t} - \tilde{E}_{\mathrm{ex}}(\mathbf{p})\right]\hat{\hat{G}}_{ij}(\mathbf{p},t,t') - \int \mathrm{d}t_1\,\hat{\Sigma}_{il}(\mathbf{p},t,t_1)\hat{\hat{G}}_{lj}(\mathbf{p},t_1,t')$$
$$- \int \mathrm{d}t_1\,\hat{\hat{\Sigma}}_{il}(\mathbf{p},t,t_1)\hat{G}_{lj}(\mathbf{p},t_1,t') = 0. \tag{6.95}$$

The normal and the anomalous self-energies $\hat{\Sigma}_{ij}$ and $\hat{\hat{\Sigma}}_{ij}$ include all the interaction processes in the system. Usually they are represented as an infinite series of Feynman graphs. The simplest, which correspond to the self-consistent field approximation, are shown in Fig. 6.8.

The macroscopic wave functions $\psi_\alpha(\mathbf{p}_0, t)$ of the coherent excitons induced by the laser radiation satisfy the following exciton-field equation:

$$\left[i\hbar\frac{\mathrm{d}}{\mathrm{d}t} - \tilde{E}_{\mathrm{ex}}(\mathbf{p}_0)\right]\psi_i(\mathbf{p}_0,t) - \int \mathrm{d}t_1\,\Sigma^r_{il}(\mathbf{p}_0,t,t_1)\psi_l(\mathbf{p}_0,t_1)$$
$$- \int \mathrm{d}t_1\,\tilde{\Sigma}^r_{il}(\mathbf{p}_0,t,t_1)\psi_l(\mathbf{p}_0,t_1) = f_i^{\mathrm{ex}}(t). \tag{6.96}$$

Here $f_i^{\mathrm{ex}}(t)$ is the source of the dipole-active coherent excitons, proportional to the external laser field $\mathbf{E}(t)$,

$$f_i^{\mathrm{ex}}(t) = \mathbf{m}_i \cdot \mathbf{E}(t), \tag{6.97}$$

where $\mathbf{m}_i$ is the dipole moment of the i exciton–photon interaction.

Equations (6.95) in the matrix representation can be transcribed in the component form. The equations for retarded components G^r_{ij} and $\tilde{G}^r_{ij}$, as well as those for advanced components G^a_{ij} and $\tilde{G}^a_{ij}$, look exactly like Eqs. (6.95) with the substitution of the proper self-energy functions. The functions $F^{\text{ex}}_{ij}(\mathbf{p},t,t')$ can be expressed in terms of the self-energies $\Omega^{\text{ex}}_{ij}(\mathbf{p},t,t')$, retarded and advanced Green's functions:

$$F^{\text{ex}}_{ij}(\mathbf{p},t,t') = \int G^r_{ik}(\mathbf{p},t,t_1)\Omega^{\text{ex}}_{kl}(\mathbf{p},t_1,t_2)G^a_{lj}(\mathbf{p},t_2,t')\,dt_1\,dt_2. \tag{6.98}$$

In what follows, the assumption is made that the self-energies Ω^{ex}_{ij} do not deviate considerably from their quasi-equilibrium values. All the retardation effects in the self-energies are neglected. The steady-state case is considered below, with the assumption that the duration of the pump pulse exceeds the relaxation time of excitons and a quasi-stationary state is established. In this case the self-energy parts become time-independent functions. Instead of the functions ψ_i of the coherent excitons, their symmetric and antisymmetric linear combinations are introduced, $\psi_\pm = (\psi_1 \pm \psi_2)/\sqrt{2}$. They lead to the substitution of the (i, j) components by other ones, as follows:

$$G^r_\pm = G^r_{11} \pm G^r_{12}, \qquad \tilde{G}^r_\pm = \tilde{G}^r_{11} \pm \tilde{G}^r_{12},$$

$$\Sigma^r_\pm = \Sigma^r_{11} \pm \Sigma^r_{12}, \qquad \tilde{\Sigma}^r_\pm = \tilde{\Sigma}^r_{11} \pm \tilde{\Sigma}^r_{12}. \tag{6.99}$$

In the case $\mathbf{p}_0 = 0$, the solutions were found as

$$G^r_\pm(\mathbf{p},E) = \frac{E + \tilde{E}_{\text{ex}}(\mathbf{p}) + \Sigma^r_\pm}{E^2 - [E_\pm(\mathbf{p})]^2}, \qquad \tilde{G}^r_\pm(\mathbf{p},E) = \frac{\tilde{\Sigma}^r_\pm}{E^2 - [E_\pm(\mathbf{p})]^2}, \tag{6.100}$$

where the energy spectra $E_\pm(\mathbf{p})$ of the different types of excitons, representing the energy spacings of the excitonic levels from the carrier frequency of the pump field, are

$$E_\pm(\mathbf{p}) = \sqrt{[\tilde{E}_{\text{ex}}(\mathbf{p}) + \Sigma^r_\pm]^2 - (\tilde{\Sigma}^r_\pm)^2}. \tag{6.101}$$

The exciton polarization ψ_+ depends on the field amplitude in the following way:

$$\psi_+ = \frac{f^{\text{ex}}_+}{\tilde{E}_{\text{ex}}(0) + \Sigma^r_+ + \tilde{\Sigma}^r_+}. \tag{6.102}$$

Instead of the usual resonance relation, the denominator in Eq. (6.102) is quite different from $E_+(\mathbf{p})$. This result is one of the manifestations of the presence of pair coherence in the system under consideration.

When the interaction constant $|V|$ is much less than U, one can write

$$\Sigma^r_+ \approx \Sigma^r_- = \Sigma, \qquad \tilde{\Sigma}^r_+ \approx -\tilde{\Sigma}^r_- = \tilde{\Sigma}.$$

The numerical results for $\tilde{\Sigma}$ and $C_{0\approx}\psi_+$ as functions of the frequency detuning $\Delta = 2\tilde{E}_{\text{ex}}(0)/|W| = 2[E_{\text{ex}}(0) - \hbar\omega_L]/|W|$ are represented in Fig. 6.9.

The zero of the frequency axis, $\Delta = 0$, corresponds in the plots to the position of the exiton level, and $\Delta = -1$ to the biexciton resonance $2\hbar\omega_L = E_{\text{biex}}(0) = 2E_{\text{ex}}(0) - |W|$, where $|W|$ is the biexciton dissociation energy. Figure 6.9. shows the resonance curves that reflect the dependences of the response ψ_+ and $\tilde{\Sigma}$ on the detuning Δ at a fixed value of the pumping field. It can be seen from these plots that in the frequency range $-1 < \Delta < 1$,

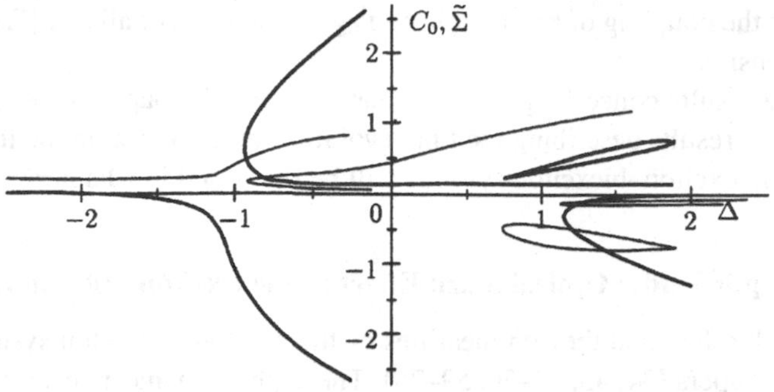

Figure 6.9. Coherent exciton amplitude C_0 (thin curves) and anomalous self-energy $\tilde{\Sigma}$ (thick curves) dependences on the frequency detunings Δ in a stationary regime for pumping-field strength $f_+^{\text{ex}} = 0.2$ and the effective repulsion coupling constant $\alpha = 0.6$ [54].

the resonance is dominated by the excitonic molecules. No resonant structure exists around $\Delta = 0$. A typical resonant structure for excitonic molecules is accompanied by a strong nonlinear resonance of exciton polarization in the vicinity of $\Delta = -1$. For frequencies above the stationary resonance $\Delta \geq -1$, the response becomes multistable. Two new states arise. Among three states with $C_0 > 0$, one is exciton dominated, that is, beyond the immediate vicinity of the resonance the majority of excitons are in the single-particle condensate, which forms the coherent polarization cloud. Two other solutions are biexciton dominated, i.e., the majority of excitons are bound into molecules at the expense of coherent polarization.

The solutions with $\Delta > 0$ and $C_0 < 0$ are unstable. The loop appearing in the interval $0.5 < \Delta < 2$ is a small remnant of a pure excitonic resonance. It disappears at a higher field. As the pumping field increases, the resonance broadens, with different branches shifting in opposite directions from those of the resonant frequency. At some critical field value for any detuning $\Delta > -1$, two positive $\tilde{\Sigma}$ solutions, one exciton dominated and the other biexciton dominated, converge and disappear while the field is still large. Only the other biexciton-dominated state, with $\tilde{\Sigma}$ negative, persists. Figure 6.9. corresponds to a relatively large value of the effective repulsion constant $\alpha = (mUa_{\text{ex}}^2)/(4\pi\hbar^2) = 0.6$.

Taking into account the interaction of the excitons with the photons of the wide-spectral-range probe light, four new polaritonlike branches are found. Their dispersion laws are

$$E^2 = \frac{1}{2}\left[E_+^2(\mathbf{p}) + (\hbar\Omega_\mathbf{p})^2 + 2\,|\mathbf{m}|^2\right]$$

$$\pm\frac{1}{2}\sqrt{[E_+^2(\mathbf{p}) - (\hbar\Omega_\mathbf{p})^2]^2 + 4|\mathbf{m}|^2[(E_+(\mathbf{p}) + \hbar\Omega_\mathbf{p})^2 + 2U_\mathbf{p}E_+(\mathbf{p})\hbar\Omega_\mathbf{p}]}, \quad (6.103)$$

where

$$U_\mathbf{p} = \frac{\tilde{E}_{\text{ex}}(\mathbf{p}) + \Sigma_+}{E_+(\mathbf{p})} - 1. \quad (6.104)$$

Each of these branches reflects the joint propagation of four particles–two excitons and two photons with complementary wave vectors. Substituting $U_\mathbf{p}$ of Eq. (6.104) into Eq. (6.103), one can see that the coupling of two polaritons is governed essentially by $|\tilde{\Sigma}_+|^2$, that is, by the biexciton density.

Besides these results concerning the stationary regime, the papers of Keldysh [52–54] also contain other results describing the time evolution of C_0 and $\tilde{\Sigma}$ in the transient stage. Self-pulsations in exciton–biexciton systems will be discussed in Chapter 9.

6.4 In-Depth Study: Optical Stark Effect in the Exciton–Biexciton System

The optical Stark effect and the nonlinearities of the exciton–biexciton system have been studied in many papers [34, 45, 48–50, 59–79]. These phenomena are characterized by the appearance of the new branches of the energy spectrum. These have the status of quasi-energy states, which were introduced in physics by Zeldovich [80] and Baz' et al. [81]. The quasienergy states are induced by the external laser radiation and lead to peculiarities in the absorption and reflection spectra.

One can explain the presence of the quasi-energy states by means of the following example. If a coherent macroscopic state with wave vector $\mathbf{k}_0$ is established in the crystal, then the creation and annihilation operators of the excitons $a^\dagger_{\mathbf{k}_0}$ and $a_{\mathbf{k}_0}$ are macroscopically large values. Besides the noncoherent excitons with energy $E_{\text{ex}}(\mathbf{k}_0 + \mathbf{q})$ described by the operators $a^\dagger_{\mathbf{k}_0+\mathbf{q}}$ and $a_{\mathbf{k}_0+\mathbf{q}}$, where $\mathbf{q} \neq 0$, quasi-energy states appear that are described by the complexes of operators of the types $(a^\dagger_{\mathbf{k}_0})^2 a_{\mathbf{k}_0-\mathbf{q}}$ and $(a_{\mathbf{k}_0})^2 a^\dagger_{\mathbf{k}_0-\mathbf{q}}$. They are characterized by the energy $[2E_{\text{ex}}(\mathbf{k}_0) - E_{\text{ex}}(\mathbf{k}_0 - \mathbf{q})]$, which under the condition of phase matching equals, or is situated near on the energy scale, the value $E_{\text{ex}}(\mathbf{k}_0 + \mathbf{q})$. In this way a new quasi-energy branch of the energy spectrum appears. This branch is virtual and exists only during the lifetime of the coherent induced state.

The macroscopic coherent states of the excitons and the biexcitons also give rise to four-wave-mixing effects, leading to the generation of new waves and the amplification of the propagating signals. Such phenomena, but without the participation of the biexcitons, have been discussed in the previous chapters. In Subsection 6.4.1 they are studied with the biexcitons taken into account. The theoretical description, as a rule, is presented in the Hartree–Fock–Bogoliubov approximation.

6.4.1 Biexcitons as Structureless Bosons

The model Hamiltonian of the excitons, photons, and the biexcitons takes into account the interaction of the excitons and the biexcitons among themselves as well as with the external laser radiation and with photons of the vacuum or the weak test beam. The laser radiation is characterized by the wave vector $\mathbf{k}_0$ and frequency ω_L. The corresponding photon creation and annihilation operators can be substituted by the C- numbers:

$$c_{\mathbf{k}_0} = \sqrt{F_{\mathbf{k}_0}} e^{-i\varphi} e^{-i\omega_L t}, \qquad c^\dagger_{\mathbf{k}_0} = (c_{\mathbf{k}_0})^*, \qquad f_{\mathbf{k}_0} = F_{\mathbf{k}_0}/V, \tag{6.105}$$

where the photon number $F_{\mathbf{k}_0}$ is proportional to the crystal volume V.

To exclude the time dependence of the Hamiltonian resulting from the time dependence of the operators $c^\dagger_{\mathbf{k}_0}$ and $c_{\mathbf{k}_0}$ of Eqs. (6.105), the double rotating reference frame is used. It rotates with the frequency ω_L for the excitons and the photons and with the frequency $2\omega_L$ for biexcitons [61]. The exciton–photon interaction is taken into account in only the rotating-wave approximation, when the antiresonant terms are neglected. The model Hamiltonian in the rotating reference frame discussed in Refs. [30, 31, and 72] has the form

$$\mathcal{H} = \sum_{\mathbf{q}} [E_{\text{ex}}(\mathbf{q}) - \hbar\omega_L]\, a_{\mathbf{q}}^{\dagger} a_{\mathbf{q}} + \sum_{\mathbf{q}} [E_m(\mathbf{q}) - 2\hbar\omega_L] b_{\mathbf{q}}^{\dagger} b_{\mathbf{q}} + \sum_{\mathbf{q}} (\hbar c q - \hbar\omega_L)\, c_{\mathbf{q}}^{\dagger} c_{\mathbf{q}}$$
$$+ \left[\varphi(\mathbf{k}_0)\sqrt{F_{\mathbf{k}_0}} e^{-i\varphi} a_{\mathbf{k}_0}^{\dagger} + \varphi^*(\mathbf{k}_0)\sqrt{F_{\mathbf{k}_0}} e^{i\varphi} a_{\mathbf{k}_0}\right] + \sum_{\mathbf{q}} [\varphi(\mathbf{q})\, a_{\mathbf{q}}^{\dagger} c_{\mathbf{q}} + \varphi^*(\mathbf{q})\, c_{\mathbf{q}}^{\dagger} a_{\mathbf{q}}]$$
$$+ \sqrt{f_{\mathbf{k}_0}} \sum_{\mathbf{q}} \left[g(\mathbf{k}_0, \mathbf{q}) e^{-i\varphi} b_{\mathbf{k}_0+\mathbf{q}}^{\dagger} a_{\mathbf{q}} + g^*(\mathbf{k}_0, \mathbf{q}) e^{i\varphi} a_{\mathbf{q}}^{\dagger} b_{\mathbf{k}_0+\mathbf{q}}\right]$$
$$+ \frac{1}{\sqrt{V}} \sum_{\mathbf{q},\mathbf{k}} [g(\mathbf{k}, \mathbf{q}) b_{\mathbf{k}+\mathbf{q}}^{\dagger} a_{\mathbf{q}} c_{\mathbf{k}} + g^*(\mathbf{k}, \mathbf{q}) a_{\mathbf{q}}^{\dagger} c_{\mathbf{k}}^{\dagger} b_{\mathbf{k}+\mathbf{q}}]$$
$$+ \frac{F_{\mathbf{k}_0}}{\sqrt{V}} \left[\mu(\mathbf{k}_0, \mathbf{k}_0)\, e^{-2i\varphi} b_{2\mathbf{k}_0}^{\dagger} + \mu^*(\mathbf{k}_0, \mathbf{k}_0)\, e^{2i\varphi} b_{2\mathbf{k}_0}\right]$$
$$+ \frac{1}{\sqrt{V}} \sum_{\mathbf{q},\mathbf{p}} [\mu(\mathbf{p}, \mathbf{q}) b_{\mathbf{p}+\mathbf{q}}^{\dagger} c_{\mathbf{q}} c_{\mathbf{p}} + \mu^*(\mathbf{p}, \mathbf{q}) b_{\mathbf{p}+\mathbf{q}} c_{\mathbf{q}}^{\dagger} c_{\mathbf{p}}^{\dagger}]$$
$$+ \frac{1}{\sqrt{V}} \sum_{\mathbf{q},\mathbf{p}} [M(\mathbf{p}, \mathbf{q}) b_{\mathbf{p}+\mathbf{q}}^{\dagger} a_{\mathbf{q}} a_{\mathbf{p}} + M^*(\mathbf{p}, \mathbf{q}) b_{\mathbf{p}+\mathbf{q}} a_{\mathbf{q}}^{\dagger} a_{\mathbf{p}}^{\dagger}]$$
$$+ \frac{1}{2V} \sum_{\mathbf{q},\mathbf{p},\mathbf{k}} U_{\text{ex}}(\mathbf{k}) a_{\mathbf{p}}^{\dagger} a_{\mathbf{q}}^{\dagger} a_{\mathbf{q}+\mathbf{k}} a_{\mathbf{p}-\mathbf{k}} + \frac{1}{2V} \sum_{\mathbf{q},\mathbf{p},\mathbf{k}} U_m(\mathbf{k}) b_{\mathbf{p}}^{\dagger} b_{\mathbf{q}}^{\dagger} b_{\mathbf{q}+\mathbf{k}} b_{\mathbf{p}-\mathbf{k}}. \qquad (6.106)$$

Here $E_{\text{ex}}(\mathbf{q})$ and $E_m(\mathbf{q})$ are the energies of the excitons and the biexcitons, respectively, and $a_{\mathbf{q}}^{\dagger}$, $a_{\mathbf{q}}$, $b_{\mathbf{q}}^{\dagger}$, $b_{\mathbf{q}}$, and $c_{\mathbf{q}}^{\dagger}$, $c_{\mathbf{q}}$ are the creation and annihilation operators of the excitons, biexcitons, and the vacuum photons, respectively. The laser photons have been taken into account already, as mentioned above. The three types of optical quantum transitions are taken into account simultaneously. They give rise to the terms linear in the operators $a_{\mathbf{k}_0}^{\dagger}$, $a_{\mathbf{k}_0}$ and $b_{2\mathbf{k}_0}^{\dagger}$, $b_{2\mathbf{k}_0}$, which can be removed from Hamiltonian (6.106) by the Bogoliubov displacement transformations, [30, 31]

$$a_{\mathbf{q}} = \sqrt{N_{\mathbf{k}_0}^{\text{ex}}} e^{-i\psi} \delta_{\mathbf{q},\mathbf{k}_0} + \alpha_{\mathbf{q}}, \qquad n_{\mathbf{k}_0}^{\text{ex}} = N_{\mathbf{k}_0}^{\text{ex}}/V,$$
$$b_{\mathbf{q}} = \sqrt{N_{2\mathbf{k}_0}^{m}} e^{-i\theta} \delta_{\mathbf{q},2\mathbf{k}_0} + \beta_{\mathbf{q}}, \qquad n_{2\mathbf{k}_0}^{m} = N_{2\mathbf{k}_0}^{m}/V, \qquad (6.107)$$

where $N_{\mathbf{k}_0}^{\text{ex}} \sim N_{2\mathbf{k}_0}^{m} \sim V$ are macroscopically large values. They can be determined by the setting to zero of the coefficients of the operators $\alpha_{\mathbf{k}_0}^{\dagger}$, $\alpha_{\mathbf{k}_0}$ and $\beta_{2\mathbf{k}_0}^{\dagger}$, $\beta_{2\mathbf{k}_0}$. These conditions determine the relations among the densities of coherent photons, excitons, and biexcitons $(f_{\mathbf{k}_0}, n_{\mathbf{k}_0}^{\text{ex}}, n_{2\mathbf{k}_0}^{m})$. These relations are

$$\sqrt{n_{\mathbf{k}_0}^{\text{ex}}}\left[E_{\text{ex}}(\mathbf{k}_0) - \hbar\omega_L + L_0^{\text{ex}}\right] - \varphi(\mathbf{k}_0)\sqrt{f_{\mathbf{k}_0}} e^{i(\varphi-\psi)}$$
$$+ \sqrt{n_{\mathbf{k}_0}^{m}}\left[2\sqrt{n_{\mathbf{k}_0}^{\text{ex}}} M(\mathbf{k}_0, \mathbf{k}_0) e^{i(\theta-2\psi)} + \sqrt{f_{\mathbf{k}_0}} g(\mathbf{k}_0, \mathbf{k}_0) e^{i(\theta-\psi-\varphi)}\right] = 0,$$
$$\sqrt{n_{2\mathbf{k}_0}^{m}}[E_m(2\mathbf{k}_0) - 2\hbar\omega_L + L_0^{m}] + \mu(\mathbf{k}_0, \mathbf{k}_0) f_{\mathbf{k}_0} e^{i(2\varphi-\theta)}$$
$$+ \sqrt{n_{\mathbf{k}_0}^{\text{ex}}}\left[\sqrt{n_{\mathbf{k}_0}^{\text{ex}}} M(\mathbf{k}_0, \mathbf{k}_0)\, e^{i(2\psi-\theta)} - \sqrt{f_{\mathbf{k}_0}} g(\mathbf{k}_0, \mathbf{k}_0)\, e^{i(\psi+\varphi-\theta)}\right] = 0. \qquad (6.108)$$

Here and in what follows below, the following notation has been introduced:

$$\tilde{E}_{\rm ex}(\mathbf{k}_0+\mathbf{k}) = E_{\rm ex}(\mathbf{k}_0+\mathbf{k}) - \hbar\omega_L + L_0^{\rm ex} + L_{\mathbf{k}}^{\rm ex},$$
$$\tilde{E}_m(2\mathbf{k}_0+\mathbf{k}) = E_m(2\mathbf{k}_0+\mathbf{k}) - 2\hbar\omega_L + L_0^m + L_{\mathbf{k}}^m,$$
$$\hbar\Omega(\mathbf{k}_0+\mathbf{k}) = \hbar c|\mathbf{k}_0+\mathbf{k}| - \hbar\omega_L,$$
$$L_{\mathbf{k}}^{\rm ex} = U_{\rm ex}(\mathbf{k})n_{\mathbf{k}_0}^{\rm ex},\ L_{\mathbf{k}}^m = U_m(\mathbf{k})n_{2\mathbf{k}_0}^m,$$
$$\mathcal{L}_{\mathbf{k}}^{\rm ex} = L_{\mathbf{k}}^{\rm ex}e^{-2i\psi} + 2\sqrt{n_{2\mathbf{k}_0}^m}M(\mathbf{k}_0+\mathbf{k},\mathbf{k}_0-\mathbf{k})e^{-i\theta}, \qquad \mathcal{L}_{\mathbf{k}}^m = L_{\mathbf{k}}^m e^{-2i\theta},$$
$$\phi(\mathbf{k}) = 2\sqrt{n_{\mathbf{k}_0}^{\rm ex}}M(\mathbf{k}_0,\mathbf{k}_0+\mathbf{k})e^{-i\psi} + \sqrt{f_{\mathbf{k}_0}}g(\mathbf{k}_0,\mathbf{k}_0+\mathbf{k})e^{-i\varphi},$$
$$\mu(\mathbf{k}) = \sqrt{n_{2\mathbf{k}_0}^m}\mu(\mathbf{k}_0+\mathbf{k},\mathbf{k}_0-\mathbf{k})e^{-i\theta},$$
$$g_1(\mathbf{k}) = \sqrt{n_{2\mathbf{k}_0}^m}g(\mathbf{k}_0+\mathbf{k},\mathbf{k}_0-\mathbf{k})e^{-i\theta},$$
$$g_2(\mathbf{k}) = \sqrt{n_{\mathbf{k}_0}^{\rm ex}}g(\mathbf{k}_0,\mathbf{k}_0+\mathbf{k})e^{-i\psi}. \tag{6.109}$$

In this notation the quadratic part of the transformed Hamiltonian has the form

$$\begin{aligned}
\tilde{\mathcal{H}}_2 &= \sum_{\mathbf{k}} \tilde{E}_{\rm ex}(\mathbf{k}_0+\mathbf{k})\alpha_{\mathbf{k}_0+\mathbf{k}}^\dagger\alpha_{\mathbf{k}_0+\mathbf{k}} + \sum_{\mathbf{k}} \tilde{E}_m(2\mathbf{k}_0+\mathbf{k})\beta_{2\mathbf{k}_0+\mathbf{k}}^\dagger\beta_{2\mathbf{k}_0+\mathbf{k}} \\
&+ \sum_{\mathbf{k}} \hbar\Omega(\mathbf{k}_0+\mathbf{k})c_{\mathbf{k}_0+\mathbf{k}}^\dagger c_{\mathbf{k}_0+\mathbf{k}} + \frac{1}{2}\sum_{\mathbf{k}}\left[\left(\mathcal{L}_{\mathbf{k}}^m\right)^*\beta_{2\mathbf{k}_0+\mathbf{k}}\beta_{2\mathbf{k}_0-\mathbf{k}} + \mathcal{L}_{\mathbf{k}}^m\beta_{2\mathbf{k}_0+\mathbf{k}}^\dagger\beta_{2\mathbf{k}_0-\mathbf{k}}^\dagger\right] \\
&+ \frac{1}{2}\sum_{\mathbf{k}}\left[\left(\mathcal{L}_{\mathbf{k}}^{\rm ex}\right)^*\alpha_{\mathbf{k}_0+\mathbf{k}}\alpha_{\mathbf{k}_0-\mathbf{k}} + \mathcal{L}_{\mathbf{k}}^{\rm ex}\alpha_{\mathbf{k}_0+\mathbf{k}}^\dagger\alpha_{\mathbf{k}_0-\mathbf{k}}^\dagger\right] \\
&+ \sum_{\mathbf{k}}\left[\phi(\mathbf{k})\beta_{2\mathbf{k}_0+\mathbf{k}}^\dagger\alpha_{\mathbf{k}_0+\mathbf{k}} + \phi^*(\mathbf{k})\alpha_{\mathbf{k}_0+\mathbf{k}}^\dagger\beta_{2\mathbf{k}_0+\mathbf{k}}\right] \\
&+ \sum_{\mathbf{k}}\left[\mu^*(\mathbf{k})c_{\mathbf{k}_0+\mathbf{k}}c_{\mathbf{k}_0-\mathbf{k}} + \mu(\mathbf{k})c_{\mathbf{k}_0+\mathbf{k}}^\dagger c_{\mathbf{k}_0-\mathbf{k}}^\dagger\right] \\
&+ \sum_{\mathbf{k}}\left[\varphi(\mathbf{k}_0+\mathbf{k})\alpha_{\mathbf{k}_0+\mathbf{k}}^\dagger c_{\mathbf{k}_0+\mathbf{k}} + \varphi^*(\mathbf{k}_0+\mathbf{k})c_{\mathbf{k}_0+\mathbf{k}}^\dagger\alpha_{\mathbf{k}_0+\mathbf{k}}\right] \\
&+ \sum_{\mathbf{k}}\left[g_1(\mathbf{k})\alpha_{\mathbf{k}_0+\mathbf{k}}^\dagger c_{\mathbf{k}_0-\mathbf{k}}^\dagger + g_1^*(\mathbf{k})\alpha_{\mathbf{k}_0+\mathbf{k}}c_{\mathbf{k}_0-\mathbf{k}}\right] \\
&+ \sum_{\mathbf{k}}\left[g_2(\mathbf{k})\,\beta_{2\mathbf{k}_0+\mathbf{k}}^\dagger c_{\mathbf{k}_0+\mathbf{k}} + g_2^*(\mathbf{k})\,c_{\mathbf{k}_0+\mathbf{k}}^\dagger\beta_{2\mathbf{k}_0+\mathbf{k}}\right].
\end{aligned} \tag{6.110}$$

The diagonalization of this quadratic form requires the solution of a system of six linear equations, which is equivalent to finding the solution of the sixth-order determinant equation. It can be obtained in an analytic form in the one- and the two-component systems when the determinant equations are reduced to quadratic or biquadratic equations.

In the case of a one-component exciton system interacting only with laser radiation, one can obtain the previously obtained result,

$$E(\mathbf{k}) = \mathcal{E}_{\rm ex}(\mathbf{k}) + \hbar\mathbf{V}_s\cdot\mathbf{k}; \qquad \mathcal{E}_{\rm ex}(\mathbf{k}) = \sqrt{\left(\tilde{\Delta}_{\rm ex} + T_{\mathbf{k}}^{\rm ex} + L_{\mathbf{k}}^{\rm ex}\right)^2 - \left(L_{\mathbf{k}}^{\rm ex}\right)^2}, \tag{6.111}$$

where

$$\tilde{\Delta}_{ex} = E_{ex}(\mathbf{k}_0) - \hbar\omega_L + L_{0,} \qquad T_{\mathbf{k}}^{ex} = \frac{\hbar^2 k^2}{2m_{ex}}, \qquad \mathbf{V}_s = \frac{\hbar \mathbf{k}_0}{m_{ex}}. \tag{6.112}$$

This was discussed in detail in Section 6.2 following Refs. 27 and 29. In the same way, one can obtain the energy spectrum in the case of a one-component biexciton system interacting only with laser radiation because of a two-photon transition from the ground state of the crystal to biexciton state. In accordance with Refs. 27 and 29, the solution is

$$E(\mathbf{k}) = \mathcal{E}_m(\mathbf{k}) + \hbar\mathbf{V}_s \cdot \mathbf{k}, \qquad \mathcal{E}_m(\mathbf{k}) = \sqrt{(\tilde{\Delta}_m + T_{\mathbf{k}}^m + L_{\mathbf{k}}^m)^2 - (L_{\mathbf{k}}^m)^2}. \tag{6.113}$$

where

$$\tilde{\Delta}_m = E_m(2\mathbf{k}_0) - 2\hbar\omega_L + L_{0,}^m \qquad T_{\mathbf{k}}^m = \frac{\hbar^2 k^2}{2m_b} = \frac{\hbar^2 k^2}{4m_{ex}}. \tag{6.114}$$

In the case of a two-component exciton–biexciton system interacting with laser radiation by means of three different quantum transitions, an analytical form of the solution may be obtained only for a special selection of the vectors $\mathbf{k}_0$ and $\mathbf{k}$ when $|\mathbf{k}_0 + \mathbf{k}| = |\mathbf{k}_0 - \mathbf{k}|$. It can be achieved for arbitrary wave vector $\mathbf{k}$ if $\mathbf{k}_0 = 0$ or for wave vector $\mathbf{k}$ perpendicular to $\mathbf{k}_0$ ($\mathbf{k} \cdot \mathbf{k}_0 = 0$). In both cases, the solutions are [31]

$$E^2 = \frac{1}{2}\{[\mathcal{E}_{ex}(\mathbf{k})]^2 + [\mathcal{E}_m(\mathbf{k})]^2 + 2\,|\phi(\mathbf{k})|^2\} \pm \frac{1}{2}\Big(\{[\mathcal{E}_{ex}(\mathbf{k})]^2 - [\mathcal{E}_m(\mathbf{k})]^2\}^2$$

$$+ 4\,|\phi(\mathbf{k})|^2\,[\tilde{E}_{ex}(\mathbf{k}_0 + \mathbf{k}) + \tilde{E}_m(2\mathbf{k}_0 + \mathbf{k})]^2 - 4\,\big|\phi \mathcal{L}_{\mathbf{k}}^{ex} - \phi^* \mathcal{L}_{\mathbf{k}}^m\big|^2\Big)^{1/2}. \tag{6.115}$$

If one substitutes $\mathcal{E}_m$ by $\hbar\Omega(\mathbf{k}_0 + \mathbf{k})$, one can see that this spectrum is of the same type as that obtained by Keldysh [54] and discussed in Section 6.3. One can be convinced of this by substituting expression (6.104) into Eq. (6.103). After some transformations, formula (6.103) will become similar to Eq. (6.115). Similar results can be obtained for other two-component systems such as exciton–photon and biexciton–photon systems. The photons in question belong to the vacuum field or to the weak probe beam.

Another way to reduce the three-component exciton–photon–biexciton system to a two-component one is to introduce the polariton operators instead of the exciton and the photon ones and to neglect the interaction of the biexcitons with the upper polariton branch. Keeping only the biexcitons and the polaritons of the lower polariton branch, one can obtain a two-component system. The converson-type elementary excitations in this system have a mixed character, as coherent superpositions of the biexciton and lower polariton states. One of them (the "biton") was studied in Ref. 82; others are discussed below.

6.4.2 Nonlinear Optical Properties of Biexcitons

The nonlinear optical properties of crystals discussed below depend on the appearance of the new quasi-energy branches of the energy spectrum of the exciton–biexciton system induced by the coherent laser radiation. Many papers have been dedicated to these questions, and the interest in them persists to the present. These investigations were initiated and carried out by different groups of investigators such as Khadzhi, Moskalenko, and Belkin [49], May, Henneberger, and Henneberger [48], Haug, März, and Schmitt-Rink [50, 59, 60], Abram [62, 63, 65], I. Abram and co-workers [61, 64, 66], Hanamura [68–70], Ivanov et al. [45],

Ivanov and Haug [73–75], Hiroshima [72], Bobrysheva, Shmiglyuk, and Russu [34], and Refs. 30 and 31. In many papers, different models and approaches have been discussed, but with common features, such as the conversion of two quasiparticles into a biexciton as a main mechanism of nonlinearity.

For example, May et al. [48] considered the biexciton–polariton system. They pointed out that if a large number of polaritons are present at frequencies near half the biexciton energy and if they form a coherent polariton mode, then renormalization of the polariton spectrum takes place because of the formation of virtual biexcitons stimulated by the coherent polariton mode. One can see that in the presence of a coherent polariton mode with energy $E_p(\mathbf{k}_0)$ and wave vector $\mathbf{k}_0$, besides the biexciton branch $E_m(\mathbf{k}_0+\mathbf{q})$, a new quasi-energy branch with energy $E_m(\mathbf{k}_0+\mathbf{q})-E_p(\mathbf{k}_0)$ appears. The new branch lies in the range of half of the biexciton energy in the vicinity of the usual lower polariton branch $E_p(\mathbf{q})$. If the biexciton and the polariton states are characterized by the annihilation operators $b^\dagger_{\mathbf{k}_0+\mathbf{q}}$ and $a_\mathbf{q}$, respectively, then the quasi-energy branch is characterized by the product of two operators $b_{\mathbf{k}_0+\mathbf{q}}a^\dagger_{\mathbf{k}_0}$, which describes the annihilation of a biexciton and the creation of one polariton in the coherent mode. The interaction of these polaritons and the quasi-energy branches gives rise to their effective splitting into two new branches. They describe the dispersion laws of the weak probe beams. May et al. [48] called the quasi-energy branch the "induced polariton" branch. An incident test photon of exactly the energy $\hbar\Omega_t=E_m(\mathbf{k}_0+\mathbf{q})-E_p(\mathbf{k}_0)$ can produce an biexciton inside the crystal by using a polariton from the induced coherent polariton mode $E_p(\mathbf{k}_0)$. The description of these processes was based on the nonequilibrium Green's function technique [48]. In contrast to this model, the three-component exciton–photon–biexciton system was considered by Khadzhi et al., who studied the mixing of the biexciton branch of the spectrum with the quasi-energy branch describing the joint state of the coherent excitons and a photon. The renormalization of the spectrum in the biexciton range is of the same type as established above in the polariton range of the spectrum [48].

The biexciton–polariton model was studied by Haug et al. [50, 59, 60] as well as by Abram and Maruani [61], who used the dielectric formalism and the electromagnetic response functions. As Haug et al. noted, the dielectric formalism allows a calculation of the electromagnetic response and the renormalized polariton spectrum in a simple way. They also confirmed that at large densities of the exciton polaritons with energies close to one half of the biexciton energy, the real or virtual formation of biexcitons is possible, which causes the renormalization of the polariton spectrum. Experimental evidence of such renormalization has been obtained by Itoh, Suzuki, and Ueta [83] in high-intensity Raman-scattering experiments in CuCl. An intensity-dependent dielectric function of several semiconductors has been observed in several studies [84–89]. These observations demand a theory of the dielectric function that goes beyond the usual linear-response theory. Haug et al. [50, 59] treated the situation in which the strong picosecond pump beam has created a high nonequilibrium polariton density $n_p(\mathbf{k})=n_p(\mathbf{k}_0)\delta_{\mathbf{k}\mathbf{k}_0}$. These polaritons interact with the excitons created by a weak test beam at $(\mathbf{k},\omega)$ and produce the biexcitons. The dielectric function in the vicinity of the exciton ground state was determined as

$$\epsilon(\mathbf{k},\omega)=\varepsilon_\infty+\frac{(\varepsilon_0-\varepsilon_\infty)\,\omega^2_{\mathrm{ex}}(\mathbf{k})}{\omega_1^2(\mathbf{k})-\omega_2^2(\mathbf{k})}\left\{\frac{\omega_1^2(\mathbf{k})-[\Delta\omega(\mathbf{k})]^2}{\omega_1^2(\mathbf{k})-\omega^2}-\frac{\omega_2^2(\mathbf{k})-[\Delta\omega(\mathbf{k})]^2}{\omega_2^2(\mathbf{k})-\omega^2}\right\}, \tag{6.116}$$

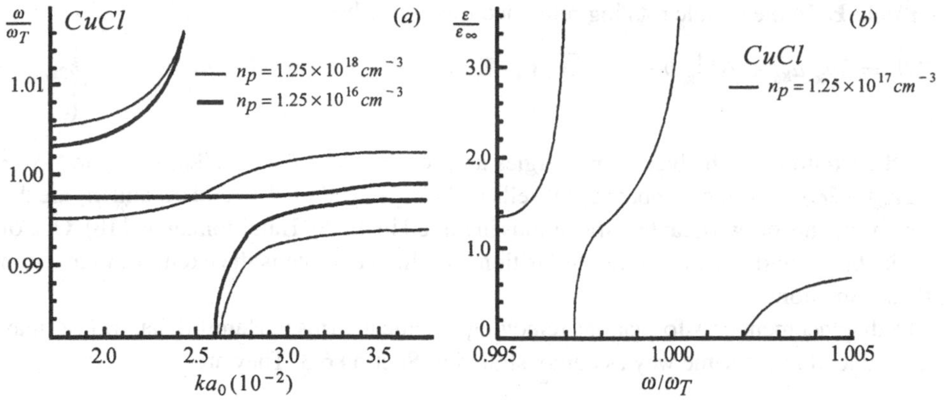

Figure 6.10. (a) Eigenmodes of the electromagnetic waves in CuCl for various polariton concentrations, (b) dielectric constant for CuCl for the polariton concentration $n_p = 1.25 \times 10^{17}\text{cm}^{-3}$. At this density $\omega_1 \approx \omega_T$ and $\omega_2 \approx \omega_m - \omega_p$ [50].

where ε_0 and ε_∞ are the zero-and the high-frequency dielectric constants, $\hbar\omega_{\text{ex}}(\mathbf{k}) = E_{\text{ex}}(\mathbf{k})$, $\hbar\Delta\omega(\mathbf{k}) = E_m(\mathbf{k}_0 + \mathbf{k}) - E_p(\mathbf{k}_0)$, and $E_p(\mathbf{k}_0) = \hbar c k_0/\sqrt{\varepsilon_0}$.

The two poles in Eq. (6.115), $\omega_{1,2}(\mathbf{k})$, are due to the excitations of an exciton and a quasi-energy state. They are

$$\omega_{1,2}(\mathbf{k}) = \frac{\omega_{\text{ex}}^2(\mathbf{k}) + [\Delta\omega(\mathbf{k})]^2}{2} \pm \frac{1}{2}\sqrt{\left\{\omega_{\text{ex}}^2(\mathbf{k}) - [\Delta\omega(\mathbf{k})]^2\right\}^2 + 16|g|^2 n_p(\mathbf{k}_0)\omega_{\text{ex}}(\mathbf{k})\Delta\omega(\mathbf{k})}, \tag{6.117}$$

where g is the constant of the two-polaritons-to-biexciton conversion. Such an expression can be obtained within the framework of the general method described in Section 6.3. The dispersion relation $\epsilon = c^2k^2/\omega^2$ together with dielectric function (6.116) yield an algebraic equation of the third order in ω^2. The solution of Eq. (6.116) gives the eigenfrequencies of the electromagnetic waves in the medium. This solution is shown in Fig. 6.10, where the material parameters for CuCl crystals have been used.

Figure 6.10 shows that the influence of the additional mode increases when the polariton density increases. If the pump beam is switched off and $n_p(\mathbf{k}_0) = 0$, the spectrum reduces to the normal polariton spectrum. In wide-gap semiconductors such as CuCl and CdS, a large nonlinearity occurs in the exciton range of the spectrum because of the two-photon generation of biexcitons [84–89]. But the biexcitons are instable against the decay into two excitonic polaritons [85, 89]. This autoionization process was also included by Schmitt-Rink and Haug [60] to determine the modifications of the dielectric function due to the Fano effect.

Abram and Maruani [61], as well as Abram [62, 63, 65], used a semiclassical nonperturbational method for the calculation of the nonlinear optical susceptibility of semiconductors in the vicinity of the two-photon–biexciton resonance by treating the electromagnetic field classically and the material excitations quantum mechanically. Their simple model Hamiltonian contains only the coherent excitons and biexcitons induced by the external laser field **E**. In the double rotating reference frame one has

$$H/\hbar = \delta a^{\dagger}_{\mathbf{k}_0} a_{\mathbf{k}_0} + \Delta b^{\dagger}_{2\mathbf{k}_0} b_{2\mathbf{k}_0} - \sqrt{N}(\mathbf{m}_1 \cdot \mathbf{E})\left(a_{\mathbf{k}_0} + a^{\dagger}_{\mathbf{k}_0}\right) - (\mathbf{m}_2 \cdot \mathbf{E})\left(b^{\dagger}_{2\mathbf{k}_0} a_{\mathbf{k}_0} + a^{\dagger}_{\mathbf{k}_0} b_{2\mathbf{k}_0}\right). \tag{6.118}$$

Here the exciton and the biexciton energies are, respectively, $\delta = \omega_{\text{ex}}(\mathbf{k}_0) - \omega_L$ and $\Delta = \omega_m(2\mathbf{k}_0) - 2\omega_L$. N is the number of unit cells in the crystal, $V = N v_0$, and $\mathbf{m}_1$ and $\mathbf{m}_2$ are the dipole moments of two quantum transitions included in model Hamiltonian (6.118). One of them is the ground-state-to-exciton transition and the other one is the exciton-to-biexciton optical transition.

The displacement transformations removing the linear terms in Hamiltonian (6.118) may be introduced in the same way as demonstrated in Section 6.3. They are

$$a_{\mathbf{k}_0} = X_{\mathbf{k}_0} + \alpha_{\mathbf{k}_0}, \qquad b_{2\mathbf{k}_0} = Y_{2\mathbf{k}_0} + \beta_{2\mathbf{k}_0},$$

$$X_{\mathbf{k}_0} = \frac{\sqrt{N}(\mathbf{m}_1 \cdot \mathbf{E})\Delta}{\Delta\delta - |\mathbf{m}_2 \cdot \mathbf{E}|^2}, \qquad Y_{2\mathbf{k}_0} = \frac{\sqrt{N}(\mathbf{m}_1 \cdot \mathbf{E})(\mathbf{m}_2 \cdot \mathbf{E})}{\Delta\delta - |\mathbf{m}_2 \cdot \mathbf{E}|^2}. \tag{6.119}$$

After these transformations, Hamiltonian (6.118) obtains a quadratic form,

$$\delta\alpha^{\dagger}\alpha + \Delta\beta^{\dagger}\beta - (\mathbf{m}_2 \cdot \mathbf{E})\,(\beta^{\dagger}\alpha + \alpha^{\dagger}\beta). \tag{6.120}$$

It may be diagonalized by the rotation transformation, and it gives rise to the energy spectrum

$$\omega_{1,2} = \frac{\delta + \Delta}{2} \pm \frac{1}{2}\sqrt{(\delta - \Delta)^2 + 4\,|\mathbf{m}_2 \cdot \mathbf{E}|^2}, \tag{6.121}$$

where the indices $\mathbf{k}_0$ and $2\mathbf{k}_0$ are omitted. The dipole operator **m** for both quantum transitions,

$$\mathbf{m} = \sqrt{N}\mathbf{m}_1 a_{\mathbf{k}_0} + \mathbf{m}_2 b_{2\mathbf{k}_0} a^{\dagger}_{\mathbf{k}_0}, \tag{6.122}$$

has to be averaged over the states in which the system is found after the field is turned on adiabatically.

The polarization of the system is

$$\mathbf{P} = \frac{\langle \mathbf{m} \rangle}{V} = \mathbf{P}_1 + \mathbf{P}_2. \tag{6.123}$$

The expectation value of the first term gives the probability that the photon emitted by the oscillating induced polarization is associated with the annihilation of an exciton, while the second term gives the probability that the emission of the photon takes place when the crystal undergoes a biexciton–exciton transition [63]. The main part of the polarization is determined by the displacement components of the operators of Eqs. (6.119). In this approximation, the polarization and the overall nonlinear susceptibility are given by

$$P = \frac{N}{V}\frac{\Delta^2\delta m_1^2 E}{(\Delta\delta - |\mathbf{m}_2 \cdot \mathbf{E}|^2)^2}, \qquad \chi = \frac{P}{E} = \frac{N}{V}\frac{\Delta^2\delta m_1^2}{(\Delta\delta - |\mathbf{m}_2 \cdot \mathbf{E}|^2)^2}. \tag{6.124}$$

If one disregards all other polarization sources, the dielectric function $\epsilon(\mathbf{k}_0,\omega_L)$ is

$$\epsilon(\mathbf{k}_0,\omega_L) = 1 + 4\pi\chi = 1 + \frac{\Delta^2\delta\omega_{LT}}{(\Delta\delta - |\mathbf{m}_2 \cdot \mathbf{E}|^2)^2}, \tag{6.125}$$

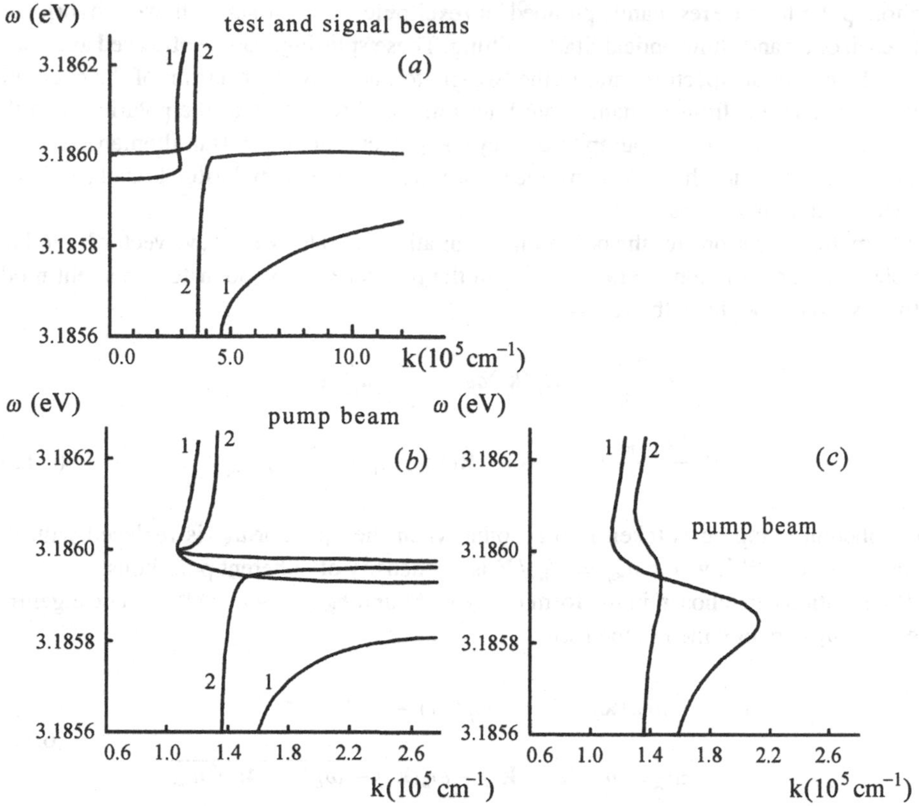

Figure 6.11. Polariton dispersion relations near the two-photon–biexciton resonance in CuCl (3.186 eV) renormalized by a strong pump beam. (a) Polariton renormalization induced on a weak test or signal beam with no damping. Curves 1 are for Rabi frequency $R = |\mathbf{m}_2\mathbf{E}| = 2$ meV; curves 2 are for $R = 0.2$ meV [65]; (b) self-renormalization of pump-beam polariton dispersion without damping; (c) the same as in (b) but with damping. Curves 1, at relativly high incident intensity $I = 30\,MW/\text{cm}^2$; curves 2, $I = 3\,MW/\text{cm}^2$ [62].

where

$$\omega_{LT} = 4\pi m_1^2(N/V), \tag{6.126}$$

is the longitudinal-transverse splitting.

The dispersion relations for the coherent modes are calculated according to Eq. (6.125), in which ϵ is substituted by c^2k^2/ω^2, ω_L by ω, and $\mathbf{k}_0$ by $\mathbf{k}$. These dispersion relations are represented in Fig. 6.11. The influence of damping on these dispersion laws is also taken into account. The energy spectrum described by formula (6.121) is the same as that for a weak test beam. This is added in Fig. 6.11.

Hanamura paid attention to the large optical nonlinearities due to excitonic molecules and dedicated a series of papers [68–70] to these phenomena. Hanamura showed theoretically the importance of the two-photon quantum transitions from the ground state to the biexciton state and showed that the nonlinear optical properties are observable under nearly two-photon resonant pumping. As in previous papers [48, 50, 65], it was stressed that when the exciton–polaritons are resonantly pumped into excitonic molecule states, they are hybridized with each other and show optical Stark splitting. These splittings can be observed as a sharp dip in the reflection spectrum and in the two-photon absorption spectrum of the excitonic molecules. These splittings change into blue shifts and red shifts of the polariton and the biexciton as functions of the pump frequency and power. The model Hamiltonian examined by Hanamura includes the biexcitons, the polaritons of the lower polariton branch, and their conversion into biexcitons.

The motion equations for the polariton annihilation operator with wave vector $\mathbf{k}_1$ and for the biexciton annihilation operator $b_{\mathbf{k}_0+\mathbf{k}_1}$ in the presence of the polariton coherent mode with wave vector $\mathbf{k}_0$ have the forms

$$
\begin{aligned}
i\hbar\frac{\mathrm{d}a_{\mathbf{k}_1}}{\mathrm{d}t} &= E_p(\mathbf{k}_1)a_{\mathbf{k}_1} + \frac{g}{\sqrt{V}}a^{\dagger}_{\mathbf{k}_0}b_{\mathbf{k}_0+\mathbf{k}_1},\\
i\hbar\frac{\mathrm{d}b_{\mathbf{k}_0+\mathbf{k}_1}}{\mathrm{d}t} &= E_m(\mathbf{k}_0+\mathbf{k}_1)\,b_{\mathbf{k}_0+\mathbf{k}_1} + \frac{g}{\sqrt{V}}a_{\mathbf{k}_1}a_{\mathbf{k}_0}.
\end{aligned}
\tag{6.127}
$$

The coherent pumping is taken into account when the operator $a_{\mathbf{k}_0}$ is replaced with the function $\sqrt{N_{\mathbf{k}_0}}e^{-i\omega_L t}$, where $n_{\mathbf{k}_0} = N_{\mathbf{k}_0}/V$ is the density of coherent polaritons.

The solutions are chosen in the form $a_{\mathbf{k}_1} \sim e^{-i\omega t}$ and $b_{\mathbf{k}_0+\mathbf{k}_1} \sim e^{-i(\omega+\omega_L)t}$. The eigenfrequencies $\omega_{1,2}$ are obtained in the form

$$
\begin{aligned}
\omega_{1,2} = {}& \frac{1}{2}[\omega_m(\mathbf{k}_0+\mathbf{k}_1) + \omega_p(\mathbf{k}_1) - \omega_L]\\
&\pm\frac{1}{2}\sqrt{[\omega_m(\mathbf{k}_0+\mathbf{k}_1) - \omega_p(\mathbf{k}_1) - \omega_L]^2 + 4|g|^2 n_{\mathbf{k}_0}}.
\end{aligned}
\tag{6.128}
$$

When the pumping $(\omega_L, \mathbf{k}_0)$ pumps the polariton resonantly into the excitonic molecule, the eigenmodes are hybridized modes of the exciton–polariton and excitonic molecule over the wave vector $\mathbf{k}_1$ and $\mathbf{k}_2 = \mathbf{k}_1 + \mathbf{k}_0$. This is the optical Stark splitting of the polariton and the biexciton, which is illustrated in Fig. 6.12. It is advantageous to observe the optical Stark effect in the two-photon absorption spectrum since this absorption line is sharp and strong because of a giant transition dipole moment.

Ivanov et al. [45] studied the optical Stark effect in the exciton–biexciton system in the electron–hole approach. The system of electrons and holes in the presence of external coherent laser radiation was considered. From the papers of Keldysh and Kozlov [51] and SCH [8], the coherent pairing of electrons and holes forming an induced Bose condensate of excitons was introduced. The new quasi-electron and quasi-hole elementary excitations were determined, as well as their energy spectrum and the interaction between them. The exciton and the biexciton operators were expressed through these new quasiparticle operators, which allowed the determination of their motion equations, and the optical Stark effect in the polariton–biexciton system was discussed in terms of the solutions of these equations. The initial and the reconstructed spectra of the polariton and the biexciton branches are presented in Fig. 6.13.

Hiroshima [72] analyzed a coherently excited exciton–biexciton system under the action of a nonresonant pump field within the framework of the Hartree–Fock–Bogoliubov approximation. The excitation spectra, which consist of optical absorption and gain bands, were determined, and it was pointed out that the optical gain is greatly enhanced through the giant oscillator strength effect of the biexcitons.

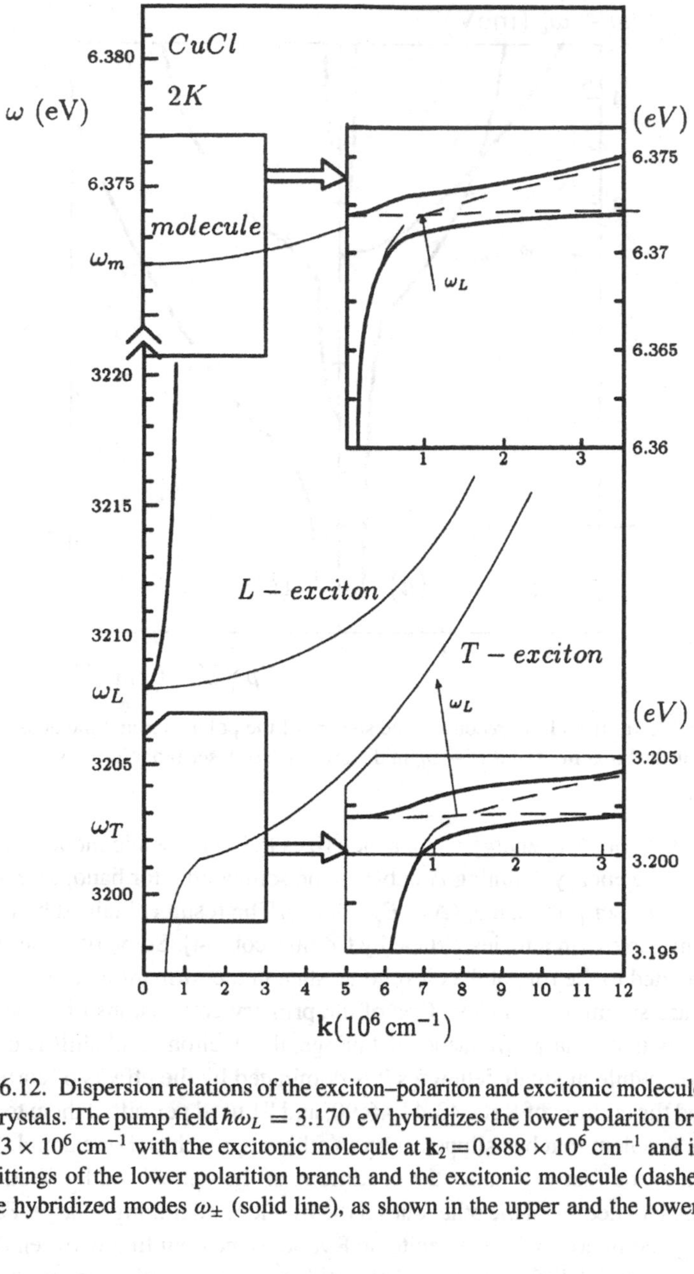

Figure 6.12. Dispersion relations of the exciton–polariton and excitonic molecule in the CuCl crystals. The pump field $\hbar\omega_L = 3.170$ eV hybridizes the lower polariton branch at $\mathbf{k}_1 = 1.3 \times 10^6$ cm^{-1} with the excitonic molecule at $\mathbf{k}_2 = 0.888 \times 10^6$ cm^{-1} and induces the splittings of the lower polarition branch and the excitonic molecule (dashed line) into the hybridized modes $\omega_\pm$ (solid line), as shown in the upper and the lower insets [68].

6.4.3 The Role of Biexcitonic States in the Excitonic Optical Stark Effect

The role of the biexcitonic states, or to be precise, of two *e–h* pairs in bound and unbound states, was clearly elucidated in the theory of the exciton optical Stark effect developed by Combescot [43, 44], Combescot and Combescot [46, 90], and Betbeder-Matibet, Combescot, and Tanguy [91]. Their approach is different from the methods described in Section 6.1 and Subsection 6.4.1 and is based on the application of perturbation theory to take into account the interaction of the electron–hole system with strong laser radiation. It is

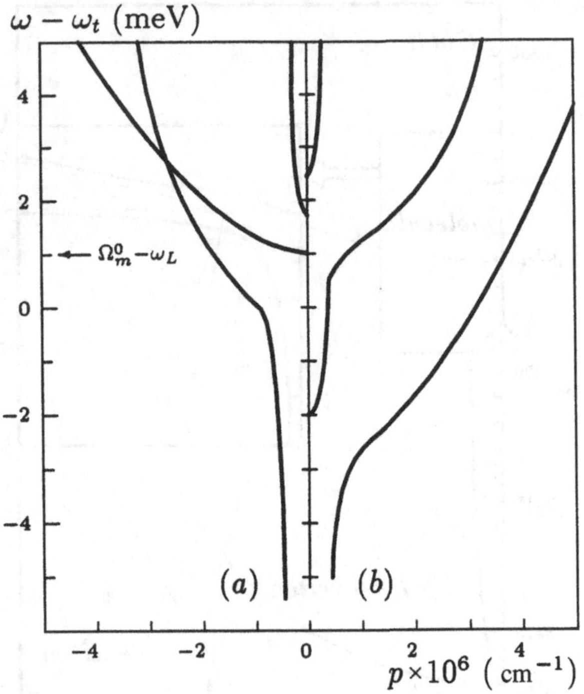

Figure 6.13. (a) initial, (b) reconstructed spectra of the polaritons and the quasi-energy branch of the biexcitons $\omega_m(\mathbf{k}) - \omega_L$ in the presence of laser radiation [45].

supposed that the coupling energy that characterizes the light–semiconductor interaction is much less than the energy detuning Δ between the semiconductor bandgap energy E_g and the energy of the laser photon $\hbar\omega_L(\Delta = E_g - \hbar\omega_L)$. The results obtained by this approach have been summarized in a review article by Combescot [44]. Some of these results, especially those related to the role of the biexcitonic state in the formation of the exciton optical Stark effect, are summarized below. One of the primary conclusions of these multilateral investigations is that at large frequency detunings, the exciton-level shift is dominated by Pauli repulsion, while at small detunings it is dominated by the effects of correlations.

The method the authors of Refs. 43,44,46,90, and 91 used permitted them to reveal some aspects of the problem that had slipped away in other approaches. For example, Combescot paid attention to the fact that under the conditions of the optical Stark effect, one exciton level cannot be isolated from the other ones if the frequency detuning is large. For frequency detunings large compared with the excitonic Rydberg, the coupling between the semiconductor ground state and different excitonic levels is as strong as with the lowest level (if one neglects the difference in the oscillator strengths), since all the exciton states are equally far in energy from the ground state. Therefore indirect links between the exciton levels through the laser radiation and the semiconductor ground state appear. The main formula describing the exciton-level shift within the framework of perturbation theory [44] reveals explicitly the decisive role of the biexcitonic bound and unbound states.

In Combescot's review article [44], the semiconductor–radiation system consists of electrons and holes, interacting among themselves by means of the Coulomb interaction, and the coherent laser radiation, which interacts with the e–h subsystem by creating and annihilating

e–h pairs. The Hamiltonian of this system is the same as that considered in Section 6.1. In the rotating reference frame, the bare states of the e–h subsystem in the zeroth-order approximation of the electron–hole–photon interaction were determined as follows.

The ground state of the semiconductor is the vacuum state for the e–h pairs. It is denoted as $|0\rangle$ and has energy $E_0 = 0$. The excited states with one e–h pair are denoted as $|x_i\rangle$ and have the eigenvalues $E_{x_i} = \Delta + \varepsilon_{x_i}$. Here ε_{x_i} can be positive as well as negative, describing not only the exciton states but also the states of the free-e–h pair. In the same way the states with two e–h pairs $|xx_n\rangle$, with the eigenenergies $E_{xx_n} = 2\Delta + \varepsilon_{xx_n}$, can describe the bound biexcitonic states and the states corresponding to two free excitons, as well as the states of two free-e–h pairs. When the electron–hole–photon interaction W is taken into consideration, the bare eigenstates of the e–h subsystem become renormalized. The new renormalized states in the second order of the perturbation theory were determined as [44]

$$E_0' = E_0 - \sum_j \frac{|\langle x_j|W|0\rangle|^2}{E_{x_j}}, \qquad E_{x_i}' = E_{x_i} + \frac{|\langle 0|W|x_i\rangle|^2}{E_{x_i}} + \sum_n \frac{|\langle xx_n|W|x_i\rangle|^2}{E_{x_i} - E_{xx_n}}. \tag{6.129}$$

The exciton-level shift δE_{x_i} was determined as the difference

$$\delta E_{x_i} = E_{x_i}' - E_0' - (E_{x_i} - E_0) = \sum_n \frac{|\langle xx_n|W|x_i\rangle|^2}{E_{x_i} - E_{xx_n}} + \frac{|\langle 0|W|x_i\rangle|^2}{E_{x_i}} + \sum_j \frac{|\langle x_j|W|0\rangle|^2}{E_{x_j}}. \tag{6.130}$$

Combescot [44] pointed out that the last two terms in Eq. (6.130) are volume dependent and must be cancelled by the volume-dependent part of the first sum. The elimination of these terms is expected physically. It is equivalent to the elimination of the disconnected diagrams in the diagrammatic calculation. This last method was also elaborated by Combescot and co-workers [44]. After the elimination of the nonphysical terms, the only remaining contribution to the exciton-level shift is related to the volume-independent part of the first sum in Eq. (6.130). Since this sum corresponds to the coupling between the exciton and all biexcitonic states (bound and unbound), the optical Stark shift can result only from the participation of these two-pair states. If there is only one exciton in the sample, the optical Stark effect cannot occur.

The review article contains a detailed comparison of the results obtained in different approaches. The main questions that appear are related to the roles of the exchange and correlation effects [92–95]. In particular, the nonequilibrium theory of the optical Stark effect elaborated by SCH [8] and reported in Section 6.1 determined the exciton-level shift in only the mean-field approximation. Combescot remarked that this approximation can give the correct leading term of the shift only in the case of large detuning compared with the excitonic Rydberg. In the case of small detuning, it is necessary to take into account the correlation effects. Combescot determined the range of application of the theory of SCH. It cannot be applied in the case of weakly bound excitons and biexciton states when $E_{x_i} \approx E_{xx_n} \approx 0$. In this case, it is necessary to take into account many-pair states, along with the exact Coulomb interaction. This calculation has not been performed, and the problem remains open [44].

As a final remark, we add that the interaction W of a strong laser field with the e–h pair can be taken into account exactly, as done by SCH [8] and discussed in Section 6.1.

References

[1] D. Fröhlich, A. Nöthe, and K. Reimann, *Phys. Rev. Lett.* **55**, 1335 (1985).

[2] A. Mysyrowicz, D. Hulin, A. Antonetti, A. Migus, W.T. Masselink, and H. Morkoc, *Phys. Rev. Lett.* **56**, 2748 (1986).

[3] A. Von Lehman, J.E. Zucker, J.P. Heritage, and D.S. Chemla, *Opt. Lett.* **11**, 609 (1986).

[4] K. Tai, J. Hegarty, and W.T. Tsang, *Appl. Phys. Lett.* **51**, 152 (1987).

[5] W.H. Knox, D.S. Chemla, D.A. B. Miller, J.B. Stark, and S. Schmitt-Rink, *Phys. Rev. Lett.* **62**, 1189 (1988).

[6] V.F. Elesin and Yu.V. Kopaev, *Zh. Eksp. Teor. Fiz.* **63**, 1447 (1972).

[7] S. Schmitt-Rink and D. S. Chemla, *Phys. Rev. Lett.* **57**, 2752 (1986).

[8] S. Schmitt-Rink, D. S. Chemla, and H. Haug, *Phys. Rev. B* **37**, 941 (1988).

[9] C. Comte and G. Mahler, *Phys. Rev. B* **38**, 10517 (1988).

[10] L.V. Keldysh, *Zh. Eksp. Teor. Fiz.* **47**, 1515 (1964) [*Sov. Phys. JETP* **20**, 1018 (1965)].

[11] E.M. Lifshitz, and L.P. Pitaevskii, *Physical Kinetics*, Vol. 10 of Theoretical Physics Series (Pergamon, New York, 1979).

[12] A.A. Abrikosov, L.P. Gor̉kov, and I.E. Dzyaloshinskii, *Methods of Quantum Field Theory in Statistical Physics* (Dover, New York, 1975).

[13] E.M. Lifshitz, and L.P. Pitaevskii, *"Statistical Physics, Part 2,"* Vol. 9 of Theoretical Physics series (Pergamon, New York, 1979).

[14] K. Henneberger and V. May, *Physica A* **138**, 537 (1986).

[15] K. Henneberger, G. Manzke, V. May, and R. Zimmermann, *Physica A* **138**, 557 (1986).

[16] W. Schäfer and J. Treuch, *Z. Phys. B* **63**, 407 (1986).

[17] C. Comte and P. Nozières, *J. Phys.* **43**, 1069 (1982).

[18] P. Nozières and C. Comte, *J. Phys.* **43**, 1083 (1982).

[19] A.S. Davydov, *Quantum Theory* (Elsevier, New York, 1976).

[20] L. Hostler and R.H. Pratt, *Phys. Rev. Lett.* **10**, 469 (1963).

[21] L. Hostler, *J. Math. Phys.* **5**, 591 (1964).

[22] M. Gavrila, *Rev. Roum. Phys.* **12**, 745 (1967).

[23] B.A. Zon, N.L. Manakov, and L.P. Rapoport, *Zh. Eksp. Teor. Fiz.* **55**, 924 (1968).

[24] G.A. Korn and T.M. Korn, *Mathematical Handbook for Scientists and Engineers* (McGraw-Hill, New York, 1961).

[25] P.W. Anderson, *Phys. Rev.* **112**, 1900 (1958).

[26] O. Betbeder-Matibet and P. Nozières, *Ann. Phys.* **51**, 392 (1969).

[27] S.A. Moskalenko and V.R. Misko, *Ukr. Fiz. Zh.* **37**, 1812 (1992).

[28] V.R. Misko and S.A. Moskalenko, *Fiz. Tverd. Tela* **35**, 606 (1993).

[29] V.R. Misko and S.A. Moskalenko, in *Nonlinear Optical Properties of Excitons in Semiconductors with Different Dimensionalities* (Shtiintsa, Kishinev, 1992), p. 24 (in Russian).

[30] S.A. Moskalenko and V.G. Pavlov, *Zh. Eksp. Teor. Fiz.* **112**, 179 (1997) [*JETP* **85**, 89 (1997)].

[31] S.A. Moskalenko, V.R. Misko and V.G. Pavlov, *Fiz. Tverd. Tela* **40**, 924 (1998).

[32] V.R. Misko, S.A. Moskalenko, and M.I. Shmiglyuk *Fiz. Tverd. Tela* **35**, 3213 (1993).

[33] V.R. Misko, S.A. Moskalenko, A.H. Rotaru, and Yu.M. Shvera, *Phys. Status Solidi B* **159**, 477 (1990).

[34] A.I. Bobrysheva, M.I. Shmiglyuk and S.S. Russu, *Proc. SPIE* **1807**, 79–88 (1993).

[35] S.A. Moskalenko, *Bose–Einstein Condensation of Excitons and Biexcitons*, (RIO, Academy of Sciences of MSSR, Kishinev, 1970) (in Russian).

[36] C. Casella, *J. Phys. Chem. Solids* **24**, 19 (1963); *J. Appl. Phys.* **34**, 1703 (1963); **36**, 2485 (1965).

[37] V.A. Gergel', R.F. Kazarinov, and R.A. Suris, *Zh. Eksp. Teor. Fiz.* **53**, 544 (1967); **54**, 298 (1968); *Proceedings of the Ninth International Conference on the Physics of Semiconductors, Moscow, 1968* (Nauka, Leningrad, 1969), Vol. 1, p. 472.

[38] A.I. Bobrysheva, *Biexcitons in Semiconductors* (Shtiintsa, Kishinev, 1979) (in Russian).

[39] F. Henneberger, K. Henneberger, and J. Voigt, *Phys. Status Solidi B* **83**, 439 (1977).
[40] F. Bechstedt and F. Henneberger, *Phys. Status Solidi B* **81**, 211 (1977).
[41] A.I. Bobrysheva, V.V. Baltaga, and M.V. Grodetskii, *Phys. Status Solidi B* **123**, 169 (1984).
[42] A.I. Bobrysheva, S.A. Moskalenko, and H.N. Kam, *Zh. Eksp. Teor. Fiz.* **103**, 301 (1993) [*JETP* **76**, 163 (1993)].
[43] M. Combescot, *Phys. Rev. B* **41**, 3517 (1990).
[44] M. Combescot, *Phys. Rep.* **221**, 167 (1992).
[45] A.L. Ivanov, L.V. Keldysh, and V.V. Panashchenko, *Zh. Eksp. Teor. Fiz.* **99**, 641 (1991) [*Sov. Phys. JETP* **72**, 359 (1991)].
[46] M. Combescot and R.Combescot, *Phys. Rev. Lett.* **61**, 117 (1988).
[47] F. Henneberger and J. Voigt, *Phys. Status Solidi B* **76**, 313 (1976).
[48] V. May, K. Henneberger, and F. Henneberger, *Phys. Status Solidi B* **94**, 611 (1979).
[49] P.I. Khadzhi, S.A. Moskalenko, and S.N. Belkin, *Sov. Phys. JETP Lett.* **29**, 200 (1979).
[50] H. Haug, R. März, and S. Schmitt-Rink, *Phys. Rev. Lett.* **77**, 287 (1980).
[51] L.V. Keldysh and A.N. Kozlov, *Zh. Eksp. Teor. Fiz. Pis'ma* **5**, 238 (1967); *Zh. Eksp. Teor. Fiz.* **54**, 978 (1968) [*Sov. Phys. JETP* **27**, 521 (1968)].
[52] L.V. Keldysh, *Solid State Commun.* **84**, 37 (1992).
[53] L.V. Keldysh, *Phys. Status Solidi B* **173**, 119 (1992).
[54] L.V. Keldysh, in *Bose–Einstein Condensation*, A. Griffin, D.W. Snoke, and S. Stringari, eds. (Cambridge U. Press, Cambridge, 1995).
[55] A.L. Ivanov and L.V. Keldysh, *Zh. Eksp. Teor. Fiz.* **84**, 404 (1983) [*Sov. Phys. JETP* **57**, 234 (1983)].
[56] L.V. Keldysh and S.G. Tikhodeev, *Zh. Eksp. Teor. Fiz.* **90**, 1852 (1986).
[57] L.V. Keldysh and S.G. Tikhodeev, *Zh. Eksp. Teor. Fiz.* **91**, 78 (1986).
[58] S.T. Beliaev, *Zh. Eksp. Teor. Fiz.* **34**, 417, 433 (1958).
[59] R. März, S. Schmitt-Rink, and H. Haug, *Z. Phys. B* **40**, 9 (1980).
[60] S. Schmitt-Rink and H. Haug, *Phys. Status Solidi B* **108**, 377 (1981); **113**, K143 (1982).
[61] I. Abram and A. Maruani, *Phys. Rev. B* **26**, 4759 (1982).
[62] I. Abram, *Phys. Rev. B* **28**, 4433 (1983).
[63] I. Abram, *Phys. Rev. B* **29**, 4480 (1984).
[64] I. Abram, A. Maruani, and S. Schmitt-Rink, *J. Phys. C* **17**, 5163 (1984).
[65] I. Abram, *J. Opt. Soc. Am. B* **2**, 1204 (1985).
[66] I. Abram, A. Maruani, D.S. Chemla, F. Bonnouvrier, and E. Batifol, *Physica B* **117/118**, 301 (1983).
[67] A. Maruani and D.S. Chemla, *Phys. Rev. B* **23**, 841 (1981).
[68] H. Hanamura, *Phys. Rev. B* **44**, 8514 (1991).
[69] H. Hanamura, *Solid State Commun.* **77**, 575 (1991).
[70] H. Hanamura, *Phys. Rev. B* **39**, 1152 (1989).
[71] I. Balslev and H. Hanamura, *Solid State Commun.* **72**, 843 (1989).
[72] T. Hiroshima, *J. Phys. C* **4**, 3847 (1992).
[73] A.L. Ivanov and H. Haug, *Phys. Rev. B* **48**, 1490 (1993).
[74] A.L. Ivanov and H. Haug, *Phys. Rev. Lett.* **74**, 438 (1995).
[75] A.L. Ivanov and H. Haug, *Phys. Status Solidi B* **188**, 61 (1995).
[76] A.L. Ivanov, M. Hasuo, N. Nagasawa, and H. Haug, *Phys. Rev. B* **52**, 11017 (1995).
[77] M. Hasuo, H. Kawano, and N. Nagasawa, *Phys. Status Solidi B* **188**, 77 (1995).
[78] M. Hasuo, H. Kawano, and N. Nagasawa, *J. Lumin.* **60/61**, 672 (1994).
[79] M. Hasuo, M. Nishino, T. Itoh, and A. Mysyrowicz, *Phys. Rev. Lett.* **70**, 1303 (1993).
[80] Ya.B. Zeldovich, *Zh. Eksp. Teor. Fiz.* **51**, 1492 (1966).
[81] A.I. Baz', Ya.B. Zeldovich, and A.M. Perelomov, *Scattering, Reactions and Decays in Nonrelativistic Quantum Mechanics* (Nauka, Moscow, 1971) (in Russian).
[82] M.I. Shmiglyuk and V.N. Pitei, *Coherent Polaritons in Semiconductors* (Shtiintsa, Kishinev, 1989) (in Russian).
[83] T. Itoh, T. Suzuki, and M. Ueta, *J. Phys. Soc. Jpn.* **44**, 345 (1978).

[84] H.M. Gibbs, S.L. McCall, T.N.C. Venkatesan, A.C.Gossard, A.Passner, and W.Wiegmann, *Appl. Phys. Lett.* **35**, 451 (1979).
[85] D.S. Chemla, A. Maruani, and E. Batifol, *Phys. Rev. Lett.* **42**, 1075 (1979).
[86] K. Kempf, G. Schmieder, G. Kurtze, and C.K. Klingshirn, *Phys. Status Solidi B* **107**, 297 (1981).
[87] G. Kurtze, W. Maier, G. Blattner, and C.K. Klingshirn, *Z. Phys. B* **39**, 95 (1980).
[88] T. Itoh and T. Suzuki, *J. Phys. Soc. Jpn.* **45**, 1939 (1978).
[89] V. May and Röseler, *Phys. Status Solidi B* **102**, 533 (1980).
[90] M. Combescot and R. Combescot, *Phys. Rev. B* **40**, 3788 (1989).
[91] O. Betbeder-Matibet, M. Combescot, and C. Tanguy, *Phys. Rev. B* **44**, 3762 (1991).
[92] I. Balslev and E. Hanamura, *Solid State Commun.* **72**, 843 (1989).
[93] A. Stahl, *Z. Phys. B* **72**, 371 (1988).
[94] M. Lindberg and S.W. Koch, *Phys. Rev. B* **38**, 3342; 7607 (1988).
[95] I. Balslev and A. Stahl, *Phys. Status Solidi B* **150**, 413 (1988).

7

Bose–Einstein Condensation of Mixed States of Excitons and Photons

7.1 Introduction. Polaritons and Semiconductor Microcavities

As we have mentioned previously, BEC of excitons or biexcitons and lasing can be seen as two limits of the same theory. Lasing occurs in the case of strong electron–photon coupling (recombination rate fast compared with interparticle-scattering rate) while excitonic BEC occurs in the case of weak electron–photon coupling (recombination rate slow compared with the interparticle-scattering rate.) A laser can be seen as a Bose condensate in which the long-range phase coherence exists in the photon states [1], while in the exciton condensate the coherence exists in the electronic states. This is one of the reasons for some of the confusing debates in the early history of excitonic Bose condensation – in many systems there is not a sharp distinction between an excitonic condensate and supperradiance, i.e., luminescence with enhanced intensity due to stimulated emission [2–4].

In the previous chapters, we have so far considered a laser light source only as a source of excitons by means of quantum transitions. In Chapter 6, the laser radiation was taken into account as a given external factor. This chapter is dedicated to phenomena related to a strong and noticeable exciton–photon interaction, when the light significantly influences the energy spectrum of the high-density excitons [5–9]. As we will see, this leads to many fascinating effects related to instabilities.

A strong exciton–photon interaction implies a strong polariton effect, in which photon and exciton states are mixed (as in Fig. 1.3). When this mixing is strong, it is not obvious how to define the ground state of the system. When the coupling is weak, one can assume that this is the $\mathbf{k} = 0$ state of the excitons, as for normal atoms, and treat quasi-equilibrium, spontaneous Bose condensation of excitons in terms of the standard weakly interacting Bose gas model, as in Chapter 2. The $\mathbf{k} = 0$ state is not necessarily the lowest energy state of the excitons in dipole-active systems, however – in a bulk crystal the exciton states transform continuously into photon states in the medium, which fall all the way to zero frequency. Therefore coherence in the polariton states analogous to Bose condensation can occur only as an inherently nonequilibrium, driven phenomenon; which state is macroscopically occupied will depend on the rate balance between generation and loss mechanisms in a given experiment. We will return to this subject of rate balance in Section 8.3 when we examine a model of the polariton distribution function based on a Fokker–Planck equation.

In this chapter, we assume that the constants of the exciton–exciton and the exciton–photon interactions are much greater than those of the exciton–phonon interaction. The exciton–photon interaction can be characterized by the splitting between the polariton branches at the point of intersection of the exciton and the photon curves. This splitting is proportional to the constants $\varphi_{\mathbf{k}}$ of the exciton–photon interaction Hamiltonian. The exciton–exciton interaction in the Bose condensed system can be characterized in terms of $L_{\mathbf{k}}$ in Eq. (2.32), which determines the shift of the exciton level. The exciton–phonon interaction strength can be measured by the exciton absorption linewidth γ. Our assumption therefore corresponds to the case in which the first two parameters are much greater than the third one:

$$L_{\mathbf{k}},\ |\varphi_{\mathbf{k}}| \gg \gamma. \tag{7.1}$$

These inequalities are fulfilled at sufficiently low temperatures and high exciton densities in such crystals as CuCl and CdS, which have strong exciton–photon interaction, which means that polariton effects are important.[a] Within the framework of these assumptions, the exciton–phonon interaction will be taken into account only afterwards or will be neglected.

The two constants $L_{\mathbf{k}}$ and $\varphi_{\mathbf{k}}$ can be taken in two different limits. One of them is

$$L_{\mathbf{k}} > |\varphi_{\mathbf{k}}|, \tag{7.2}$$

which means that the exciton–exciton interaction prevails over the exciton–photon interaction. In this case one can treat first the formation of an excitonic Bose condensate with wave vector **k**. This condensate then gives rise to a coherent electromagnetic wave with the same wave vector. The interaction of exciton and photon macroscopic waves will result in the formation of one of two possible condensate-photon modes [5,6]. In the opposite limit, when

$$|\varphi_{\mathbf{k}}| > L_{\mathbf{k}}, \tag{7.3}$$

it is necessary from the beginning to start with the polariton picture and to study the collective properties of polaritons. Both variants are studied in this chapter.

To obtain a system with a given number of excitons and photons, one can enclose the crystal in a high-Q Fabry–Perot cavity that prevents the escape of photons that have frequency resonant with the exciton level. One can accomplish this by means of modern crystal-growth technologies by sandwiching a semiconductor layer between two high-reflection Bragg mirrors made of multiple layers such as those used in vertical-cavity surface-emitting lasers, in what is called a "semiconductor microcavity." These microcavities allow control of the photon densities in the optical frequency range, similar to traditional microwave cavities. If the resonance of the cavity is chosen near to the excitonic transition level, strong coupling between excitons and photons occurs [17, 18]. This system has been experimentally realized, with a clear suppression of the rate of spontaneous emission of the excitons [19]. In recent years, an enormous amount of experimental work has been performed on polaritons in semiconductor microcavities (e.g., Refs. 20–26; for useful reviews see Refs. 27 and 28), including recent work that has given evidence for the stimulated scattering of polaritons [29, 30]. Since the photonlike polariton branch now exists in a cavity, it does not have zero energy at zero crystal momentum, as in Fig. 1.3; instead it has finite energy at zone center, leading to two distinct peaks in the reflectivity and photoluminecence spectra. These two adjacent states can be coherently excited, leading to Rabi oscillations and quantum beats.

[a] Along with the dipole-active exciton transitions, it is also possible to consider quadrupole-active transitions and quadrupole polaritons [10–16] when the constant γ is very small. Such a case is realized for the $n = 1$ exciton level of the yellow series of the Cu_2O crystal. The quadrupole-active exciton level has the symmetry Γ_5^+ and is known as the orthoexciton (see Appendix A).

Experimental work in recent years has accurately determined the dispersion curves of the two polariton branches in various semiconductor microcavities.

These new possibilities stimulated further theoretical investigations of the coherent macroscopic states of excitons and photons [5–9]. At the same time, new questions arise about matters such as the effects of increasing the exciton–photon coupling in the resonator and the quantization of the excitonic center-of-mass motion [17]. In the same way, one can discuss the quantization of the polariton dispersion curve. In general, we will not treat such delicate details here.

Taking into account the parameters of the excitons and crystals as well as those of the optical light sources and resonators, one can imagine three different possible methods of achieving BEC of excitons and photons. One of them was considered above, within the framework of condition (7.2). Another case can be expected when exciton accumulation takes place along with frequent exciton–photon transformations. The mirrors of the microcavity can facilitate the accumulation of excitons in the bulk crystal embedded in the resonator. Because of repeated reflections from the mirrors, the photons with a selected wave vector will contribute to the formation of the exciton Bose condensate. The spontaneous accumulation of excitons will transform gradually into an induced Bose condensation stimulated by the interior photons of the microcavity, which are different from the photons of the nonresonant external pumping source. The coherence property of the pump radiation is not important in this case. This case is the same as that in the usual laser. The difference consists of the appearance of coherent excitons and photons with the same wave vector simultaneously. A third method involves the excitation of crystals without microcavities by coherent external laser beams. If the excitons are excited directly by resonant coherent light, a real, nonequilibrium, induced exciton BEC can be realized, as suggested in Ref. 31.

Using lasers with different frequency detunings opens many possibilities. In crystals in which inequality (7.3) can be fulfilled at moderate exciton densities, the polariton state with wave vector **k** and frequency $\omega_p(\mathbf{k})$ can be excited coherently by one- or two-photon absorptions as well as by photon–phonon transitions. For example, the polaritons in CuCl crystals were detected in the two-photon absorption of two laser beams with the frequencies ω_1 and ω_2 and wave vectors $\mathbf{k}_1$ and $\mathbf{k}_2$ satisfying the conditions $\omega_1 + \omega_2 = \omega_p(\mathbf{k})$ and $\mathbf{k}_1 + \mathbf{k}_2 = \mathbf{k}$ [32]. Induced Bose condensation of the polaritons at the bottom of the upper polariton branch (UPB) was observed by Brodin, Goer, and Matsko, [33–35]. They excited ZnTe crystals by laser radiation with frequencies ω greater than the frequency of the longitudinal exciton $\omega_{\parallel}(0)$ by the frequencies $n\omega_{LO}$ of a few optical phonons. This implies the relation $\omega = \omega_{\parallel}(0) + n\omega_{LO}$, where $n = 1, 2, 3$. The scattering of this radiation with the emission of n optical phonons will give rise to the coherent induced state, which was identified by the appearance of a very sharp luminescence line at the frequency $\omega_{\parallel}(0)$. More detailed discussions of these and other experiments are given in the monograph by Shmiglyuk and Pitei [36]. The discovery of the optical Stark effect in the excitonic spectral range resulted in many experimental and theoretical investigations along these lines, some aspects of which were discussed in Chapter 6.

While the induced BEC of dipole-active excitons and photons has been achieved experimentally, up to now the spontaneous BEC of these quasiparticles has not been evidenced. In Section 7.2 we discuss the influence of a Bose condensate of the excitons and the photons on the noncondensed quasiparticles. The condensate properties depend on the polariton structure. When the frequencies of the exciton and the photon $\omega_{ex}(k)$ and ck do not coincide, one of the two polariton components is real, but the other is virtual. Their reciprocal transformations are governed by a conservation law for momentum but not energy. In the first method of excitation, the accumulated dipole- or quadrupole-active excitons are real, while the photons are virtual. In the third method, the external photons are real but the excitons are virtual. When $\omega_{ex}(k)$ equals ck, both components are real. But in all cases, these components form a unique polariton state, and so for these reasons the methods of excitation are not so important. The main factor is the presence of a coherent macroscopic mode, which is nothing but a reservoir filled with energy. This reservoir influences all the quasiparticles of the system.

7.2 Condensate-Photon Modes and Giant Polaritons

7.2.1 The Keldysh Equations Describing the Coherent Excitons and Photons

The way in which dipole-active excitons give rise to an electric field was explicitly demonstrated by Hall [37]. Here we undertake a similar task in the case of BEC of dipole-active excitons. Our treatment of mixed states of coherent excitons and photons follows that of Keldysh [7], who started from the *e–h* Hamiltonian and found two equations, a nonlinear Schrödinger equation for the coherent excitons and a linear wave equation for the coherent electromagnetic-field strength, which describe the time- and the space-varying amplitudes of the coherent excitons and photons. We present similar equations in the momentum representation, describing spatially homogeneous states of coherent photons and excitons regarded as true bosons. Reference 7 did not include a detailed deduction of the Keldysh equations, however, and so here we present one possible derivation, suggested by Moskalenko, Kiselyova, and Pavlov [38], which will permit us to understand better the properties of coherent dipole-active excitons.

The starting point of this derivation here, as well as in the paper of Hall [37], is the electron–hole exchange Coulomb interaction. Its long-range part can be represented as a dipole–dipole interaction, leading to the appearance of the electric-field strength. This term is absent in the case of scalar excitons and has not been considered up till now, for example in electron–hole Hamiltonian equation (2.93).

To understand better the origin of the dipole–dipole interaction and to deduce from first principles the equations describing the coherent excitons and photons, we need to step back to review the interaction of the band electrons in a crystal. The full Hamiltonian, which describes the Coulomb interaction of the electrons belonging to the valence shells of the crystal atoms, has the form

$$\frac{1}{2}\sum_{\substack{n_1,n_2,n_1',n_2',\\ \mathbf{p},\mathbf{q},\mathbf{k},\sigma_1,\sigma_2}} F(n_1,\mathbf{p};n_2,\mathbf{q};n_1',\mathbf{p}-\mathbf{k};n_2',\mathbf{q}+\mathbf{k})a^{\dagger}_{n_1,\mathbf{p},\sigma_1}a^{\dagger}_{n_2,\mathbf{q},\sigma_2}a_{n_2',\mathbf{q}+\mathbf{k},\sigma_2}a_{n_1',\mathbf{p}-\mathbf{k},\sigma_1}. \tag{7.4}$$

Here $a^{\dagger}_{n,\mathbf{p},\sigma}$ and $a_{n,\mathbf{p},\sigma}$ are Bose creation and the annihilation operators, respectively, of an electron in a single-electron state characterized by the number of the electron band n, momentum $\mathbf{k}$, and spin projection σ. We assume simple bands and neglect the spin-orbit interaction; for the more complicated case of Cu_2O, see Appendix A. The Coulomb interaction terms kept in *e–h* Hamiltonian (2.93) correspond to definite components of Hamiltonian (7.4), namely, the numbers of bands n_i, n_i' take the notations of conduction (c) or valence (v) bands in the following ways: $n_1 = n_2 = n_1' = n_2' = c\,(v)$, $n_1 = n_1' = c\,(v)$, and $n_2 = n_2' = v\,(c)$. The approximate expressions of these integrals coincide with

$$F(c, \mathbf{p}; c, \mathbf{q}; c, \mathbf{p}-\mathbf{k}; c, \mathbf{q}+\mathbf{k}) = F(v, \mathbf{p}; v, \mathbf{q}; v, \mathbf{p}-\mathbf{k}, v, \mathbf{q}+\mathbf{k})$$

$$= F(c, \mathbf{p}; v, \mathbf{q}; c, \mathbf{p}-\mathbf{k}; v, \mathbf{q}+\mathbf{k}) = V_k = \frac{4\pi e^2}{V k^2 \varepsilon_\infty}. \tag{7.5}$$

This expression contains the high-frequency dielectric constant ε_∞ instead of the static dielectric constant ε_0, which was introduced phenomenologically in Chapter 2 in the direct Coulomb *e–h* interaction (2.94). ε_0 is affected by all the exciton levels, whereas ε_∞ is due to all these levels except for the given dipole-active exciton level, which is considered explicitly. The polarization determined by this quantum transition and its contribution to the dielectric constant is considered separately and directly. For this reason it cannot be added again to the background dielectric constant ε_∞.

In this background medium the light velocity c differs from the value c_0 in vacuum ($c = c_0/\sqrt{\varepsilon_\infty}$). This fact was taken into account in the review article by Poluektov, Popov, and Roitberg [39]. The question arises of how one can obtain the resulting *e–h* potential (2.94) starting with initial potential (7.5). This was discussed in the review article by Haken [40] regarding the exciton–phonon interaction. It turns out that it is necessary to take into account the interaction of the electrons and the holes with the longitudinal optical phonons. This interaction is proportional to the difference $\varepsilon_\infty^{-1} - \varepsilon_0^{-1}$ of the inverse values of the high-frequency ε_∞ and the low-frequency ε_0 dielectric constants.

Along with the components of Eq. (7.5), there also exist Coulomb exchange interaction terms, for which the band numbers take the values $n_1 = n_2' = c\,(v)$ and $n_2 = n_1' = v\,(c)$. One of these integrals has the expression

$$F(c, \mathbf{p}; v, \mathbf{q}; v, \mathbf{p}-\mathbf{k};\ c, \mathbf{q}+\mathbf{k})$$

$$= \int d\mathbf{r}_1 \int d\mathbf{r}_2 \psi_{c\mathbf{p}}^*(\mathbf{r}_1)\psi_{v,\mathbf{p}-\mathbf{k}}(\mathbf{r}_1)\frac{e^2}{\varepsilon_\infty r_{12}}\psi_{v\mathbf{q}}^*(\mathbf{r}_2)\psi_{c,\mathbf{q}+\mathbf{k}}(\mathbf{r}_2). \tag{7.6}$$

Here $\psi_{c\mathbf{p}}(\mathbf{r})$ and $\psi_{v\mathbf{p}}(\mathbf{r})$ are the Bloch electron wave functions belonging to the conduction and the valence bands. They can be represented as a product of plane waves and periodic parts $u_{n\mathbf{p}}(\mathbf{r})$ as follows:

$$\psi_{n\mathbf{p}}(\mathbf{r}) = \frac{1}{\sqrt{V}} e^{i\mathbf{p}\cdot\mathbf{r}} u_{n\mathbf{p}}(\mathbf{r}), \quad n = c, v. \tag{7.7}$$

The periodic parts have the property

$$u_{n\mathbf{p}}(\mathbf{R}+\mathbf{s}) = u_{n\mathbf{p}}(\mathbf{s}),$$

where $\mathbf{R}$ is a lattice vector, while the vector $\mathbf{s}$ changes within the volume v_0 of the elementary lattice cell. By making the substitutions

$$\mathbf{r}_1 = \mathbf{R}_1 + \mathbf{s}_1, \qquad \mathbf{r}_2 = \mathbf{R} + \mathbf{R}_1 + \mathbf{s}_2, \qquad |\mathbf{r}_{12}| = |\mathbf{R} + \mathbf{s}_2 - \mathbf{s}_1|,$$

$$\int d\mathbf{r}_1 \int d\mathbf{r}_2 = \sum_{\mathbf{R}_1}\sum_{\mathbf{R}_2}\int_{v_0} d\mathbf{s}_1 \int_{v_0} d\mathbf{s}_2,$$

$$\frac{1}{|\mathbf{R}+\mathbf{s}_2-\mathbf{s}_1|} \cong \begin{cases} \frac{1}{|\mathbf{s}_2-\mathbf{s}_1|}, & \mathbf{R}=0, \\ \frac{1}{R} + \frac{\mathbf{s}_1\cdot\mathbf{s}_2}{R^3} - \frac{3(\mathbf{s}_1\cdot\mathbf{R})(\mathbf{s}_2\cdot\mathbf{R})}{R^5}, & \mathbf{R}\neq 0 \end{cases} \tag{7.8}$$

one can obtain the following expression where we have kept only terms contributing to the exchange integral,

$$F(c, \mathbf{p}; v, \mathbf{q}; v, \mathbf{p}-\mathbf{k}; c, \mathbf{q}+\mathbf{k}) = \frac{1}{N}\{A_{cv}(\mathbf{p}, \mathbf{q}, \mathbf{k}) + V[\mathbf{k}, \mathbf{d}_{cv}(\mathbf{p}), \mathbf{d}^*_{cv}(\mathbf{q})]\}. \quad (7.9)$$

Here N is the number of lattice sites ($V = Nv_0$) and $A_{cv}(\mathbf{p}, \mathbf{q}, \mathbf{k})$ is the contact exchange Coulomb integral with $\mathbf{R} = 0$,

$$A_{cv}(\mathbf{p}, \mathbf{q}, \mathbf{k}) = \frac{1}{v_0}\int_{v_0} d\mathbf{s}_1 \frac{1}{v_0}\int_{v_0} d\mathbf{s}_2 e^{i\mathbf{k}\cdot(\mathbf{s}_2-\mathbf{s}_1)} \times u^*_{c\mathbf{p}}(\mathbf{s}_1)u_{v,\mathbf{p}-\mathbf{k}}(\mathbf{s}_1)\frac{e^2}{|\mathbf{s}_2-\mathbf{s}_1|\,\varepsilon_\infty}u^*_{v\mathbf{q}}(\mathbf{s}_2)u_{c,\mathbf{q}+\mathbf{k}}(\mathbf{s}_2). \quad (7.10)$$

It determines the difference between the orthoexciton and the paraexciton levels. The long-range part of the exchange Coulomb integral with $\mathbf{R} \neq 0$ has the form

$$\begin{aligned} V[\mathbf{k}, \mathbf{d}_{cv}(\mathbf{p}), \mathbf{d}^*_{cv}(\mathbf{q})] &= \sum_{\mathbf{R}\neq 0} e^{i\mathbf{k}\cdot\mathbf{R}}\frac{1}{v_0}\int_{v_0} ds_1 \frac{1}{v_0}\int_{v_0} d\mathbf{s}_2 \\ &\quad \times u^*_{c\mathbf{p}}(\mathbf{s}_1)u_{v,\mathbf{p}-\mathbf{k}}(\mathbf{s}_1)\frac{e^2}{\varepsilon_\infty\,|\mathbf{R}+\mathbf{s}_2-\mathbf{s}_1|}u^*_{v\mathbf{q}}(\mathbf{s}_2)u_{c,\mathbf{q}+\mathbf{k}}(\mathbf{s}_2) \\ &\approx \sum_{\mathbf{R}\neq 0} e^{i\mathbf{k}\cdot\mathbf{R}}\left\{\frac{[\mathbf{d}_{cv}(\mathbf{p})\cdot\mathbf{d}^*_{cv}(\mathbf{q})]}{\varepsilon_\infty R^3} - \frac{3\,[\mathbf{d}_{cv}(\mathbf{p})\cdot\mathbf{R}]\,[\mathbf{d}^*_{cv}(\mathbf{q})\cdot\mathbf{R}]}{\varepsilon_\infty R^5}\right\}. \end{aligned} \quad (7.11)$$

Here the following relations were used:

$$\frac{1}{v_0}\int_{v_0} d\mathbf{s}\, u_{n\mathbf{p}}(\mathbf{s})u_{n',\mathbf{p}-\mathbf{k}}(\mathbf{s}) \cong \delta_{n,n'}, \qquad (n, n') = (c, v), \quad \mathbf{k} \ll \mathbf{p},$$

$$\frac{e}{v_0}\int_{v_0} d\mathbf{s}\, u^*_{c\mathbf{p}}(\mathbf{s})\mathbf{s}u_{v,\mathbf{p}-\mathbf{k}}(\mathbf{s}) \approx \frac{e}{v_0}\int_{v_0} d\mathbf{s}\, u^*_{c\mathbf{p}}(\mathbf{s})\mathbf{s}u_{v\mathbf{p}}(\mathbf{s}) = \mathbf{d}_{cv}(\mathbf{p}) = \mathbf{d}^*_{vc}(\mathbf{p}). \quad (7.12)$$

It is possible to neglect $\mathbf{k}$ compared with the momentum $\mathbf{p}$. This is justified by the fact that $\mathbf{p}$ ranges over the whole Brillouin zone, while the wave vector $\mathbf{k}$ equals the wave vector of the exciton and coincides with the wave vector of light.

The term $V[\mathbf{k}, \mathbf{d}_{cv}(\mathbf{p}), \mathbf{d}^*_{cv}(\mathbf{q})]$ is known as the long-range dipole–dipole interaction. The electron and hole operators used in Hamiltonian (2.93) are connected with the electron-band operators by the relations

$$a^\dagger_{\mathbf{p}\sigma} = a^\dagger_{c\mathbf{p}\sigma}, \qquad b^\dagger_{\mathbf{p}\sigma} = a_{v,-\mathbf{p},-\sigma}. \quad (7.13)$$

Writing the Coulomb exchange term of Hamiltonian (7.4) in normal order relative to the electron and hole operators, one can obtain the expression

$$\frac{1}{N}\sum_{\substack{\mathbf{p},\mathbf{q},\mathbf{k},\\ \sigma,\sigma'}} \{A_{cv}(\mathbf{p}, \mathbf{q}, \mathbf{k}) + V[\mathbf{k}, \mathbf{d}_{cv}(\mathbf{p}), \mathbf{d}^*_{cv}(\mathbf{q})]\}a^\dagger_{\mathbf{p}\sigma}b^\dagger_{\mathbf{k}-\mathbf{p},-\sigma}b_{-\mathbf{q},-\sigma'}a_{\mathbf{q}+\mathbf{k},\sigma'}, \quad (7.14)$$

which describes the contact and the long-range exchange interactions that exist in addition to the direct Coulomb interaction.

The difference between the two normal-ordered terms (one represented in band-electron operators and the other in e–h operators) is a quadratic term in e–h operators. It determines, along with other similar terms, the effective periodic potential created in a self-consistent way by all the valence electrons and by the remaining parts of the crystal atoms.

One can see that the creation of an electron–hole pair is accompanied by the appearance of a dipole moment $\mathbf{d}_{cv}(\mathbf{p})$ characterizing the interband quantum transition. Its properties are determined by the band structure and the crystal symmetry. If the crystal-symmetry group has a center of inversion and the bands c and v have different parities at the point $\mathbf{k} = 0$, then the transition dipole moment $\mathbf{d}_{cv}(\mathbf{p})$ is nonzero even for the momentum $\mathbf{p} = 0$. For this reason it is called an allowed band-to-band transition [41]. In the case in which $\mathbf{d}_{cv}(\mathbf{p})$ is proportional to $\mathbf{p}$, the band-to-band transition is called a forbidden transition.

As discussed below, the exciton transition dipole moment depends on the band-to-band dipole moment $\mathbf{d}_{cv}(\mathbf{p})$ as well as on the wave function of the relative e–h motion in the exciton. Thus the appearance of an electron–hole pair is also characterized by a virtual band-to-band transition dipole moment $\mathbf{d}_{cv}(\mathbf{p})$. Such dipole moments appear simultaneously at the each site of the excited crystal lattice, giving rise to the crystal polarization. In its turn this polarization gives rise to the electromagnetic field.

In the case of BEC of dipole-active excitons, the polarization of the crystal will become macroscopically large, as will the amplitude of the condensed excitons. To consider this effect one needs to add term (7.14) to Hamiltonian (2.93) and to subject the completed Hamiltonian to u, v transformation (2.108). The additional term (7.14) will give rise to supplementary terms in the compensation condition (2.113). They are

$$\frac{1}{N}\sum_{\mathbf{pq}}\sum_{\sigma'}\{A_{cv}(\mathbf{p},\mathbf{q},0) + V[\mathbf{k}\to 0, \mathbf{d}_{cv}(\mathbf{p}), \mathbf{d}^*_{cv}(\mathbf{q})]\}u_{\mathbf{q}}v_{\mathbf{q}}\left(u^2_{\mathbf{p}} - v^2_{\mathbf{p}}\right)$$

$$+\frac{2}{N}\sum_{\mathbf{pq}}\{A_{cv}(\mathbf{p},\mathbf{q},\mathbf{p}-\mathbf{q}) + V[\mathbf{p}-\mathbf{q}, \mathbf{d}_{cv}(\mathbf{p}), \mathbf{d}^*_{cv}(\mathbf{q})]\}v^2_{\mathbf{q}}u_{\mathbf{p}}v_{\mathbf{p}}. \quad (7.15)$$

In accordance with the selection rule, the wave vector $\mathbf{k}$ in the first term of formula (7.15) must be set equal to the wave vector of the single-exciton state in which the BEC occurred. In Chapter 2, excitonic BEC was considered, for simplicity, in the single-particle state with wave vector $\mathbf{k} = 0$. But in the case of dipole-active excitons, the long-range dipole–dipole interaction $V(\mathbf{k}, \mathbf{d}_{cv}, \mathbf{d}^*_{cv})$ is a nonanalytical function that depends on the vector $\mathbf{k}$. At small values of the modulus $|\mathbf{k}|$ it depends on the direction in which the wave vector $\mathbf{k}$ tends to zero. As shown below, for $\mathbf{k} \parallel \mathbf{d}_{cv}$ and $\mathbf{k} \perp \mathbf{d}_{cv}$ the values of $V(\mathbf{k}, \mathbf{d}_{cv}, \mathbf{d}^*_{cv})$ are different. For this reason it is necessary to keep the direction $\mathbf{k}\to 0$, to calculate first the sum over the lattice sites in Eq. (7.11) for a given direction of $\mathbf{k}$, and only after that to put $|\mathbf{k}| = 0$. In the case of the contact exchange interaction $A_{cv}(\mathbf{p}, \mathbf{q}, 0)$, such precaution is excessive. To remember that particularity of the long-range Coulomb exchange interaction, we will write $\mathbf{k}\to 0$, instead of $\mathbf{k}=0$. At the same time, such notation opens the possibility of generalizing the obtained formula from the case $\mathbf{k}=0$ to $\mathbf{k}\neq 0$. Probably it would be better to obtain from the beginning the results concerning BEC of e–h pairs in a state with $\mathbf{k} \neq 0$. This program was realized by Ivanov, Keldysh, and Panashchenko [42], without taking into account the dipole–dipole interaction. But it is possible to consider the case $\mathbf{k} \to 0$, to introduce in the final result the kinetic energy of the exciton translational motion $\hbar^2k^2/2m_{\text{ex}}$, that determines the spatial dispersion, and to add the index $\mathbf{k}$ at the amplitude of the exciton condensate. This simple method is presented below. After that the limit of the wave vector $\mathbf{k}$ to zero is omitted. This method takes into account the results of Refs. 7 and 42. The second term in expression (7.15) contains the rapidly oscillating factors $\exp[i(\mathbf{p}-\mathbf{q}, \mathbf{R})]$ and $\exp[i(\mathbf{p}-\mathbf{q}, \mathbf{s}_2-\mathbf{s}_1)]$. One of them leads to the exponential decrease of the interaction of two dipoles with distance $\mathbf{R}$. The second factor diminishes the value of the contact exchange integral. The second term in expression (7.15) is dropped.

Compensation condition (2.113), supplemented by the first term of expression (7.15) for the spin-oriented e–h pairs, has the form

$$\left[\mathcal{E}(\mathbf{p}) - \mu - 2\sum_{\mathbf{q}} V_{\mathbf{p}-\mathbf{q}} v_{\mathbf{q}}^2\right] u_{\mathbf{p}} v_{\mathbf{p}} - \left(u_{\mathbf{p}}^2 - v_{\mathbf{p}}^2\right)\sum_{\mathbf{q}} V_{\mathbf{p}-\mathbf{q}} u_{\mathbf{q}} v_{\mathbf{q}}$$
$$+ \frac{1}{N}\sum_{\mathbf{q}} \{A_{cv}(\mathbf{p},\mathbf{q},0) + V[\mathbf{k}\to 0, \mathbf{d}_{cv}(\mathbf{p}), \mathbf{d}_{cv}^*(\mathbf{q})]\} u_{\mathbf{q}} v_{\mathbf{q}}\left(u_{\mathbf{p}}^2 - v_{\mathbf{p}}^2\right) = 0. \quad (7.16)$$

The physically meaningful quantity is the dipole moment of the exciton transition from the ground state of the crystal. It is determined as follows:

$$\mathbf{d}_{nl} = \frac{\sqrt{v_0}}{V}\sum_{\mathbf{p}} \varphi_{nl}(\mathbf{p})\mathbf{d}_{cv}(\mathbf{p}),$$
$$\frac{1}{V}\sum_{\mathbf{p}} |\varphi_{nl}(\mathbf{p})|^2 = 1, \qquad \frac{1}{V}\sum_{\mathbf{p}} \varphi_{nl}(\mathbf{p}) = \psi_{nl}(\mathbf{r}=0). \quad (7.17)$$

Here $\varphi_{nl}(\mathbf{p})$ is the wave function of the relative e–h motion in the exciton in the momentum representation. The same function in the real-space representation is written as $\psi_{nl}(\mathbf{r})$. The quantum numbers n and l determine the exciton level of the hydrogenlike series. In the case of an allowed band-to-band transition [41], the exciton dipole moment is

$$\mathbf{d}_{nl} = \mathbf{d}_{cv}(0)\sqrt{v_0}\,\psi_{nl}(0), \qquad |\mathbf{d}_{nl}|^2 = |\mathbf{d}_{cv}|^2\, v_0\, |\psi_{nl}(0)|^2. \quad (7.18)$$

It is nonzero for the exciton s states. Forbidden band-to-band transitions, when $\mathbf{d}_{cv}(\mathbf{p}) = A\mathbf{p}$, lead to exciton dipole moments

$$|\mathbf{d}_{nl}|^2 = |A|^2\, v_0 |[\nabla\varphi_{nl}(\mathbf{r})]_{\mathbf{r}=0}|^2, \quad (7.19)$$

which are nonzero for the p states of the excitonic hydrogenlike series.

If we confine ourselves to a BEC of excitons in the ground $1s$ state and suppose that the transition is dipole active, then we must assume that the crystal has an allowed first-order band-to-band transition, $\mathbf{d}_{cv}(\mathbf{p}) = \mathbf{d}_{cv}(0)$. To express the compensation condition through the exciton dipole moment $\mathbf{d}$ (the index $1s$ is dropped) one needs to average the left-hand side of Eq. (7.16) with the help of the $\varphi_{1s}(\mathbf{p})$ exciton wave function. Then Eq. (7.16) can be rewritten in the form

$$\mu\frac{1}{\sqrt{V}}\sum_{\mathbf{p}} u_{\mathbf{p}} v_{\mathbf{p}} \varphi_{1s}(\mathbf{p}) = \frac{1}{\sqrt{V}}\sum_{\mathbf{p}} \mathcal{E}(\mathbf{p}) u_{\mathbf{p}} v_{\mathbf{p}} \varphi_{1s}(\mathbf{p}) - \frac{2}{\sqrt{V}}\sum_{\mathbf{pq}} V_{\mathbf{p}-\mathbf{q}} v_{\mathbf{q}}^2 u_{\mathbf{p}} v_{\mathbf{p}} \varphi_{1s}(\mathbf{p})$$
$$- \frac{1}{\sqrt{V}}\sum_{\mathbf{pq}} V_{\mathbf{p}-\mathbf{q}} u_{\mathbf{q}} v_{\mathbf{q}}\left(u_{\mathbf{p}}^2 - v_{\mathbf{p}}^2\right)\varphi_{1s}(\mathbf{p}) + \frac{1}{N\sqrt{V}}\sum_{\mathbf{pq}} \{A_{cv}(\mathbf{p},\mathbf{q},0)$$
$$+ V[\mathbf{k}\to 0, \mathbf{d}_{cv}(\mathbf{p}), \mathbf{d}_{cv}^*(\mathbf{q})]\} u_{\mathbf{q}} v_{\mathbf{q}}\left(u_{\mathbf{p}}^2 - v_{\mathbf{p}}^2\right)\varphi_{1s}(\mathbf{p}). \quad (7.20)$$

Solutions (2.114)–(2.116) mean that $v_{\mathbf{p}}^0 = \sqrt{n_{ex}}\varphi_{1s}(\mathbf{p})$, $u_{\mathbf{p}}^2 = 1 - v_{\mathbf{p}}^{02}$, $u_{\mathbf{p}} = 1 - v_{\mathbf{p}}^{02}/2$, and $N_{ex} = Vn_{ex}$. After their substitution into Eq. (7.20), this equation transforms into the equality

$$\mu\sqrt{N_{ex}} = (-\mathrm{Ry}_{ex} + A)\sqrt{N_{ex}} + \frac{26\pi}{3}a_{ex}^3\mathrm{Ry}_{ex}\frac{N_{ex}}{V}\sqrt{N_{ex}} + V(\mathbf{k}\to 0, \mathbf{d}, \mathbf{d}^*)\left(1 - \frac{35}{4}\pi a_{ex}^3\frac{N_{ex}}{V}\right)\sqrt{N_{ex}}. \quad (7.21)$$

The expression $(-\mathrm{Ry}_{ex} + A)$ is the energy of the exciton level $E_{ex}(0)$ measured relative to the bottom of the conduction band in the same manner as the chemical potential μ. A is the shift of the exciton level due the contact exchange interaction. Its density correction is dropped. If one wishes to measure the exciton creation energy $E_{ex}(0)$ and the chemical potential μ from the top of the valence band, it is necessary only to add the expression $E_g\sqrt{N_{ex}}$ to the left and the right sides of Eq. (7.21). The long-range exchange interaction can be represented in the form

$$V(\mathbf{k}, \mathbf{d}, \mathbf{d}^*)\sqrt{N_{ex}} = \sqrt{N_{ex}}\sum_{\mathbf{R}\neq 0} e^{i\mathbf{k}\cdot\mathbf{R}}\left[\frac{|\mathbf{d}|^2}{\varepsilon_\infty R^3} - \frac{3\,|\mathbf{d}\cdot\mathbf{R}|^2}{\varepsilon_\infty R^5}\right] = -(\mathbf{d}e^{i\mathbf{k}\cdot\mathbf{n}})^*\cdot\mathbf{E}_{\mathrm{eff}}(\mathbf{n}). \quad (7.22)$$

Here $\mathbf{R}$ has been set equal to the difference of two integer vectors of the lattice sites $\mathbf{m}$ and $\mathbf{n}$ $(\mathbf{R} = \mathbf{m} - \mathbf{n} \neq 0)$. The effective electric-field strength $\mathbf{E}_{\mathrm{eff}}(\mathbf{n})$ is

$$\mathbf{E}_{\mathrm{eff}}(\mathbf{n}) = -\sqrt{N_{ex}}\sum_{\mathbf{m}\neq\mathbf{n}}\left[\frac{\mathbf{d}e^{i\mathbf{k}\cdot\mathbf{m}}}{\varepsilon_\infty R^3} - \frac{3\mathbf{R}(e^{i\mathbf{k}\cdot\mathbf{m}}\mathbf{d}\cdot\mathbf{R})}{\varepsilon_\infty R^5}\right]. \quad (7.23)$$

Following Hall [37], one can say that a BEC of excitons with wave vector $\mathbf{k}\to 0$ induces the appearance of a virtual macroscopic dipole moment equal to $\mathbf{d}\sqrt{N_{ex}}e^{i\mathbf{k}\cdot\mathbf{m}}$ at each site $\mathbf{m}$ of the lattice, modulated along the lattice as a plane wave. These macroscopic dipole moments from all the sites $\mathbf{m}$ except the given site $\mathbf{n}$ give rise in the last site to the effective macroscopic electric field $\mathbf{E}_{\mathrm{eff}}(\mathbf{n})$ of Eq. (7.23). It can be expressed through the mean electric field $\mathbf{E}(\mathbf{n})$ created by all the macroscopic dipoles, including that of the site $\mathbf{n}$, by the help of Lorentz–Lorenz formula

$$\mathbf{E}_{\mathrm{eff}}(\mathbf{n}) = \mathbf{E}(\mathbf{n}) + \frac{4\pi}{3}\mathbf{P}(\mathbf{n}). \quad (7.24)$$

The macroscopic polarization $\mathbf{P}(\mathbf{n})$ can be determined as a macroscopic dipole moment of the given site referred to the volume v_0 of the elementary cell:

$$\mathbf{P}(\mathbf{n}) = \frac{\mathbf{d}\sqrt{N_{ex}}e^{i\mathbf{k}\cdot\mathbf{n}}}{\varepsilon_\infty v_0}. \quad (7.25)$$

In what follows, the electric-field strength and polarization are considered as continuous functions $\mathbf{E}(\mathbf{r})$ and $\mathbf{P}(\mathbf{r})$ that obey the Maxwell equations for a condensed medium:

$$\nabla\times\mathbf{E} = -\frac{1}{c}\frac{\partial\mathbf{H}}{\partial t}, \qquad \nabla\times\mathbf{H} = \frac{1}{c}\frac{\partial\mathbf{D}}{\partial t},$$
$$\mathbf{D} = \mathbf{E} + 4\pi\mathbf{P}, \qquad \nabla\cdot\mathbf{D} = 0, \qquad \nabla\cdot\mathbf{H} = 0. \quad (7.26)$$

As mentioned above in relation to formula (7.5), the background medium has the dielectric constant ε_∞. The light velocity in this medium is $c = c_0/\sqrt{\varepsilon_\infty}$ and differs from the light velocity c_0 in vacuum. If one multiplies both parts of Eq. (7.21) by the factor $\exp(-i\mu t/\hbar + i\mathbf{k}\cdot\mathbf{r})$, substituting the site vector $\mathbf{n}$ by continuous vector $\mathbf{r}$, then the exciton amplitude $a_{\mathbf{k}}(\mathbf{r}, t)$ appears, which has the properties

$$a_{\mathbf{k}}(\mathbf{r}, t) = \sqrt{N_{\rm ex}}e^{-\frac{i}{\hbar}\mu t + i\mathbf{k}\cdot\mathbf{r}}, \qquad \frac{|a_{\mathbf{k}}(\mathbf{r}, t)|^2}{V} = \frac{N_{\rm ex}}{V},$$

$$\mu a_{\mathbf{k}}(\mathbf{r}, t) = i\hbar\frac{da_{\mathbf{k}}(\mathbf{r}, t)}{dt}, \tag{7.27}$$

which are in accordance with formulas (2.69) for the condensate amplitude in Beliaev's technique.

Finally, the kinetic energy of the exciton translation motion can be introduced as an effect of spatial dispersion. Taking into account Eqs. (7.22), (7.24), and (7.27), one can transform Eq. (7.21) as follows:

$$i\hbar\frac{da_{\mathbf{k}}(\mathbf{r}, t)}{dt} = \left[E_{\rm ex}(0) - \frac{\hbar^2}{2m_{\rm ex}}\Delta\right]a_{\mathbf{k}}(\mathbf{r}, t) + \frac{26\pi a_{\rm ex}^3}{3}\mathrm{Ry}_{\rm ex}\frac{|a_{\mathbf{k}}|^2}{V}a_{\mathbf{k}}(\mathbf{r}, t)$$

$$-\mathbf{d}^*\left[\mathbf{E}(\mathbf{r}, t) + \frac{4\pi}{3}\mathbf{P}(\mathbf{r}, t)\right]\left(1 - \frac{35\pi}{4}a_{\rm ex}^3\frac{|a_{\mathbf{k}}|^2}{V}\right),$$

$$\mathbf{P}(\mathbf{r}, t) = \frac{\mathbf{d}a_{\mathbf{k}}(\mathbf{r}, t)}{v_0\varepsilon_\infty}. \tag{7.28}$$

Equations (7.28) and (7.26) solve the question in view. One possible solution is the electrostatic case, when $\mathbf{H} = 0$ and $\{d\mathbf{D}(\mathbf{r}, t)/(dt)\} = 0$. Then Maxwell's equations become

$$\mathbf{k}\times\mathbf{E} = 0, \qquad \mathbf{k}\cdot(\mathbf{E} + 4\pi\mathbf{P}) = 0, \tag{7.29}$$

and their solution is

$$\mathbf{E} = -4\pi\frac{(\mathbf{P}\cdot\mathbf{k})\mathbf{k}}{k^2}. \tag{7.30}$$

It leads to the following values of the dipole–dipole interaction for two different orientations:

$$V(\mathbf{k}\parallel\mathbf{d}, \mathbf{d}, \mathbf{d}^*) = \frac{8\pi\,|\mathbf{d}|^2}{3v_0\varepsilon_\infty},$$

$$V(\mathbf{k}\perp\mathbf{d}, \mathbf{d}, \mathbf{d}^*) = -\frac{4\pi\,|\mathbf{d}|^2}{3v_0\varepsilon_\infty}, \tag{7.31}$$

which exhibit the nonanalytical behavior of the dipole–dipole interaction mentioned above. Here it is appropriate to make one remark. In the spin-oriented model we have chosen, the e–h pair has the spin structure $(e\uparrow, h\downarrow)$. Its wave function differs from the eigenfunctions with definite total spin S and its projection S_z. In a non-spin-oriented case, such eigenfunctions have the structures $1/\sqrt{2}[(e\uparrow, h\downarrow) \pm (e\downarrow, h\uparrow)]$. One of them, with the upper sign, corresponds to the paraexciton state with $S = 0$ and $S_z = 0$. The lower sign agrees with the orthoexciton state with $S = 1$ and $S_z = 0$. The exchange Coulomb interaction will be doubled for the paraexciton state and will vanish for all orthoexciton states.

For transverse coherent photons and excitons when $\mathbf{E}\parallel\mathbf{P}\parallel\mathbf{d}\perp\mathbf{k}$, the term

$$-\frac{4\pi}{3}\mathbf{d}\cdot\mathbf{P} = -\frac{4\pi\,|\mathbf{d}|^2}{3v_0\varepsilon_\infty}a_{\mathbf{k}} \tag{7.32}$$

can be added to the first term of Eqs. (7.28). It gives rise to the limiting frequency of the transverse exciton,

$$\hbar\omega_{\perp}(0) = E_{\text{ex}} - \frac{4\pi}{3v_0}\frac{|\mathbf{d}|^2}{\varepsilon_\infty}.$$

If one uses the condition $\nabla \cdot \mathbf{E} = 0$ and the equality $\nabla \times \nabla \times \mathbf{E} = -\Delta\mathbf{E} + \nabla(\nabla \cdot \mathbf{E})$, the two equations describing the coherent transverse excitons and photons will obtain the forms

$$i\hbar\frac{da_{\mathbf{k}}}{dt} = \left[\hbar\omega_{\perp}(0) - \frac{\hbar^2}{2m_{\text{ex}}}\Delta\right]a_{\mathbf{k}} + \frac{26\pi}{3}a_{\text{ex}}^3\text{Ry}_{\text{ex}}\frac{|a_{\mathbf{k}}|^2}{V}a_{\mathbf{k}} - \mathbf{d}\cdot\mathbf{E}\left(1 - \frac{35\pi}{4}a_{\text{ex}}^3\frac{|a_{\mathbf{k}}|^2}{V}\right),$$

$$\left(\frac{\varepsilon_\infty}{c_0^2}\frac{\partial^2}{\partial t^2} - \Delta\right)\mathbf{E} = -\frac{4\pi}{c_0^2}\frac{\mathbf{d}}{v_o}\frac{\partial^2 a_{\mathbf{k}}}{\partial t^2}. \tag{7.33}$$

The first of Eqs. (7.33) is the nonlinear Schrödinger equation describing the coherent transverse excitons. The second equation is the wave equation for the coherent electromagnetic field. These equations describe the amplitudes of the dipole-active transverse excitons and will be used in Chapter 9 to describe coherent nonlinear light propagation in crystals in the excitonic range of the spectrum.

Haken and Schenzle [8] pointed out that the theory of polaritons breaks down at high light intensities. They presented nonlinear equations for the exciton–photon coupling and a polaritonlike solution that may contain a macroscopically large, or “giant,” number of excitons bound to photons. Searching for solutions that resemble conventional polaritons, they treated the exciton and the photon operators as classical quantities. This approach is well justified for high field amplitudes. The dispersion relation for giant polaritons is power dependent. In the limit $N_{\mathbf{k}_0} \to 0$, which implies that $F_{\mathbf{k}_0} \to 0$, the conventional polariton dispersion relation in the rotating-wave approximation was obtained. Besides the plane-wave solutions, qualitatively new solutions were found, including undamped oscillations described by the elliptic functions and undistorted solitonlike solutions. These will be discussed in Chapter 9.

7.2.2 The Excitons as Simple Bosons

For simplicity in this subsection we will consider dipole-active excitons and photons with only one transverse polarization [43]. A second simplification is related to use of the exciton–photon interaction Hamiltonian in the rotating-wave approximation, in which the antiresonant terms are neglected. These terms are important for obtaining a correct polariton dispersion relation in the range of small wave vectors. In the range of wave vectors near the intersection of the photon and the exciton dispersion curves, the main role is played by the resonant terms. If the resonator Fabry–Perot or the λ cavity is tuned in such a way to maintain a high photon density, one can suppose conservation of the total number of excitons and photons. This assumption permits us to introduce into the Hamiltonian a unique chemical potential μ for excitons and photons. The Hamiltonian then has the form [43, 44]

$$H = \sum_{\mathbf{q}}(\Delta + T_{\mathbf{q}} - \mu)a_{\mathbf{q}}^\dagger a_{\mathbf{q}} + \sum_{\mathbf{q}}(\hbar cq - \mu)\,c_{\mathbf{q}}^\dagger c_{\mathbf{q}}$$
$$+ \sum_{\mathbf{q}}\varphi_{\mathbf{q}}(a_{\mathbf{q}}^\dagger c_{\mathbf{q}} - c_{\mathbf{q}}^\dagger a_{\mathbf{q}}) + \frac{1}{2V}\sum_{\mathbf{pqk}}U(k)a_{\mathbf{p}}^\dagger a_{\mathbf{q}}^\dagger a_{\mathbf{q}+\mathbf{k}}a_{\mathbf{p}-\mathbf{k}}. \tag{7.34}$$

Here $a_{\mathbf{q}}^{\dagger}$, $a_{\mathbf{q}}$ and $c_{\mathbf{q}}^{\dagger}$, $c_{\mathbf{q}}$ are Bose creation and the annihilation operators of exciton and photons, respectively. Δ is the exciton creation energy at the bottom of the exciton band. $\hbar\omega_{\perp}(q) = \Delta + T_q$, where T_q is the exciton kinetic energy and $U(k)$ is the coefficient of the exciton–exciton interaction. The coefficient of the exciton–photon interaction $\varphi_{\mathbf{q}}$ in model Hamiltonian (7.34) is assumed to be zero at the point $\mathbf{q} = 0$. In this model the longitudinal-transverse polariton splitting Δ_{LT} equals zero. The antiresonant terms are also taken into account below.

The choice of the optimum resonator mode will determine the wave vector of the Bose condensate. Bose condensation of excitons and photons with wave vector $\mathbf{k}_0$ can be taken into account by the Bogoliubov displacement procedure (introduced in Chapter 2) for the operators $a_{\mathbf{q}}$ and $c_{\mathbf{q}}$ as follows:

$$a_{\mathbf{q}} = \sqrt{N_{\mathbf{k}_0}} e^{-i\psi} \delta_{\mathbf{q},\mathbf{k}_0} + \alpha_{\mathbf{q}},$$
$$c_{\mathbf{q}} = \sqrt{F_{\mathbf{k}_0}} e^{-i\theta} \delta_{\mathbf{q},\mathbf{k}_0} + \xi_{\mathbf{q}}, \tag{7.35}$$

where the macroscopic amplitudes $N_{\mathbf{k}_0}$ and $F_{\mathbf{k}_0}$ are proportional to the volume V.

Up to terms quadratic in the operators $\alpha_{\mathbf{q}}^{\dagger}$, $\alpha_{\mathbf{q}}$, $\xi_{\mathbf{q}}^{\dagger}$, and $\xi_{\mathbf{q}}$, the transformed Hamiltonian has the form

$$H = E_0 + \sum_{\mathbf{q}} [\hbar\omega_{\perp}(q) + L_0 + L_{\mathbf{q}-\mathbf{k}_0} - \mu(k_0)]\alpha_{\mathbf{q}}^{\dagger}\alpha_{\mathbf{q}} + \sum_{\mathbf{q}} [\hbar c q - \mu(k_0)]\xi_{\mathbf{q}}^{\dagger}\xi_{\mathbf{q}}$$
$$+ \sum_{\mathbf{q}} \varphi_{\mathbf{q}}(\alpha_{\mathbf{q}}^{\dagger}\xi_{\mathbf{q}} - \xi_{\mathbf{q}}^{\dagger}\alpha_{\mathbf{q}}) + \sum_{\mathbf{q}} \frac{L_{\mathbf{q}-\mathbf{k}_0}}{2}\left(e^{2i\psi}\alpha_{\mathbf{q}}\alpha_{2\mathbf{k}_0-\mathbf{q}} + e^{-2i\psi}\alpha_{\mathbf{q}}^{\dagger}\alpha_{2\mathbf{k}_0-\mathbf{q}}^{\dagger}\right), \tag{7.36}$$

where

$$L_{\mathbf{q}} = U(q)\frac{N_{\mathbf{k}_0}}{V}. \tag{7.37}$$

The terms linear in the operators $\alpha_{\mathbf{k}_0}^{\dagger}$, $\alpha_{\mathbf{k}_0}$, $\xi_{\mathbf{k}_0}^{\dagger}$, and $\xi_{\mathbf{k}_0}$ that appear after the substitution of Eqs. (7.35) into Eq. (7.34) are canceled when the corresponding coefficients are set equal to zero, as follows:

$$[\hbar\omega_{\perp}(k_0) + L_0 - \mu(k_0)]\sqrt{N_{\mathbf{k}_0}}e^{i(\psi-\theta)} - i|\varphi_{\mathbf{k}_0}|\sqrt{F_{\mathbf{k}_0}} = 0,$$
$$[\hbar c k_0 - \mu(k_0)]\sqrt{F_{\mathbf{k}_0}} + i|\varphi_{\mathbf{k}_0}|\sqrt{N_{\mathbf{k}_0}}e^{i(\psi-\theta)} = 0. \tag{7.38}$$

Here it has been taken into account that $\varphi_{\mathbf{k}_0}$ is a purely imaginary number, i.e., $\varphi_{\mathbf{k}_0} = i|\varphi_{\mathbf{k}_0}|$. Equations (7.38) and their complex conjugates determine the dispersion relation $\mu(\mathbf{k}_0)$, the ratio $N_{\mathbf{k}_0}/F_{\mathbf{k}_0}$, and the difference $(\psi - \theta)$ of the two phases. The dispersion equation has the form

$$[\hbar c k_0 - \mu(k_0)]\,[\hbar\omega_{\perp}(k_0) + L_0 - \mu(k_0)] = |\varphi_{\mathbf{k}_0}|^2. \tag{7.39}$$

The dispersion relation represents two polaritonlike branches but with the polariton gap Δ_{LT} equal to zero. The upper $\mu_{\text{upb}}(\mathbf{k}_0)$ and lower $\mu_{\text{lpb}}(\mathbf{k}_0)$ polaritonlike branches are represented in Fig. 7.1(a). The phase difference $(\psi - \theta)$ equals $+\pi/2$ on the $\mu_{\text{lpb}}(\mathbf{k}_0)$ branch and $-\pi/2$ on the $\mu_{\text{upb}}(\mathbf{k}_0)$ branch. The ratio $N_{\mathbf{k}_0}/F_{\mathbf{k}_0}$ is

$$\frac{N_{\mathbf{k}_0}}{F_{\mathbf{k}_0}} = \frac{|\varphi_{\mathbf{k}_0}|^2}{[\hbar\omega_{\perp}(k_0) + L_0 - \mu(k_0)]^2} = \frac{[\hbar c k_0 - \mu(k_0)]^2}{|\varphi_{\mathbf{k}_0}|^2}. \tag{7.40}$$

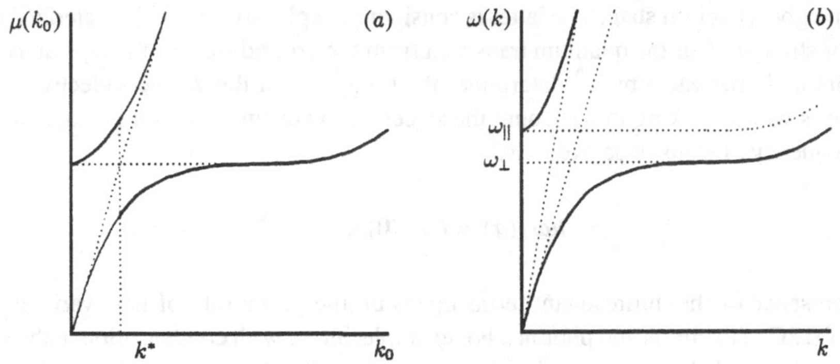

Figure 7.1. Frequency spectrum of the condensate-photon modes, (a) with the antiresonance terms neglected, and (b) with them taken into account [43, 44].

Taking into account these results, one can determine the ground-state energy E_0, which is

$$E_0 = [\hbar\omega_\perp(k_0) + L_0 - \mu(k_0)]N_{\mathbf{k}_0} + [\hbar c k_0 - \mu(k_0)]F_{\mathbf{k}_0} - \frac{1}{2}L_0 N_{\mathbf{k}_0}$$
$$+ \sqrt{F_{\mathbf{k}_0}N_{\mathbf{k}_0}}|\varphi_{\mathbf{k}_0}|i\left[e^{i(\psi-\theta)} - e^{-i(\psi-\theta)}\right] = -\frac{1}{2}L_0 N_{\mathbf{k}_0}. \tag{7.41}$$

When repulsion between the excitons prevails and $U(0) > 0$, minimal values of the ground-state energy E_0 are realized on one of the two branches of the chemical potential $\mu_{\mathrm{lpb}}(\mathbf{k}_0)$ and $\mu_{\mathrm{upb}}(\mathbf{k}_0)$, where the values $N_{\mathbf{k}_0}$ are maximal. When at a given $\mathbf{k}_0$ the exciton number $N_{\mathbf{k}_0}$ is greater than the photon number $F_{\mathbf{k}_0}$, these regions are called excitonlike regions and are represented in Fig. 7.1(a) by thick curves. On these thick branches, the ground-state energy with a given value $\mathbf{k}_0$ obtains a minimal value. The influence of the antiresonant terms can be taken into account by use of the exciton–photon interaction Hamiltonian proposed by Hopfield [45]. Instead of Eq. (7.34), the new Hamiltonian describing the exciton–photon and the exciton–exciton interactions has the form

$$H = \sum_{\mathbf{q}} \hbar\omega_\perp(q)a_{\mathbf{q}}^\dagger a_{\mathbf{q}} + \sum_{\mathbf{q}}(\hbar c q + \psi_{\mathbf{q}})c_{\mathbf{q}}^\dagger c_{\mathbf{q}} + \frac{1}{2}\sum_{\mathbf{q}} \psi_{\mathbf{q}}(c_{\mathbf{q}}c_{-\mathbf{q}} + c_{\mathbf{q}}^\dagger c_{-\mathbf{q}}^\dagger)$$
$$+ \sum_{\mathbf{q}} \varphi_{\mathbf{q}}(a_{\mathbf{q}}^\dagger - a_{-\mathbf{q}})(c_{\mathbf{q}} + c_{-\mathbf{q}}^\dagger) + \frac{1}{2V}\sum_{\mathbf{pqk}} U(k)a_{\mathbf{p}}^\dagger a_{\mathbf{q}}^\dagger a_{\mathbf{q}+\mathbf{k}}a_{\mathbf{p}-\mathbf{k}}. \tag{7.42}$$

The sum is taken over all $\mathbf{q}$ space and over only one polarization perpendicular to $\mathbf{q}$. The interaction constants $\psi_{\mathbf{q}}$ and $\varphi_{\mathbf{q}}$ are

$$\psi_{\mathbf{q}} = \frac{f(\hbar\omega_p)^2}{2\hbar c q}, \qquad \varphi_{\mathbf{q}} = \frac{i}{2}\sqrt{f}\hbar\omega_p\sqrt{\frac{\omega_\perp(q)}{cq}},$$

with

$$\omega_p = \frac{\omega_p^0}{\sqrt{\varepsilon_\infty}}, \qquad c = \frac{c_0}{\sqrt{\varepsilon_\infty}}, \qquad \omega_p^{02} = \frac{4\pi e^2}{m_0 v_0}. \tag{7.43}$$

Here the plasmon frequency ω_p and the light velocity in the crystal c are less than ω_p^0 and c_0 by the value $\sqrt{\varepsilon_\infty}$. The dielectric constant ε_∞ is determined by all the possible exciton states except the given exciton state. The latter is considered explicitly and is characterized by the oscillator strength f of the quantum transition from the ground state of the crystal. c_0 is the light velocity in the vacuum. ω_p^0 determines the frequency of the valence–electron-plasma oscillations without taking into account the screening by excited exciton states, and $\omega_\perp(q)$ is the frequency of transverse excitons:

$$\hbar\omega_\perp(q) = \hbar\omega_\perp(0) + T_{\mathbf{q}}.$$

The presence of the antiresonant terms opens up the possibility of new variants in the theory of BEC of excitons and photons. For example, along with condensation in the single-particle exciton and photon states with wave vectors $\mathbf{k}$, BEC also occurs in the single-particle states with the opposite wave vector $-\mathbf{k}$. In other words, backwards-running waves exist simultaneously with the forward-running coherent exciton and photon waves. Together, these form standing waves that are characterized by wave vector $\pm\mathbf{k}$. Their appearance is a linear optical effect and is determined by the antiresonant terms in the exciton–photon interaction in (7.42).

The nonlinearity in this system is due to the exciton–exciton interaction. It generates higher-order harmonics of the condensate standing waves, which are characterized by the wave vectors $\pm 3\mathbf{k}$, $\pm 5\mathbf{k}$, and so on. For example, consider the product of four operators $a^\dagger_{\mathbf{p}} a^\dagger_{\mathbf{q}} a_{\mathbf{q}+\mathbf{k}'} a_{\mathbf{p}-\mathbf{k}'}$, which is present in Hamiltonian (7.42). If one supposes that the two creation operators have the same wave vectors as one of the condensate waves, for example $\mathbf{p} = \mathbf{q} = \pm\mathbf{k}$, and at the same time one of the annihilation operators has the wave vector $\mathbf{q} + \mathbf{k}' = \mp\mathbf{k}$, then the difference wave vector $\mathbf{p} - \mathbf{k}'$ will be equal to $\pm 3\mathbf{k}$. Because the three first operators are macroscopically large numbers by supposition, this term of the Hamiltonian will induce the appearance of a macroscopically large occupation number of the single-particle exciton states with wave vectors $\pm 3\mathbf{k}$. The existence of the Bose condensates with wave vectors $\pm\mathbf{k}$ and $\pm 3\mathbf{k}$ induces, in its turn, the appearance of the Bose condensates with wave vectors $\pm 5\mathbf{k}$, etc. The existence of the higher-order exciton-condensate harmonics gives rise inevitably to the appearance of the higher-order photon-condensate harmonics. This fact significantly complicates the determination of the condensate-photon mode parameters.

Below only the first-order harmonic with wave vectors $\pm\mathbf{k}$ are kept. With the higher-order harmonics neglected, the terms in Hamiltonian (7.42) with the greatest magnitude are

$$\begin{aligned}
&\hbar\omega_\perp(k)(a^\dagger_{\mathbf{k}} a_{\mathbf{k}} + a^\dagger_{-\mathbf{k}} a_{-\mathbf{k}}) + (\hbar ck + \psi_{\mathbf{k}})(c^\dagger_{\mathbf{k}} c_{\mathbf{k}} + c^\dagger_{-\mathbf{k}} c_{-\mathbf{k}}) + \psi_{\mathbf{k}}(c_{\mathbf{k}} c_{-\mathbf{k}} + c^\dagger_{\mathbf{k}} c^\dagger_{-\mathbf{k}}) \\
&+ \varphi_{\mathbf{k}}[(c_{\mathbf{k}} + c^\dagger_{-\mathbf{k}})(a^\dagger_{\mathbf{k}} - a_{-\mathbf{k}}) + (c_{-\mathbf{k}} + c^\dagger_{\mathbf{k}})(a^\dagger_{-\mathbf{k}} - a_{\mathbf{k}})] \\
&+ \frac{U(0)}{2V}(a^\dagger_{\mathbf{k}} a^\dagger_{\mathbf{k}} a_{\mathbf{k}} a_{\mathbf{k}} + a^\dagger_{-\mathbf{k}} a^\dagger_{-\mathbf{k}} a_{-\mathbf{k}} a_{-\mathbf{k}}) + \frac{[U(0) + U(2k)]}{V} a^\dagger_{\mathbf{k}} a^\dagger_{-\mathbf{k}} a_{-\mathbf{k}} a_{\mathbf{k}}.
\end{aligned} \tag{7.44}$$

The motion equations for the operators describing the condensed excitons and photons can be represented in the form

$$i\hbar\frac{\mathrm{d}}{\mathrm{d}t}\hat{A}_{\mathbf{k}} = \hat{L}_{\mathbf{k}}\hat{A}_{\mathbf{k}}, \tag{7.45}$$

where the 4 × 4 matrices are

$$\hat{A}_{\mathbf{k}} = \begin{vmatrix} (a_{\mathbf{k}} + a^{\dagger}_{-\mathbf{k}}) & 0 & 0 & 0 \\ (a_{\mathbf{k}} - a^{\dagger}_{-\mathbf{k}}) & 0 & 0 & 0 \\ (c_{\mathbf{k}} + c^{\dagger}_{-\mathbf{k}}) & 0 & 0 & 0 \\ (c_{\mathbf{k}} - c^{\dagger}_{-\mathbf{k}}) & 0 & 0 & 0 \end{vmatrix},$$

$$\hat{L}_{\mathbf{k}} = \begin{vmatrix} 0 & (\hbar\omega_{\perp}(k) + \mathrm{L}_{\mathbf{k}}) & 2\varphi_{\mathbf{k}} & 0 \\ [\hbar\omega_{\perp}(k) + \mathrm{L}_{\mathbf{k}}] & 0 & 0 & 0 \\ 0 & 0 & 0 & \hbar ck \\ 0 & -2\varphi_{\mathbf{k}} & (\hbar ck + 2\psi_{\mathbf{k}}) & 0 \end{vmatrix},$$

$$L_{\mathbf{k}} = \frac{N_{\mathbf{k}}}{V}\,[2U(0) + U(2k)]\,, \qquad \hbar\tilde{\omega}_{\perp}(k) = \hbar\omega_{\perp}(k) + L_{\mathbf{k}}. \tag{7.46}$$

In the stationary state, when

$$i\hbar\frac{\mathrm{d}}{\mathrm{d}t}\hat{A}_{\mathbf{k}} = \hbar\omega\hat{A}_{\mathbf{k}}, \tag{7.47}$$

the dispersion equation has the form

$$\omega^4 - \omega^2\left[\tilde{\omega}_{\perp}^2(k) + (ck)^2 + f\omega_p^2\right] + \tilde{\omega}_{\perp}^2(k)(ck)^2 = 0. \tag{7.48}$$

Its solution gives the dispersion relation

$$\omega^2 = \frac{\tilde{\omega}_{\perp}^2(k) + (ck)^2 + f\omega_p^2}{2} \pm \frac{1}{2}\sqrt{\left[\tilde{\omega}_{\perp}^2(k) - (ck)^2 - f\omega_p^2\right]^2 + 4\tilde{\omega}_{\perp}^2(k) f\omega_p^2}, \tag{7.49}$$

which is illustrated in Fig. 7.1(b). The two branches coincide completely with the polariton dispersion curves.

7.3 The Energy Spectrum of the Noncondensed Excitons and Photons in the Presence of a Condensate

7.3.1 The Absolute and Convective Instabilities. Sturrock's Rules

We now mention some elements of instability theory. These are discussed in detail in the basic papers by Sturrock [46] and Arecchi, Masserini, and Schwendimann [47], as well as in the textbooks, Refs. 48 and 49. This preliminary knowledge is necessary to understand better the properties of the energy spectrum of the elementary excitations in the presence of the Bose condensate formed by the dipole-active excitons and photons.

Following the procedure of Ref. 48, the main rules can be demonstrated in the simple example of two initial branches of the spectrum $\omega_1 = v_1 k$ and $\omega_2 = v_2 k$, which interact weakly between themselves with the interaction constant m. The dispersion equation for the interacting branches has the form

$$(\omega - v_1 k)(\omega - v_2 k) - m = 0. \tag{7.50}$$

It can be solved for ω as a function of k or for k as a function of ω. The two solutions, $\omega(k)$ and $k(\omega)$, are

$$\omega_{1,2}(k) = \frac{(v_1 + v_2)k}{2} \pm \frac{1}{2}\sqrt{(v_1 - v_2)^2 k^2 + 4m},$$

$$k_{1,2}(\omega) = \frac{(v_1 + v_2)\omega}{2v_1 v_2} \pm \frac{1}{2v_1 v_2}\sqrt{(v_1 - v_2)^2 \omega^2 + 4m v_1 v_2}. \tag{7.51}$$

The character of these solutions depends on the signs of the interaction constant m and of the product $v_1 v_2$ of the two initial velocities.

One can consider four different cases,

$$\begin{aligned}
&(a) \quad m > 0, \quad v_1 v_2 > 0, \quad m v_1 v_2 > 0;\\
&(b) \quad m > 0, \quad v_1 v_2 < 0, \quad m v_1 v_2 < 0;\\
&(c) \quad m < 0, \quad v_1 v_2 > 0, \quad m v_1 v_2 < 0;\\
&(d) \quad m < 0, \quad v_1 v_2 < 0, \quad m v_1 v_2 > 0.
\end{aligned} \tag{7.52}$$

In cases (a) and (b), the interaction constant m is positive and so is the expression under the square root in the functions $\omega_{1,2}(k)$. Then for all the real values of k, the eigenfrequencies of the system are real. No nonzero imaginary parts of the solutions of the type Im $\omega(k) \neq 0$ and no instabilities appear–the system is stable. These two cases are represented in Figs. 7.2(a) and 7.2(b). The velocities with the same signs are directed upwards, but those with different signs are directed in the following way: one of them upwards and the other downwards. There is one difference between these cases in regard to the properties of the corresponding functions $k_{1,2}(\omega)$. In case (a) the functions $k_{1,2}(\omega)$ are real for all real values of ω. For this reason simple stable waves can exist in the system and can propagate without amplification for all values of ω. In case (b) there are some values of ω in the frequency gap

$$0 \leq \omega^2 \leq \frac{4\,|m v_1 v_2|}{(v_1 - v_2)^2}$$

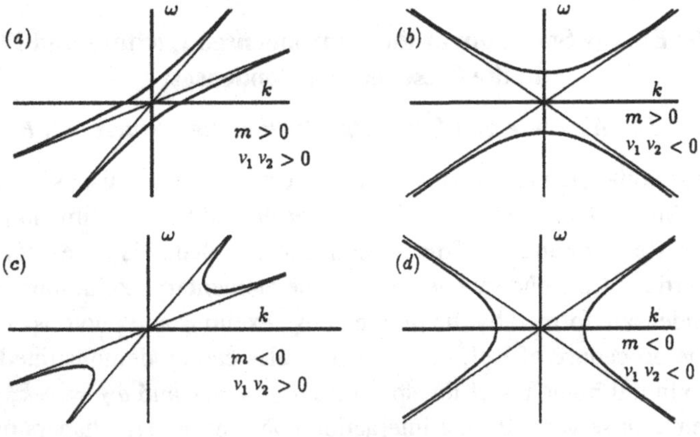

Figure 7.2. The dispersion curves of two weakly interacting initial branches. The initial branches $\omega_1 = v_1 k$ and $\omega_2 = v_2 k$ are represented by thin lines. The resulting frequency spectra are represented by thick curves (from Ref. 48).

for which the functions $k_{1,2}(\omega)$ are complex. When k is complex, the function e^{ikx} becomes an increasing or decreasing function of x. But amplification of the wave amplitude must be excluded because the system is stable. The only possible variant remains the decreasing amplitude of the evanescent wave. In case (b) the system is stable, but evanescent waves can exist in some frequency regions.

In cases (c) and (d), one deals with interaction constants $m < 0$. The functions $\omega_{1,2}(k)$ become complex in the region of the real values of wave numbers k determined by the relation

$$(v_1 - v_2)^2 k^2 \leq 4\,|m|\,.$$

One of two functions $\omega_{1,2}(k)$ has a positive imaginary part $\mathrm{Im}\ \omega(k) > 0$ in this region. This fact leads to an unlimited increase of the corresponding wave amplitude $e^{-i\omega t}$ in the limit $t \to \infty$. It shows that the system is unstable in both cases (c) and (d) in some regions of the real wave numbers k. Between them there also exists a difference, as in cases (a) and (b) discussed above, namely, in case (c) the functions $k_{1,2}(\omega)$ are complex functions for some real values of ω. In case (c), unlike case (b), amplification of the propagating wave amplitude is possible. The amplification takes place at the expense of the energy supply in the system. This instability is closely related to the amplification of the propagating wave and is called "convective." In case (d), the increase of the wave amplitude is not related to its propagation because the functions $k_{1,2}(\omega)$ are real and the factor $e^{ik(\omega)x}$ remains restricted. The unrestricted growth of the disturbance appears because of the evolution of the amplitude in time, which does not depend on the wave propagation. This type of instability is called "nonconvective" or "absolute." The two cases of convective and absolute instabilities are represented in Figs. 7.2(c) and 7.2(d).

Some concluding remarks of Lifshitz and Pitaevskii are instructive [48]. If the dispersion equation has roots in the upper part of the complex ω plane, i.e., if $\mathrm{Im}\,\omega$ is positive, then a small initial perturbation in the form of a plane wave will increase, which means that the system is unstable with respect to such perturbations. In reality, any initial perturbation represents a wave packet with a finite size in space. Eventually the packet propagates, spreads, and increases in amplitude in an unstable system. Here there are two possibilities. In the former case, in spite of the wave-packet displacement, the perturbation continues to grow in all the places where it arises. In the latter case, the amplification is accompanied by the propagating wave packet, which takes with it the growing perturbation. This difference is relative, as pointed out in Ref. 48, because the character of the instability depends on the reference frame. If one switches to the reference frame moving with the growing wave packet, then the convective instability is transformed into an absolute instability, and vice versa.

Some of Sturrock's rules are comprehensive [46]. If ω is real for all real k, then any complex k for real ω denotes an evanescent wave. Conversely, if k is real for all real ω, then any complex ω for real k denotes nonconvective instability. If the function $k(\omega)$ is complex for some values ω of interest, there are two possibilities: either the wave is evanescent or the wave is amplified. If a system is unstable and the growing disturbance is propagated away from the point of origin, this instability is convective. If the disturbance grows in amplitude and in extent, but the instability embraces the original point of origin, this instability is called nonconvective or absolute. These types of intersections and instabilities will be met frequently below.

7.3.2 *Energy Spectrum of the Elementary Excitations in the Presence of the Condensate-Photon Mode*

Our aim in this subsection is to describe in a single formalism the energy spectrum of the elementary excitations for both cases of spontaneous and induced BEC of dipole-active excitons and photons. These results were obtained by Moskalenko et al. [43], as well as by Misko, Moskalenko, and Shmiglyuk [50].

In this approach, the formulas for the energy spectrum $\omega(q)$ are the same in both cases, with one exception. In the case of a spontaneous appearance of BEC, they will contain the chemical potential $\mu(k_0)$, while in the case of BEC induced by external laser radiation, they will contain the laser frequency $\omega_L = ck_0$. Another parameter of the theory is the number $N_{\mathbf{k}_0}$ of the coherent excitons. For the spontaneous BEC, this is present as a free parameter and is related to the number of coherent photons $F_{\mathbf{k}_0}$ by formula (7.40). For the induced BEC, the number $F_{\mathbf{k}_0}$ of the laser beam photons determines the number $N_{\mathbf{k}_0}$ through relation (6.63).

The subject of this subsection has much in common with the determination of the energy spectrum of a semiconductor with a saturated interband transition [51]. In contrast, the saturation of the exciton band leads to Bose degeneracy and to the formation of a Bose condensate. The level of the chemical potential plays the role of the reference level from which the levels of the noncondensate quasiparticles are measured. Exactly the same role is played by the laser frequency ω_L when the Hamiltonian is written in the reference frame rotating with this frequency.

To determine the energy spectrum of the elementary excitations, it is necessary to diagonalize the quadratic Hamiltonian (7.36). The equation for the operators $\alpha^{\dagger}_{\mathbf{k}_0\pm\mathbf{q}}$, $\alpha_{\mathbf{k}_0\pm\mathbf{q}}$, $c^{\dagger}_{\mathbf{k}_0\pm\mathbf{q}}$, and $c_{\mathbf{k}_0\pm\mathbf{q}}$ are represented in a matrix form,

$$i\frac{\mathrm{d}}{\mathrm{d}t}\begin{vmatrix}\alpha_{\mathbf{k}_0+\mathbf{q}}\\ \alpha^{\dagger}_{\mathbf{k}_0-\mathbf{q}}\\ c_{\mathbf{k}_0+\mathbf{q}}\\ c^{\dagger}_{\mathbf{k}_0-\mathbf{q}}\end{vmatrix} = \hat{L}_{\mathbf{q}}\begin{vmatrix}\alpha_{\mathbf{k}_0+\mathbf{q}}\\ \alpha^{\dagger}_{\mathbf{k}_0-\mathbf{q}}\\ c_{\mathbf{k}_0+\mathbf{q}}\\ c^{\dagger}_{\mathbf{k}_0-\mathbf{q}}\end{vmatrix} = \hat{L}_{\mathbf{q}}\hat{A}_{\mathbf{q}}. \tag{7.53}$$

The 4×4 matrix $\hat{L}_{\mathbf{q}}$ is

$$\hat{L}_{\mathbf{q}} = \begin{vmatrix}\sigma_{\mathbf{q}}(+) & l_{\mathbf{q}}e^{2i\psi} & \chi_{\mathbf{q}}(+) & 0\\ -l_{\mathbf{q}}e^{-2i\psi} & -\sigma_{\mathbf{q}}(-) & 0 & \chi_{\mathbf{q}}(-)\\ -\chi_{\mathbf{q}}(+) & 0 & \Omega_{\mathbf{q}}(+) & 0\\ 0 & -\chi_{\mathbf{q}}(-) & 0 & -\Omega_{\mathbf{q}}(-)\end{vmatrix}. \tag{7.54}$$

Here the following notations were used:

$$\sigma_{\mathbf{q}}(\pm) = \omega_{\perp}(\mathbf{k}_0 \pm \mathbf{q}) + l_0 + l_{\mathbf{q}} - \mu(k_0)/\hbar,$$
$$\omega_{\mathbf{q}}(\pm) = c|\mathbf{k}_0 \pm \mathbf{q}| - \mu(k_0)/\hbar,$$
$$l_{\mathbf{q}} = L_{\mathbf{q}}/\hbar, \quad \chi_{\mathbf{q}}(\pm) = \varphi_{\mathbf{k}_0\pm\mathbf{q}}/\hbar. \tag{7.55}$$

$\mu(\mathbf{k}_0)/\hbar$ will be substituted sometimes by ω_L. The stationary solution of the motion equation

$$i\frac{d}{dt}\hat{A}_{\mathbf{q}} = \omega\hat{A}_{\mathbf{q}}$$

gives the energy spectrum $\omega(\mathbf{q})$. This function obeys the fourth-order algebraic equation

$$\omega^4 + A_{\mathbf{q}}\omega^3 + B_{\mathbf{q}}\omega^2 + C_{\mathbf{q}}\omega + D_{\mathbf{q}} = 0. \tag{7.56}$$

The coefficients $A_{\mathbf{q}}$, $B_{\mathbf{q}}$, $C_{\mathbf{q}}$, and $D_{\mathbf{q}}$ are given by the following expressions (in which the index $\mathbf{q}$ is omitted):

$$\begin{aligned}
A &= \Omega(-) - \Omega(+) + \sigma(-) - \sigma(+),\\
B &= l^2 - \sigma(+)\sigma(-) - [|\chi(+)|^2 + |\chi(-)|^2]\\
&\quad - \Omega(+)\Omega(-) + [\sigma(-) - \sigma(+)]\,[\Omega(-) - \Omega(+)]\,,\\
C &= [\Omega(+) - \Omega(-)]\,\sigma(+)\sigma(-) + [\sigma(+) - \sigma(-)]\,\Omega(+)\Omega(-) - |\chi(+)|^2\,[\sigma(-)\\
&\quad + \Omega(-)] + |\chi(-)|^2\,[\sigma(+) + \Omega(+)] + l^2\,[\Omega(-) - \Omega(+)]\,,\\
D &= -\,|\chi(-)|^2\,\sigma(+)\Omega(+) - |\chi(+)|^2\,\sigma(-)\Omega(-)\\
&\quad + |\chi(-)|^2\,|\chi(+)|^2 - l^2\Omega(+)\Omega(-) + \sigma(+)\sigma(-)\Omega(+)\Omega(-).
\end{aligned}$$

Here the equality $\chi^2(\pm) = -|\chi(\pm)|^2$ was taken into account. In the general case of an arbitrarily oriented wave vector $\mathbf{q}$ relative to the wave vector $\mathbf{k}_0$, it is impossible to find an analytical expression of the desired energy spectrum. Such possibilities appear only in the particular cases when $\mathbf{k}_0 = 0$ or when $\mathbf{k}_0 \neq 0$, but $\mathbf{q} \perp \mathbf{k}_0$. Then the following simplifications appear:

$$\begin{aligned}
&\Omega_{\mathbf{q}}(+) = \Omega_{\mathbf{q}}(-) = \Omega_{\mathbf{q}}, \qquad \sigma_{\mathbf{q}}(+) = \sigma_{\mathbf{q}}(-) = \sigma_{\mathbf{q}},\\
&\chi_{\mathbf{q}}(+) = \chi_{\mathbf{q}}(-) = \chi_{\mathbf{q}}, \qquad A_{\mathbf{q}} = C_{\mathbf{q}} = 0.
\end{aligned} \tag{7.57}$$

They permit us to reduce Eq. (7.56) to a biquadratic one with the solutions

$$\begin{aligned}
\omega_{1,2}^2(q) = \omega_{\pm}^2(q) &= \frac{\sigma_{\mathbf{q}}^2 - l_{\mathbf{q}}^2 + \Omega_{\mathbf{q}}^2 + 2|\chi_{\mathbf{q}}|^2}{2}\\
&\quad \pm \frac{1}{2}\sqrt{\left(\sigma_{\mathbf{q}}^2 - l_{\mathbf{q}}^2 - \Omega_{\mathbf{q}}^2\right)^2 + 4|\chi_{\mathbf{q}}|^2[(\sigma_{\mathbf{q}} + \Omega_{\mathbf{q}})^2 - l_{\mathbf{q}}^2]}.
\end{aligned} \tag{7.58}$$

But the introduction of the chemical potential μ in Hamiltonian (7.34) is not obvious. It is of interest to confirm result (7.58) by an independent method by use of Hamiltonian (7.34) without the introduction of μ. Here we demonstrate this for the simpler case $\mathbf{k}_0 = 0$, when the exciton operators $a_0^\dagger$ and a_0 are macroscopically large compared with the operators $a_{\mathbf{q}}^\dagger$,

$a_{\mathbf{q}}$, $c^{\dagger}_{\mathbf{q}}$, and $c_{\mathbf{q}}$, where $\mathbf{q} \neq 0$. At the same time, because of the lack of photons in the state $\mathbf{k}_0 = 0$, the operators $c^{\dagger}_0$ and c_0 are negligibly small.

By using the perturbation theory of the previous chapters, one can obtain the motion equation for the operator a_0. It has the form

$$i\hbar \frac{\mathrm{d}a_0}{\mathrm{d}t} = (\Delta + L_0)\, a_0, \qquad L_0 = U(0)\frac{N_0}{V},$$
$$N_0 = a^{\dagger}_0 a_0 \approx a_0 a^{\dagger}_0. \tag{7.59}$$

It follows from Eqs. (7.59) that the products of two operators of the type a_0^2 and $a_0^{\dagger 2}$ are time dependent. The motion equations for the operators $a_{\mathbf{q}}$ and $c_{\mathbf{q}}$, as well as for the complexes of operators $[(a_0^2 a^{\dagger}_{-\mathbf{q}})/N_0]$ and $[(a_0^2 c^{\dagger}_{-\mathbf{q}})/N_0]$ are

$$i\hbar \frac{\mathrm{d}a_{\mathbf{q}}}{\mathrm{d}t} = (\Delta + L_0 + T_{\mathbf{q}} + L_{\mathbf{q}})a_{\mathbf{q}} + L_{\mathbf{q}} \frac{a_0^2 a^{\dagger}_{-\mathbf{q}}}{N_0} - \varphi_{\mathbf{q}} c_{\mathbf{q}},$$
$$i\hbar \frac{\mathrm{d}c_{\mathbf{q}}}{\mathrm{d}t} = \hbar c q c_{\mathbf{q}} + \varphi_{\mathbf{q}} a_{\mathbf{q}},$$
$$i\hbar \frac{\mathrm{d}}{\mathrm{d}t}\left(\frac{a_0^2 a^{\dagger}_{-\mathbf{q}}}{N_0}\right) = (\Delta + L_0 - T_{\mathbf{q}} - L_{\mathbf{q}})\left(\frac{a_0^2 a^{\dagger}_{-\mathbf{q}}}{N_0}\right) - L_{\mathbf{q}} a_{\mathbf{q}} - \varphi_{\mathbf{q}}\left(\frac{a_0^2 c^{\dagger}_{-\mathbf{q}}}{N_0}\right),$$
$$i\hbar \frac{\mathrm{d}}{\mathrm{d}t}\left(\frac{a_0^2 c^{\dagger}_{-\mathbf{q}}}{N_0}\right) = [2(\Delta + L_0) - \hbar c q]\left(\frac{a_0^2 c^{\dagger}_{-\mathbf{q}}}{N_0}\right) + \varphi_{\mathbf{q}}\left(\frac{a_0^2 a^{\dagger}_{-\mathbf{q}}}{N_0}\right). \tag{7.60}$$

The appropriate Green's functions are

$$G_{11}(\mathbf{q}) = -i\langle T[a_{\mathbf{q}}(t) a^{\dagger}_{\mathbf{q}}(0)]\rangle,$$
$$G_{21}(\mathbf{q}) = -i\langle T[c_{\mathbf{q}}(t) a^{\dagger}_{\mathbf{q}}(0)]\rangle,$$
$$G_{31}(\mathbf{q}) = -i\left\langle T\left[\frac{a_0^2(t) a^{\dagger}_{-\mathbf{q}}(t)}{N_0} a^{\dagger}_{\mathbf{q}}(0)\right]\right\rangle,$$
$$G_{41}(\mathbf{q}) = -i\left\langle T\left[\frac{a_0^2(t) c^{\dagger}_{-\mathbf{q}}(t)}{N_0} a^{\dagger}_{\mathbf{q}}(0)\right]\right\rangle. \tag{7.61}$$

Their Fourier transforms $G_{i1}(q, E)$, where $i = 1, 2, 3, 4$, obey the equations

$$EG_{11}(\mathbf{q}, E) = \frac{1}{2\pi} + (\Delta + L_0 + L_{\mathbf{q}} + T_{\mathbf{q}})G_{11}(\mathbf{q}, E) + L_{\mathbf{q}} G_{31}(\mathbf{q}, E) - \varphi_{\mathbf{q}} G_{21}(\mathbf{q}, E),$$
$$EG_{21}(\mathbf{q}, E) = \hbar c q G_{21}(\mathbf{q}, E) + \varphi_{\mathbf{q}} G_{11}(\mathbf{q}, E),$$
$$EG_{31}(\mathbf{q}, E) = (\Delta + L_0 - L_{\mathbf{q}} - T_{\mathbf{q}})G_{31}(\mathbf{q}, E) - L_{\mathbf{q}} G_{11}(\mathbf{q}, E) - \varphi_{\mathbf{q}} G_{41}(\mathbf{q}, E),$$
$$EG_{41}(\mathbf{q}, E) = [2(\Delta + L_0) - \hbar c q]\, G_{41}(\mathbf{q}, E) + \varphi_{\mathbf{q}} G_{31}(\mathbf{q}, E). \tag{7.62}$$

The energy spectrum can be determined by the solution of the determinant equation:

$$\begin{vmatrix} E-\Delta-L_0-T_{\mathbf{q}}-L_{\mathbf{q}} & -L_{\mathbf{q}} & \varphi_{\mathbf{q}} & 0 \\ -L_{\mathbf{q}} & -E+\Delta+L_0-T_{\mathbf{q}}-L_{\mathbf{q}} & 0 & -\varphi_{\mathbf{q}} \\ -\varphi_{\mathbf{q}} & 0 & E-\hbar cq & 0 \\ 0 & \varphi_{\mathbf{q}} & 0 & -E+2(\Delta+L_0)-\hbar cq \end{vmatrix} = 0.$$

If one introduces the designation $\mu(0) = \Delta + L_0$ and

$$E = \mu(0) + \hbar\omega(q),$$

the expression for $\omega(q)$ will be exactly the same as in formula (7.58) when $\mathbf{k}_0 = 0$,

$$\hbar\omega(q) = \Bigg[\frac{1}{2}\{T_{\mathbf{q}}^2 + 2T_{\mathbf{q}}L_{\mathbf{q}} + [\hbar cq - \mu(0)]^2 + 2|\varphi_{\mathbf{q}}|^2\} \pm \frac{1}{2}\big(\{T_{\mathbf{q}}^2 + 2T_{\mathbf{q}}L_{\mathbf{q}} - [\hbar cq - \mu(0)]^2\}^2 + 4|\varphi_{\mathbf{q}}|^2\{[\hbar cq - \mu(0) + T_{\mathbf{q}} + L_{\mathbf{q}}]^2 - L_{\mathbf{q}}^2\}\big)^{\frac{1}{2}}\Bigg]^{\frac{1}{2}}. \quad (7.63)$$

The same method can be used when the condensation takes place in states of dipole-active excitons and photons with $\mathbf{k}_0 \neq 0$.

We now discuss the energy spectrum in the different cases. First, we consider spontaneous BEC in the exciton state with wave vector $\mathbf{k}_0 = 0$ when the chemical potential is $\mu(0) = \hbar\omega_\perp(0) + L_0$. The energy spectrum is shown in Fig. 7.3(a) for the case of no exciton–photon interaction. When the exciton–photon interaction is taken into account, the spectrum changes significantly, as one can see in Fig. 7.3(b). The dotted lines show, as in Fig. 7.3(a), the functions $(T_{\mathbf{q}}^2 + 2T_{\mathbf{q}}L_{\mathbf{q}})^{1/2}$ and $|\hbar cq - \mu(0)|$, which intersect at the points $q_\pm$:

$$\hbar cq_\pm = \mu(0) \pm \left(T_{\mathbf{q}_\pm}^2 + 2T_{\mathbf{q}_\pm}L_{\mathbf{q}_\pm}\right)^{1/2}.$$

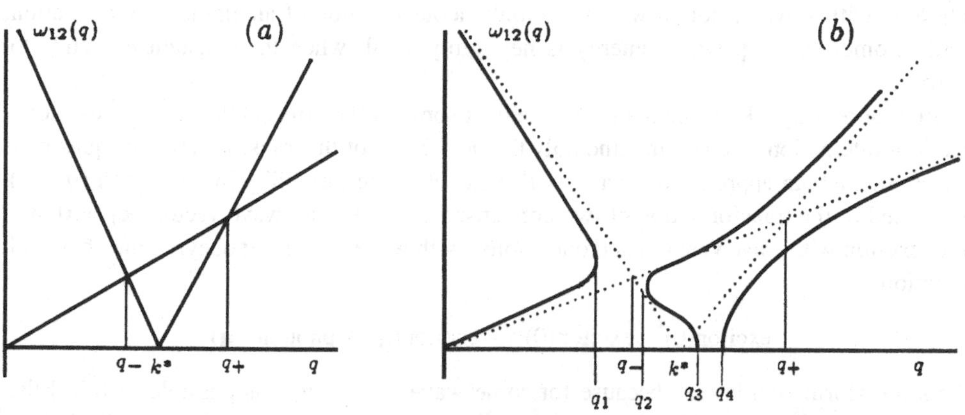

Figure 7.3. The spectrum of the elementary excitations at the BEC of excitons into a state with vector $\mathbf{k}_0 = 0$; (a) without exciton–photon interaction, (b) with this interaction taken into account when $q_+ > q_4$ [43, 44].

In the vicinity of these points the most important changes of the spectrum occur. Let us consider the neighborhood of the point q_-, the boundaries of which are determined by the limit wave numbers q_1 and q_2. At the limit points, the expression in the interior square root of Eq. (7.63) becomes zero and the two branches of the spectrum coincide,

$$\omega_1^2(q_i) = \omega_2^2(q_i); \quad i = 1, 2.$$

In the small region $q_1 < q < q_2$, the energy spectrum is complex and the system is unstable. This is to be expected, since for $\hbar cq < \mu(0)$ the branch $|\hbar cq - \mu(0)| = \mu(0) - \hbar cq$ has negative group velocity, whereas the group velocity of the branch $(T_{\mathbf{q}}^2 + T_{\mathbf{q}} L_{\mathbf{q}})^{1/2}$ is positive. It is a case similar to that shown in Fig. 7.2(d). In this region, the condensed excitons with $\mathbf{k}_0 = 0$ can be transformed into photons with wave vector $+\mathbf{q}$, emitting sound vibration quanta ("hydrons") with wave vector $-\mathbf{q}$. For such a process, the energy and the momentum conservation laws are fulfilled simultaneously. As mentioned above, this type of instability is called an "absolute" or "nonconvective" instability. The perturbation remains in the same region where it appeared, growing slowly to include more of the crystal. Its amplitude increases in time as long as the damping in the system is not important. Thus the system of Bose condensed excitons can simultaneously generate both light and sound with macroscopic amplitudes, i.e., it can operate as a laser–phaser.

In the region q_+ the group velocities of the initial waves have the same signs. Here the condensed excitons cannot be transformed into photons and the system is stable, similar to the case of Fig. 7.2(a). The elementary excitation spectrum consists of two simple waves. At the points q_3 and q_4, defined by the conditions

$$\hbar c q_3 = \mu(0) + \frac{|\varphi_{\mathbf{q}_3}|^2}{(T_{\mathbf{q}_3} + 2L_{\mathbf{q}_3})}, \qquad \hbar c q_{q_4} = \mu(0) + \frac{|\varphi_{\mathbf{q}_4}|^2}{T_{\mathbf{q}_4}},$$

the lower branch $\omega_-(q)$, corresponding to a minus sign before the interior square root in Eq. (7.63), equals zero. In the region $q_3 < q < q_4$ this branch does not exist. This absolute instability is associated with the possibility of transforming condensed excitons into photons with wave vectors $\mathbf{q}$ with the simultaneous creation of an elementary excitation with momentum $-\mathbf{q}$, whose energy is negligibly small when the constant φ_q tends to zero.

In a more rough description, one can neglect some of the effects related to the exciton–exciton interaction, keeping in mind only the possibility of the transformation of quasiparticles. In such an approach one can say that the absolute instability in the region (q_1, q_2) is related to the transformation of two condensed excitons with wave vectors $\mathbf{k}_0 = 0$ into one exciton with wave vector $\mathbf{q}$ and one photon with wave vector $-\mathbf{q}$ obeying the chemical reaction

$$\text{exciton}(0) + \text{exciton}(0) \rightarrow \text{exciton}(\mathbf{q}) + \text{photon}(-\mathbf{q}).$$

This transformation is real, because for some wave vectors $\mathbf{q}$, it is possible to fulfill the energy conservation law. For these wave numbers the energy of the exciton at $\mathbf{q}$ is slightly more than the energy of the exciton with zero wave vector, while the energy of the photon at $-\mathbf{q}$ is less than the energy of the second condensed exciton at zero wave vector by the same amount.

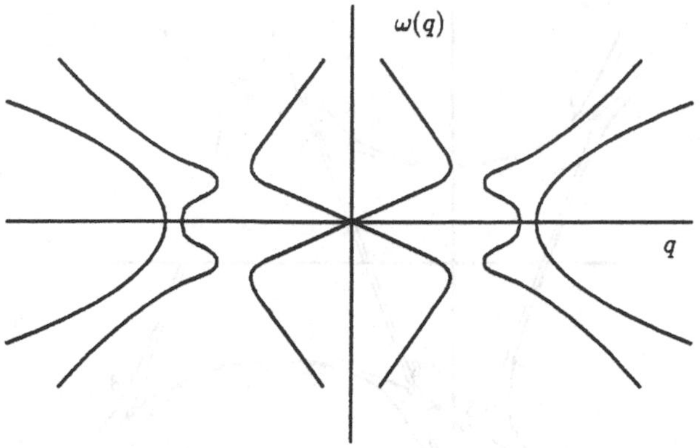

Figure 7.4. The full picture of the energy spectrum of the noncondensed excitons and photons when the BEC of excitons takes place in the state $\mathbf{k}_0 = 0$ [44].

The second absolute instability in the region (q_3, q_4) is related to the transformation of two condensed excitons into two noncondensate photons in accordance with the chemical reaction

$$\text{exciton}(0) + \text{exciton}(0) \rightarrow \text{photon}(\mathbf{q}) + \text{photon}(-\mathbf{q}).$$

Taking into account the positive and negative values of the frequencies $[\pm\omega_{1,2}(q)]$ in Eq. (7.63), one can construct the full picture of the energy spectrum for positive as well as for negative wave numbers $\pm q$. This is illustrated in Fig. 7.4.

Now we examine the case $\mathbf{k}_0 \neq 0$. Solutions (7.58) are real if two conditions are fulfilled,

$$\Omega_{\mathbf{q}}^2\left(\sigma_{\mathbf{q}}^2 - l_{\mathbf{q}}^2\right) + |\chi_{\mathbf{q}}|^4 \geq 2|\chi_{\mathbf{q}}|^2\sigma_{\mathbf{q}}\Omega_{\mathbf{q}},$$

$$\left(\sigma_{\mathbf{q}}^2 - l_{\mathbf{q}}^2 - \Omega_{\mathbf{q}}^2\right) + 4|\chi_{\mathbf{q}}|^2\left[\left(\sigma_{\mathbf{q}}^2 + l_{\mathbf{q}}^2\right)^2 - l_{\mathbf{q}}^2\right] \geq 0. \tag{7.64}$$

For $q = 0$, the left-hand side of the first of conditions (7.64) equals exactly the right-hand side, whereas the second expression is positive. Therefore one can expand the coefficients $\Omega_{\mathbf{q}}$, $\sigma_{\mathbf{q}}$, and $\chi_{\mathbf{q}}$ in series on $\mathbf{q}$ to find dispersion laws in the forms

$$\hbar\omega_+(q) = \mathcal{E}_0(k_0) + \frac{\hbar^2 q^2}{2M(k_0)}, \qquad \omega_-(q) = u(k_0)q. \tag{7.65}$$

The expressions for $\mathcal{E}_0(k_0)$, $M(k_0)$, and $u(k_0)$ were determined and reported in Ref. 44. After some simplifications for $k_0 = k^*$, $u(k^*)$ and $\mathcal{E}_0(k^*)$ are

$$u(k^*) = c\left|\frac{\hbar c k^* - \mu(k^*)}{2\hbar c k^*}\right|^{1/2},$$

$$\mathcal{E}_0(k^*) = (2L_0\,|\varphi_{\mathbf{k}^*}|)^{1/2}.$$

They determine the q dependences only in the direction $\mathbf{q} \perp \mathbf{k}_0$. If one sets $L_0 \approx |\varphi_{\mathbf{k}^*}| \approx |\hbar c k^* - \mu(k^*)| \approx 10^{-3}$ eV and $k^* \approx 10^5$ cm^{-1} then the values $\mathcal{E}_0(k^*) \approx 10^{-3}$ eV and $u(k^*) = 5 \times 10^8$ cm/s are obtained.

For the direction of $\mathbf{q}$ parallel to $\mathbf{k}_0$ ($\mathbf{q} \parallel \mathbf{k}_0$), the energy spectrum can be represented only qualitatively, with Sturrock's rules, mentioned above. The illustration below refers

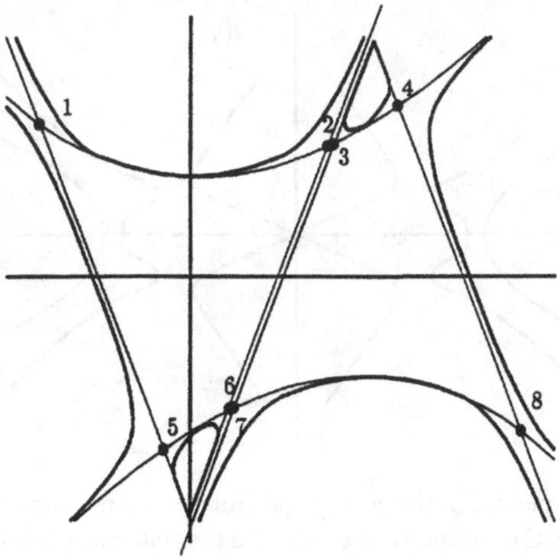

Figure 7.5. The energy spectrum in the vicinities of the eight points of wave synchronization. At points 1 and 8 the two polariton-type branches occur, at points 4 and 5 the absolute instabilities occur, and at points 2, 3, 6, and 7 the convective instabilities occur.

to the case of induced BEC, when the role of chemical potential is played by the laser frequency ω_L. The real quasiparticle transformations are shown in Fig. 6.5. The fourth order of dispersion equation (7.56) is related to the fact that along with the usual passive branches of the spectrum $\sigma_{\mathbf{q}}(+)$ and $\Omega_{\mathbf{q}}(+)$, there are also new quasi-energetical active branches $-\sigma_{\mathbf{q}}(-)$ and $-\Omega_{\mathbf{q}}(-)$. The first two branches increase when the wave vector $\mathbf{q}$ increases, whereas the last two branches decrease with increasing q. The intersections of the four branches of the initial spectrum take place at eight points of phase matching. They are determined by the equalities

$$2\omega_L = \omega_\perp(\mathbf{k}_0+\mathbf{q}) + \omega_\perp(\mathbf{k}_0-\mathbf{q}) + 2l_0 + 2l_{\mathbf{q}},$$
$$2\omega_L = \omega_\perp(\mathbf{k}_0+\mathbf{q}) + c\,|\mathbf{k}_0-\mathbf{q}| + l_0 + l_{\mathbf{q}},$$

as follows from Fig. 6.5.

The eight points are shown in Fig. 7.5. At points 1 and 8, the polariton-type reconstruction of the spectrum takes place. Because of the spatial dispersion effect, the polariton spectrum is gapless, and two simple waves can propagate. This is the case in Fig. 7.2(a). At points 4 and 5 absolute instabilities occur, while at points 2, 3, 6, and 7, convective instabilities take place. These correspond to Figs. 7.2(d) and 7.2(c), respectively.

Here it is appropriate to note that similar instabilities were evidenced in another range of the spectrum. For example, Elesin and Kopaev [52] considered the theory of the semiconductor laser, which lases because of the interband transitions. They pointed out that the electromagnetic field generated by the laser above the threshold is nothing more than a Bose condensate of photons with nonzero frequency ω_L. The analogy between the phase transition in a laser and the phase transitions in systems in thermodynamic equilibrium was noted earlier by Graham and Haken [53]. Elesin and Kopaev showed that the photons of the nonequilibrium laser Bose condensate interact indirectly among themselves through their direct interaction with the electrons, which effectuate the band-to-band transitions. As a result of that indirect interaction between the condensed photons, the noncondensate photons appear. This effect is analogous to the removal of particles from the Bose condensate in the Bogoliubov theory of the equilibrium weakly nonideal gas. But unlike it, the number of noncondensate photons grows in time if the damping in the system is not taken into account.

The condensate of photons without losses is unstable. The instability appears because the squeezing out of condensed photons is real, not virtual. It is accompanied by the simultaneous fulfillment of the energy and the momentum conversation laws. This process, as noted Elesin and Kopaev [52], is equivalent to the parametric transformation of two photons of the same frequency into two photons of different frequencies.

7.3.3 The Instabilities in a System with a Strong Polariton Effect

Now we consider another limiting case [inequality (7.3)], in which the exciton–photon interaction, i.e., the polariton effect, prevails over the exciton–exciton and the exciton–phonon interactions. In this case it is necessary to switch at the outset from the exciton and the photon operators to polariton operators. After such a transformation, the problem will look similar to those considered above. In this subsection, therefore, we will deal with a system of high-density polaritons interacting among themselves. Their interaction with phonons is discussed in Subsection 7.3.4.

Many aspects related to coherent polaritons at high density were studied by Shmiglyuk and Pitei in a special monograph [36]. Much of the complexity arises because the polariton–polariton interaction looks much more complicated than the exciton–exciton interaction. In most cases, one can assume that excitons interact in real space through a central force. As a result, the interaction constant in the momentum representation depends on only the momentum transfer between the two excitons involved in a collision. In general, however, two-body scattering between excitons involves four wave vectors for the four exciton operators describing the two incoming and two outgoing momenta. Because of momentum conservation, this means that scattering of two excitons can be characterized by three independent wave vectors. In moving to the polariton picture, one introduces polariton operators written in terms of the exciton operators, with transformation coefficients that depend on the wave vectors of the individual exciton operators. For this reason, the constant of the polariton–polariton interaction depends on three wave vectors and can be written in the form $\nu(\mathbf{p}, \mathbf{q}, \mathbf{k})$. In general, unlike the case of excitons, none of these dependences can be ignored. This dependence makes numerical calculations of the kinetic processes difficult.

The determination of the energy spectrum in the mean-field Hartree–Fock–Bogoliubov approximation considerably simplifies the problem because one deals with only a quadratic Hamiltonian. Following Ref. 36, we consider induced BEC in a single-polariton state with wave vector $\mathbf{k}_0$ in the lower polariton branch (LPB) in the region of the "bottleneck" (discussed briefly in Section 5.1). The energy of this state lies below the bottom of the upper polariton branch (UPB) and below the limiting frequency $\omega_\perp$ of the transverse exciton. In this case the interaction of the polaritons belonging to lower and upper branches can be neglected because it does not lead to the appearance of new instabilities. Two condensed polaritons from the region of the LPB bottleneck can only virtually transform themselves into two UPB quasiparticles. The real process is forbidden by energy conservation. These virtual processes give rise to the shifts of the polariton branches but do not cause qualitative changes.

After some simplifications, the Hamiltonian describing the polaritons of the LPB can be written as

$$H = \sum_{\mathbf{p}} \mathcal{E}(\mathbf{p}) a_{\mathbf{p}}^{\dagger} a_{\mathbf{p}} + \frac{1}{2V} \sum_{\mathbf{p},\mathbf{q},\mathbf{k}} \nu(\mathbf{p},\mathbf{q},\mathbf{k}) a_{\mathbf{p}}^{\dagger} a_{\mathbf{q}}^{\dagger} a_{\mathbf{q}+\mathbf{k}} a_{\mathbf{p}-\mathbf{k}},$$

$$\nu(\mathbf{p},\mathbf{q},\mathbf{k}) = \nu(\mathbf{p},\mathbf{q},-\mathbf{k}), \qquad \nu^{*}(\mathbf{p},\mathbf{q},\mathbf{k}) = \nu(\mathbf{p}-\mathbf{k},\mathbf{q}+\mathbf{k},-\mathbf{k}). \tag{7.66}$$

Here $a_{\mathbf{p}}^{\dagger}$ and $a_{\mathbf{p}}$ are the boson-type polariton creation and annihilation operators, $\mathcal{E}(\mathbf{p})$ is the bare dispersion relation of the LPB, and $\nu(\mathbf{p},\mathbf{q},\mathbf{k})$ is the constant of the polariton–polariton interaction. The index of the LPB is dropped.

If one supposes that in some way the induced BEC in the single-particle state $\mathbf{k}_0$ has occurred, then the dominant terms in Hamiltonian equation (7.66) will be

$$\tilde{H} = \mathcal{E}(\mathbf{k}_0) a_{\mathbf{k}_0}^{\dagger} a_{\mathbf{k}_0} + \frac{\nu(\mathbf{k}_0,\mathbf{k}_0,0)}{2V} a_{\mathbf{k}_0}^{+2} a_{\mathbf{k}_0}^{2} + \sum_{\mathbf{p}\neq\mathbf{k}_0} \tilde{\mathcal{E}}(\mathbf{p}) a_{\mathbf{p}}^{\dagger} a_{\mathbf{p}}$$
$$+ \frac{1}{2} \sum_{\mathbf{q}\neq 0} \left[\frac{\nu(\mathbf{k}_0,\mathbf{k}_0,\mathbf{q})}{V} a_{\mathbf{k}_0}^{+2} a_{\mathbf{k}_0+\mathbf{q}} a_{\mathbf{k}_0-\mathbf{q}} + \frac{\nu^{*}(\mathbf{k}_0,\mathbf{k}_0,\mathbf{q})}{V} a_{\mathbf{k}_0}^{2} a_{\mathbf{k}_0+\mathbf{q}}^{\dagger} a_{\mathbf{k}_0-\mathbf{q}}^{\dagger} \right]. \tag{7.67}$$

The terms contain either two or four operators describing the condensed polaritons. The dressed dispersion relation $\tilde{\mathcal{E}}(\mathbf{p})$ is

$$\tilde{\mathcal{E}}(\mathbf{p}) = \mathcal{E}(\mathbf{p}) + \frac{N_{\mathbf{k}_0}}{V} \nu(\mathbf{p},\mathbf{k}_0,0) + \frac{N_{\mathbf{k}_0}}{V} \nu(\mathbf{p},\mathbf{k}_0,\mathbf{p}-\mathbf{k}_0). \tag{7.68}$$

Here the macroscopic number of condensed polaritons $N_{\mathbf{k}_0} = a_{\mathbf{k}_0}^{\dagger} a_{\mathbf{k}_0}$ is treated as a c number $N_{\mathbf{k}_0} \sim V$, $n_{\mathbf{k}_0} = N_{\mathbf{k}_0}/V$.

In the zeroth-order approximation, the condensate amplitude $a_{\mathbf{k}_0}$ obeys the motion equation

$$i\hbar \frac{da_{\mathbf{k}_0}}{dt} = \left[\mathcal{E}(\mathbf{k}_0) + \frac{\nu(\mathbf{k}_0,\mathbf{k}_0,0)}{V} a_{\mathbf{k}_0}^{\dagger} a_{\mathbf{k}_0} \right] a_{\mathbf{k}_0} = \mu(\mathbf{k}_0) a_{\mathbf{k}_0}, \tag{7.69}$$

where $\mu(\mathbf{k}_0)$ plays the role of the chemical potential for a nonequilibrium state of the polariton system:

$$\mu(\mathbf{k}_0) = \mathcal{E}(\mathbf{k}_0) + \frac{\nu(\mathbf{k}_0,\mathbf{k}_0,0) N_{\mathbf{k}_0}}{V}. \tag{7.70}$$

At the same time, $\mu(\mathbf{k}_0)$ will be used as a reference point for measurement of the energy spectrum of the elementary excitations. In the case of repulsion between the polaritons ($\nu > 0$), which is the only case considered below, the value of $\mu(\mathbf{k}_0)$ is less than the value of the dressed function $\tilde{\mathcal{E}}(\mathbf{k}_0)$. As noted by Shmiglyuk and Pitei, this fact is important.

The motion equations for the operator $a_{\mathbf{k}_0+\mathbf{q}}$ and for the complex of operators of the type $(a_{\mathbf{k}_0}^{2} a_{\mathbf{k}_0-\mathbf{q}}^{\dagger}/N_{\mathbf{k}_0})$ can be obtained with the Hamiltonian $\tilde{H}$ of Eq. (7.67) and motion equation (7.69). It is reasonable to introduce the damping of the noncondensate polaritons phenomenologically in these motion equations in the following way:

$$i\hbar \frac{\mathrm{d}a_{\mathbf{k}_0+\mathbf{q}}}{\mathrm{d}t} = [\tilde{\mathcal{E}}(\mathbf{k}_0+\mathbf{q}) - i\gamma(\mathbf{k}_0+\mathbf{q})]a_{\mathbf{k}_0+\mathbf{q}} + L^*_{\mathbf{q}}\left(\frac{a^2_{\mathbf{k}_0}a^\dagger_{\mathbf{k}_0-\mathbf{q}}}{N_{\mathbf{k}_0}}\right),$$

$$i\hbar \frac{\mathrm{d}}{\mathrm{d}t}\left(\frac{a^2_{\mathbf{k}_0}a^\dagger_{\mathbf{k}_0-\mathbf{q}}}{N_{\mathbf{k}_0}}\right) = [2\mu(\mathbf{k}_0) - \tilde{\mathcal{E}}(\mathbf{k}_0-\mathbf{q}) - i\gamma(\mathbf{k}_0-\mathbf{q})]\left(\frac{a^2_{\mathbf{k}_0}a^\dagger_{\mathbf{k}_0-\mathbf{q}}}{N_{\mathbf{k}_0}}\right) - L_{\mathbf{q}}a^\dagger_{\mathbf{k}_0+\mathbf{q}},$$

$$L_{\mathbf{q}} = \frac{\nu(\mathbf{k}_0, \mathbf{k}_0, \mathbf{q})N_{\mathbf{k}_0}}{V}. \tag{7.71}$$

The eigenvalues E of these motion equations can be represented in the form

$$E = \mathcal{E} + \mu(\mathbf{k}_0), \tag{7.72}$$

where $\mathcal{E}$ obeys the determinant equation

$$\begin{vmatrix} [\tilde{\mathcal{E}}(\mathbf{k}_0+\mathbf{q}) - \mu(\mathbf{k}_0) - i\gamma(\mathbf{k}_0+\mathbf{q}) - \mathcal{E}] & L^*_q \\ -L_q & [\mu(\mathbf{k}_0) - \tilde{\mathcal{E}}(\mathbf{k}_0-\mathbf{q}) - i\gamma(\mathbf{k}_0-\mathbf{q}) - \mathcal{E}] \end{vmatrix} = 0$$

and equals

$$\mathcal{E}(\mathbf{q}) = \frac{\tilde{\mathcal{E}}(\mathbf{k}_0+\mathbf{q}) - \tilde{\mathcal{E}}(\mathbf{k}_0-\mathbf{q}) - i[\gamma(\mathbf{k}_0+\mathbf{q}) + \gamma(\mathbf{k}_0-\mathbf{q})]}{2}$$
$$\pm\frac{1}{2}\sqrt{\{\tilde{\mathcal{E}}(\mathbf{k}_0+\mathbf{q}) + \tilde{\mathcal{E}}(\mathbf{k}_0-\mathbf{q}) - 2\mu(\mathbf{k}_0) - i\,[\gamma(\mathbf{k}_0+\mathbf{q}) - \gamma(\mathbf{k}_0-\mathbf{q})]\}^2 - 4|L_q|^2}. \tag{7.73}$$

Let us first analyze the energy spectrum in the case of no damping [$\gamma(\mathbf{k}_0 \pm \mathbf{q}) = 0$] following Ref. 36. It is represented in Fig. 7.6 for the case $\mathbf{q} \parallel \mathbf{k}_0$. The wave vector $\mathbf{k}_0$ and the chemical potential $\mu(\mathbf{k}_0)$ are determined by the external laser radiation $\mu(\mathbf{k}_0) = \hbar\omega_L$. The solid curves represent the dispersion curve $\tilde{\mathcal{E}}(\mathbf{k}_0+\mathbf{q})$ and the quasi-energy curve $[2\mu(\mathbf{k}_0) - \tilde{\mathcal{E}}(\mathbf{k}_0-\mathbf{q})]$, which appear in motion equations (7.71). The intersections of these curves determine the

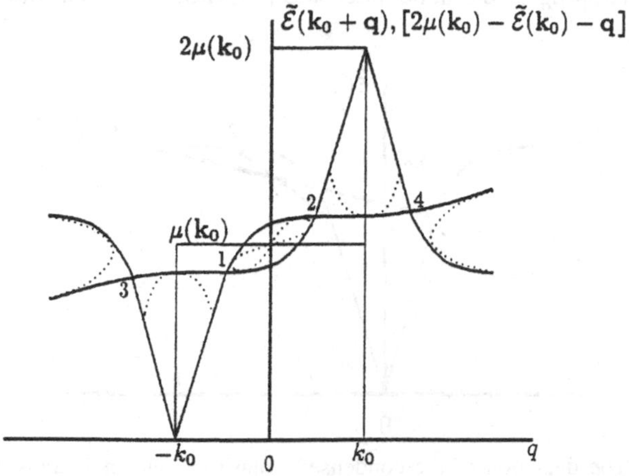

Figure 7.6. The energy spectrum of the noncondensed polaritons [36].

points of phase matching. They obey the equation

$$2\mu(\mathbf{k}_0) = \tilde{\mathcal{E}}(\mathbf{k}_0 + \mathbf{q}) + \tilde{\mathcal{E}}(\mathbf{k}_0 - \mathbf{q}) \tag{7.74}$$

and are indicated by the numbers 1, 2, 3, and 4 in Fig. 7.6. The dotted curves show the renormalized spectrum. When the expression under the square root in Eq. (7.73) is negative for some values of q, which satisfy the relation

$$[\tilde{\mathcal{E}}(\mathbf{k}_0 + \mathbf{q}) + \tilde{\mathcal{E}}(\mathbf{k}_0 - \mathbf{q}) - 2\mu(\mathbf{k}_0)]^2 \leq 4|L_{\mathbf{q}}|^2, \tag{7.75}$$

then the renormalized spectrum is complex for these real values of q. Conditions (7.75) are fulfilled in the vicinity of the four points of phase matching. The boundaries of these regions can also be determined by the conditions (7.75). This case completely coincides with the classical examples demonstrated in Section 7.3. Therefore at points 1–4 in Fig. 7.6 the instabilities of the spectrum take place. Following Sturrock's rules, the convective instability occurs in regions 1 and 2. For real values of q there are the complex values of $\mathcal{E}$, and for real values of $\mathcal{E}$ there are complex values of q.

The system can be used as an amplifier of polariton waves. The region of amplification 1 can lie in the photonlike part of the LPB for a suitable choice of the laser frequency ω_L. It should be possible to observe experimentally the amplification of a weak probe-light signal propagating through the crystal in this spectral region. Naturally this amplification is possible because of the external laser beam, which coherently excites the polariton mode $\mathbf{k}_0$. In regions 3 and 4 the system undergoes absolute instability and can generate the photonlike and the excitonlike polaritons, respectively. The source of such generation is the same as that given earlier.

Regions 1 and 2 as well as 3 and 4 are pair conjugated because of the pair annihilation of two condensed polaritons and their resulting transformation into two scattered polaritons, and vice versa. This depletion of two condensed polaritons and their transformation into two noncoherent quasiparticles is represented in Fig. 7.7. The condensed polaritons have wave vector $\mathbf{k}_0$ and energy $\mu(\mathbf{k}_0)$, which is lower than $\tilde{\mathcal{E}}(\mathbf{k}_0)$. This fact permits real quantum transitions to take place. Regions 1 and 2 in Fig. 7.6 are situated not so far from the value $q \approx 0$, while regions 3 and 4 are located near the values $q \approx \pm 2k_0$. One can see that when q tends to zero, the two energy branches $\mathcal{E}_{1,2}(q)$ also tend to zero. It means that energy branches $E_{1,2}(q)$ of Eq. (7.72) coincide at this point with $\mu(\mathbf{k}_0)$. This fact is depicted in Fig. 7.6. When damping of the noncondensate polaritons is taken into account, solution

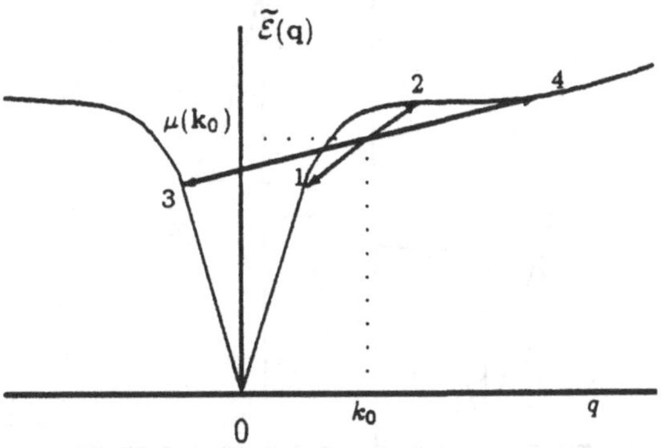

Figure 7.7. The depletion of the condensed polaritons and their transformation into scattered quasiparticles [36].

(7.73) has a common negative imaginary part in first term, outside of the square root. It is more difficult to obtain a net positive imaginary part for at least one of the two solutions. This can occur when the square-root term is taken in account.

The resulting imaginary parts Im $\mathcal{E}_{1,2}(q)$ can be easily obtained at the points of phase matching of Eq. (7.74). They are

$$\text{Im } \mathcal{E}_{1,2}(q) = \frac{-[\gamma(\mathbf{k}_0+\mathbf{q})+\gamma(\mathbf{k}_0-\mathbf{q})]}{2} \pm \frac{1}{2}\sqrt{[\gamma(\mathbf{k}_0+\mathbf{q})-\gamma(\mathbf{k}_0-\mathbf{q})]^2+|L_{\mathbf{q}}|^2}. \tag{7.76}$$

One of them can be positive for values of $|L_q|^2$ and $n_{\mathbf{k}_0}$ which obey the conditions

$$|L_q|^2 \geq \gamma(\mathbf{k}_0+\mathbf{q})\gamma(\mathbf{k}_0-\mathbf{q}),$$
$$n_{\mathbf{k}_0} \geq n_{\text{th}} = \frac{\sqrt{\gamma(\mathbf{k}_0+\mathbf{q})\gamma(\mathbf{k}_0-\mathbf{q})}}{|\nu(\mathbf{k}_0,\mathbf{k}_0,\mathbf{q})|}. \tag{7.77}$$

The threshold condensate density n_{th} is directly proportional to the root mean square of the damping constants $\gamma(\mathbf{k}_0 \pm \mathbf{q})$ and inversely proportional to the constant $|\nu(\mathbf{k}_0,\mathbf{k}_0,\mathbf{q})|$ of the polariton–polariton interaction. When the density of condensed polaritons $n_{\mathbf{k}_0}$ exceeds the threshold value n_{th}, the instability in the system can appear.

Another effect of the damping is its influence on the spectral width of the generated and amplified waves, reducing the regions of the frequencies and wave vectors where the instabilities can appear. The evolution of the absolute instability will give rise to the appearance of two new condensates symmetrically situated on the energy scale relative to the initial condensate. A similar effect was observed in studies of light generation in the semiconductor laser. Along with the main generation mode, the appearance of accompanying radiation modes was observed with similar properties. Elesin and Kopaev [52] attributed their appearance to the development of an instability in the semiconductor laser and to the depletion of the condensed photons. Unlike the band-to-band transition considered by Elesin and Kopaev, in the case of polaritons studied by Shmiglyuk and Pitei [36], the structure of the two growing modes is different. At point 4 the polaritons are excitonlike, whereas at point 3 they are photonlike quasiparticles.

7.3.4 In-Depth Study: *The Elementary Excitations in the Three-Component System of Excitons, Photons, and Phonons*

In Chapter 3, we studied the energy spectrum of the elementary excitations in the system of Bose condensed excitons interacting with phonons. In that case, the exciton–photon interaction was taken into account only as a cause of band-to-band transitions generating the population of excitons. On the other hand, in Subsection 7.3.3 we have considered only the energy spectrum of the Bose condensed excitons interacting with photons but without the participation of phonons. Of course, it would be of interest to join together both of these variants. Although this joint variant is more cumbersome than each of the two preceding ones, some qualitative results can be obtained. This program was initiated by Belkin, Miglei, and Moskalenko [54] within the framework of inequality (7.2), considering the excitons and the photons separately without using polariton operators. These results were completed by Kiselyova and Moskalenko by introduction of mixed-type photon–phonon Green's functions and self-energy parts. As a result, one can generalize the Beliaev technique to discuss the three-component system and the energy spectrum of the noncondensate quasiparticles. For simplicity the antiresonant terms in the exciton–photon interaction are neglected in the discussion below.

The Hamiltonian of the three-component system can be constructed from the Hamiltonians (3.35) and (7.34) and can be written in the form

$$H = \sum_{\mathbf{q}}(\Delta + T_{\mathbf{q}} - \mu)a^{\dagger}_{\mathbf{q}}a_{\mathbf{q}} + \sum_{\mathbf{q}}(\hbar cq - \mu)\,c^{\dagger}_{\mathbf{q}}c_{\mathbf{q}} + \sum_{\mathbf{q}} B_{\mathbf{q}}b^{\dagger}_{\mathbf{q}}b_{\mathbf{q}} + \sum_{\mathbf{q}} \varphi_{\mathbf{q}}(a^{\dagger}_{\mathbf{q}}c_{\mathbf{q}} - c^{\dagger}_{\mathbf{q}}a_{\mathbf{q}})$$
$$+ \frac{i}{\sqrt{N_a}}\sum_{\mathbf{pq}}\theta(\mathbf{p}-\mathbf{q})a^{\dagger}_{\mathbf{p}}a_{\mathbf{q}}(b_{\mathbf{p}-\mathbf{q}} - b^{\dagger}_{\mathbf{q}-\mathbf{p}}) + \frac{1}{2V}\sum_{\mathbf{pqk}} \nu(k)a^{\dagger}_{\mathbf{p}}a^{\dagger}_{\mathbf{q}}a^{\dagger}_{\mathbf{q}+\mathbf{k}}a_{\mathbf{p}-\mathbf{k}}. \tag{7.78}$$

The meaning of the letter symbols and designations are the same as in original expressions (3.35) and (7.34). The introduction of the chemical potential μ reflects the assumption that the total number of excitons and photons is given. At the same time, the total number of phonons is not conserved.

Following Eqs. (3.7) and (3.46), one can introduce a 5×5 matrix of the type

$$\hat{A}_{\mathbf{p}} = \begin{vmatrix} \cdots \\ \cdots \\ A_{i\mathbf{p}} \\ \cdots \\ \cdots \end{vmatrix} = \begin{vmatrix} \alpha_{\mathbf{p}} & 0 & 0 & 0 & 0 \\ \alpha^{\dagger}_{-\mathbf{p}} & 0 & 0 & 0 & 0 \\ (b_{\mathbf{p}} - b^{\dagger}_{-\mathbf{p}}) & 0 & 0 & 0 & 0 \\ c_{\mathbf{p}} & 0 & 0 & 0 & 0 \\ c^{\dagger}_{-\mathbf{p}} & 0 & 0 & 0 & 0 \end{vmatrix},$$
$$\hat{A}^{+}_{\mathbf{p}} = |\cdots \quad \cdots \quad A^{+}_{i\mathbf{p}} \quad \cdots \quad \cdots|, \tag{7.79}$$

and the Green's functions in a matrix form similar to that of Eq. (3.49),

$$\hat{G}(\mathbf{p}, E) = \langle\langle \hat{A}_{\mathbf{p}} \mid \hat{A}^{+}_{\mathbf{p}} \rangle\rangle_E = \|G_{ij}(\mathbf{p}, E)\|. \tag{7.80}$$

These can be represented graphically as in Subsection 3.2.3. The matrix elements $G_{i1}(\mathbf{p}, E)$ (where $i = 1, 2, \ldots 5$) are

$$\begin{aligned} G_{11}(\mathbf{p}, E) &= \langle\langle a_{\mathbf{p}} | a^{\dagger}_{\mathbf{p}} \rangle\rangle_E = \\ G_{21}(\mathbf{p}, E) &= \langle\langle a^{\dagger}_{-\mathbf{p}} | a^{\dagger}_{\mathbf{p}} \rangle\rangle_E = \\ G_{31}(\mathbf{p}, E) &= \langle\langle b_{\mathbf{p}} - b^{\dagger}_{-\mathbf{p}} | a^{\dagger}_{\mathbf{p}} \rangle\rangle_E = \\ G_{41}(\mathbf{p}, E) &= \langle\langle c_{\mathbf{p}} | a^{\dagger}_{\mathbf{p}} \rangle\rangle_E = \\ G_{51}(\mathbf{p}, E) &= \langle\langle c^{\dagger}_{-\mathbf{p}} | a^{\dagger}_{\mathbf{p}} \rangle\rangle_E = \end{aligned} \tag{7.81}$$

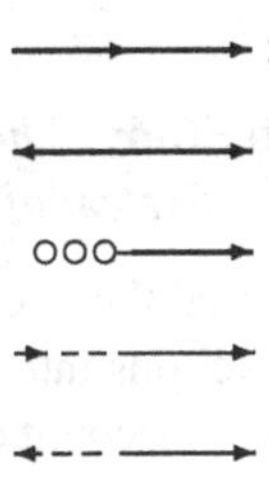

The full Green's functions are represented by thick lines with arrows at the ends in most cases. When the Green's functions consist of two exciton operators, they are represented by solid lines. The lines are dashed when two photon operators are used and dotted when two phonon operators are used. The mixed-type Green's functions contain the operators of different quasiparticles. There are mixed exciton–phonon, exciton–photon, and phonon–photon Green's functions. They are represented by thick lines that are semisolid–semidotted, semisolid–semidashed, and semidotted–semidashed, respectively. The arrows denote the creation and annihilation operators. They are directed toward the exterior of the diagram when the quasiparticles are created and toward the interior when the annihilation of quasiparticles takes place. For the phonon operator $(b_{\mathbf{p}} - b^{\dagger}_{-\mathbf{p}})$, no arrow with definite direction can be introduced. Its closed motion equation is second order. Instead of this linear combination, it is possible to use the phonon creation and annihilation operators separately. This will increase the number of Green's functions and the corresponding number of degrees of freedom in the first-order motion equations, but will not change the results concerning the energy spectrum of the noncondensate quasiparticles. This fact was demonstrated in Chapter 3 with different approaches.

Following the structure of the Green's functions, the corresponding self-energy parts can be introduced in the same manner:

$$\Sigma_{11}(\mathbf{p}, E) = \quad ;$$

$$\Sigma_{12}(\mathbf{p}, E) = \quad ;$$

$$\Sigma_{13}(\mathbf{p}, E) = \quad ;$$

$$\Sigma_{14}(\mathbf{p}, E) = \quad ;$$

$$\Sigma_{15}(\mathbf{p}, E) = \quad ;$$

$$\Sigma_{21}(\mathbf{p}, E) = \quad ;$$

$$\Sigma_{22}(\mathbf{p}, E) = \quad ;$$

$$\Sigma_{23}(\mathbf{p}, E) = \quad ;$$

$$\Sigma_{24}(\mathbf{p}, E) = \quad ;$$

$$\Sigma_{25}(\mathbf{p}, E) = \quad ;$$

$$\Sigma_{31}(\mathbf{p}, E) = \quad ;$$

$$\Sigma_{32}(\mathbf{p}, E) = \quad ;$$

$$\Sigma_{33}(\mathbf{p}, E) = \quad ;$$

$$\Sigma_{34}(\mathbf{p}, E) = \quad ;$$

$$\Sigma_{35}(\mathbf{p}, E) = \quad ;$$

$$\Sigma_{41}(\mathbf{p}, E) = \quad ;$$

$$\Sigma_{42}(\mathbf{p}, E) = \quad ;$$

$$\Sigma_{43}(\mathbf{p}, E) = \quad ;$$

$$\Sigma_{44}(\mathbf{p}, E) = \quad ;$$

$$\Sigma_{45}(\mathbf{p}, E) = \quad ;$$

$$\Sigma_{51}(\mathbf{p}, E) = \quad ;$$

$$\Sigma_{52}(\mathbf{p}, E) = \quad ;$$

$$\Sigma_{53}(\mathbf{p}, E) = \quad ;$$

$$\Sigma_{54}(\mathbf{p}, E) = \quad ;$$

$$\Sigma_{55}(\mathbf{p}, E) = \quad ;$$

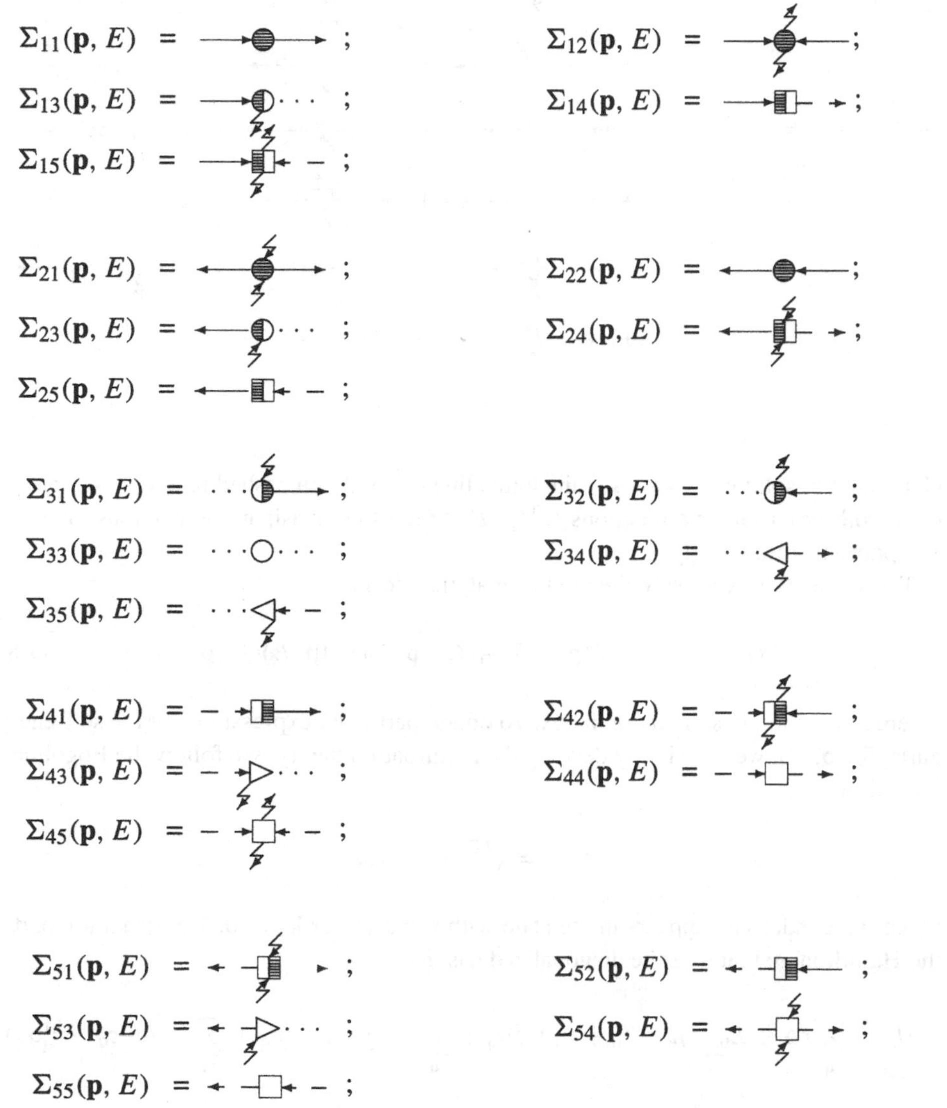

$$(7.82)$$

The closed set of graphical equations for the five Green's functions is the following:

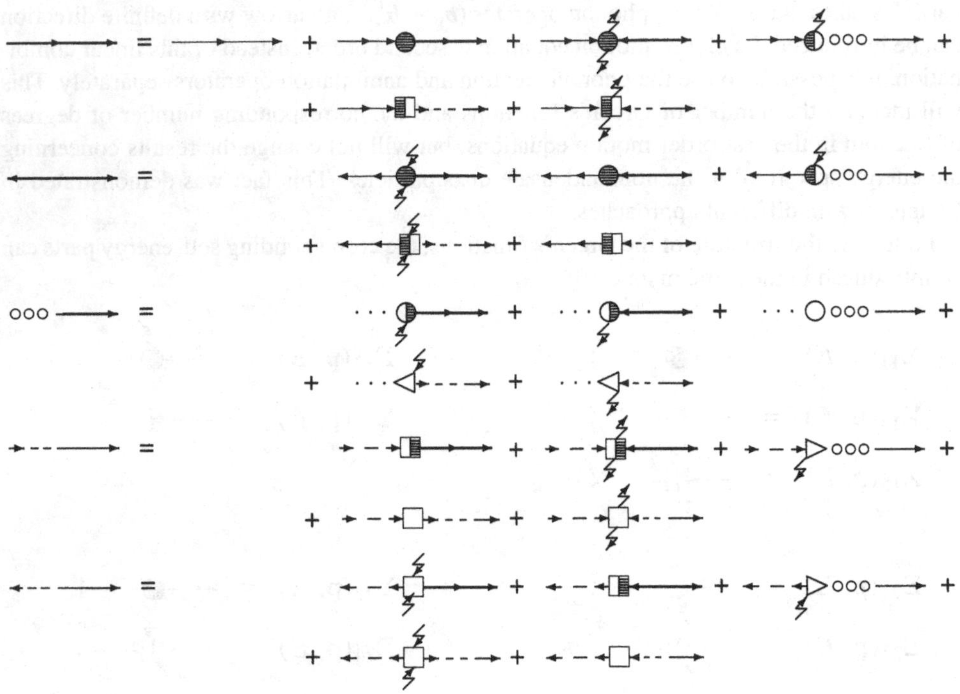

(7.83)

Here the thin solid line, thin dotted line $\cdots$, and thin dashed line represent the zeroth-order Green's functions $G^0_{ii}(\mathbf{p}, E)$ of the free-quasiparticle excitons, phonons, and photons, respectively.

These equations can be written in the analytical form as

$$G_{ij}(\mathbf{p}, E) = G^0_{ii}(\mathbf{p}, E)\delta_{ij} + G^0_{ii}(\mathbf{p}, E)\Sigma_{il}(\mathbf{p}, E)G_{lj}(\mathbf{p}, E) \tag{7.84}$$

where l is a dummy summation index. To obtain particular expressions for the self-energy parts $\Sigma_{ij}(\mathbf{p}, E)$, we can simply develop the perturbation theory. We follow the Bogoliubov procedure

$$a_\mathbf{q} = \sqrt{N_0}\delta_{\mathbf{q},0} + \alpha_\mathbf{q}$$

when the condensate appears in the state with wave vector $\mathbf{k}_0 = 0$. The quadratic part of the Hamiltonian that must be diagonalized has the form

$$H_2 = \sum_\mathbf{q}(\Delta + L_0 - \mu + T_\mathbf{q} + L_\mathbf{q})\alpha^\dagger_\mathbf{q}\alpha_\mathbf{q} + \sum_\mathbf{q}(\hbar c q - \mu)c^\dagger_\mathbf{q}c_\mathbf{q} + \sum_\mathbf{q}\varphi_q(\alpha^\dagger_\mathbf{q}c_\mathbf{q} - c^\dagger_\mathbf{q}\alpha_\mathbf{q})$$
$$+ \frac{1}{2}\sum_\mathbf{q} L_\mathbf{q}(\alpha_\mathbf{q}\alpha_{-\mathbf{q}} + \alpha^\dagger_\mathbf{q}\alpha^\dagger_{-\mathbf{q}}) + i\sqrt{\frac{N_0}{N_a}}\sum_\mathbf{q}\theta(\mathbf{q})(\alpha_{-\mathbf{q}} + \alpha^\dagger_\mathbf{q})(b_\mathbf{q} - b^\dagger_{-\mathbf{q}}). \tag{7.85}$$

The motion equations, obtained with the help of Hamiltonian defined in Eq. (7.85), are

$$i\hbar\frac{d\alpha_{\mathbf{p}}}{dt} = \tilde{\Delta}_{\mathbf{p}}\alpha_{\mathbf{p}} + L_{\mathbf{p}}\alpha^{\dagger}_{-\mathbf{p}} + \varphi_{\mathbf{p}}c_{\mathbf{p}} + i\theta(\mathbf{p})\sqrt{\frac{N_0}{N_a}}(b_{\mathbf{p}} - b^{\dagger}_{-\mathbf{p}}),$$

$$i\hbar\frac{d\alpha^{\dagger}_{-\mathbf{p}}}{dt} = -\tilde{\Delta}_{\mathbf{p}}\alpha^{\dagger}_{-\mathbf{p}} - L_{\mathbf{p}}\alpha_{\mathbf{p}} + \varphi_{\mathbf{p}}c^{\dagger}_{-\mathbf{p}} - i\theta(\mathbf{p})\sqrt{\frac{N_0}{N_a}}(b_{\mathbf{p}} - b^{\dagger}_{-\mathbf{p}}),$$

$$(i\hbar)^2\frac{d^2}{dt^2}(b_{\mathbf{p}} - b^{\dagger}_{-\mathbf{p}}) = B^2_{\mathbf{p}}(b_{\mathbf{p}} - b^{\dagger}_{-\mathbf{p}}) - 2iB_{\mathbf{p}}\theta(\mathbf{p})\sqrt{\frac{N_0}{N_a}}(\alpha_{\mathbf{p}} + \alpha^{\dagger}_{-\mathbf{p}}),$$

$$i\hbar\frac{dc_{\mathbf{p}}}{dt} = (\hbar cp - \mu)c_{\mathbf{p}} - \varphi_{\mathbf{p}}\alpha_{\mathbf{p}},$$

$$i\hbar\frac{dc^{\dagger}_{-\mathbf{p}}}{dt} = -(\hbar cp - \mu)c^{\dagger}_{-\mathbf{p}} - \varphi_{\mathbf{p}}\alpha^{\dagger}_{-\mathbf{p}},$$

$$\tilde{\Delta}_{\mathbf{p}} = \left(\Delta + L_0 - \mu + T_{\mathbf{p}} + L_{\mathbf{p}}\right). \tag{7.86}$$

Here it is necessary to set $\mu = \Delta + L_0$ for the quasi-equilibrium state, while the condensed photons with $\mathbf{k}_0 = 0$ do not exist. In the nonequilibrium state induced by laser radiation, μ must be substituted by the photon energy $\hbar\omega_L$. The wave vector $\mathbf{k}_0$ can be only approximately set to zero, keeping in mind that the number of coherent laser photons is macroscopically large and determines the number of condensed excitons $N_{\mathbf{k}_0}$.

In the stationary state, when all the operators A_i have the time dependence $\exp(-iEt/\hbar)$, the condition for solution of Eqs. (7.86) can be written in the determinant form

$$\begin{vmatrix} (\tilde{\Delta}_{\mathbf{p}} - E) & L_{\mathbf{p}} & i\theta(\mathbf{p})\sqrt{\frac{N_0}{N_a}} & \varphi_{\mathbf{p}} & 0 \\ L_{\mathbf{p}} & (\tilde{\Delta}_{\mathbf{p}} + E) & i\theta(\mathbf{p})\sqrt{\frac{N_0}{N_a}} & 0 & -\varphi_{\mathbf{p}} \\ 2iB_{\mathbf{p}}\theta(\mathbf{p})\sqrt{\frac{N_0}{N_a}} & 2iB_{\mathbf{p}}\theta(\mathbf{p})\sqrt{\frac{N_0}{N_a}} & (E^2 - B^2_{p}) & 0 & 0 \\ -\varphi_{\mathbf{p}} & 0 & 0 & (\hbar cp - \mu - E) & 0 \\ 0 & \varphi_{\mathbf{p}} & 0 & 0 & (\hbar cp - \mu + E) \end{vmatrix} = 0. \tag{7.87}$$

This implies that in the first order of the perturbation theory, the self-energy parts introduced in Eqs. (7.82) are

$$\Sigma^{(1)}_{11}(\mathbf{p}, E) = \Sigma^{(1)}_{22}(\mathbf{p}, E) = L_0 + L_{\mathbf{q}},$$

$$\Sigma^{(1)}_{12}(\mathbf{p}, E) = \Sigma^{(1)}_{21}(\mathbf{p}, E) = L_{\mathbf{q}},$$

$$\Sigma^{(1)}_{13}(\mathbf{p}, E) = \Sigma^{(1)}_{23}(\mathbf{p}, E) = i\theta(\mathbf{p})\sqrt{\frac{N_0}{N_a}},$$

$$\Sigma_{14}(\mathbf{p}, E) = -\Sigma^{(1)}_{25}(\mathbf{p}, E) = \varphi_p,$$

$$\Sigma^{(1)}_{15}(\mathbf{p}, E) = \Sigma^{(1)}_{24}(\mathbf{p}, E) = 0.$$

The dispersion equation has a bicubic form

$$AE^6 + BE^4 + CE^2 + D = 0. \tag{7.88}$$

The coefficients A, B, C, and D in the case when $\mu = \Delta + L_0$ are given by

$$A = 1,$$

$$B = -\left[(\hbar cp - \mu)^2 + B_{\mathbf{p}}^2 + T_{\mathbf{p}}^2 + 2T_{\mathbf{p}}L_{\mathbf{p}} + 2|\varphi_{\mathbf{p}}|^2\right],$$

$$C = (\hbar cp - \mu)^2 \left(B_{\mathbf{p}}^2 + T_{\mathbf{p}}^2 + 2T_{\mathbf{p}}L_{\mathbf{p}}\right) - 2|\varphi_{\mathbf{p}}|^2 \left[(\hbar cp - \mu)(T_{\mathbf{p}} + L_{\mathbf{p}}) - B_{\mathbf{p}}^2\right]$$
$$+ |\varphi_p|^4 - 4\frac{N_0}{N_a}\theta^2(p)T_{\mathbf{p}}B_{\mathbf{p}} + B_{\mathbf{p}}^2\left(T_{\mathbf{p}}^2 + 2T_{\mathbf{p}}L_{\mathbf{p}}\right),$$

$$D = -2|\varphi_p|^2 B_{\mathbf{p}}(\hbar cp - \mu)\left[2\frac{N_0}{N_a}\theta^2(p)B_{\mathbf{p}}(T_{\mathbf{p}} + L_{\mathbf{p}})\right]$$
$$+ (\hbar cp - \mu)^2\left[4\frac{N_0}{N_a}\theta^2(p)T_{\mathbf{p}}B_{\mathbf{p}} - B_{\mathbf{p}}^2\left(T_{\mathbf{p}}^2 + 2T_{\mathbf{p}}L_{\mathbf{p}}\right)\right] - |\varphi_{\mathbf{p}}|^4 B_{\mathbf{p}}^2. \tag{7.89}$$

The energy spectrum described by dispersion equation (7.87) is much more complicated than the simple variants of Subsection 7.3.1. For this reason, only qualitative considerations can be suggested. Below, we regard the different intersection points as a problem of two interacting waves, neglecting the influence of the third one. In this approach, the interaction constant of dispersion equation (7.50) must be determined at each intersection point.

At points 3, 4, and 5 in Fig. 7.8, the self-energy parts determined in the first order of the perturbation theory are sufficient to solve the problem. But for points 1 and 2, where the photon branches intersect the dispersion line $B_{\mathbf{p}}$ of the acoustic phonon, this approximation is insufficient, because Hamiltonian defined in Eq. (7.78) does not contain the direct phonon–photon interaction. At these points, as well as at point 6, a higher-order perturbation theory is needed. Figure 7.8 shows one possible variant of the energy spectrum of a three-component system, which takes into account the results already discussed in the two limiting cases. It

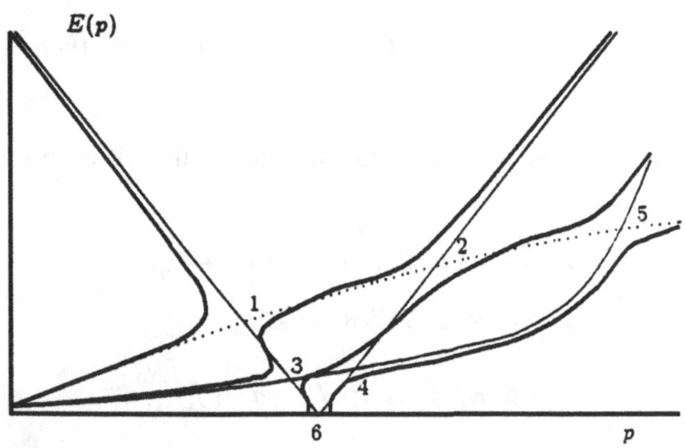

Figure 7.8. The expected variant of the energy spectrum of the elementary exitations in the three-component system [54].

is assumed that the sound velocity u that determines the dispersion line $B_\mathbf{p} = \hbar u p$ is greater than the critical velocity u^* of the hydron's dispersion curve.

7.4 In-Depth Study: "Phonoritons" and Nonlinear Absorption of Polaritons

In Section 7.3 the possibility of obtaining a complicated energy spectrum in a three-component system was studied, but sufficient information about this spectrum was not obtained. Moreover, so far we have not investigated the polariton–phonon and its quasi-energy spectrum at all. In a series of papers, Ivanov and Keldysh [55–58] and Keldysh and Tikhodeev [59–61] studied in detail the dynamic and kinetic processes that accompany the propagation of an intense electromagnetic wave in direct-gap semiconductors with a strong polariton effect and strong exciton–phonon interaction. In these works, Ivanov and Keldysh introduced the concept of the "phonoriton."

An intense electromagnetic wave can excite a macroscopically large polariton wave with wave vector $\mathbf{k}_0$ and number $N_{\mathbf{k}_0}$ proportional to the volume V of the system ($N_{\mathbf{k}_0} \propto V$). If the phase of this wave is not well determined, one deals with a polariton noise wave. If the phase of the wave amplitude is well defined, a coherent macroscopic polariton wave appears. It is nothing more than a BEC of polaritons induced by a coherent external electromagnetic field. Ivanov and Keldysh proposed the description of the dynamic and kinetic processes in both cases. To simplify the problem, some restrictions were introduced, namely, the intensity of the polariton wave is assumed to be low enough that the polariton–polariton interaction can be ignored. The coherent wave frequency is taken as lying on the LPB in the region of the bottleneck. Only the phonon–polariton interaction is taken into account.

Ivanov and Keldysh began with the anti-Stokes scattering of an initial macroscopically large (coherent or incoherent) polariton wave. This process occurs with the absorption of the acoustic phonons and gives rise to the up-conversion of the initial polaritons into the excitonlike polaritons of the LPB. Their density of states is greater than that of the photonlike region of the LPB. For this reason, the probability of the anti-Stokes scattering processes involving the absorption of phonons can be greater than the probability of Stokes scattering involving the emission of phonons. The anti-Stokes-scattering processes need the existence of a sufficient number of equilibrium phonons, however. Their depletion during the polariton up-conversion process gives rise to a new phonon mechanism of nonlinear absorption of the initial polariton wave. The strong decrease of the occupation numbers of the phonon modes leads to a decrease of the absorption coefficient of the initial polariton wave, yielding the new mechanism of phonon nonlinearity. On the other hand, the increase of the occupation numbers of the scattered polaritons in the anti-Stokes region, which takes place until these polariton occupation numbers reach the equilibrium values of the phonon occupation numbers, leads to an effective quenching of the scattering process for the initial polaritons. The scattered polaritons begin to return to the initial mode $\mathbf{k}_0$ with high probability. As a result of these processes, the damping coefficient of the initial noise polariton wave becomes lower by some orders of magnitude than in the linear theory.

These theoretical investigations were effectuated within the framework of the Keldysh diagram technique for nonequilibrium processes. The variant of the coherent polariton wave is described with a wave equation for its photon component and a nonlinear Schrödinger equation for its exciton component. The up-scattered polaritons are, in fact, considered to be excitonlike quasiparticles. They and the phonons are described by kinetic equations. The dynamic mixing and superposition of the states of scattered polaritons and phonons in the presence of a coherent polariton mode give rise to the reconstruction of their quasi-energy branches. We discuss this effect in more detail in Subsection 7.4.1.

7.4.1 Acoustic and Optical Phonoritons

The concept of the phonoriton can be introduced starting from different Hamiltonians [55, 56]. The variant proposed by Ivanov [57] is discussed here because it allows several specific physical explanations as well as numerical estimates. Based on these estimates, it was shown that the optical phonoriton is of experimental interest.

In Ref. 57 a coherent macroscopic polariton wave with wave vector $\mathbf{k}_0$ was considered. Such a description is valid as long as the polariton damping is small and the imaginary part of the wave vector $\mathbf{k}_0$ is much less than its real part. At the same time, it is necessary to choose the initial polariton state in such a way that its exciton component is not too small, because the polaritons interact with other quasiparticles only through their exciton components. These two apparently contradictory conditions can be satisfied simultaneously when the frequency $\omega_{\mathbf{k}_0}$ of the initial polariton wave belongs to the LPB, falls inside a region not far from the transverse exciton frequency $\omega_t \equiv \omega_\perp$, and obeys the inequalities

$$\Omega_c \gg \omega_\perp - \omega_{\mathbf{k}_0} \gg \gamma. \tag{7.90}$$

Here Ω_c characterizes the polariton splitting and equals the root mean square of the frequency $\omega_\perp$ and the polariton gap ω_{lt} ($\Omega_c = \sqrt{2\omega_t\omega_{lt}}$). The polariton gap, in its turn, equals the difference between the longitudinal and the transverse exciton frequencies ($\omega_{lt} = \omega_l - \omega_t$). γ is the damping of the polariton wave. For example, in a CdS crystal, where $\Omega_c = 100$ meV and $\omega_{lt} = 2$ meV, the frequency $\omega_{\mathbf{k}_0}$ can be chosen in the region of crystal transparency near the exciton absorption line. The initial coherent polariton wave, as mentioned above, can be regarded as a nonequilibrium Bose condensate of excitons and photons with wave vector $\mathbf{k}_0$, induced by an external electromagnetic field. The density of excitons $n_{\mathbf{k}_0}$ in the coherent polariton mode can be determined by the formula

$$n_{\mathbf{k}_0} = |A(\mathbf{k}_0)|^2 \frac{I}{\hbar\omega_{\mathbf{k}_0} v_g(\mathbf{k}_0)}, \tag{7.91}$$

where I is the power of the external pumping wave, $v_g(\mathbf{k}_0)$ is the polariton group velocity, and $|A(\mathbf{k}_0)|^2$ is a weighting factor that determines the excitonic part of the polariton. In the excitonlike region of the LPB, it can be approximated as

$$|A(\mathbf{k}_0)|^2 \approx \frac{1}{\left[1 + 4\frac{(\omega_t - \omega_{\mathbf{k}_0})^2}{\Omega_c^2}\right]}. \tag{7.92}$$

The Hamiltonian (7.78) for the exciton–photon–phonon system was used but without the exciton–exciton interaction. Because the aim of the paper was the induced BEC of polaritons, the chemical potential μ in Eq. (7.78) was dropped. Instead, the Bogoliubov displacement transformation for the Bose condensate was used, with the exciton operator $a_{\mathbf{k}_0}$ substituted by a macroscopically large amplitude $a_{\mathbf{k}_0} = \sqrt{N_{\mathbf{k}_0}} e^{-i\omega_{\mathbf{k}_0}t}$. It explicitly depends on time with the frequency of the coherent polariton mode $\omega_{\mathbf{k}_0}$ lying on the LPB.

An essential simplification of the calculations was achieved in Ref. 57 in the case of a strong exciton–photon interaction $|\varphi_{\mathbf{q}}|^2 \gg \frac{N_{\mathbf{k}_0}}{N_a}|\theta|^2 \equiv |Q|^2$. In this case, instead of treating the exciton and the photon branches of the spectrum separately, one can introduce the LPB with frequencies ω_p. This means that the three-component system can be reduced to a two-component one. Another simplification in Ref. 57 was the neglect of the Stokes-scattering terms, with only the anti-Stokes process taken into account. Given these approximations, the dispersion equation can be represented in the simple form,

$$(E - \hbar\omega_{\mathbf{p}})(E - B_{\mathbf{p}-\mathbf{k}_0} - \hbar\omega_{\mathbf{k}_0}) = |A(\mathbf{p})|^2\,|Q|^2\,. \tag{7.93}$$

Equation (7.93) characterizes the alteration of the polariton and the phonon spectra. In this two-component model, investigated earlier by Ivanov and Keldysh [55], the exciton component of the polariton plays the role of the representative of this polariton. The excitons have the polariton frequencies $\omega_{\mathbf{p}}$ and the weight factors of the order of unity ($|A(\mathbf{p})|^2 \approx 1$).

The phonoritons can be acoustic or optical, depending on the type of the longitudinal phonons that interact with the excitons. The interaction with optical phonons is of special interest. The optical-phonoriton gap has the dependence

$$\Delta(\mathbf{p} - \mathbf{k}_0) \approx |Q|^2 \approx n_{\mathbf{k}_0}|\mathbf{p} - \mathbf{k}_0|^2.$$

This anisotropic value depends on the angle between the vectors $\mathbf{p}$ and $\mathbf{k}_0$ and achieves its maximum value in a backscattering geometry. The oscillator strength of the induced anti-Stokes phonoriton resonance at energy $\hbar\omega_{\mathbf{k}_0} + B_{\mathbf{p}-\mathbf{k}_0}$ is also proportional to $|Q|^2$.

Estimates of the phonoriton gap were made for the parameters of the CdS crystal: $\hbar\omega_\perp = 2.52$ eV, the energy of the optical phonon $B_{LO} = 38$ meV, $a_{\text{ex}} = 27$ Å, $\epsilon_0 = 8.87$, $\epsilon_\infty = 5.10$, $m_e = 0.2m_0$, $m_h = 1.35m_0$. The energy of the coherent polariton mode was chosen as 2.47 eV, and the exciton density was chosen as $n_{\mathbf{k}_0} = 10^{17}$ cm^{-3}. The phonoriton splitting, determined as $\sqrt{2B_{\mathbf{p}-\mathbf{k}_0}\Delta(\mathbf{p} - \mathbf{k}_0)}$ has a maximum value of 0.3 meV, for the optical phonons and 0.02–0.03 meV for the acoustic phonons.

The main condition for observing the splitting experimentally lies in the relation

$$\sqrt{2B_{\mathbf{p}-\mathbf{k}_0}\Delta(\mathbf{p} - \mathbf{k}_0)} \geq \gamma_{\text{pol}}(\mathbf{p}) + \gamma_{\text{ph}}(\mathbf{p} - \mathbf{k}_0). \tag{7.94}$$

This means that the splitting must prevail over the summary damping of polariton and phonon branches. For an acoustic phonoriton in a CdS crystal at $T = 4.2$ K, $\gamma_{\text{ph}}(\mathbf{p} - \mathbf{k}_0) < \gamma_{\text{pol}}(\mathbf{p}) \leq 0.01$ meV. For an optical phonoriton, the damping is greater, $\gamma_{\text{pol}}(\mathbf{p}) < \gamma_{\text{ph}}(\mathbf{p} - \mathbf{k}_0) \sim 0.1$ meV, but at the same time the phonoriton splitting is also greater. Thus, in both cases, the phonoriton splitting can be measured.

In addition to condition (7.94), the spectral width of the initial coherent polariton mode γ_{imp} must be less than the value of the phonoriton gap,

$$\Delta(\mathbf{p} - \mathbf{k}_0) \geq \gamma_{\text{imp}}. \tag{7.95}$$

In the CdS crystal, the polariton splitting $\hbar\Omega_c$ is greater than the phonon energy B_{LO}. This fact is also in favor of phonoriton observation. The full dispersion equation must also take into account the Stokes terms, which give rise to the instabilities in the system. These issues are discussed in the Subsection 7.4.2 with a simplified two-component model.

7.4.2 Stokes- and Anti-Stokes-Scattering Processes in a Two-Component Model

Shmiglyuk and Pitei [36, 62] and Shmiglyuk, Pitei, and Miglei [63] proposed another variant, which on the one hand confirms the above results, but on the other hand adds some new details to the above picture. They began with the polaritons on the LPB and pointed out that when one takes into account both the Stokes- and the anti-Stokes-scattering processes, it leads to an indirect polariton–polariton interaction by means of the phonon system. Using a single method, they described the quasi-energy spectrum not only in the excitonic range of the spectrum, but also in the range of the phonon frequencies. As in Subsection 7.4.1, the polariton density is assumed to be low enough that the direct interaction between the polaritons can be ignored. The coherent polariton wave with wave vector $\mathbf{k}_0$ is also considered as a nonequilibrium Bose condensate whose frequency lies in the region of the point of inflection on the LPB. The noncondensate polaritons and phonons interact with each other parametrically through the Bose condensate.

The model Hamiltonian they used was chosen in the form

$$H = \hbar\omega(\mathbf{k}_0)a^\dagger_{\mathbf{k}_0}a_{\mathbf{k}_0} + \sum_{\mathbf{p}\neq\mathbf{k}_0}\hbar\omega(\mathbf{p})a^\dagger_{\mathbf{p}}a_{\mathbf{p}} + \sum_{\mathbf{p}} B_{\mathbf{p}}b^\dagger_{\mathbf{p}}b_{\mathbf{p}}$$
$$+ i\frac{1}{\sqrt{N_a}}\sum_{\mathbf{p}\neq\mathbf{k}_0}\Theta(\mathbf{p}-\mathbf{k}_0)\big[A_{\mathbf{p},\mathbf{k}_0}a^\dagger_{\mathbf{p}}a_{\mathbf{k}_0}\big(b_{\mathbf{p}-\mathbf{k}_0} - b^\dagger_{\mathbf{k}_0-\mathbf{p}}\big)$$
$$- A_{\mathbf{k}_0,\mathbf{p}}a^\dagger_{\mathbf{k}_0}a_{\mathbf{p}}\big(b^\dagger_{\mathbf{p}-\mathbf{k}_0} - b_{\mathbf{k}_0-\mathbf{p}}\big)\big],$$

with

$$A_{\mathbf{p},\mathbf{k}_0} = A^*_{\mathbf{k}_0,\mathbf{p}}. \tag{7.96}$$

Here the polariton operators are denoted by $a^\dagger_{\mathbf{p}}$ and $a_{\mathbf{p}}$, while the phonon operators are denoted by $b^\dagger_{\mathbf{p}}$ and $b_{\mathbf{p}}$. The polariton–phonon interaction is characterized by the factors $A_{\mathbf{p},\mathbf{k}_0} = A^*_{\mathbf{k}_0,\mathbf{p}}$ in addition to the constant $\Theta(\mathbf{p}-\mathbf{k}_0) = \Theta(\mathbf{k}_0-\mathbf{p})$. The operators $a^\dagger_{\mathbf{k}_0}$ and $a_{\mathbf{k}_0}$ are treated as a macroscopically large amplitudes, which obey the zeroth-order motion equations

$$i\frac{\mathrm{d}}{\mathrm{d}t}a_{\mathbf{k}_0} = a_{\mathbf{k}_0}\omega(\mathbf{k}_0), \qquad i\frac{\mathrm{d}}{\mathrm{d}t}N_{\mathbf{k}_0} = 0.$$

The noncondensate quasiparticles, corresponding to the interaction of one energy and three quasi-enenergy branches, can be described by the simple and compound operators $a_{\mathbf{p}}$, $a^2_{\mathbf{k}_0}a^\dagger_{2\mathbf{k}_0-\mathbf{p}}/N_{\mathbf{k}_0}$, $a_{\mathbf{k}_0}b_{\mathbf{p}-\mathbf{k}_0}/\sqrt{N_{\mathbf{k}_0}}$, and $a_{\mathbf{k}_0}b^\dagger_{\mathbf{k}_0-\mathbf{p}}/\sqrt{N_{\mathbf{k}_0}}$. Their motion equations can be represented in a matrix form,

$$i\hbar\frac{\mathrm{d}}{\mathrm{d}t}\hat{A}_{\mathbf{p}} = \hat{L}_{\mathbf{p}}\hat{A}_{\mathbf{p}}, \qquad \hat{A}_{\mathbf{p}} = \begin{vmatrix} a_{\mathbf{p}} \\ \dfrac{a^2_{\mathbf{k}_0}a^\dagger_{2\mathbf{k}_0-\mathbf{p}}}{N_{\mathbf{k}_0}} \\ \dfrac{a_{\mathbf{k}_0}b_{\mathbf{p}-\mathbf{k}_0}}{\sqrt{N_{\mathbf{k}_0}}} \\ \dfrac{a_{\mathbf{k}_0}b^\dagger_{\mathbf{k}_0-\mathbf{p}}}{\sqrt{N_{\mathbf{k}_0}}} \end{vmatrix}, \tag{7.97}$$

where the 4×4 matrix $\hat{L}_{\mathbf{p}}$ has the form

$$\hat{L}_{\mathbf{p}} = \begin{vmatrix} \varepsilon_1 & 0 & Q_{AS}A_{\mathbf{p},\mathbf{k}_0} & -Q_SA_{\mathbf{p},\mathbf{k}_0} \\ 0 & \varepsilon_2 & -Q_SA^*_{2\mathbf{k}_0-\mathbf{p},\mathbf{k}_0} & Q_{AS}A^*_{2\mathbf{k}_0-\mathbf{p},\mathbf{k}_0} \\ -Q_{AS}A^*_{\mathbf{p},\mathbf{k}_0} & -Q_SA_{2\mathbf{k}_0-\mathbf{p},\mathbf{k}_0} & \varepsilon_3 & 0 \\ -Q_SA^*_{\mathbf{p},\mathbf{k}_0} & -Q_{AS}A_{2\mathbf{k}_0-\mathbf{p},\mathbf{k}_0} & 0 & \varepsilon_4 \end{vmatrix}. \tag{7.98}$$

Here the following notations have been introduced:

$$\begin{aligned}
\varepsilon_1 &= \hbar\omega(\mathbf{p}) - i\gamma(\mathbf{p}),\\
\varepsilon_2 &= 2\hbar\omega(\mathbf{k}_0) - \hbar\omega(2\mathbf{k}_0 - \mathbf{p}) - i\gamma(2\mathbf{k}_0 - \mathbf{p}),\\
\varepsilon_3 &= \hbar\omega(\mathbf{k}_0) + B_{\mathbf{p}-\mathbf{k}_0} - i\gamma_{\text{ph}}(\mathbf{p} - \mathbf{k}_0),\\
\varepsilon_4 &= \hbar\omega(\mathbf{k}_0) - B_{\mathbf{k}_0-\mathbf{p}} - i\gamma_{\text{ph}}(\mathbf{k}_0 - \mathbf{p}),\\
Q &= i\sqrt{\frac{\mathbf{N}_{\mathbf{k}_0}}{N_a}}\,\Theta(\mathbf{p} - \mathbf{k}_0).
\end{aligned} \tag{7.99}$$

All four energy and quasi-energy values ε_i in Eqs. (7.99) are supplemented by the corresponding dampings $\gamma(\mathbf{p})$ and $\gamma_{\text{ph}}(\mathbf{p})$. The latter values were introduced into the motion equations phenomenologically. The exciton–phonon interaction constants Q are labeled formally by the indices S and AS in order to denote those which describe the Stokes- or the anti-Stokes-scattering processes.

The steady-state solutions of Eqs. (7.97) can be obtained with the determinant equation

$$|\hat{L}_p - \hat{I}E| = 0, \tag{7.100}$$

where $\hat{I}$ is the 4×4 unit matrix and E is the desired eigenvalue. The 4×4 determinant equation can be analyzed qualitatively. To obtain the analytical solutions it is necessary to keep the anti-Stokes terms Q_{AS}, neglecting the Stokes ones Q_S, or vice versa. In the former case, the energy spectrum is

$$\begin{aligned}
E = {} & \frac{1}{2}\left\{\hbar\omega(\mathbf{p}) + \hbar\omega(\mathbf{k}_0) + B_{\mathbf{p}-\mathbf{k}_0} - i[\gamma(\mathbf{p}) + \gamma_{\text{ph}}(\mathbf{p} - \mathbf{k}_0)]\right\}\\
& \pm \frac{1}{2}\sqrt{\left\{\hbar\omega(\mathbf{p}) - \hbar\omega(\mathbf{k}_0) - B_{\mathbf{p}-\mathbf{k}_0} - i[\gamma(\mathbf{p}) - \gamma_{\text{ph}}(\mathbf{p} - \mathbf{k}_0)]\right\}^2 + 4|Q|^2|A_{\mathbf{p},\mathbf{k}_0}|^2}.
\end{aligned} \tag{7.101}$$

If one compares Eq. (7.101) with Eqs. (7.51), one can conclude that two simple propagating waves exist. The phonoriton splitting at the point of phase matching $\hbar\omega(\mathbf{p}) = \hbar\omega(\mathbf{k}_0) + B_{\mathbf{p}-\mathbf{k}_0}$ for small dampings is

$$2|QA_{\mathbf{p},\mathbf{k}_0}| \sim \sqrt{n_{\mathbf{k}_0}}|\Theta(\mathbf{p} - \mathbf{k}_0)|. \tag{7.102}$$

If the damping $|\gamma(\mathbf{p}) - \gamma_{\text{ph}}(\mathbf{p} - \mathbf{k}_0)|$ exceeds value (7.102), the phonoriton splitting does not appear. When only the Stokes-scattering processes are taken into account, the energy spectrum has a similar form, but with one important distinction:

$$\begin{aligned}
E = {} & \frac{1}{2}\left\{\hbar\omega(\mathbf{p}) + \hbar\omega(\mathbf{k}_0) - B_{\mathbf{k}_0-\mathbf{p}} - i[\gamma(\mathbf{p}) + \gamma_{\text{ph}}(\mathbf{k}_0 - \mathbf{p})]\right\}\\
& \pm \frac{1}{2}\sqrt{\left\{\hbar\omega(\mathbf{p}) - \hbar\omega(\mathbf{k}_0) + B_{\mathbf{k}_0-\mathbf{p}} - i[\gamma(\mathbf{p}) - \gamma_{\text{ph}}(\mathbf{k}_0 - \mathbf{p})]\right\}^2 - 4|QA_{\mathbf{p},\mathbf{k}_0}|^2}.
\end{aligned} \tag{7.103}$$

The interaction constant $4|QA_{\mathbf{p},\mathbf{k}_0}|^2$ enters with a minus sign, unlike the anti-Stokes case, in which it was present with a plus sign. If one compares it with examples (7.51), one can conclude that an instability in the system exists. The character of this instability depends on the dispersion law of the bare phonon branch $B_{\mathbf{k}_0-\mathbf{p}}$.

In the case of Stokes phase matching,

$$\hbar\omega(\mathbf{p}) + B_{\mathbf{k}_0-\mathbf{p}} = \hbar\omega(\mathbf{k}_0),$$

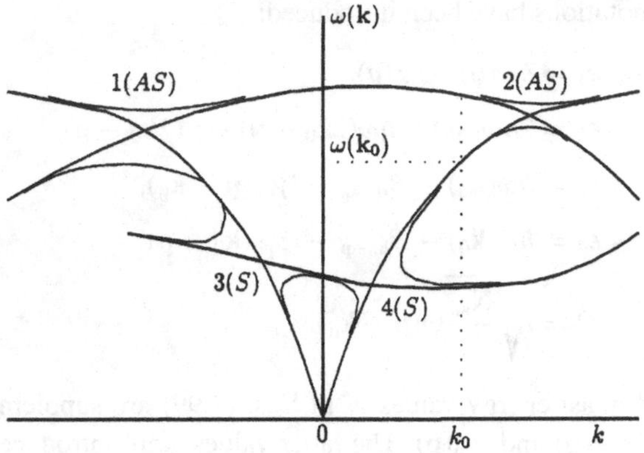

Figure 7.9. The anti-Stokes-scattering (points 1 and 2) and Stokes-scattering (points 3 and 4) processes on the optical-longitudinal phonons of the coherent polariton wave [36].

the imaginary parts of solutions (7.103) are

$$\operatorname{Im} E = -\frac{\gamma(\mathbf{p})+\gamma_{ph}(\mathbf{k}_0-\mathbf{p})}{2} \pm \frac{1}{2}\sqrt{[\gamma(\mathbf{p})-\gamma_{ph}(\mathbf{k}_0-\mathbf{p})]^2+4\left|QA_{\mathbf{p},\mathbf{k}_0}\right|^2}. \qquad (7.104)$$

One of two solutions corresponding to the minus sign before the square root in Eq. (7.104) remains negative at all values of the condensate density $n_{\mathbf{k}_0}$. This solution is stable. The second solution, corresponding to the plus sign, begins to grow when $n_{\mathbf{k}_0}$ increases. At some threshold value $n_{\mathbf{k}_0} = n_c$,

$$n_c = \frac{\gamma(\mathbf{p})\gamma_{ph}(\mathbf{k}_0-\mathbf{p})}{\Theta^2(\mathbf{p}-\mathbf{k}_0)v_0}, \qquad (7.105)$$

the resulting imaginary part becomes equal to zero. Below the threshold point, the system remains stable. Above it the resulting imaginary part turns out to be positive and the system becomes unstable.

In Figure 7.9 the optical phonoriton branches are depicted. At points 1 and 2 the anti-Stokes-scattering processes take place. At these points the phonoriton gaps appear. For real values of the frequency ω lying within the gaps, the wave vectors are complex, and evanescent waves can be observed. At point 1 the phonoriton gap is much larger than in region 2 because backscattering occurs with greater momentum transfer than forward scattering. The existence of evanescent waves is demonstrated by an anomalously high reflectivity of the crystal for a probe signal propagating in the direction $\mathbf{k}_0$ for frequencies lying in range 2 and propagating in the opposite direction for the frequencies belonging to range 1. The phonoriton energy spectrum for the Stokes-scattering processes is also represented in Fig. 7.9. At point 3 the convective instability occurs, whereas at point 4 the absolute instability takes place. For the frequency and wave vectors corresponding to point 3, the system is an amplifier of the propagating waves. For the same parameters corresponding to point 4, the system will act as a generator of new waves, and the phenomenon of induced two-body scattering will be observed. The system will generate light and nonequilibrium optical phonons. Brodin, Kadan, and Matsko [64] demonstrated coherent modes of nonequilibrium optical phonons whose appearance was due to the induced two-body scattering of the initial coherent polariton modes in HgI_2 and PbI_2 crystals.

If one takes into account both the Stokes and the anti-Stokes processes simultaneously, as described by Eq. (7.100), a more complicated energy spectrum can be obtained. There are four bare initial energy and three quasi-energy branches that intersect each other at ten points of phase matching. Four points among them correspond to evanescent waves. One can expect that the other four points represent regions with absolute instabilities, while at the remaining two points the convective instabilities occur. At this point, however, we leave off our discussion of phonoritons.

7.5 In-Depth Study: High-Intensity Polariton Wave near the Stimulated-Scattering Threshold

In this section we consider the Stokes-scattering process of a coherent polariton wave with acoustic phonons, which gives rise in some cases to an instability in the polariton–phonon system. The results reviewed below come entirely from the papers of Keldysh and Tikhodeev [59–61], who proposed a quantum-statistical description of those many-body phenomena that accompany the appearance of the instability near its threshold. On the basis of this work and related papers, it is possible to understand better the intrinsic quantum-statistical processes that take place in the system and manifest themselves as the precursors of the instability phenomenon. Such investigations give a quantum-statistical foundation of the phenomenological theory of instabilities expounded above.

As in the papers by Ivanov and Keldysh [55, 56] dedicated to the anti-Stokes-scattering processes, Keldysh and Tikhodeev considered a high-intensity electromagnetic wave that excites a coherent polariton wave on the LPB in a direct-gap semiconductor with a strong exciton–photon interaction. This coherent polariton wave can be regarded as an induced Bose condensed state of polaritons with wave vector $\mathbf{k}_0$ lying in the region of the LPB bottleneck. The amplitude of this wave is assumed to be a macroscopically large value $a_{\mathbf{k}_0} \sim \sqrt{V}$. It is considered as a given magnitude that is not changed under the influence of scattering processes. Unlike anti-Stokes processes, through which up-conversion of the coherent polaritons and the absorption of the phonons take place, the Stokes-scattering processes transform the coherent polaritons with wave vector $\mathbf{k}_0$ into the scattered polaritons with wave vectors $\mathbf{k}_-$, as shown in Fig. 7.10. This polariton downconversion is accompanied by the emission of an acoustic phonon with wave vector $\mathbf{k}_0 - \mathbf{k}_-$. Although both types of anti-Stokes- and

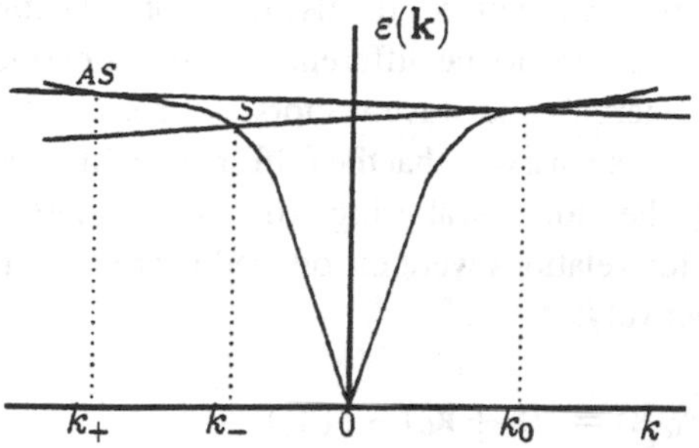

Figure 7.10. The LPB and the spectra of absorbed and emitted acoustic phonons. The Stokes (S) and anti-Stokes (AS) processes of the coherent wave $\mathbf{k}_0$ are denoted [59].

Stokes-scattering processes take place in a real system simultaneously, nevertheless their separate investigation is justified because it permits one to understand the underlying physics better. As in the anti-Stokes case, the correlation between the scattered polaritons $\mathbf{k}_-$ with the emitted phonons $(\mathbf{k}_0 - \mathbf{k}_-)$ in the presence of the condensed polaritons $\mathbf{k}_0$ results in the formation of mixed quasi-energetical polariton–phonon modes. But unlike the anti-Stokes case, in which two simple propagating waves arise in the region $\mathbf{k}_+$, the instability in the renormalized energy spectrum and the growing waves can appear in the region $\mathbf{k}_-$ of the Stokes-scattering process. If the dampings of the bare scattered polaritons and emitted phonons are neglected, they will appear at all densities of the coherent polaritons $n_{\mathbf{k}_0} = a^\dagger_{\mathbf{k}_0} a_{\mathbf{k}_0}/V$. Taking into account these dampings, one can see that the system remains stable until the density $n_{\mathbf{k}_0}$ reaches a critical value n_c [Eq. (7.105)]. The tendency to instability in the system persists, however, and manifests itself when the stable system approaches the threshold of the instability. This instability is nothing other than a growing Mandelstam–Brillouin stimulated process. As shown above, as the density $n_{\mathbf{k}_0}$ increases gradually, the imaginary part of one of the two renormalized quasi-energy branches, being negative, tends to zero. It becomes zero when $n_{\mathbf{k}_0}$ reaches the value n_c. When $n_{\mathbf{k}_0}$ exceeds n_c, the system becomes unstable.

Keldysh and Tikhodeev also took into account the reverse process of down-scattered polaritons $\mathbf{k}_-$ returning into the region near wave vector $\mathbf{k}_0$. These polaritons with $\mathbf{k} \approx \mathbf{k}_0$ are called transmitted polaritons, in contrast to the coherent ones. The down-scattered polaritons with $\mathbf{k} \approx \mathbf{k}_-$ are called scattered polaritons. Thus the physical model contains three types of polaritons (coherent, scattered, and transmitted) as well as the phonons with wave vectors $\mathbf{k} \approx \mathbf{k}_0 - \mathbf{k}_-$. The transmitted polaritons form the fluctuating part that accompanies the given coherent wave with wave vector $\mathbf{k}_0$. This model is described by Keldysh and Tikhodeev in terms of the Keldysh nonequilibrium diagram technique, whose basis was expounded in Chapter 6. The Dyson equations for the nonequilibrium Green's functions were solved in the so-called τ approximation. The self-energy parts are also called polarization operators, because they are determined by the polariton–phonon interaction. The τ approximation means that the normal polarization operators are assumed to be independent of the coherent wave density $n_{\mathbf{k}_0}$, whereas the anomalous polarization operators are linear with respect to the amplitude $a^\dagger_{\mathbf{k}_0}$, i.e., proportional to $\sqrt{n_{\mathbf{k}_0}}$. Keldysh and Tikhodeev showed that the τ approximation is inadequate near the threshold, because of particle accumulation in the weakly damped mode. Formally, it leads to the growth of the diagrams with increasing order of the perturbation theory. But this effect can be avoided if large-loop diagrams of the polarization operators are taken into account. In this new approach, near the threshold all the diagrams are found to be of the same order of magnitude. Just as in the phase-transition theory, the complete solution of the problem requires the use of renormalization group methods.

Below, the polaritons belonging to the different wave-vector regions are treated as different quasiparticles. Their energies and wave vectors are measured from different reference points, but they are chosen in such a way that the fulfillment of the energy and the momentum conservation laws during the Stokes-scattering processes is guaranteed.

The following dispersion relations were introduced for transmitted and scattered polaritons and phonons, respectively:

$$\begin{aligned} \varepsilon_0(\mathbf{k}) &= \varepsilon(\mathbf{k} + \mathbf{k}_0) - \varepsilon(\mathbf{k}_0), \\ \varepsilon_-(\mathbf{k}) &= \varepsilon(\mathbf{k} + \mathbf{k}_-) - \varepsilon(\mathbf{k}_-), \\ \hbar\omega_-(\mathbf{k}) &= \hbar u|\mathbf{k} + \mathbf{k}_0 - \mathbf{k}_-| - [\varepsilon(\mathbf{k}_0) - \varepsilon(\mathbf{k}_-)]. \end{aligned} \tag{7.106}$$

Here u is the sound velocity. As shown in Chapter 6, the Green's functions and self energy parts in the Keldysh diagram technique are 2×2 matrices that depend on two real-time indices. They have the triangular representation

$$\hat{G}(k) = \begin{vmatrix} 0 & G^q(k) \\ G^r(k) & F(k) \end{vmatrix}, \qquad \hat{\Sigma}(k) = \begin{vmatrix} \sigma(k) & \hat{\Sigma}^r(k) \\ \Sigma^q(k) & 0 \end{vmatrix}. \tag{7.107}$$

The τ approximation used in Refs. [59] and [61] supposes that near the surfaces $\varepsilon = \varepsilon_-(\mathbf{k})$ and $\varepsilon = \omega_-(\mathbf{k})$ it is possible to express the components of the normal polariton and phonon self-energy matrices through the corresponding relaxation times or the quasiparticle dampings $i\gamma_p(k)$ and $i\gamma_{\text{ph}}(k)$.

At the same time, the anomalous self-energy parts are taken in the first order of the perturbation theory. In this approximation, which is equivalent to the Bogoliubov approach in the theory of the nonideal Bose gas, the poles of the Green's function describe the energy spectrum of phonoritons in the Stokes-scattering region:

$$\begin{aligned} \varepsilon = {} & \frac{1}{2}\{\varepsilon_-(\mathbf{k}) - \hbar\omega_-(\mathbf{k}) - i[\gamma_p(k) + \gamma_{\text{ph}}(k)]\} \\ & \pm \frac{1}{2}\sqrt{\{\varepsilon_-(\mathbf{k}) + \hbar\omega_-(\mathbf{k}) - i[\gamma_p(k) - \gamma_{\text{ph}}(k)]\}^2 - 4n_{\mathbf{k}_0} v_0 \Theta^2}. \end{aligned} \tag{7.108}$$

One can see that at small values of k, especially at the point $k = 0$, where $\varepsilon_-(0) = \hbar\omega_-(0) = 0$, one of two solutions of Eq. (7.108) has a finite, negative imaginary part and describes a stable state. The imaginary part of the other solution, being negative at small values of $n_{\mathbf{k}_0}$, tends to zero as the density $n_{\mathbf{k}_0}$ increases. It achieves the value of zero at the threshold value n_c, which is

$$n_c = \frac{\gamma_p(k)\gamma_{\text{ph}}(k)}{v_0 \Theta^2}. \tag{7.109}$$

This result correlates with one obtained by Shmiglyuk and Pitei. When the density of coherent polaritons $n_{\mathbf{k}_0}$ exceeds the threshold value n_c, the imaginary part of this solution becomes positive and the corresponding state becomes unstable. The generation of a growing wave takes place at the expense of the coherent polariton mode. For small deviations from the threshold value, when $n_{\mathbf{k}_0}$ can be represented in a form

$$n_{\mathbf{k}_0} = n_c(1 - \lambda), \qquad 0 < \lambda \ll 1, \tag{7.110}$$

the polariton–phonon system remains stable. But in the region of small values of ε and $|\mathbf{k}|$, namely for $|\varepsilon| \le \lambda$ and $|\mathbf{k}| \le \lambda^{1/2}$, the Green's functions become large, being proportional to the matrix $R(k)$ that is divergent at $\lambda \to 0$:

$$R(k) = \begin{vmatrix} 0 & 1/\lambda \\ 1/\lambda & 1/\lambda^2 \end{vmatrix}. \tag{7.111}$$

Keldysh and Tikhodeev [59] noted that the τ approximation became insufficient in the neighborhood of the threshold because there were divergent diagrams that were not taken into account in that approach. It is a specific feature of the Stokes-scattering process and it is related to the possible appearance of the instability in the system. The diagrams of higher-order perturbation theory contain several Green's functions of the type of Eq. (7.111), as well as integrations over the variables ε and $|\mathbf{k}|$. Each integration on the small regions of ε

and k gives a small factor $\lambda^{1+n/2}$, where n is the dimensionality of the system. Nevertheless, the divergence of such a diagram calculated separately grows becuase of the increasing number of the Green's functions (7.111). But it is possible to group these higher-order divergent diagrams in such a way as to compensate for these higher-order divergencies. This compensation relation is similar to the Hugenholtz–Pines formula in the theory of a nonideal Bose gas [65]. After that, the higher-order diagrams have divergence similar only to one of the lower-order diagrams. As Keldysh and Tikhodeev pointed out, the situation near the threshold is like the one in the vicinity of the phase transition. The full solution of the problem requires the use of the renormalization group methods.

In addition, the total spatial density of the down-scattered polaritons near the threshold diverges as

$$n_{-} = \frac{n_c}{\lambda^{1/2}}. \tag{7.112}$$

It grows at the expense of the coherent polariton mode. In this case, the assumption of a fixed coherent wave amplitude becomes invalid. At the end of their paper, Keldysh and Tikhodeev [59] proposed a self-consistent approximation for a simplified one-dimensional model, making allowance for only single-loop diagrams for the polarization operators. In their second paper on these problems, Keldysh and Tikhodeev [60] studied nonstationary Manldelstam–Brillouin scattering of the intense polariton wave. They showed that after the intense coherent polariton wave is switched on, the amplitude of the anti-Stokes line in the Mandelstam–Brillouin spectrum may oscillate in time. The oscillation frequency is determined by the splitting of the phonoriton branches. But this splitting must be more than the total damping of the scattered quasiparticles. When the amplitude of the coherent mode and the phonoriton effect are small, the process of establishing stationary occupation numbers of the scattered polaritons and phonons reveals the character of the relaxation.

In the down-scattering region, the oscillations do not appear in the central part of the Stokes line. They can exist at only the boundaries of this line below the threshold. When the amplitude of the coherent wave exceeds the threshold value, then the stimulated scattering appears and the numbers of scattered quasiparticles grow exponentially in time. The mutual influence of coherently excited and scattered polaritons and their quantum fluctuations in a stationary state will be discussed in Chapter 8 for a slightly different model.

References

[1] The analogy between a laser and superfluid helium is reviewed by P.C. Martin in *Low Temperature Physics - LT9*, J.E. Daunt, D.O. Edwards, F.J. Milford, and M. Yaqub, eds. (Plenum, New York, 1965), p. 9.

[2] N. Nagasawa, N. Nakata, Y. Doi, and M. Ueta, *J. Phys. Soc. Jpn.* **38**, 593 (1975); **39**, 987 (1975).

[3] J.B. Grun, R. Levy, E. Ostertag, and D.P. Vu, in *Proceedings of the Oji Seminar on Physics of Highly Excited Solids, (Japan, 1976)* (Springer, Berlin, 1976), p. 49; R. Levy, C. Klingshirn, E. Ostertag, D.P. Vu, and J.B. Grun, *Phys. Status Solidi B* **77**, 381 (1977).

[4] M. Ojima, T. Kushida, Y. Tanaka, and S. Shionaya, *Solid State Commun.* **20**, 847 (1976).

[5] A.V. Lelyakov and S.A. Moskalenko, *Fiz. Tverd. Tela* **11**, 3260 (1969), [*Sov. Phys. Solid State* **11**, 2642 (1969).

[6] S.A. Moskalenko, *Bose-Einstein Condensation of Excitons and Biexcitons* (RIO, Academy of Sciences of MSSR, Kishinev, 1970) (in Russian).

[7] L.V. Keldysh, in *Problems of Theoretical Physics* (Nauka, Moscow, 1972), pp. 433–444 (in Russian).
[8] H. Haken and A. Schenzle, *Phys. Lett. A* **41**, 405 (1972).
[9] V. Langer, H. Stolz, and W. von der Osten, *Phys. Rev. B* **51**, 2103 (1995).
[10] E.F. Gross, *Usp. Fiz. Nauk* **63**, 575 (1957); **76**, 433 (1962).
[11] V.M. Agranovich and V.L. Ginzburg, *Crystal Optics with Spatial Dispersion and the Theory of Excitons* (Springer, Berlin, 1984).
[12] S.I. Pekar, *Crystal Optics and Additional Light Waves* (Naukova Dumka, Kiev, 1982) (in Russian).
[13] S.A. Moskalenko, *Introduction to the Theory of High Density Excitons* (Shtiintsa, Kishinev, 1983) (in Russian).
[14] D. Frohlich, A. Kulik, B. Uebbing, V. Langer, H. Stolz, and W. von der Osten, *Phys. Status Solidi B* **173**, 31 (1992).
[15] D. Frölich, A. Kulik, B. Uebbing, A. Mysyrowicz, V. Langer, H. Stolz, and W. von der Osten, *Phys. Rev. Lett.* **67**, 2343 (1991).
[16] S.I. Pekar, V.N. Piskovoi, and B.E. Tsekvava, *Fiz. Tverd. Tela* **23**, 1905 (1981); [*Sov. Phys. Solid State* **23**, 1113 (1981)].
[17] Y. Chen, A. Tredicucci and F. Bassani, *Phys. Rev. B* **52**, 1800 (1995).
[18] For a recent review, see P.R. Berman, ed., *Cavity Quantum Electrodynamics* (Academic, Boston, 1994).
[19] See, e.g., Y. Yamamoto, S. Machidwa, and G. Björk, in *Coherence, Amplification, and Quantum Effects in Semiconductor Lasers*, Y. Yamamoto, ed. (Wiley, New York, 1991).
[20] C. Weisbuch, M. Nishioka, A. Ishakawa, and Y. Arakawa, *Phys. Rev. Lett.* **69**, 3314 (1992).
[21] R. Houdre et al., *Phys. Rev. Lett.* **73**, 2043 (1994).
[22] H. Wang, J. Shah, T.C. Damen, W.Y. Jan, J.E. Cunningham, M. Hong, and J.P. Mannaerts *Phys. Rev. B* **51**, 14713 (1995).
[23] T. Amand, X. Marie, P. Le Jeune, M. Brosseau, D. Robart, and J. Barrau, *Phys. Rev. Lett.* **78**, 1355 (1997).
[24] H. Cao, S. Pau, J.M. Jacobson, G. Björk, Y. Yamamoto, and A. Imamoglu, *Phys. Rev. A* **55**, 4632 (1997).
[25] M. Kira et al., *Phys. Rev. Lett.* **79**, 5170 (1997).
[26] C. Ell et al., *Phys. Rev. Lett.* **80**, 4795 (1998).
[27] J. Bloch, in *Proceedings of the Twenty-Fourth Conference on the Physics of Semiconductors* (World Scientific, Singapore, 1999).
[28] M.S. Skolnick, in *Proceedings of the Twenty-Fourth Conference on the Physics of Semiconductors* (World Scientific, Singapore, 1999).
[29] M.D. Martin, G. Aichmayr, L. Vina, J.K. Son, and E.E. Mendez, in *Proceedings of the Twenty-Fourth Conference on the Physics of Semiconductors* (World Scientific, Singapore, 1999).
[30] R. Andre, F. Boeuf, D. Heger, R. Romastain, and L.-S. Dang, in *Proceedings of the Twenty-Fourth Conference on the Physics of Semiconductors* (World Scientific, Singapore, 1999).
[31] S.A. Moskalenko, *Fiz. Tverd. Tela* **4**, 276 (1962).
[32] D. Frolich, E. Mohler and P. Weisner, *Phys. Rev. Lett.* **26**, 554 (1971).
[33] M.C. Brodin, D.B. Goer, and M.G. Matzko, *Pis'ma Zh. Eksp. Teor. Fiz.* **20**, 300 (1974).
[34] M.G. Matzko, *Pis'ma Zh. Eksp. Teor. Fiz.* **21**, 281 (1975).
[35] M.S. Brodin, M.G. Matzko, *Izv. Akad. Nauk USSR Ser. Fiz* **45**, 1567 (1981).
[36] M.I. Shmiglyuk and V.N. Pitei, *Coherent Polaritons in Semiconductors* (Shtiintsa, Kishinev, 1989) (in Russian).
[37] G.G. Hall, *Proc. R. Soc. London Ser. A* **270**, 285 (1962).
[38] S.A. Moskalenko, E.S. Kiselyova, and V.G. Pavlov, *Conference Handbook, ICL'96*, Jan Hala, ed. (International Conference on Luminescence and Optical Spectroscopy of Condensed Matter, 18–23 August 1996, Prague, Czech Republic) (Publisher's name and location, 1996), pp. 12–17.

[39] I.A. Poluektov, Yu.M. Popov, and V.S. Roitberg, *Usp. Fiz. Nauk* **114**, 97 (1974).
[40] H. Haken, *Fortschr. Phys.* **6**, 271 (1958); *Usp. Fiz. Nauk* **68**, 4 (1959).
[41] R.J. Elliott, *Phys. Rev.* **124**, 340 (1961).
[42] A.L. Ivanov, L.V. Keldysh, and V.V. Panashchenko, *Zh. Eksp. Teor. Fiz.* **99**, 641 (1991) [*Sov. Phys. JETP* **72**, 359 (1991)].
[43] S.A. Moskalenko, M.F. Miglei, M.I. Shmiglyuk, P.I. Khadzhi, and A.V. Lelyakov, *Zh. Eksp. Teor. Fiz.* **64**, 1786 (1973) [*Sov. Phys. JETP* **37**, 902 (1973)].
[44] S.A. Moskalenko, A.I. Bobrysheva, A.V. Lelyakov, M.F. Miglei, P.I. Khadzhi, and M.I. Shmiglyuk, *Interaction of Excitons in Semiconductors* (Shtiintsa, Kishinev, 1974) (in Russian).
[45] J.J. Hopfield, *Phys. Rev.* **112**, 1555 (1958).
[46] P.A. Sturrock, *Phys. Rev.* **112**, 1488 (1958).
[47] F.T. Arecchi, G.L. Masserini, and P. Schwendimann, *Riv. Nouvo Cimento* **1**, 181 (1969).
[48] E.M. Lifshitz and L.P. Pitaevskii, *Physical Kinetics*, Vol. 10 of Theoretical Physics Series (Pergamon, New York, 1979), §62–64.
[49] A.M. Fedorchenko and N.Ya. Kotsarenko, *Absolute and Convective Instabilities in Plasmas and Solids* (Nauka, Moscow, 1981), pp. 42–44.
[50] V.R. Misko, S.A. Moskalenko, and M.I. Shmiglyuk, *Fiz. Tverd. Tela* **35**, 3213 (1993).
[51] V.M. Galitskii, S.P. Goreslavskii, and V.F. Elesin, *Zh. Eksp. Teor. Fiz.* **57**, 207 (1969).
[52] V.F. Elesin and Yu.V. Kopaev, *Zh. Eksp. Teor. Fiz.* **72**, 334 (1977).
[53] R. Graham and H. Haken, *Z. Phys.* **237**, 31 (1970).
[54] S.N. Belkin, M.F. Miglei, and S.A. Moskalenko, in *Theoretical Spectroscopy* (Academy of Sciences of the USSR, Moscow, 1977), p. 207 (in Russian); in *Intrinsic Semiconductors at High Levels of Excitation* (Shtiintsa, Kishinev, 1978), p. 124 (in Russian).
[55] A.L. Ivanov and L.V. Keldysh, *Zh. Eksp. Teor. Fiz.* **84**, 404 (1983); [*Sov. Phys. JETP* **57**, 234 (1983)].
[56] A.L. Ivanov and L.V. Keldysh, *Dok. Akad. Nauk SSSR* **264**, 1363 (1982).
[57] A.L. Ivanov, *Zh. Eksp. Teor. Fiz.* **90**, 158 (1986).
[58] A.L. Ivanov, *Dok. Akad. Nauk SSSR* **283**, 99 (1985).
[59] L.V. Keldysh and S.G. Tikhodeev, *Zh. Eksp. Teor. Fiz.* **90**, 1852 (1986).
[60] L.V. Keldysh and S.G. Tikhodeev, *Zh. Eksp. Teor. Fiz.* **91**, 78 (1986).
[61] L.V. Keldysh and S.G. Tikhodeev, *Preprints of the P.N. Lebedev Physics Institute*, No. 331, (1985).
[62] M.I. Shmiglyuk and V.N. Pitei, *Ukr. Fiz. Zh.* **30**, 56 (1985); **31**, 1670 (1986).
[63] M.I. Shmiglyuk, V.N. Pitei and M.F. Miglei, *Ukr. Fiz. Zh.* **32**, 1033 (1987).
[64] M.S. Brodin, V.N. Kadan and M.G. Matsko, *Ukr. Fiz. Zh.* **30**, 1465 (1985); *Fiz. Tverd. Tela* **27**, 776 (1985).
[65] N. Hugenholtz and D. Pines, *Phys. Rev.* **116**, 489 (1959).

8

Nonequilibrium Kinetics of High-Density Excitons

8.1 Condensate Formation in the Bose Gas

An important question in connection with the experiments aimed at achieving BEC of a weakly interacting Bose gas with a finite lifetime is the time needed for the formation of the condensate. It is not at all evident that the condensation can always take place within the lifetime of the system. For the case of liquid He, the assumption of equilibrium seems satisfied because ^{4}He atoms, which do not decay, undergo roughly 10^{11} interparticle-scattering events per second, assuming classical hard-sphere scattering. In other boson systems, however, nonequilibrium effects may have a significant effect. For excitons in a semiconductor, the lifetime of the particle may be only a few hundred scattering times. Spin-polarized hydrogen also has a finite lifetime to nonpolarized states [1], and alkali atoms can recombine into molecules [2]. Naively judging from classical behavior, one might assume that hundreds of scattering times may seem more than adequate time to establish equilibrium. The case of Bose condensation is unique, however, since a macroscopic number of particles must enter a single quantum state that by random-scattering processes alone is highly improbable. Therefore we need not assume that the time for Bose condensation is short.

Early experiments with orthoexcitons in Cu_2O seemed to indicate that the excitons might not have enough time to condense within their lifetime [3]. As discussed in Subsection 1.4.3, the orthoexciton density increased to approach the BEC boundary given in Eq. (1.18) and highly degenerate Bose–Einstein momentum distributions were observed, but a significant condensate fraction did not appear. Instead the gas increased its temperature to accommodate extra particles, reaching temperatures well above the lattice temperature. Later experimental work [4, 5], as well as a theoretical work by Kavoulakis, Baym, and Wolfe [6] has suggested that the temperature of the excitons at high densities can be understood primarily as a result of the balance of a number of processes, including the acoustic- and the optical-phonon emission rates and the Auger recombination rate [7–9], and that the paraexcitons can indeed exceed the critical density for condensation. Neverthless, the fact that at the time of the early experiments in Cu_2O, there was no well-established theoretical answer to the question of how long was needed for condensation, pointed out how little was known about this phase transition that had already been studied for 50 years. Until experiments were performed with finite-lifetime particles, no one had addressed this issue.

Several papers have contributed to the progress on this complicated problem, in which different models of Bose gas and various approaches have been used. Some of the earliest work was done by Levich and Yakhot [10–12] in the context of the theory of superfluid ^{4}He; they put forward the question of what would happen to a system that initially was at temperature $T > T_c$ and through subsequent evolution comes to equilibrium at temperature $T < T_c$. Levich and Yakhot investigated several models in different approaches, beginning with the kinetic theory of Bose condensation of a gas of hard spheres coupled to a thermal bath, without taking into account the transformation of the energy spectrum of the system below the critical point. Earlier, the BEC of a gas of noninteracting photons was studied by Zeldovich and Levich [13]. Using the random-phase approximation, i.e., ignoring coherent effects, Levich and Yakhot found that if no condensate is initially present, then BEC would need an infinite interval of time to be completed if the initial temperature of the system is above T_c. On the other hand, if a negligibly small fraction of the particles has already formed a condensate, this new phase will grow continuously until it reaches its macroscopic level. Levich and Yakhot realized that the Boltzmann equation cannot take into account the coherent effects and noted that the derivation of the Boltzmann equation contains an inconsistency, namely, that the dispersion relation is left unchanged, as that of the free particles. A loss of information about the evolution of the dispersion relation due to interaction occurs in the transition from a reversible Hamiltonian to an irreversible Boltzmann equation. To avoid this irreversibility and to take into account the coherent effects, Levich and Yakhot proposed a reversible quasi-Boltzmann equation, which led to an explosive growth of the condensate. Although they felt that this effect could be an artifact, nevertheless their idea of an instability in the system has something in common with the instability discussed by Stoof [14] as a necessary step of the Bose condensate nucleation, discussed below.

Later papers have continued to wrestle with this issue, e.g., works by Snoke and Wolfe [15], Tikhodeev [16], Eckern [17], and Semikoz and Tkachev [18], and comprehensive approaches by Yu, Kagan, Svistunov, and Shlyapnikov [19–25] and Stoof [14, 26, 27]. The picture that emerges from these papers is that the full process of the phase transition can be divided into three well-defined stages. First is the kinetic stage, in which the gas is quenched into the region near the critical point of the phase transition. The second stage is the coherent stage, in which the proper phase transition takes place and the nonequilibrium condensate is nucleated and built up. The third stage is also a kinetic stage, in which quasi-equilibrium between the condensate formed in the previous stage and the noncondensate quasiparticles is established. In this stage, the final distribution of the Bogoliubov quasiparticles is achieved and the superfluidity of the condensed gas appears.

All these papers contain interesting and instructive results. But the fundamental question of the initial appearance of broken symmetry is addressed most directly in the papers by Stoof [14, 26, 27]. For this reason we review these papers first and return to the question of the initial, kinetic stage of the evolution afterwards. The new approaches proposed by Stoof are based on the Keldysh nonequilibrium Green's function technique developed in the functional variant, as well as on the field theoretical methods in the theory of the Bose gas elaborated by Beliaev [28] and Hugenholtz and Pines [29]. The formulation of the Ginzburg–Landau theory of superfluidity in the coherent state representation was proposed by Langer [30]. Stoof [14, 26, 27] generalized the Landau theory of the second-order phase transition and proposed its nonequilibrium time-dependent variant. This permitted Stoof to describe the coherent stage of Bose condensation. A nonlinear Schrödinger equation for the order parameter was used by Kagan and co-workers in their theory of the appearance of the quasi-condensate. To begin, we discuss the Langer paper [30].

8.1.1 Ginzburg–Landau Theory of Superfluidity in the Coherent State Representation

Langer [30] showed that the coherent state representation of the many-boson wave function can be identified with the complex-valued order-parameter function in the Ginzburg–Landau phenomenological theory of superfluidity [31]. This theory was used to describe the superfluidity of He by Ginzburg and Pitaevskii [32], Pitaevskii [33], and Gross [34]. Langer [30] proposed the quantum-statistical derivation of the Ginzburg–Landau free-energy functional that plays an important role in the recent papers dedicated to the problem of BEC formation mentioned above. In this subsection the main results of the paper by Langer [30] are summarized.

The boson field operator $\Psi(\mathbf{r})$ was introduced in Eq. (2.24) and has the form

$$\Psi = \frac{1}{\sqrt{V}} \sum_{\mathbf{k}} a_{\mathbf{k}} e^{i\mathbf{k}\cdot\mathbf{r}}, \tag{8.1}$$

where $a_{\mathbf{k}}$ are the Bose annihilation operators. The coherent states were introduced by formulas (2.17) and (2.18). They are most useful for many-body systems in which the boson modes are highly occupied and behave in some sense classically. The coherent states, as Langer noted, give the systematic quantum-mechanical description of intense radiation fields for which classical electrodynamics provides a valid, although incomplete, description. The set of all the states of the form

$$\prod_{\mathbf{k}} |\alpha_{\mathbf{k}}\rangle \equiv |\{\alpha_{\mathbf{k}}\}\rangle = |\{\psi\}\rangle \tag{8.2}$$

is a complete set for a many-boson system. Here $\psi(\mathbf{r})$ is the order parameter, which is determined as the diagonal matrix element of the boson field operator of Eq. (8.1) in the basis of states defined in Eq. (8.2):

$$\psi(\mathbf{r}) = \langle\{\alpha\}|\Psi(\mathbf{r})|\{\alpha\}\rangle = \frac{1}{\sqrt{V}} \sum_{\mathbf{k}} \alpha_{\mathbf{k}} e^{i\mathbf{k}\cdot\mathbf{r}}. \tag{8.3}$$

$\psi(\mathbf{r})$ is assumed to have the time dependence $\exp(-i\mu t/\hbar)$, where μ is the chemical potential of the system.

Langer noted that the function $\psi(\mathbf{r},t)$ is the classical Schrödinger field, which describes the many-boson system in the same way the Maxwell field describes the classical limit of quantum electrodynamics. For many-particle Bose systems, as opposed to the many-photon system, the validity of the classical description implies superfluidity. The space of all functions $\psi(\mathbf{r},t)$ is appropriate for the representation of an isothermal canonical ensemble. The statistical fluctuations of the system can be represented as a continuous random motion of the system point $\psi(\mathbf{r},t)$ in the function space $\{\mathrm{Re}\,\psi, \mathrm{Im}\,\psi\}$. Each point of this space is visited by the system with a frequency proportional to the Boltzmann factor $\exp(-F\{\psi\}/k_B T)$, where $F\{\psi\}$ is the Ginzburg–Landau free-energy functional. The grand canonical partition function for the ideal Bose gas Z_0 is

$$Z_0 = \mathrm{Tr}\left(e^{-(H_0-\mu N)/k_B T}\right), \tag{8.4}$$

where, as in Eq. (2.1),

$$H_0 - \mu N = \sum_{\mathbf{k}} \tilde{\mathcal{E}}_{\mathbf{k}} a_{\mathbf{k}}^{\dagger} a_{\mathbf{k}}, \qquad \tilde{\mathcal{E}}_{\mathbf{k}} = \frac{\hbar^2 k^2}{2m} - \mu.$$

In the coherent state representation, Z_0 has the form

$$Z_0 = \prod_{\mathbf{k}} \left(\int \frac{\mathrm{d}^2 \alpha_{\mathbf{k}}}{2\pi} \right) \langle \{\alpha\} | e^{-(H_0 - \mu N)/(k_B T)} | \{\alpha\} \rangle. \tag{8.5}$$

The free-energy functional $F\{\alpha\}$ is determined as

$$F_0\{\alpha\} = -k_B T \ln \langle \{\alpha\} | e^{-(H_0-\mu N)/(k_B T)} | \{\alpha\} \rangle,$$
$$\langle \{\alpha\} | e^{-(H_0-\mu N)/(k_B T)} | \{\alpha\} \rangle = \prod_{\mathbf{k}} \langle \alpha_{\mathbf{k}} | e^{-(\tilde{\mathcal{E}}_{\mathbf{k}})/(k_B T) a_{\mathbf{k}}^{\dagger} a_{\mathbf{k}}} | \alpha_{\mathbf{k}} \rangle. \tag{8.6}$$

Inserting representation (2.19), which expresses the coherent state $|\alpha\rangle$ through the Fock states $|n\rangle$, and taking into account that

$$\langle n | e^{-(\tilde{\mathcal{E}})/(k_B T) a^{\dagger} a} | n' \rangle = \delta_{nn'} e^{-(\tilde{\mathcal{E}})/(k_B T) n},$$

one can obtain the formula

$$\langle \alpha_{\mathbf{k}} | e^{-(\tilde{\mathcal{E}}_{\mathbf{k}})/(k_B T) a_{\mathbf{k}}^{\dagger} a_{\mathbf{k}}} | \alpha_{\mathbf{k}} \rangle = \exp\left[|\alpha_{\mathbf{k}}|^2 \left(e^{-(\tilde{\mathcal{E}}_{\mathbf{k}})/(k_B T)} - 1 \right) \right]. \tag{8.7}$$

This leads to the following expression of the free-energy functional $F\{\alpha\}$ of the ideal Bose gas:

$$\begin{aligned} F_0\{\alpha\} &= -k_B T \sum_{\mathbf{k}} |\alpha_{\mathbf{k}}|^2 \left(e^{-(\tilde{\mathcal{E}}_{\mathbf{k}})/(k_B T)} - 1 \right) \\ &= F_0\{\psi\} = -k_B T \int \mathrm{d}\mathbf{r}\, \psi^*(\mathbf{r}) \left[\exp\left(\frac{\hbar^2 \nabla^2}{2m k_B T} + \frac{\mu}{k_B T} \right) - 1 \right] \psi(\mathbf{r}) \\ &\approx \int \mathrm{d}\mathbf{r} \left[\frac{\hbar^2}{2m} |\nabla \psi(\mathbf{r})|^2 - \mu\, |\psi(\mathbf{r})|^2 \right]. \end{aligned} \tag{8.8}$$

Here the order parameter $\psi(\mathbf{r})$ was used. The functional is quadratic in ψ and is the correct result for noninteracting bosons. Since for the ideal Bose gas, one has $\mu = -|\mu| \le 0$, the functional $F_0\{\psi\}$ does not predict any metastable states except the one at $\psi = 0$. Therefore no superfluidity exists according to above-formulated criterion. To improve Eq. (8.8) and to obtain a description of the system with metastable superfluidity, it is necessary to derive a higher-order term in the expression of $F\{\psi\}$ of the form $|\psi|^4$. Such a term with a positive sign will be generated by the repulsive two-body interaction, which must be added in the full Hamiltonian H of the type of Eq. (2.29). Then the full free-energy functional $F\{\alpha\}$ may be expressed as

$$F\{\alpha\} = F_0\{\alpha\} + \sum_{\Gamma_c} W_{\Gamma_c}\{\alpha\},$$

where only the connected two-body ladder diagrams are included. They are represented in Fig. 8.1. The wavy lines represent the interaction $U(\mathbf{k})$ of Eq. (2.29).

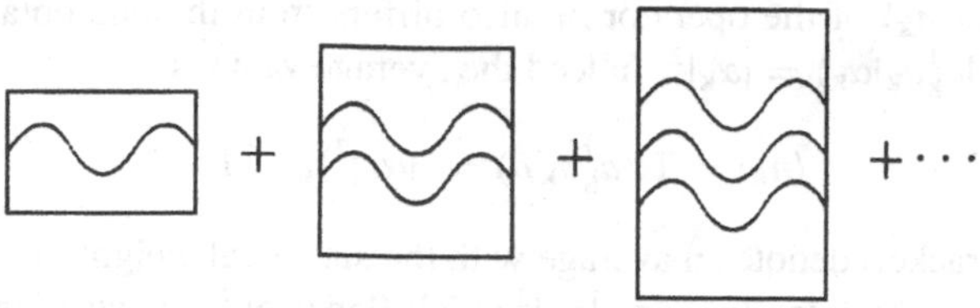

Figure 8.1. The sequence of the two-body ladder diagrams giving the contribution to $F\{\alpha\}$ of the order of $|\alpha|^4$ [30].

The formalism is similar to that developed by Lee and Yang [35]. In this way Langer [30] constructed a free-energy functional of Ginzburg–Landau form,

$$F\{\psi\} = \int d\mathbf{r} \left[\frac{\hbar^2}{2m} |\nabla\psi|^2 - A(T)\,|\psi|^2 + \frac{B\,(T)}{2}\,|\psi|^4 \right] = \int d\mathbf{r}\, f\{\psi\}, \tag{8.9}$$

where $f\{\psi\}$ is the density of the functional. The coefficients A and B depend on the temperature T. The quartic term $|\psi|^4$ will be correct if the ψ function varies slowly over distances of the order of the two-body-scattering length. The crucial temperature dependence of the coefficient $A(T)$ is essentially the same as the temperature dependence of the chemical potential μ. For noninteracting particles, μ is negative at high temperatures and goes to zero at the BEC point. When the repulsive interaction is added and the term $|\psi|^4$ appears with $B > 0$, then μ may be expected to change sign and to become positive below some critical temperature, as required by the phenomenological model of the Landau theory of the second-order phase transition. The functional

$$\rho\{\psi\} = \frac{1}{Z} \exp\left(-\frac{F\{\psi\}}{k_B T} \right) \tag{8.10}$$

is the diagonal element of the density matrix ρ in the coherent state representation. It gives the probability that the system will be found in the state ψ. One can suppose that the statistical weight $\rho\{\psi\}$ has a most probable state $\psi = \psi_s$ in its part of the function space ψ and that ψ_s can describe a stable or metastable state of the system. In the particular case of BEC in the single-particle state $\mathbf{k}_s = 0$, the function ψ_s coincides with ψ_0. In the general case, one can associate ψ_s with a superfluid density n_s and supercurrent $\mathbf{j}_s$ as follows:

$$n_s = |\psi_s|^2, \qquad \mathbf{j}_s = \frac{1}{2i}(\psi_s^* \nabla \psi_s - \psi_s \nabla \psi_s^*). \tag{8.11}$$

For these quantities to be meaningful, ψ_s must be of the order of unity and not of the order of $N^{-1/2}$, where the number of particles N is proportional to the volume V. ψ_s must have the properties of a classical field and must obey the superfluid equation of motion. But the most probable values of ψ_s or ψ_0 do not coincide with the average value ψ of the Bose field operator Ψ, just as n_s and $\mathbf{j}_s$ are not the same as the expected average values of n and $\mathbf{j}$. To evaluate n and $\mathbf{j}$, one must take proper account of the width of the peak $\rho\,\{\psi\}$ near the point $\psi = \psi_s$. In another words, one must include the fluctuations. In such a way the normal components n_n and $\mathbf{j}_n$ of the full density and current appear alongside the superfluid ones:

$$n = n_s + n_n, \qquad \mathbf{j} = \mathbf{j}_s + \mathbf{j}_n. \tag{8.12}$$

The average value $\langle n_{\mathbf{k}} \rangle$ of the operator $n_{\mathbf{k}}$ also differs from the one obtained in the pure coherent state $\langle \alpha_{\mathbf{k}} | a_{\mathbf{k}}^{\dagger} a_{\mathbf{k}} | \alpha_{\mathbf{k}} \rangle = |\alpha_{\mathbf{k}}|^2$. Indeed the average value is

$$\langle n_{\mathbf{k}} \rangle = \mathrm{Tr}(a_{\mathbf{k}}^{\dagger} a_{\mathbf{k}} \hat{\rho}) = \langle |\alpha_{\mathbf{k}}|^2 \rangle_{\psi} - 1, \tag{8.13}$$

where the angle brackets denote an average with the statistical weight of functional (8.10). The expected identity between $\langle n_{\mathbf{k}} \rangle$ and $|\alpha_{\mathbf{k}}|^2$ is fulfilled in only the classical limit $\langle n_{\mathbf{k}} \rangle \gg 1$ and becomes exactly correct only when the mode $\mathbf{k}$ is occupied macroscopically, i.e., when $\langle n_{\mathbf{k}} \rangle$ is of the order of $N \sim V$.

So far, we have discussed the stationary variant of the Ginzburg–Landau theory in the coherent state representation discussed by Langer [30, 36]. Stoof [14, 26, 27] elaborated the time-dependent generalization of the Ginzburg–Landau theory of the second-order phase transition, which allows the possibility of solving the problem of the time needed for formation of a Bose condensate. This theory is discussed in Subsection 8.1.2.

8.1.2 The Nucleation and Buildup of the Bose Condensate

Stoof argued that the solution of the problem of the nucleation of BEC does not lie in the study of the kinetic stages of the phase transition, since the Boltzmann equation describing the kinetic stages is unable to treat the buildup of the coherence. It cannot lead to a macroscopic occupation of the single-particle ground state, since the production rate of the condensate fraction in the thermodynamic limit $N \to \infty$, $V \to \infty$ is nonzero only if a condensate already exists. If the gas is quenched sufficiently far into the critical region of the phase transition, the first stage of its evolution is the kinetic stage discussed in Refs. 10, 12, and 15–18. It leads to the thermalization of the gas, but does not lead to phase coherence; rather, the kinetic stage is related to the transformation of the initially highly nonequilibrium distribution into a quasi-equilibrium distribution. This process is characterized by the typical time scale τ_{el}, which is the average time between two incoherent elastic collisions and has the order of magnitude

$$\tau_{\mathrm{el}} = \frac{1}{n\sigma v} = O\left[\frac{\hbar}{k_B T}\left(\frac{\Lambda}{a}\right)^2\right], \qquad \Lambda^3 n = O(1), \qquad \Lambda = \left(\frac{2\pi\hbar^2}{m k_B T}\right)^{1/2},$$
$$v = \left(\frac{3k_B T}{m}\right)^{1/2}, \qquad \sigma = \pi a^2, \qquad n = \frac{N}{V}, \qquad \frac{a}{\Lambda} \ll 1. \tag{8.14}$$

Here σ is the elastic cross section, a is the scattering length, v is thermal relative velocity of colliding particles, and Λ is the thermal de Broglie wavelength. The scattering length a is positive in the repulsive Bose gas and determines the interaction constant between the particles in Eq. (2.29), $U(0) \approx 4\pi\hbar^2 a/m$.

In the region of the phase transition, the degeneracy parameter $n\Lambda^3$ is $O(1)$ and the ratio a/Λ is much less than unity. Thus the phase transition depends, to a great extent, on the way in which the gas is quenched into the critical region, where the dominant time scale of the kinetic processes is determined by τ_{el}.

The preceding and ensuing kinetic stages of the phase transition have been studied elsewhere [10, 12, 15–18] and were not treated by Stoof. Stoof's main objective [14, 26, 27] was the accurate discussion of the nucleation and the buildup of the condensate. To this end Stoof used a functional approach to the Keldysh nonequilibrium Green's function formalism. On this basis, he elaborated the time-dependent nonequilibrium description of the dynamics of a weakly interacting Bose gas. The theory was used to show how the system initially enters the critical region and to determine the critical temperature T_c of the nonideal Bose gas. Attention was concentrated mainly on the time evolution of the condensate density after the quench of the system into the critical region.

Below, we discuss only qualitative results of the Stoof's papers and the physics involved. The accurate discussion of the nucleation question requires the evaluation of the interaction between the particles within the T matrix approximation, which means the summation of the ladder-type diagrams shown in Fig. 8.1.

The time-dependent Ginzburg–Landau theory elaborated by Stoof [14, 26, 27] contains a motion equation for the order parameter $\psi(\mathbf{r}, t)$ of the form

$$\left(i\hbar \frac{\mathrm{d}}{\mathrm{d}t} + \frac{\hbar^2 \nabla^2}{2m} \right) \psi(\mathbf{r}, t) = [S^+(t, T) + T^+ |\psi(\mathbf{r}, t)|^2]\, \psi(\mathbf{r}, t). \tag{8.15}$$

Here $S^+(t, T)$ is the time-dependent coefficient that plays the crucial role. Taken with the minus sign, it is similar to the chemical potential μ in Eq. (8.8) and is equivalent to the coefficient A in Eq. (8.9). $S^+(t, T)\,\delta(t - t')$ is a simple approximation for the retarded self-energy $\hbar\Sigma^+(0, t, t')$ of a Bose particle with zero momentum. The coefficient $T^+ = 4\pi\hbar^2 a/m$, where a is the scattering length. Only in the case of a positive scattering length will the above Ginzburg–Landau theory be stable and thus useful. T^+ plays the role of the effective interaction constant and corresponds to the coefficient B in Eq. (8.9).

Equation of motion (8.15) was found by Stoof by means of a variational method that uses the action $S[\psi(\mathbf{r}, t), t]$. This action is the desired generalization of the Ginzburg–Landau free-energy functional $F\{\psi\}$. The time dependence of the coefficient $S^+(t, T)$ for different initial temperatures of the Bose gas are represented in Fig. 8.2. Here T_c is the critical temperature of the weakly interacting Bose gas, which differs slightly from the critical temperature of the ideal Bose gas T_0.

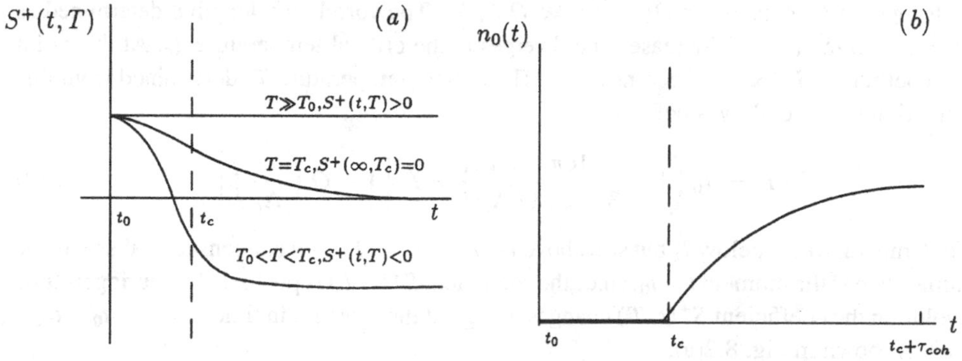

Figure 8.2. (a) Time dependence of the coefficient $S^+(t, T)$ for three different temperatures, (b) time evolution of the condensate density $n_0(t)$ during the coherent stage of the phase transition [14].

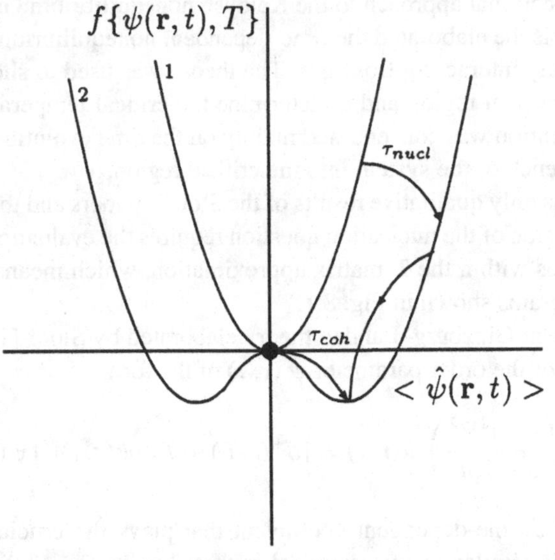

Figure 8.3. Visualizations of the time scale τ_{nucl} associated with the appearance of the instability and the time scale τ_{coh} for the relaxation of the order parameter to its equilibrium value. Curve 1 corresponds to the coefficient $S^+(t, T) > 0$ and curve 2 to $S^+(t, T) < 0$ [14].

Stoof showed that for $T \gg T_0$ the coefficient $S^+(t, T) = 2nT^+ > 0$ is positive. In this case the free-energy functional density $f\,[\psi\,(\mathbf{r},\, t)\,,T]$, determined as

$$f\,[\psi(\mathbf{r},t),\,T] = \frac{\hbar^2 |\nabla\psi(\mathbf{r},t)|^2}{2m} + S^+(t,\,T)\,|\psi(\mathbf{r},t)|^2 + \frac{1}{2}T^+|\psi(\mathbf{r},t)|^4, \tag{8.16}$$

is represented by the curve 1 in Fig. 8.3.

In this region of temperatures the evolution of the gas is described by the quantum Boltzmann equation. When the temperature is lowered, the occupation numbers for momenta $\hbar k < O(\hbar/\Lambda)$ increase and lead to an enhancement of the coherent population of the states with momenta $\hbar k < O(\hbar\sqrt{na}) \ll O(\hbar/\Lambda)$. The correlation length ξ determined as $\xi = \hbar/\sqrt{2mS^+(\infty, T)}$ increases and diverges at the critical temperature T_c. At this point the coefficient $S^+(\infty, T)$ becomes zero. The critical temperature T_c determined from this condition was found by Stoof to be

$$T_c = T_0\left(1 + \frac{16\pi}{3 \times 2.612}\frac{a}{\Lambda_c}\right) = T_0\left[1 + O\left(\frac{a}{\Lambda_c}\right)\right]. \tag{8.17}$$

For temperatures T below T_c but still above T_0 ($T_0 < T < T_c$), the condensate is absent in the initial state at the moment $t = t_0$, since the coefficient $S^+(t, T)$ is positive. But during its time evolution the coefficient $S^+(t, T)$ changes its sign at the moment in time $t = t_c = t_0 + \tau_{\text{nucl}}$. This is shown in Fig. 8.2(a).

At the beginning of this period of time evolution ($t = t_0$), the system is represented by a point ψ lying on curve 1 in Fig. 8.3, whereas at the end of this period ($t = t_c$) the self-consistent state of the system has to be represented by the point ψ lying on the curve 2 in Fig. 8.3. Because of the time dependence of the coefficient $S^+(t, T)$, which differs from the stationary value of the coefficient A in Eq. (8.9), the possibility of having an intermediate non-self-consistent state appears. In this intermediate state the free-energy develops a double-well structure, whereas the coherent state is still zero. This state is unstable and gives rise to the instability leading to the phase transition of the point ψ from one free-energy curve to another. This phase transition giving rise to the nucleation of the condensate is represented in Fig. 8.3 by the nucleation time τ_{nucl}. As Stoof pointed out, the gas has to develop the instability associated with the phase transition by itself.

The time scale τ_{nucl} of the Bose condensate nucleation was determined by Stoof as

$$\tau_{\text{nucl}} = O\left[\frac{\hbar}{k_B(T_c - T)}\frac{a}{\Lambda_c}\right] = O\left(\frac{\hbar}{k_B T_c}\right), \quad T_c - T = O\left(T_c\frac{a}{\Lambda_c}\right). \tag{8.18}$$

Expression (8.18) is not true at temperatures very close to T_c, when $(T_c - T)$ is much less than $T_c a/\Lambda_c$. Surprisingly, it is independent of the density – the nucleation time scale (8.18) is much less than the time of the elastic collisions τ_{el} of Eq. (8.14). Therefore the nucleation process does not impede the phase transition. Comparing the kinetic stage and the nucleation period, one can conclude that the phase transition in this interval of its evolution is determined by the way in which the gas is quenched into the critical region where $T_0 < T < T_c$. But the full course of the phase transition is not complete, since the nucleation is only the beginning of the coherent stage, and it is necessary to study its subsequent evolution.

The nucleus of the condensate that appears as a result of the instability has a small density at the moment $t = t_c$ of the order of

$$n_0(t_c) = O\left[n\left(\frac{a}{\Lambda_c}\right)^2\right]. \tag{8.19}$$

The subsequent buildup of the condensate density is determined by equation of motion (8.15). This stage of evolution is called the coherent stage. The time scale involved in this process has a typical value of

$$\tau_{\text{coh}} = \frac{\hbar}{n_0 T^+} = O\left(\frac{\hbar}{k_B T_c}\frac{1}{n_0 a \Lambda_c^2}\right), \qquad T^+ = \frac{4\pi\hbar^2 a}{m}. \tag{8.20}$$

During the coherent stage the density of the condensate $n_0(t)$ gradually increases, as shown in Fig. 8.2(b), and achieves the value

$$n_0(t_c + \tau_{\text{coh}}) = O\left[n\left(\frac{a}{\Lambda_c}\right)\right]. \tag{8.21}$$

Substituting the value of n_0 of Eq. (8.21) into (8.20), one can estimate the time scale τ_{coh}, which turns out to be of the same order of magnitude as the elastic collision time τ_{el}:

$$\tau_{\text{coh}} = O\left[\frac{\hbar}{k_B T_c}\left(\frac{\Lambda_c}{a}\right)^2\right] = \tau_{\text{el}}, \qquad n\Lambda_c^3 = O(1). \tag{8.22}$$

In this way, for the time t slightly larger than t_c the nucleation of condensate takes place by means of the coherent population of the single-particle ground state. However, by this mechanism only a small condensate density of the order of Eq. (8.19) is formed. At the end of the coherent stage, the density n_0 reaches the value of Eq. (8.21), which is represented in Fig. 8.2(b). The source of this increase is the depopulation of the noncondensate quasiparticle states, the time evolution of which remains to be discussed. Their evolution in time is governed by the conservation law of the full number of particles, which is enforced by the $U(1)$ gauge symmetry of the Hamiltonian.

Stoof arrived at the conclusion that the phase of the order parameter never has a fixed value and the $U(1)$ symmetry is not really broken dynamically. This is, of course, expected, since the system evolves according to a symmetric Hamiltonian [14]. This assertion needs a comment, since the existence of the ground state and the order parameter with a fixed phase is associated with the breaking of the $U(1)$ gauge symmetry of the Hamiltonian. These are two apparently contradictory assertions related to the gauge symmetry of the system. The solution of this paradox can be found when the difference between the most probable value ψ_0 of the order parameter and its average value ψ is taken into account. The consequence of this difference was discussed in Subsection 8.1.2, following the papers of Langer [30, 36], which give the clue to solve the paradox. Taking into account this difference, Stoof represented the order parameter at $T \neq 0$ as a sum of the most probable value ψ_0 and of the fluctuating part $\psi'(\psi = \psi_0 + \psi')$. Considering the fluctuations around ψ_0 as a small perturbation $|\psi'| \ll |\psi_0|$, he obtained the linearized equation of motion for the fluctuating part ψ'. For the time-independent value $n_0 = |\psi_0|^2$, the linearized equation of motion leads to the Bogoliubov dispersion relation for the noninteracting quasiparticles,

$$\hbar\omega(k) = \sqrt{\left(\frac{\hbar^2 k^2}{2m}\right)^2 + 2T^+ n_0 \left(\frac{\hbar^2 k^2}{2m}\right)}. \tag{8.23}$$

But Stoof remarked that, following the work by Lee and Yang [37], near the critical temperature of the phase transition the interaction between the quasiparticles is important and cannot be neglected. Taking this interaction into consideration in a mean-field-like manner, Stoof [26, 27] indicated another dispersion relation different from relation (8.23):

$$\hbar\omega(k) = \sqrt{\left(\frac{\hbar^2 k^2}{2m}\right)^2 - (T^+ n_0)^2}. \tag{8.24}$$

From this dispersion relation, the modes for which $\hbar^2 k^2/2m < T^+ n_0$ are unstable and show an exponential decay of their population. This decay results in the buildup of the condensate during the coherent stage.

The coherent stage does not include the subsequent kinetic stage of the phase transition. During this last stage the condensate and the quasiparticles equilibrate. The characteristic time scale for this stage τ_{rel} was determined by Eckern [17] as

$$\tau_{\text{rel}} = O\left[\frac{\hbar}{k_B T_c}\left(\frac{\Lambda_c}{a}\right)^3\right]. \tag{8.25}$$

Therefore there is a sequence of time scales,

$$\tau_{\text{nucl}} \ll \tau_{\text{el}} \simeq \tau_{\text{coh}} \ll \tau_{\text{rel}} \leq \tau_{\text{inel}},$$

where τ_{inel} is the time scale of the inelastic processes determining the lifetime of the system, which will be the recombination lifetime in the case of excitons. Stoof concluded that the most important requirement for the achievement of the phase transition is

$$\tau_{\text{el}} \ll \tau_{\text{inel}},$$

which is relatively mild requirement.

A somewhat different picture is given by the work of Kagan et al. [19–25], who divided the evolution of the gas into a kinetic stage and a subsequent coherent stage. They pointed out that the initial nonequilibrium state of the gas without a condensate seed gives rise to a particle flux in energy space toward low energies. This stage is described by a nonlinear Boltzmann equation with a characteristic time given by the interparticle collisions. Kagan et al. introduced the notion of the "quasi-condensate," which has a fixed density but a fluctuating phase. When the particles are quenched into a region of low kinetic energies that are less than the potential energy, a quasi-condensate starts to form. In this stage, in their opinion, the fluctuations of the density are suppressed but not of the phase. The short-range-order coherent correlation is established. But the formation of the long-range topological order and the genuine condensate needs a much longer time, which depends on the system size. The off-diagonal long-range order arising after the attenuation of the long-wave phase fluctuations has a size-dependent relaxation time.

The detailed study of the kinetic stage of evolution developed by Kagan et al. confirms the conjecture presented in Subsection 8.1.1. But their investigations of the coherent stage of the evolution lead to the result that complete BEC cannot occur in a finite amount of time in the thermodynamic limit. In the opinion of Stoof [14], the physical picture of two different time scales for the amplitude and for phase fluctuations of the order parameter, which has been put forward by Kagan and co-workers [19–25], is applicable only if these fluctuations exist independently of each other. Stoof argued in favor of a strong coupling between the amplitude and the phase fluctuations. He paid attention to the fact that for the order parameter ψ to be nonzero at the initial moment of the time evolution, one must show that the system has an instability corresponding to the formation of the condensate [14]. Such a step in the full case of the phase transition elaborated by Kagan et al. has not been provided at the present time.

8.2 Population Dynamics of a Bose Gas

In Ref. 15 the population dynamics of a Bose gas near condensation was investigated, but in the normal state. This work showed by numerical calculation that for a weakly interacting Bose gas the number of characteristic scattering times required for achieving an equilibrium Bose–Einstein distribution increases with increasing degeneracy. At the same time, the characteristic scattering time for the degenerate gas is significantly shortened by stimulated emission, with the net result that the equilibrium times may be comparable with the classical case. We review the results of Ref. 15 here. As in Refs. 10, 12, and 38, these calculations used the random-phase approximation, which is valid for a weakly interacting, noncondensed Bose gas. Of course, this approximation means that the model cannot incorporate the transition to phase coherence, but, as we will see, the calculations successfully model the evolution of a Bose gas very far from equilibrium up to the moment just before condensation.

As is well known [39], the basic result of the Bose–Einstein statistics for the occupation number for bosons in equilibrium,

$$N_{\mathbf{k}} = N[E(\mathbf{k}); \mu, T] = \frac{1}{e^{[E(\mathbf{k})-\mu]/k_B T} - 1}, \tag{8.26}$$

follows from postulating a process of stimulated scattering, i.e., that the probability of a particle scattering into a state $\mathbf{k}$ is proportional to $P = (1 + N_{\mathbf{k}})$, in contrast to the

Fermi–Dirac case of repressed scattering or Pauli exclusion $P = (1 - N_{\mathbf{k}})$. Both of these results follow from the random-phase approximation. In a nonequilibrium Bose gas, this corresponds to writing the matrix element for two bosons with wave vectors $\mathbf{k}_1$ and $\mathbf{k}_2$ scattering to $\mathbf{k}_1'$ and $\mathbf{k}_2'$ as

$$\langle \mathbf{k}_1', \mathbf{k}_2' | M a^\dagger_{\mathbf{k}_2'} a^\dagger_{\mathbf{k}_1'} a_{\mathbf{k}_2} a_{\mathbf{k}_1} | \mathbf{k}_1, \mathbf{k}_2 \rangle^2 = M^2(k) N_{\mathbf{k}_1} N_{\mathbf{k}_2} \left[1 + N_{\mathbf{k}_1'}\right] \left[1 + N_{\mathbf{k}_2'}\right], \tag{8.27}$$

where $M(k)$ is the Fourier transform of the particle–particle interaction potential for momentum exchange $\mathbf{k} = \mathbf{k}_1 - \mathbf{k}_1'$. In these calculations, M was assumed to be constant, equal to the standard interaction potential for a hard-sphere gas, $M = (4\pi\hbar^2 a/mV)$, where a is the scattering length, but the exact form of M turns out to be unimportant as long as it remains finite at all k. The total scattering rate into state $\mathbf{k}_1$ can then be expressed as

$$\Gamma_i(\mathbf{k}_1) = \frac{2\pi}{\hbar} \left[\frac{V}{(2\pi)^3}\right]^2 \int \mathrm{d}^3k_2\, \mathrm{d}^3k_1'\, \mathrm{d}^3k_2'\, S(\mathbf{k}_1', \mathbf{k}_2'; \mathbf{k}_1, \mathbf{k}_2) \tag{8.28}$$

where

$$S(\mathbf{k}_1, \mathbf{k}_2; \mathbf{k}_1', \mathbf{k}_2') = \delta(\mathbf{k}_1 + \mathbf{k}_2 - \mathbf{k}_1' - \mathbf{k}_2')\, \delta(E_1 + E_2 - E_1' - E_2') \times M^2(k) N_{\mathbf{k}_1'} N_{\mathbf{k}_2'} \left[1 + N_{\mathbf{k}_1}\right] \left[1 + N_{\mathbf{k}_2}\right].$$

The scattering rate out of state $\mathbf{k}_1$, $\Gamma_o(\mathbf{k}_1)$, is the same expression, but with the integrand $S(\mathbf{k}_1', \mathbf{k}_2'; \mathbf{k}_1, \mathbf{k}_2)$. One can check that the distribution function given in Eq. (8.26) is the equilibrium solution simply by substituting it into the equation for equilibrium, $\Gamma_i(\mathbf{k}) = \Gamma_o(\mathbf{k})$.

The ninth-order integrals for Γ_i and Γ_o can be simplified by integration over all the delta functions and angles, leaving only the integral over energies. When the work is completed, one obtains [40, 41] the following Boltzmann equation for the rate of change of the number of particles per volume with energies in the interval $(E_1, E_1 + \mathrm{d}E_1)$:

$$\begin{aligned} \frac{\mathrm{d}n(E_1)\mathrm{d}E_1}{\mathrm{d}t} &= \frac{2\pi}{\hbar} g^2 \left[\frac{V}{(2\pi)^3}\right]^2 \frac{1}{16}\, \mathrm{d}E_1 \int \mathrm{d}E_2\, \mathrm{d}E_1' \int_{k_{\mathrm{lo}}}^{k_{\mathrm{hi}}} M^2(k)\, \mathrm{d}k \\ &\times \left[\left.\frac{\mathrm{d}E_1}{\mathrm{d}(k^2)}\right|_{k_1}\right]^{-1} \left[\left.\frac{\mathrm{d}E}{\mathrm{d}(k^2)}\right|_{k_2}\right]^{-1} \left[\left.\frac{\mathrm{d}E}{\mathrm{d}(k^2)}\right|_{k_1'}\right]^{-1} \left[\left.\frac{\mathrm{d}E}{\mathrm{d}(k^2)}\right|_{k_2'}\right]^{-1} \\ &\times \{N(E_1')N(E_2')[1 + N(E_1)][1 + N(E_2)] \\ &- N(E_1)N(E_2)[1 + N(E_1')][1 + N(E_2')]\}, \end{aligned} \tag{8.29}$$

where $k_{\mathrm{lo}} = \max(|k_1 - k_1'|, |k_2 - k_2'|)$ and $k_{\mathrm{hi}} = \min(k_1 + k_1', k_2 + k_2')$, $N(E)$ is the average occupation number of states with energy E, and g is the spin degeneracy of the gas. Interactions will, in general, change the form of the dispersion relation $E = \hbar^2 k^2/2m_{\mathrm{ex}}$, and in principle one could use a density-dependent dispersion relation without greatly increasing the difficulty of the calculation. The calculations of Ref. 15 used a constant mass dispersion relation with no renormalization; this is the weak interaction assumption.

In the simulations, an initial nonequilibrium distribution of particles $N(E)$ was represented as a set of floating-point numbers on a grid of discrete energy points, $(N_0, N_{E_1}, \ldots, N_{E_i}, \ldots)$, and the total rate given by Eq. (8.29) was calculated at each energy. A small time step δt was chosen such that the total change of the distribution per iteration would be small, and the distribution was updated according to $N(E) \rightarrow N(E) + \frac{\mathrm{d}N(E)}{\mathrm{d}t}\delta t$. Then the total rate of Eq. (8.29) was recalculated, etc., to evolve the distribution in time.

8.2.1 Evolution of the Particle Distribution Function

Figure 8.4 shows the time evolution of a low-density, nearly classical hard-sphere gas, thermally isolated from its surroundings. An interesting result is that regardless of the initial distribution, the particles move to the classical Maxwell–Boltzmann distribution in less than five particle-scattering times, in which the scattering time is defined as the average time for all N of the particles to scatter once.

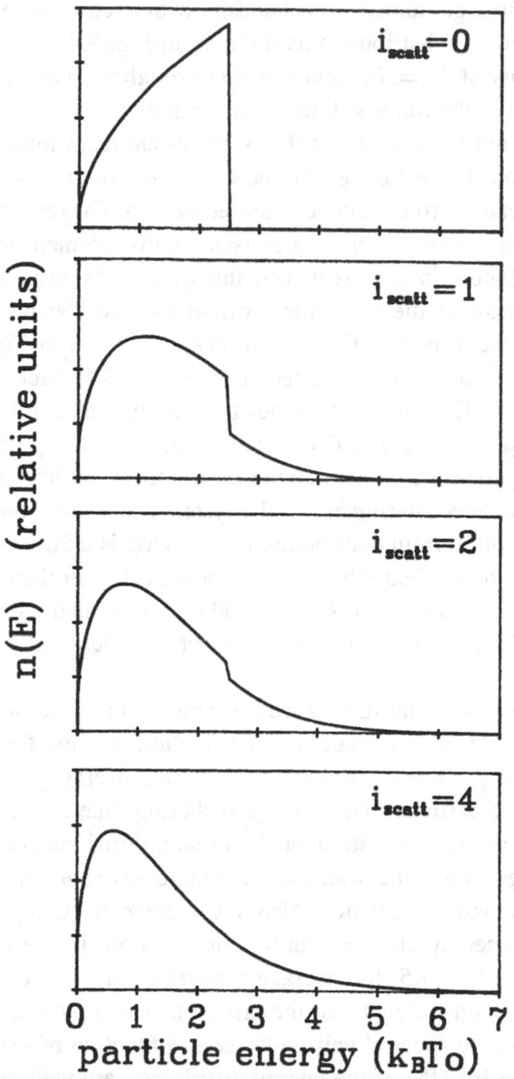

Figure 8.4. Particle distribution for boson gas undergoing only interparticle scattering for the low-density case, $n = 0.00125n_c^0$, where n_c^0 is the critical density at temperature T_0 and T_0 is defined by $\bar{E} = 1.5k_BT_0$. The top curve is the initial distribution, and the following curves correspond to the distribution after all the one, two, and four characteristic scattering times. The index i_{scatt} is the number of scattering events per particle since the initial creation. The last curve is within 1% of a Maxwell–Boltzmann distribution (from Ref. 15).

In this particular case, the initial distribution corresponds to a uniform occupation number for k states up to E_0 and zero occupation number of k states above this energy. The average particle energy of this system, $\bar{E} = 0.6E_0$, gives the final equilibrium temperature, $\bar{E} = 1.5k_B T_0$. Although the full quantum statistics are included in this calculation, the particle density for this case is only $n = 0.00125n_c^0$, where n_c^0 is the critical density at temperature T_0. Therefore quantum effects are not observed.

As the density is increased into the quantum regime, one finds that the system takes many more scattering events to come into equilibrium. Figure 8.5 shows the energy distributions at different times for a Bose gas above critical density with fixed volume, energy, and density, for the same initial energy distribution as the case in Fig. 8.4. The transients caused by the initial discontinuity at $E = E_0$ continue up to roughly 20 scattering times. After 80 scattering times, the distribution is still far from condensation.

On reflection, it is not surprising that the Bose system takes many scattering events to approach condensation. From Eq. (8.27) one can show that the scattering rate into the ground state is identically zero for a noncondensed system. The rate into nearby states goes as the magnitude of the wave vector k for constant matrix element M. Terms of M^2 with higher orders of k will contribute to scattering into these states even less.

After the initial transients die away, the distribution is reasonably well fit by a single-chemical-potential Bose–Einstein distribution $f(E)$, as given in Eq. (8.26); a fit to the distribution at $t = 80$ is shown as the dotted curve in Fig. 8.5. Each fit has two free parameters, the temperature T_s, which determines the energy scale, and the effective unitless chemical potential $\alpha_{\text{eff}} = -\mu_{\text{eff}}/k_B T_{\text{eff}}$, which determines the spectral shape. The fitted values of α_{eff} and T_{eff} change in time, moving toward their equilibrium values. In the case shown in Fig. 8.5, after 80 scattering events the system has reached only $\alpha_{\text{eff}} = 0.024$, with $T_{\text{eff}} = 1.625T_0$. The equilibrium distribution for this case is a 50% condensate ($\alpha = 0$) at a temperature of $3T_0$ (the final equlibrium temperature is higher than T_0, since the average energy of a Bose gas is less than $3/2k_B T$.) In other words, as the gas is equilibrating, its energy spectrum will look similar to that of a much less dense gas in equilibrium, at an elevated temperature.

So far we have described the thermalization process in terms of the number of scattering events per particle. The average time per scattering event for a particle obviously depends on the actual particle density $n = N/V$, the average energy, and the scattering cross section of the particles. The average scattering time also depends on the degree of degeneracy, since the $(1 + N)$ stimulated emission terms enhance the scattering rate. At high quantum degeneracy, the scattering rate increases as n^3, instead of linearly with n, as expected from classical statistics. Also, the average scattering rate changes in time as the degree of degeneracy changes during equilibration. In the example of the high-density case shown in Fig. 8.5, the average scattering time decreases by almost an order of magnitude between the first and the last plot. If one plots the values of the fitted $\alpha_{\text{eff}} \equiv |\mu_{\text{eff}}/k_B T|$ and the spectral half-width Δ as a function of real time, then both the fit value of α_s and the FWHM of the energy distribution are well described by an exponential decay in time. According to these calculations, if a weakly interacting gas has a density above the critical density for condensation at the temperature T_0 defined by the average energy, then the effective chemical potential α_s will reach 0.01 in less than one classical scattering time, in which the classical scattering time is defined as $\tau_0 = 1/n\sigma\bar{v}$ and σ is the s-wave-scattering cross section, nominally equal to $4\pi a^2$ for a hard-sphere Bose gas.

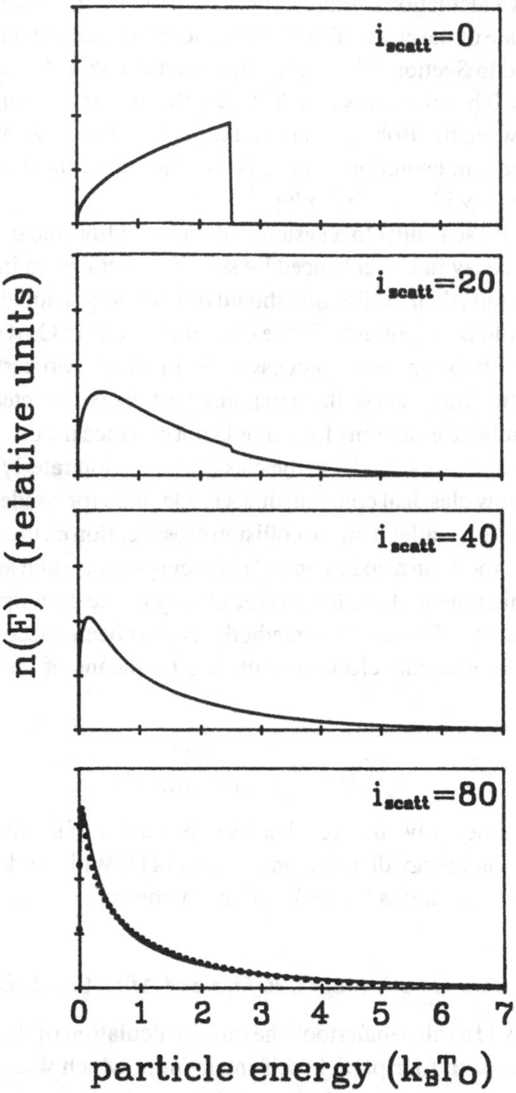

Figure 8.5. Particle distribution for boson gas undergoing only interparticle scattering for the high-density case, $n = 12.5n_c^0$, where n_c^0 is the critical density at temperature T_0 and T_0 is defined by $\bar{E} = 1.5k_BT_0$. The top curve is the initial distribution, and the following curves correspond to the distribution after all the particles have scattered 20, 40, and 80 times. The curve at $i_{\text{scatt}} = 80$ is best fit by a Bose–Einstein distribution of the form of relation (1.46) with $T_s = 1.625T_0$ and $\alpha_s = 0.024$, shown as the dotted curve (from Ref. 15).

At first, it may seem that in the thermodynamic limit, the system can never condense. If one recalls, however, that the occupation number (2.2) gives approximately

$$N_0 = \frac{1}{e^{\alpha_{\text{eff}}} - 1} \sim \frac{1}{\alpha_{\text{eff}}},$$

then the exponential decay of the effective chemical potential α_{eff} means that the condensate fraction grows exponentially in time, on a time scale given by the classical collision time.

Extrapolation of this calculation therefore implies that the gas does condense, with the characteristic time scale of the classical collision time, in agreement with the results of Stoof [14, 26, 27], discussed in Section 8.1. Of course, when the FWHM of the peak of the energy distribution near $E = 0$ becomes less than $\hbar^2/ma^2$, then the random-phase approximation, which allows one to write the Boltzmann equation, will break down and this model cannot account for the subsequent evolution of the gas. At that point, the considerations of Stoof, reviewed in Section 8.1, will come into play.

The application of these results to a system with a finite lifetime depends on the mechanism of decay. If the decay rate is enhanced by stimulated emission in the same way as the interparticle collision rate, then the lifetime should remain proportional to the scattering time and condensation would be impossible. In the case of excitons in Cu_2O, the dominant decay mechanism is believed to be an Auger process [7–9] in which two particles collide and one is annihilated. Since the final state of the remaining particle is expected to be a high-energy ionized state, the stimulated emission effect should not enhance the decay rate, and the Auger process should simply be proportional to the classical collision rate $1/\tau_0 = \sigma n v$. As shown above, a lifetime of a few classical collision times is adequate for condensation. Therefore, if the Auger cross section is smaller than the collision cross section σ, the system can condense.

We note that this method can also be applied to other systems, such as a weakly interacting Bose gas in a harmonic potential. As long as the energy of the particles increases monotonically with k, Eq. (8.29) will hold. This method has also been successfully applied to the case of nonequilibrium electron–electron scattering by means of the long-range screened Coulomb interaction,

$$M_{121'2'} = \frac{1}{V}\frac{4\pi e^2}{\epsilon(K,\omega)K^2}, \tag{8.30}$$

reproducing the experimentally observed density dependence of the rate of disappearance of transients in the electron energy distribution in GaAs [41]. With modern computing power, the simulations of Figs. 8.4 and 8.5 take less than an hour on a typical workstation.

8.2.2 Two Kinetic Stages, Before and After Condensation

Semikoz and Tkachev [18] also undertook the direct calculation of the distribution function of the scalar Bose gas with four-particle self-interaction, which was in contact with a heat bath.

Their investigation is similar to the approach of the Ref. 15 reported in Subsection 8.2.1, but Semikoz and Tkachev performed the numerical integration of the Boltzmann equation over a much wider dynamical range of relative energies and densities of particles and directly analyzed the distribution function. In agreement with our comments above, they noted that it is impossible to observe the buildup of coherence within the framework of the Boltzmann kinetic equation, which is valid only to describe the stage before and after the moment of the condensate formation. The condensation itself involves spontaneous symmetry breaking and the formation of long-range phase coherence, while the kinetic equation describing the time evolution before condensation does not contain any mechanism of symmetry breaking by which the proper phase transition can take place.

The most striking feature of the results of Semikoz and Tkachev [18] is the self-similar character of the distribution function $f(\varepsilon, t)$ at different moments of time as it evolves. The value $f(0, t)$ at zero momentum and single-particle energy $\varepsilon = 0$ reaches infinity at the moment of time $t = t_c$ [$f(0, t_c) = \infty$], which means the onset of condensation. The stationary solution of the kinetic equation has a tendency to the limiting dependence $f(\varepsilon) \sim \varepsilon^{-7/6}$ at small ε [22, 42], but never reaches it.

Adding the condensate "by hand," Semikoz and Tkachev also considered the second kinetic stage taking place after the condensation. Solving the kinetic equation, Semikoz and Tkachev assumed that the initial distribution function had a power dependence at small energies of the type $f(\varepsilon) \sim \varepsilon^{-7/6}$, and an exponential tail at $\varepsilon > 1$. They found that the time evolution leads to a change of the initial power law and to its transformation into the power law $f(\varepsilon) \sim 1/\varepsilon$. Later on, this power law stays at the equilibrium value $f(\varepsilon)$, but the amplitude of $f(\varepsilon)$ gradually decreases. During the period of time before the change of the distribution function reaches the exponential tail of the initial distribution function, the density of the condensate grows linearly with time. As in the previous kinetic stage, the time evolution is self-similar. Semikoz and Tkachev concluded that the picture they observed differs qualitatively from previous work. In the first kinetic stage, the distribution function of excess particles, which eventually has to form the condensate, does not narrow in time, gradually approaching a δ function in an infinite time, as found by Levich and Yakhot [10, 12]. Instead, a singular power-law profile, $f \sim \varepsilon^{-7/6}$, that corresponds to a constant flux of particles in momentum space toward the condensate [42], tends to be established during a finite period of time.

In the second kinetic stage, after the condensate formation, instead of this steady flow, particles from all energy levels throughout the whole energy interval jump directly into the condensate, retaining the equilibrium shape of the distribution function $f(\varepsilon) \sim 1/\varepsilon$ during most of the time. The constant of proportionality in this law gradually decreases until it reaches the equilibrium value [18].

8.2.3 Bose Condensation by means of Particle–Phonon Interaction

In the case of excitons in a solid, another issue arises that has no analog in the case of atoms in a gas. Since the excitons are coupled to the phonon field at all times, one must ask how the existence of exciton–phonon interactions changes the above picture, if at all.

The behavior of the exciton–phonon-scattering rate is different from that of elastic exciton–exciton scattering discussed so far. Comparing it with the rate of thermalization by means of exciton–exciton scattering depends on the density and the temperature regime. If the average energy of the excitons is low compared with the zone-center optical-phonon energy $\hbar\omega_0$ and low compared with the energy $2m_{ex}u^2$, where u is the acoustic sound velocity, then scattering by phonons will be exponentially suppressed. For parabolic exciton bands, only states with energy $2m_{ex}u^2$ couple to the exciton ground state, while excitons with less than the energy $(1/2)m_{ex}u^2$ cannot emit acoustic phonons at all, an effect that has been observed experimentally to lead to extremely rapid diffusion at low temperatures [43].

On the other hand, since exciton–exciton scattering depends linearly on the exciton density, at very low density the thermalization of the exciton gas will always depend on the exciton–phonon-scattering rate. This is quite different from the case of alkali atoms in a trap, in which the rate of thermalization always depends on the particle density.

A Boltzmann equation can be written for the nonequilibrium Bose gas involving exciton–phonon scattering, just as in the case of exciton–exciton scattering, discussed in Subsections 8.2.1 and 8.2.2. With the total rate of scattering written as in Eq. (8.28), the instantaneous net rate of scattering into state **k** is

$$\Gamma_i(\mathbf{k}) = \frac{2\pi}{\hbar}\frac{V}{(2\pi)^3}\int d^3q M^2(\mathbf{q})\{N(E_{\mathbf{k}+\mathbf{q}})[1+N(E_{\mathbf{k}})]$$
$$\times[\delta(E_{\mathbf{k}+\mathbf{q}} - E_{\mathbf{k}} - \hbar\omega_{\mathbf{q}})(1+F_{\mathbf{q}}) + \delta(E_{\mathbf{k}+\mathbf{q}} - E_{\mathbf{k}} + \hbar\omega_{\mathbf{q}})F_{\mathbf{q}}]$$
$$-[1+N(E_{\mathbf{k}-\mathbf{q}})]N(E_{\mathbf{k}})$$
$$\times[\delta(E_{\mathbf{k}} - E_{\mathbf{k}-\mathbf{q}} - \hbar\omega_{\mathbf{q}})(1+F_{\mathbf{q}}) + \delta(E_{\mathbf{k}} - E_{\mathbf{k}-\mathbf{q}} + \hbar\omega_{\mathbf{q}})F_{\mathbf{q}}]\}, \quad (8.31)$$

where $F_{\mathbf{q}}$ is the occupation number of the phonons with wave vector **q**. Acoustic exciton–phonon interaction potential (3.2) can be written as $M^2(\mathbf{q}) = \hbar\Xi^2 q/2\rho u V$, where Ξ is an effective deformation potential for the excitons, ρ is the crystal mass density, and u is the average acoustic sound velocity. Early work by Inoue and Hanamura [38] used this equation to study the approach to Bose condensation of excitons by means of exciton–phonon scattering. Using several approximations to reduce the Boltzmann equation to a Fokker–Planck equation that does not involve integration, they then modeled the evolution of the nonequilibrium exciton gas. As in the exciton–exciton-scattering calculations discussed above and as found by Levich and Yakhot [10] and Tikhodeev [16], since they started with a Boltzmann equation they could not model the buildup of the true condensate; instead, Inoue and Hanamura introduced the condensate by hand at some point in time. It is a general feature of all these Boltzmann equation models that at some point in time, number conservation is violated, as the total rate of scattering out of does not equal the total rate of scattering into excited particle states. This point in time was taken by Inoue and Hanamura as the time of onset of condensation.

Snoke, Braun, and Cardona [44] later used direct numerical solution of the Boltzmann equation for exciton–phonon scattering, treating the phonon population as an unchanging bath, to model the approach to Bose condensation. After all the angles are integrated out, the Boltzmann equation for the change of the exciton population due to phonon emission, analogous to Eq. (8.29), becomes

$$\frac{dn(E_k)dE_k}{dt} = \frac{1}{32\pi^3}\frac{g\Xi^2}{\rho u}dE_k\left[\left.\frac{dE}{d(k^2)}\right|_{E_k}\right]^{-1}\int q^2\,dq$$
$$\times\left\{\left[\left.\frac{dE}{d(k^2)}\right|_{E_k+\hbar uq}\right]^{-1} N(E_k+\hbar uq)[1+N(E_k)](1+F_q)\right.$$
$$\times\Theta(E_{k+q} - E_k - \hbar uq)\Theta(E_k + \hbar uq - E_{k-q})$$
$$-\left[\left.\frac{dE}{d(k^2)}\right|_{E_k-\hbar uq}\right]^{-1} N(E_k)[1+N(E_k-\hbar uq)](1+F_q)$$
$$\left.\times\Theta(E_{k+q} - E_k + \hbar uq)\Theta(E_k - \hbar uq - E_{k-q})\right\} \quad (8.32)$$

where $\Theta(E)$ is the Heavyside function.

One writes a similar equation for the rate of change due to phonon absorption. The distribution of the excitons can then be evolved by the iterative discussed for Eq. (8.29).

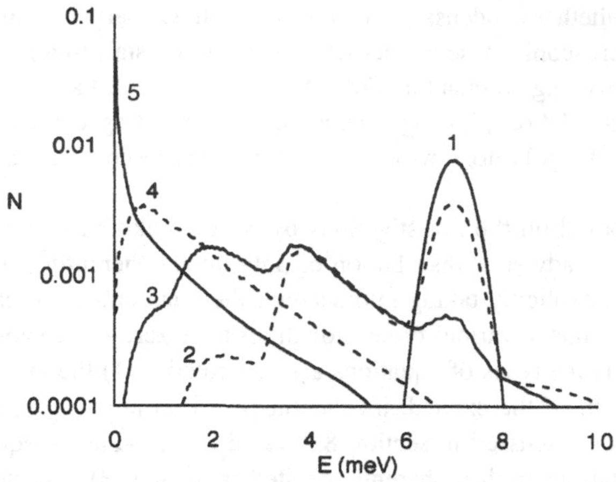

Figure 8.6. Evolution of excitons by means of scattering with acoustic phonons. The system consists of a gas of excitons in Cu_2O at density $n = 7 \times 10^{18}$ cm^{-3} and a lattice-phonon temperature of 16 K, with exciton–exciton scattering "turned off." The curves correspond to the instantaneous distribution $N(E)D(E)\,dE$, where $D(E) \propto \sqrt{E}$ is the density of states, at times (1) $t = 0$, (2) $t = 11.4$ ps, (3) $t = 42$ ps, (4) $t = 120$ ps, and (5) $t = 832$ ps.

As seen in Fig. 1.8, these calculations give an excellent fit to time-resolved experimental data of excitons at low density. The entire time series of Fig. 1.8 is fit with a single parameter, which is the effective deformation potential Ξ. The fit value of the deformation potential was then compared with measurements of the shifts of the exciton line under hydrostatic pressure and static uniaxial stress and was found to agree within a few percent [44].

Figure 8.6 shows a solution of this model for excitons in Cu_2O, in which the exciton–exciton interactions have been turned off, so that the gas evolves only by means of exciton–phonon scattering. As seen in this figure, at late times, a sharp peak occurs at $E = 0$; the subsequent breakdown of number conservation in the calculations indicates that a coherent model of Bose condensation is necessary at this point. The time scales for exciton–phonon scattering found with these calculations are long compared with typical exciton–exciton-scattering times at experimental densities, e.g., in Cu_2O, in which typical elastic-scattering times are a few picoseconds, so that in most experiments the exciton–exciton scattering will control the evolution to Bose condensation. Nevertheless, the time scale for Bose condensation by means of phonon scattering can be significantly less than the exciton lifetime and may be important in experiments at low exciton density and low temperatures.

8.3 In-Depth Study: Quantum Fluctuations and Induced Bose–Einstein Condensation of Polaritons – Squeezed States of Polaritons in Quantum Dots

In the rest of this chapter, we now turn to an inherently nonequilibrium system, which is the induced polariton condensate. Since the ground state of the polariton is poorly defined (the true ground state is the absence of any excitations), the state in which Bose condensation occurs, if it occurs at all, is determined by the rate balance of the various scattering, generation, and decay processes. The issue in this case is not the time scale for condensation, but whether condensation can occur at all. Of particular interest is the transition from the macroscopic Bose condensate (in a bulk crystal) to a microscopic system of low dimensionality (e.g., a quantum dot.) At what point will the coherence disappear? As seen in these calculations, having a coherent laser exciting the states does not necessarily ensure that the polaritons will also exist in a macroscopically occupied coherent state.

This section is based on the investigations by Moskalenko and co-workers [45–52], who solved for the steady-state distribution of polaritons coherently generated in a state $\mathbf{k}_0$ by a pump laser, explicitly taking into account the quantum-statistical fluctuations of the polaritons, in crystals of various dimensionalities from zero-dimensional (0D) to three-dimensional (3D). Three types of equations are generated – (1) the kinetic equations describing the evolution of the excited, incoherent polariton population, analogous to the Boltzmann equations discussed in section 8.2, (2) a Fokker–Planck equation describing the statistical fluctuations in the coherently excited state, and (3) a master equation relating the two populations. Polaritons are assumed to leave the coherently excited state and enter the incoherent population by means of a two-body, polariton–polariton-scattering process [53]. It is assumed that the $\mathbf{k}_0$ polaritons belong to the lower polariton branch (LPB) and that they are converted into two scattered polaritons on the same branch. For simplicity, we do not consider the upper polariton branch.

As a model system we choose a crystallite of finite volume V, which is embedded in a glassy matrix located in an optical resonator. External laser radiation with frequency ω_L and energy flux S_L directly excites a single polariton mode of this crystallite with wave vector $\mathbf{k}_0$. The polariton–polariton interaction leads to the removal of two polaritons from the mode $\mathbf{k}_0$ and the creation of two scattered polaritons with wave vectors $\mathbf{k}_0 \pm \Delta\mathbf{k}$. This process in turn influences the original induced mode $\mathbf{k}_0$, so that the steady-state solutions to the Fokker–Planck equation for the $\mathbf{k}_0$ mode polaritons and the kinetic equations for the scattered polaritons in the LPB must be obtained in a self-consistent fashion. As density increases, the nonequilibrium distribution function of the scattered polaritons gradually begins to be affected by stimulated two-body scattering because of the $(1 + N)$ terms, as discussed in regard to Eq. (8.27). At the same time, the coherently excited polaritons can exhibit such quantum-statistical properties as bunching, antibunching, and squeezed states. These effects occur for small values of the product $S_L V$ for certain detunings from resonance. As $S_L V$ increases, the coherently excited polaritons become completely coherent. In the limit of a bulk crystal they acquire the properties of a driven macroscopically occupied Bose condensate. One of these properties is optical bistability, which will be studied as a function of S_L.

As mentioned above, a specific polariton mode $\mathbf{k}_0$ is coherently excited by external laser radiation in crystals of various dimensionality, including bulk samples and crystallites embedded in a glassy matrix placed in a resonator. The glassy matrix serves as a heat bath, while the resonator selects a single mode with wave vector $\mathbf{k}_0$ to be excited. It is assumed that the pump laser radiation has a power flux S_L and the frequency ω_L, which in general is not the same as the bare frequency $\omega(\mathbf{k}_0)$ of the selected polariton mode, leading to a detuning from resonance $\Delta\omega = \omega(\mathbf{k}_0) - \omega_L$. The evolution of the polariton distribution will in general depend on all of the quantities S_L and $\Delta\omega$, the polariton–polariton interaction constant g, and the crystal volume $V = d^3$.

As the dimensionality of the crystal changes from 3D to 0D, there is a nontrivial change in the role played by quantum fluctuations in the number of polaritons, along with the changes in the corresponding diffusion terms of the Fokker–Planck equation for the reduced density matrix of the selected mode. When the volume is large, i.e., $V \to \infty$, even small coherent pumping amplitudes can give rise to a macroscopically occupied state. A large volume V suppresses the diffusion term in the Fokker–Planck equation and favors drift, i.e., a deterministic description of the induced Bose condensate. For small finite volume V, a minimum pumping intensity is necessary to suppress the fluctuations. Under these conditions, new states and processes appear, which are described below. Instead of a deterministic description of the condensed mode, which is valid for bulk crystals [54–56], one should use a more general quantum-statistical description that will also be correct when the approximation of a fixed, externally applied field is not applicable.

Polaritons in 0D systems can exist only when their wave vector $\mathbf{k}$ and frequency ω satisfy the inequality $k = \omega n(\omega)/c \geq 2\pi/d$, where $n(\omega)$ is the index of refraction; in other words, the wavelength must be less than the dimension of the crystallite. In general, for the linear photonlike branch of the polariton, this cutoff implies an energy much larger than the gap energy, lying on the upper polariton branch – typical dimensions at which size-quantization effects of polaritons are observed are 100–200 nm, e.g., in CdS and GaAs [57, 58] while the wavelength resonant with a typical semiconductor gap of 1–2 eV falls in the range 600–1200 nm. On the LPB, however, $n(\omega)$ increases rapidly as the frequency approaches the polariton gap from below, so that this condition can be met for wave vectors on the LPB near the gap and above. Below this region, it is possible to speak only of the constituent parts of the polariton individually, i.e., an exciton component that undergoes size quantization, leading to a discrete set of confined states, and a photon component with a continuous energy spectrum. In principle, then, we are dealing with particles that have a mixed discrete-continuous energy spectrum. Since we are primarily interested in crystals having dimensions $d \approx$ 60–200 nm, however, we do not explicitly take into account the discreteness of the spectrum, since this becomes important primarily when d becomes comparable with the exciton diameter, of the order of 50 nm or less. In this range of d, the polariton description is entirely replaced by the exciton–photon description [59].

Strictly speaking, the momentum conservation law does not hold in crystallites – it serves to indicate only the most probable quantum transitions. In Subsection 8.3.1, this fact will be built explicitly into the polariton interaction Hamiltonian. This fact naturally gives rise to inhomogeneous broadening of the two-particle quantum transitions involved in the scattering process, which has an important role. If one does not include either homogeneous or inhomogeneous broadening of the polariton spectrum and the quantum transitions, then the constraints imposed by conservation of energy and momentum imply that real polariton conversion is possible only for well-defined initial and final states. This leads to singularities in the kinetic equations, which must be removed by phenomenologically introducing a homogeneous broadening of the polariton states.

The work presented here proposes a quantum-statistical description of the process by which an induced coherent state is established in macroscopic systems with various volumes V and numbers $N_{\mathbf{k}_0} \sim V$ of coherently excited particles in a selected mode $\mathbf{k}_0$. The number $N_{\mathbf{k}_0}$ depends not only on V and power flux S_L of the laser radiation, but also on the distribution of the scattered particles. Although this system is more complicated than the model investigated in Ref. 60, in which a single nonlinear oscillator interacts with a heat bath, the two models are nevertheless similar in regard to the behavior of the selected mode. The difference between them arises because in Ref. 60 the dimensionless volume of the mode is fixed, $N = V/v_0 = 1$, whereas in this model one can track the way in which N influences the processes occurring in the system as it varies from 1 to ∞. The detailed description of the nonequilibrium distribution function of the scattered polaritons on an equal footing with the polaritons of the selected mode gives the possibility of tracking how the real excitation of the mode $\mathbf{k}_0$ takes place under conditions in which the detuning from resonance $\Delta\omega$ is nonzero. The existence of nonequilibrium-scattered particles under steady-state conditions ensures that the energy conservation law can be fulfilled for each microscopic quantum-transition event, because of the lifetime broadening of states, leading to the classical picture of forced oscillations of a damped oscillator under the action of an external periodic force. This investigation showed that the mode volume V and the average number of coherently excited particles $N_{\mathbf{k}_0}$ affect the evolution of their statistical properties in different ways. Thus, for example, the diffusion terms in the Fokker–Planck equation are proportional to V^{-1}, while the statistical properties of the coherently excited polaritons depend on the product $S_L V$.

Coherent pumping can be introduced into the equation of motion by means of an average value $\langle a_{\mathbf{k}_0}\rangle$ of the annihilation operator $a_{\mathbf{k}_0}$ for polaritons. To do this one must relate the intensity of the external laser radiation field to the photon component of the polariton mode $\mathbf{k}_0$ by using the Maxwell–Fresnel boundary conditions at the surface of the resonator mirror [61, 62]. This source of pumping is introduced into the Hamiltonian in a way that is equivalent to the boundary conditions.

8.3.1 The Model Hamiltonian, Master, and Kinetic Equations

The system consisting of the polaritons (S), the heat bath (T), and the source of coherent laser radiation (L) is described by the following Hamiltonian, which is written in the rotating-wave approximation with a coordinate system rotating with frequency ω_L:

$$\mathcal{H} = \mathcal{H}_{S,0} + \mathcal{H}_{S,L} + \mathcal{H}_{S,\text{int}} + \mathcal{H}_T + \mathcal{H}_{ST}. \tag{8.33}$$

The free-polariton Hamiltonian $\mathcal{H}_{S,0}$ has the form

$$\mathcal{H}_{S,0} = \sum_{\mathbf{k}} \hbar[\omega(\mathbf{k}) - \omega_L] a_{\mathbf{k}}^{\dagger} a_{\mathbf{k}}, \tag{8.34}$$

where $a_{\mathbf{k}}^{\dagger}$ and $a_{\mathbf{k}}$ are creation and annihilation operators. The Hamiltonian for the interaction of polaritons of the $\mathbf{k}_0$ mode with the external laser radiation $\mathcal{H}_{S,L}$ and among themselves $\mathcal{H}_{S,\text{int}}$ are as follows:

$$\begin{aligned}
\mathcal{H}_{S,L} &= i\left[(dE_0)a_{\mathbf{k}_0}^{\dagger} - (dE_0)^* a_{\mathbf{k}_0}\right], \\
\mathcal{H}_{S,\text{int}} &= \frac{1}{2V} \sum_{\mathbf{p},\mathbf{q},\mathbf{k},\mathbf{s}} g\phi(\mathbf{k},\mathbf{s}) a_{\mathbf{p}}^{\dagger} a_{\mathbf{q}}^{\dagger} a_{\mathbf{q}+\mathbf{k}} a_{\mathbf{p}-\mathbf{s}} \\
&= U + L + F^{+} + F + P^{+} + P + \cdots \\
&= U + L + \mathcal{H}_{S,\text{int}}' + \cdots,
\end{aligned} \tag{8.35}$$

where E_0 is the amplitude of the driving field within the resonator. Its relation to the external laser radiation follows the notation of Refs. 61 and 62; here d is the dipole moment of the transition. The interaction parameter is chosen to be a constant g for wave vectors that do not exceed the inverse exciton radius a_{ex}^{-1} and to vanish for all other ranges of wave vector. The function $\phi(\mathbf{k}, \mathbf{s})$ takes into account the deviation from the law of momentum conservation; as $V \to \infty$ it reduces to a Kronecker δ symbol. For simplicity it is assumed that the smeared-out function $\phi(\mathbf{k}, \mathbf{s})$ retains the property $\phi(\mathbf{k}, \mathbf{s}) = \phi(\mathbf{p} - \mathbf{s}, \mathbf{p} - \mathbf{k})$.

It is expedient to write out those parts of the Hamiltonian $\mathcal{H}_{S,\mathrm{int}}$ that correspond to the processes that are interesting:

$$
\begin{aligned}
U &= \frac{g\phi(0,0)}{2V} A^+ A, \\
L &= \frac{2N_{\mathbf{k}_0}}{V} \sum_{\mathbf{q},\Delta\mathbf{k}} g\phi(\Delta\mathbf{k}, 0) a^\dagger_{\mathbf{k}_0+\mathbf{q}} a_{\mathbf{k}_0+\mathbf{q}+\Delta\mathbf{k}}, \\
F^+ &= \frac{A^+}{2V} \sum_{\mathbf{k},\Delta\mathbf{k}} g\phi(\mathbf{k}, \Delta\mathbf{k}) a_{\mathbf{k}_0+\mathbf{k}} a_{\mathbf{k}_0-\Delta\mathbf{k}}, \\
P^+ &= \frac{a^\dagger_{\mathbf{k}_0}}{V} \sum_{\mathbf{p},\mathbf{q},\Delta\mathbf{k}} g\phi(\mathbf{q}, \Delta\mathbf{k}) a^\dagger_{\mathbf{k}_0+\mathbf{p}} a_{\mathbf{k}_0+\mathbf{p}+\mathbf{q}} a_{\mathbf{k}_0-\Delta\mathbf{k}}, \\
A^+ &= (a^\dagger_{\mathbf{k}_0})^2, \qquad N_{\mathbf{k}_0} = a^\dagger_{\mathbf{k}_0} a_{\mathbf{k}_0}, \\
B &= A^+ A, \qquad C = A A^+.
\end{aligned}
\tag{8.36}
$$

The operators F^+ and F describe the conversion of two polaritons in the state $\mathbf{k}_0$ into two scattered polaritons with momenta $\mathbf{k}_0 + \mathbf{k}$ and $\mathbf{k}_0 - \Delta\mathbf{k}$. According to Ref. 53, for a bulk crystal points 1 and 2 shown in Fig. 8.7 are located at a distance of the order of $\mathbf{k}_0$ from the

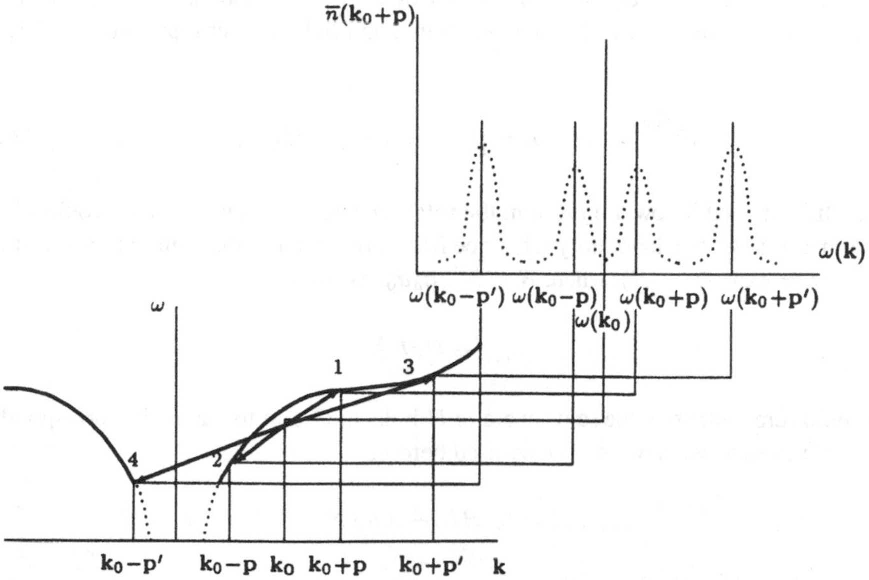

Figure 8.7. Dispersion law for the LPB in crystallites and the distribution function of scattered polaritons, found as the solution of self-consistent kinetic equations (8.55), described in Subsection 8.3.3 [46].

point $\mathbf{k}_0$, while points 3 and 4 are at a distance of the order of $2\mathbf{k}_0$, away. The operators P^+ and P describe scattering processes with the participation of one polariton in the mode $\mathbf{k}_0$. The explicit forms of the terms $\mathcal{H}_T$ and $\mathcal{H}_{ST}$ are not presented here. They are included in the standard way and give rise to phenomenological constants in the master and the kinetic equations [50, 63].

Here one remark is needed. In what follows, the terms F^+ and F of the Hamiltonian $\mathcal{H}'_{S,\mathrm{int}}$ are considered within the framework of perturbation theory. They give rise to the conversion of the coherently excited polaritons into the scattered ones. The terms F^+ and F are not taken into account exactly, as is done in the Bogoliubov theory of the weakly nonideal Bose gas, because the coherently exited polaritons in crystallites, whose dimensionalities change from 0D to 3D, in general do not form a true Bose condensate with a macroscopically large, fixed amplitude. Only in the limiting case $V \to \infty$ is it possible to substitute the operators $a^\dagger_{\mathbf{k}_0}$ and $a_{\mathbf{k}_0}$ by c numbers. In objects such as crystallites with a small volume V, even under the action of the external coherent laser radiation, the polaritons of the mode $\mathbf{k}_0$ are not purely coherent. In spite of their high density, their total number in a small-volume crystallite is low. In this case, the fluctuations of the operators $a^\dagger_{\mathbf{k}_0}$ and $a_{\mathbf{k}_0}$ are important and must be studied explicitly.

In some respects, the properties of the coherently excited polaritons in the crystallite mode $\mathbf{k}_0$ are similar to the properties of the Bose condensate at nonzero temperatures T not far from the critical temperature T_c. In this case, the density of the condensate is small and its influence on the noncondensate particles is not as important as at $T = 0$. Therefore one cannot apply the Bogoliubov procedure for the terms F^+ and F. Our aim in this section is to track the gradual alterations of the quantum-statistical properties of the coherently excited polaritons and to describe by a single approach the different limiting cases of their existence in stationary conditions. For this purpose, the proposed method is more adequate than the assumption of a given coherent condensate.

The quantum Liouville equation for the density matrix or statistical operator ρ of the polariton system S in the rotating reference frame and Schrödinger representation has the form

$$i\hbar\frac{\mathrm{d}\rho}{\mathrm{d}t} = [\mathcal{H}_S, \rho] = \left[\mathcal{H}_{S,0} + \mathcal{H}_{S,L} + \mathcal{H}_{S,\mathrm{int}}, \rho\right]. \tag{8.37}$$

The results obtained below are given in the rotating reference frame. If one would like to pass from a rotating to a laboratory reference frame, one can use the unitary transformation operator $U = \exp(-i\omega_L t N)$, where $N = \sum_{\mathbf{p}} a^\dagger_{\mathbf{p}} a_{\mathbf{p}}$, as follows:

$$\rho_{\mathrm{Lab}} = U\rho U^\dagger.$$

One should remember that the same operator U links the Hamiltonian in the corresponding reference frames as we have repeatedly used before,

$$\mathcal{H} = U^\dagger H U - i\hbar\omega_L N.$$

The density matrix in the interaction representation is

$$\rho_i(t) = e^{(i/\hbar)t\mathcal{H}_{S,0}}\rho(t)e^{-(i/\hbar)t\mathcal{H}_{S,0}}, \tag{8.38}$$

which obeys the equation

$$i\hbar\frac{d\rho_i(t)}{dt} = [\mathcal{H}_{SL,i}(t) + \mathcal{H}_{S,\text{int},i}(t), \rho_i(t)], \tag{8.39}$$

where the operators $\mathcal{H}_{SL,i}(t)$ and $\mathcal{H}_{S,\text{int},i}(t)$ are also taken in the interaction representation (8.38).

Following the method elaborated in the quantum theory of lasers [63], the rate of change of the density matrix can be divided into two parts. One of them is the coherent change under the influence of the laser radiation,

$$i\hbar\frac{d\rho_i(t)}{dt}\bigg|_{\text{coh}} = [\mathcal{H}_{SL,i}(t), \rho_i(t)], \tag{8.40}$$

and the other one is due to incoherent scattering processes under influence of the interaction Hamiltonian $\mathcal{H}_{S,\text{int},i}(t)$,

$$i\hbar\frac{d\rho_i(t)}{dt}\bigg|_{\text{incoh}} = [\mathcal{H}_{S,\text{int},i}(t), \rho_i(t)]. \tag{8.41}$$

The Liouville equation in the Schrödinger representation can be rewritten in the form

$$i\hbar\frac{d\rho}{dt} = \left[\mathcal{H}_{S,0} + \mathcal{H}_{S,L}, \rho\right] + e^{-(i/\hbar)t\mathcal{H}_{S,0}}\left[i\hbar\frac{d\rho_i(t)}{dt}\bigg|_{\text{incoh}}\right]e^{(i/\hbar)t\mathcal{H}_{S,0}}. \tag{8.42}$$

Incoherent contribution (8.41) can be represented in the form

$$\rho_i(t) = \rho_i(0) - \frac{i}{\hbar}\int_0^t dt'[U_i(t') + L_i(t') + \mathcal{H}'_{S,\text{int},i}(t'), \rho_i(t')]. \tag{8.43}$$

The approximate solution of this equality is obtained by means of first-order iteration for the operators U and L and second-order iteration for the operator $\mathcal{H}'_{S,\text{int}}$. Thus the incoherent evolution can be approximated by the formula

$$\begin{aligned}\rho_i(t) \simeq \rho_i(0) &- \frac{i}{\hbar}\int_0^t dt'[U_i(t') + L_i(t'), \rho_i(0)] \\ &- \frac{1}{\hbar^2}\int_0^t dt_1 \int_0^{t_1} dt_2[\mathcal{H}'_{S,\text{int},i}(t_1), [\mathcal{H}'_{S,\text{int},i}(t_2), \rho_i(0)]],\end{aligned} \tag{8.44}$$

which substitutes the small-scale variations by the large-scale variations. The first-order term in the interaction operator $\mathcal{H}'_{S,\text{int},i}$ was dropped because it is nondiagonal in the operators of the scattered polaritons and because its average value calculated with the density matrix of these polaritons taken in the diagonal representation is zero, as will be clear from the discussion below. For the same reasons, one can drop the mixed term of the second-order interaction,

$$[U_i(t_1) + L_i(t_1), [\mathcal{H}'_{S,\text{int},i}(t_2), \rho_i(0)]],$$

and can simplify the second-order term of Eq. (8.44), keeping only a few combinations of the operators F^+, F, P^+, and P, as follows:

$$\begin{aligned}&[F_i^\dagger(t_1), [F_i(t_2), \rho_i(0)]], \qquad [F_i(t_1), [F_i^\dagger(t_2), \rho_i(0)]], \\ &[P_i^\dagger(t_1), [P_i(t_2), \rho_i(0)]], \qquad [P_i(t_1), [P_i^\dagger(t_2), \rho_i(0)]].\end{aligned} \tag{8.45}$$

Here one remark should be made. Approximate expression (8.44) contains the density matrix $\rho_i(t')$ in the limit $t' = 0$. To the accuracy of the highest order of the perturbation theory, $\rho_i(0)$ in this formula can be substituted by the operator $\rho_i(t)$ at the other limit of the integration. It is possible to pull out the function $\rho_i(t')$ from the integrand and to substitute it by its value at one of the integration limits, if the function $\rho_i(t')$ changes in time much more slowly than the other part of the integrand, which has the form $\exp(i\varepsilon t')$. The latter factor can be considered as a rapidly varying function in the limit $|\varepsilon|\, t \gg 1$. The value ε determines the energy transfer during the scattering process. Its inverse absolute value $|\varepsilon|^{-1}$ determines the collision or memory time τ_m. The upper integration limit t has the meaning of the free path time or relaxation time τ_{rel} and is supposed to be much more than the collision time ($\tau_{\text{rel}} \gg \tau_m$).

The approach in which one extracts the density matrix from the integrand and expands the integration limit to infinity ($t \to \infty$) is known as a Markov approximation [65, 66]. The extension of the time t to infinity means the introduction of irreversibility in the scattering process, because the time needed for the system to return back in the initial state at $t = 0$ becomes infinite.

In the calculations performed below, the density matrix $\rho_i(0)$ is used. After that, the reduced density matrix $\rho_{\mathbf{k}_0,i}(0)$ of the coherently excited polaritons and the mean occupation numbers $\bar{n}_{\mathbf{k}_0+\Delta\mathbf{k}}(0)$ of the scattered polaritons is substituted by the values $\rho_{\mathbf{k}_0,i}(t)$ and $\bar{n}_{\mathbf{k}_0+\Delta\mathbf{k}}(t)$. The total number of polaritons is divided into two parts. One of them belongs to the mode $\mathbf{k}_0$ and represents the coherently excited polaritons, whereas the other part is referred to as the scattered or the incoherent polaritons. These are distributed among all the modes $\mathbf{k}_0 + \Delta\mathbf{k}$, where $\Delta\mathbf{k} \neq 0$.

The density matrix $\rho_i(0)$ is represented in a factorized form in which the coherently excited mode $\mathbf{k}_0$ is distinguished in the following way:

$$\rho_i(0) = \rho_{\mathbf{k}_0,i}(0) \prod_{\Delta\mathbf{k}\neq 0} \rho_{\mathbf{k}_0+\Delta\mathbf{k},i}(0) = \rho_{\mathbf{k}_0,i}(0)\rho_{\{\Delta\mathbf{k}\},i}(0) \tag{8.46}$$

The statistical operator $\rho_{\mathbf{k}_0,i}(0)$ is chosen in the representation of the coherent states $|\alpha\rangle$, which are more appropriate for the description of the excited polaritons. The coherent states are the eigenfunctions of the annihilation operator $a_{\mathbf{k}_0}$,

$$a_{\mathbf{k}_0}|\alpha\rangle = \alpha|\alpha\rangle.$$

The index $\mathbf{k}_0$ of the eigenvalue α is dropped.

The diagonal representation of the density matrix $\rho_{\mathbf{k}_0,i}(0)$ in terms of the coherent states has the form

$$\rho_{\mathbf{k}_0,i}(0) = \int \mathrm{d}^2\alpha\, P(\alpha, \alpha^*, 0)|\alpha\rangle\langle\alpha|,$$

$$\alpha = \alpha' + i\alpha'', \qquad \mathrm{d}^2\alpha = \mathrm{d}\alpha'\mathrm{d}\alpha'', \tag{8.47}$$

which in the case of a well-defined coherent state transforms into

$$\rho_{\mathbf{k}_0,i}(0) = \frac{1}{\pi}|\alpha\rangle\langle\alpha|.$$

But these diagonal representations are not sufficient to take into account all the quantum fluctuations in some models. This fact was established by Drummond and Walls [60] for a model of a nonlinear oscillator interacting with a thermal bath. Below, the nondiagonal representation in terms of the coherent states proposed by Drummond and Gardiner [67] are used.

The statistical operators of incoherent polaritons with wave vectors $\mathbf{k}_0 + \Delta\mathbf{k}$ are chosen in the diagonal representation of the Fock states, which are eigenstates of the number operators $a^\dagger_{\mathbf{k}_0+\Delta\mathbf{k}} a_{\mathbf{k}_0+\Delta\mathbf{k}}$;

$$a^\dagger_{\mathbf{k}_0+\Delta\mathbf{k}} a_{\mathbf{k}_0+\Delta\mathbf{k}}|m_{\Delta\mathbf{k}}\rangle = m_{\Delta\mathbf{k}}|m_{\Delta\mathbf{k}}\rangle.$$

In the particular case

$$P(m_{\Delta\mathbf{k}}, \Delta\mathbf{k}, 0) \simeq \exp\left[-\frac{\hbar\,|\omega(\mathbf{k}_0+\Delta\mathbf{k})-\omega_L|}{k_B T} m_{\Delta\mathbf{k}}\right], \tag{8.48}$$

the statistical operator

$$\rho_{\mathbf{k}_0+\Delta\mathbf{k},i}(0) = \sum_{m_{\Delta\mathbf{k}}=0}^{\infty} P(m_{\Delta\mathbf{k}}, \Delta\mathbf{k}, 0)|m_{\Delta\mathbf{k}}\rangle\langle m_{\Delta\mathbf{k}}| \tag{8.49}$$

transforms into the quasi-equilibrium statistical operator.

Relation (8.44) gives rise to the approach on which basis the master equation for the reduced density matrix $\rho_{\mathbf{k}_0,i}(0)$ and the kinetic equations for the mean occupation numbers $\bar{n}_{\mathbf{k}_0+\Delta\mathbf{k}}(t)$ are deduced. These equations must be self-consistent. That means that in both parts of each equation, the values to be found must be present at the same moment t. But in basic equation (8.44), $\rho_i(t)$ is present on the left side, whereas on the right side $\rho_i(0)$ is present. This will lead to the same situation in the equations we are seeking. The removal of this deficiency can be achieved by substitution of the values $\rho_{\mathbf{k}_0,i}(0)$ and $\bar{n}_{\mathbf{k}_0+\Delta\mathbf{k}}(0)$ on the right sides of the equations for their values $\rho_{\mathbf{k}_0,i}(t)$ and $\bar{n}_{\mathbf{k}_0+\Delta\mathbf{k}}(t)$, as was argued above. Such substitutions are justified with the accuracy of the highest order of the perturbation theory. This program of deductions follows the well-established methods of the quantum theory of lasers [63].

Following these prescriptions, instead of the small-scale incoherent variation of the reduced density matrix, one can introduce the large-scale derivative

$$\left.\frac{\partial\rho_{\mathbf{k}_0,i}(t)}{\partial t}\right|_{\text{incoh}} \cong \frac{\rho_{\mathbf{k}_0,i}(t)-\rho_{\mathbf{k}_0,i}(0)}{t}. \tag{8.50}$$

The large-scale derivative can be transformed from the interaction representation to the Schrödinger representation with Eq. (8.42). After that, the master equation for the reduced density matrix in the Schrödinger representation and in the rotating reference frame has the form

$$\begin{aligned}
\frac{\partial\rho_{\mathbf{k}_0}(t)}{\partial t} &= \left[\frac{dE_0}{\hbar}a^{\dagger}_{\mathbf{k}_0} - \left(\frac{dE_0}{\hbar}\right)^{*} a_{\mathbf{k}_0}, \rho_{\mathbf{k}_0}(t)\right] \\
&\quad - i\left\{\left[\Delta\omega + \frac{l(t)}{\hbar} + p_1(t) - q_1(t) + m(\mathbf{k}_0)\right]\left[N_{\mathbf{k}_0}, \rho_{\mathbf{k}_0}(t)\right]\right. \\
&\quad \left. + \frac{1}{2V}\left[\frac{g\phi(0,0)}{\hbar} + f_1(t)\right]\left[B, \rho_{\mathbf{k}_0}(t)\right] - \frac{g_1}{2V}\left[C, \rho_{\mathbf{k}_0}(t)\right]\right\} \\
&\quad - \left(\frac{f_2(t)}{2V}\{[A^{+}, A\rho_{\mathbf{k}_0}(t)] + [\rho_{\mathbf{k}_0}(t)A^{+}, A]\}\right. \\
&\quad + \frac{g_2}{2V}\{[A, A^{+}\rho_{\mathbf{k}_0}(t)] + [\rho_{\mathbf{k}_0}(t)A, A^{+}]\} \\
&\quad + \left[p_2(t) + \frac{\gamma_1(\mathbf{k}_0)}{2}\right]\{[a^{\dagger}_{\mathbf{k}_0}, a_{\mathbf{k}_0}\rho_{\mathbf{k}_0}(t)] + [\rho_{\mathbf{k}_0}(t)a^{\dagger}_{\mathbf{k}_0}, a_{\mathbf{k}_0}]\} \\
&\quad \left. + \left[q_2(t) + \frac{\gamma_2(\mathbf{k}_0)}{2}\right]\{[a_{\mathbf{k}_0}, a^{\dagger}_{\mathbf{k}_0}\rho_{\mathbf{k}_0}(t)\} + [\rho_{\mathbf{k}_0}(t)a_{\mathbf{k}_0}, a^{\dagger}_{\mathbf{k}_0}]\}\right).
\end{aligned} \tag{8.51}$$

Here the coefficients $l(t)$, $f_i(t)$, $g_i(t)$, $p_i(t)$, and $q_i(t)$, as well as the sought-after reduced density matrix $\rho_{\mathbf{k}_0}(t)$, depend self-consistently on the same moment of time t.

The coefficients $f_i(t)$, $g_i(t)$, $p_i(t)$, and $q_i(t)$ represent the real and the imaginary self-energy terms generated by the operators F^+ and P^+, whereas the coefficient $l(t)$ is generated by the operator L. They depend on the mean occupation numbers $\bar{n}_{\mathbf{k}_0+\Delta\mathbf{k}}(0)$ in the limit $t = 0$ as follows:

$$\left|\begin{matrix} f_1(0) \\ g_1(0) \\ \text{---} \\ f_2(0) \\ g_2(0) \end{matrix}\right| = \frac{1}{\hbar^2 V}\sum_{\mathbf{p},\mathbf{q}} g^2\phi^2(\mathbf{p},\mathbf{q}) \left|\begin{matrix} \mathrm{P}\dfrac{1}{\Omega(\mathbf{p},\mathbf{q})} \\ \hline \pi\delta[\Omega(\mathbf{p},\mathbf{q})] \end{matrix}\right| \times \left|\begin{matrix} \left[1+\bar{n}_{\mathbf{k}_0+\mathbf{p}}(0)\right]\left[1+\bar{n}_{\mathbf{k}_0-\mathbf{q}}(0)\right] \\ \bar{n}_{\mathbf{k}_0+\mathbf{p}}(0)\bar{n}_{\mathbf{k}_0-\mathbf{q}}(0) \end{matrix}\right|,$$

$$\left|\begin{matrix} p_1(0) \\ q_1(0) \\ \text{---} \\ p_2(0) \\ q_2(0) \end{matrix}\right| = \frac{2}{\hbar^2 V^2}\sum_{\mathbf{p},\mathbf{q},\Delta\mathbf{k}} g^2\phi^2(\mathbf{q},\Delta\mathbf{k}) \left|\begin{matrix} \mathrm{P}\dfrac{1}{\theta(\mathbf{p},\mathbf{q},\Delta\mathbf{k})} \\ \hline \pi\delta\left[\theta(\mathbf{p},\mathbf{q},\Delta\mathbf{k})\right] \end{matrix}\right|$$

$$\times \left|\begin{matrix} \bar{n}_{\mathbf{k}_0+\mathbf{p}}(0)\left[1+\bar{n}_{\mathbf{k}_0+\mathbf{p}+\mathbf{q}}(0)\right]\left[1+\bar{n}_{\mathbf{k}_0-\Delta\mathbf{k}}(0)\right] \\ \bar{n}_{\mathbf{k}_0+\mathbf{p}+\mathbf{q}}(0)\bar{n}_{\mathbf{k}_0-\Delta\mathbf{k}}(0)\left[1+\bar{n}_{\mathbf{k}_0+\mathbf{p}}(0)\right] \end{matrix}\right|, \tag{8.52}$$

$$l(0) = \frac{2}{V}\sum_{\mathbf{q}} g\phi(0,0)\bar{n}_{\mathbf{k}_0+\mathbf{q}}(0). \tag{8.53}$$

The quantities $m(\mathbf{k}_0)$ and $\gamma_i(\mathbf{k}_0)$ are determined by the interactions of the polaritons with the heat bath and do not depend on the state of the system S. The frequencies $\Omega(\mathbf{p},\mathbf{q})$ and $\Theta(\mathbf{p},\mathbf{q},\Delta\mathbf{k})$ determine the energy transfers during the scattering processes and are written as

$$\Omega(\mathbf{p},\mathbf{q}) = 2\omega(\mathbf{k}_0) - \omega(\mathbf{k}_0+\mathbf{p}) - \omega(\mathbf{k}_0-\mathbf{q})$$
$$\Theta(\mathbf{p},\mathbf{q},\Delta\mathbf{k}) = \omega(\mathbf{k}_0) + \omega(\mathbf{k}_0+\mathbf{p}) - \omega(\mathbf{k}_0+\mathbf{p}+\mathbf{q}) - \omega(\mathbf{k}_0-\Delta\mathbf{k}).$$

Master equation (8.51) depends on the mean occupation numbers $\bar{n}_{\mathbf{k}_0+\Delta\mathbf{k}}(t)$ of the scattered polaritons through the the functions $f_i(t)$, $g_i(t)$, $p_i(t)$, $q_i(t)$, and $l(t)$. In their turn, the mean occupation numbers obey the kinetic equations, which can be obtained with the definitions

$$\bar{n}_{\mathbf{k}_0+\Delta\mathbf{k}}(t) = \mathrm{Tr}\left[a^\dagger_{\mathbf{k}_0+\Delta\mathbf{k}}a_{\mathbf{k}_0+\Delta\mathbf{k}}\rho(t)\right] = \mathrm{Tr}\left[a^\dagger_{\mathbf{k}_0+\Delta\mathbf{k}}a_{\mathbf{k}_0+\Delta\mathbf{k}}\rho_i(t)\right],$$
$$\frac{\partial\bar{n}_{\mathbf{k}_0+\Delta\mathbf{k}}(t)}{\partial t} = \mathrm{Tr}\left[a^\dagger_{\mathbf{k}_0+\Delta\mathbf{k}}a_{\mathbf{k}_0+\Delta\mathbf{k}}\frac{\partial\rho(t)}{\partial t}\right] = \mathrm{Tr}\left[a^\dagger_{\mathbf{k}_0+\Delta\mathbf{k}}a_{\mathbf{k}_0+\Delta\mathbf{k}}\left.\frac{\partial\rho_i(t)}{\partial t}\right|_S\right]$$
$$= \mathrm{Tr}\left[a^\dagger_{\mathbf{k}_0+\Delta\mathbf{k}}a_{\mathbf{k}_0+\Delta\mathbf{k}}\left.\frac{\partial\rho_i(t)}{\partial t}\right|_{\mathrm{incoh}}\right]. \tag{8.54}$$

Here the notation $\frac{\partial\rho_i(t)}{\partial t}|_S$ means the change of the density matrix resulting only from the influence of the polariton system without the participation of the laser radiation and the thermal bath. The laser radiation does not affect the scattered polaritons, whereas the influence of the thermal bath is taken into account phenomenologically. As done above, the incoherent variation $\frac{\partial \rho_i(t)}{\partial t}|_{\text{incoh}}$ is replaced with the large-scale variation, determined by relation (8.44). The self-consistent kinetic equations are then

$$\begin{aligned}
\frac{\partial \bar{n}_{\mathbf{k}_0+\Delta\mathbf{k}}(t)}{\partial t} &= -\gamma(\mathbf{k}_0+\Delta\mathbf{k})\bar{n}_{\mathbf{k}_0+\Delta\mathbf{k}}(t) + \frac{2\pi}{\hbar^2 V^2}\sum_{\mathbf{q}} g^2\phi^2(\mathbf{q},\Delta\mathbf{k})\delta\left[\Omega(\Delta\mathbf{k},\mathbf{q})\right] \\
&\quad\times \left\{\langle B(t)\rangle\left[1+\bar{n}_{\mathbf{k}_0+\Delta\mathbf{k}}(t)\right]\left[1+\bar{n}_{\mathbf{k}_0-\mathbf{q}}(t)\right] - \langle C(t)\rangle \bar{n}_{\mathbf{k}_0+\Delta\mathbf{k}}(t)\bar{n}_{\mathbf{k}_0-\mathbf{q}}(t)\right\} \\
&\quad+ \frac{4\pi}{\hbar^2 V^2}\sum_{\mathbf{q},\mathbf{s}} g^2\phi^2(\mathbf{q},\mathbf{s})\delta\left[\theta(\Delta\mathbf{k},\mathbf{q},\mathbf{s})\right] \\
&\quad\times \left\{\left[1+\langle N_{\mathbf{k}_0}(t)\rangle\right]\left[1+\bar{n}_{\mathbf{k}_0+\Delta\mathbf{k}}(t)\right]\bar{n}_{\mathbf{k}_0+\Delta\mathbf{k}+\mathbf{q}}(t)\bar{n}_{\mathbf{k}_0-\mathbf{s}}(t)\right. \\
&\quad\left. - \langle N_{\mathbf{k}_0}(t)\rangle \bar{n}_{\mathbf{k}_0+\Delta\mathbf{k}}(t)\left[1+\bar{n}_{\mathbf{k}_0+\Delta\mathbf{k}+\mathbf{q}}(t)\right]\left[1+\bar{n}_{\mathbf{k}_0-\mathbf{s}}(t)\right]\right\} \\
&\quad+ \frac{8\pi}{\hbar^2 V^2}\sum_{\mathbf{p},\mathbf{q}} g^2\phi^2(\mathbf{q},-\Delta\mathbf{k})\delta\left[\theta(\mathbf{p},\mathbf{q},-\Delta\mathbf{k})\right] \\
&\quad\times \left\{\langle N_{\mathbf{k}_0}(t)\rangle \bar{n}_{\mathbf{k}_0+\mathbf{p}}(t)\left[1+\bar{n}_{\mathbf{k}_0+\Delta\mathbf{k}}(t)\right]\left[1+\bar{n}_{\mathbf{k}_0+\mathbf{p}+\mathbf{q}}(t)\right]\right. \\
&\quad\left. - \left[1+\langle N_{\mathbf{k}_0}(t)\rangle\right]\left[1+\bar{n}_{\mathbf{k}_0+\mathbf{p}}(t)\right]\bar{n}_{\mathbf{k}_0+\mathbf{p}+\mathbf{q}}(t)\bar{n}_{\mathbf{k}_0+\Delta\mathbf{k}}(t)\right\}. \qquad (8.55)
\end{aligned}$$

The first term on right side of Eq. (8.55) contains the damping constant $\gamma(\mathbf{k}_0+\Delta\mathbf{k})$, of the same type as $\gamma(\mathbf{k}_0)$. The second term describes the conversion of polaritons of the $\mathbf{k}_0$ mode into two scattered polaritons. The last two terms correspond to scattering processes between the four polaritons involved in the scattering process – two in the initial state and two in the final state, with only one belonging to the $\mathbf{k}_0$ mode.

As one can see, the kinetic equations contain the averages of the operators belonging to the coherently excited mode. The mean occupation numbers $\bar{n}_{\mathbf{k}_0+\Delta\mathbf{k}}(t)$ depend on the average values $\langle N_{\mathbf{k}_0}(t)\rangle$, $\langle B(t)\rangle$ and $\langle C(t)\rangle$. To determine them it is necessary to find the solution $\rho_{\mathbf{k}_0}(t)$ of master equation (8.51). In its turn master equation (8.51) depends on the mean occupation numbers $\bar{n}_{\mathbf{k}_0+\Delta\mathbf{k}}(t)$ through the functions $f_i(t)$, $g_i(t)$, $p_i(t)$, and $q_i(t)$. Master equation (8.51) and the set of kinetic equations (8.55) form a complete set of equations describing the polariton system.

8.3.2 The Fokker–Planck Equation and Squeezed States

To find the explicit expression of the mean values of the selected mode operators,

$$\langle a^\dagger_{\mathbf{k}_0}\cdots a_{\mathbf{k}_0}\rangle = \mathrm{Tr}\left[a^\dagger_{\mathbf{k}_0}\cdots a_{\mathbf{k}_0}\rho_{\mathbf{k}_0}(t)\right],$$

the explicit form of the reduced density matrix $\rho_{\mathbf{k}_0}(t)$ is needed. As pointed out by Drummond and Walls [60], the Glauber–Sudarshan diagonal representation is insufficient in some cases, because the full set of coherent states is overcomplete and the probability function $P(\alpha, \alpha^*, t)$ in Eq. (8.47) does not always exist. The nondiagonal representation introduced by Drummond and Gardiner [67] has the form

$$\rho_{\mathbf{k}_0}(t) = \int_C\int_{C'} d\alpha\, d\beta\, P(\alpha,\beta,t)\frac{|\alpha\rangle\langle\beta^*|}{\langle\beta^*|\alpha\rangle}, \qquad (8.56)$$

where the independent variables α and β vary along the independent Hankel contours C and C' in the complex plane. The Hankel path of integration goes from $-\infty$ on the real axis around the origin in an anticlockwise direction back to $-\infty$ [60].

The properties of coherent states were discussed in Chapter 2, introduced by formulas (2.17)–(2.23). Along with the one-photon coherent states $|\alpha\rangle$ there also exist two-photon coherent states called "squeezed" states. They were introduced by Caves and Shumaker [68] and play the same role in two-photon quantum optics as Glauber coherent states play in one-photon optics. In the general case, two photons can belong to two different modes of the electromagnetic field. But the particular case of one selected mode is of special interest for us, called the degenerate case.

The degenerate squeezed state $|\alpha, y\rangle$ is

$$|\alpha, y\rangle = D(\alpha)S(y)|0\rangle, \tag{8.57}$$

where $D(\alpha)$ is the displacement operator, whereas $S(y)$ is the degenerate squeezing operator that has the form

$$S(y) = \exp\left[\frac{y}{2}([a^\dagger]^2 - a^2)\right], \qquad S^+(y) = S(-y), \tag{8.58}$$

where y is a real value.

One can introduce the degenerate squeezed annihilation operator b in the following way:

$$b = SaS^+. \tag{8.59}$$

The squeezed state $|\alpha, y\rangle$ is the eigenstate of the squeezed annihilation operator b of Eq. (8.59),

$$b|\alpha, y\rangle = \beta|\alpha, y\rangle, \tag{8.60}$$

with the eigenvalue β, where

$$\beta = \alpha \cosh y - \alpha^* \sinh y. \tag{8.61}$$

By expanding $S(y)$ in series, one can prove the relations

$$S(y)\begin{vmatrix} a \\ a^\dagger \end{vmatrix} S^+(y) = \cosh y \begin{vmatrix} a \\ a^\dagger \end{vmatrix} - \sinh y \begin{vmatrix} a^\dagger \\ a \end{vmatrix}. \tag{8.62}$$

They lead to the following changes of the degenerate quadrature phase operators:

$$\begin{gathered} x_1 = a + a^\dagger, \qquad x_2 = i\left(a - a^\dagger\right), \\ S(y)\begin{vmatrix} x_1 \\ x_2 \end{vmatrix} S^+(y) = \begin{vmatrix} x_1 e^{-y} \\ x_2 e^{y} \end{vmatrix}. \end{gathered} \tag{8.63}$$

One of them is squeezed by the factor e^{-y} at the expense of the other one, which is changed by the factor e^{y}. This fact is important for the calculation of the mean-square deviations of the quadrature phase operators x_i,

$$\langle(\Delta x_i)^2\rangle = \langle x_i{}^2\rangle - \langle x_i\rangle^2. \tag{8.64}$$

Equation (8.60) can be verified as follows:

$$\begin{aligned} b|\alpha, y\rangle &= S(y)aS^+(y)D(\alpha)S(y)|0\rangle, \\ S^+(y)D(\alpha)S(y) &= \exp\left[S^+\left(\alpha a^\dagger - \alpha^* a\right)S\right] = D(\beta), \\ b|\alpha, y\rangle &= S(y)aD(\beta)|0\rangle = \beta|\alpha, y\rangle. \end{aligned} \tag{8.65}$$

Now we return to one-photon coherent states, which are widely used in this section.

The unity operator I for the coherent states equals

$$I = \frac{1}{\pi}\int d^2\gamma|\gamma\rangle\langle\gamma|, \qquad \langle\alpha|I|\alpha\rangle = 1. \tag{8.66}$$

The projection operator $\Lambda(\alpha,\beta)$, which is one part of the integrand of nondiagonal representation (8.56), can be transformed by use of Eqs. (2.21) and (2.23) as follows:

$$\Lambda(\alpha,\beta) \equiv \frac{|\alpha\rangle\langle\beta^*|}{\langle\beta^*|\alpha\rangle} = e^{-\alpha\beta}e^{\alpha a^\dagger}|0\rangle\langle 0|e^{\beta a}. \tag{8.67}$$

This representation permits one to express the action on the projection operator $\Lambda(\alpha,\beta)$ of the creation and annihilation operators $a^\dagger$ and a and of their different combinations through the algebraic and differential operations as follows:

$$\begin{aligned}
&a\Lambda(\alpha,\beta) = \alpha\Lambda(\alpha,\beta), \quad \Lambda(\alpha,\beta)a^\dagger = \beta\Lambda(\alpha,\beta),\\
&a^\dagger\Lambda(\alpha,\beta) = \left(\beta + \frac{\partial}{\partial\alpha}\right)\Lambda(\alpha,\beta), \quad \Lambda(\alpha,\beta)a = \left(\alpha + \frac{\partial}{\partial\beta}\right)\Lambda(\alpha,\beta),\\
&a^\dagger\Lambda(\alpha,\beta)a = \left(\beta + \frac{\partial}{\partial\alpha}\right)\left(\alpha + \frac{\partial}{\partial\beta}\right)\Lambda(\alpha,\beta), \quad \text{and so on.}
\end{aligned} \tag{8.68}$$

These rules lead to the evolution of the distribution function $P(\alpha,\beta,t)$ induced by different terms (8.68), which gives rise to the Fokker–Planck equation that determines the evolution of the distribution function $P(\alpha,\beta,t)$:

$$\begin{aligned}
\frac{\partial P(\alpha,\beta,t)}{\partial t} = &\left(-\frac{dE_0}{\hbar}\frac{\partial}{\partial\alpha} - \left(\frac{dE_0}{\hbar}\right)^*\frac{\partial}{\partial\beta}\right.\\
&+\left\{i\left[\omega(\mathbf{k}_0) - \omega_L + \frac{l}{\hbar} + p_1 - q_1 - \frac{2g_1}{V}\right] + \left(p_2 - q_2 - \frac{2g_2}{V}\right)\right\}\frac{\partial}{\partial\alpha}\alpha\\
&+\left\{-i\left[\omega(\mathbf{k}_0) - \omega_L + \frac{l}{\hbar} + p_1 - q_1 - \frac{2g_1}{V}\right] + \left(p_2 - q_2 - \frac{2g_2}{V}\right)\right\}\frac{\partial}{\partial\beta}\beta\\
&+\frac{1}{V}\left\{i\left[\frac{2g\phi(0,0)}{\hbar} + f_1 - g_1\right] + (f_2 - g_2)\right\}\frac{\partial}{\partial\alpha}\alpha^2\beta\\
&+\frac{1}{V}\left\{-i\left[\frac{2g\phi(0,0)}{\hbar} + f_1 - g_1\right] + (f_2 - g_2)\right\}\frac{\partial}{\partial\beta}\alpha\beta^2\\
&-\frac{1}{2V}\left\{i\left[\frac{2g\phi(0,0)}{\hbar} + f_1 - g_1\right] + (f_2 - g_2)\right\}\frac{\partial^2}{\partial\alpha^2}\alpha^2\\
&-\frac{1}{2V}\left\{-i\left[\frac{2g\phi(0,0)}{\hbar} + f_1 - g_1\right] + (f_2 - g_2)\right\}\frac{\partial^2}{\partial\beta^2}\beta^2\\
&+\left(2q_2 + \frac{4g_2}{V}\right)\frac{\partial^2}{\partial\alpha\partial\beta} + \frac{4g_2}{V}\frac{\partial^2}{\partial\alpha\partial\beta}\alpha\beta - \frac{2g_2}{V}\frac{\partial^3}{\partial\alpha^2\partial\beta}\alpha\\
&\left.-\frac{2g_2}{V}\frac{\partial^3}{\partial\beta^2\partial\alpha}\beta + \frac{g_2}{V}\frac{\partial^4}{\partial\alpha^2\partial\beta^2}\right)P(\alpha,\beta,t).
\end{aligned} \tag{8.69}$$

The values $\pm i\,[\omega(\mathbf{k}_0) - \omega_L + l/\hbar + p_1 - q_1]$ and $\pm i\,[2g\phi(0,0)/\hbar + f_1 - g_1]$ determine the dispersive components whereas $(p_2 - q_2)$ and $(f_2 - g_2)$ determine the absorptive components of the self-energy parts describing the coherently excited polaritons. Their interaction with the scattered polaritons is represented by the terms $(p_i - q_i)$. The influence of the conversion processes, which lead to the transformations of two polaritons of the selected mode into two scattered ones and vice versa, is reflected by the terms$(f_i - g_i)$. These terms describe the indirect interaction of the coherently excited polaritons with each other through the creation and annihilation of pairs of scattered polaritons. The direct interaction of the polaritons in the selected mode is represented by the term $g\phi(0,0)/\hbar$, whereas the influence on them of the external laser radiation is represented by the expressions $dE_0/\hbar$ and $(dE_0/\hbar)^*$.

The Fokker–Planck equation contains first-order derivatives with respect to the variables α and β, which determine the drift or deterministic behavior of the distribution function $P(\alpha, \beta, t)$, that is, of the coherently excited polaritons. The second- and higher-order derivatives describe their quantum fluctuations and the deviation from the deterministic evolution. This stochastic behavior is also affected by the same interaction processes discussed above. The terms containing the second-order derivatives are called diffusive terms. There are two diagonal terms and one nondiagonal diffusive term. An important property regarding all the stochastic terms makes it advantageous to divide them into two groups, one that contains the diagonal diffusive terms and another that contains the nondiagonal diffusive term and the higher-order derivatives. This property is related to the dependences of the corresponding coefficients on the density of the scattered polaritons n_1, defined as

$$n_1(t) = \frac{1}{V}\sum_{\Delta\mathbf{k}} \bar{n}_{\mathbf{k}_0+\Delta\mathbf{k}}(t). \tag{8.70}$$

If this density is much smaller than the density of the coherently excited polaritons $n_{\mathbf{k}_0}$,

$$n_{\mathbf{k}_0}(t) = \frac{N_{\mathbf{k}_0}}{V},$$

then the nondiagonal diffusive term and the higher-order derivatives can be neglected compared with the diagonal diffusive terms.

The coefficients of the diagonal second-order derivatives contain the dispersive and absorptive terms $\{\pm i\,[2g\phi(0,0)/\hbar + f_1 - g_1] + f_2 - g_2\}$ that are nonzero in the quantum limit when the mean occupation numbers $\bar{n}_{\mathbf{k}_0+\Delta\mathbf{k}}(t)$ and the density of scattered polaritons n_1 tend to zero. In contrast to this, the coefficients of the nondiagonal second-order derivative and of the higher-order derivatives are determined by the coefficients g_2 and q_2, which, following formulas (8.52), have quadratic dependences on the mean occupation numbers $\bar{n}_{\mathbf{k}_0+\Delta\mathbf{k}}$. In the case of small mean occupation numbers $\bar{n}_{\mathbf{k}_0+\Delta\mathbf{k}}(t)$, which is the only case considered below, when the inequality $n_1 \ll n_{\mathbf{k}_0}$ is fulfilled, one can simplify the Fokker–Planck equation by neglecting the nondiagonal diffusive term and the higher-order derivatives. In this limit the functions f_i can be considered as having zeroth-order magnitude in the density n_1, the functions p_i as having first-order magnitude, and the functions g_i and q_i as having second-order magnitude and much smaller than f_i. As the density of scattered polaritons increases, i.e., as the laser radiation intensity is increased, the values f_i and g_i in pairs, as well as p_i and q_i in pairs, become approximately equal, and the above distinction becomes invalid.

In our case α and β are macroscopic variables proportional to the volume V of the crystallite. It is reasonable to indicate their dependence on the volume explicitly, as

$$\alpha = \sqrt{N}\xi, \quad \beta = \sqrt{N}\eta,$$
$$E_0 = \sqrt{N}\varepsilon_0, \quad N = Vn_c. \tag{8.71}$$

The dimensionless volume N can be introduced in various ways. We use the density n_c, which will appear below in the course of finding $\bar{n}_{\mathbf{k}_0+\Delta\mathbf{k}}$. Finally, after the above-mentioned simplifications, we find the following Fokker–Planck equation:

$$\frac{\partial P(\xi,\eta,t)}{\partial t} = \left\{\frac{\partial}{\partial \xi}\left[\kappa\xi + 2\chi\xi^2\eta - \frac{d\varepsilon_0}{\hbar}\right] + \frac{\partial}{\partial \eta}\left[\kappa^*\eta + 2\chi^*\eta^2\xi - \left(\frac{d\varepsilon_0}{\hbar}\right)^*\right]\right.$$
$$\left. - \frac{\chi}{N}\frac{\partial^2}{\partial\xi^2}\xi^2 - \frac{\chi^*}{N}\frac{\partial^2}{\partial\eta^2}\eta^2\right\}P(\xi,\eta,t). \tag{8.72}$$

The basis of this simplified approach is the approximation that the density of scattered polaritons n_1 is small compared with the density of coherently excited polaritons $n_{\mathbf{k}_0}$, i.e., $n_1 < n_{\mathbf{k}_0}$.

There is also one more limitation to values $N \geq 1$. It arises from the fact that the neglected terms containing higher-order derivatives have dependences on N of N^{-2} and N^{-3}, which become important at the values $N < 1$. This leads to lower bounds on the sizes of the crystallites. The square brackets on the right side of the Fokker–Planck equation contain drift terms, which specify the degree of deterministic behavior of the $\mathbf{k}_0$ mode polaritons. The remaining two terms are diagonal diffusion terms, which describe the quantum fluctuations of the polaritons of this mode. The coefficients κ and χ that enter into the Fokker–Planck equation are related to the previously defined coefficients by the relations

$$\kappa = i\overline{\Delta\omega} + \sigma, \qquad \chi = i\mu + \delta, \tag{8.73}$$

where

$$\overline{\Delta\omega} = \Delta\omega + \frac{l}{\hbar} + p_1 - q_1 + m(\mathbf{k}_0), \qquad \Delta\omega = \Delta\omega = \omega(\mathbf{k}_0) - \omega_L,$$
$$\sigma = \frac{\gamma(\mathbf{k}_0)}{2} + (p_2 - q_2), \qquad \delta = \frac{\mathrm{n_c}}{2}(\mathrm{f}_2 - \mathrm{g}_2);$$
$$\mu = n_c\left[\frac{g\phi(0,0)}{\hbar} + \frac{1}{2}(f_1 - g_1)\right], \qquad \gamma(\mathbf{k}_0) = \gamma_1(\mathbf{k}_0) - \gamma_2(\mathbf{k}_0) > 0,$$
$$m(\mathbf{k}_0) = m_1(\mathbf{k}_0) - m_2(\mathbf{k}_0). \tag{8.74}$$

Kinetic equations (8.55) and Fokker–Planck equation (8.72) form a new full set of equations that self-consistently describe the polariton system coherently driven by laser radiation.

To conclude this section, we specify the relation between the constant $d\varepsilon_0/\hbar$ for the source of coherent pumping and the power flux of the laser radiation S_L by using the example of a ring resonator with two semitransparent mirrors with reflection coefficients R and two opaque mirrors. In the space between the two mirrors we place a glassy matrix in the form of a film of thickness L containing microcrystallites. Following Ref. 61, we then write boundary conditions that relate the electromagnetic field outside the resonator to the field inside it. We depart from Ref. 61 only in our replacement of the field within the resonator by the photon component of the mode $\mathbf{k}_0$ polaritons. The relation we are looking for is

$$Y = \left|\frac{d\varepsilon_0}{\hbar\gamma_{\text{eff}}}\right|^2 = \frac{S_L \Omega_{\text{res}}^2}{S_c \gamma_{\text{eff}}^2}\xi. \tag{8.75}$$

Here γ_{eff} has the sense of the effective attenuation for the polariton level, and the following notations are introduced:

$$S_c = \frac{\hbar\omega_L^2 n_c \varepsilon_\infty}{2k_0}, \qquad \Omega_{\text{res}}^2 = \frac{c_0^2(1-R)}{L^2\varepsilon_\infty},$$

$$\xi = \frac{[\omega_{\text{ex}}(\mathbf{k}_0) - \omega(\mathbf{k}_0)]^2}{[\omega_\perp(\mathbf{k}_0) - \omega(\mathbf{k}_0)]^2 + \left|\frac{\varphi_{\mathbf{k}_0}}{\hbar}\right|^2},$$

$$\left|\frac{\varphi_{\mathbf{k}_0}}{\hbar}\right|^2 = \frac{\omega_\perp \omega_{LT}}{2}\varepsilon_\infty. \tag{8.76}$$

The factor ξ indicates the fraction of the photon component that enters into a polariton of frequency $\omega(\mathbf{k}_0)$; $\omega_\perp(\mathbf{k}_0)$ is the transverse exciton frequency, ε_∞ is the high-frequency dielectric constant of the crystallite, and c_0 is the velocity of light in vacuum.

Let us now turn to the investigation of the stationary states of this system.

8.3.3 Stationary Self-Consistent Solution

Integrodifferential equations like Eq. (8.55) can be solved only approximately. In Eq. (8.55) we have explicitly taken into account the inhomogeneous broadening of the energy spectrum of scattered polaritons associated with the loss of the quasi-momentum conservation in finite-volume crystallites.

One can also investigate a variant of Eq. (8.55) that takes into account homogeneous broadening of the polariton energy due to two-particle losses from the mode $\mathbf{k}_0$ phenomenologically. The treatment of this equation is described in full in Ref. 45. Despite the differences in the two approaches, the qualitative properties of the solutions obtained and the conclusions that follow from them coincide.

The stationary solution of Eq. (8.55) for the distribution function of the scattered quasiparticles is shown in Fig. 8.7. This stationary solution consists of two terms. The first term leads to a peaked structure in the nonequilibrium **k**-space distribution, while the second term leads to a continuous background. The difference between them is the same as the difference between the shapes of exciton absorption bands for direct-gap and indirect-gap semiconductors. As seen in Fig. 8.7, four peaks occur, which correspond to points 1, 2, 3, and 4 on the dispersion curve.

The peaked part of the nonequilibrium distribution function for scattered quasiparticles that rises above the continuous background is found to be

$$\bar{n}_{\mathbf{k}_0+\Delta\mathbf{k}} = \frac{\pi}{2}\frac{\left(b + 4n_{\mathbf{k}_0}n_1\right)\gamma_{\text{eff}}\tilde{\delta}\left[\Omega(\Delta\mathbf{k})\right]}{n_c^2 + \sigma n_{\mathbf{k}_0}n_1 + \varepsilon n_c n_1/N - b}, \tag{8.77}$$

where

$$n_c = \frac{\hbar \gamma_{\text{eff}}}{2\,|g|}, \tag{8.78}$$

$$b = \frac{\langle B \rangle}{V^2}, \tag{8.79}$$

while the unknown constants σ and ε are positive. The term $\varepsilon n_c n_1/N$ has an additional dependence on $1/N$, which was omitted in Ref. 45.

The denominator of expression (8.77) equals $(n_c^2 - b)$ if $\sigma = \varepsilon = 0$. In this case it is possible for the nonequilibrium distribution function $\bar{n}_{\mathbf{k}_0+\Delta\mathbf{k}}$ to exhibit a threshold dependence on the value of b, which is an average over the state of the coherently excited mode. A dependence of this sort is analogous to the threshold dependence encountered in stimulated Brillouin scattering, as described in Refs. 55, 56, and 69. In our case, an increase in b is accompanied by simultaneous increases in the quantities $n_{\mathbf{k}_0}$ and n_1. Thus the effective threshold value, which increases with increasing $n_{\mathbf{k}_0}$, b, and n_1 in the case $\sigma \neq 0$ and $\varepsilon \neq 0$, is smeared out to the point of being inaccessible in practice. The system is continuous at the point $b = n_c^2$, because the denominator is finite there.

The smoothing function $\tilde{\delta}[\Omega(\Delta\mathbf{k})]$ was chosen in the form of a Lorentzian:

$$\tilde{\delta}[\Omega(\Delta\mathbf{k})] = \frac{1}{\pi} \frac{q(\Delta\mathbf{k})}{\Omega^2(\Delta\mathbf{k}) + q^2(\Delta\mathbf{k})}. \tag{8.80}$$

The quantity $q(\Delta\mathbf{k})$ is determined by the inhomogeneous broadening in this variant or by the homogeneous $\gamma_{\text{eff}}(\Delta\mathbf{k})$ in Ref. 45. In practice it is necessary to choose the larger of the two broadenings.

To determine the average values of $n_{\mathbf{k}_0}$ and b, we must find a stationary solution of Fokker–Planck equation (8.72). For the special case $N = 1$, such a solution was found by Drummond and Walls [60]. This solution was generalized to arbitrary finite values of N, as

$$P_{ss}(\xi, \eta) = \frac{1}{I(c, z, N)} \xi^{Nc-2} \eta^{Nc^*-2} \exp\left(\frac{Nz}{\xi} + \frac{Nz^*}{\xi} + 2N\xi\eta \right),$$

$$c = \frac{\kappa}{\chi}, \qquad z = \frac{\mathrm{d}\varepsilon_0}{\hbar \gamma}, \qquad 1 \le N < \infty. \tag{8.81}$$

The normalization constant entering into Eq. (8.81) has the form

$$I(c, z, N) = -4\pi^2 (Nz)^{Nc} (Nz^*)^{Nc^*} \frac{{}_0F_2(Nc, Nc^*, 2N\,|zN|^2)}{|zN|^2\, \Gamma(Nc)\Gamma(Nc^*)}. \tag{8.82}$$

Here the generalized hypergeometric function ${}_0F_2(Nc, Nc^*, 2N\,|zN|^2)$ is expressed by the series

$${}_0F_2\,(a, b, x) = \sum_{n=1}^{\infty} \frac{x^n}{n!} \frac{\Gamma(a)\Gamma(b)}{\Gamma(a+n)\Gamma(b+n)}, \tag{8.83}$$

where $\Gamma(x)$ is the gamma function. By using Eqs. (8.81) and (8.82) one can find the average values $\langle (a_{\mathbf{k}_0}^\dagger)^p (a_{\mathbf{k}_0})^q \rangle$ from the operators for the coherently excited mode. These quantities can be expressed in terms of the generalized hypergeometric series ${}_0F_2\,(a, b, x)$ and the gamma function $\Gamma(x)$ in the following way:

$$\left\langle (a^\dagger_{\mathbf{k}_0})^p (a_{\mathbf{k}_0})^q \right\rangle = N^{(p+q)/2}(Nz)^q(Nz^*)^p \times \frac{\Gamma(Nc)\Gamma(Nc^*)_0F_2(Nc+q, Nc^*+p, 2N|zN|^2)}{\Gamma(Nc+q)\Gamma(Nc^*+p)_0F_2(Nc, Nc^*, 2N|zN|^2)}. \tag{8.84}$$

The second-order correlation function $g^{(2)}(t)$ at the instant of time $t = 0$ equals

$$g^{(2)}(0) = \frac{\left\langle (a^\dagger_{\mathbf{k}_0})^2 (a_{\mathbf{k}_0})^2 \right\rangle}{\left\langle a^\dagger_{\mathbf{k}_0} a_{\mathbf{k}_0} \right\rangle^2} = \frac{|Nc|^2 \; {}_0F_2(Nc+2, Nc^*+2, 2N\,|zN|^2)\,{}_0F_2(Nc, Nc^*, 2N\,|zN|^2)}{|Nc+1|^2\,[{}_0F_2(Nc+1, Nc^*+1, 2N\,|zN|^2)]^2}. \tag{8.85}$$

This function determines the required statistical properties of the coherently excited polaritons for various values of the dimensionless volume of the crystallite N. Expression (8.85) depends on the coefficients c and z. It follows from Eqs. (8.74) and (8.81) that these in turn depend on the nonequilibrium distribution function $\bar{n}_{\mathbf{k}_0+\Delta\mathbf{k}}$. Thus the problem is self-consistent and can in principle be solved with Eqs. (8.72)–(8.74), (8.51), and (8.81)–(8.84).

Further analysis is not possible without a number of simplifications and approximations. The analytic expression for the functions ${}_0F_2\,(a, b, x)$ can be used in two limiting cases – for $x < ab$ and $x \to \infty$ [70]. The case $x < ab$, where Eq. (8.83) is approximated by two terms,

$${}_0F_2\,(a, b, x) \approx 1 + \frac{x}{ab}, \tag{8.86}$$

has particular interest. Since $x = 2N\,|zN|^2$ and one of the values ab equals $|c|^2$, we can find a restriction on the product NY of the two dimensionless quantities – the volume of the crystallite and the density of the power of the coherent pump Y – of the form

$$NY < \frac{1}{2}\left[\frac{1}{4} + \left(\frac{\Delta\omega}{\gamma_{\text{eff}}}\right)^2\right]. \tag{8.87}$$

From this restriction, we can obtain an analytic result within a rather small region of Y, for which the maximum value Y_{max} is inversely proportional to N. To appreciate better the information supplied by the second-order correlation function $g^{(2)}(0)$, one can remember that it equals 2 in the thermal equilibrium state, it equals 1 in a pure coherent state $|\alpha\rangle$, and it vanishes in a quantum Fock state with $n = 1$. Such quantum-statistical properties are well known in quantum optics and can be measured by coincidence counters, e.g., in the experiment of Hanbury-Brown and Twiss (for a review, see, e.g., Ref. 71). This experiment reveals the characteristic feature of thermal light consisting of the tendency of the photons to arrive in pairs, called "photon bunching." Laser radiation is characterized by a coincidence counting rate that does not depend on the delay time. An effect that is opposite to the bunching phenomenon is called the antibunching effect, which is characterized as a deficit, rather than an excess, of nondelayed coincidences with respect to the random ones [71]. The quantum-mechanical Fock states, corresponding to definite values of the photon numbers, exhibit such an antibunching effect. Antibunching has no analog in classical optics and shows the intrinsic quantum nature of the radiation field.

Analysis shows that for $(c + c^*)N + 1 > 0$, the correlation function $g^{(2)}(0) < 1$ and the polaritons of the mode $\mathbf{k}_0$ exhibit the property of antibunching. When the further inequality $(c + c^*)N + 3 < 0$ is fulfilled, we have $g^{(2)}(0) > 1$, which is a sign that the coherently excited polaritons are bunched when condition (8.87) holds. With Eq. (8.84), it is easy to calculate the mean-square deviations of Eq. (8.64) of the quadrature phase operator $\hat{x}_i$ of Eqs. (8.63). The mean-square deviations are subject to the uncertainty relation

$$\langle(\Delta x_1)^2\rangle\langle(\Delta x_2)^2\rangle \geq 1. \tag{8.88}$$

In single-photon quantum optics, the minimum noise moment satisfies $\langle(\Delta x_i)^2\rangle = 1$. This moment, which is referred to as the zero point, is achieved in a coherent state. In two-photon quantum optics, the squeezed states of Eq. (8.57) are possible. In these states, the noise moment in one of the quadrature phases is smaller than the zero point noise because of relations (8.63). For example, in the case of Eqs. (8.63), when $y > 0$ the deviation $\langle(\Delta x_1)^2\rangle$ becomes less than the zero point noise $\langle(\Delta x_1)^2\rangle < 1$, because of the increased moment in the conjugate operator $\langle(\Delta x_2)^2\rangle > 1$ [68]. Analogous states of coherently excited polaritons are possible because the two particles from the $\mathbf{k}_0$ mode behave like a two-photon transition in the case of degenerate modes. Analysis shows that squeezed states occur in the presence of both bunching and antibunching when certain relations between the phases of the complex quantities c and z are satisfied. These novel statistical properties of polaritons in crystallites have not often been discussed in the literature. They are sensitive to increases in N and Y and disappear as $NY \to \infty$ when condition (8.87) is violated.

Numerical estimates based on Eqs. (8.84)–(8.86) were made for three values $N = 1, 10, 50$ and five values $\Delta\omega/\gamma_{\text{eff}} = 0, \pm 3, \pm 5$, for $p = q = 10^{-1}$, $\sigma = 10^{-4}$–10^{-5}, and $\varepsilon = 0$, and the constants $g = 10^{-32}$ erg cm^3, $\gamma_{\text{eff}} = 10^{11}$ s^{-1}, and $n_c = 5 \times 10^{15}$ cm^{-3}, which are close to those of the crystal CdS. These particular values of N and n_c correspond to crystallite dimensions $d \approx 600$–2200 Å, which fall within the restrictions presented above. The calculation shows that for $N = 1$ the effects of bunching and antibunching are such that $g^{(2)}(0)$ deviates from 1 by 20%–30%. For $N = 10$ the difference comes to 2%–3%, while for $N = 50$ it is only 0.3%–0.6% in all. For $N \geq 100$ the effects under discussion are vanishingly small and polaritons of the mode $\mathbf{k}_0$ are found to be in a purely coherent state. At the boundaries of the crystallite, the polaritons can be converted into light, which can be subjected to analysis by the method of photon counting. The extent of the light bunching and antibunching can be evaluated following Ref. 72. The degree of squeezing of the fluctuations, i.e., the departure of $\langle(\Delta x_i)^2\rangle$ from 1 for $N = 1$, comes to 10% under the condition of antibunching and 26% under bunching conditions. For $N = 10$, the degree of squeezing decreases by an order of magnitude and comes to 2.4%–2.6%.

The connection between these effects and the amount of detuning from resonance is the following. Let us assume that the frequency of the laser radiation ω_L is larger than the bare frequency of the polariton mode $\omega(\mathbf{k}_0)$, such that $\Delta\omega/\gamma_{\text{eff}} = -3, -5$ and its intensity is small and bounded by condition (8.87). Under these conditions, the coherent pumping is not capable of creating purely coherent polaritons of a selected mode in a crystallite of finite volume. The excess frequency of the photons above the frequency of polaritons $\omega(\mathbf{k}_0)$ causes a partial randomization of the latter, whose statistical properties are now reminiscent of the properties of thermal radiation, and a tendency toward bunching is observed. When the excitation frequency of the radiation does not exceed the bare frequency of the mode

$\omega(\mathbf{k}_0)$, a deficit of photon energy below that required for exciting the $\mathbf{k}_0$ polaritons arises. Therefore, when $\Delta\omega/\gamma_{\text{eff}} = 0, 3, 5$, the quantum effects of the forced oscillations are more evident. As we noted previously, this rather low-intensity excitation radiation cannot excite purely coherent polaritons of the mode $\mathbf{k}_0$. However, in this case the polaritons are observed to exhibit the phenomenon of antibunching, which is a characteristic of quantum states of the Fock type [71]. The different behavior of coherently excited polaritons as a function of the sign of the detuning from resonance $\Delta\omega$ for rather small pumping intensities correlates with the properties of an exciton absorption band when the fundamental mechanism for energy dissipation is exciton–exciton and exciton–phonon interactions [73].

Let us now estimate the laser radiation energy S_L required for observing these phenomena. According to condition (8.87), Y varies in the interval $(0, Y_{\max})$, where $Y_{\max} \approx [(\Delta\omega/\gamma_{\text{eff}})^2 + 1/4]/2N$ as $n_1 \to 0$. For all the cases that we investigated $Y \leq 10$ holds. We chose the following parameters: for the photons, $\omega_L = 3 \times 10^{15}\ \text{s}^{-1}$, for the resonator, $L = 1\ \mu m$ and $R = 0.5$; for the excitons, polaritons, and exciton–polariton interactions, $\omega_{\text{ex}}(\mathbf{k}_0) = ck_0$, $\omega_{\text{ex}}(\mathbf{k}_0) - \omega(\mathbf{k}_0) = 1/2|\varphi_{\mathbf{k}_0}/\hbar|$, $\gamma_{\text{eff}} = 10^{11}\ \text{s}^{-1}$, and $n_c = 5 \times 10^{15}\ \text{cm}^{-3}$, respectively. According to Eqs. (8.75) and (8.76) we find that $\Omega^2_{\text{res}}/\gamma^2_{\text{eff}} = 10^6$, $S_c = 10^{15}\ \text{erg/cm}^2\ \text{s}$ and $\xi = 0.2$.

The maximum laser radiation power needed to observe these effects is found to be $5\ \text{kW/cm}^2$ for resonant excitation of crystallites of CdS. In all these cases $N_{\mathbf{k}_0}$ does not exceed 1, and Vn_1 is at least 1 order of magnitude less than $N_{\mathbf{k}_0}$, as supposed at the beginning. One can see that restriction (8.87) is very tight, strongly limiting the total number of excited polaritons.

Let us briefly discuss the case of a bulk crystal, i.e., $V \to \infty$. The asymptotic expressions of Ref. 70 are not sufficient to describe the case of a bulk crystal for low coherent pump intensities. A more general solution to the Fokker–Planck equation as $N \to \infty$ in steady state is

$$P_{ss}(\xi, \eta) \cong \text{const}\, \delta(\xi - \xi_1)\, \delta(\eta - \eta_1), \tag{8.89}$$

where ξ_1 and η_1 satisfy the equations

$$\eta_1 = \xi_1^*, \qquad |\xi|^2 = u, \qquad w = u^2, \qquad g^{(2)}(0) = 1, \qquad c\xi_1 + 2\,|\xi_1|^2\,\xi_1 = z,$$
$$4u^3 + 2\,(c + c^*)\,u^2 + |c|^2\,u = |z|^2$$
$$u = \frac{n_{\mathbf{k}_0}}{n_c}, \qquad w = \frac{b}{n_c^2}. \tag{8.90}$$

The diffusive terms disappear and the description becomes deterministic. Using the expressions for $(c + c^*)$ and $|c|^2$ for the coefficient c defined in Eq. (8.81) along with Eqs. (8.73) and (8.74), one finds the function $u(Y)$ shown in Fig. 8.8. The coherent macroscopic state possesses the property of optical bistability, with a hysteresis loop that is not pronounced. The curve $u(Y)$ increases very slowly in the quasi-threshold region $u = 1$. Many aspects of the optical-bistability effect in the excitonic range of the spectrum were studied in detail by Khadzhi [74] and Khadzhi, Shibarshina and Rotaru [75]. These issues will be discussed in Chapter 9.

Similar problems for high-density polaritons in a direct-gap semiconductor excited by laser radiation were investigated by Beloussov and Shvera [76, 77], who used the Keldysh diagram technique for nonequilibrium processes formulated in terms of functional derivatives. The Dyson equations for the normal and anomalous Green's functions were obtained

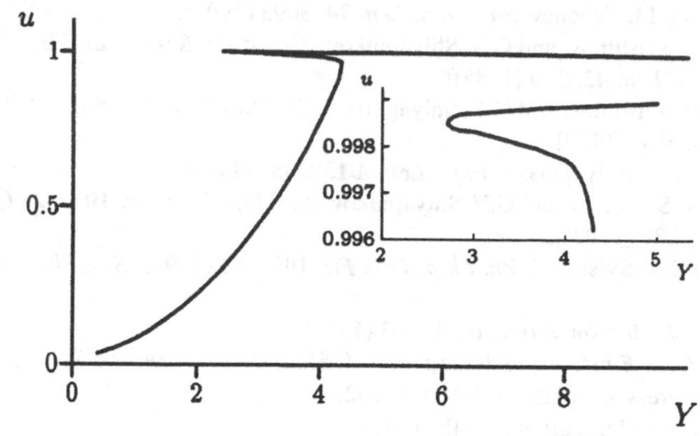

Figure 8.8. Density of coherent polaritons u as a function of pump intensity Y for a detuning from resonance $\Delta\omega/\gamma_{\text{eff}} = -3$ [46].

with the functional Legendre transformation. In the Born approximation the Dyson equations were transformed into Boltzmann-type kinetic equations, which were supplemented by the motion equation for the average value of the operator describing the coherent part of the polariton field. The full set of equations allowed the study of the quantum fluctuations of the system in the stationary regime. Within the framework of these equations, Beloussov and Frolov [78, 79] determined the parameters of the condensate time evolution that exhibits proper nutation and time decay.

A self-consistent quantum-statistical theory for Bose systems excited by a powerful external pump was also developed by Safonov [80] and Kalafati and Safonov [81]. In that work it was also shown that the nonequilibrium distribution function consists of a condensate part, coherent with the external pump, and a thermal part, describing the deviations from the forced oscillations of the system.

References

[1] I. Silvera and J.T.M. Walraven, *Prog. Low Temp. Phys.* **10**, 139 (1986).
[2] For example, M.H. Anderson et al., *Science* **269**(7), 198 (14 July 1995).
[3] D.W. Snoke, J.P. Wolfe, and A. Mysyrowicz, *Phys. Rev. Lett.* **59**, 827 (1987).
[4] D.W. Snoke, J.P. Wolfe, and A. Mysyrowicz, *Phys. Rev. Lett.* **64**, 2543 (1990).
[5] D.W. Snoke, J.P. Wolfe and A. Mysyrowicz, *Phys. Rev. B* **41**, 11171 (1990).
[6] G.M. Kavoulakis, G. Baym, and J.P. Wolfe, *Phys. Rev. B* **53**, 1 (1996).
[7] A. Mysyrowicz, D. Hulin and C. Benoit a la Guillaume, *J. Lumin.* **24/25**, 629 (1981).
[8] D.P. Trauernicht, J.P. Wolfe, and A. Mysyrowicz, *Phys. Rev. B* **34**, 2561 (1986).
[9] D.W. Snoke and J.P. Wolfe, *Phys. Rev. B* **42**, 7876 (1990).
[10] E. Levich and V. Yakhot, *Phys. Rev. B* **15**, 243 (1977).
[11] E. Levich and V. Yakhot, *J. Low Temp. Phys.* **28**, 197 (1977).
[12] E. Levich and V. Yakhot, *J. Phys. A* **11**, 2237 (1978).
[13] Ya.B. Zeldovich and E. Levich, *Sov. Phys. JETP* **28**, 1287 (1969).
[14] H.T.C. Stoof, in *Bose-Einstein Condensation*, A. Griffin, D.W. Snoke, and S. Stringari, eds. (Cambridge U. Press, Cambridge, 1995), p. 226.
[15] D.W. Snoke and J.P. Wolfe, *Phys. Rev. B* **39**, 4030 (1989).
[16] S.G. Tikhodeev, *Zh. Eksp. Teor. Fiz.* **97**, 681 (1990).
[17] U. Eckern, *J. Low Temp. Phys.* **54**, 333 (1984).

[18] S.V. Semikoz and I.I. Tkachev, *Phys. Rev. Lett.* **74**, 3093 (1995).
[19] Yu. Kagan, B.V. Svistunov, and G.V. Shlyapnikov, *Pis'ma Zh. Eksp. Teor. Fiz.* **42**, 169 (1985) [*Sov. Phys. JETP Lett.* **42**, 209 (1985)].
[20] Yu. Kagan, B.V. Svistunov, and G.V. Shlyapnikov, *Zh. Eksp. Teor. Fiz.* **93**, 552 (1987) [*Sov. Phys. JETP* **66**, 314 (1987)].
[21] Yu. Kagan and G.V. Shlyapnikov, *Phys. Lett. A* **130**, 483 (1988).
[22] Yu. Kagan, B.V. Svistunov and G.V. Shlyapnikov, *Zh. Eksp. Teor. Fiz.* **101**, 528 (1992) [*Sov. Phys. JETP* **75**, 387 (1992)].
[23] Yu. Kagan and B.V. Svistunov, *Zh. Eksp. Teor. Fiz.* **105**, 353 (1994) [*Sov. Phys. JETP* **75**, 387 (1992)].
[24] B.V. Svistunov, *J. Moscow Phys. Soc.* **1**, 363 (1991).
[25] Yu. Kagan, in *Bose-Einstein Condensation*, A. Griffin, D.W. Snoke, and S. Stringari, eds. (Cambridge U. Press, Cambridge, 1995), p. 202.
[26] H.T.C. Stoof, *Phys. Rev. Lett.* **66**, 3148 (1991).
[27] H.T.C. Stoof, *Phys. Rev. A* **45**, 8398 (1992).
[28] S.T. Beliaev, *Zh. Eksp. Teor. Fiz.* **34**, 417, 433 (1958).
[29] N. Hugenholtz and D.Pines, *Phys. Rev.* **116**, 489 (1959).
[30] J.S. Langer, *Phys. Rev.* **167**, 183 (1968).
[31] V.L. Ginzburg and L.D. Landau, *Zh. Eksp. Teor. Fiz.* **20**, 1064 (1950).
[32] V.L. Ginzburg and L.P. Pitaevskii, *Zh. Eksp. Teor. Fiz.* **34**, 1240 (1958) [*Sov. Phys. JETP* **7**, 858 (1958)].
[33] L.P. Pitaevskii, *Zh. Eksp. Teor. Fiz.* **40**, 646 (1961); *Usp. Fiz. Nauk* **90**, 623 (1966).
[34] E.P. Gross, *Nuovo Cimento* **20**, 454 (1961); *J. Math. Phys.* **4**, 147 (1963); *Ann. Phys.* **9**, 292 (1960).
[35] T.D. Lee and C.N. Yang, *Phys. Rev.* **113**, 1165 (1959).
[36] J.S. Langer, *Phys. Rev.* **184**, 219 (1969).
[37] T.D. Lee and C.N. Yang, *Phys. Rev.* **112**, 1419 (1958); **113**, 1406 (1959).
[38] M. Inoue and E. Hanamura, *J. Phys. Soc. Jpn.* **41**, 771 (1976).
[39] D.I. Blokhintsev, *Quantum Mechanics* (Reidel, Dordrecht, 1964), pp. 493–500.
[40] D.W. Snoke, W.W. Rühle, Y.-C. Lu, and E. Bauser, *Phys. Rev. B* **45**, 10979 (1992).
[41] D.W. Snoke, *Phys. Rev. B* **50**, 11583 (1994).
[42] V.E. Zakharov, *Zh. Eksp. Teor. Fiz.* **51**, 688 (1966) [*Sov. Phys. JETP* **24**, 455 (1966)]; **62**, 1745 (1972) [**35**, 908 (1972)].
[43] D.P. Trauernicht, J.P. Wolfe, and A. Mysyrowicz, *Phys. Rev. Lett.* **52**, 855 (1984).
[44] D.W. Snoke, D. Braun, and M. Cardona, *Phys. Rev. B* **44**, 2991 (1991).
[45] V.R. Misko, S.A. Moskalenko, A.H. Rotaru, and Yu.M. Shvera, *Phys. Status Solidi B* **159**, 477 (1990).
[46] V.R. Misko, S.A. Moskalenko, A.H. Rotaru, and Yu.M. Shvera, *Zh. Eksp. Teor. Fiz.* **99**, 1215 (1991).
[47] S.A. Moskalenko, A.H. Rotaru, Yu.M. Shvera, and V.A. Zalozh, *Phys. Status Solidi B* **149**, 187 (1988).
[48] S.A. Moskalenko, A.H. Rotaru, and Yu.M. Shvera, *Teor. Mater. Fiz.* **75**, 295 (1988).
[49] S.A. Moskalenko, A.H. Rotaru, and Yu.M. Shvera, in *Quantum Statistical Properties of High Density Excitons* (Shtiintsa, Kishinev, 1988), pp. 3–42 (in Russian).
[50] S.A. Moskalenko, A.H. Rotaru, and Yu.M. Shvera, in *Cooperative and Nonequilibrium Processes in a System of High Density Excitons* (Shtiintsa, Kishinev, 1989), pp. 169–198 (in Russian).
[51] S.A. Moskalenko, A.H. Rotaru, and Yu.M. Shvera, in *Excitons and Biexitons in Size Restricted Systems* (Shtiintsa, Kishinev, 1990), pp. 3–29 (in Russian).
[52] S.A. Moskalenko, A.H. Rotaru, and Yu.M. Shvera, in *Interaction of Excitons with Laser Radiation* (Shtiintsa, Kishinev, 1991), pp. 32–57 (in Russian).
[53] M.I. Shmiglyuk and V.N. Pitei, *Coherent Polaritions in Semiconductors* (Shtiintsa, Kishinev, 1989) (in Russian).

[54] A.L. Ivanov and L.V. Keldysh, *Zh. Eksp. Teor. Fiz.* **84**, 404 (1983) [*Sov. Phys. JETP* **57**, 234 (1983)].

[55] L.V. Keldysh and S.G. Tikhodeev, *Zh. Eksp. Teor. Fiz.* **90**, 1852 (1986).

[56] L.V. Keldysh and S.G. Tikhodeev, *Zh. Eksp. Teor. Fiz.* **91**, 78 (1986).

[57] M.I. Strashnikova, V.Y. Reznichenko, and V.V. Cherniy, *Ukr. Fiz. Zh.* **32**, 187 (1985).

[58] J. Kusano et al., *Solid State Commun.* **72**, 215 (1989).

[59] A.D'Andrea and R. del Sole, *Phys. Rev. B* **41**, 1413 (1990).

[60] P.D. Drummond and D.F. Walls, *J. Phys. A* **13**, 725 (1980).

[61] R. Bonifacio and L.A. Lugiato, *Lett. Nuovo Cimento* **21**, 517 (1987).

[62] V.A. Zalozh, S.A. Moskalenko, and A.H. Rotaru, *Zh. Eksp. Teor. Fiz.* **95**, 601 (1989) [*Sov. Phys. JETP* **68**, 338 (1989)].

[63] F. Arecchi, M. Scully, H. Haken, and W. Weidlich, *Quantum Fluctuations in Laser Radiation* (Mir, Moscow, 1974), p.143.

[64] F. Bassani and M. Rovere, *Solid State Commun.* **19**, 887 (1976).

[65] K.-H. -Li, *Phys. Rep.* **134**, 1 (1986).

[66] For a general review of nonequilibrium theory, see, e.g., J. Rammer and H. Smith, *Rev. Mod. Phys.* **58**, 323 (1986).

[67] P.D. Drummond and C.V. Gardiner, *J. Phys. A* **13**, 2353 (1980).

[68] C.M. Caves and B.L. Schumaker, *Phys. Rev. A* **31**, 3068 (1985).

[69] L.V. Keldysh and S.G. Tikhodeev, *Preprints of the P.N. Lebedev Physics Institute*, No. 331, (1985).

[70] O.I. Marichev, *Handbook of Integral Transforms of Higher Transcedental Functions* (Halsted, New York, 1985).

[71] H. Paul, *Rev. Mod. Phys.* **54**, 1061 (1982).

[72] S. Fridberg and L. Mandel, *Opt. Commun.* **46**, 141 (1983).

[73] A.V. Lelyakov and S.A. Moskalenko, *Fiz. Tverd. Tela* **11**, 3260 (1969) [*Sov. Phys. Solid State* **11**, 2642 (1969)].

[74] P.I. Khadzhi, *Nonlinear Optical Processes in the System of Excitons and Biexcitons in Semiconductors* (Shtiintsa, Kishinev, 1985) (in Russian).

[75] P.I. Khadzhi, G.D. Shibarshina, and A.H. Rotaru, *Optical Bistability in the System of Coherent Excitons and Biexcitons in Semiconductors* (Shtiintsa, Kishinev, 1988) (in Russian).

[76] I.V. Beloussov and Yu.M. Shvera, *Z. Phys. B* **90**, 51 (1993).

[77] I.V. Beloussov and Yu.M. Shvera, *Teor. Mat. Fiz.* **85**, 237 (1991).

[78] I.V. Beloussov and V.V. Frolov, *Zh. Eksp. Teor. Fiz.* **109**, 1806 (1996) [*Sov. Phys. JETP* **82**, 974 (1996)].

[79] I.V. Beloussov and V.V. Frolov, *Phys. Rev. B* **54**, 2523 (1996).

[80] V.L. Safonov, *Physica A* **188**, 675 (1992).

[81] Yu.D. Kalafati and V.L. Safonov, *Zh. Eksp. Teor. Fiz.* **95**, 2009 (1989) [*Sov. Phys. JETP* **68**, 1162 (1989)].

9

Coherent Nonlinear Optics with Excitons

Turbulence and chaos in a semiconductor crystal? These are just some of the intriguing predictions of nonlinear optics with coherent excitons. In this chapter we discuss some aspects of nonlinear coherent optics with the participation of the excitons, photons, and biexcitons. As in many nonlinear systems, these effects arise because of feedback mechanisms, i.e., the light impinging on the system affects the electronic states, which in turn affect the dielectric constant of the medium through which the light passes. These effects are closely related to the phenomena of self-organization [1–8] and optical bistability [7–10] and self-pulsations and the appearance of chaos [2–6]. The cooperative nonlinear coherent processes in optical systems have recently attracted much attention [7–20]. These investigations have stimulated the development of new mathematical methods [21–31] and have opened up possibilities for many new applications. The theoretical and the experimental investigations in these directions give rise to the theory of solitons [32–42] and lead to the new developments in the optics of ultrashort pulses [43–58].

Ultrafast phenomena such as self-induced transparency and nutation are related to the propagation of solitons and pulse trains, whereas the transient stages of the ultrashort light-pulse penetration in the media are described by area theorems. These coherent ultrafast phenomena typically take place during intervals of time less than the relaxation time, i.e., femtoseconds to picoseconds in typical solids. Collisions and incoherent scattering processes destroy the coherent evolution of the excited particles. But if coherent macroscopic states of the excitons, photons, and biexcitons have been formed because of spontaneous or induced BEC, in this case one can consider their time evolution over intervals of time much greater than the relaxation time. If one is dealing with spontaneous BEC, the interval of the time evolution will be determined by the recombination lifetime instead of the relaxation time. In the case of induced BEC, the coherence time will be determined by the temporal width of the coherent pumping source.

These peculiarities give rise to essential differences between the system of coherent excitons, photons, and biexcitons and the well-known model of two-level atoms, which is usually discussed in quantum optics and optoelectronics. In the case of two-level atoms, all the atoms are assumed independent of each other except for the coupling through the photon field. The exciton in this case is just the excitation of a single atom, which cannot move from that atom. In the case of excitons and biexcitons in a solid medium, collective effects due to the coherent motion of excitons can arise.

The theory of coherent nonlinear light propagation in the excitonic range of the spectrum, which is reported below, is based on some of the fundamental works and achievements mentioned above. In its turn it leads to predictions of new and interesting effects.

9.1 Optical Bistability Effects and Self-Pulsations

Risken and Numendal [59] analyzed self-oscillations in a laser by using the Maxwell–Bloch equations that describe the interaction of an electromagnetic field with two-level atoms. It was shown that for a certain set of parameters, steady-state continuous generation becomes unstable and self-oscillations in the lasing intensity occur. The self-oscillations in the geometry of a ring resonator and a Fabry–Perot resonator filled with two-level absorbers were studied in Refs. 60–62.

Haken [3–5] and Oraevskii [63] showed that the system of Maxwell–Bloch equations is homologous to the system of Lorenz equations [64]. In addition to ordinary attractors, the Lorenz equations have some singular strange attractors. Their presence indicates that in dissipative systems there are processes of a dynamic stochastic nature. The onset of dynamic chaos in optics has been the subject of many studies.

It was shown in Refs. 65–67 that the amplitude of the field that is transmitted through a ring resonator with a nonlinear medium undergoes a sequence of period-doubling bifurcations that result in the appearance of optical turbulence. This effect was observed experimentally [68]. The transition from periodic to chaotic behavior in bistable optical devices as well as the dynamic stochastic behavior in quantum generators was also studied.

In this section, we review the research on these phenomena in a system of excitons and biexcitons in a condensed medium. The biexciton–exciton conversion process and the transitions between the different excitonic levels exhibit many similarities to the model of two-level atoms. There are, on the other hand, some substantial differences. Specifically, a system of excitons and biexcitons differs from a disordered set of atoms or impurity centers in a crystal by virtue of its translational and quantum-statistical properties. It also differs in the method by which the initial state is prepared.

Rotaru and Shibarshina [69], Rotaru [70, 71], and Rotaru and Zalozh [72] studied the optical turbulence accompanying exciton–exciton and exciton–biexciton transitions in semiconductors. It was shown that these transitions are described by a generalized system of Lorenz equations in a four-dimensional phase space, and the conditions were found under which that system of equations can be reduced and converted into the ordinary system of Lorenz equations. To begin, we consider the pure exciton transition from the ground state of the crystal to the exciton state, following the work of Zalozh, Moskalenko, and Rotaru [73].

9.1.1 Self-Pulsations in the Excitonic Range of the Spectrum

The conditions for the appearance of various temporal structures in a system of coherent excitons and photons can be found starting with the Keldysh equations (7.33). They describe coherent excitons and photons with the amplitudes a and E, which are assumed to be slightly nonuniform in space and time. These equations contain neither an external pump nor dissipation effects.

For waves that are propagating along the x axis and that correspond to the positive-frequency part of the oscillatory electromagnetic field, the Keldysh equations (7.33) take the form

$$i\frac{\mathrm{d}a}{\mathrm{d}t} = \left(\Omega_{\perp} - \frac{\hbar}{2m_{\mathrm{ex}}}\frac{\mathrm{d}^2}{\mathrm{d}x^2} + \frac{g\,|a|^2}{\hbar V}\right)a - \frac{\mathrm{d}}{\hbar}E^+, \qquad c^2\frac{\mathrm{d}^2E^+}{\mathrm{d}x^2} - \frac{\mathrm{d}^2E^+}{\mathrm{d}t^2} = \frac{4\pi\mathrm{d}}{v_0}\frac{\mathrm{d}^2a}{\mathrm{d}t^2}. \tag{9.1}$$

Here only the dependences on x and t of the amplitudes $a\,(x,t)$ and $E^+\,(x,t)$ are taken into account. $E^{+}(x, t)$ is the positive-frequency part of the electric strength, g is the constant of the exciton–exciton interaction, d is the dipole moment of the transition from the ground state of the crystal to the exciton state, m_{ex} is the translational exciton mass, and $\Omega_{\perp}$ is the limiting frequency of a transverse exciton. The background dielectric constant ε_{∞} that was introduced in Eqs. (7.33) is set to unity here. V and v_0 are the volumes of the crystal and the unit cell, respectively.

The macroscopic amplitudes of the excitons and the field are represented as plane waves with a frequency ω and a wave vector **k**, multiplied by the envelope functions

$$a(x, t) = \sqrt{V}\tilde{A}e^{(-i\omega t + ikx)},$$
$$E^{+}(x, t) = \sqrt{V/v_0}\tilde{\varepsilon}e^{(-i\omega t + ikx)}. \tag{9.2}$$

The envelope functions $\tilde{A}$ and $\tilde{\varepsilon}$ of the corresponding wave packets are assumed to be slowly varying functions. They obey the conditions

$$\left|\frac{d\tilde{A}}{dt}\right| \ll \omega|\tilde{A}|, \qquad \left|\frac{d\tilde{A}}{dx}\right| \ll k|\tilde{A}|, \quad \text{etc.}, \tag{9.3}$$

which means that the envelopes are fairly smooth functions compared with the rapidly oscillating carrier waves. The envelopes change only slightly over a wavelength and over a period of the light incident upon the crystal.

Substituting Eqs. (9.2) into Eqs. (9.1), ignoring the spatial dispersion effects ($m_{ex} \to \infty$), in inequalities (9.3) one finds the following simplified equations for $\tilde{A}$ and $\tilde{\varepsilon}$, for coherent excitons and photons with uniform spatial distributions:

$$\frac{d\tilde{A}}{dt} = i\left(\omega - \Omega_{\perp} - \frac{g}{\hbar}|\tilde{A}|^2\right)\tilde{A} - \gamma_{ex}\tilde{A} + i\frac{\mathrm{d}}{\hbar v_0^{1/2}}\tilde{\varepsilon}, \tag{9.4}$$

$$\frac{d\tilde{\varepsilon}}{dt} = i\frac{2\pi\omega d}{v_0^{1/2}}|\tilde{A}| + i\frac{\omega^2 - c^2k^2}{2\omega}\tilde{\varepsilon} - \gamma\tilde{\varepsilon} + \varepsilon_0. \tag{9.5}$$

Here γ_{ex} and γ are the decay constants of the excitons and the photons, respectively, whereas ε_0 is the amplitude of the coherent external pump, introduced phenomenologically. These equations give a comprehensive description of the dynamic evolution of coherent excitons and photons that are distributed uniformly in a crystal when an external pump is acting and with decay. In the most general case, amplitudes $\tilde{A}$ and $\tilde{\varepsilon}$ are complex quantities, so that the system of Eqs. (9.4) and (9.5) consists of four independent ordinary differential equations. These equations are the same as those in Ref. 74, in which they were deduced rigorously on the basis of the quantum theory of fluctuations.

The following dimensionless quantities are introduced:

$$x = \frac{\tilde{A}}{\tilde{A}_0}, \qquad y = \frac{\tilde{\varepsilon}}{\tilde{\varepsilon}_0}, \qquad \tilde{A}_0 = \left(\frac{\hbar\gamma_{ex}}{|g|}\right)^{1/2}, \qquad \tilde{\varepsilon}_0 = (2\pi\hbar\omega)^{1/2}\tilde{A}_0$$
$$\Delta = \frac{\omega^2 - c^2k^2}{2\omega\gamma_{ex}}, \qquad \delta = \frac{\omega - \Omega_{\perp}}{\gamma_{ex}}, \qquad \sigma = \frac{\gamma}{\gamma_{ex}},$$
$$\alpha = \left(\frac{\omega\Omega_0}{2\gamma_{ex}^2}\right)^{1/2}, \qquad \Omega_0 = \frac{4\pi d^2}{v_0\hbar}, \qquad P = \frac{\varepsilon_0}{\gamma_{ex}\tilde{\varepsilon}_0},$$
$$T = t\gamma_{ex}, \qquad \nu = g/|g|. \tag{9.6}$$

Below, the value $\nu = 1$ is selected, which corresponds to repulsion between the excitons. When Eqs. (9.6) are used, Eqs. (9.4) and (9.5) take the forms

$$\frac{dx}{dT} = i(\delta - \nu|x|^2)x - x + i\alpha y, \tag{9.7}$$

$$\frac{dy}{dT} = i\alpha x + i\Delta y - \sigma y + P. \tag{9.8}$$

Equations (9.7) and (9.8) fall into the class of nonlinear ordinary differential equations, which describe open dynamic systems. Several steady-state solutions x, y are possible for such equations. Not all of them, however, will be stable, and the question of their stability is discussed below. The steady-state solutions are found from the conditions $\dot{x} = \dot{y} = 0$. They lead to the expressions

$$x_s = -\alpha P \frac{\sigma(\delta - n) + i(\sigma + \alpha^2)}{\sigma^2(\delta - n)^2 + (\sigma + \alpha^2)^2},$$

$$y_s = -\frac{x_s}{\alpha}(\delta - n + i) = P \frac{[\sigma(\delta - n) + i(\sigma + \alpha^2)]\,(\delta - n + i)}{\sigma^2\,(\delta - n)^2 + (\sigma + \alpha^2)^2}, \tag{9.9}$$

where $n = |x_s|^2$ is the steady-state density of coherent excitons. Equations (9.9) lead to the relation

$$I = |P|^2 = \frac{n}{\alpha^2}[\sigma^2(\delta - n)^2 + (\sigma + \alpha^2)^2], \tag{9.10}$$

where I is the dimensionless intensity of the coherent pumping source.

Expression (9.10), which contains the nonlinearity due to the exciton–exciton interaction, relates the exciton density n to the intensity I and is the typical equation of the theory of optical bistability in the excitonic range of the spectrum [9, 10, 75]. If the complex amplitudes x and y are written in the form $y = x_1 + ix_2$, $x = x_3 + ix_4$, instead of Eqs. (9.7) and (9.8), the following system of equations is found:

$$\frac{dx_1}{dT} = -\sigma x_1 - \Delta x_2 - \alpha x_4 + P, \tag{9.11}$$

$$\frac{dx_2}{dT} = -\sigma x_2 + \Delta x_1 + \alpha x_3, \tag{9.12}$$

$$\frac{dx_3}{dT} = -x_3 - \alpha x_2 - (\delta - n)\,x_4, \tag{9.13}$$

$$\frac{dx_4}{dT} = -x_4 + \alpha x_1 + (\delta - n)\,x_3, \tag{9.14}$$

$$n = \left(x_3^2 + x_4^2\right). \tag{9.15}$$

These nonlinear differential equations form the basis for analyzing the possible occurrence of self-pulsations in the excitonic range of the spectrum. The effect of saturation of the exciton dipole moment is taken into account in Subsection 9.1.2.

An important property of the exciton–photon system is the fact that for low-density n, the amplitude x of the exciton wave is proportional to the amplitude y of the electromagnetic field. The polarization of the medium is determined by the amplitude of the exciton wave. By contrast, the polarization of two-level atoms depends not only on the amplitude of the electromagnetic field, but also on the difference between the occupation numbers of the levels involved in the quantum transition.

We find the stationary solutions of Eqs. (9.11)–(9.15) by setting the derivatives $(\mathrm{d}x_i/\mathrm{d}T)$ equal to zero; we denote these as $x_{i,s}$. The stability of the stationary solutions $x_{i,s}$ can be analyzed with Lyapunov's small-perturbation method. This consists of the supposition that in the vicinity of the stationary points, the solutions $x_i(T)$ can be represented in the form

$$x_i(T) = x_{i,s} + \xi_i(T), \qquad \xi_i(T) \sim e^{\lambda T}, \tag{9.16}$$

where $\xi_i(T)$ are small perturbations compared with $x_{i,s}$. In this case, differential equations (9.11)–(9.15) can be linearized in the small values $\xi_i(T)$, as follows:

$$\frac{\mathrm{d}}{\mathrm{d}T}\begin{vmatrix} \cdot \\ \xi_i(T) \\ \cdot \\ \cdot \end{vmatrix} = \hat{J}\begin{vmatrix} \cdot \\ \xi_i(T) \\ \cdot \\ \cdot \end{vmatrix}, \tag{9.17}$$

where the matrix $\hat{J}$ has the form

$$\hat{J} = \begin{Vmatrix} (-1+2x_{3,s}x_{4,s}) & \left(-\delta + x_{3,s}^2 + 3x_{4,s}^2\right) & 0 & -\alpha \\ \left(\delta - 3x_{3,s}^2 + x_{4,s}^2\right) & (-1-2x_{3,s}x_{4,s}) & \alpha & 0 \\ 0 & -\alpha & -\sigma & -\Delta \\ \alpha & 0 & \Delta & -\sigma \end{Vmatrix}. \tag{9.18}$$

Equation (9.17) gives rise to the characteristic equation in the determinant form:

$$\left|\hat{J} - \lambda\hat{I}\right| = 0, \tag{9.19}$$

where $\hat{I}$ is the four-dimensional identity matrix. This equation determines the exponents λ of the small perturbations. Characteristic equation (9.19) takes the form

$$\lambda^4 + a_1\lambda^3 + a_2\lambda^2 + a_3\lambda + a_4 = 0. \tag{9.20}$$

For simplicity, we consider the case in which the frequency ω of the external field coincides with the resonator mode frequency ck. For $\omega = ck$ and $\Delta = 0$, the coefficients a_i are

$$\begin{gathered} a_1 = 2(1+\sigma), \qquad a_2 = 2(1+\sigma)^2 + 2(\sigma+\alpha^2) + D, \\ a_3 = 2[(1+\sigma)(\sigma+\alpha^2) + 6D], \qquad a_4 = (\sigma+\alpha^2) + \sigma^2 D, \\ D = (\delta - n)(\delta - 3n), \qquad n = \left(x_{3,s}^2 + x_{4,s}^2\right). \end{gathered} \tag{9.21}$$

Following Lyapunov's criteria, the stationary solutions are stable if all four roots λ of Eq. (9.20) have negative real parts, i.e.,

$$\mathrm{Re}\,(\lambda_i) < 0, \qquad i = 1, 2, 3, 4. \tag{9.22}$$

The necessary and sufficient Routh–Hurwitz conditions for this negativity resolve into the requirements that all the main diagonal minors Δ_i of the Hurwitz matrix $\|\Delta_4\|$ must be positive. The matrix $\|\Delta_4\|$ is constructed from the coefficients a_i in the following way:

$$\|\Delta_4\| = \begin{Vmatrix} a_1 & 1 & 0 & 0 \\ a_3 & a_2 & a_1 & 1 \\ 0 & a_4 & a_3 & a_2 \\ 0 & 0 & 0 & a_4 \end{Vmatrix}. \tag{9.23}$$

The diagonal minors Δ_i of the corresponding determinant Δ_4 are

$$\Delta_1 = a_1, \qquad \Delta_2 = \begin{vmatrix} a_1 & 1 \\ a_3 & a_2 \end{vmatrix}, \qquad \Delta_3 = \begin{vmatrix} a_1 & 1 & 0 \\ a_3 & a_2 & a_1 \\ 0 & a_4 & a_3 \end{vmatrix}, \qquad \Delta_4 = a_4\Delta_3. \tag{9.24}$$

The Routh–Hurwitz conditions,

$$\Delta_i > 0, \qquad i = 1, 2, 3, 4, \tag{9.25}$$

mean that the following inequalities must hold:

$$D > D_1 = -(1+\sigma)^2, \qquad D > D_2 = -(1+\frac{\alpha^2}{\sigma})^2. \tag{9.26}$$

If the exponents λ_i have the properties $\mathrm{Re}(\lambda_i) < 0$ and $\mathrm{Im}(\lambda_i) = 0$ for all values of i, then the corresponding stationary solution represents a stable node. If the real parts are negative but the imaginary parts are nonzero, $\mathrm{Im}(\lambda_i) \neq 0$ for at least one value of i, then the given stationary solution represents a stable focus. If the real parts of the exponents λ_i are equal to zero or positive ($\mathrm{Re}(\lambda_i) \geq 0$) for at least one value of i, then the given stationary solution is unstable and the phase trajectories assume the shapes of a limiting cycle or a strange attractor.

But all these trajectories, no matter what their concrete forms, possess one common property, namely, that the volume V_0 of the four-dimensional phase space occupied by the points determined by the coordinates $x_i(T)$ and belonging to a given trajectory evolves in time in a similar way. This evolution is determined by the sum of the diagonal elements of the matrix $\hat{J}$ of Eq. (9.18) and obeys the equation of motion

$$\frac{\mathrm{d}V_0}{\mathrm{d}T} = -2(1+\sigma)V_0, \qquad V_0(T) \sim \exp[-2(1+\sigma)T]. \tag{9.27}$$

It means that the volume $V_0\,(T)$ tends to zero when T tends to infinity, with the characteristic dimensionless time $1/2\,(1+\sigma)$. All the discussed trajectories concentrate into subspaces with zero phase volumes. Regarding I of Eq. (9.10) as a function of n, one can see that under the condition $0 < \delta < \delta_2 = \sqrt{3}(\alpha^2/\sigma + 1)$, this function is a single-valued, monotonically increasing function and has an inflection point at $n_p = 2\delta/3$. At this point, I equals

$$I_p = \frac{2\delta^3\sigma^2}{27\alpha^2}\left[1 + \frac{9}{\delta^2}\left(1+\frac{\alpha^2}{\sigma}\right)^2\right].$$

If $\delta > \delta_2$, the function $I\,(n)$ has two extrema at $n = n_3$ and $n = n_4$, given by the values

$$n_{3,4} = \frac{2\delta}{3} \mp \left[\frac{\delta^2}{9} - \frac{1}{3}\left(1+\frac{\alpha^2}{\sigma}\right)^2\right]^{1/2}. \tag{9.28}$$

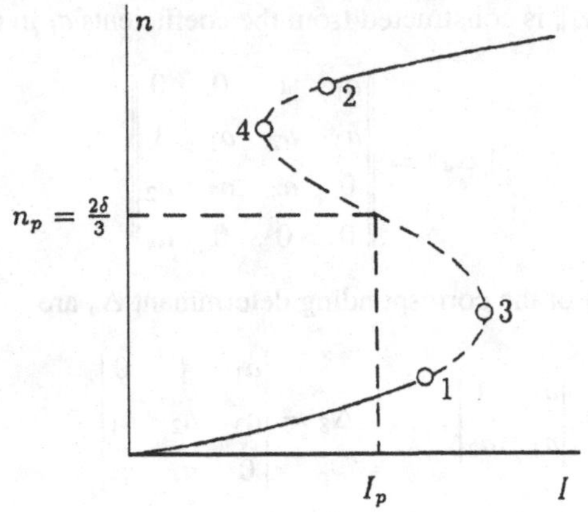

Figure 9.1. The exciton density $n = |x|^2$ versus the intensity of the electromagnetic field incident upon the crystal, $I = |P|^2$.

The inverse function $n(I)$ is triple valued at $\delta > \delta_2$ in some region of intensity I; in other words, there are three values of the exciton density in the crystal that correspond to a given value of the amplitude of the external field. This is illustrated in Fig. 9.1. It is easy to see that the region $3 \to 4$ of the functional dependence $n(I)$ is unstable for $\delta > \delta_2$, since the second of inequalities (9.26) does not hold on this part of the curve.

The optical-bistability region in Fig. 9.1 at $\delta > \delta_2$ and $\alpha > \sigma$ arises in the pump interval $I_4 < I < I_3$, where

$$I_{3,4} = \frac{2\delta^3\sigma^2}{27\alpha^2}[1 + 3v \pm (1-v)^{3/2}], \qquad v = 3(\frac{\sigma + \alpha^2}{\sigma\delta})^2. \tag{9.29}$$

On the lower and the upper branches, however, instabilities arise in a certain interval of the pump intensity. The lower branch of the curve is stable for $0 < I < I_1$ and unstable for $I_1 < I < I_3$. The upper branch is stable for $I > I_2$ and unstable for $I_4 < I < I_2$. In the case $\delta > \delta_2$ but $\alpha < \sigma$, a bistability arises in the system as before, but both the upper and the lower branches of the curve are stable. For resonance detuning δ such that the relations $\delta < \delta_1 = \sqrt{3}(\sigma + 1)$ and $\alpha > \sigma$ hold, the functional dependence $n(I)$ is single valued and a steady state is stable for all values of the external pump.

Although there is no bistability in the system in the case $\delta_1 < \delta < \delta_2$ and $\alpha > \sigma$, the part of the curve $n(I)$ in the interval $I_1 < I < I_2$ becomes unstable. Here I_1 and I_2 are given by

$$I_{1,2} = \frac{2\delta^3\sigma^2}{27\alpha^2}\left[1 + 3v \mp \left(1 + \frac{w - 3v}{2}\right)(1-w)^{1/2}\right],$$
$$w = 3\left(\frac{\sigma+1}{\delta}\right)^2. \tag{9.30}$$

In the case $\delta < \delta_2$ and $\alpha < \sigma$, there will be no bistability either, and the entire $n(I)$ curve will be stable.

At present there is no standard algorithm for solving general nonlinear differential equations, and it is difficult to find analytic solutions of Eqs. (9.11)–(9.15). Numerical solutions

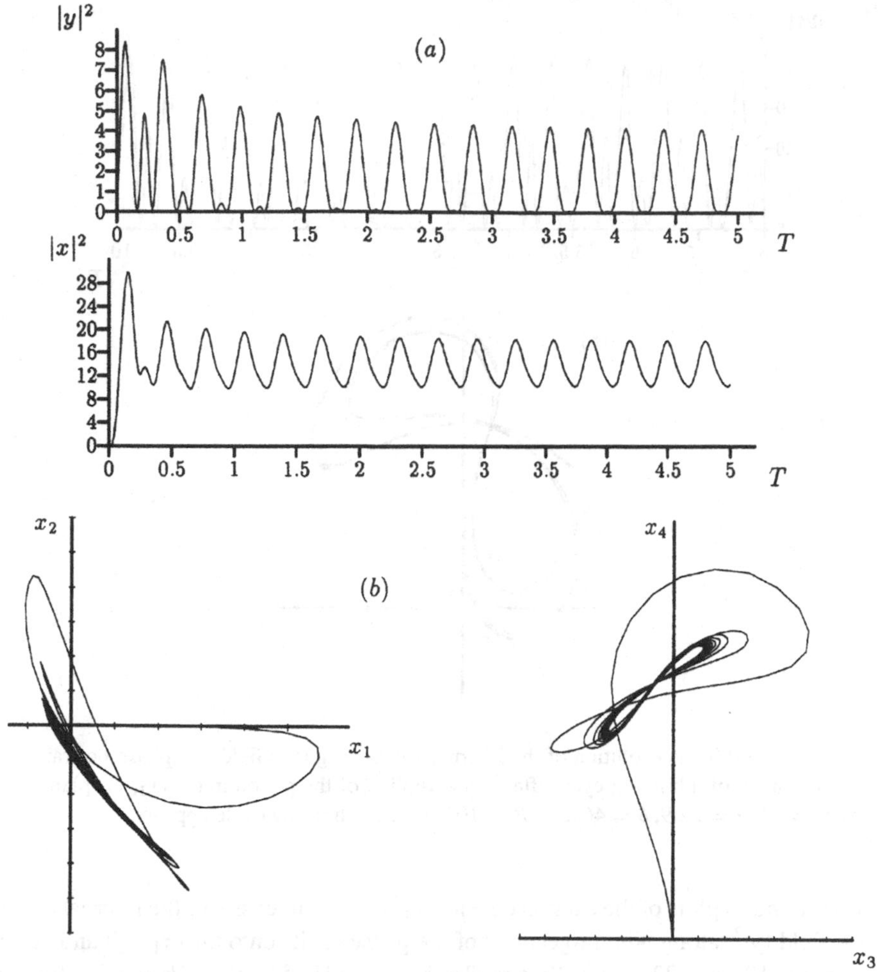

Figure 9.2. (a) Time evolution of the exciton density $|x|^2$ and the intensity of the electromagnetic field $|y|^2$, (b) projections of the phase orbits onto the (x_1, x_2) and (x_3, x_4) planes for $\sigma = 10$, $\alpha = 22.3$, $\delta = 20$, and $P = 85$. A stable limiting cycle appears.

are discussed below, which give results such as those shown in Figs. 9.2 and 9.3. The numerical results obtained are similar to the properties of the famous Lorenz system [64], which plays a special role in modern physics primarily in connection with the problem of turbulence. The "strange" behavior of the solutions of the deterministic nonlinear equations has been reported independently by Grasyuk and Oraevskii [76].

If the steady-state solutions are unstable, the attractors in phase space may be one of several things: a limit cycle, a torus, or a strange attractor. These entities correspond to nonlinear periodic, quasi-periodic, and stochastic self-oscillations in the system. A characteristic property of the latter is that the random self-oscillations do not arise in the system because of the introduction of random forces in the initial conditions or because of the action of random external forces. Their appearance is due to an internal property of the system and is related to the complex motion of the orbits in phase space.

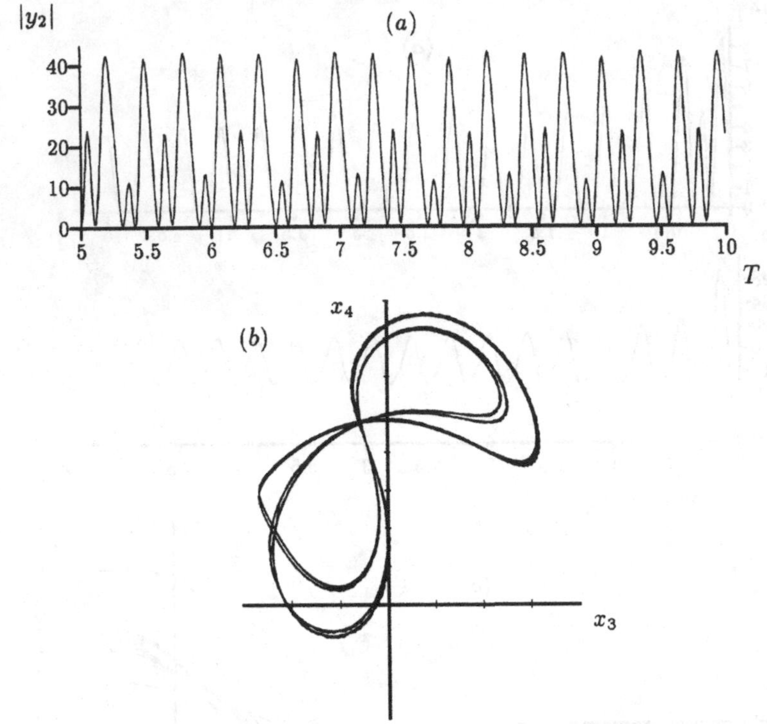

Figure 9.3. (a) Time evolution of the internal electromagnetic field, (b) phase portrait of bifurcation of a limiting cycle after a quadrupling of the period in the (x_3, x_4) plane for $\sigma = 10$, $\alpha = 22.9$, $\delta = 46$, and $P = 107$. A stable limiting cycle appears.

Figure 9.2 shows plots of the exciton density $|x|^2$ and the intensity of the internal electromagnetic field $|y|^2$, along with projections of the phase orbits onto the (x_1, x_2) and (x_3, x_4) planes for $\sigma = 10$, $\alpha = 22.3$, $\delta = 20$, and $P = 85$ ($\delta_1 = 19$, $\delta_2 = 98$). We see that for these parameter values, nonlinear periodic self-oscillations arise in the system, and as time passes a phase orbit becomes a stable limit cycle. Many other numerical investigations of this system have been performed, which show all the beautiful phenomena of random self-oscillations, chaos, Hopf bifurcations, strange attractors, etc. [69–73]. So far, these effects have not been observed experimentally; the numerical estimates indicate that the most likely material for observing these effects is CdS, in which the polariton effect is quite pronounced. Experimental evidence has been seen for optical bistability resulting from the exciton–exciton interaction in a GaSe crystal [77], but without self-oscillations.

9.1.2 Self-Pulsating Laser Radiation due to Exciton–Biexciton Conversion

The conditions for lasing by means of one- and two-photon biexciton–exciton conversion were investigated by Moskalenko, Rotaru, and Shvera [78], following the papers by Wang and Haken [79]. Of course, these quantum transitions must be allowed by the selection rules. In this Subsection we discuss the self-pulsation of laser generation during single-photon conversion, based on the paper by Moskalenko, Rotaru, and Zalozh [80]. One has three macroscopically large amplitudes, namely, the amplitude of the coherent excitons α_1, the amplitude of the coherent biexcitons α_2, and the amplitude of the photons α ($\alpha_1 \sim \alpha_2 \sim \alpha \sim \sqrt{V}$). The proper frequencies of these free quasiparticles are ω_1, ω_2, and ω, respectively.

The densities of the two populations n_1 and n_2, as well as the coherent polarization of the crystal Q, are

$$n_i = \frac{|\alpha_i|^2}{V}, \qquad i = 1, 2, \qquad Q = \frac{\alpha_2 \alpha_1^*}{V}. \tag{9.31}$$

The Hamiltonian describing the coherent excitons and biexcitons with single-photon conversion taken into account was considered in Eq. (6.105). On this basis the motion equations for the densities n_1 and n_2, the amplitude α, and polarization Q can be deduced. They contain the decay constants γ_1, γ_2, and γ and the pumping rates p_1 and p_2 of the exciton and the biexciton levels. These equations are

$$\begin{aligned}
\frac{dn_1}{dt} &= -2\gamma_1 n_1 - \frac{f}{\sqrt{N}}(Q\alpha^* + Q^*\alpha) + p_1, \\
\frac{dn_2}{dt} &= -2\gamma_2 n_2 - \frac{f}{\sqrt{N}}(Q\alpha^* + Q^*\alpha) + p_2, \\
\frac{d\alpha}{dt} &= -i\omega\alpha - \gamma\alpha - fQ\sqrt{N}v_0, \\
\frac{dQ}{dt} &= -i\left(\omega_2 - \omega_1\right) Q - \frac{f}{\sqrt{N}}\left(n_2 - n_1\right)\alpha - \left(\gamma_2 + \gamma_1\right) Q.
\end{aligned} \tag{9.32}$$

Here f is the constant of the one-photon exciton–biexciton conversion and $V = Nv_0$, where v_0 is the volume of the unit cell. Instead of the variables n_i, α, Q, and t, the dimensionless values x, y, z, F, and τ are introduced:

$$\begin{aligned}
&\alpha = \sqrt{N}x\alpha_0 e^{-i\Omega t}, \qquad Q = -2s_0 y e^{-i\Omega t} \\
&z = R - \frac{(n_2 - n_1)}{s_0}, \qquad F = \frac{(n_2 + n_1)}{F_0}, \qquad \tau = t\left(\gamma_2 + \gamma_1\right).
\end{aligned} \tag{9.33}$$

The units are then

$$\alpha_0 = \frac{(\gamma_2 + \gamma_1)}{2f}, \qquad s_0 = \frac{2\alpha_0}{f v_0}, \qquad F_0 = \frac{(p_1\gamma_2 + p_2\gamma_1)}{2\gamma_2\gamma_1 v_0}. \tag{9.34}$$

The following parameters are also introduced:

$$\begin{aligned}
&\sigma = \frac{\gamma}{(\gamma_2 + \gamma_1)}, \qquad R = \frac{f^2\left(p_2\gamma_1 - p_1\gamma_2\right)}{2\gamma_2\gamma_1\gamma\left(\gamma_2 + \gamma_1\right)}, \\
&P = R\frac{1 + ab}{a + b}, \qquad a = \frac{(\gamma_1 - \gamma_2)}{(\gamma_2 + \gamma_1)}, \qquad b = \frac{(p_1 - p_2)}{(p_1 + p_2)}.
\end{aligned} \tag{9.35}$$

In what follows, the variables x and y are considered to be real, and the frequencies are assumed to obey the equality

$$\Omega = \omega = \left(\omega_2 - \omega_1\right). \tag{9.36}$$

When the new variables and parameters are used, the motion equations have the form

$$\begin{aligned}
\frac{dx}{d\tau} &= -\sigma x + \sigma y, \\
\frac{dy}{d\tau} &= -y + (R - z)\,y, \\
\frac{dz}{d\tau} &= -z + xy - aP\,(F - 1), \\
\frac{dF}{d\tau} &= 1 - F + \frac{a}{P} z.
\end{aligned} \tag{9.37}$$

The parameters σ and R characterize the resonator quality and effective inversion population, respectively.

In Eqs. (9.37) the variables x and y depend on the inversion z of the occupation numbers of two levels, just as in the case of two-level atoms. System (9.37) is closely related to the Lorenz equations [64]. The two systems of equations differ in their dimensionalities, however. In the exciton–biexciton case, the phase space is four dimensional, in contrast to the three-dimensional Lorenz phase space. The additional fourth equation in Eqs. (9.37) determines the time evolution of the total number of coherent excitons and biexcitons F. This is different from the case of two-level atoms, in which the number of two-level atoms in the ground and the excited states remains unchanged, given by the integral of motion. If one could introduce a similar restriction on the variable F, system of equations (9.37) will transform into the Lorenz equations.

As in Subsection 9.1.1, Lyapunov's small-perturbation method and the Routh–Hurwitz criteria can be used to determine the stability of the stationary states. Numerical simulations have also been performed that permit one to draw the phase trajectories for some concrete parameter, and the events of Hopf bifurcation have been elucidated.

Analysis of Eqs. (9.37) shows that stable laser generation due to single-photon biexciton–exciton conversion takes place when the parameters σ and R obey the restrictions

$$\sigma < 2, \quad R > 1 \text{ or } \sigma > 2, \quad 1 < \mathrm{R} < \mathrm{R}_{\mathrm{cr}}. \tag{9.38}$$

Here the critical value of R is introduced. In the case $\sigma > 2$ and $R > R_{cr}$, the trajectory becomes unstable and is attracted to the strange attractor, as one can see in Fig. 9.4.

Investigations in this direction by Zalozh, Rotaru, and Tronchu concern the suppression of stochastic self-pulsations under the influence of an external periodic force and the nonstationary optical switchovers between different branches of the optical-bistability curve [81, 82].

9.2 Self-Induced Transparency, Nutation, and Quantum Beats

The phenomenon of self-induced transparency (SIT) in optics was predicted theoretically and was demonstrated experimentally by McCall and Hahn [51]. It was understand on the basis of a model of light propagation through a system of two-level atoms. It proved to be similar to the case of a solitary wave packet propagating in shallow water, observed and reported for the first time by Russell [83].

The SIT phenomenon in optics described by McCall and Hahn [51] is one of the interesting manifestations of the coherent nonlinear phenomena in a system of noninteracting two-level atoms. Other such phenomena are supperradiance, discussed, e.g., by Dicke [84], the

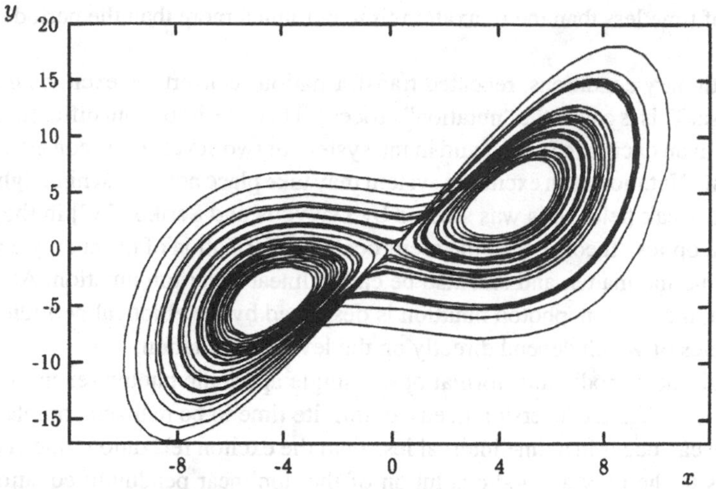

Figure 9.4. The strange-attractor phase picture on the plane (field amplitude – medium polarization) $\leftrightarrow (x, y)$ for the values $R = 55$ ($R > R_{cr} = 51.58$), $\sigma = 3$, and $\alpha = 0.8$.

photon-echo effect, investigated, e.g., by Kopvillem and Naghibarov [52] as well as by Abella, Kurnit, and Hartmann [53], and the nutation effect studied by Tang and Statz [85], Hocker and Tang [86], and Burshtein and Pusep [57]. All of these have an analog in the exciton–biexciton system, as we will see below. The excitation of the atoms of a crystal, which have a density much greater than that of a gas, leads to new peculiarities of the coherent nonlinear processes. These are closely related to the crystalline structure of the matter, namely, the strong interaction between the atoms and the fact that the number of photons and excited atoms is much less than the number of unexcited atoms. This differs essentially from the case of a rarefied gas of two-level atoms, in which the density of photons can be much greater than the density of the gas, and in many cases the photons can be considered as a fixed electromagnetic field.

In crystals, the excitation energy of a single atom passes quickly to the neighboring atoms and forms a collective excited state, namely, the exciton. If the exciting laser power is not too high and exciton metallization does not take place, then coherent excitons are formed under the action of the coherent, resonant laser radiation. These are characterized by the same wave vector, a definite phase, and macroscopic occupation of a single-particle state. The coherent excitons are characterized by a sharper spontaneous recombination line than that of incoherent excitons. But it is important to note that in both the coherent and the incoherent cases, the line intensities are proportional to the densities of the excitons, if one neglects density-dependent lifetime effects such as Auger recombination, unlike the system of two-level atoms with a coherent polarization, in which formation of a supperradiant state is characterized by a spontaneous radiation probability proportional to the square of the atom concentration.

The stationary state of an exciton–photon system is well known in linear crystal optics, namely, the polariton branches of the energy spectrum, which appear as a result of the exciton–photon interaction. If one additionally takes into account the exciton–exciton interaction, then one deals with nonlinear optics. In this section, we discuss the time evolution of the system of coherent excitons and photons with the same wave vector **k** during an interval of time less than the relaxation time but much more than the period of the light wave.

In nonstationary conditions, repeated transformations convert the excitons into photons and vice versa. This is called the "nutation" process. There are important differences between the nutation in an excitonic system and in the system of two-level atoms considered in Refs. 57, 85, and 86. Nutation in an excitonic system may take place not only at high light intensity, but also with weak fields; this was studied by Davydov and Serikov [87] in the context of linear crystal optics. It consists of the reciprocal transformation of the energy accumulated in the field and the matter, and can also be called linear polariton nutation. At high levels of excitation, the exciton–photon nutation is described by the elliptical periodic functions, the frequencies of which depend directly on the level of excitation.

Along with the periodic transformations, a single aperiodic photon–exciton conversion may take place. Such a conversion needs an infinite time to be realized completely, but its main feature can occur in a time interval less than the exciton relaxation time. This process is analogous to the only aperiodic solution of the nonlinear pendulum equation. Exactly this kind of aperiodic process is responsible for the SIT phenomenon in the excitonic range of the spectrum. The SIT phenomenon reduces to the propagation of solitary wave packets through the crystal with anomalously small losses, involving the coherent excitons and photons. A stable pulse profile occurs because the wave-packet dispersion is compensated for by the nonlinearity of the system. Such a wave packet, also known as the soliton, has a spatial size that is greater than the wavelength of the light. The above-mentioned aperiodic transformation of photons into excitons, as well as the subsequent returning of energy to the photons, takes place in the running reference frame, which moves together with the center of the soliton.

Haken and Schenzle [88] were the first to investigate this effect in excitonic systems. They found a soliton-type solution for the ultrashort laser pulses propagating through a crystal. But in this and other papers, the soliton parameters were not found in a self-consistent way [88–92]. Moreover, Hanamura [91] and Inoue [92] concluded that soliton formation for excitons is impossible because of the polariton effect. Moskalenko, Sinyak and Khadzhi, however, showed that the polariton effect does not prevent soliton formation in any way, but quite the reverse: it favors the formation of a polariton soliton [93, 94]. Agranovich and Rupasov [95] arrived at the same conclusions by studying Frenkel excitons, as did Akimoto and Ikeda [96], taking into account the polariton effect in the system of two-level atoms.

The starting point of this work is the approximation of small phase modulation [97, 98]. In this context, exact solutions of the SIT theory equations in the slowly varying amplitude approximation were obtained independently by Moskalenko and co-workers [9, 41, 99–101] and by Goll and Haken [102, 103]. In the following sections, we review the results of the two groups of investigators, comparing the solutions obtained.

Hanamura [91] noted that the effect of saturation of the dipole moment must be taken into account and therefore included these terms in the material and wave equations. In Subsection 9.2.1 we discuss this effect following Hanamura's paper. The presence of such a term in the wave equation considerably complicates the solution of the nonlinear equations. Saturation of the dipole moment leads to an increase of the soliton width and to a decrease of its amplitude. Taking into account the phase modulation leads to the appearance of a branch with anomalous dispersion for the soliton carrier wave frequency. This unexpected result was obtained first by Akimoto and Ikeda [96] and by Goll and Haken [102, 103] and is discussed below.

9.2.1 The Polariton Soliton

The system of Keldysh equations describing coherent excitons and photons was derived in Chapter 7. Following Hanamura's paper [91], these equations can be generalized to include the effect of saturation of the dipole moment. For waves traveling in the x direction in the medium with $\varepsilon_\infty \to 1$, the equations are

$$i\hbar\frac{\mathrm{d}a}{\mathrm{d}t} = \left(\hbar\Omega_\perp - \frac{\hbar^2}{2m_{\mathrm{ex}}}\frac{\mathrm{d}^2}{\mathrm{d}x^2}\right)a + g\frac{|a|^2}{V}a - \mathrm{d}E^+ + \frac{\mathrm{d}\chi}{V}(aaE^- + 2a^\dagger aE^+), \quad (9.39)$$

$$\left(\frac{\mathrm{d}^2}{\mathrm{d}x^2} - \frac{1}{c^2}\frac{\mathrm{d}^2}{\mathrm{d}t^2}\right)E^+ = \frac{4\pi}{c^2}\frac{\mathrm{d}}{v_0}\frac{\mathrm{d}^2}{\mathrm{d}t^2}\left[a\left(1 - \frac{\chi}{V}a^\dagger a\right)\right]. \quad (9.40)$$

As in Chapter 7, here $a(x,t)$ is the amplitude of the coherent excitons, $E^+(x,t)$ is the positive-frequency part of the electromagnetic field $[E^- = (E^+)^*]$, g is the constant of exciton–exciton interaction, χ is the constant of the dipole-moment saturation, d is the value of the dipole moment for the transition from the ground state of the crystal to the exciton state, $\Omega_\perp$ is the frequency of the transverse exciton at the point $k = 0$, m_{ex} is its mass, v_0 is the elementary cell volume, and V is the crystal volume. The expressions for $a(x,t)$ and $E^+(x,t)$ are chosen as

$$a(x,t) = \sqrt{V/2}\,(u + iv)\exp(-i\omega t + ikx), \quad (9.41)$$

$$E^+(x,t) = \sqrt{V/2}\,(\varepsilon + iF)\exp(-i\omega t + ikx). \quad (9.42)$$

Here the envelope functions u, v, ε, and F of the wave packet are slowly varying functions that depend on the running variable $\tau = t - x/s$. The value $(u^2 + v^2)/2$ is equal to the exciton density.

Equations of motion (9.39) and (9.40) are valid only in the limit $a_{\mathrm{ex}}^3(u^2 + v^2)/2 \ll 1$, where a_{ex} is the exciton radius. This means that the exciton density must be far from the metallization boundary. The term $[1 - \chi(u^2 + v^2)/2]$ does not change its sign, remaining positive at all exciton densities, and does not play the same role as a difference between the population numbers in the model of two-level atoms. For this reason, the solutions of Eqs. (9.39) and (9.40) are similar to the solutions of the nonlinear Schrödinger equation and do not contain as a special case the soliton-type solutions of the Maxwell–Bloch equations proposed by McCall and Hahn [51]. In this way the solution with $g \neq 0$ and different values of χ describing the SIT phenomenon in an excitonic system is quite different from the analogous solution in the case of two-level atoms. In both cases, however, they consist of aperiodic energy transformations between the photon field and the matter.

In the excitonic range of the spectrum, following an incident pulse, the light transforms into excitons in the first half of the full aperiodic transformation, and then the energy of the crystal is returned to the light in the second half of the process. A soliton-type wave-packet envelope is ultimately established because of the existence of the nonlinearity in the system, which compensates for the spread of the wave packet that is due to the polariton dispersion law.

The slowly varying amplitude approximation supposes that the derivatives of the envelope functions obey the conditions

$$\left|\frac{du}{dt}\right| \ll \omega\,|u|\,, \qquad \left|\frac{dv}{dt}\right| \ll \omega\,|v|\,, \qquad \dot{u} = \frac{du}{d\tau},$$
$$\left|\frac{du}{dx}\right| \ll k\,|u|\,, \qquad \left|\frac{dv}{dx}\right| \ll k\,|v|\,, \qquad \dot{v} = \frac{dv}{d\tau}. \tag{9.43}$$

Substituting expressions (9.41) and (9.42) into Eqs. (9.39) and (9.40), keeping only the first-order derivatives $\dot{u}$, $\dot{v}$, $\dot{\varepsilon}$, and $\dot{F}$, which is the case in inequalities (9.43), neglecting the second-order derivatives $\ddot{u}$, $\ddot{v}$, $\ddot{\varepsilon}$, and $\ddot{F}$, and setting the real and the imaginary parts of the resulting expressions to zero, one obtains the following four first-order differential equations:

$$\frac{d}{\hbar}\varepsilon[1-2\chi(u^2+v^2)] = \left[\omega_{\text{ex}}(k) + \frac{g}{2\hbar}(u^2+v^2) - \omega\right]\left\{\chi uv^2 + \left[1-\frac{\chi}{2}(u^2+3v^2)\right]u\right\}$$
$$+\left(1-\frac{\hbar k}{ms}\right)\left\{\left[1-\frac{\chi}{2}(u^2+3v^2)\right]\dot{v} - \chi uv\dot{u}\right\}, \tag{9.44}$$

$$\frac{d}{\hbar}F[1-2\chi(u^2+v^2)] = \left[\omega_{\text{ex}}(k) + \frac{g}{2\hbar}(u^2+v^2) - \omega\right]\left\{\chi u^2v + \left[1-\frac{\chi}{2}(3u^2+v^2)\right]v\right\}$$
$$+\left(1-\frac{\hbar k}{ms}\right)\left\{\chi uv\dot{v} - \left[1-\frac{\chi}{2}(3u^2+v^2)\right]\dot{u}\right\}, \tag{9.45}$$

$$\left(\frac{\omega^2}{c^2}-k^2\right)\varepsilon - 2\left(\frac{\omega}{c^2}-\frac{k}{s}\right)\dot{F} = -\frac{4\pi d\omega^2}{c^2v_0}\left[1-\frac{\chi}{2}(u^2+v^2)\right]u$$
$$+\frac{8\pi d\omega}{c^2v_0}\left[1-\frac{\chi}{2}(u^2+v^2)\right]\dot{v}, \tag{9.46}$$

$$\left(\frac{\omega^2}{c^2}-k^2\right)F + 2\left(\frac{\omega}{c^2}-\frac{k}{s}\right)\dot{\varepsilon} = -\frac{4\pi d\omega^2}{c^2v_0}\left[1-\frac{\chi}{2}(u^2+v^2)\right]v$$
$$-\frac{8\pi d\omega}{c^2v_0}\left[1-\frac{\chi}{2}(u^2+v^2)\right]\dot{u}. \tag{9.47}$$

These four first-order differential equations can be transformed into two differential equations that are second-order in the functions u and v:

$$\chi_k u + G_k(u^2+v^2)u + M_k\ddot{u} - P_k\dot{v} - Q_k\left[\frac{1}{2}(u^2+3v^2)\dot{v} + uv\dot{u}\right]$$
$$+\Gamma_k\left[\frac{1}{2}(3u^2+v^2)\dot{v} - uv\dot{u}\right] - R_k\left[u(\dot{v}^2-\dot{u}^2) - 2v\dot{u}\dot{v} + uv\ddot{v} - \frac{1}{2}(u^2+3v^2)\ddot{u}\right] = 0, \tag{9.48}$$

$$\chi_k v + G_k(u^2+v^2)v + M_k\ddot{v} + P_k\dot{u} + Q_k\left[\frac{1}{2}(3u^2+v^2)\dot{u} + uv\dot{v}\right]$$
$$-\Gamma_k\left[\frac{1}{2}(u^2+3v^2)\dot{u} - uv\dot{v}\right] + R_k\left[v(\dot{v}^2-\dot{u}^2) + 2u\dot{u}\dot{v} + uv\ddot{u} + \frac{1}{2}(3u^2+v^2)\ddot{v}\right] = 0, \tag{9.49}$$

where

$$\chi_k = \left(k^2 - \frac{\omega^2}{c^2}\right)[\omega_{\text{ex}}(k) - \omega] - \Omega_0 \frac{\omega^2}{c^2},$$

$$G_k = \frac{g}{2\hbar}\left(k^2 - \frac{\omega^2}{c^2}\right) + \frac{\chi}{2}\left\{\Omega_0 \frac{\omega^2}{c^2} + 3[\omega_{\text{ex}}(k) - \omega]\left(k^2 - \frac{\omega^2}{c^2}\right)\right\},$$

$$M_k = 2\left(\frac{k}{s} - \frac{\omega}{c^2}\right)\left(1 - \frac{\hbar k}{ms}\right),$$

$$\Gamma_k = \chi\left(k^2 - \frac{\omega^2}{c^2}\right)\left(1 - \frac{\hbar k}{ms}\right),$$

$$P_k = 2\left(\frac{k}{s} - \frac{\omega}{c^2}\right)[\omega_{\text{ex}}(k) - \omega] - \left(k^2 - \frac{\omega^2}{c^2}\right)\left(1 - \frac{\hbar k}{ms}\right) - 2\Omega_0 \frac{\omega}{c^2},$$

$$Q_k = \frac{2g}{\hbar}\left(\frac{k}{s} - \frac{\omega}{c^2}\right) + 6\chi\left(\frac{k}{s} - \frac{\omega}{c^2}\right)[\omega_{\text{ex}}(k) - \omega] + 2\chi\Omega_0 \frac{\omega}{c^2},$$

$$R_k = 2\chi\left(\frac{k}{s} - \frac{\omega}{c^2}\right)\left(1 - \frac{\hbar k}{ms}\right),$$

$$\Omega_0 = \frac{4\pi \mathrm{d}^2}{\hbar v_0}, \quad \omega_{\text{ex}}(k) = \Omega_\perp + \frac{\hbar k^2}{2m}. \tag{9.50}$$

We first find the solution of Eqs. (9.48) and (9.49), neglecting the dipole-moment saturation effect, that is, setting $\chi = 0$. In terms of the new complex function $z = u + iv$, the system of two equations may be joined into one nonlinear equation:

$$M_k \ddot{z} + \chi_k z + G_k |z|^2 z + i P_k \dot{z} + i Q_k |z|^2 \dot{z} + i Q_k z(\dot{z} z^* + z \dot{z}^*) = 0. \tag{9.51}$$

The slowly varying amplitude z is decomposed into the modulus $\rho(\tau)$ and the phase factor $\varphi(\tau)$:

$$z = \rho(\tau) e^{i\varphi(\tau)}. \tag{9.52}$$

The differential equations for $\rho\,(\tau)$ and $\varphi\,(\tau)$ are

$$\begin{aligned} &\ddot{\rho} + \alpha\rho + \beta\rho^3 - \gamma\rho\dot{\varphi} - \delta\rho^3\dot{\varphi} - \rho\dot{\varphi}^2 = 0, \\ &\rho\ddot{\varphi} + 2\dot{\rho}\dot{\varphi} + \gamma\dot{\rho} + 3\delta\rho^2\dot{\rho} = 0, \end{aligned} \tag{9.53}$$

where

$$\alpha = \frac{\chi_k}{M_k}, \qquad \beta = \frac{G_k}{M_k}, \qquad \gamma = \frac{P_k}{M_k}, \qquad \delta = \frac{Q_k}{M_k}. \tag{9.54}$$

The general solution of Eqs. (9.53) was found by Khadzhi and is given by

$$\dot{\varphi} = \frac{C}{\rho^2} - \frac{3}{4}\delta\rho^2 - \frac{1}{2}\gamma. \tag{9.55}$$

It has physical meaning when $\rho \to 0$ only for $C = 0$. The occurrence of the term $-1/2\gamma$ in Eq. (9.55) means the appearance of a linear part $-1/2\gamma\tau$ in the expression $\varphi(\tau)$. If one assumes that ω is the central frequency of the pulse, the constant γ must be set to zero. This condition was used in Refs. 99–103 as well as in Refs. 9 and 41 and may be regarded as an equation for the determination of the soliton velocity as a function of the other parameters. The modulus $\rho(\tau)$ obeys a differential equation for a nonlinear oscillator,

$$\ddot{\rho} + \alpha\rho + \beta\rho^3 + \frac{3}{16}\delta^2\rho^5 = 0, \tag{9.56}$$

with the first integral of motion

$$\dot{\rho}^2 + \alpha\rho^2 + \frac{1}{2}\beta\rho^4 + \frac{1}{16}\delta^2\rho^6 = 2D. \tag{9.57}$$

The expression

$$W(\rho) = \alpha\rho^2 + \frac{1}{2}\beta\rho^4 + \frac{1}{16}\delta^2\rho^6 \tag{9.58}$$

plays the role of the potential energy of the nonlinear oscillator, and D has the sense of its total energy. Formula (9.57) expresses the energy conservation law. The soliton-type solution takes place only in the case in which $\alpha = -|\alpha|$.

Figure 9.5 shows $W(\rho)$ and the phase trajectories for $\alpha < 0$ in coordinates $\dot{\rho}$, ρ. The function $W(\rho)$ attains its extrema at the three points

$$\rho_1 = 0,$$
$$\rho_{2,3} = \rho_\pm = \pm\frac{2}{\delta}\sqrt{\frac{2}{3}\left(-\beta + \sqrt{\beta^2 + \frac{3}{4}\delta^2|\alpha|}\right)}. \tag{9.59}$$

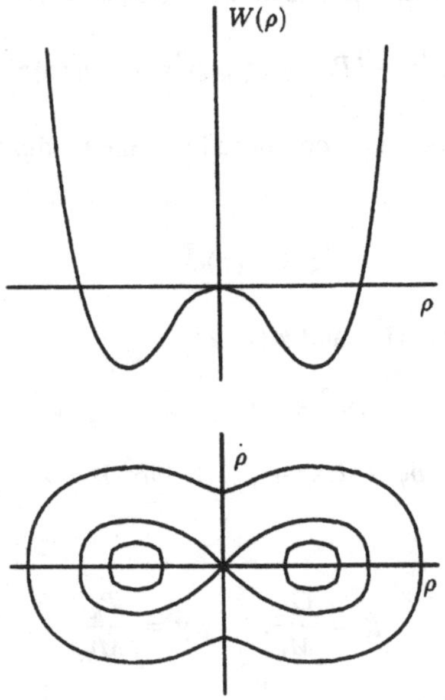

Figure 9.5. The plot of the potential energy of the nonlinear oscillator and the phase trajectories for the condition $\alpha < 0$.

If $D < 0$, then nonlinear oscillations relative to the two symmetric equilibrium positions ρ_- and ρ_+ take place. The case $D > 0$ also corresponds to oscillations but with large amplitudes. The motion on the separatrix is determined by the value $D = 0$. The solution of Eq. (9.56) in this case is a soliton with its center at the point $\tau = 0$. This was also found by Khadzhi and co-workers [9, 41, 99, 100]:

$$\rho(\tau) = \frac{\sqrt{\frac{2|\alpha|}{\sqrt{\beta^2+|\alpha|\delta^2}}}}{\sqrt{\cosh^2\left(\sqrt{|\alpha|}\tau\right) + \frac{\beta-\sqrt{\beta^2+|\alpha|\delta^2}}{2\sqrt{\beta^2+|\alpha|\delta^2}}}}. \tag{9.60}$$

The soliton parameters are defined by Eq. (9.60). Its amplitude A is

$$A = \frac{2\sqrt{|\alpha|}}{\sqrt{\beta + \sqrt{\beta^2 + |\alpha|\,\delta^2}}}.$$

The FWHM value σ is determined as $\rho(\sigma/2) = A/2$, from which it follows that

$$\sigma = \frac{2}{\sqrt{|\alpha|}} \ln\left(\sqrt{3\frac{\beta + 5\sqrt{\beta^2 + |\alpha|\,\delta^2}}{\sqrt{\beta^2 + |\alpha|\,\delta^2}}} + \sqrt{3\frac{\beta + \sqrt{\beta^2 + |\alpha|\,\delta^2}}{\sqrt{\beta^2 + |\alpha|\,\delta^2}}}\right). \tag{9.61}$$

The soliton area S is evaluated from the integral

$$S = \int_{-\infty}^{\infty} \rho(\tau)\mathrm{d}\tau,$$

which equals

$$S = \frac{8F(\psi, K)}{\sqrt{\beta + \sqrt{\beta^2 + |\alpha|\,\delta^2} + \sqrt{2\sqrt{\beta^2 + |\alpha|\,\delta^2}}}}, \tag{9.62}$$

where $F(\psi, K)$ is the full elliptic integral of the first kind [104, 105] with the parameters

$$\psi = \frac{\sqrt{2\sqrt{\beta^2 + |\alpha|\,\delta^2}} + \sqrt{\sqrt{\beta^2 + |\alpha|\,\delta^2} - \beta}}{\sqrt{\beta + \sqrt{\beta^2 + |\alpha|\,\delta^2}}}, \tag{9.63}$$

$$K = 2\frac{\sqrt[4]{\frac{1}{2}\left(1 - \frac{\beta}{\sqrt{\beta^2+|\alpha|\delta^2}}\right)}}{1 + \sqrt{\frac{1}{2}\left(1 - \frac{\beta}{\sqrt{\beta^2+|\alpha|\delta^2}}\right)}}. \tag{9.64}$$

With the aid of Eq. (9.60), the function $\varphi(\tau)$ is determined as

$$\varphi(\tau) = \varphi_0 + 3\,\mathrm{sgn}(\tau)\sqrt{|\alpha|}\ln\left|\frac{\beta + \sqrt{\beta^2 + |\alpha|\,\delta^2} + \delta\sqrt{|\alpha|}\tanh\sqrt{|\alpha|}\tau}{\beta + \sqrt{\beta^2 + |\alpha|\,\delta^2} - \delta\sqrt{|\alpha|}\tanh\sqrt{|\alpha|}\tau}\right|, \tag{9.65}$$

where φ_0 is a constant of the integration.

At small values of δ^2, the exact values for $u = \rho\cos\varphi$ and $v = \rho\sin\varphi$ practically coincide with the approximate solutions within the framework of perturbation theory, taking into account the phase modulation [94, 98]. A complete description of the soliton requires knowledge of its five parameters, namely the frequency ω and the wave vector $\mathbf{k}$ of the carrier wave, the soliton velocity s, the amplitude at the maximum A, and the FWHM σ. These last two quantities can be expressed in terms of ω, k, and s.

The condition $\gamma = 0$ permits one to express the velocity s in terms of ω and k. As a consequence, one more condition is needed to find the dispersion relation $\omega(k)$. Akimoto and Ikeda [96] and Goll and Haken [102, 103] used the condition $|\alpha(k)| = 1/T^2$, which suggests that the constant $\alpha(k)$ does not depend on k. By using two conditions,

$$\gamma = 0, \quad |\alpha(k)| = 1/T^2, \tag{9.66}$$

one can determine $\omega(k)$ and $s(k)$ to be

$$2\left(\frac{k}{s} - \frac{\omega}{c^2}\right)[\omega_{\mathrm{ex}}(k) - \omega] = \left(k^2 - \frac{\omega^2}{c^2}\right)\left(1 - \frac{\hbar k}{ms}\right) + 2\Omega_0 \frac{\omega}{c^2}, \tag{9.67}$$

$$\frac{1}{T^2} = \frac{\{\Omega_0 \omega^2 - (c^2k^2 - \omega^2)[\omega_{\mathrm{ex}}(k) - \omega]\}}{2c^2\left(\frac{k}{s} - \frac{\omega}{c^2}\right)\left(1 - \frac{\hbar k}{ms}\right)}. \tag{9.68}$$

One can see that the dispersion law contains a polariton gap at large values of T. But simultaneously with the lower and the upper polaritonlike branches, a new extensive branch appears, nearly horizontal with an anomalous dispersion. As the soliton width T shortens, the polariton gap disappears, and the branch with the anomalous dispersion becomes shorter and wider, while the dispersion curve tends as a whole to that of the photon.

The ratio $s(\omega)/c$ decreases quickly with the decrease of the frequency detuning $\omega - ck_0$ and depends on the parameter $\Omega_0 T$. When the dispersion law $\omega(k)$ approaches the photon line, then the soliton velocity $s(\omega)$ approaches the light velocity c, and vice versa. In the range of the polariton gap, when it exists, the difference between $s(\omega)$ and c is very large.

We have seen that the polariton effect not only does not prevent soliton formation, but on the contrary, leads to new, fascinating soliton behavior. Further investigations of polariton solitons, as well as solitons in the system of coherent excitons, photons, and biexcitons ("simultons") were pursued in Refs. 106–110. Phonon-type polariton solitons in solids [111, 112] as well as in biological systems [113] have also been investigated.

9.2.2 Exciton–Photon Nutation

In the system of two-level atoms, nutation is a nonstationary phenomenon related to the coherent superposition of two quantum states in an external laser radiation field. One of them is the excited atomic state, and the other is the renormalized state formed from the atomic ground state in the presence of coherent laser radiation. As a result, the atom is found alternately in the ground and the excited states, and coherent transformations are induced by the external laser field. When the frequency detuning between the frequencies of the laser radiation and the atom excitation equals zero, then the nutation frequency coincides with the Rabi frequency, which is proportional to the amplitude of the external laser field. The nutation process in the two-level atom implies the coherent periodic alternation of the electron population between two levels, induced by the external laser radiation.

In contrast to this picture, the nutation effect in an excitonic system is a much more simple process. It can take place in linear crystal optics in the absence of an external radiation source and is related to the coherent superposition of the exciton and photon states, which arises because of the polariton effect. In stationary conditions the polariton effect gives rise to the formation of the polariton branches, whereas nonstationary conditions [41, 87] lead to the nutation effect. In this way the nutation effect in the excitonic range of the spectrum is a nonstationary process that consists of multiple reciprocal transformations of coherent excitons into coherent photons repeated over time. The nutation transformation takes place during an interval of time less than the relaxation time.

In linear crystal optics the exciton–photon nutation effect was studied for the first time by Davydov and Serikov [87] and in the more general nonlinear case by Moskalenko et al. [9, 41]. The nutation oscillations differ essentially from the self-pulsations discussed in Section 9.1. Indeed, the latter oscillations appear on the background of the stationary states of the exciton–photon system and are the manifestations of their instability. In this Subsection we review some results obtained by Khadzhi and coworkers [9, 41]. Since the nutation process deals with only the time evolution of the quasiparticles, their spatial distribution is assumed homogeneous.

The Hamiltonian of the coherent excitons and photons with the same wave vector, taking into account only the polariton effect and the main term of the exciton–exciton interaction, has the form

$$\frac{H}{\hbar} = \omega_{\mathrm{ex}} a^\dagger a + ckc^\dagger c + \frac{i\varphi}{\hbar}(a^\dagger c - c^\dagger a) + \frac{g}{2\hbar V} a^\dagger a^\dagger a a. \tag{9.69}$$

Here the wave vectors of the coherent state operators are dropped. The exciton operators $a^\dagger$ and a and the photon operators $c^\dagger$ and c can be considered as macroscopically large amplitudes that depend on the time t.

The frequency detuning Δ between the exciton frequency ω_{ex} and the photon frequency ck is defined as $\Delta = \omega_{\mathrm{ex}} - ck$. The constants φ and g determine the polariton effect and the exciton–exciton interaction. With the exciton and photon creation and annihilation operators, the operators R_1, R_2, and R_3 that represent the components of the Bloch pseudovector can be constructed. They are

$$\begin{aligned} &R_1 = a^\dagger c + c^\dagger a, \qquad R_2 = i(a^\dagger c - c^\dagger a),\\ &R_3 = a^\dagger a - c^\dagger c, \qquad N = a^\dagger a, \qquad F = c^\dagger c, \end{aligned} \tag{9.70}$$

where R_3 is the difference between the occupation number operators N and F of the coherent excitons and photons. The motion equations for these operators are

$$\begin{aligned} \dot{R}_1 &= (\Delta + \bar{g}n)\, R_2 + 2\bar{\varphi} R_3,\\ \dot{R}_2 &= -\,(\Delta + \bar{g}n)\, R_1,\\ \dot{R}_3 &= -2\bar{\varphi} R_1,\\ \dot{N} &= -\bar{\varphi} R_1,\\ \dot{F} &= \bar{\varphi} R_1, \end{aligned}$$

$$\text{where} \quad n = \frac{N}{V}, \quad f = \frac{F}{V}, \quad \bar{\varphi} = \frac{\varphi}{\hbar}, \quad \bar{g} = \frac{g}{\hbar}. \tag{9.71}$$

The operators of Eqs. (9.70), being macroscopically large values proportional to the crystal volume V, can be treated as functions of time t, for which there are two integrals of motion,

$$N + F = \text{const}, \qquad R_1^2 + R_2^2 + R_3^2 = \text{const}. \tag{9.72}$$

The solution of nonlinear equations (9.71) was found for the following initial conditions:

$$N(0) = 0, \qquad F(0) = F_0, \qquad \dot{N}(0) = \dot{F}(0) = 0. \tag{9.73}$$

Equations (9.71) and requirements (9.73) lead to the integral of motion for the coherent exciton density $n(t)$:

$$\dot{n}^2 + W(n) = 0. \tag{9.74}$$

It expresses the energy conservation law for the nonlinear oscillator with total kinetic and potential energy equal to zero. The double potential energy of the nonlinear oscillator has the form

$$W(n) = n\left[-4\bar{\varphi}^2 f_0 + (\Delta^2 + 4\bar{\varphi}^2)n + \Delta\bar{g}n^2 + \frac{1}{4}\bar{g}^2 n^3\right],$$
$$f_0 = \frac{F_0}{V}. \tag{9.75}$$

To simplify the next calculations, a particular case $\Delta = 0$ is considered. The curve $W(n)$ is depicted in Fig. 9.6. It represents a potential well in which periodic motion of the oscillator is possible. The maximum density $n_{\max}$ of the coherent excitons that may be achieved during their time evolution is determined by the requirement

$$\frac{1}{4}g^2 n^3 + 4\varphi^2 n - 4\varphi^2 f_0 = 0, \tag{9.76}$$

which has an approximate solution

$$n_{\max} \simeq f_0\left(1 - \frac{g^2 f_0^2}{16\varphi^2}\right), \qquad |g|\, n \ll 4\,|\varphi|\,. \tag{9.77}$$

When the level of excitation f_0 increases, the value $n_{\max}$ differs increasingly from f_0.

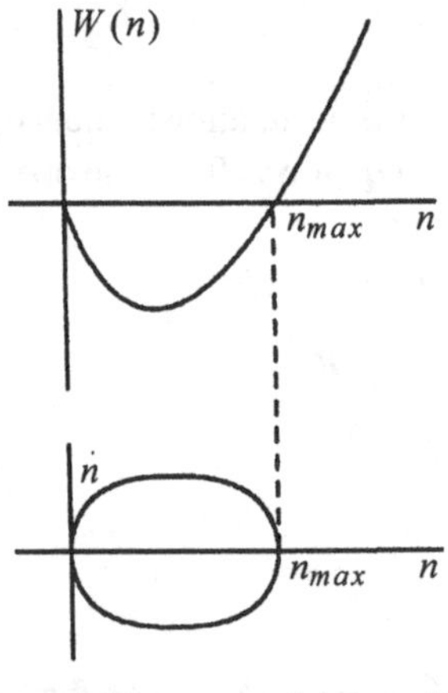

Figure 9.6. The plots of the potential energy of the nonlinear oscillator and the phase trajectory in the case $\Delta = 0$ [9, 41].

The classical phase trajectory in the plane $(\dot{n}, n)$ has two points of return, $n = 0$ and $n = n_{\max}$. Between these points the finite motion of the nonlinear oscillator with the full energy equal to zero takes place. The phase trajectory is represented in Fig. 9.6. The solution of Eqs. (9.71) and (9.73) at $\Delta = 0$ has the form of a pulse train:

$$n(t) = n_{\max} \frac{1 - \mathrm{cn}\left(\frac{1}{2}\sqrt{l_1 l_2}\, t\right)}{\left(1 + \frac{l_1}{l_2}\right) + \left(1 - \frac{l_1}{l_2}\right) \mathrm{cn}\left(\frac{1}{2}\sqrt{l_1 l_2}\, t\right)}. \tag{9.78}$$

Here the following notations were introduced:

$$l_1 = \sqrt{16\bar{\varphi}^2 + \frac{3}{4}\bar{g}^2 f_0^2 + \frac{1}{4}\bar{g}^2 f_0^2 \left(3 - \frac{\bar{g}^2 f_0^2}{8\bar{\varphi}^2}\right)^2},$$

$$l_2 = \sqrt{16\bar{\varphi}^2 + \bar{g}^2 f_0^2}. \tag{9.79}$$

The elliptic cosine cn(z) [104, 113, 114] has a modulus k that is given by

$$k = \sqrt{\frac{1}{2}\left(1 - \frac{16\bar{\varphi}^2 + \frac{3}{2}\bar{g}^2 f_0^2 - \frac{1}{32}\frac{\bar{g}^4 f_0^4}{\bar{\varphi}^2}}{l_1 l_2}\right)}. \tag{9.80}$$

The nutation frequency ω_n is

$$\omega_n = \frac{\pi\sqrt{l_1 l_2}}{4K(k)}, \tag{9.81}$$

where $K(k)$ is the full elliptic integral of the first kind [104, 114, 115]. One can conclude that the coherent exciton density n at $\Delta = 0$ changes periodically in the interval $(0, n_{\max})$, where $n_{\max}$ is less than f_0. The density of the coherent photons $f = f_0 - n$ changes in the interval $(f_0 - n_{\max}, f_0)$. This behavior is represented in Fig. 9.7.

The nutation frequency at $\Delta = 0$ increases monotonically with the increasing initial photon density f_0. At a high exciton level the nutation frequency is greater than in the linear

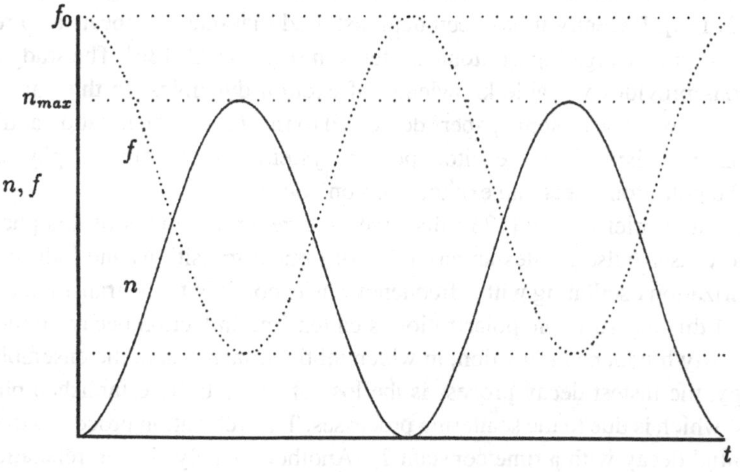

Figure 9.7. The time evolution of the coherent excitons and photons in the case $\Delta = 0$ [9, 41].

limit. In the linear case one has the parameters

$$\bar{g} = 0, \qquad l_1 = l_2 = 4\bar{\varphi}, \qquad k = 0, \qquad K(0) = (\pi/2), \tag{9.82}$$

which lead to the linear nutation frequency

$$\omega_{\rm ln} = \frac{2|\varphi|}{\hbar}, \tag{9.83}$$

and to linear oscillations of the types [87]

$$n = f_0 \sin^2\left(\frac{\varphi t}{\hbar}\right), \qquad f = f_0 \cos^2\left(\frac{\varphi t}{\hbar}\right). \tag{9.84}$$

Khadzhi and coworkers also determined the nutation frequency in the case of nonzero frequency detuning ($\Delta \neq 0$). Expanded in a series in terms of the photon density f_0, in the first order of perturbation theory this expression takes the form

$$\omega_n = \omega_{\rm ln}\left[1 + \frac{\bar{\varphi}^2 \bar{g}\Delta}{(\Delta^2 + 4\bar{\varphi}^2)^2} f_0\right], \tag{9.85}$$

which means that in the case $\bar{g}\Delta > 0$ the nutation frequency increases monotonically proportional to the level of excitation [9, 41].

9.2.3 Quantum Beats of Excitons

Time-resolved, high-resolution quantum-beat spectroscopy is a powerful method in atomic and molecular spectroscopy [116–119], as well as in solid-state physics [120–136]. It is useful for determining the small energy splittings due to fine and hyperfine interactions, for studying the relaxation processes, for determining the dephasing times and lifetimes of the excited electronic states, and for distinguishing between Raman scattering and hot luminescence. The essence of this method, including its enormous potential for applications, has been discussed in detail in several review articles [116–118], monographs [15, 119], and papers [122–136]. Recently it has been demonstrated that quantum beats also represent a powerful tool for investigating excitonic systems in solids [122–136]. The study of beating from excitons provides valuable knowledge of exciton dynamics. In this subsection, we present a short overview of some papers dedicated to this topic. The nutation and quantum-beat phenomena arising in the exciton–photon system have a common physical origin, related to the polariton effect and exciton–photon conversion.

Leo et al., in a brief review [125], discussed the general features of this phenomenon. An ultrashort laser pulse excites an ensemble of optical transitions and induces a macroscopic polarization oscillating with a frequency corresponding to the transition energy. The relaxation of this macroscopic polarization is called free-induction decay. In the case of a homogeneously broadened transition, in which all the transitions in the ensemble have the same energy, the fastest decay process is the loss of the initially established phase of the excitations, which is due to the scattering processes. This relaxation process is described by an exponential decay with a time constant T_2. Another, usually slower, relaxation process is the relaxation of the excitations back to the ground state, called the T_1 time. As pointed out by Stolz [123], T_2 is the decay time of the coherence terms represented by off-diagonal elements of the density matrix, whereas T_1 is the decay time of the population terms represented by diagonal elements of the density matrix.

In Ref. 125 the model of three-level atoms was first considered. Two quantum transitions, from the ground state of the atom to two excited states with nearly the same energy, were considered. If they are excited with an ultrashort laser pulse having a sufficient spectral width to cover both transition energies, then a coherent superposition of the two wave functions is created. The slightly different oscillation frequencies of the two transitions will then lead to a beating of the polarization. This evolves in the form of a beating oscillation that has a central frequency and a slowly varying beating amplitude. The period of the beating amplitude is

$$T_{\mathrm{QB}} = \frac{h}{\Delta E} = \frac{2\pi}{\Delta\omega}, \quad \Delta E \equiv \hbar\Delta\omega. \tag{9.86}$$

Here ΔE is the difference in the transition energies of the two excited levels. The subsequent time-resolved fluorescence signal exhibits typical quantum-beat oscillations originating from the superposition of the wave functions of two excited quantum states.

Along with quantum beats of this type, there is also the possibility of forming a coherent polarization in a system of two-level atoms with an inhomogeneous broadening of the transition frequency. Interference phenomena can be observed if energetically closely spaced two-level transitions are excited coherently. In this case, the quantum beats are not caused by interference of the wave functions, but by a direct polarization interference.

As an example of the first type of system, one can consider the heavy-hole (hh) and the light-hole (lh) exciton states in a homogeneous quantum-well structure. The two exciton states plus the ground state of the quantum-well structure form the three-level model. The second type of system can be modeled by hh exciton states in quantum-well structures with spatially varying confinement, which gives rise to inhomogeneous broadening, a situation that is typical in semiconductor quantum-well structures because of surface roughness created during the fabrication process. The hh exciton state and the ground state form the two-level system. Unlike the free-carrier extended states, the excitons have dephasing times T_2 much longer than the electron intraband-scattering time. The excitation of the closely spaced exciton states gives rise to quantum beats with dephasing time constant T_2 in the picosecond range. Reference 125 presented an extensive experimental investigation of quantum beats originating from excitonic states in quantum wells. These structures hold particular interest because of their strong excitonic transitions and interesting applications in modulators. It was observed that if the laser beam is tuned between the hh and the lh exciton transitions and if it has sufficient spectral width, the two states are excited coherently. The beat period of 960 fs is in excellent agreement with the energy splitting of 4.2 meV. The polarization decay can be described by a single exponential with time constant T_2 of ~2 ps. Quantum beating of a second type, arising from the interference of excitons with slightly different quantum-confinement energies because of well-width fluctuations, were also evidenced [124, 125]. Schmitt-Rink et al. [134] showed the polarization dependence of hh and lh quantum beats.

The Zeeman splitting of the hh and the lh exciton states in GaAs/Al$_x$Ga$_{1-x}$As quantum wells and the time evolution of the excitonic signal arising in four-wave-mixing experiments were observed by Carmel, Shtrikman, and Bar-Joseph [131, 132]. The excitonic signal is strongly modified because of quantum beats and the rotation of the echo polarization. An increase in the dephasing times in the magnetic field is observed [132].

Magnetoquantum beats and coherence in resonant light scattering from quadrupole polaritons in Cu_2O were investigated by Langer et al. [126]. Using the picosecond time-resolved light-scattering method, they observed quantum beats of the $1s$ quadrupole-active Γ_5^+ exciton level, split by a magnetic field. The observations were made for both one-phonon- and two-phonon-scattering processes. From these experiments the Raman and hot-luminescence contributions could be discriminated, and exciton–polariton coherence times of up to 800 ps at 2K were measured. The coherence times strongly decrease at larger wave vectors and higher temperatures because of phonon scattering. To obtain the coherence time of the intermediate quadrupole-polariton states, two different light-scattering processes were investigated, with the participation of either one odd-parity Γ_3^- phonon with energy $\hbar\omega(\Gamma_3^-) = 13.6$ meV or with the participation of two such phonons. The one-phonon-scattering process becomes resonant in two cases. One of them occurs if the incident laser is tuned directly into the quadrupole transition of the Γ_5^+ orthoexciton level with energy $E_{\mathrm{or}} = 2.0329$ eV and is called the "incoming" resonance. The second "outgoing" resonance takes place when the laser frequency coincides with the energy of the indirect exciton–phonon absorption energy $E_{\mathrm{or}} + \hbar\omega(\Gamma_3^-)$. In both cases, the electric quadrupole and phonon-assisted electric dipole transitions are involved in the scattering process. Because of the extremely small oscillator strength of the quadrupole transition ($f = 3.7 \times 10^{-9}$), the signals can be distinguished from the nonresonant background only very near to the two resonance poles.

Quantum beats and exciton coherence in time-resolved resonant light scattering were investigated by Stolz [123]. It was shown that the coherent nature of the states involved in the scattering process can be directly revealed by quantum beats in the time-dependent scattering spectrum if the resonant states are energetically split. The exciton dephasing time for the lowest exciton polariton in Cu_2O was found to be surprisingly long, up to 1 ns, determined by the energy relaxation time T_1 due to inelastic phonon scattering. Pure dephasing due to elastic scattering gives only a minor contribution.

Langer, Stolz, and von der Osten [120] carried out subsequent investigations of the quantum beats in picosecond time-resolved resonance fluorescence in Cu_2O for the same $1s$ quadrupole-polariton states. The experiments allowed a clear distinction between the Raman and the hot-luminescence-like contributions to the scattered-light intensity. The strong effect of the spectral width of the excitation pulse on the temporal behavior of the scattering and the temperature dependence of the coherence time indicate the importance of the polariton character of the excited state.

In the quantum-beat experiments discussed so far, propagation effects were not relevant. These effects were seen in quantum-beat spectroscopy for the first time by Fröhlich et al. [122]. They considered only one excited electronic state, namely, the $1s$ quadrupole-active orthoexciton level in Cu_2O. Its coupling to the electromagnetic field leads to the formation of two polariton branches, as was pointed out by Pekar, Piskovoi and Tsekvava [137]. The rather weak quadrupole exciton–photon interaction leads to a very small splitting of the two polariton branches and therefore to a narrow region of strong dispersion in (ω, k) space. The dispersion law is described by the formula

$$\varepsilon(\omega) = \frac{c^2k^2}{\omega^2} = \varepsilon_\infty + \frac{f_{lmn}c^2k^2}{\omega_{\mathrm{or}}^2(0) + \omega_{\mathrm{or}}(0)\frac{\hbar k^2}{m_{\mathrm{ex}}} - \omega^2 - i\omega\gamma}, \tag{9.87}$$

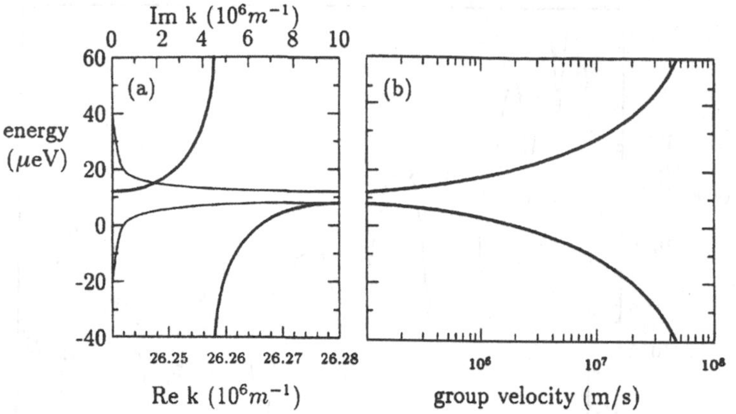

Figure 9.8. Calculated dispersion of a 1s exciton polariton in Cu_2O and the corresponding group velocity (from Ref. 121).

where $\varepsilon_\infty = 6.5$ is the background dielectric constant, $\hbar\omega_{or}(0) = 2.0329$ eV is the energy of the quadrupole-active orthoexciton level, and f_{lmn} is the oscillator strength, which depends on the directions of the wave vector **k** and the light polarization **e**. It has a value of approximately $f = 3.7 \times 10^{-9}$. γ is the damping constant, introduced phenomenologically and taken as approximately equal to $\hbar\gamma = 0.9\ \mu$eV. The term $\omega_{or}(0)\hbar k^2/m_{ex}$ describing the spatial dispersion effects depends on the translational-exciton mass $m_{ex} = 2.7m_0$.

The dispersion law of the quadrupole polariton in a narrow region around the resonance is represented in Fig. 9.8(a). The zero of the energy scale refers to the energy of the orthoexciton. The dependences of Re $k(\omega)$ and Im $k(\omega)$ are depicted by thick and thin solid curves, respectively. In Fig. 9.8(b), the energy dependence of the group velocity $v_g = d\omega/dk$ is shown for both branches.

In Fig. 9.9, the delayed signal measured on the rear side of the sample is shown. The incident pulse on the illuminated side of the crystal excites both quadrupole-polariton branches resonantly because the pulse spectral width is larger than the polariton dispersion region. The delayed signal on the rear side exhibited a pronounced modulation, the amplitude of which depended critically on the fine tuning of the laser to the 1s resonance. The period of the modulation increased with time delay. The thicker samples exhibited shorter periods of beats for a given delay time. All these characteristics were explained in Ref. 122 on the basis of propagating quantum beats. The authors supposed that the superposition of two polariton wave packets with the same group velocities takes place, one constructed from states lying on the upper polariton branch and one constructed from states in the lower polariton branch. As one can see in Fig. 9.8(b), the states with higher velocities lie on the photonlike parts of the two polariton branches. The difference between their energies is large, and therefore the quantum-beat period is short. Having a high velocity, these wave packets arrive at the rear side of the sample with a small time delay. For this reason the quantum beats at a small delay time have a short period, whereas at a larger delay time the quantum beats have longer period.

In addition to these properties, at higher excitation intensities new peculiarities were observed. For example, the polarization decay becomes faster and the quantum beats disappear. The experimental curves cannot be explained by the theory of linear pulse propagation.

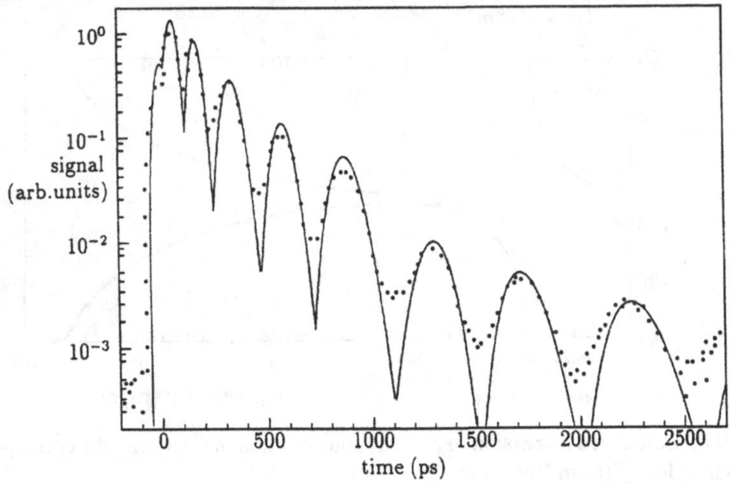

Figure 9.9. Propagation beats signal from a Cu_2O crystal of thickness 0.91 mm. The $1s$ exciton was excited resonantly by a 55-ps laser pulse. The solid curve represents the numerical fit, the points are experimental data (from Ref. 121).

A possible mechanism is polariton–polariton scattering that leads to additional energy and phase relaxation.

In an earlier paper, Ulbrich and Fehrenbach [138] studied the time of flight of polaritons with group velocity $v_g = d\omega/dk$ in a GaAs crystal. But in their paper, the coherence properties were neglected, since the spectral width $\Delta\omega$ of the laser was smaller than the splitting of two polariton branches for dipole-active exciton transitions.

The quantum beats of excitons in a Cu_2O crystal were also studied in the condition of a double optical resonance. Shmiglyuk, Bobrysheva, and Pavlov [139] considered the case when stationary pumping performed by the infrared laser couples dynamically the $1s(\Gamma_5^+)$ and $2p(\Gamma_2^- + \Gamma_3^- + \Gamma_4^- + \Gamma_5^-)$ excitonic states of the yellow series in the Cu_2O crystal. It leads to the formation of the new quasi-energy levels in the vicinities of the initial $1s$ and $2p$ levels. An ultrashort laser pulse with a frequency lying in the region of the $1s(\Gamma_5^+)$ exciton resonance simultaneously excites the $1s(\Gamma_5^+)$ exciton state and one or two new quasi-energy levels arising in its vicinity. It leads to the appearance of quantum beats of the crystal fluorescence and permits one to determine the quasi-energy spectrum and the matrix elements of the $1s$–$2p$ transitions.

Not only excitonic states but also biexcitonic effects can be revealed by quantum-beat spectroscopy. The temporal oscillations originating in biexciton pairing appear in short-pulse optical experiments in strong magnetic fields in GaAs quantum wells. Finkelstein et al. [135] and Bar-Ad et al. [136] showed that the biexcitonic states introduce a channel for coupling between the σ^+ and σ^- exciton states. The manifestation of this coupling is the appearance of oscillations whose frequency is the binding energy of the biexciton.

9.3 Photon-Echo and Transient Effects

The phenomenon of the photon echo, completely analogous to the nuclear-magnetic-spin-resonance echo effect, was predicted and investigated by Kopvillem and Naghibarov [52] and by Abella et al. [53]. It gives rise to the new field of photon-echo spectroscopy [46]. The interest in this phenomenon came about because of the possibility of influencing the free evolution of the dipole moments of two-level atoms.

The states of the dipoles can be represented by Bloch pseudovectors, just as in the case of nuclear spins. In this picture, the Bloch vectors representing these states undergo precessions on the surface of the Bloch sphere. The Bloch vector consists of the two components (real and imaginary) of the dipole moment and has a third component equal to the difference of the population numbers of the two levels of the atom. The action of ultrashort laser pulses on these states may cause changes of the Bloch vector precessions that are equivalent to the inversion in time of their preceding evolution.

The photon-echo phenomenon can be observed only in systems with a significant inhomogeneous broadening of the energy spectrum and can be observed only during intervals of time less than the polarization relaxation time. The photon echo appears as a response of the system to the successive action of two ultrashort pulses having well-defined "areas." The pulse area is defined as the integral over time of the envelope $\varepsilon(x, t)$ of the electric-field strength $E = \varepsilon(x, t)\exp(-i\omega t + ikx)$ and is denoted by $\theta(x)$:

$$\theta(x) = \frac{\mathrm{d}}{\hbar}\int_{-\infty}^{\infty} \varepsilon(x, t)\mathrm{d}t. \tag{9.88}$$

Here d is the dipole moment of the quantum transition in the system. The light is assumed to be propagating in the x direction.

The first strong light pulse, with the area $\pi/2$, excites the two-level atoms in the states with nonzero transition dipole moments, which have the same phases and together form a superradiant state of the system. The superradiant state is characterized by a probability of the spontaneous photon emission proportional to the square of the density of two-level atoms, in contrast to the linear dependence on the density in the usual radiant state.

The subsequent free evolution of the wave functions of the atoms takes place with slightly different frequencies of the excitations of the atoms, because of the inhomogeneous broadening in the system. The different phases of the transition dipoles destroy the superradiant state of the system, which begins to emit photons in the usual regime. During this period of time, between the action of the two ultrashort pulses, the Bloch vectors describing the states of the atoms undergo precession in the equatorial plane of the Bloch sphere.

The action of the second ultrashort pulse, with area π, changes the precessions of the Bloch vectors in such a way that their subsequent motions lead to the restoration of the identical phases of all the dipoles, which occurs at some moment of time after the action of the second pulse. This interval is equal exactly to the time interval between the two ultrashort pulses. The equal phases of the dipoles that occur at this moment temporarily restore the superradiant state and lead to the appearance of a bright, superradiant echo signal. The familiar picture applies here, which is often given in the context of nuclear-spin resonance, of runners in a race all starting at the same point, running at different speeds, who then receive a signal to run back to the starting line. Even though they have all run different distances, since they each run at the same speed on the way back as they ran on the way out, they all arrive at the starting line again at the same time.

In all the foundational papers on this phenomenon, the theoretical investigations were based on the approximation of a given photon field, when the photon density of the external laser pulses is much greater than the density of the two-level atoms. Khadzhi [140] elaborated for the first time a theory of the photon echo beyond this approximation, which permits one to consider arbitrary relationships between the photon and atom densities. These results are reported below.

Another topic of this section is the description of the transition stage of the ultrashort light-pulse propagation in the media. This stage concerns the first part of the penetration depth of the pulse into the medium. During this time, the envelope shape undergoes a change until it achieves a stationary temporal profile. In the subsequent propagation, the pulse envelope remains unchanged until the effects of the relaxation processes appear.

Because this stage is described by nonlinear differential equations with partial derivatives depending on two variables x and t, it is reasonable at least to choose an analytical description for the pulse area $\theta(x)$. The usual nonlinear differential equation of this type,

$$\frac{d\theta(x)}{dx} = -\frac{\alpha}{2}\sin\theta(x), \tag{9.89}$$

was obtained by McCall and Hahn [51] and was called the "area theorem." Here α is the coefficient of the linear absorption of the medium. If the area $\theta(x)$ of the pulse equals 2π, then $d\theta/dx = 0$ and the propagating pulse does not change its area. In the case of small area $\sin\theta(x) \approx \theta(x)$, the solution of this equation will be

$$\theta(x) = \theta_0 e^{-\alpha x/2}. \tag{9.90}$$

In this case, the pulse area will decrease exponentially, which corresponds to the Bouger–Lambert–Beer law. The area theorems in the case of coherent excitons and biexcitons, taking into account their initial densities and different quantum transitions, were derived by Khadzhi and are reviewed below, following Refs. 141–143.

9.3.1 Photon-Echo Theory Beyond the Fixed-Field Approximation

Following the work of Khadzhi [140], we now consider the system of coherent excitons and biexcitons in semiconductor quantum-well structures and examine the quantum transition in the range of the M band due to the one-photon exciton–biexciton conversion, discussed in Subsection 4.1.1. In typical quantum-well structures, inhomogeneous broadening of the transition line arises because of the fluctuations of the width of the quantum wells. This broadening will be characterized by a Gaussian function with a half-width Δ_0. We have already presented in Subsection 9.1.2 the system of nonlinear differential equations describing the time evolution of the coherent excitons, photons, and biexcitons, in connection with optical-bistability and self-pulsation phenomena. But in contrast to the stationary case, in which the dampings and coherent stationary pumpings are important, in the case of the photon-echo description these terms must be excluded from the initial equations. Another difference between the two cases is related to the role played by the frequency detuning Δ. In the previous cases, we considered systems of spatially homogeneous coherent excitons and biexcitons with a single value Δ for all quasiparticles. In the present case, the frequency detuning Δ for different groups of coherent excitons and biexcitons can be quite different. Moreover, this difference is the cause of the photon-echo effect. For these reasons, the detuning Δ in the nonlinear equations must be kept nonzero. In some cases it will exceed the nonlinear terms, which will permit us to simplify the nonlinear equations.

The polarization of the medium, calculated for a group of coherent excitons and biexcitons with a well-defined detuning Δ, will be averaged with the help of the Gaussian-type distribution function for the inhomogeneous broadening. There are also some small differences between the variables used in Subsection 9.1.2 and those used in the present case. For example, the polarization is represented in the form of real and imaginary parts, and the difference z between the occupation numbers of the excitons and biexcitons, as well as the envelope function ε of the electric-field strength, are used:

$$Q = u + iv, \qquad z = n_2 - n_1. \tag{9.91}$$

The starting equations of Ref. 140 are

$$\dot{u} = \Delta v, \qquad \dot{v} = -\Delta u - \frac{1}{2} f z \varepsilon,$$
$$\dot{\varepsilon} = -\alpha f v, \qquad \dot{z} = 2 f v \varepsilon. \tag{9.92}$$

Here f is the constant of the exciton–biexciton conversion, $\alpha = 2\pi\hbar\omega$, and $\Delta = \omega - \omega_0$ is the detuning between the laser pulse frequency ω and the M-band frequency ω_0. In the fixed-field approximation, the equation containing the field derivative $\dot{\varepsilon}$ was neglected, and the amplitude ε was substituted by the amplitude of the external laser pulse. The theoretical approach proposed by Khadzhi allows one to take into account the coherent photon field ε generated by coherent excitons and biexcitons that coexist with the external laser pulses. The latter are characterized by amplitudes ε_{0i}, time widths τ_i, photon densities f_{0i}, and pulse areas θ_i:

$$\theta_i = f \varepsilon_{0i} \tau_i, \qquad f_{0i} = \frac{\varepsilon_{0i}^2}{4\pi\hbar\omega}. \tag{9.93}$$

It is assumed that the initial density of coherent excitons is n_0, which determines the total density of coherent excitons and biexcitons at subsequent times. The biexcitons appear because of the action of the ultrashort laser pulses and because of the conversion of excitons into biexcitons.

As mentioned above, a simplification of the nonlinear equations (9.92) is possible, which concerns the intervals of time when the external laser pulses do not act. In this case, the main role is played by the terms containing the frequency detuning Δ, whereas the nonlinear terms $f z \varepsilon$ and $f v \varepsilon$ play a secondary role.

The nonlinear differential equations (9.92) in the intervals of time τ_1 and τ_2 were solved exactly. The external ultrashort laser pulses with square forms were taken into account by means of the initial conditions. The time- and Δ-dependent components of the polarization, $u(\Delta, t)$ and $v(\Delta, t)$, were averaged over the distribution function of the inhomogeneous broadening. It gives rise to the polarization of the medium

$$P(t) = \frac{1}{\sqrt{\pi}\,\Delta_0} \int_{-\infty}^{\infty} v(\Delta, t) e^{-\Delta^2/\Delta_0^2} \mathrm{d}\Delta. \tag{9.94}$$

The function $u(\Delta, t)$, being an odd function Δ, vanishes after averaging. The polarization $P(t)$ determines the response of the exciton–biexciton system to the action of two successive ultrashort laser pulses acting during the intervals of time τ_1 and τ_2, respectively, with the interval τ between the end of the first pulse and the beginning of the second pulse. All these parameters, as well as the others mentioned in Eq. (9.93), determine the final expression of the polarization $P(t)$ of Eq. (9.94). Its time dependence can be expressed in terms of the time T from the end of the second pulse or in terms of the time t from the beginning of the action of the first pulse. These two time variables are related by the equality $t = \tau_1 + \tau + \tau_2 + T$.

Khadzhi considered four possible relations between the densities of photons f_{0i} and excitons n_0 and found the analytical expressions for each case separately [142]. The first case corresponds to the conditions

$$f_{01} > n_0, \quad f_{02} > n_0 cn^2(\theta_1/2) \tag{9.95}$$

and is close to the limit considered in the fixed-field approximation. The polarization has the form

$$P(\tau) = \hbar f n_0 \,\mathrm{sn}\left(\frac{\theta_1}{2}\right) \mathrm{cn}\left(\frac{\theta_1}{2}\right) \left\{ 1 + \mathrm{dn}(\psi_2) \left[\frac{\frac{f_{02}}{n_0} \mathrm{sn}^2(\psi_2)}{\frac{f_{02}}{n_0} + \mathrm{sn}^2\left(\frac{\theta_1}{2}\right)} - \mathrm{cn}^2(\psi_2) \right] \right.$$
$$\left. \times \left[1 - \frac{\mathrm{sn}^2(\frac{\theta_1}{2})\mathrm{sn}^2(\psi_2)}{\frac{f_{02}}{n_0} + \mathrm{sn}^2(\frac{\theta_1}{2})} \right]^{-2} \right\} e^{-\frac{\Delta_0^2 (T-\tau)^2}{4}} \sin \omega t. \tag{9.96}$$

Here sn(x), dn(x), and cn(x) are the well-known elliptic functions [104, 113, 114, 143]. The moduli k_1 and k_2 of the elliptic functions with variables θ_1 and ψ_2 are

$$k_1^2 = \frac{n_0}{f_{01}}, \qquad k_2^2 = \left[\frac{f_{02}}{n_0} + \mathrm{sn}^2\left(\frac{\theta_1}{2}\right) \right]^{-1},$$

$$\psi_2 = \frac{\theta_2}{2} \left[1 + \frac{n_0}{f_{02}} + \mathrm{sn}^2\left(\frac{\theta_1}{2}\right) \right]^{1/2}. \tag{9.97}$$

One can see that the echo signal has the intensity $I \sim |P(\tau)|^2$, proportional to the square of the exciton density n_0, and coincides with the result obtained in the fixed-field approximation. But even in this limit, new properties of the echo signal are revealed. For example, the polarization depends not only on the areas θ_1 and θ_2, but also on the amplitudes ε_{0i} and the densities f_{0i}. At the moment of time when the second pulse arrives, the medium is already prepared by the first pulse and has the parameters determined by it. This fact explains the appearance of memory.

Another limit, the opposite of the previous one, corresponds to

$$f_{01} < n_0, \quad f_{02} < n_0 \,\mathrm{dn}^2(\varphi_1). \tag{9.98}$$

In this case the polarization equals

$$P(\tau) = \hbar f \sqrt{f_{01} n_0}\, \mathrm{sn}(\varphi_1)\mathrm{dn}(\varphi_1) \left\{ 1 + \mathrm{dn}(\varphi_2) \left[\frac{\frac{f_{02}}{n_0} \mathrm{sn}^2(\varphi_2)}{\frac{f_{02}}{n_0} + k_1^2 \,\mathrm{sn}^2(\varphi_1)} - \mathrm{cn}^2(\varphi_2) \right] \right.$$
$$\left. \times \left[1 - \frac{k_1^2 \,\mathrm{sn}^2(\varphi_1)\,\mathrm{sn}^2(\varphi_2)}{\frac{f_{02}}{n_0} + k_1^2 \,\mathrm{sn}^2(\varphi_1)} \right]^{-2} \right\} e^{-\frac{\Delta_0^2 (T-\tau)^2}{4}} \sin \omega t, \tag{9.99}$$

where

$$k_1^2 = \frac{f_{01}}{n_0}, \qquad k_2^2 = \frac{f_{02}}{n_0} + k_1^2 \,\mathrm{Sn}^2(\varphi_1), \qquad \varepsilon_{ex}^2 = 4\pi\hbar\omega n_0,$$

$$\varphi_1 = \sqrt{\frac{\alpha n_0}{2}} f\tau_1 = \frac{1}{2} f\varepsilon_{ex}\tau_1, \qquad \varphi_2 = \sqrt{\alpha n_0/2} f\tau_2 = \frac{1}{2} f\varepsilon_{ex}\tau_2. \tag{9.100}$$

The polarization of the medium does not depend on the areas of the two external pulses θ_1 and θ_2. It is determined exclusively by the areas of the material field, φ_1 and φ_2, which are proportional to the product of the exciton wave amplitude ε_{ex} and the length of the corresponding external light pulse.

The intensity of the echo signal is proportional to $n_0 f_{01} \sim n_0\varepsilon_{01}^2$. This fact is extremely important for an experimental study of optical echo effect in the cases in which the density of the two-level atoms is higher than the photon density of the first laser pulse. In both cases the echo signals appear at a time $T = \tau$ after the end of the second pulse.

Under conditions $f_{01} > n_0$ but $f_{02} < n_0 \,\mathrm{cn}^2(\theta_1/2)$ the polarization is proportional to n_0 as in the first case considered above.

On the other hand, for the conditions $f_{01} < n_0$ and $f_{02} > n_0 \,\mathrm{dn}^2(\varphi_2)$ the dependence $P \sim \sqrt{n_0 f_{01}}$ was obtained. As mentioned above, the system of coherent excitons and biexcitons, as regards the quantum transition in the range of the M band of the exciton–biexciton conversion, is completely equivalent to the system of two-level atoms. But if one considers the one-photon transitions from the ground state of the crystal to the exciton state and the two-photon transitions from the ground state of the crystal into the biexciton state, one can see that the problem becomes much more complicated.

9.3.2 The Transient Stage and the Area Theorems

Area theorem (9.89) was deduced by McCall and Hahn [51] for the case of two-level atoms. It was assumed that all the atoms are initially in their ground states when the propagating ultrashort laser pulse impinges on the system. But it is possible that some of the atoms are excited at the moment the pulse arrives. Then the McCall and Hahn theorem does not work and must be generalized.

The same task arises when an ultrashort pulse impinges on the system of coherent excitons and biexcitons, giving rise to exciton–biexciton conversion. Both cases were considered by Khadzhi and co-workers [9, 141–143], who derived the corresponding area theorems. It was supposed that the coherent excitons and biexcitons with densities n_{ex} and n_b were created in advance, before the arrival of the ultrashort laser pulse, which has a frequency that lies in the range of the M band. For biexcitons in CuCl-type crystals with a large dissociation energy, assuming that the pulse spectral width is not too large, the exciton–biexciton conversion takes place alone, without the accompaniment of ground-state to exciton-state transitions. In CdS-type crystals with a comparatively weak biexciton binding energy, both quantum transitions must be taken into account simultaneously. In the first case, the Khadzhi area theorem has the form

$$\frac{d\theta}{dx} = -\frac{\alpha}{2}[(n_{ex} - n_b)\sin\theta + 2\sqrt{n_{ex}n_b}\,(1 - \cos\theta)]. \tag{9.101}$$

Here α is the coefficient of the light absorption due to exciton–biexciton conversion. In the analogy with two-level atoms, n_{ex} gives the density in the ground state and n_b denotes the

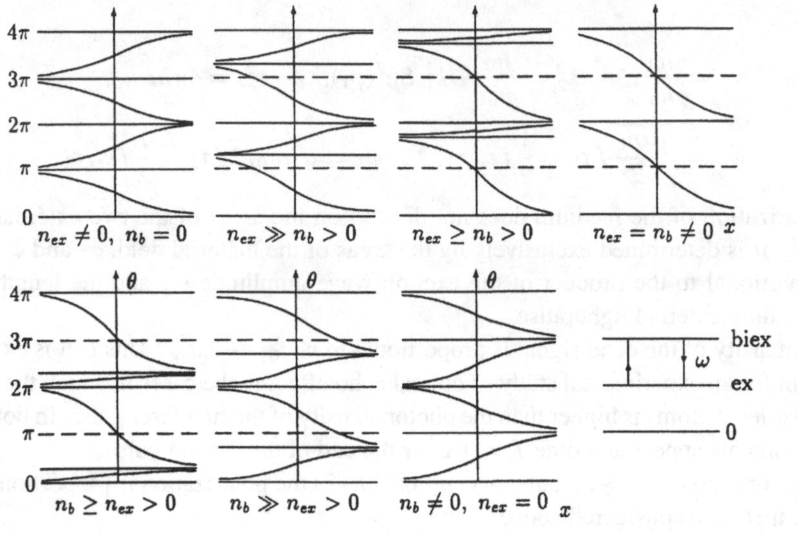

Figure 9.10. The spatial area evolution of the ultrashort laser pulse in the range of the M band.

density of excited atoms. The system in which the inequality $n_{ex} > n_b$ holds is an absorbing medium, whereas the inverse relation $n_{ex} < n_b$ implies an amplifying medium.

Stationary pulses in a totally absorbing medium $n_{ex} \neq 0, n_b = 0$ have areas equal to $2\pi n$, whereas in a completely amplifying medium, $n_{ex} = 0, n_b \neq 0$, they are equal to $(2n+1)\pi$. The spatial evolution of the area $\theta(x)$ of the pulse in media with different relations between n_{ex} and n_b is represented in Fig. 9.10. In partially absorbing or in partially amplifying media, the stationary pulses were determined by Khadzhi to be

$$\theta = 2\pi n - 2\arctan\left(\frac{n_{ex} - n_b}{2\sqrt{n_{ex} n_b}}\right). \tag{9.102}$$

In the case $n_{ex} = n_b$, the area plot is characterized only by decreasing branches, and the SIT phenomenon cannot take place.

If the laser pulse acts in the second type of medium, it can create coherent excitons and at the same time convert them into coherent biexcitons. In this case one does not need to prepare the system of coherent excitons and biexcitons in advance, since the laser pulse does it. In the case $n_{ex} = n_b = 0$, Khadzhi and co-workers obtained the area theorem [9, 141–143]

$$\frac{d\theta}{dx} = -k_1 \sin\frac{\theta}{2} - k_2 \sin\frac{\theta}{2}\left(1 - \cos\frac{\theta}{2}\right), \tag{9.103}$$

where the coefficients k_1 and k_2 are related to the parameters of the medium. This spatial evolution is presented in Fig. 9.11.

One can see that the stationary pulses have areas of $2\pi n$. If the initial area θ_0 is greater than 2π, then the area of the transmitted pulse rises to the value 4π. When θ_0 is less then 2π, it decreases to zero during its propagation. A pulse with $\theta_0 = 2\pi$ propagates with the same area. But the evolution of the initial pulse area $\theta_0 \neq 2\pi$ toward 2π is impossible. This result differs essentially from the case of quantum transitions in the range of the M band alone. Khadzhi generalized the area theorem (9.103) to the case in which the density

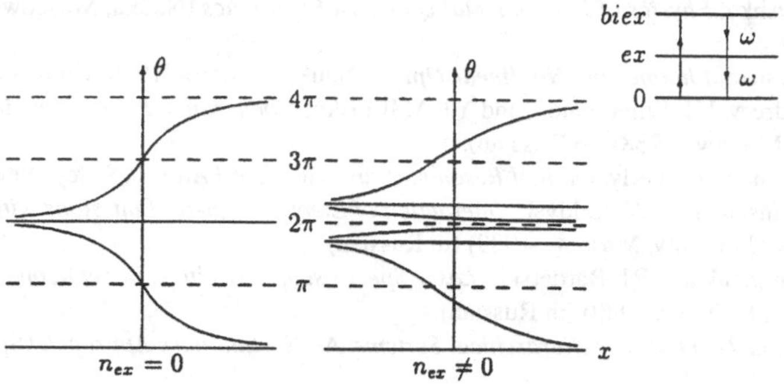

Figure 9.11. The spatial evolution of the pulse area when two quantum transitions take place simultaneously.

of excitons n_{ex} before the pulse penetration into crystal is nonzero. In this case, two new stationary area values appear in the vicinity of the value 2π, denoted as $2\pi \pm \theta_1$. In the interval of areas $0 \leq \theta \leq 4\pi$, there are four stationary values: 2π, 4π, and $2\pi \pm \theta_1$. The stationary values $2\pi \pm \theta_1$ are unstable, however. The area $2\pi - \theta_1$ evolves either toward the value zero or toward 2π. The area $2\pi + \theta_1$ evolves either toward the value 2π or toward 4π.

These results allow the possibility of influencing the propagation of the ultrashort laser pulses in the crystals by changing the exciton density n_{ex} by means of a weak probe signal. This effect can be used for discrimination of ultrashort laser pulses, depending on their area, and for development of new methods of active spectroscopy.

References

[1] I. Prigogine, *From Being to Becoming: Time and Complexity in the Physical Sciences* (Freeman, San Francisco, 1980).

[2] I. Prigogine and I. Stengers, *Order Out of Chaos: Man's New Dialogue with Nature* (Heinemann, London, 1984).

[3] H. Haken, *Synergetics: An Introduction. Nonequilibrium Phase Transitions and Self-Organization in Physics, Chemistry and Biology* (Springer, Berlin, New York, 1978).

[4] H. Haken, *Advanced Synergetics. Instability Hierarchies of Self-Organizing Systems and Devices* (Springer, Berlin, 1983).

[5] H. Haken, *Laser Light Dynamics*, Vol.2 of Light Series (North-Holland, Amsterdam,1985).

[6] B.B. Kadomtsev, ed., *Synergetics* (Mir, Moscow, 1984).

[7] H.M. Gibbs, *Optical Bistability: Controlling Light with Light* (Academic, New York, 1985).

[8] A.H. Rotaru and V.A. Zalozh, *Optical Self-Organization of Excitons and Biexcitons in Semiconductors* (Shtiintsa, Kishinev, 1990) (in Russian).

[9] P.I. Khadzhi, *Nonlinear Optical Processes in the System of Excitons and Biexcitons in Semiconductors* (Shtiintsa, Kishinev, 1985) (in Russian).

[10] P.I. Khadzhi, G.D. Shibarshina, and A.H. Rotaru, *Optical Bistability in the System of Coherent Excitons and Biexcitons in Semiconductors* (Shtiintsa, Kishinev, 1988) (in Russian).

[11] S. Leibovich and A.R. Seebass, eds., *Nonlinear Waves* (Cornell U. Press, Ithaca, NY, 1974).

[12] A.A. Andronov, A.A. Vitt, and S.E. Khaikin, *Theory of Oscillations* (Nauka, Moscow, 1981) (in Russian).

[13] H.A. Haus, *Waves and Fields in Optoelectronics* (Prentice-Hall, Englewood Cliffs, NJ, 1984).

[14] A.V. Gaponov-Grekhov and M.I. Rabinovich, eds., *Nonlinear Waves. Self-Organization* (Nauka, Moscow, 1983) (in Russian).

[15] D.N. Klyshko, *Physics of Fundamental Quantum Electronics* (Nauka, Moscow, 1986) (in Russian).
[16] D.N. Klyshko, *Photons and Nonlinear Optics* (Nauka, Moscow, 1980) (in Russian).
[17] A.V. Andreev, V.I. Emel'yanov, and Yu. A. Il'inskii, *Cooperative Phenomena in Optics* (Nauka, Moscow, 1988) (in Russian).
[18] L. Allen and J.H. Eberly, *Optical Resonance and Two-Level Atoms* (Wiley, New York, 1975).
[19] Yu.A. Il'inskii and L.V. Keldysh, *Interaction of Electromagnetic Emissions with materials* (Moscow University, Moscow, 1989) (in Russian).
[20] M.I. Shmiglyuk and P.I. Bardetskii, *Laser Spectroscopic Excitons in Semiconductors* (Shtiintsa, Kishinev, 1980) (in Russian).
[21] J.S. Nicolis, *Dynamics of Hierarchical Systems. An Evolutionary Approach* (Springer, Berlin, 1986).
[22] A.H. Nayfeh, *Introduction to Perturbation Techniques* (Wiley, New York, 1981); *Perturbation Methods* (wiley, New York, 1973).
[23] J.S. Keller and S. Antman, eds., *Bifurcation Theory and Nonlinear Eigenvalue Problems* (Benjamin, New York, 1969).
[24] T. Poston and I. Stewart, *Catastrophe Theory and its Applications* (Pitman, London, 1978).
[25] V.I. Arnold, *Catastrophe Theory* (Nauka, Moscow, 1990) (in Russian).
[26] J.M.T. Thompson, *Instabilities and Catastrophes in Science and Engineering* (Wiley, New York, 1982).
[27] G.B. Whitham, *Linear and Nonlinear Waves* (Wiley, New York, 1974).
[28] J.D. Murray, *Lectures on Nonlinear Differential Equations. Models in Biology* (Clarendon, Oxford, 1977).
[29] Yu.I. Neimark and P.S. Landa, *Stochastic and Chaotic Oscillations* (Nauka, Moscow, 1987) (in Russian).
[30] G. Iooss and D.D. Joseph, *Elementary Stability and Bifurcation Theory* (Springer, Berlin, 1980).
[31] R. Gilmore, *Catastrophe Theory for Scientists and Engineers* (Wiley, New York, 1981).
[32] V.E. Zakharov, S.V. Manakov, S.P. Novikov, and L.P. Pitaevskii, *Soliton Theory. Counterpoint Method* (Nauka, Moscow, 1980) (in Russian).
[33] G.L. Lamb Jr., *Elements of Soliton Theory* (Wiley, New York, 1980).
[34] R.K. Bullough and P.J. Caudrey, eds., *Solitons* (Springer, Berlin, 1980).
[35] M.J. Ablowitz and H. Segur, *Solitons and the Inverse Scattering Transform* (Society for Industrial and Applied Mathematics, Philadelphia, 1981).
[36] R.K. Dodd, J.C. Eilbeck, J.D. Gibbon, and H.C. Morris, *Solitons and Nonlinear Wave Equations*, 2nd ed. (Academic, London, 1984).
[37] A.C. Newell, *Solitons in Mathematics and Physics* (Society for Industrial and Applied Mathematics, Philadelphia, 1985).
[38] A.S. Davydov, *Solitons in Molecular Systems* (Naukova Dumka, Kiev, 1988) (in Russian).
[39] A.S. Davydov, *Solitons in Bioenergetics* (Naukova Dumka, Kiev, 1986) (in Russian).
[40] K. Lonngren and A. Scott, eds., *Solitons in Action* (Academic, New York, 1978).
[41] S.A. Moskalenko, P.I. Khadzhi, and A.H. Rotaru, *Solitons and Nutation in Exciton Spectra* (Shtiintsa, Kishinev, 1980) (in Russian).
[42] A.T. Filippov, *Many-Sided Solitons* (Nauka, Moscow, 1986) (in Russian).
[43] S.L. Shapiro, ed., *Ultrashort Light Pulses. Picosecond Techniques and Applications* (Springer, Berlin, 1977).
[44] S.A. Akhmanov, V.A. Visloukh, and A.S. Chirkin, *Optics of Femtosecond Laser Impulses* (Nauka, Moscow, 1988) (in Russian).
[45] A.P. Sukhorukov, *Nonlinear Wave Interactions in Optics and Radiophysics* (Nauka, Moscow, 1988) (in Russian).
[46] E.A. Manykin and V.V. Samartsev, *Optical Echo Spectroscopy* (Nauka, Moscow, 1984) (in Russian).
[47] P.I. Khadzhi, *Pis'ma Zh. Eksp. Teor. Fiz.* **60**, 85 (1994).
[48] P.I. Khadzhi and S.L. Gaivan, *Zh. Eksp. Teor. Fiz.* **108**, 1813 (1995).

[49] P.I. Khadzhi and S.L. Gaivan, *Opt. Spektrosk.* **81**, 333 (1996).
[50] P.I. Khadzhi and S.L. Gaivan, *Kvantovaya Elektron.* **23**, 451; 837; 1009 (1996).
[51] S.L. McCall and E.L. Hahn, *Phys. Rev.* **183**, 457 (1969).
[52] U.H. Kopvillem and V.P. Naghibarov, *Fiz. Met. Metalloved.* **15**, 313 (1963).
[53] I.D. Abella, N.A. Kurnit and S.R. Hartmann, *Phys. Rev.* **141**, 391 (1966).
[54] P.G. Kryukov and V.S. Letokhov, *Usp. Fiz. Nauk* **99**, 169 (1969).
[55] A.N. Oraevskii, *Usp. Fiz. Nauk* **91**, 181 (1973).
[56] R.F. Bullough, P.J. Caudrey, J.C. Eilbeck, and J.D. Gibbon, *Opto-electronics* **6**, 121 (1974).
[57] A.I. Burshtein and A.Yu. Pusep, *Zh. Eksp. Teor. Fiz.* **69**, 6 (1972).
[58] V.S. Dneprovskii, *Usp. Fiz. Nauk* **145**, 149 (1985).
[59] H. Risken and K. Numendal, *J. Appl. Phys.* **39**, 4622 (1968).
[60] R. Bonifacio and L.A. Lugiato, *Lett. Nuovo Cimento* **21**, 510 (1978).
[61] R. Bonifacio, M. Gronci, and L.A. Lugiato, *Opt. Commun.* **30**, 129 (1979).
[62] F. Casagrande, L.A. Lugiato, and M.L. Asquini, *Opt. Commun.* **32**, 492 (1980).
[63] A.N. Oraevskii, *Kvantovaya Elektron.* **8**, 130 (1981) [*Sov. Phys. J. Quantum Electronics* **11**, 71 (1981)].
[64] E.N. Lorenz, *J. Atmos. Sci.* **20**, 130 (1963).
[65] L.A. Lugiato, L.M. Narducci, D.K. Bandy, and C.A. Pennise, *Opt. Commun.* **43**, 281 (1982).
[66] K. Ikeda, *Opt. Commun.* **30**, 257 (1979).
[67] K. Ikeda and O. Akimoto, *Phys. Rev. Lett.* **48**, 617 (1981).
[68] H.M. Gibbs, F.A. Hopf, D.L. Kaplan, and R.L. Shoemaker, *Phys. Rev. Lett.* **46**, 474 (1981).
[69] A.H. Rotaru and G.D. Shibarshina, *Phys. Lett. A* **109**, 292 (1985).
[70] A.H. Rotaru, *Fiz. Tverd. Tela* **28**, 2492 (1986) [*Sov. Phys. Solid State* **28**, 1393 (1986)].
[71] A.H. Rotaru, *Fiz. Tverd. Tela* **29**, 3282 (1987) [*Sov. Phys. Solid State* **29**, 1883 (1987)].
[72] A.H. Rotaru and V.A. Zalozh, *Fiz. Tverd. Tela* **29**, 3483 (1987) [*Sov. Phys. Solid State* **29**, 1969 (1987)].
[73] V.A. Zalozh, S.A. Moskalenko, and A.H. Rotaru, *Zh. Eksp. Teor. Fiz.* **95**, 601 (1989) [*Sov. Phys. JETP* **68**, 338 (1989)].
[74] S.A. Moskalenko, A.H. Rotaru, and Yu.M. Shvera, *Teor. Mater. Fiz.* **75**, 295 (1988).
[75] V.F. Elesin and Yu.V. Kopaev, *Zh. Eksp. Teor. Fiz.* **63**, 1447 (1972).
[76] A.Z. Grasyuk and A.N. Oraevskii, *Radiotekhnika i electronika* **9**, 527 (1964).
[77] G.P. Golubev, V.S. Dneprovskii and E.A. Kiselev, *Dokl. Akad. Nauk SSSR* **280**, 591 (1985) [*Sov. Phys. Dokl.* **30**, 71 (1985)].
[78] S.A. Moskalenko, A.H. Rotaru, and Yu.M. Shvera, *Fiz. Tverd. Tela* **29**, 2396; 3474 (1987).
[79] Z.C. Wang and H. Haken, *Z. Phys. B* **55** 364; **56**, 77; **56**, 83 (1984).
[80] S.A. Moskalenko, A.H. Rotaru, and V.A. Zalozh, *Phys. Status Solidi B* **150**, 401 (1988).
[81] V.A. Zalozh, A.H. Rotaru and V.Z. Tronchu, *Zh. Eksp. Teor. Fiz.* **103**, 994 (1993); **105**, 260 (1994) [*JETP* **76**, 487 (1993); **78**, 138 (1994)].
[82] A.H. Rotaru and V.Z. Tronchu, *Fiz. Tverd. Tela* **36**, 20 (1994) [*Sov. Phys. Solid State* **36**, 10 (1994)].
[83] J.S. Russell, in *Reports of the 14th Meeting of the British Association for the Advancement of Science* (JMurray, London, 1844), p. 311.
[84] R.H. Dicke, *Phys. Rev.* **93**, 99 (1954).
[85] C.L. Tang and H. Statz, *Appl. Phys. Lett.* **183**, 457 (1968).
[86] G.B. Hocker and C.L. Tang, *Phys. Rev. Lett.* **21**, 591 (1968).
[87] A.S. Davydov and A.A. Serikov, *Phys. Status Solidi B* **56**, 351, (1973).
[88] H. Haken and A. Schenzle, *Phys. Lett. A* **41**, 405 (1972).
[89] V. Krishan and S. Krishan, *Can. J. Phys.* **52**, 2127, (1974).
[90] V.V. Samartsev, A.I. Siraziev, and Yu.E. Sheibut, *Spectrosc. Lett.* **6**, 659 (1973).
[91] E. Hanamura, *J. Phys. Soc. Jpn.* **37**, 1553 (1974).
[92] M. Inoue, *J. Phys. Soc. Jpn.* **37**, 1560 (1974).
[93] S.A. Moskalenko, in *Molecular Spectroscopy of Dense Phases*, M. Grosmann, S.G. Elkomoss, and J. Ringeissen, eds. (Elsevier, Amsterdam, 1976), p. 45.
[94] S.A. Moskalenko, V.A. Sinyak, and P.I. Khadzhi, *Kvantovaya Electron.* **3**, 852, (1976).

[95] V.M. Agranovich and V.N. Rupasov, *Fiz. Tverd. Tela* **18**, 801, (1976).
[96] O. Akimoto and K. Ikeda, *J. Phys. A* **10**, 425 (1977).
[97] S.A. Moskalenko, A.H. Rotaru, and P.I. Khadzhi, *Optics Commun.* **23**, 367, (1977).
[98] S.A. Moskalenko, A.H. Rotaru, V.A. Sinyak, and P.I. Khadzhi, *Fiz. Tverd. Tela* **19**, 2172 (1977).
[99] S.A. Moskalenko, P.I. Khadzhi and A.H. Rotaru, in *Intrinsic Semiconductors at High Levels of Excitation* (Shtiintsa, Kishinev, 1978), p. 3; *Proceedings of the All-Union Conference on Coherent and Nonlinear Optics* (Chast'I Leningrad, 1978), p. 240.
[100] S.N. Belkin, S.A. Moskalenko, A.H. Rotaru, and P.I. Khadzhi, *Izv. Akad. Nauk SSSR*, **43**, 355 (1979).
[101] S.N. Belkin, P.I. Khadzhi, S.A. Moskalenko, and A.H. Rotaru, *J. Phys. C* **14**, 4109 (1981).
[102] J. Goll and H. Haken, *Phys. Rev. A* **18**, 2241 (1978).
[103] J. Goll and H. Haken, *Opt. Commun.* **21**, 1, (1978).
[104] Yu.S. Sikorsky, *Basic Theories of Elliptical Functions with Applications to mechanics* (NKTP, Moscow, 1936) (in Russian)
[105] I.S. Gradshteyn and I.M. Ryzhik, *Tables of Integrals, Sums, Series and Products* (Academic, New York, 1965).
[106] P.I. Khadzhi, S.A. Moskalenko, A.H. Rotaru, and G.D. Shibarshina, *Phys. Status Solidi B* **144**, K25 (1982).
[107] S.A. Moskalenko, P.I. Khadzhi, G.D. Shibarshina, A.H. Rotaru, and F.N. Georghitsa, *Fiz. Tverd. Tela* **25,** 678 (1983).
[108] G.D. Shibarshina, S.A. Moskalenko, and P.I. Khadzhi, *Phys. Status Solidi B* **119**, 153 (1983).
[109] S.A. Moskalenko, P.I. Khadzhi, A.H. Rotaru, E.S. Kiselyova, and G.D. Shibarshina, *Izv. Akad. Nauk SSSR* Ser. Fiz **47**, 1971 (1983).
[110] S.A. Moskalenko, P.I. Khadzhi, A.H. Rotaru, and E.S. Kiselyova, *Izv. Akad. Nauk SSSR* Ser. Fiz **46**, 609 (1982).
[111] E.S. Moskalenko, *Fiz. Tverd. Tela* **20**, 2246 (1978).
[112] E.S. Kiselyova, *Phys. Status Solidi B* **104**, 497 (1981).
[113] S.A. Moskalenko, P.I. Khadzhi, A.H. Rotaru, V.V. Baltaga, E.S. Kiselyova, S.S. Russu, and G.D. Shibarshina, in *The Living State II*, R.K. Mishra, ed. (World Scientific, Singapore, 1985), p. 340.
[114] G.A. Korn and T.M. Korn, *Mathematical Handbook for Scientists and Engineers* (McGraw-Hill, New York, 1961).
[115] A.M. Zhuravskii, *Handbook on Elliptical Functions* (Academy of Sciences, SSSR, Moscow 1941) (in Russian).
[116] S. Haroche, in *High Resolution Spectroscopy*, K. Shimoda, ed. (Springer, Heidelberg, 1976), p. 253.
[117] E.B. Alexandrov, N.I. Kaliteevskii, and M.P. Chaika, *Usp. Fiz. Nauk* **129**, 155 (1979); [*Sov. Phys. Usp.* **22**, 760, 1979].
[118] S. Stenholm, in *Les Houches Session XXVII, 1975. Aux Frontieres de la Spectroscopie Laser (Frontiers in Laser Spectroscopy)* R. Balian, S. Haroche, and S. Liberman, eds. (North-Holland, Amsterdam, 1977), Vol. I, p. 400.
[119] V.S. Letokhov and V.P. Chebotaev, *Principles of Nonlinear Laser Spectroscopy* (Nauka, Moscow, 1975).
[120] V. Langer, H. Stolz, and W. von der Osten, *Phys. Rev. B* **51**, 2103 (1995).
[121] D. Frölich, A. Kulik, B. Uebbing, A. Mysyrowicz, V. Langer, H. Stolz, and W. von der Osten, *Phys. Rev. Lett.* **67**, 2343 (1991).
[122] D. Fröhlich, A. Kulik, B. Uebbing, V. Langer, H. Stolz, and W. von der Osten, *Phys. Status Solidi B* **173**, 31, (1992).
[123] H. Stolz, *Phys. Status Solidi B* **173**, 99 (1992).
[124] E.O. Göbel, K. Leo, T.C. Damen, J. Shah, S. Schmitt-Rink, W. Schäfer, J.F. Müller, and K. Köhler, *Phys. Rev. Lett.* **64**, 1801 (1990).

[125] K. Leo, J. Shah, E.O. Göbel, J.C. Damen, S. Schmitt-Rink, W. Schäfer, J.F. Müller, and K. Köhler, *Mod. Phys. Lett. B* **5**, 87 (1991).

[126] V. Langer, H. Stolz, W. von der Osten, D. Fröhlich, A. Kulik, and B. Uebbing, *Europhys. Lett.* **18**, 723 (1992).

[127] H. Stolz, V. Langer, E. Schreiber, S. Permogorov, and W. van der Osten, *Phys. Rev. Lett.* **67**, 679 (1991).

[128] K.H. Pantke, P. Schillak, B.S. Razbirin, V.G. Lyssenko, and J.M. Hvam, *Phys. Rev. Lett.* **70**, 327 (1993).

[129] R. Shimano and M. Kuwata-Gonokami, *Phys. Rev. Lett.* **72**, 530 (1994).

[130] S. Bar-Ad and I. Bar-Joseph, *Phys. Rev. Lett.* **66**, 2491 (1991).

[131] O. Carmel, H. Shtrikman, and I. Bar-Joseph, *Phys. Rev. B* **48**, 1955 (1993).

[132] O. Carmel and I. Bar-Joseph, *Phys. Rev. B* **47**, 7606 (1993).

[133] M. Shmiglyuk, A. Bobrysheva, and V. Pavlov, *Phys. Status Solidi B* **199**, 427 (1997).

[134] S. Schmitt-Rink et al., *Phys. Rev. B* **46**, 10460 (1992).

[135] G. Finkelstein, S. Bar-Ad, O. Carmel, I. Bar-Joseph, and Y. Levinson, *Phys. Rev. B* **47**, 12964 (1993).

[136] S. Bar-Ad, I. Bar-Joseph, G. Finkelstein, and Y. Levinson, *Phys. Rev. B* **50**, 18375 (1994).

[137] S.I. Pekar, V.N. Piskovoi, and B.E. Tsekvava, *Fiz. Tverd. Tela* **23**, 1905 (1981) [*Sov. Phys. Solid State* **23**, 1113 (1981)].

[138] R.G. Ulbrich and G.W. Fehrenbach, *Phys. Rev. Lett.* **43**, 963 (1979).

[139] M.I. Shmiglyuk, A.I. Bubrysheva and V.G. Pavlov, *phys. Stat. Sol. B* **199**, 427 (1997).

[140] P.I. Khadzhi, *Pis'ma Zh. Exp. Teor. Fiz.* **60**, 85 (1994) [*JETP Lett.* **60**, 89 (1994)].

[141] S.A. Moskalenko, P.I. Khadzhi , A.H. Rotaru, E.S. Kiselyova, and G.D. Shibarshina, in *Proceedings of the International Conference and School on Lasers and Applications, Bucharest, 1982*, Part 1, I. Ursu and A.M. Prokhorov, eds. (CIP, Bucharest, 1983), p. 629.

[142] A.I. Bobrysheva, S.A. Moskalenko, and P.I. Khadzhi, *Bul. Acad. shtiintse Repub. Moldova* **2**, 77 (1990).

[143] S.A. Moskalenko and P.I. Khadzhi, in *Interaction of Excitons with Laser Radiation* (Shtiintsa, Kishinev, 1991), pp. 3–31 (in Russian).

10

New Directions

In the previous chapters, we have surveyed a number of physical systems in which either induced or spontaneous coherence of excitonic complexes leads to novel optical effects. Most of the systems we have discussed have been bulk semiconductors, however, or in some cases a single two-dimensional quantum well or a zero-dimensional quantum dot. Much of the recent and ongoing work in Bose condensation of excitons involves much more exotic band structures, however. It is beyond the scope of this book to give an in-depth treatment of these ideas, but in this chapter we give a brief survey of some of the fascinating recent proposals and experiments.

10.1 Trapping Excitons with Stress

The recent successes [1–3] at creating a Bose condensate of alkali atoms in a magneto–optical trap have spawned a tremendous amount of theoretical work on the properties of a Bose condensate in a harmonic potential (e.g., Refs. 4–8). In a harmonic potential, Bose condensation leads to a dramatic "smoking gun" associated with Bose condensation, namely, a two-component spatial distribution inside the harmonic potential-well trap, with a sharp peak in the center that corresponds to the ground state of the harmonic oscillator, in which a macroscopic number of particles accumulate. It is also possible to trap excitons in a three-dimensional harmonic potential, in which case the same theory would apply.

We consider an ideal gas of bosons trapped in a parabolic potential of the form

$$V(r) = \frac{1}{2}\alpha r^2, \tag{10.1}$$

where r is the distance from the center of the well and α is the force constant. In the low-density classical case, the chemical potential is the sum of the internal chemical potential

$$\mu_{\text{int}} = -k_B T \ln(n_Q/n) \tag{10.2}$$

and an external chemical potential given by $\mu_{\text{ext}} = \frac{1}{2}\alpha r^2$. In equilibrium, $\mu_{\text{int}} + \mu_{\text{ext}} =$ const. over the well, leading to the spatial distribution of the density,

$$n(r) = n(0)e^{-\alpha r^2/2k_B T}, \tag{10.3}$$

i.e., a Gaussian profile. This expression is based on Maxwell–Boltzmann statistics. The spatial distribution of noncondensed particles exhibiting Bose–Einstein statistics is not much different from this form [9].

The single-particle states of the harmonic-oscillator potential given above comprise an energy ladder, $E_s = (s + 3/2)\hbar\omega$, with $\omega = (\alpha/m)^{1/2}$ and $s = 1, 2, \ldots$. Each level has a degeneracy of $(s+1)(s+2)/2$, implying that the density of states for $s \gg 1$ is proportional to E^2. The full expression for the density of states in this limit is

$$D(E) = \frac{gE^2}{2(\hbar\omega)^3}, \tag{10.4}$$

where g is the spin multiplicity of the particle. The chemical potential is determined by the condition $N(E) = \int D(E) f(E)\,\mathrm{d}E$, where $f(E)$ is distribution function (1.24). Setting $\mu = 0$ in this equation yields the critical condition for BEC,

$$N_c = 1.202g(k_B T/\hbar\omega)^3. \tag{10.5}$$

For a given temperature, when the total number of particles in the well exceeds N_c , the excess particles occupy the ground state. Since the ground state of the harmonic oscillator is a narrow Gaussian wave function centered at the bottom of the well, BEC in a harmonic potential is qualitatively different from the uniform-potential case – BEC corresponds to a spatial condensation rather than a k-space condensation.[a]

The spatial extent of the ground state for an ideal gas, calculated simply as the wave function of a single particle in the ground state of the well, is so small that condition (1.19) for the excitons to remain individual bosons, $na^3 \ll 1$, would not hold if a significant fraction of the gas were condensed in this state. In this case, one cannot ignore the effect of interactions among the particles in the condensate. As shown by Hijmans et al. [11], in the limit of weak interaction, assuming a condensate density much greater than noncondensate density near the center of the well, the spatial extent of the condensate is fixed by the relation

$$n_0(r)U + V(r) = \mu, \tag{10.6}$$

where U is the interaction vertex (proportional to the scattering length), $V(r)$ is the external potential, and μ is the chemical potential determined by $n'(r_c)U + V(r_c) = \mu$, where n' is the excited-state density. The radius r_c of the region in which the condensate density is nonzero is determined by the condition that the integral over the condensate density profile over this region is equal to the number of condensate particles. The volume of the condensate therefore depends on the condensate fraction. Although the volume of the condensate taking into account interactions is much larger than the single-particle (noninteracting) ground state, so that the condition $na^3 \ll 1$ holds, it should still be much smaller than the excited-state spatial distribution and therefore should give a clear telltale for the presence of a condensate.

A way of creating a three-dimensional parabolic potential for excitons has long been known [9]. This method uses the fact that stress applied to a crystal shifts the energy bands by means of the deformation-potential interaction. When a nonuniform stress is used, a point of high shear stress can be created inside a crystal, which corresponds to a potential-energy minimum for the excitons, since the shear-stress deformation potential is negative. Figure 10.1(a) shows a schematic of the equipotential curves caused by a Hertzian contact geometry, which is defined as the contact between a curved surface with a plane surface.

[a] As noted by Goble and Trainor [10] and others, this approximation breaks down in the limit of low particle density, since in that case the $k_B T_c$ becomes comparable to the level spacing $\hbar\omega$. In other words, the thermodynamic limit cannot be assumed.

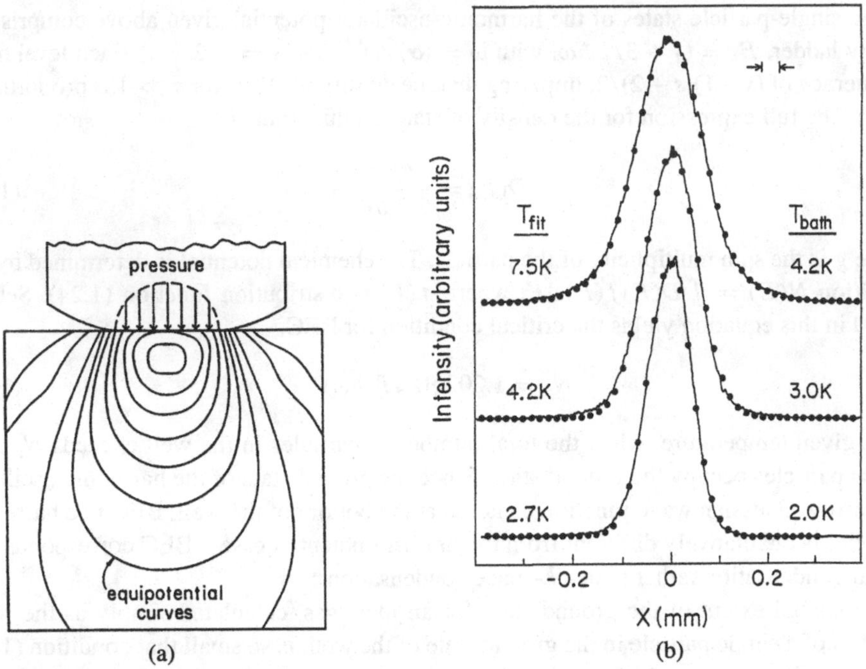

Figure 10.1. (a) Equipotential curves for a Hertzian contact stress, (b) spatial profiles of the exciton luminescence in the three-dimensional parabolic well created in a Cu_2O crystal with Hertzian contact stress at various temperatures. At higher temperatures, the excitons occupy a greater volume (from Ref. 9).

The shear-stress maximum is formed just below the contact area of the rounded stress rod. Figure 10.1(b) shows the spatial profiles of the recombination luminescence of excitons in Cu_2O in the region of the stress maximum. Since the spatial extent of a classical gas depends on the temperature according to

$$<r> = \frac{\int r^3 \, dr \, e^{-\alpha r^2/2k_BT}}{\int r^2 \, dr \, e^{-\alpha r^2/2k_BT}} = (\mathrm{const})\sqrt{k_BT}, \qquad (10.7)$$

a fit of the spatial profiles gives a measurement of the temperature. As seen here, the fit temperature exceeds the bath temperature in each case, presumably because the laser that generates the excitons heats the crystal.

In early experiments at the University of Illinois [12, 13], although the excitons in Cu_2O were successfully trapped, the density of the excitons was not high enough for the Bose condensation phase transition. At that time, a cw laser source was used with an average power of a few hundred milliwatts. As laser power was increased, the density of the excitons increased sublinearly because of the two-body Auger recombination process. There is no fundamental reason why excitons in this kind of trap cannot exceed the critical density for Bose condensation, however, if the instantaneous laser power is high enough to create enough excitons and observations are made on time scales that are short compared with the exciton lifetime.

This strain-confinement technique has also previously been exploited for studying excitonic phases in Ge and Si [14]. As mentioned in Chapter 4, from the analogy to alkaki vapors used by Keldysh, one could also look for BEC of excitons in Si and Ge on short time scales. In steady state, excitons at high density in these semiconductors turn into the metallic electron–hole liquid (EHL), just as alkali atoms at low temperature and high density eventually undergo a phase transition to a metallic state. The transition requires the excitons to get rid of energy, however, and therefore it cannot occur simply through elastic two-body collisions. Depending on the experimental conditions, the transition to the metallic state can take a long time compared with the interparticle collision time. The recent successes in observing BEC of alkali atoms (e.g. Refs. 1–3) take advantage of this fact – the alkali atoms are observed on time scales comparable with the interparticle collision time, but short compared with the time for three-body collisions. During this time, the atoms act as a weakly interacting gas and undergo BEC. In the same way, one would expect that observations of indirect excitons in Si and Ge on time scales comparable with the exciton–exciton collision time but short compared with the exciton–phonon-scattering time could lead to observation of BEC of excitons in these materials.

10.2 Two-Dimensional Systems

As is well known, free particles in two dimensions cannot undergo BEC since the density of states in two dimensions allows a macroscopic number of particles in the excited states at all densities. It turns out that Bose particles in two dimensions can still become a superfluid, however, by undergoing a Kosterlitz–Thouless transition with a power-law decrease of the particle–particle correlation function at long distances [15].

Two-dimensional physics can be naturally studied in semiconductor systems by use of layered structures, e.g., quantum wells. One ingenious way of making excitons in two dimensions is to put the electrons and the holes into two different, adjacent planes, as illustrated in Fig. 10.2. All the excitons are then polarized perpendicular to the planes. These are called "dipole excitons," "indirect excitons," or "barbell excitons."

10.2.1 Dipole Excitons in Coupled Quantum Wells

If a DC electric field is applied perpendicularly to two coupled quantum wells, then the electrons and the holes will feel a force in opposite directions, leading them to accumulate in separate two-dimensional planes. The excitons formed from these spatially separated

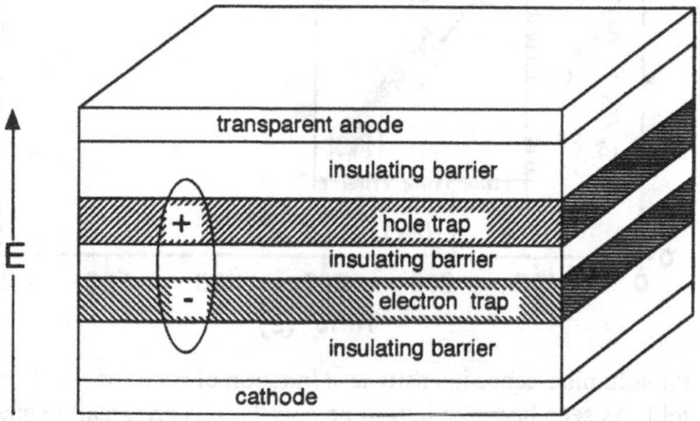

Figure 10.2. Schematic of a two-dimensional layered structure with spatially separated electrons and holes.

electrons and holes are then polarized perpendicularly to the planes. This has two benefits. First, the overlap of the wave functions of the electron and the hole is greatly reduced, which can lead to much longer lifetimes for excitons that have dipoled-allowed transitions. Dramatically longer exciton lifetimes have been seen in coupled GaAs wells when an electric field is applied [16, 17]. Second, since all the excitons are polarized in the same direction, the interaction between the excitons is predominantly repulsive; although a liquid state has been predicted [18, 19] under certain conditions, as discussed below.

Although the authors of early experiments [16, 20] aimed at producing a Bose condensate of excitons in coupled quantum wells succeeded in observing long lifetimes, the interface roughness created during the fabrication process led to a large inhomogeneous broadening of the luminescence, and the excitons became trapped in local energy minima instead of becoming superfluid [20–22]. This situation resembles quite closely the case of the destruction of superfluidity in a random potential considered by Huang [15].

In more recent experiments, Butov et al. [23, 24] used a similar structure and applied a magnetic field in addition to the electric field. As in previous experiments [25], the magnetic field has the effect of enhancing the repulsion between the excitons, since the Pauli exclusion effect is enhanced when the excitons are spin aligned, as discussed in Chapter 4.

Butov and co-workers found novel effects by examining the recombination luminescence in new ways. First, they examined the temporal correlations of the exciton recombination luminescence [23]. They found that, at a critical density, these correlations became very long, of the order of minutes (see Fig. 10.3). They interpreted this as evidence of a phase transition, since critical behavior near a phase transition can lead to large correlations.

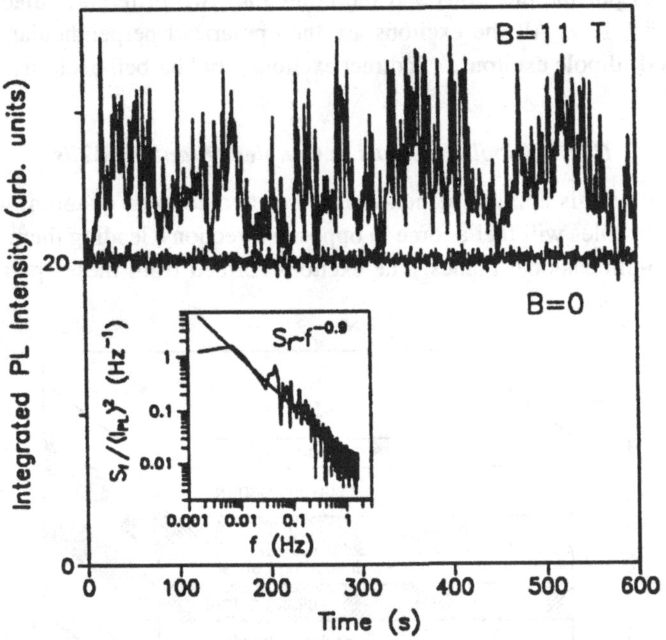

Figure 10.3. Photoluminescence intensity as a function of time for two different values of magnetic field. As seen here, the long-time correlations are greatly enhanced in the case of a high magnetic field. Inset: the frequency spectrum of the photoluminescence (from Ref. 23).

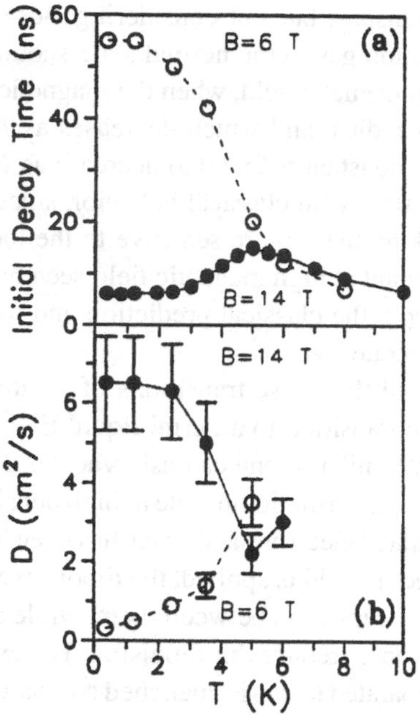

Figure 10.4. (a) Decay time of dipole exciton luminescence as a function of temperature for two different magnetic fields. Comparison of the decay time for a sample that is partially masked and a sample that is unmasked gives a measure of the lateral diffusion constant since excitons that move out of the excitation region and under the masked area do not contribute to the detected luminescence, thereby causing faster decay. (b) Diffusion constant deduced from these measurements, as a function of temperature, for two different magnetic fields (from Ref. 24).

In later experiments [24], Butov et al. indirectly examined the transport of the dipole excitons. In these experiments, most of the quantum-well area was masked, and only small stripes were left unmasked, for optical access. The excitons were created in these regions by a laser, and then the subsequent temporal evolution of the exciton luminescence was recorded and compared with the case in which the sample was not masked. This gives a measurement of the diffusion constant of the particles, because the particles that diffuse into the region behind the mask will not contribute to the recorded luminescence. The authors found that above a critical threshold of magnetic field and below a critical temperature, the diffusion constant of the excitons dramatically increased, as illustrated in Fig. 10.4.

The result of fast diffusion above a critical threshold is indicative of superfluidity, but at present, there are no quantitative theoretical predictions for the behavior of Bose-condensed dipole excitons in a magnetic field in the presence of weak localization, to which the experiments may be compared. Lozovik and Yudson [26] predicted that a Bose condensate of two-dimensional dipole excitons should become superfluid, but they did not include localization or magnetic field in their model. Lerner and Lozovik [27] and Paquet, Rice, and Ueda [18] predicted that a two-dimensional exciton gas in high magnetic field (such that only the lowest Landau state is occupied) should be superfluid, but they did not treat the case of spatial separation of the electrons and holes nor the effect of localization in a random potential. Recently, Arseev and Dzyubenko [28] studied the diffusion of dipole excitons in the presence of magnetic field and weak localization, taking into account quantum mechanical corrections in the exciton transport but not considering Bose condensation of excitons. They found that a dipole exciton gas in the normal state should have a diffusion constant which increases as B^2 in low magnetic field, when the magnetic length $(\hbar c/eB)^{1/2}$ is much larger than the exciton Bohr radius, and which decreases as B^{-1} at high magnetic field. Experimentally, the diffusion constant is found to decrease as B increases at low magnetic field [24, 29], which is consistent with classical behavior, since the magnetic field shrinks the size of the excitons, making them more sensitive to the local potential. The dramatic increase of the diffusion constant at high magnetic field seen by Butov et al. [24] therefore stands in sharp contrast to both the classical prediction and the prediction of Arseev and Dzyubenko for the nondegenerate gas.

In section 1.3, we discussed the phase transitions of excitons in 3D and noted that if the excitons undergo a phase transition to a Fermi liquid EHL state at high density, Bose condensation will not occur. Similarly, one can ask whether 2D dipole excitons remain a nearly ideal gas or transform to a Fermi liquid state at high density. In general, the transition to EHL for 2D spatially separated electrons and holes has been less well studied than the 3D EHL transition. When an electric field is applied, the dipole excitons are all aligned, so that there is a strong dipole–dipole repulsion between them, while exchange and correlation of the electrons and of the holes act to reduce this repulsion. The exchange interaction between the carriers in the spatially separated layers is quenched compared to the monolayer 2D case, due to the small penetration of the dipole–dipole interaction through the barriers. Lozovik and Yudson [26] and other authors [30, 31] have argued that no EHL state is possible for spatially separated electrons and holes, but more recently, Lozovik and Berman [19] have studied this question in detail and have predicted that under certain conditions, the system of spatially separated electrons and holes can both form a dielectric liquid, which could still be a Bose condensate, and a metallic EHL.

Lozovik and Berman [19] used the Keldysh–Kozlov–Kopaev approach [32, 33] and its generalization proposed by Noziéres and Comte [34, 35] (discussed in Chapter 2) for a model of spatially separated electrons and holes in isotropic, parallel planes at $T = 0$. Figure 10.5 shows the calculation of Lozovik and Berman for the energy per *e–h* pair as a function of the pair density for several well separation values, relative to the single exciton ground state energy $E_{ex}(D)$, which also depends on the well separation. As seen in this figure, a minimum in the energy per pair occurs at nonzero pair density. In the interval of values $0 < D < D_{cr} \approx 1.1a$, the energy of the *e–h* pair E is negative, and is greater in magnitude than the biexciton binding energy, which means that the excitons will form a stable "excitonic liquid" with fixed density. As the interlayer distance D increases, the depth of the minimum decreases, and the liquid density decreases. At the interlayer distance $D = D_{cr}$, the binding energy of the excitonic liquid is comparable with the binding energy of a single exciton. At this point a phase transition from the excitonic liquid to the excitonic gas takes place. However, in a sharp region $D_{cr} < D < D_{sp} \approx 1.9a$, the relative minimum of the liquid phase still exists. This means that the liquid phases remain metastable. At larger values of $D > D_{sp}$, the liquid phase is absolutely unstable and only the excitonic or biexcitonic phases remain stable, in which case the exciton densities will be determined only by the external conditions, e.g., the excitation intensity. This instability of the liquid phase at $D > D_{sp}$ is a specific feature of excitons with spatially separated electrons and holes and is related to the enhanced dipole–dipole repulsion at large distances.

Lozovik and Berman found that for an istropic 2D system, the energy spectrum of the single-particle elementary excitations at all pair densities n and all interwell separations D

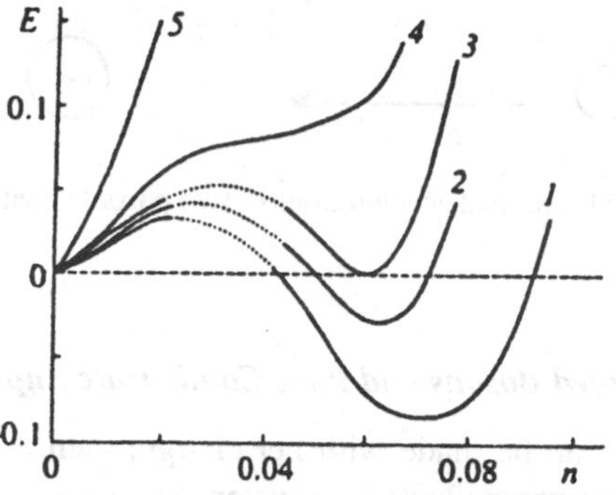

Figure 10.5. Energy $E = E_{eh}(D, n) - E_{ex}(D)$ of a system of spatially separated electrons and holes as a function of the surface density. E is given in units of the two-dimensional excitonic Rydberg, $Ry_{ex2D} = 2me^4/\hbar 2\epsilon$, and the surface density n is given in units of $1/a^2$, where $a = \hbar^2\epsilon/2me^2$ is the 2D excitonic Bohr radius. The labeled curves correspond to (1) $D = 0$; (2) $D = 0.5a$; (3) $D = 1.1a$; (4) $D = 1.9a$; and (5) $D = 5a$ (from Ref. 19).

is characterized by an energy gap at $T = 0$, even in the excitonic liquid state, and therefore both the exciton gas and the excitonic liquid state should be a Bose superfluid. This superfluidity will persist up to the Kosterlitz–Thouless transition temperature, given by

$$T_c = \frac{0.45\pi\hbar^2 n_s}{k_B m}, \tag{10.8}$$

where n_s is the superfluid density. Lozovik and Berman argued, however, that if the system is anisotropic in the plane of the layers, a transition can occur at high pair density from the excitonic dielectric phase into a metallic EHL phase which is not superfluid.

In principle, the value of D can be continuously varied by varying the external electric field in a monolayer 2D exciton system. In this case, one can imagine studying the phase transition as a function of D by observing the spectral shift of the photoluminescence line. Lozovik and Berman suggested that the movement of the liquid excitonic droplets will lead to large fluctuations in the photoluminescence intensity.

When a strong magnetic field is applied normal to the planes such that only the lowest Landau level is occupied, two dimensional excitons transform into "magnetoexcitons," which have the interesting properties that no free electron and hole states exist, only excitonic bound states, and that the exciton mass depends only on the applied magnetic field and not on the original 2D band structure [18]. Since the application of magnetic field generally causes the interaction of the excitons to become weaker and more repulsive [19, 27, 36], one generally does not expect a liquid state for 2D magnetoexcitons, but Paquet, Rice and Ueda [18] argued that a 2D system with asymmetric electron and hole states (e.g. spatially separated electrons and holes) can form a liquid state with droplets of filling factor $v = 1$ when the magnetic fields exceeds a critical threshold. At low magnetic fields (but still large enough that the cyclotron energy is large compared to k_BT and the excitonic Rydberg) the ground state of the 2D magnetoexciton system is expected to always remain a nearly ideal gas [18, 27].

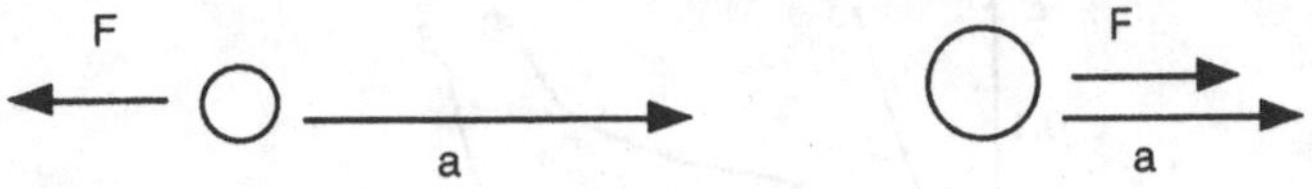

Figure 10.6. Relative forces and accelerations of two particles with the same charge and opposite mass.

10.2.2 Charged Bosons and Bose Condensate Superconductors

If an excitonic complex can be made with net charge, then a Bose condensate of such particles would also be a superconductor – a system known as a Schafroth superconductor [37]. How can such complexes be produced?

One fascinating proposal for charged bosons relies on the existence of states with negative mass in the valence-band structure of many semiconductor heterostructures. It is well known [38] that the mixing and the splitting of the Γ_8 valence-band states in GaAs quantum wells can lead to negative mass at zone center in the first excited hole band, depending on the well thickness, barrier height, stress, and other material parameters. Shvartsman and Golub [39] proposed that charged, bound complexes of two holes can exist when the reduced mass of the pair is negative, which is the case when the absolute value of the mass of the negative-mass particle is less than the mass of the positive-mass particle. This kind of complex had been proposed earlier by Gross, Perel', and Shekhmamet'ev [40] as an explanation for an unusual, apparently reversed excitonic spectrum in bulk BiI_3 [41]; a similar spectrum has also been reported in bulk ZnP_2 [42]. The bihole complex is similar to the bipolaron invoked in some theories of high-T_c superconductivity [43–46], but relies on Coulomb interaction rather than phonon deformations for the binding mechanism.

Figure 10.6 illustrates how these bihole states can be bound. Since the negative-mass particle accelerates in a direction opposite to the force it feels, the Coulomb repulsion between the two particles causes the negative-mass particle to accelerate toward the positive-mass particle. If the magnitude of its mass is less than that of the positive-mass particle, it will catch up and pass the positive-mass particle. The direction of the forces then reverses. This leads to bound orbits exactly analogous to the Rydberg states of an exciton.

For the bihole system to be a superconductor, one needs a quasi-permanent population of biholes. Figure 10.7 gives one possible way of doing this, using three valence subbands (Shvartsman and Golub considered only two subbands.) The hole band with negative-mass states can be filled with holes (e.g., by modulation doping) up to a level that prevents the holes from leaving the zone-center maximum. The hole can then be excited into the higher-lying states by interband optical excitation. The metastable biholes can decay only if the holes recombine with a conduction-band electron, emitting a photon.

There is no firm experimental evidence for biholes to date. Since band-structure engineering is progressing to such a high art, however, there seems to be no reason why biholes cannot be created.

As mentioned in Chapter 4, Yudson [47] suggested the possibility of BEC of charged bosonic complexes such as three electrons and one spatially separated hole (*eee–h*) in a double quantum well with a parallel metal plate (see Fig. 10.8). As Yudson showed, the image charges from a perfectly conducting surface affect the interparticle Coulomb forces in such a way that an overall attractive force can arise. This effect relies on the fact that the carriers are constrained within two-dimensional planes at different distances from the

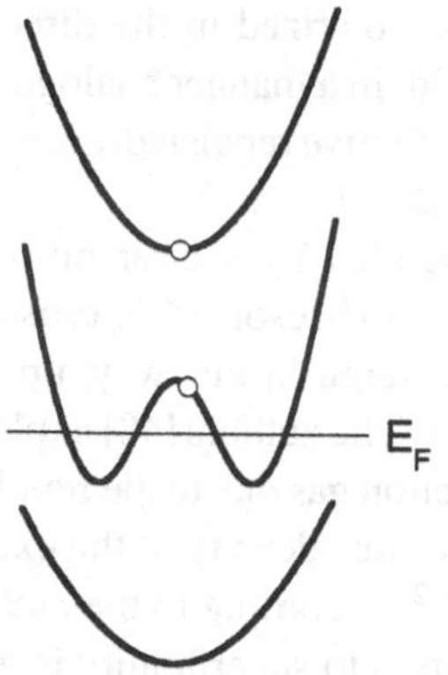

Figure 10.7. Schematic of a possible hole subband structure in a quantum well that would allow stable charged bosons in the form of biholes. The first excited subband has negative effective mass at zone center, while the second excited subband has positive mass at zone center.

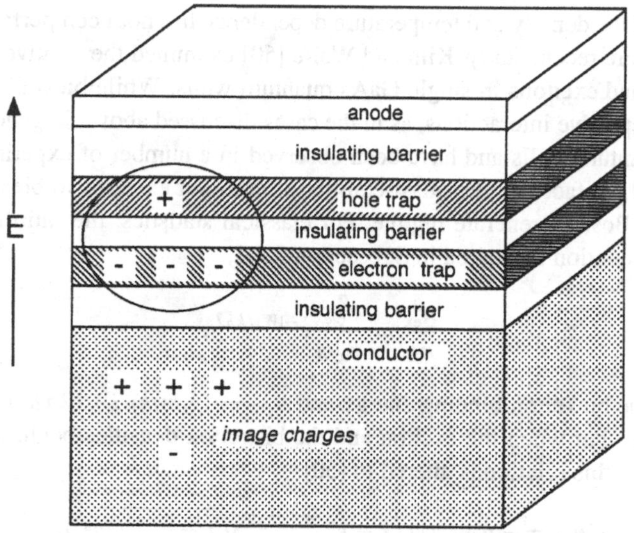

Figure 10.8. Schematic of a two-dimensional layered structure with spatially separated electrons and holes, with a metal plate to moderate the interparticle Coulomb forces.

metal sheet. When the metal plate is closer to the *e* layer, for example, the *e*–*e* Coulomb repulsion is suppressed considerably by the image-charge polarization of the metal plate. The hole on the opposite side, by contrast, feels much less effect of the image charge.

10.2.3 Bose Statistics of Excitons and Biexcitons in Single Quantum Wells

Besides the above studies, recent experiments have also examined the behavior of excitons and biexcitons in single quantum wells when their densities reach the Bose regime.

In one study [48], the band bending at the interface of a single, wide (300-nm) *n*-type GaAs quantum well effectively created two parallel, two-dimensional sheets of excitons.

The excitons in these sheets were polarized in the direction perpendicular to the planes because of the intrinsic electric field, in a manner analogous to that discussed in Subsection 10.2.2. This polarization led to an effective repulsive force between the excitons, as discussed above, and record lifetimes, up to 2.5 μs.

After these excitons were generated by a laser pulse, their diffusion rate out of the excitation area was measured by time-resolved microscopy (see Fig. 10.9.) Amazingly high diffusion constants were measured in this way, up to 3600 cm^2/s, corresponding to supersonic transport of the excitons. The authors [48] explained the anomalously fast expansion in terms of pressure in the exciton gas due to the repulsive forces between the excitons. Alternatively, one can note that the area density of the exciton gas when these effects were seen was approximately 10^{11} cm^{-2}, according to the authors, while the critical density σ_c for the Kosterlitz–Thouless transition to superfluidity is given approximately by [49]

$$\sigma_c = \frac{2k_B T m}{\pi \hbar^2}, \tag{10.9}$$

for which is equal to 10^{11} cm^{-2} for exciton mass $m = 0.6m_0$ and $T = 1.8$ K. The authors showed that the anomalously fast diffusion disappeared at lower densities; unfortunately, a careful study of the density and temperature dependence has not been performed.

In another more recent study, Kim and Wolfe [50] examined the relative populations of the biexcitons and excitons in single GaAs quantum wells. While biexcitons do not exist in the case of repulsive interactions, as in the cases discussed above, in general, biexcitons can exist in quantum wells and have been observed in a number of experiments [51–57]. This study used the fact that the quasi-equilibrium ratio of excitons to biexcitons changes strongly in the Bose degenerate regime. For classical statistics, the ratio is given by the standard Saha equation discussed in Subsection 5.2.3,

$$\frac{n_{\rm ex}^2}{n_B} = \frac{n_{Q\rm ex}^2}{n_{QB}} e^{-|W|/k_B T}, \tag{10.10}$$

where $|W|$ is the biexciton binding energy and $n_{Qi} = (g_i m_i k_B T)/(2\pi\hbar^2)$ is the quantum density of states, but when the exciton and the biexciton densities begin to approach the quantum regime, this relation is altered to [50]

$$n_{\rm ex} = -n_{Q\rm ex} \log\left(1 - e^{-|W|/2k_B T}\sqrt{1 - e^{-n_B/n_{QB}}}\right), \tag{10.11}$$

assuming $m_B = 2m_{\rm ex}$. Essentially this means that the ratio of biexcitons to excitons will greatly increase in the Bose–Einstein regime, since the two-particle chemical potential cannot increase above the biexciton ground state, so that eventually all new particles must be added to the biexciton condensate as density increases at the same temperature. This allows for a test of Bose–Einstein statistics even when the photoluminescence line shapes are too broad to allow examination of the biexciton ground state directly. The total luminescence intensity of the two populations can be measured and compared to the prediction of Eq. (10.11). This is shown in Fig. 10.10.

As seen in this figure, the behavior of the biexcitons and the excitons closely follows the predictions of the quantum-statistical theory. Since no test has been made of the momentum distribution or long-range phase coherence of the biexcitons, it is not possible to say conclusively that the biexcitons are condensed. These measurements are strong evidence of the Bose statistics of the system, however.

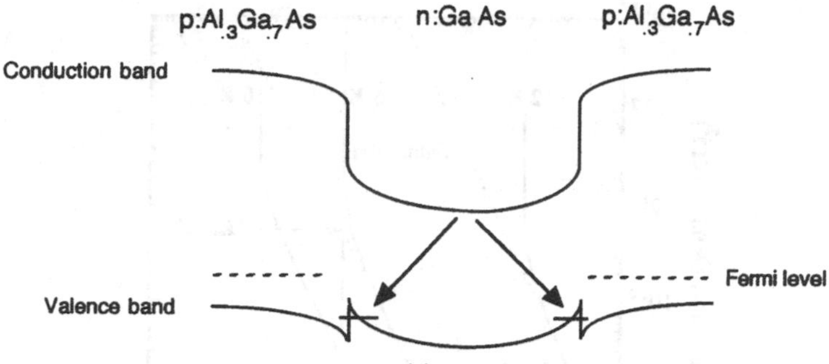

(a)

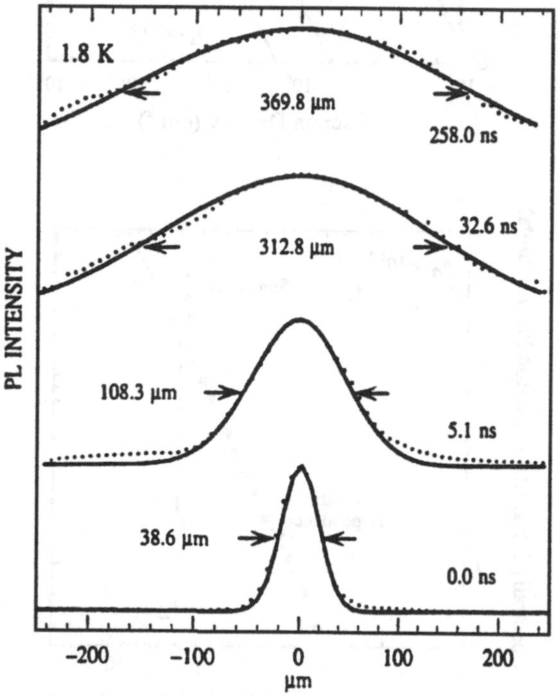

(b)

Figure 10.9. (a) Band structure of the polarized excitons in a single, wide n-doped quantum well surrounded by p-type barriers. Spatially polarized excitons are created from electrons in the center of the well and holes in localized states at the edges of the well. (b) Spatially and temporally resolved photoluminescence profiles measured from two-dimensional excitons in GaAs quantum wells at 1.8 K. The expansion from the initial distribution to the 100-μm distribution within 5 ns corresponds to an average velocity of 7×10^5 cm/s, i.e., ballistic expansion faster than the speed of sound in the lattice (from Ref. 48).

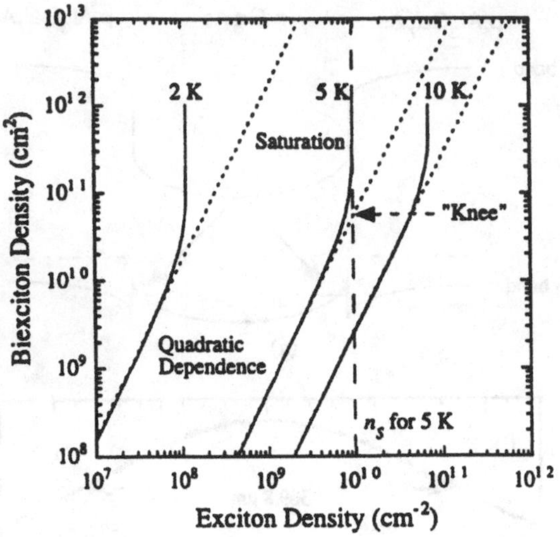

(a)

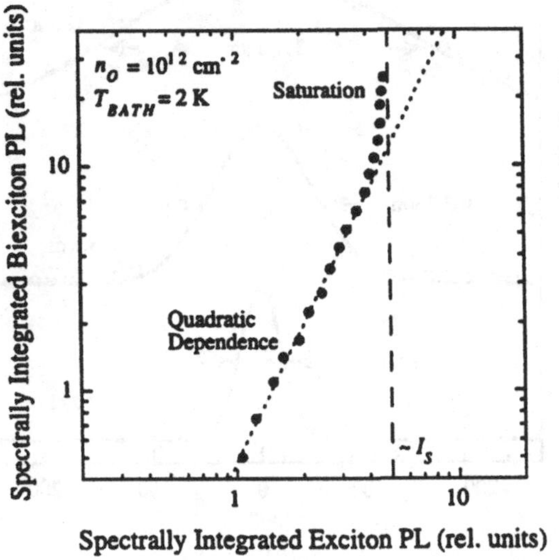

(b)

Figure 10.10. (a) Spectrally integrated exciton and biexciton luminescence intensities in a GaAs quantum well, (b) predictions of Eq. (10.11) (from Ref. 50).

10.2.4 Trapping Excitons in Two Dimensions

As discussed above, at least two experiments have shown an increase of diffusion of excitons at high density and low temperature in two-dimensional structures, which might be associated with superfluidity of the excitons. In general, however, it is difficult to distinguish between fast diffusion due to superfluidity and fast expansion due to a pressure gradient or phonon wind, as in the experiments with Cu_2O discussed in Chapter 3 (although a theoretical study by Link and Baym [58] has indicated that it may be possible to distinguish between the two based on their spatial profiles). Therefore experiments that show fast transport in quantum wells such as those discussed above do not necessarily demand the

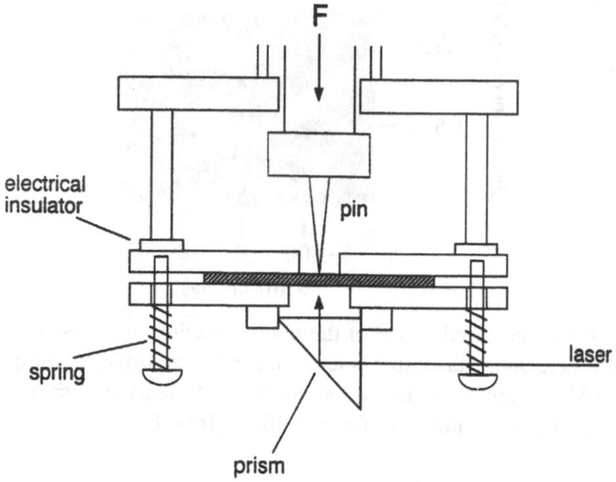

Figure 10.11. Geometry of the stressor that creates a shear-stress maximum in the quantum-well sample with nearly zero hydrostatic stress (from Ref. 17).

interpretation of superfluidity. An alternative method of showing superfluidity is to introduce a confining force that causes the excitons to be attracted inward, opposite of the direction of motion caused by the pressure gradient. As pointed out by Nozières [59] and others, even a weak trapping potential that destroys the translational symmetry of a two-dimensional system will allow true Bose condensation in two dimensions, analogous to Bose condensation in a three-dimensional harmonic potential, discussed at the beginning of this chapter. This two-dimensional Bose condensate would have a distinctive spatial profile as in the three-dimensional case discussed in Section 10.1.

One way to create a trapping potential in a semiconductor heterostructure is to deliberately introduce a well-width variation [30]. Since a wider well width corresponds to lower energy of the band-to-band transition in a quantum-confined system, a localized spot with wider width will correspond to a trap for the dipole excitons discussed above. Another approach is to use inhomogenous stress, an inhomogeneous magnetic field, or an inhomogeneous electric field to trap the excitons. These latter have the advantage in that the trapping field can be applied externally and varied continously. Recent experiments [17] show that this is possible in quantum heterostructures. Figure 10.11 shows a stress geometry that creates a point of high shear stress in a quantum well. This shear-stress maximum corresponds to an energy minimum in this case just as in the bulk semiconductors discussed in Section 10.1. In addition, when the pin is held at a fixed voltage, a current flows through the substrate from the pin to the metal holder, so that the voltage across the sample drops to zero far away from the pin. If the sample is a coupled quantum-well structure, as shown in Fig. 10.2, a voltage maximum corresponds to an energy minimum for the dipole excitons. These two effects combine to create an harmonic potential trap for dipole excitons in two dimensions, as shown in Fig. 10.12.

In a two-dimensional harmonic potential, the critical number for Bose condensation, from a calculation analogous to that in Section 10.1, is $1.6g(k_BT)^2/(\hbar\omega_0)^2$. For the well shown in Fig. 10.11, $\alpha = 65$ meV/mm^2, which implies a critical number at a temperature of 2 K of approximately 5×10^7 for an exciton mass of the order of the electron mass.

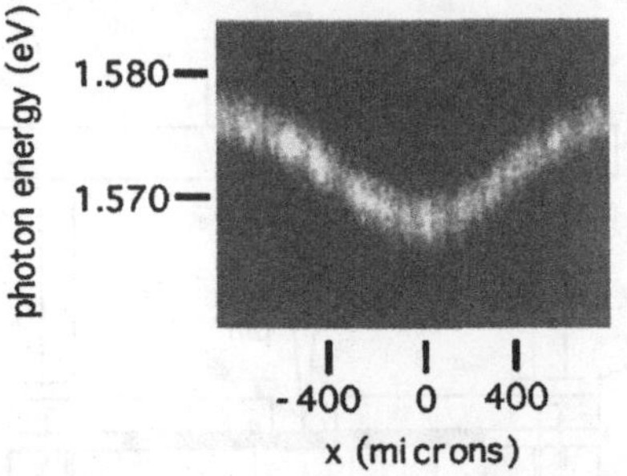

Figure 10.12. Time-integrated image of the indirect exciton luminescence through an imaging spectrometer as the laser spot is scanned across the surface of the quantum-well sample with 43-kV/cm applied field. The point of lowest energy corresponds to the point directly below the tip of the pin shown in Fig. 10.10 (from Ref. 17).

This is well within the number of photons in a typical laser pulse. So far, however, the effective temperature of the excitons in the well in these experiments has been too high for Bose condensation [17]. Nevertheless, this is a promising avenue that may reveal the two-component distribution that is the telltale of Bose condensation in a harmonic potential.

10.3 The Excitonic Insulator, or "Excitonium"

We have already briefly discussed the excitonic-insulator state in Chapter 5 in the context of Keldysh's theory of a nonequilibrium excitonic-insulator state. Much of the theory has centered on the nonequilibrium excitonic-insulator state, which occurs when an intense laser pulse creates a high density of electrons and holes in a normal semiconductor. A different, fascinating proposal would produce permanent excitons, i.e., an excitonic ground state. The theory of this state, also called "excitonium," was elaborated, starting in the 1960's, in a number of papers [33, 60–75].

One can discuss this transition in terms of three relevant energies: the energy gap between the single-particle electron and hole states E_{gap}, the excitonic Rydberg Ry_{ex}, and the thermal energy k_BT. Figure 10.13 shows a model in which E_{gap} is continuously tuned, while Ry_{ex} is assumed constant. If $E_{\text{gap}} \gg \text{Ry}_{\text{ex}} \gg k_BT$, one has the case of a semiconductor such as Cu_2O at room temperature or GaAs at low temperature. Excitons exist in this case only when the crystal is excited by a photon source, in which case the exciton gas (including the Bose condensed exciton gas) can exist as a metastable quasi-equilibrium state. If $E_{\text{gap}} \gg k_BT \gg \text{Ry}_{\text{ex}}$, then the excitons can spontaneously ionize when the crystal is excited by a photon source, and the carriers will form an electron–hole plasma, as in the case of GaAs at room temperature. The crossover from exciton gas to electron–hole plasma is fairly complicated, however, as discussed in the first part of Chapter 5, and does not depend simply on k_BT. Other states such as the nonequilibrium, BCS-like excitonic insulator or the Fermi EHL state can also exist, depending on the exciton–exciton interactions and whether the photon intensity is high enough.

In the case $E_{\text{gap}} \sim k_BT \gg \text{Ry}_{\text{ex}}$, one has a narrow-gap semiconductor, in which a thermal population of excited electrons and holes exists in equilibrium, which would be a conductor. If $E_{\text{gap}} - \text{Ry}_{\text{ex}} \sim k_BT$ but $\text{Ry}_{\text{ex}} \gg k_BT$, then one has the case of a thermal-equilibrium population of excitons, which is an insulating state. BEC of excitons cannot occur in this

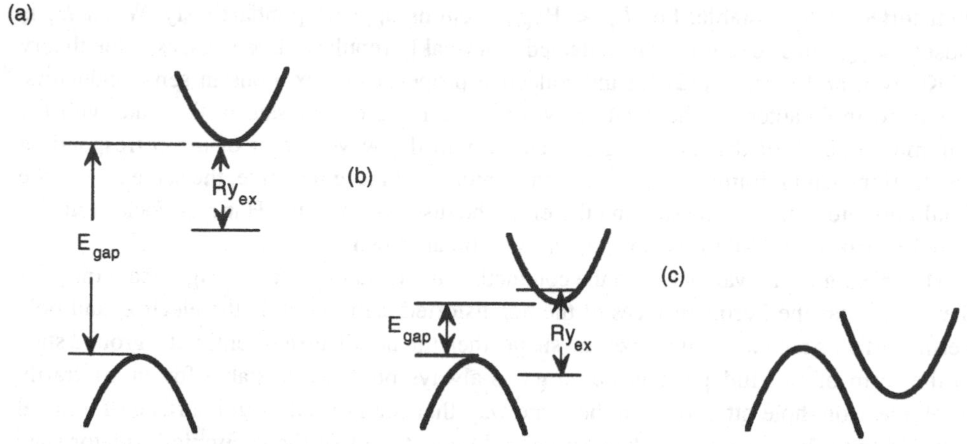

Figure 10.13. Relevant energies in the excitonic-insulator transition in three cases: (a) a normal semiconductor, $E_{gap} > \mathrm{Ry}_{ex}$; (b) the excitonic insulator, $\mathrm{Ry}_{ex} > E_{gap}$; (c) a metal, $E_{gap} < 0$. In case (b), the bands will not remain unaltered, since an instability exists for spontaneous creation of excitons.

case as an equilibrium phenomenon, since the equilibrium ground state is a state in which no excitons exist.

If $E_{gap} < 0$, then one has a metal. What about the intermediate case, when $E_{gap} - \mathrm{Ry}_{ex} < 0$? As noted by Mott [60] and Knox [61], in this case, excitons can form spontaneously even at $T = 0$. Therefore a macroscopic number of carriers will pair into permanent excitons, leading to a complete reconstruction of the bands.

This effect relies on the possibility of the excitonic binding energy's remaining large even while the bandgap decreases. To first order, this is not impossible to imagine, since the excitonic Rydberg depends on the masses of the carriers and the dielectric constant, not directly on the bandgap energy.

10.3.1 Long-Range Order in the Excitonic Insulator

If the energy to create an indirect exciton becomes negative, then the ordinary ground state of the crystal built up in the Hartree–Fock approximation by use of the single electron states of the valence band becomes unstable with respect to the formation of excitons [71]. The theory of this instability was first developed by Keldysh and Kopaev [33] and Des Cloizeaux [62]. Further details of this state, named "excitonium" in Ref. 71, have been worked out by Arkhipov [63], Kozlov and Maksimov [64, 65], Kohn [66, 70], Baklanov and Chaplik [67], Kopaev [68], Jérome, Rice, and Kohn [69], Halperin and Rice [71, 73], Guseinov and Keldysh [72], Zittartz [74] and others. In these papers the electrons and the holes were treated in an effective-mass approach, and only the long-range Coulomb interaction between electrons and holes was considered. In this Subsection we follow the theory elaborated in the review article by Halperin and Rice [71], drawing on these other papers as well. As realized by Keldysh and Kopaev [33] and Des Cloizeaux [62], the theory of an excitonic insulator in this case is mathematically similar to the BCS theory of superconductivity.

Halperin and Rice considered the case of a narrow-gap semiconductor or semimetal at $T = 0$. For $E_g > \mathrm{Ry}_{ex}$, no excitons are present in the ground state of the crystal, and the undistorted state is stable. For $E_g < \mathrm{Ry}_{\mathrm{ex}}$, excitons appear spontaneously. When E_g is close to $\mathrm{Ry}_{\mathrm{ex}}$, these excitons can be treated as a weakly repulsive Bose gas, as in the theory of Keldysh and Kozlov [32] for the collective properties of excitons in semiconductors, discussed in Chapter 2. The excitons will form a Bose condensate in the state with the minimum energy of the indirect-gap exciton, with the wave vector of the corresponding conduction-band minimum. As E_{ex} becomes more and more negative, the screening of the Coulomb interaction increases, and therefore the discussion of the Hartree–Fock excitonic state in terms of a dilute Bose exciton gas will break down.

The mixing of the valence- and the conducting-band states becomes significant only for the states near the Fermi surfaces of the nondistorted semimetal. If the electron and hole Fermi surfaces are identical in size and shape, then the nondistorted semimetal ground state in the form of two independent Fermi gases always becomes unstable for an arbitrarily weak electron–hole attraction, in the same way that the normal Fermi surface of a metal is unstable to the formation of Cooper pairs. Many studies of the excitonic-insulator state have been carried out in the model of semimetal with isotropic bands. As shown by Kopaev [68], anisotropy substantially complicates the picture. Anisotropy of the electron bands enhances the stability of the semimetal state, and a significant anisotropy will destroy the excitonic-insulator state even at very low temperatures. A difference of the densities of electrons and holes will also have the same effect.

The Hamiltonian describing the electrons of the conduction band (c) and the valence band (v) and their Coulomb interaction in an indirect-gap semiconductor can be written as [69]

$$H = \sum_{\mathbf{k}} \varepsilon_c(\mathbf{k}) a^\dagger_{c\mathbf{k}} a_{c\mathbf{k}} + \sum_{\mathbf{k}} \varepsilon_v(\mathbf{k}) a^\dagger_{v\mathbf{k}} a_{v\mathbf{k}} + \frac{1}{2} \sum_{\mathbf{q}} V_{\mathbf{q}} \rho(\mathbf{q}) \rho(-\mathbf{q}), \tag{10.12}$$

where the creation and annihilation operators $a^\dagger_{c\mathbf{k}}$ and $a_{c\mathbf{k}}$ create and destroy the electrons in the band c with wave vectors $\mathbf{w} + \mathbf{k}$, while the operators $a^\dagger_{v\mathbf{k}}$ and $a_{v\mathbf{k}}$ create and destroy electrons in the valence band with wave vectors $\mathbf{k}$. The valence band is assumed to have a maximum at the point $\mathbf{k} = 0$, and the conduction band a minimum at the point $\mathbf{w}$ of the Brillouin zone. The energies of the electron in two bands $\varepsilon_c(\mathbf{k}_c)$ and $\varepsilon_v(\mathbf{k}_v)$ are measured relative to the center of the indirect bandgap, and the wave vectors $\mathbf{k}_c$ and $\mathbf{k}_v$ are taken relative to the respective band extrema. The single-particle energies are

$$\varepsilon_c(\mathbf{k}_c) = \frac{1}{2} E_g + \frac{\hbar^2 k_c^2}{2m_e},$$

$$\varepsilon_v(\mathbf{k}_v) = -\left(\frac{1}{2} E_g + \frac{\hbar^2 k_v^2}{2m_h}\right). \tag{10.13}$$

Here m_e and m_h are the electron and the hole masses and E_g may be positive or negative. The charge-density operator $\rho(\mathbf{q})$ and the Coulomb interaction constant $V_{\mathbf{q}}$ have the expressions

$$\rho(\mathbf{q}) = \sum_{\mathbf{k}} (a^\dagger_{c,\mathbf{k}+\mathbf{q}} a_{c\mathbf{k}} + a^\dagger_{v,\mathbf{k}+\mathbf{q}} a_{v,\mathbf{k}}),$$

$$V_{\mathbf{q}} = \frac{4\pi e^2}{\varepsilon V q^2}, \tag{10.14}$$

where ε is the dielectric constant and V is the volume of the system. The spin degeneracy of the electrons is neglected.

In the absence of the Coulomb interaction, for negative values of E_g we have a semimetal consisting of two Fermi gases with the Fermi energies μ_e and μ_h and Fermi wave vectors $\mathbf{k}_F$ given by

$$\mathbf{k}_F^2 = 2\mu|E_g|, \qquad \mu = \frac{m_e m_h}{m_e + m_h},$$
$$\mu_i = \frac{\hbar^2 k_F^2}{2m_i}, \qquad i = e, h, \qquad |E_g| = \mu_e + \mu_h. \tag{10.15}$$

The conventional insulating ground state of the semiconductor is given by the function

$$|\varphi\rangle = \prod_{\mathbf{k}} a^\dagger_{v,\mathbf{k}} |\mathrm{vac}\rangle \tag{10.16}$$

where $|\mathrm{vac}\rangle$ is the state without electrons and $\mathbf{k}$ runs over the whole Brillouin zone. The possible instability of this state against bound-pair formation suggests that one try a new Hartree–Fock ground state of the form

$$|\psi\rangle = \prod_{\mathbf{k}} \alpha^\dagger_{\mathbf{k}} |\mathrm{vac}\rangle, \tag{10.17}$$

where $\alpha^\dagger_{\mathbf{k}}$ creates an electron in a linear combination of band c and v states according to

$$\alpha_{\mathbf{k}} = u_{\mathbf{k}} a_{v,\mathbf{k}} - v_{\mathbf{k}} a_{c,\mathbf{k}},$$
$$|u_{\mathbf{k}}|^2 + |v_{\mathbf{k}}|^2 = 1. \tag{10.18}$$

The state $|\psi\rangle$ can be also written in the equivalent form,

$$|\psi\rangle = \prod_{\mathbf{k}} (u^*_{\mathbf{k}} - v^*_{\mathbf{k}} a^\dagger_{c\mathbf{k}} a_{v\mathbf{k}})\, |\varphi\rangle, \tag{10.19}$$

which shows in parallel with BCS theory, since in this state a hole in a state $(v, \mathbf{k})$ and an electron in a state $(c, \mathbf{k} + \mathbf{w})$ are either both present or both absent.

To establish the connection of formulas (10.16)–(10.19) with the ones in Chapter 2, in which the electron–hole representation was used, it is sufficient to pass from the operators describing the conduction- and the valence-band electrons to the electron–hole operators by the usual substitutions

$$a_{c\mathbf{k}} = a_{\mathbf{k}}, \qquad a_{v\mathbf{k}} = b^\dagger_{-\mathbf{k}}, \tag{10.20}$$

and to remember that a valence band filled with electrons and an empty conduction band is the vacuum state for electron–hole pairs. If the function $|\varphi\rangle$ is substituted by the electron–hole vacuum function $|0\rangle$, then the function $|\psi\rangle$ obtains the form

$$|\psi\rangle = \prod_{\mathbf{k}} (u^*_{\mathbf{k}} - v^*_{\mathbf{k}} a^\dagger_{\mathbf{k}} b^\dagger_{-\mathbf{k}})\, |0\rangle, \tag{10.21}$$

It is mathematically analogous to the wave function of Bose condensed correlated electron–hole pairs.

Minimization of the total energy with respect to $u_{\mathbf{k}}$ and $v_{\mathbf{k}}$ gives the following results:

$$u^2_{\mathbf{k}} = \frac{1}{2}\left(1 + \frac{\xi_{\mathbf{k}}}{E_{\mathbf{k}}}\right), \qquad v^2_{\mathbf{k}} = \frac{1}{2}\left(1 - \frac{\xi_{\mathbf{k}}}{E_{\mathbf{k}}}\right), \tag{10.22}$$

where

$$\xi_{\mathbf{k}} = \frac{1}{2}\left(E_g + \frac{\hbar^2 k^2}{2\mu}\right), \qquad E_{\mathbf{k}}^2 = \xi_{\mathbf{k}}^2 + \Delta_{\mathbf{k}}^2. \tag{10.23}$$

The gap function $\Delta_{\mathbf{k}}$ is determined by

$$\Delta_{\mathbf{k}} = \sum_{\mathbf{q}} V_{\mathbf{k}-\mathbf{q}} \frac{\Delta_{\mathbf{q}}}{2E_{\mathbf{q}}}. \tag{10.24}$$

The wave function $\psi_{\mathbf{q}} = (\Delta_{\mathbf{q}}/2E_{\mathbf{q}})$ obeys the equation

$$\left[\left(E_g + \frac{\hbar^2 k^2}{2\mu}\right)^2 + |\Delta_{\mathbf{k}}|^2\right]^{1/2} \psi_{\mathbf{k}} = \sum_{\mathbf{q}} V_{\mathbf{k}-\mathbf{q}} \psi_{\mathbf{q}}. \tag{10.25}$$

Equation (10.24) can be compared with the elementary equation for the exciton wave function $\varphi_{\mathbf{k}}$,

$$\left(\frac{\hbar^2 k^2}{2\mu} + E_B\right)\varphi_{\mathbf{k}} = \sum_{\mathbf{q}} V_{\mathbf{k}-\mathbf{q}} \varphi_{\mathbf{k}}, \qquad E_B = \mathrm{Ry}_{\mathrm{ex}}. \tag{10.26}$$

This comparison [64] confirms that $|\Delta_{\mathbf{k}}| = 0$ when $E_g \geq \mathrm{Ry}_{\mathrm{ex}}$. The nontrivial solution $|\Delta_{\mathbf{k}}| > 0$ exists for all values of E_g, positive or negative, but obeys the restriction $|E_g| < \mathrm{Ry}_{\mathrm{ex}}$ [64].

Jérome et al. [69] showed that in spite of the formal similarity with the BCS theory, the nature of the order in the excitonic-insulator state is entirely different from that of superfluid systems. In particular, the excitonic-insulator state exhibits no superfluid transport properties [71]. These conclusions are supported by the most of early papers, excluding Refs. 64, and 65.

As shown by Yang [76], a superconductor is characterized by off-diagonal long-range order (ODLRO), which is manifested in the properties of the two-particle density matrix

$$\langle \mathbf{r}_1', \mathbf{r}_2' |\, \rho_2 \,| \mathbf{r}_1, \mathbf{r}_2 \rangle \equiv \frac{\mathrm{Tr}[e^{-\beta H} \Psi^\dagger(\mathbf{r}_2)\Psi^\dagger(\mathbf{r}_1)\Psi(\mathbf{r}_1')\Psi(\mathbf{r}_2')]}{\mathrm{Tr}(e^{-\beta H})}, \tag{10.27}$$

where $\Psi(\mathbf{r})$ is the electron-field operator. In the case of superconductor this matrix element remains finite in the limit $\mathbf{r}_1 \approx \mathbf{r}_2$, $\mathbf{r}_1' \approx \mathbf{r}_2'$, and $|\mathbf{r}_1 - \mathbf{r}_1'| \to \infty$. As shown in papers by Jérome et al. [69] and Kohn and Sherrington [77], in the case of an excitonic dielectric there is no ODLRO. As pointed out by Keldysh [78], this result assumes thermodynamic equilibrium in the system, when all electrons of valence and conduction bands are characterized by one and the same chemical potential. We add that the two-electron-field operators can be chosen of the type $\Psi_v(\mathbf{r}_1)\Psi_c(\mathbf{r}_2)$, which leads to a two-particle density matrix [63] in the form

$$\langle \Psi_c^\dagger(\mathbf{r}_2)\Psi_v^\dagger(\mathbf{r}_1)\Psi_v(\mathbf{r}_1')\Psi_c(\mathbf{r}_2')\rangle.$$

Here the average is made over function (10.19). This expression does not have ODLRO, as seen by a comparison of ground-state wave function (10.19) and the corresponding wave function of the superconductor. As shown by Keldysh [78], the tendency to extrapolate this result to the case of high-density nonequilibrium excitons in semiconductors [77] is not justified. The difference between these cases is discussed below. However, in the excitonic insulator, there is an additional diagonal long-range order [69]. This means that in the particular case of $\mathbf{r}_1 = \mathbf{r}_1'$, $\mathbf{r}_2 = \mathbf{r}_2'$, and $|\mathbf{r}_1 - \mathbf{r}_2| \to \infty$, the two-particle density matrix is finite and has a periodic dependence on $|\mathbf{r}_1 - \mathbf{r}_2|$. This corresponds to a diagonal long-range order of the crystal, i.e., a new real-space ordering. This new periodicity is characterized by the wave vector $\mathbf{w}$ of the electron band minimum. The translational symmetry of the ground state is broken by the introduction of this wave vector $\mathbf{w}$ characterizing the Bose condensate of the indirect-bandgap excitons [69].

When spin is taken into account, the picture becomes somewhat more complicated. Halperin and Rice [71] introduced the creation operator $c_{\mathbf{w}}^\dagger$ for an exciton with wave vector $\mathbf{w}$ in the form

$$c_{\mathbf{w}}^\dagger = \sum_{\mathbf{k}} \sum_{\sigma,\sigma'} f_{\sigma,\sigma'}(\mathbf{k})\, a_{c,\mathbf{k}+\mathbf{w},\sigma}^\dagger a_{v,\mathbf{k},\sigma'}. \tag{10.28}$$

Unlike the $a_{c,\mathbf{k}}$ operator used above, which used $\mathbf{k}$ measured relative to the band extremum, here the dependence on the wave vector $\mathbf{k}+\mathbf{w}$ is shown explicitly, and the electron and hole can exist in either a singlet or a triplet state. The properties of the excitonic-insulator phase depend on whether the macroscopically occupied exciton state is a singlet or triplet, leading to additional magnetic effects [71]. Bose condensation of indirect-bandgap excitons with wave vector $\mathbf{w}$ means that $\langle c_{\mathbf{w}}^\dagger \rangle \neq 0$, and hence the averages $\langle a_{c,\mathbf{k}+\mathbf{w},\sigma}^\dagger a_{v,\mathbf{k},\sigma'} \rangle$ are nonzero for suitable choices of σ and σ'. Halperin and Rice [71] pointed out that in the Hartree–Fock picture of the new distorted crystal phase, the one-electron states are made up from linear combinations of states of wave vectors $\mathbf{k}$ and $\mathbf{k}+\mathbf{w}$ from the valence and conducting bands of the old nondistorted crystal, as shown in formulas (10.18). For the distorted state to have a lower energy than the undestorted one, the exchange potential must be strong enough to produce the required admixture of valence- and conduction-band states.

Halperin and Rice [71] discussed the nature of the distortions that can arise, using a model without spin-orbit coupling. The nature of the distortion depends on whether the expectation value $\langle c_{\mathbf{w}}^\dagger \rangle$ is real or imaginary as well as on whether the macroscopically occupied exciton state is a singlet or a triplet. Halperin and Rice enumerated four possibilities:

- Singlet state with real phase, characterized by a charge-density oscillation.
- Triplet state with real phase, characterized by antiferromagnetic spin-density oscillations.
- Singlet state with imaginary phase, characterized by transverse currents that change sign from one unit cell to the next, i.e., orbital antiferromagnetism.
- Triplet state with imaginary phase, characterized by transverse spin currents.

Halperin and Rice [71] also considered the possibility of a crystalized excitonium state. They pointed out that at some values of the parameters, the excitonic-insulator phase does not exist at all. For example, at small deviations from isotropy, at low density of electrons and holes and with a large difference in masses ($m_h \gg m_e \approx m_o$, where m_o is the free electron mass), the excitons will not form a Bose fluid, but instead will condense to form a periodic array of excitonic molecules analogous to solid hydrogen. For $m_h \sim m_e \approx m_o$, such a solid array will be unstable because of zero point motion [78]. In the limit $m_h \gg m_e \approx m_o$, when the density of e–h pairs increases, one might expect a Wigner crystallization of the holes in a more or less uniform background of light electrons. This state would be conducting and analogous to a metallic state. At higher densities of e–h pairs, the Wigner array of holes would melt and the crystal would become a normal semimetal, with well-defined Fermi surfaces of the electrons and holes.

In regard to the phase transition from the semimetal to the excitonic-insulator state, a warning suggested in Ref. [71] is relevant here. Even if the Hartree–Fock calculation leads to a second-order phase transition, there are other interactions that have been neglected, which might lead to a first-order phase transition of the gas–liquid type. Such a result was obtained by Guseinov and Keldysh [72]. They added supplementary terms to Hamiltonian (10.12) that simultaneously create and destroy two electron–hole pairs with total wave vector equal to zero. These terms, which remove the analogy between the excitonic insulator and a superfluid Bose gas, play the role of a source of excitons and give rise to an energy gap in the dispersion law of the collective elementary excitations, leading to a first-order phase transition instead of to a second-order one. The excitonic insulator then becomes no different from any other usual dielectric. Similar conclusions were reported in Ref. 71.

As noted above and at the beginning of Chapter 5, the theory of the equilibrium excitonic-insulator state differs essentially from the theory of nonequilibrium excitons in semiconductors, which has been the subject of most of this book. We return again here to discuss some of the main differences, following the discussion of Keldysh [78].

As Keldysh pointed out, the excitonic-insulator state is a thermodynamic equilibrium state, in which all the electrons of the valence and the conduction bands have one and the same chemical potential [78]. In fact, only the Coulomb attraction of the pairs and the condition of electroneutrality, i.e., equal numbers of electrons and holes, make the description of this system similar to the case of nonequilibrium electron–hole pairs in semiconductors. In the latter case, the electron and the hole populations each have their own, separate chemical potentials. The establishment of quasi-equilibrium among the electrons, holes, and excitons leads the relation among their chemical potentials, taking into account the fact that the total number of quasiparticles is determined by the external source of excitation and not by the condition of thermodynamic equilibrium. This fact means that one cannot extrapolate the results concerning the absence of superfluidity in the excitonic insulator to the case of nonequilibrium excitons in semiconductors. In addition, as discussed in Section 2.2, moving excitons do not transfer mass and electric charge, but only excitation energy and dipole moments, if they exist. Superfluidity of excitons implies, for example, the nondissipative transfer of energy during a period of time determined by the exciton lifetime [78].

To put it another way, in the electron–hole description of nonequilibrium excitons, one deals with the electron–hole field operators $\Psi_e(\mathbf{r}_1)\Psi_h(\mathbf{r}_2)$, which are equivalent to the product $\Psi_c(\mathbf{r}_1)\Psi_v^\dagger(\mathbf{r}_2)$ and differ from the two-fermion operators $\Psi_c(\mathbf{r}_1)\Psi_v(\mathbf{r}_2)$ used above for the equilibrium excitonic insulator. One can prove the existence of the ODLRO in the case of correlated electron–hole pairs in a semiconductor by considering the two-particle density matrix of the type

$$\langle\Psi_e^\dagger(\mathbf{x}')\Psi_h^\dagger(\mathbf{y}')\Psi_h(\mathbf{y})\Psi_e(\mathbf{x})\rangle, \tag{10.29}$$

where the electron and hole field operators have the forms

$$\Psi_e(\mathbf{x}) = \frac{1}{\sqrt{V}}\sum_{\mathbf{p}} a_{\mathbf{p}}e^{i\mathbf{p}\cdot\mathbf{x}}, \qquad \Psi_h(\mathbf{y}) = \frac{1}{\sqrt{V}}\sum_{\mathbf{q}} b_{\mathbf{q}}e^{i\mathbf{q}\cdot\mathbf{y}}, \tag{10.30}$$

and the average is made over the ground state, which appears in the Bogoliubov–Hartree–Fock approximation after the Bogoliubov u, v transformation:

$$\begin{aligned} a_{\mathbf{p}} &= u_{\mathbf{p}}\alpha_{\mathbf{p}} - v_{\mathbf{p}}\beta_{-\mathbf{p}}^\dagger, \\ b_{\mathbf{p}} &= u_{\mathbf{p}}\beta_{\mathbf{p}} + v_{\mathbf{p}}\alpha_{-\mathbf{p}}^\dagger. \end{aligned} \tag{10.31}$$

The new ground state in the presence of BEC of correlated *e–h* pairs obeys the conditions

$$\alpha_{\mathbf{p}}\,|0\rangle = \beta_{\mathbf{p}}\,|0\rangle = 0. \tag{10.32}$$

At $T = 0$, matrix element (10.29) takes exactly the same form as that used by Jérome et al. [69] for the case of BCS theory, namely,

$$\begin{aligned}&\langle 0|\,\Psi_e^\dagger(\mathbf{x}')\Psi_h^\dagger(\mathbf{y}')\Psi_h(\mathbf{y})\Psi_e(\mathbf{x})\,|0\rangle\\&\quad= g(\mathbf{x}-\mathbf{x}')g(\mathbf{y}-\mathbf{y}') + f^*(\mathbf{x}'-\mathbf{y}')f(\mathbf{x}-\mathbf{y}),\end{aligned} \tag{10.33}$$

where

$$\begin{aligned}g\,(\mathbf{x}) &= \frac{1}{V}\sum_{\mathbf{p}} |v_{\mathbf{p}}|^2 e^{i\mathbf{p}\cdot\mathbf{x}},\\ f\,(\mathbf{x}) &= \frac{1}{V}\sum_{\mathbf{q}} v_{\mathbf{q}}u_{\mathbf{q}}e^{i\mathbf{q}\cdot\mathbf{x}}.\end{aligned} \tag{10.34}$$

As in the case of BCS theory, the second term in Eq. (10.30) leads to ODLRO because it remains finite when $|\mathbf{x}-\mathbf{x}'| \to 0$.

The above considerations show that the equilibrium excitonic insulator has fundamental differences from the nonequilibrium exciton system, even though their mathematical descriptions both are based on BCS-type ground-state wave functions and the Bogoliubov *u–v* transformations.

10.3.2 Experimental Evidence for the Excitonic-Insulator Phase

Although the above theoretical considerations indicate that ODLRO, and therefore superfluidity, should be absent in the case of an equilibrium excitonic insulator, the other fascinating predictions of this state, e.g., new types of charge-density and spin-density waves, make experimental investigations of this system worthwhile.

Of course, the most direct evidence for this new state is a sharp transition from a conductor to insulator as the gap is tuned. Recent experiments with bulk narrow-gap semiconductors seem to show the expected behavior of the permanent excitonic-insulator state. If one can continuously tune the bandgap of a semiconductor while keeping the excitonic Rydberg constant, one would expect to see the following behavior: First, as the gap decreases, one would see the resistivity decrease, as more thermally excited carriers occupy the conduction band. If the excitonic-insulator state occurs, one would expect to see a sharp increase in the resistivity. Then as the conduction band is lowered even further, one would expect to see decreasing resistivity again, as the material turned metallic.

This behavior has been seen in materials in which stoichiometry was used to create a narrow-gap semiconductor and then pressure was used to lower the gap further [79,80]. As seen in Fig. 10.13, as the gap energy is varied, at one point the resistance of the material increases sharply, and then decreases again. This is consistent with the interpretation of the intermediate state as an excitonic insulator. While the change of resistivity is dramatic, it is not a direct test of long-range order, any more than a decrease of resistivity is alone a test for a purported superconductor. Portengen, Östreich, and Sham [81,82] have recently proposed an experiment which could be used to detect spontaneous coherence of the excitonium state directly. Noting that the spontaneous symmetry breaking which occurs in Bose condensation of excitons leads to a macroscopic polarization, they showed that this built-in polarization should lead to a nonvanishing second-order nonlinear optical response, which could be detected in a two-beam mixing experiment. This test may apply to a broad class of mixed-valent systems. Portengen, Östreich, and Sham noted that the self-consistent mean-field solution of the Falicov–Kimball model [83] for mixed-valent materials corresponds to a Bose condensation of excitons [81]. Therefore a number of these materials may be intrinsic excitonic insulators which have a built-in polarization, allowing second-harmonic generation to take place.

Another recent proposal is to search for a stable excitonic insulator in a two-dimensional system [84–86]. In this system, a double quantum well like that shown in Fig. 10.2 is produced that is also a narrow-bandgap semiconductor. As is well known [87, 88], $InAs/Al_xGa_{1-x}Sb$ heterostructures make an ideal narrow-gap system with a tunable bandgap. Recent experiments [89, 90] have reported evidence for an excitonic-insulator ground state in this system, based on far-infrared absorption and cylotron resonance measurements. A basic problem for interpreting these measurements, however, is that the low-energy excitation spectrum of the excitonic insulator, which should show up in far-infrared measurements, has never really been calculated for a realistic system. Presumably, comparison of such a calculation to the experimental excitation spectrum should show strong evidence of the long-range order of the system in the same way that the excitation spectrum deduced from neutron-scattering data gives evidence of long-range order in the case of liquid He [91].

10.4 Optical Coherence Without Lasing

We have already discussed many times in this volume the fact that Bose condensation of excitons corresponds to a coherent, macroscopic polarization of the electronic states. This means that, in principle, the light emitted from such a state can also be coherent. In the case of a driven Bose condensate, e.g., as in the optical phase conjugation experiments in CuCl discussed in Chapter 4 or in the cases envisioned in the previous few chapters, the coherence can lead to optical four-wave mixing. In that case, the coherence of the emitted light does not seem surprising, since a coherent laser source was used to generate the exciton state in the first place. In the case of spontaneous, or thermodynamic, condensation, however, one has the interesting possibility of optical coherence in an experiment in which no optical state is amplified, i.e., in which no stimulated emission of photons has occurred.

An illustration of such an experiment is shown in Fig. 10.15. If an incoherent pump creates excitons at a high enough density and a low enough temperature in a trap, then the excitons can undergo spontaneous Bose condensation, i.e., spontaneous symmetry breaking, leading to macroscopic phase coherence. If the coherence of the excitons is then transmitted to the photons emitted when they recombine, then in principle, if one collects the light emitted from the ground state of this system, the light collected from one region of the condensate will have a definite phase relative to light emitted from another part of the condensate. Photons can be collected from two different solid angles from the same source and then combined into a single path that then ends on the entrance slit of the spectrometer. The spontaneous coherence of the excitons should produce the uncanny result of a "missing beam-splitter interferometer." Although the geometry resembles that of a Michelson interferometer, which normally would have two beam splitters, in this case the noncollinear light from the source should already be coherent. A single beam splitter is used to put light from two directions back together. When photons emitted by direct recombination of excitons in the ground state are examined within the proper time gate, the intensity should vary sinusoidally as

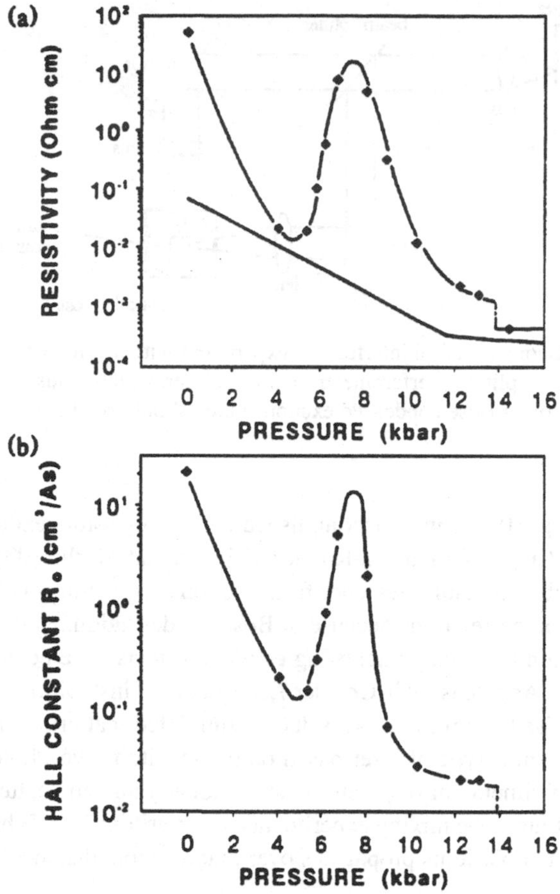

Figure 10.14. Resistivity and Hall constant of the narrow-gap semiconductor $TmSe_{0.45}Te_{0.55}$ as the gap is tuned lower by means of increasing pressure (from Ref. 79).

the path length difference of the two legs is varied, because of constructive and destructive interference of the phase-coherent photons. Similar effects can be seen by a two-detector analysis of the temporal correlations of the emitted photons [92].

The form of the exciton–photon coupling will determine whether this is possible. As discussed in Chapter 7, if the exciton states are dipole coupled to the photon states, then one must deal with polariton states, and the ground state of the system will not be the zero-momentum state. It would seem difficult to distinguish between a spontaneous polariton condensate and lasing in such a system. In the case of weak coupling, however, as in the case of the semiconductor Cu_2O or the two-dimensional dipole excitons discussed in Subsection 10.2.1, then spontaneous condensation can occur in the excitonic states.

In the case of the phonon-assisted luminescence process discussed in Subsection 1.4.1, the participation of a phonon from the incoherent phonon population would destroy the phase coherence of the luminescence. Single-photon direct luminescence from the exciton ground state is not allowed, by momentum conservation. Two-photon luminescence from the ground state, however, can satisfy energy and momentum conservation as well as preserve the phase coherence of the exciton states. Therefore it seems that an experiment such as

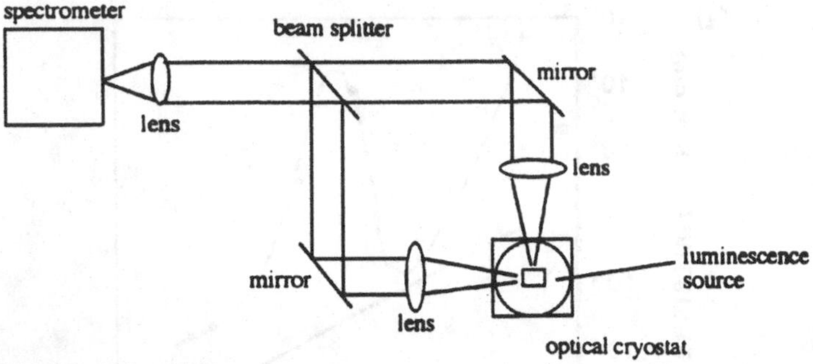

Figure 10.15. Geometry of an interference experiment with a coherent exciton source, or "missing beam-splitter interferometer." Unlike a laser, which sends coherent light in only one direction, a Bose condensed exciton source should send coherent photons in all directions.

that envisioned in Fig. 10.14 can be accomplished with two-photon–exciton recombination luminescence from Cu_2O. Fernández-Rossier, Tejedor, and Merlin [93] have also argued that the direct recombination luminescence from excitons in quantum-well structures should show coherent photon correlations because of Bose condensation.

Finally, we mention that many interesting coherent effects have been proposed and observed for polaritons. As discussed in Chapter 7, the theory of instabilities of Bose condensed polaritons can give rise to amplified waves due to stimulated scattering of polaritons, which can be the basis of a new type of laser based on two-exciton, two-photon transitions [94, 95], and recent experiments have given evidence for stimulated scattering of polaritons [96–99]. Recently, four-wave mixing experiments on polaritons in BiI_3 have shown that the coherent phase of the polaritons propagates over macroscopic distances at high excitation densities [100].

As seen in this brief survey, many new possibilities exist for novel excitonic effects. Work on Bose–Einstein condensation of excitons has moved beyond the realm of simply asking whether the effect can occur, and into new physical effects that not only show the quantum character of these systems, but also may lead to applications.

References

[1] M.H. Anderson et al., *Science* **269** (7), 198 (1995).
[2] K.B. Davis et al., *Phys. Rev. Lett.* **75**, 3969 (1995).
[3] M.R. Andrews et al., *Science* **273** (7), 84 (1996).
[4] S. Stringari, *Phys. Rev. Lett.* **76**, 1405 (1996); F. Dalfovo and S. Stringari, *Phys. Rev. A* **53**, 2477 (1996).
[5] M. Holland and J. Cooper, *Phys. Rev. A* **53**, R1954 (1996).
[6] M. Edwards et al., *Phys. Rev. A* **53**, R1950 (1996).
[7] S. Grossman and M. Holthaus, *Phys. Lett. A* **208**, 188 (1996).
[8] P.A. Ruprecht et al., *Phys. Rev. A* **51**, 4704 (1995).
[9] D.P. Trauernicht, J.P. Wolfe, and A. Mysyrowicz, *Phys. Rev. B* **34**, 2561 (1986).
[10] D. F. Goble and L. E. H. Trainor,*Phys. Lett.* **18**,122 (1965).
[11] T.W. Hijmans, Yu. Kagan, G.V. Shlyapnikov, and J.T.M. Walraven, to be published; see also D.A. Huse and E.D. Siggia, *J. Low Temp. Phys.* **46**, 137 (1982) and V.V. Goldman, I. Silvera, and A.J. Leggett, *Phys. Rev. B* **24**, 2870 (1981).

[12] J.P. Wolfe, J.L. Lin, and D.W. Snoke, in *Bose–Einstein Condensation*, A. Griffin, D.W. Snoke, and S. Stringari, eds. (Cambridge U. Press, Cambridge, 1995).
[13] D.P.Trauernicht, Ph.D. dissertation, (University of Illinois, Urbana-Champaign, 1986).
[14] J.P. Wolfe and C.D. Jeffries, in *Electron–Hole Droplets in Semiconductors*, C.D. Jeffries and L.V. Keldysh, eds. (North-Holland, Amsterdam, 1987).
[15] See, e.g., K. Huang, in *Bose–Einstein Condensation*, A. Griffin, D.W. Snoke, and S. Stringari, eds. (Cambridge U. Press, Cambridge, 1995).
[16] T. Fukuzawa, E.E. Mendez, and J.M. Hong, *Phys. Rev. Lett.* **64**, 3066 (1990).
[17] V. Negoita, D.W. Snoke, and K. Eberl, in press.
[18] D. Paquet, T. M. Rice, and K. Ueda, *Phys. Rev. B* **32**, 5208 (1985).
[19] Yu. E. Lozovik and O. L. Berman, *JETP* **84**, 1027 (1997).
[20] J.A. Kash, M. Zachau, E.E. Mendez, J.M. Hong, and T. Fukuzawa, *Phys. Rev. Lett.* **66**, 2247 (1991).
[21] S. D. Baranovskii, R. Eichmann, and P. Thomas, *Phys. Status Solidi B* **205**, R19 (1998).
[22] V. B. Timofeev et al., *Europhys. Lett.* **41**, 535 (1998).
[23] L.V. Butov, A. Zrenner, G. Abstreiter, G. Böhm, and G. Weimann, *Phys. Rev. Lett.* **73**, 304 (1994).
[24] L.V. Butov et al., *Surf. Sci.* **361/362**, 243 (1996).
[25] V.B. Timofeev, V.D. Kulakovskii, and I.V. Kukushkin, *Physica B+C* **117/118**, 327 (1983).
[26] Yu. E. Lozovik and V.I. Yudson, *JETP Lett.* **22**, 274 (1976).
[27] I.V. Lerner and Yu. E. Lozovik, *Sov. Phys. JETP* **53**, 763 (1981).
[28] P.I. Arseev and A.B. Dzyubenko, *JETP* **87**, 200 (1998).
[29] H. Akiyama, H. Sakaki, and T. Matsusue, *Appl. Phys. Lett.* **67**, 2037 (1995).
[30] X. Zhu, P.B. Littlewood, M. Hybertson, and T.M. Rice , *Phys. Rev. Lett.* **74**, 1633 (1995).
[31] J. Fernandez-Rossier and C. Tejedor, *Phys. Rev. Lett.* **78**, 4809 (1997).
[32] L. V. Keldysh and A.N. Kozlov, *Zh. Eksp. Teor. Fiz.* **54**, 978 (1968) [*Sov. Phys. JETP* **27**, 521 (1968)].
[33] L. V. Keldysh and Yu. V. Kopaev, *Fiz. Tverd. Tela* **6**, 2791 (1964) [*Sov. Phys. Solid State* **6**, 2219 (1965)].
[34] C. Comte and P. Nozières, *J. Physique* **43**, 1069 (1982).
[35] P. Nozières and C. Comte, *J. Physique* **43**, 1083 (1982).
[36] A.V. Korolev and M.A. Liberman, *Phys. Rev. Lett.* **72**, 270 (1994).
[37] M.R. Schafroth, *Phys. Rev.* **100**, 463 (1955).
[38] For example, Y.-C. Chang and R.B. James, *Phys. Rev. B* **39**, 12672 (1989); B. Laikhtman, R.A. Kiehl, and D.J. Frank, *J. Appl. Phys.* **70**, 1531 (1991).
[39] L. Shvartsman and J.E. Golub, in *Bose–Einstein Condensation*, A. Griffin, D.W. Snoke, and S. Stringari, eds. (Cambridge U. Press, Cambridge, 1995).
[40] E.F. Gross, V.I. Perel', and R.I. Shekhmamet'ev, *Sov. Phys. JETP Lett.* **13**, 229 (1971).
[41] E.F. Gross, N.V. Starostin, M.P. Shepilov, and R.I. Shekhmamet'ev, *Proc. USSR Acad. Sci.* **37**, 885 (1973).
[42] A.V. Sel'kin, I.G. Stamov, N.N. Syrbu, and A.G. Umanets, *Sov. Phys. JETP Lett.* **35**, 57 (1982).
[43] V.A. Kovarskii and A.A. Golub, *Fiz. Tver. Tela* **16**, 617 (1974).
[44] V.A. Kovarskii, *Phys. Status Solidi B* **151**, K35, (1989); *Fiz. Tver. Tela* **32**, 867 (1990).
[45] J. Ranninger, in *Bose–Einstein Condensation*, A. Griffin, D.W. Snoke, and S. Stringari, eds. (Cambridge U. Press, Cambridge, 1995).
[46] A.S. Alexandrov, in *Bose–Einstein Condensation*, A. Griffin, D.W. Snoke, and S. Stringari, eds. (Cambridge U. Press, Cambridge, 1995).
[47] V.I. Yudson, *Phys. Rev. Lett.* **77**, 1564 (1996).
[48] G.D. Gilliland, D.J. Wolford, G.A. Northrop, M.S. Petrovic, T.F. Kuech, and J.A. Bradley, *J. Vac. Sci. Technol. B* **10**, 1959 (1992).
[49] See, for example, I.F. Silvera, in *Bose–Einstein Condensation*, A. Griffin, D.W. Snoke, and S. Stringari, eds. (Cambridge U. Press, Cambridge, 1995); D.O. Edwards, *Physica B* **109/110**, 1531 (1982).

[50] J.C. Kim and J.P. Wolfe, *Phys. Rev. B* **57**, 9861 (1998).
[51] B.F. Feuerbacher, J. Kuhl, and K. Ploog, *Phys. Rev. B* **43**, 2439 (1991).
[52] D.J. Lovering, R.T. Phillips, G.J. Denton, and G.W. Smith, *Phys. Rev. Lett.* **23**, 1880 (1992).
[53] D. Oberhauser et al., *Phys. Status Solidi B* **173**, 53 (1992); K.H. Pantke et al., *Phys. Rev. B* **47**, 2413 (1993).
[54] G. Finkelstein, S. Bar-Ad, O. Carmel, I. Bar-Joseph, and Y. Levinson, *Phys. Rev. B* **47**, 12964 (1993).
[55] E.J. Mayer et al., *Phys. Rev. B* **51**, 10909 (1995).
[56] D. Birkedal et al., *Phys. Rev. Lett.* **76**, 672 (1996).
[57] K.B. Ferrio and D.G. Steele, *Phys. Rev. B* **54**, (1996).
[58] B. Link and G. Baym, *Phys. Rev. Lett.* **69**, 2959 (1992).
[59] P. Nozières, in *Bose–Einstein Condensation*, A. Griffin, D.W. Snoke, and S. Stringari, eds. (Cambridge U. Press, Cambridge, 1995).
[60] N.F. Mott, *Philos. Mag.* **6**, 287 (1961).
[61] R.S. Knox, *Theory of Excitons* (Academic, New York, 1963), p. 100.
[62] J. Des Cloizeaux, *J. Phys. Chem. Solids* **26**, 259 (1965).
[63] R.G. Arkhipov, *Zh. Exp. Teor. Fiz.* **43**, 349 (1962) [*Sov. Phys. JETP* **16**, 251 (1962)].
[64] A.N. Kozlov and L.A. Maksimov, *Zh. Exp. Teor. Fiz.* **48**, 1184 (1965) [*Sov. Phys. JETP* **21**, 790 (1965)].
[65] A.N. Kozlov and L.A. Maksimov, *Zh. Exp. Teor. Fiz.* **49**, 1284 (1965) [*Sov. Phys. JETP* **22**, 889 (1966)].
[66] W. Kohn, in *Physics of Solids at High Pressures*, C.T. Temiznka and R.M. Emrick, eds. (Academic, New York, 1965).
[67] E.V. Baklanov and A.V. Chaplik, *Fiz. Tverd. Tela* **7**, 2768 (1965) [*Sov. Phys. Solid State* **7**, 2240 (1966)].
[68] Yu.V. Kopaev, *Fiz. Tverd. Tela* **8**, 223 (1966) [*Sov. Phys. Solid State* **8**, 175 (1966)].
[69] D. Jérome, T.M. Rice, and W. Kohn, *Phys. Rev.* **158**, 462 (1967).
[70] W. Kohn, *Phys. Rev. Lett.* **19**, 439 (1967).
[71] B.I. Halperin and T.M. Rice, *Rev. Mod. Phys.* **40**, 755 (1968).
[72] R.R. Guseinov and L.V. Keldysh, *Zh. Exp. Teor. Fiz.* **63**, 2255 (1972).
[73] B.I. Halperin and T.M. Rice, in *Solid State Physics*, F. Seitz, D. Turnbull, and H. Ehrenreich, eds. (Academic, New York, 1968), Vol. 21.
[74] J. Zittartz , *Phys. Rev.* **162**, 752 (1967); **164**, 575 (1967); **165**, 605 (1968).
[75] W. Kohn, in *Many-Body Physics*, C. de Witt and R. Balian, eds. (Gordon and Breach, New York, 1968).
[76] C.N.Yang, *Rev. Mod. Phys.* **34**, 694 (1962).
[77] W. Kohn and D. Sherrington, *Rev. Mod. Phys.* **42**, 1 (1970).
[78] L.V. Keldysh, *Problems of Theoretical Physics* (Nauka, Moscow, 1972), p. 433.
[79] B. Bucher, P. Steiner, and P. Wachter, *Phys. Rev. Lett.* **67**, 2717 (1991).
[80] P. Wachter, A. Jung, and P. Steiner, *Phys. Rev. B* **51**, 5542 (1995).
[81] T. Portengen, Th. Östreich, and L. J. Sham, *Phys. Rev. Lett.* **76**, 3384 (1996).
[82] Th. Östreich, T. Portengen, and L. J. Sham,*Solid State Commun.,* **100**,325 (1996).
[83] L.M. Falicov and J.C. Kimball, *Phys. Rev. Lett.* **22**, 997 (1969).
[84] X. Zhu, J.J. Quin, and G. Gumbs, *Solid State Commun.* **75**, 595 (1990).
[85] E.G. Wang, Y. Zhou, C.S. Ting, J. Zhang, T. Pang, and C. Chen, *J. Appl. Phys.* **78**, 7099 (1995).
[86] Y. Naveh and B. Laikhtman, *Phys. Rev. Lett.* **77**, 900 (1996).
[87] Y. Naveh and B. Laikhtman, *Appl. Phys. Lett.* **66**, 1980 (1995).
[88] L.M. Claessen, J.C. Maan, M. Altarelli, P. Wyder, L.L. Chang, and L. Esaki, *Phys. Rev. Lett.* **57**, 2556 (1986).
[89] J.P. Cheng, J. Kono, B.D. McCombe, I. Lo, W.C. Mitchelm, and C.E. Stutz, *Phys. Rev. Lett.* **74**, 450 (1995).
[90] T.P. Marlow et al., in *Proceedings of the Twenty-Fourth conference on the Physics of Semiconductors* (World Scientific, Singapore, 1998).

[91] A. Griffin, *Excitations in a Bose–Condensed Liquid* (Cambridge U. Press, Cambridge, 1993).
[92] B. Laikhtman, *Europhys. Lett.* **43**, 53 (1998).
[93] J. Fernández-Rossier, C. Tejedor, and R. Merlin, *Solid State Commun.* **108**, 473 (1998).
[94] S.A. Moskalenko, *Bose-Einstein Condensation of Excitons and Biexcitons*, (Academy of Sciences of the MSSR, Kishinev, 1970) (in Russian.)
[95] V.F. Elesin and Yu.V. Kopaev, *Zh. Eksp. Teor. Phys.* **72**, 334 (1977).
[96] J. Bloch, in *Proc. 24th Int. Conf. Phys. Semiconductors*, (World, Singapore, 1999).
[97] M.S. Skolnick, in *Proc. 24th Int. Conf. Phys. Semiconductors*, (World, Singapore, 1999).
[98] M.D. Martin, G. Aichmayr, L. Vina, J.K. Son, and E.E. Mendez, in *Proc. 24th Int. Conf. Phys. Semiconductors*, (World, Singapore, 1999).
[99] R. Andre, F. Boeuf, D. Heger, R. Romastain, and L.-S. Dang, in *Proc. 24th Int. Conf. Phys. Semiconductors*, (World, Singapore, 1999).
[100] T. Karasawa, H. Mino, and M. Yamamoto, in *Proceedings of the 1999 International Conference on Luminescence and Optical Spectroscopy* (Osaka, Japan, August 1999), in press.

Appendix A: Properties of Excitons in Cu_2O

Because in many of the experiments discussed in this volume the semiconductor Cu_2O was used, and because this material is less well known than other semiconductors such as Si and GaAs, we review here some of the basic properties of Cu_2O.

Cu_2O is actually one of the earliest semiconductors studied [1], and the proof of the existence of excitons in this crystal [2, 3], which confirmed the theories of Frenkel [4], Wannier [5], and Mott [6], led to an exciting period of expansion in the field of excitonic physics in the late 1950s and early 1960s, during which time many of the optical properties of this intriguing crystal were established [7–19]. (For general reviews, see Refs. 20 and 21.) The same property that makes it an excellent material for nonlinear optical effects, namely, a very large excitonic binding energy (150 meV), makes it a poor electrical conductor even at room temperature. Therefore Cu_2O has not been studied or fabricated extensively by the electronics industry, and high-quality samples are still not widely available.

Band Structure. Most of the complexities of the band structure of Cu_2O stem from the *d* orbitals of the Cu atoms in the valence band. Cu_2O forms into a cubic lattice with inversion symmetry, which is the O_h symmetry group [22]. In the O_h symmetry, the five 3*d* orbitals are split by means of the crystal field into a higher threefold degenerate Γ_5^+ level and a lower twofold degenerate Γ_3^+ level. The lowest conduction band, on the other hand, is formed from the 4*s* orbitals that have Γ_1^+ symmetry in the O_h group.

When electron spin is taken into account, the lowest conduction band has the doubly degenerate Γ_6^+ symmetry, while the upper Γ_5^+ valence level splits by means of spin-orbit interaction into a higher doubly degenerate Γ_7^+ level and a lower fourfold degenerate Γ_8^+ level. These two bands are the highest occupied valence states. When a conduction electron forms an exciton with the lower Γ_8^+ level, the energy gap is roughly 2.3 eV, leading these excitons to be called the "green" exciton series, while the excitons formed from the Γ_6^+ conduction electrons and Γ_7^+ holes form the "yellow" series, with a gap energy of approximately 2.173 eV at low temperature (and 2.096 eV at room temperature [23]). These "yellow" excitons, with a binding energy of 0.15 eV, are the lowest electronic excited state of the crystal. In addition, there are "blue" and "indigo" excitons formed from the higher-lying Γ_8^- conduction band and the Γ_7^+ and Γ_8^+ valence bands. The yellow and the green excitons are energetically distant from other electronic states, so that one can generally ignore the effects of mixing with the blue and the indigo excitons and other higher- or lower-lying states. In general, one can also ignore the mixing of yellow states with green states, but studies of the shifts of the yellow exciton states with pressure [24, 25], which yield the deformation potentials for the exciton–phonon interaction as well as the static energy of the stress trap discussed in Chapter10, can be understood only in terms of mixing of all 12 yellow and green exciton states.

The four exciton states of the Γ_6^+ conduction electrons and the Γ_7^+ holes are split by electron-hole exchange into a higher triplet Γ_5^+ level and a singlet Γ_2^+ level. The upper triplet state is usually called the orthoexciton and the lower singlet state is usually called the paraexciton, but these are not pure spin states as in the case of orthohydrogen and parahydrogen, since the Γ_7^+ hole states are not pure hole-spin states, but are rather total angular momentum states, given by

$$\begin{aligned} |\uparrow_H\rangle &= -\frac{i}{\sqrt{3}}|yz\rangle|\downarrow_h\rangle + \frac{1}{\sqrt{3}}|xz\rangle|\downarrow_h\rangle - \frac{i}{\sqrt{3}}|xy\rangle|\uparrow_h\rangle, \\ |\downarrow_H\rangle &= -\frac{i}{\sqrt{3}}|yz\rangle|\uparrow_h\rangle - \frac{1}{\sqrt{3}}|xz\rangle|\uparrow_h\rangle + \frac{i}{\sqrt{3}}|xy\rangle|\downarrow_h\rangle, \end{aligned} \tag{A.1}$$

where the states with lower case h are pure hole spin states. The excitonic basis states are then given by

$$\begin{aligned} &|J=1, J_z=1\rangle = |\uparrow_e, \uparrow_H\rangle, \\ &|J=1, J_z=0\rangle = \frac{1}{\sqrt{2}}\left(|\uparrow_e, \downarrow_H\rangle + |\downarrow_e, \uparrow_H\rangle\right), \\ &|J=1, J_z=-1\rangle = |\downarrow_e, \downarrow_H\rangle, \\ &|J=0, J_z=0\rangle = \frac{1}{\sqrt{2}}\left(|\uparrow_e, \downarrow_H\rangle - |\downarrow_e, \uparrow_H\rangle\right), \end{aligned} \tag{A.2}$$

which equal

$$\begin{aligned} |J=1, J_z=1\rangle &= \frac{|0\ 0\rangle + |1\ 0\rangle}{\sqrt{6}}\left(-i|yz\rangle + |xz\rangle\right) - \frac{i}{\sqrt{3}}|1\ 1\rangle|xy\rangle, \\ |J=1, J_z=0\rangle &= \frac{-1}{\sqrt{6}}|1\ 1\rangle\left(i|yz\rangle + |xz\rangle\right) + \frac{i}{\sqrt{3}}|0\ 0\rangle|xy\rangle \\ &\quad -\frac{1}{\sqrt{6}}|1\ -1\rangle\left(i|yz\rangle - |xz\rangle\right), \\ |J=1, J_z=-1\rangle &= \frac{|0\ 0\rangle - |1\ 0\rangle}{\sqrt{6}}\left(i|yz\rangle + |xz\rangle\right) + \frac{i}{\sqrt{3}}|1\ -1\rangle|xy\rangle, \\ |J=0, J_z=0\rangle &= \frac{-1}{\sqrt{6}}|1\ 1\rangle\left(i|yz\rangle + |xz\rangle\right) + \frac{i}{\sqrt{3}}|1\ 0\rangle|xy\rangle \\ &\quad +\frac{1}{\sqrt{6}}|1\ -1\rangle\left(i|yz\rangle - |xz\rangle\right), \end{aligned} \tag{A.3}$$

and where

$$\begin{aligned} &|1\ 1\rangle = |\uparrow_e, \uparrow_h\rangle, \\ &|1\ 0\rangle = \frac{1}{\sqrt{2}}\left(|\uparrow_e, \downarrow_h\rangle + |\downarrow_e, \uparrow_h\rangle\right), \\ &|1\ -1\rangle = |\downarrow_e, \downarrow_h\rangle, \\ &|0\ 0\rangle = \frac{1}{\sqrt{2}}\left(|\uparrow_e, \downarrow_h\rangle - |\downarrow_e, \uparrow_h\rangle\right) \end{aligned} \tag{A.4}$$

are the pure spin triplet and singlet states. Since the electron–hole exchange interation affects only states with opposite electron and hole spin (since in the virtual annihilation process the electron must be able to go into a vacant state of the same spin), only the $|0\,0\rangle$ terms in states (A.3) contribute. This implies that the triplet state of the yellow exciton in Cu_2O feels a positive exchange energy, while the singlet is unaffected by electron–hole exchange, in contrast to the case of simple spins, i.e., excitons created from Γ_6^+ electrons and Γ_6^+ holes, in which the triplet state feels no exchange interaction and the singlet state does.

The electron–hole Coulomb exchange term was discussed in Section 7.2 in regard to Keldysh's equations for coherent excitons and photons, for a simple model with one doubly degenerate conduction band and one doubly degenerate valence band. In the notation of that section, the exchange energy for Cu_2O is written as

$$H_{\rm exch} = \frac{1}{2}\sum_{\mathbf{p},\mathbf{q},\mathbf{k}}\sum_{\sigma_1,\sigma_2}\sum_{i,j} F(c,\sigma_1,\mathbf{p};v,i,\mathbf{q};v,j,\mathbf{p}-\mathbf{k};c,\sigma_2,\mathbf{q}+\mathbf{k})$$
$$\times a^\dagger_{c,\sigma_1,\mathbf{p}} a^\dagger_{v,i,\mathbf{q}} a_{c,\sigma_2,\mathbf{q}+\mathbf{k}} a_{v,j,\mathbf{p}-\mathbf{k}}$$
$$+\frac{1}{2}\sum_{\mathbf{p},\mathbf{q},\mathbf{k}}\sum_{\sigma_1,\sigma_2}\sum_{i,j} F(v,i,\mathbf{p};c,\sigma_1,\mathbf{q};c,\sigma_2,\mathbf{p}-\mathbf{k};v,j,\mathbf{q}+\mathbf{k})$$
$$\times a^\dagger_{v,i,\mathbf{p}} a^\dagger_{c,\sigma_1,\mathbf{q}} a_{v,j,\mathbf{q}+\mathbf{k}} a_{c,\sigma_2,\mathbf{p}-\mathbf{k}}, \tag{A.5}$$

where σ_1 and σ_2 are the two pure spin states of the Γ_6^+ conduction band and i and j are the two Γ_7^+ hole states given by Eqs. (A.1). The coefficients $F(\cdots)$ are defined by the relation

$$F\,(c,\sigma_1,\mathbf{p};v,i,\mathbf{q};v,j,\mathbf{p}-\mathbf{k};c,\sigma_2,\mathbf{q}+\mathbf{k}) \tag{A.6}$$
$$= \int d\mathbf{r}_1 \int d\mathbf{r}_2 \langle \psi_{\Gamma_6^+,\sigma_1,\mathbf{p}}(\mathbf{r}_1)|\langle \psi_{\Gamma_7^+,i,\mathbf{q}}(\mathbf{r}_2)|\frac{e^2}{\varepsilon_\infty \mathbf{r}_{12}}|\psi_{\Gamma_7^+,j,\mathbf{p}-\mathbf{k}}(\mathbf{r}_1)\rangle|\psi_{\Gamma_6^+,\sigma_2,\mathbf{q}+\mathbf{k}}(\mathbf{r}_2)\rangle.$$

When the standard electron and hole operators

$$a_{c,\sigma,\mathbf{p}} = a_{\mathbf{p},\sigma}, \qquad a_{v,i,\mathbf{q}} = b^\dagger_{-\mathbf{q},i}, \tag{A.7}$$

are used, the normal-ordered part $\tilde{H}_{\rm exch}$ of Hamiltonian (A.5) takes the form

$$\tilde{H}_{\rm exch} = \sum_{\mathbf{p},\mathbf{q},\mathbf{k}}\sum_{\sigma_1,\sigma_2}\sum_{i,j} F(c,\sigma_1,\mathbf{p};v,i,\mathbf{q};v,j,\mathbf{p}-\mathbf{k};c,\sigma_2,\mathbf{q}+\mathbf{k})$$
$$\times a^\dagger_{\mathbf{p},\sigma_1} b^\dagger_{-\mathbf{p}+\mathbf{k},j} b_{-\mathbf{q},i} a_{\mathbf{q}+\mathbf{k},\sigma_2}. \tag{A.8}$$

For the exciton wave functions

$$|\psi\rangle = \frac{1}{\sqrt{2V}}\sum_{\mathbf{q}} \varphi(\mathbf{q})|J\,J_z\rangle,$$

Table A.1. *Physical parameters for Cu_2O*

Name	Symbol	Value	Reference
Dielectric constant	ϵ_∞	7	[27]
Electron mass	m_e	$1.0 m_0$	[28, 29]
Hole mass	m_h	$0.7 m_0$	[29]
$n = 1$ exciton mass	m_{ex}	$2.7 m_0$	[30, 31]
Rydberg ($n > 1$)	Ry_{ex}	97 meV	[32]
Bohr radius ($n > 1$)	a_{ex}	11.1 Å	[26]
$n = 1$ binding energy	$\mathrm{Ry}_{ex}^{(1)}$	150 meV	[20]
$n = 1$ Bohr radius	$a_{ex}^{(1)}$	5.0 Å	[26]
Bandgap	E_{gap}	2.17 eV (4 K)	[32]
Ortho–para exchange splitting	Δ	12 meV	[33]

where $\varphi(\mathbf{q})$ describes the relative electron–hole motion and $|J J_z\rangle$ are the basis states given in Eqs. (A.2), the average values of the exchange interaction $\tilde{H}_{exch}$ are found to be

$$E_{exch}(\mathbf{k}, \Gamma_5^+) = \frac{2}{3V} \sum_{\mathbf{p},\mathbf{q}} \varphi^*(\mathbf{p}-\alpha\mathbf{k})\varphi(\mathbf{q}+\beta\mathbf{k}) \times F(c, \mathbf{p}; 3d, xy, \mathbf{q}; 3d, xy, \mathbf{p}-\mathbf{k}; c, \mathbf{q}+\mathbf{k}),$$

$$E_{exch}(\mathbf{k}, \Gamma_2^+) = 0. \tag{A.9}$$

In Chapter 1, we wrote simply that the exciton mass is $m_{ex} = m_e + m_h$ and $\mathrm{Ry}_{ex} = e^2/2\epsilon a_{ex}$, where $a_{ex} = \hbar^2\epsilon/e^2 m_r$. In Cu_2O, neither of these approximations is quite true for the ground state of the exciton, because of central-cell corrections [26]; essentially, because the radius of the exciton is so small, the Wannier approximation breaks down and one nearly has the case of a Frenkel excitons. The $n = 1$ exciton in Cu_2O has a mass of 2.7 m_0, where m_0 is the vacuum electron mass, while the free electron and hole have masses of 1.0 and 0.7 m_0, respectively. Kavoulakis, Chang, and Baym [26] have shown that the mass correction is the same for both the orthoexcitons and the paraexcitons. Also, the $n = 1$ state has a binding energy of 150 meV and a Bohr radius of 5 Å, compared with $\mathrm{Ry}_{ex} = 97$ meV and $a_{ex} = 11$ Å for the excited $n \geq 2$ states. Table A.1 summarizes relevant parameters for the Cu_2O band structure.

Phonons. Since the unit cell of Cu_2O has six atoms, there are $3 \times 6 = 18$ phonon modes at zone center. Three of these are the two transverse-acoustic and longitudinal-acoustic-phonon mode, and the remaining 15 are optical phonons, which have the symmetries and energies listed in Table A.2. The optical phonons therefore cannot be simply discussed in terms of longitudinal-optical and transverse-optical modes.

Time-resolved luminescence experiments [31] showed that the emission of optical phonons by the excitons is extremely slow, of the order of tens of picoseconds compared with hundreds of femtoseconds for free carriers in GaAs. This is not surprising, since the Fröhlich interaction is forbidden for excitons with electrons and holes of equal mass; the low-energy optical phonons in Cu_2O also all have negative parity, while the exciton states have positive parity, leading to supression of phonon emission near zone center.

Table A.2. *Zone-center optical phonons in Cu_2O*

Designation	energy (meV)	Relative radiative efficiency Ortho	Para
$\Gamma_5^-(^3\Gamma_{25}^-)$	11.0	17 (Ref. 34)	1 (Ref. 35)
$\Gamma_3^-(^2\Gamma_{12}^-)$	13.8	500	≈0
$\Gamma_4^-(^3\Gamma_{15}^-)$	18.7	17 (Ref. 36)	≈0
$\Gamma_2^-(^1\Gamma_2^-)$	43	3	
$\Gamma_5^+(^3\Gamma_{25}^+)$	64	3	
$\Gamma_4^-(^3\Gamma_{15}^-)$	79	8	

Selection rules. The symmetry properties of Cu_2O give it one of its most remarkable properties, which is an extremely long exciton lifetime, despite the fact that Cu_2O is a direct-gap semiconductor. Based on absorption measurements [33], the intrinsic radiative lifetime of paraexcitons in Cu_2O is 8 ms [37]; experimentally, lifetimes up to 10 μs have been measured [38] at low temperature. Orthoexciton lifetimes as long as 300 ns have been measured at room temperature [31].

Since both the conduction band and the hole band of the yellow exciton in Cu_2O at zero stress have positive parity, direct dipole recombination (which has a matrix element with Γ_4^- symmetry in the O_h group; see Table A.3) is forbidden for both the orthoexciton and the paraexciton. Direct recombination is quadrupole allowed for the orthoexciton, while for the paraexciton, direct recombination is forbidden to all orders at zero stress, but becomes allowed under uniaxial stress [25].

The phonon-assisted recombination therefore usually dominates the luminescence. As seen in Subsections 1.4.1 and 2.1.5, the $^2\Gamma_{12}^-$ phonon-assisted recombination from the orthoexcitons particularly dominates the luminescence spectrum. Birman [39] has explained this by noting that transitions involving the $^2\Gamma_{12}^-$ phonon can involve the nearby Γ_8^- conduction-band states as intermediate states (which together with the Γ_7^+ valence states form the blue series of excitons in Cu_2O), while transitions involving all other phonons must use the $^3\Gamma_{15}^-$ valence-band states, which are energetically much further away. The

Table A.3. *Symmetry notations in the O_h group*

Koster notation [22]	Common notation	Basis functions
Γ_1^+	$^1\Gamma_1^+$	$x^2+y^2+z^2$
Γ_2^+	$^1\Gamma_2^+$	$(x^2-y^2)(y^2-z^2)(z^2-x^2)$
Γ_3^+	$^2\Gamma_{12}^+$	$(2z^2-x^2-y^2), \sqrt{3}(x^2-y^2)$
Γ_4^+	$^3\Gamma_{15}^+$	S_x, S_y, S_z
Γ_5^+	$^3\Gamma_{25}^+$	yz, xz, xy
Γ_1^-	$^1\Gamma_1^-$	$\Gamma_2^- \times \Gamma_2^+$
Γ_2^-	$^1\Gamma_2^-$	xyz
Γ_3^-	$^2\Gamma_{12}^-$	$\Gamma_3^+ \times \Gamma_2^-$
Γ_4^-	$^3\Gamma_{15}^-$	x, y, z
Γ_5^-	$^3\Gamma_{25}^-$	$\Gamma_5^+ \times \Gamma_1^-$

paraexciton phonon-assisted transitions are weak compared with those of the orthoexcitons, so that the paraexcitons in Cu_2O were not definitively observed until the late 1970s [33,40].

Many papers [7,9,13,15,17,41–45] have explored the effects of magnetic and electric fields on the exciton states in Cu_2O. The new lines that appear after the symmetry of the crystal is lowered lead to a complicated overall picture. Recently, studies by Matsumoto et al. [46] and Naka et al. [47], who used two-photon excitation and photovoltaic spectroscopy, have revived interested in the level structure of Cu_2O.

References

[1] M. Hayashi and K. Katsuki, *J. Phys. Soc. Jpn.* **5**, 381 (1950).
[2] E.F. Gross and A.N. Karryev, *Dokl. Akad. Nauk SSSR* **84**, 261; 471 (1952).
[3] M. Hayashi and K. Katsuki, *J. Phys. Soc. Jpn.* **7**, 599 (1952).
[4] Ya.I. Frenkel, *Phys. Rev.* **37**, 17, 1276 (1931).
[5] G.A.Wannier, *Phys. Rev.* **52**, 191 (1937).
[6] N.F. Mott, *Trans. Faraday Soc.* **34**, 500 (1938).
[7] E.F. Gross, *Usp. Fiz. Nauka* **63**, 575 (1957); **76**, 433 (1962).
[8] E.F. Gross, B.S. Razbirin, and M.A. Yakobson, *Zh. Eksp. Teor. Fiz.* **27**, 1149 (1957).
[9] E.F. Gross, B.P. Zakharchenya, and N.M. Reinov, *Dokl. Akad. Nauk SSSR* **97**, 57, 221 (1957).
[10] R.J. Elliott, *Phys. Rev.* **108**, 1384 (1957).
[11] I.S. Gorban' and V.B. Timofeev, *Zh. Opt. Spektrosk.* **9**, 482 (1960).
[12] S.I. Pekar and B.E. Tsekvava, *Fiz. Tverd. Tela* **2**, 261 (1960).
[13] S.A. Moskalenko, *Fiz. Tverd. Tela* **2**, 1755 (1960); *Zh. Opt. Specktrosk.* **9**, 369 (1960).
[14] J.B. Grun, M. Sieskind, and S. Nikitine, *J. Phys. Chem. Solids* **19**, 189 (1961).
[15] E.F. Gross, B.P. Zakharchenya and L.M. Kanskaya, *Fiz. Tverd. Tela* **3**, 972 (1961).
[16] E.F. Gross, A.A. Kaplyanskii, V.T. Agekyan, and D.S.Bulyanitsa, *Fiz. Tverd. Tela* **4**, 1660 (1962).
[17] S.A. Moskalenko and A.I. Bobrysheva, *Fiz. Tverd. Tela* **4**, 2169 (1962).
[18] E.F. Gross, A.A. Kaplyanskii, and V.T. Agekyan, *Fiz. Tverd. Tela* **4**, 2169 (1962).
[19] M.I. Shmiglyuk and S.A. Moskalenko, *Fiz. Tverd. Tela* **6**, 2729 (1964).
[20] V.T. Agekyan, *Phys. Status Solidi A* **43**, 11 (1977).
[21] S. Nikitine, in *Optical Properties of Solids* (Plenum, New York, 1969), Chap. 9.
[22] We use the notation of G. Koster, J. Dimmock, R. Wheeler, and H. Statz, *Properties of the Thirty-Two Point Groups* (MIT, Cambridge, MA, 1963).
[23] D.W. Snoke, A.J. Shields, and M. Cardona, *Phys. Rev. B* **45**, 11693 (1992).
[24] R.G. Waters, F.H. Pollack, R.H. Bruce, and H.Z. Cummins, *Phys. Rev. B* **21**, 1665 (1980).
[25] A. Mysyrowicz, D.P. Trauernicht, J.P. Wolfe, and H.-R. Trebin, *Phys. Rev. B* **27**, 2562 (1983).
[26] G.M. Kavoulakis, Y.-C. Chang, and G. Baym, *Phys. Rev. B* **55**, 7593 (1997).
[27] M. O'Keefe, *J. Chem. Phys.* **39**, 1789 (1963).
[28] A. Gotzene and C. Schwab, *Solid State Commun.* **18**, 1565 (1976).
[29] J.W. Hodby et al., *J. Phys. C* **9**, 1429 (1976).
[30] P.Y. Yu and Y.R. Shen, *Phys. Rev. Lett.* **32**, 939 (1974); *Phys. Rev. B* **12**, 1377 (1975).
[31] D.W. Snoke, D. Braun, and M. Cardona, *Phys. Rev. B* **44**, 2991 (1991).
[32] S. Nikitine, J.B. Grun, and M. Sieskind, *J. Phys. Chem. Solids* **17**, 292 (1961).
[33] P.D. Bloch and C. Schwab, *Phys. Rev. Lett.* **41**, 514 (1978).
[34] D.W. Snoke, J.P. Wolfe, and A. Mysyrowicz, *Phys. Rev. B* **41**, 11171 (1990).
[35] The measured relative absorption ratio of 50 in Ref. 33 is reduced by a factor of 3 to take into account the orthoexciton spin degeneracy.
[36] K. Reimann and K. Syassen, *Phys. Rev. B* **39**, 11113 (1989).
[37] K. O'Hara, Ph. D. thesis (University of Illinois at Urbana-Champaign, 1999).

[38] A. Mysyrowicz, D. Hulin, and A. Antonetti, *Phys. Rev. Lett.* **43**, 1123 (1979).
[39] J.L. Birman, *Solid State Commun.* **13**, 1189 (1978).
[40] G. Kuwabar, M. Tanaka, and H. Fukutani, *Solid State Commun.* **21**, 599 (1977).
[41] A.A. Kaplyanskii and A.K. Prjevudskii, in *Spectroscopy of Crystals* (Nauka, Moscow, 1966) p. 135.
[42] R.P. Seisyan, *Diamagnetic Exciton Spectroscopy* (Nauka, Moscow, 1984).
[43] R.P. Seisyan and B.P. Zakharchenya, in *Landau Level Spectroscopy*, G. Landwehr and E. Rashba, eds. (North-Holland, Amsterdam, 1981).
[44] S.A. Moskalenko, *Introduction to the Theory of High-Density Excitons* (Shtiintsa, Kishinev, 1983), Chap. 2 (in Russian).
[45] A.I. Bobrysheva and S.A.Moskalenko, *Fiz. Tekh. Poluprovodn.* **2**, 438 (1968).
[46] H. Matsumoto, K. Saito, M. Hasuo, S. Kono, and N. Nagasawa, *Solid State Commun.* **97**, 125 (1996).
[47] N. Naka, M. Hasuo, S. Kono and N. Nagasawa, in *Proceedings of the Twenty-Third International Conference on the* Physics of Semiconductors (Springer, Berlin, 1996), Vol.1, p. 273.

Author Index

Subject Index